D1223265

Operator Theory: Advances and Applications
Vol. 119

Optimization in Elliptic Problems with Applications to Mechanics of Deformable Bodies and Fluid Mechanics

William G. Litvinov

Birkhäuser Verlag
Basel · Boston · Berlin

Author:

William G. Litvinov
Institute of Statics and Dynamics of Aero-Space Structures
University of Stuttgart
Pfaffenwaldring 27
D-70550 Stuttgart

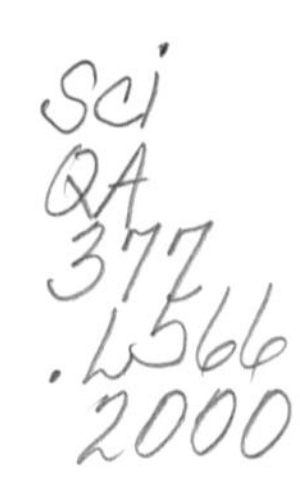

2000 Mathematics Subject Classification 49K20; 35J45, 73Kxx

A CIP catalogue record for this book is available from the
Library of Congress, Washington D.C., USA

Deutsche Bibliothek Cataloging-in-Publication Data

Litvinov, Vil'iam G.:
Optimization in elliptic problems with applications to mechanics of deformable bodies and fluid mechanics / William G. Litvinov. - Basel ; Boston ; Berlin : Birkhäuser, 2000
(Operator theory ; Vol. 119)
ISBN 3-7643-6199-9

ISBN 3-7643-6199-9 Birkhäuser Verlag, Basel – Boston – Berlin

Printed on acid-free paper produced from chlorine-free pulp. TCF ∞
Cover design: Heinz Hiltbrunner, Basel
Printed in Germany
ISBN 3-7643-6199-9

9 8 7 6 5 4 3 2 1

For my wife Tatiana and my son Eugene

Contents

Preface xv

Introduction xix

1 Basic Definitions and Auxiliary Statements

1.1 Sets, functions, real numbers 1
1.1.1 Notations and definitions 1
1.1.2 Real numbers 2
1.2 Topological, metric, and normed spaces 4
1.2.1 General notions 4
1.2.2 Metric spaces 5
1.2.3 Normed vector spaces 6
1.3 Continuous functions and compact spaces 10
1.3.1 Continuous and semicontinuous mappings 10
1.3.2 Compact spaces 12
1.3.3 Continuous functions on compact spaces 13
1.4 Maximum function and its properties 14
1.4.1 Discrete maximum function 14
1.4.2 General maximum function 16
1.5 Hilbert space 17
1.5.1 Basic definitions and properties 17
1.5.2 Compact and selfadjoint operators in a Hilbert space . . . 20
1.5.3 Theorem on continuity of a spectrum 25
1.5.4 Embedding of a Hilbert space in its dual 31
1.5.5 Scales of Hilbert spaces and compact embedding 33
1.6 Functional spaces that are used in the investigation of boundary value and optimal control problems 36
1.6.1 Spaces of continuously differentiable functions 36
1.6.2 Spaces of integrable functions 37
1.6.3 Test and generalized functions 37
1.6.4 Sobolev spaces 39
1.7 Inequalities of coerciveness 44
1.7.1 Coercive systems of operators 44
1.7.2 Korn's inequality 48

1.8 Theorem on the continuity of solutions of functional equations . . . 50
1.9 Differentiation in Banach spaces and the implicit function theorem . . . 51
1.9.1 Fréchet derivative and its properties . . . 51
1.9.2 Implicit function . . . 52
1.9.3 The Gâteaux derivative and its connection with the Fréchet derivative . . . 53
1.10 Differentiation of the norm in the space $W_p^m(\Omega)$ . . . 54
1.10.1 Auxiliary statement . . . 54
1.10.2 Theorem on differentiability . . . 55
1.11 Differentiation of eigenvalues . . . 58
1.11.1 The eigenvalue problem . . . 58
1.11.2 Differentiation of an operator-valued function . . . 60
1.11.3 Eigenspaces and projections . . . 61
1.11.4 Differentiation of eigenvalues . . . 64
1.12 The Lagrange principle in smooth extremum problems . . . 70
1.13 G-convergence and G-closedness of linear operators . . . 72
1.14 Diffeomorphisms and invariance of Sobolev spaces with respect to diffeomorphisms . . . 73
1.14.1 Diffeomorphisms and the relations between the derivatives . . . 73
1.14.2 Sequential Fréchet derivatives and partial derivatives of a composite function . . . 75
1.14.3 Theorem on the invariance of Sobolev spaces . . . 76
1.14.4 Transformation of derivatives under the change of variables . . . 78

2 Optimal Control by Coefficients in Elliptic Systems
2.1 Direct problem . . . 81
2.1.1 Coercive forms and operators . . . 81
2.1.2 Boundary value problem . . . 82
2.2 Optimal control problem . . . 86
2.2.1 Nonregular control . . . 86
2.2.2 Regular control . . . 88
2.2.3 Regular problem and necessary conditions of optimality . . . 90
2.2.4 Nonsmooth (discontinuous) control . . . 97
2.2.5 Some remarks on the use of regular and discontinuous controls . . . 102
2.3 The finite-dimensional problem . . . 103

2.4 The finite-dimensional problem (another approach) 105
2.4.1 The set $U^{(t)}$ 105
2.4.2 Approximate solution of the problem (2.2.22) 107
2.4.3 Approximate solution of the optimal control problem when the set $\overset{\circ}{U}_{\text{ad}}$ is empty 109
2.4.4 On the computation of the functional $h \to \Psi_k(h, u_h)$ . . . 110
2.4.5 Calculation and use of the Fréchet derivative of the functional $h \to \Psi_m(h, u_h)$ 113
2.5 Spectral problem 117
2.5.1 Eigenvalue problem 117
2.5.2 On the continuity of the spectrum 118
2.6 Optimization of the spectrum 120
2.6.1 Formulation of the problem and the existence theorem 120
2.6.2 Finite-dimensional approximation of the optimal control problem 122
2.6.3 Computation of eigenvalues 127
2.7 Control under restrictions on the spectrum 129
2.7.1 Optimal control problem 129
2.7.2 Approximate solution of the problem (2.7.7) 131
2.7.3 Second method of approximate solution of the problem (2.7.7) 132
2.7.4 Differentiation of the functionals $h \to A_i\mu(h)$ and necessary conditions of optimality 135
2.8 The basic optimal control problem 138
2.8.1 Setting of the problem. Existence theorem 138
2.8.2 Approximate solution of the problem (2.8.6) 140
2.9 The combined problem 142
2.10 Optimal control problem for the case when the state of the system is characterized by a set of functions 145
2.10.1 Setting of the problem 145
2.10.2 The existence theorem 146
2.11 The general control problem 149
2.11.1 Bilinear form a_q and the corresponding equation 150
2.11.2 Bilinear form b_r and the spectral problem 153
2.11.3 Basic control problem 154
2.11.4 Application of the basic control problem (combined problem) 157
2.12 Optimization by the shape of domain and by operators 159
2.12.1 Domains and bilinear forms 159

2.12.2 Optimization problem connected with solution of an operator equation 160
2.12.3 Eigenvalue optimization problem 162
2.12.4 Some realizations of the spaces M_l and N_l 164
2.13 Optimization problems with smooth solutions of state equations 168
2.13.1 Systems of elliptic equations 168
2.13.2 Elliptic problems in domains and in a fixed domain 170
2.13.3 The problem of domain shape optimization 173
2.13.4 Approximate solution of the direct problem ensuring convergence in the norm of a space of smooth functions 174

3 Control by the Right-hand Sides in Elliptic Problems
3.1 On the minimum of nonlinear functionals 177
3.1.1 Setting of the problem. Auxiliary statements 177
3.1.2 The existence theorem 179
3.1.3 Characterization of a minimizing element 181
3.1.4 Functionals continuous in the weak topology 182
3.2 Approximate solution of the minimization problem 183
3.2.1 Inner point lemma 183
3.2.2 Finite-dimensional problem 185
3.3 Control by the right-hand side in elliptic problems provided the goal functional is quadratic 191
3.3.1 Setting of the problem 191
3.3.2 Existence of a solution. Optimality conditions 192
3.3.3 An example of a system described by the Dirichlet problem 194
3.4 Minimax control problems 198
3.5 Control of systems whose state is described by variational inequalities 201
3.5.1 Setting of the problem 201
3.5.2 The existence theorem 203
3.5.3 An example of control of a system described by a variational inequality 205

4 Direct Problems for Plates and Shells
4.1 Bending and free oscillations of thin plates 209
4.1.1 Basic relations of the theory of bending of thin plates 209
4.1.2 Orthotropic plates 211
4.1.3 Bilinear form corresponding to the strain energy of the plate 212
4.1.4 Problem of bending of a plate 215

4.1.5 Problem of free oscillations of a plate 221
4.2 Problem of stability of a thin plate 223
4.2.1 Stored energy of a plate 223
4.2.2 Conditions of stationarity 226
4.2.3 Auxiliary statements 228
4.2.4 Transformation of the problem (4.2.27), (4.2.28) 231
4.2.5 Stability of a plate and bifurcation 235
4.2.6 An example of nonexistence of stable solutions 239
4.3 Model of the three-layered plate ignoring shears in the middle layer 242
4.3.1 Basic relations 242
4.3.2 Problems of the bending and of the free flexural oscillations 244
4.4 Model of the three-layered plate accounting for shears in the middle layer 246
4.4.1 Basic relations 246
4.4.2 Bilinear form corresponding to the three-layered plate 250
4.4.3 Bending of the three-layered plate 253
4.4.4 Natural oscillations of three-layered plate 255
4.5 Basic relations of the shell theory 257
4.6 Shells of revolution 260
4.6.1 Deformations and functional spaces 260
4.6.2 The bilinear form a_h 262
4.6.3 The subspace of functions with zero-point strain energy 264
4.7 Shallow shells 265
4.8 Problems of statics of shells 267
4.9 Free oscillations of a shell 268
4.10 Problem of shell stability 270
4.10.1 On some approaches to stability problems 270
4.10.2 Reducing of the stability problem to the eigenvalue problem 271
4.10.3 Spectral problem (4.10.12) 272
4.11 Finite shear model of a shell 274
4.11.1 Strain energy of an elastic shell 274
4.11.2 Shallow shell 276
4.11.3 A relation between the Kirchhoff and Timoshenko models of shell 278
4.12 Laminated shells 282
4.12.1 The strain energy of a laminated shell 282
4.12.2 Shell of revolution 284
4.12.3 Shallow shells 286

5 Optimization of Deformable Solids

5.1 Settings of optimization problems for plates and shells . . . 287
5.1.1 Goal functional and a function of control . . . 287
5.1.2 Restrictions . . . 289
5.2 Approximate solution of direct and optimization problems for plates and shells . . . 291
5.2.1 Direct problems and spline functions . . . 291
5.2.2 The spaces V_m for plates . . . 292
5.2.3 The spaces V_m for shells . . . 294
5.2.4 Direct problems for nonfastened plates and shells . . . 297
5.2.5 Solution of optimization problems . . . 298
5.3 Optimization problems for plates (control by the function of the thickness) . . . 300
5.3.1 Optimization under restrictions on strength . . . 300
5.3.2 Stability optimization problem . . . 305
5.3.3 Optimization of frequencies of free oscillations . . . 311
5.3.4 Combined optimization problem and optimization for a class of loads . . . 312
5.4 Optimization problems for shells (control by functions of midsurface and thickness) . . . 312
5.4.1 Problem of optimization of a shell of revolution with respect to strength . . . 313
5.4.2 Optimization according to the stability of a cylindrical shell subject to a hydrostatic compressive load . . . 316
5.5 Control by the shape of a hole and by the function of thickness for a shallow shell . . . 319
5.5.1 Problem of optimization according to strength . . . 319
5.5.2 Approximate solution of the optimization and direct problems . . . 322
5.5.3 Problem of optimization of eigenvalues . . . 324
5.5.4 Approximate solution of the eigenvalue problem . . . 325
5.6 Control by the load for plates and shells . . . 326
5.6.1 General problem of control by the load . . . 326
5.6.2 Optimization problems for plates . . . 327
5.7 Optimization of structures of composite materials . . . 333
5.7.1 Concept of a composite material . . . 333
5.7.2 Homogenization (averaging) of a periodical structure based on G-convergence . . . 334
5.7.3 Effective elasticity characteristics of granule and fiber reinforced composites . . . 343
5.7.4 Optimization of the effective elasticity constants of a composite . . . 348

5.7.5 Optimization of a granule reinforced composite 354
5.7.6 Optimization of composite laminate shells 357
5.7.7 Optimization of the composite structure 367
5.8 Optimization of laminate composite covers according to mechanical and radio engineering characteristics 373
5.8.1 Propagation of electromagnetic waves through a laminated medium 373
5.8.2 Optimization problems 380
5.9 Shape optimization of a two-dimensional elastic body 383
5.9.1 Sets of controls and domains in the optimization problem 383
5.9.2 Problems of elasticity in domains 384
5.9.3 The optimization problem 386
5.10 Optimization of the internal boundary of a two-dimensional elastic body 388
5.11 Optimization problems on manifolds and shape optimization of elastic solids 391
5.11.1 Optimization problem for an elastic solid 392
5.11.2 Spaces and operators on $\mathbb{R}/2\pi\mathbb{Z}$, auxiliary statements 398
5.11.3 Optimization problem on $\mathbb{R}/2\pi\mathbb{Z}$ 405
5.12 Optimization of the residual stresses in an elastoplastic body 409
5.12.1 Force and thermal loading of a nonlinear elastoplastic body 410
5.12.2 Residual stresses and deformations 421
5.12.3 Temperature pattern in a medium 424
5.12.4 Optimization problem 426

6 Optimization Problems for Steady Flows of Viscous and Nonlinear Viscous Fluids
6.1 Problem of steady flow of a nonlinear viscous fluid 431
6.1.1 Basic equations and assumptions 431
6.1.2 Formulation of the problem 434
6.1.3 Existence theorem 439
6.2 Theorem on continuity 443
6.3 Continuity with respect to the shape of the domain 446
6.3.1 Formulation of the problem 446
6.3.2 Lemmas on operators $\widetilde{L}_q$ and $\widetilde{B}_q$ 448
6.3.3 Theorem on continuity 451
6.4 Control of fluid flows by perforated walls and computation of the function of filtration 454

6.4.1 The problem of flow in a circular cylinder and the function of filtration 455
6.4.2 The passage factor for the power model 458
6.4.3 Control of the surface forces at the inlet by the perforated wall 459
6.5 The flow in a canal with a perforated wall placed inside 460
6.5.1 Basic equations 460
6.5.2 Generalized solution of the problem 462
6.6 Optimization by the functions of surface forces and filtration 463
6.6.1 Formulation of the problem and the existence theorem 463
6.6.2 On the differentiability of the function $T \to (v(T), p(T))$ 465
6.6.3 Differentiability of the functionals Φ_i and necessary optimality conditions 470
6.7 Problems of the optimal shape of a canal 471
6.7.1 Set of controls and diffeomorphisms 472
6.7.2 Optimization problems 474
6.8 A problem of the optimal shape of a hydrofoil 478
6.8.1 State equation for a moving hydrofoil 478
6.8.2 Fixed-domain problem and Fréchet differentiability of the functionals 485
6.8.3 Optimization problem 493
6.9 Direct and optimization problems with consideration for the inertia forces 495
6.9.1 Setting and solution of the direct problem 495
6.9.2 Approximation of the problem (6.9.10)–(6.9.12) 500
6.9.3 Some remarks on models, optimization problems, and existence results 501

Bibliography 503

Index 519

Preface

This book is intended to be both a thorough introduction to contemporary research in optimization theory for elliptic systems with its numerous applications and a textbook at the undergraduate and graduate level for courses in pure or applied mathematics or in continuum mechanics.

Various processes of modern technology and production are described by elliptic partial differential equations. Optimization of these processes reduces to optimization problems for elliptic systems. The numerical solution of such problems is associated with the solution of the following questions.

1. The setting of the optimization problem ensuring the existence of a solution on a set of admissible controls, which is a subset of some infinite-dimensional vector space.
2. Reduction of the infinite-dimensional optimization problem to a sequence of finite-dimensional problems such that the solutions of the finite-dimensional problems converge, in a sense, to the solution of the infinite-dimensional problem.
3. Numerical solution of the finite-dimensional problems.

The book is devoted to these questions. Attention is focused on the setting of a problem, on the proof of an existence theorem, and on methods of approximate solution of optimization problems, the existence theorems being the base for construction of approximate solutions. General systems of elliptic equations are considered. For such systems, we study optimization problems in which the coefficients of equations, the shapes of domains, and the right-hand sides of equations are considered to be controls. General results are applied to various optimization problems of mechanics of deformable bodies, plates, shells, composite materials, and structures made of them, as well as to optimization problems of mechanics of viscous and nonlinear viscous fluids.

In order to read the book it is necessary to know some basic topics of topology, function theory, and functional analysis. Nevertheless, all the necessary mathematical topics are set forth in Chapter 1.

I hope the book will be useful not only for students and scientists in mathematics and mechanics, but also for researching engineers and various application-minded readers. It will be possible for them to see correct formulations of direct and optimization problems, methods of approximation, and possibly they will be convinced that modern mathematical analysis is an indispensable tool for the setting, genuine understanding, and solution of the direct and optimization problems arising in applications. Mathematically oriented readers will see many open mathematical problems generated by technology and production.

Chapters 1 and 4 of this book were the basis of my lectures to undergraduate students of Kiev University.

I must acknowledge the contributions of the many people who aided, either directly or indirectly, in the conception and writing of the manuscript. The seminars of Professors Yu. M. Berezansky, S. D. Eidelman, and M. L. Gorbachuk attached to the Institute of Mathematics of National Academy of Sciences of Ukraine have exerted an essential influence on the creation of this book. The attention and encouragement of the leaders and participants of the seminars stimulated my activity.

Some parts of the book entered into my lectures to graduate students of the University of Jyväskylä. I very sincerely thank Professor P. Neittaanmäki for the invitation and hospitality.

I had the pleasure of delivering lectures on topics of the book in the Voronezh Winter Mathematical Schools, whose organizer, life, and soul was Professor S. G. Krein, and in the Crimean Fall Mathematical Schools, organized by Professor N. D. Kopachevsky. Many famous mathematicians participated in these schools, and I had a good chance to discuss various problems with them. Discussions with Professors S. G. Krein and P. E. Sobolevsky were especially helpful to me.

Some of the problems considered in the book have appeared as a result of discussions with Professors N. V. Banichuk and K. A. Lurie.

Professors N. S. Bakhvalov, A. V. Fursikov, G. M. Kobelkov, V. B. Kolmanovsky, R. Mennicken, Z. Mróz, J. Nečas, E. Schnack, I. V. Skrypnik, D. Tiba, V. M. Tikhomirov offered me opportunities for delivering lectures, which gave me a clearer view on the subject, and I could test various forms of the presentation of the material of the book.

I am gratefully indebted to Professors Yu. Deynekin, Ya. Rojtberg, A. Seyranian, and Z. Sheftel for many helpful discussions.

During many years I had the good fortune of working in the Institute of Mechanics of National Academy of Sciences of Ukraine with my colleagues Professors I. Babich, A. Panteleev, Yu. Rubezhanky, N. Semeniuk, Doctors M. Belonosov, N. Ivanova, T. Zarutskaia. Each of them facilitated the creation of the book.

Since May 1997, I have lived in Germany where I have had plenty of useful discussions with Professors K.-H. Hoffmann, R. Hoppe, B. Kröplin, E. Meister, and W. Wendland. Moreover, these professors, especially Prof. B. Kröplin, have done many good deeds for me.

I am very thankful to the editor of *Operator Theory: Advances and Applications*, Prof. I. Gohberg, for his amiable consent to publish the book in the series.

I am also indebted to Dr. A. Butyrin and my son, Dr. E. Lytvynov, for the translation of the larger part of the manuscript from Russian into English, for significant improvements of my English in the other part, and for capturing the TEX files for formatting the book. I very sincerely thank Dr. N. Botkin, Dr. G. Shchepan'uk, and Eng. T. Maier for the help in the preparation of the book for publication.

Stuttgart, April 1999 William G. Litvinov

Introduction

"This, O my Best Beloved, is a story — a new and a wonderful story — a story quite different from the other stories — a story about the Most Wise Sovereign Suleiman-bin-Daoud-Solomon the Son of David.

"There are three hundred and fifty-five stories about Suleiman-bin-Daoud; but this is not one of them."

— *R. Kipling*
"The Butterfly that Stamped"

Optimization problems for elliptic equations are met with in various applications. A great number of such problems arise in continuum mechanics, in the optimization of structures, processes, and so forth.

The book is devoted to the optimization problems for elliptic equations and systems of equations. In the majority of cases, we apply the generalized (weak) solution of a boundary value problem, which is defined as an element u such that

$$u \in V, \quad a(u,v) = (f,v), \qquad \text{for all } v \in V. \tag{1}$$

Here, V is a Hilbert space, a is a bilinear form generated by an elliptic equation or by a system of elliptic equations, and f is the right-hand side, which is considered to be an element of the dual V^* of the space V. In problems of mechanics of deformable bodies, the bilinear form a is generated by the strain energy, and the right-hand side f by the applied load. From the physical point of view, the equation (1) is the principle of virtual work.

Analogously, a generalized solution of an eigenvalue problem is defined as a set of pairs $(u_i, \lambda_i) \subset V \times \mathbb{R}$ such that

$$a(u_i, v) = \lambda_i b(u_i, v), \qquad \text{for all } v \in V. \tag{2}$$

Here, b is a bilinear form on $V \times V$, $\mathbb{R}$ is the set of real numbers, b is generated by the kinetic energy of the deformable body, the equation (2) is obtained from the principle of virtual work in which the inertial forces are considered to be applied loads, λ_i being connected with the frequencies of free vibrations.

The bilinear forms a and b depend on a control, e.g., in the case of a shell, the functions of thickness and curvatures of the midsurface may be considered as a control, the bilinear forms a, b depending on these functions.

In order to cover a wide range of optimization problems, we consider the following abstract bilinear form a given on the space $W = \prod_{i=1}^{\nu} W_2^{l_i}(\Omega)$, where $W_2^{l_i}(\Omega)$ is the Sobolev space, $l_i \geq 1$, $V \subset W$, and

$$a(u, v) = \int_\Omega \sum_{i,j=1}^{k} a_{ij}(P_i u)(P_j v)\, dx.$$

Here, P_i's are linear continuous mappings of W into $L_2(\Omega)$, and the variable coefficients a_{ij} depend on a control. In some cases, the operators P_i may also depend on a control. In problems of mechanics of deformable bodies, $P_i u$'s are the components of the strains generated by a function of displacement u. The goal and the restriction functionals, the sets of controls and admissible controls are introduced. The goal and restriction functionals are defined on the set of the pairs (h, u_h), where h is a control and u_h a solution of the problem (1) or (2) for the control h. Usually, the set of controls is supposed to be a convex set in a Sobolev space or in the product of Sobolev spaces, and these spaces are chosen so that the goal and restriction functionals are continuous with respect to the topology generated by the weak one of the corresponding Sobolev space (or of the product of Sobolev spaces). The set of admissible controls is given by putting restrictions on the values of the restriction functionals and on some characteristics of h and u_h.

The optimization problem consists in finding an element that minimizes or maximizes the goal functional on the set of admissible controls. For problems under consideration, we prove existence theorems and study some methods of the construction of approximate solutions. We also consider the domain shape optimization problems in which a control defines the shape of the domain. Here, we apply smooth diffeomorphisms of domains defined by the set of controls onto a fixed domain, and after the replacement of variables corresponding to the diffeomorphisms, the domain shape optimization problem reduces to a problem of control by coefficients in a fixed domain. We also study optimization problems in which the right-hand sides of equations are considered to be controls. Here, existence theorems are proved, necessary optimality conditions are established, and methods of approximation are studied.

General results are applied to exploring and solving various optimization problems of mechanics of deformable bodies and mechanics of viscous and nonlinear viscous fluids, including various problems of technology and production.

A typical peculiarity of almost all problems considered in the book is that the set of admissible controls is nonconvex. Moreover, the goal functional is also nonconvex in some problems. These circumstances create considerable difficulties for the numerical solution of optimization problems. Besides, the direct problems of deformable bodies and viscous and nonlinear viscous fluids reduce to complicated systems of elliptic equations. This is why only few numerical solutions of nonconvex optimization problems are presented in the book, and the solutions found are locally optimal or the best of some locally optimal solutions.

It should be noted that, when the coefficients of equations or the shape of a domain are considered to be a control, we apply, as a rule, controls of Sobolev spaces, which possess some smoothness. In many cases, smooth controls comply with the physical and engineering points of view. For example, the models of plates and shells are valid if the derivatives of the function of thickness are not great and the functions of curvatures of a shell are smooth. Moreover, in case of a nonsmooth function of thickness, the concentration of stresses, leading to the failure of a structure, arises, and this concentration is not taken into account in the models of plates and shells. Similar situations take place in optimization of various processes of technology and production. Discontinuous controls were calculated for the control of flight of rockets. However, the engineers rejected these, as such controls might lead to dangerous effects, which were not taken into account in the mathematical models. "The control as well as the management have to be gentle," as the saying goes.

We note also that, in the majority of the general optimization problems studied in the book, the boundary value problems are considered in bounded domains in $\mathbb{R}^2$. Nevertheless, almost all the results obtained may be extended to bounded domains in $\mathbb{R}^n$ with $n > 2$.

The book consists of six chapters. In Chapter 1, we set forth some topics from functional analysis needed to read the book, and establish some results which are used in the following chapters for the investigation and solution of the optimization problems. We introduce the concepts of topological, metric, normed, and Hilbert spaces, linear and k-linear continuous mappings. Spectral problems associated with bilinear continuous forms in a Hilbert space are considered. The Sobolev spaces are defined, and the embedding theorem, the theorem on the coerciveness of a system of operators, the implicit function theorem, and others are adduced. The problem of the differentiability of the eigenvalues of (2) with respect to the control that determines the forms a and b is studied.

Chapter 2 is devoted to the optimization by coefficients of elliptic systems. We consider optimization problems in which the state of a system is defined by one function or by a set of functions. The latter is an abstract analog of the problem of optimization of a structure subject to a class of loads. Various problems of eigenvalue and domain shape optimization are studied.

In Chapter 3, we study optimization problems in which the right-hand sides of equations are controls. Existence theorems and the necessary optimality conditions are established, and the convergence of approximate solutions is studied. We also consider the minimax optimization problems and problems of optimization of systems described by variational inequalities.

In Chapter 4, we study direct problems for various models of plates and shells, problems of stress-strain state, of stability, and of free vibrations. The bilinear forms associated with the potential (strain) and kinetic energies of plates and shells are defined and studied. Existence results for the problems under consideration are established.

The reader interested in the models and mathematical theory of plates and shells can read this chapter after looking through Sections 1.1–1.9.

Chapter 5 is devoted to the optimization of structures. We consider various settings of optimization problems and numerical methods for solving direct and optimization problems. Optimization problems for plates and shells under restrictions on strength, stiffness, stability, and frequencies of free vibrations are studied. We also consider various problems of optimization of composite materials and structures made of them.

In Chapter 6, we formulate and study various optimization problems for steady flows of viscous and nonlinear viscous fluids. In particular, we consider the optimization by body and surface forces, by the distribution of velocities on the inlet, by perforated walls, and by the shape of domains. We also study some engineering problems, such as problems of the optimal shape of a canal and of the optimal shape of the hydrofoil and the problem of optimization of the header of a paper machine.

Chapter 1

Basic Definitions and Auxiliary Statements

"All this recitative by the chorus is only to bring us to the point where you may be told why Dry Valley Johnson shook up the insoluble sulphur in the bottle"

– *O. Henry*
"The Indian Summer of Dry Valley Johnson"

1.1 Sets, functions, real numbers

1.1.1 Notations and definitions

We suppose the reader to be familiar with notions connected with sets and functions; nevertheless, we state the terminology and introduce notations that will be used throughout the book.

Let X be a set; the writing $x \in X$ means that x is an element of the set X; if the writing $x \in X$ follows some equality (or inequality) and no additional conditions are imposed on x, then it should be read that the equality (inequality) holds for an arbitrary x from X. The sign $\forall$ means "for all." In the case when it is clear which set X we bear in mind, we simply write $\forall x$ instead of $\forall x \in X$. The writing $x \notin X$ means that x does not belong to X. The writing $\{x \mid P(x)\}$ denotes the set of all the x satisfying the condition (or conditions) $P(x)$. If A is a subset of a set X, which is denoted by $A \subset X$, then $X \setminus A$ is the complement of the subset A to the whole set X, i.e.,

$$X \setminus A = \{\, x \mid x \in X, \ x \notin A \,\}.$$

The union of sets A, B is denoted by $A \cup B$, and the intersection of A, B by $A \cap B$.

The product $E \times F$ of two sets E and F is the set of all ordered pairs (x, y) such that x is an element of the set E and y an element of the set F. The word "ordered" means that, if E and F coincide, then the pairs (x, y) and (y, x) with $x \neq y$ are considered as different ones.

In the same way, one defines the product of several sets, or of an arbitrary collection of sets.

Let sets E and F be given. A map (mapping) of E into F, or a function defined on E and taking values in F, or an operator acting from E into F, is a correspondence f that assigns to every element x from E some element from F, which is denoted by $f(x)$ (or fx) and is called the image of x under the mapping f. In particular, if F is the set of real numbers, then f is called a functional. The notation $f\colon E \to F$ means that f is a mapping of E into F.

A mapping f is said to be injective, or an injection, if the images of two different elements of E under the mapping f are different elements of F. A mapping f is called surjective, or a surjection, or a mapping "onto" if every element of F is the image under the mapping f of some element of E.

A mapping $f\colon E \to F$ is called bijective, or a bijection, or a one-to-one mapping, if every element of F is the image of a unique element of E. A mapping is one-to-one if and only if it is both injective and surjective.

Let $f\colon E \to F$ be a one-to-one mapping and let y belong to F. Denote by $f^{-1}(y)$ the unique element x of E such that $f(x) = y$. So, the mapping f^{-1} from F into E is well-defined. It is also a one-to-one mapping. This mapping is called the inverse of f.

1.1.2 Real numbers

The set of real numbers is denoted by $\mathbb{R}$, the set of integers by $\mathbb{Z}$, and the set of natural numbers by $\mathbb{N}$. The set of real x satisfying $a < x < b$ (a, b are real numbers, or it may be that $a = -\infty$, $b = \infty$) is called an open interval and is denoted by (a, b). The set of those x which satisfy the inequalities $a \leq x \leq b$ is called a segment, or a closed interval, and is denoted by $[a, b]$. We will use also the following notations: $[a, b) = \{x \mid a \leq x < b\}$, $(a, b] = \{x \mid a < x \leq b\}$, $\mathbb{R}_+ = [0, \infty)$. Sets $[a, b)$, $(a, b]$, where $a, b < \infty$, are called half-intervals.

Let A be a subset of $\mathbb{R}$. The set A is called bounded above (below) if there exists a number M (m) that is not less (greater) than all the numbers x from A, M called a majorant of the set A, and m a minorant.

A set A is called bounded if it is both bounded above and below. A set in $\mathbb{R}$ is said to have maximum if there exists a majorant that belongs to this set. A set need not have maximum, but if it does have, then this maximum is unique. For if a and b are two maximums of the same set, then we have both $a \leq b$ and $b \leq a$, so $a = b$. The maximum of a set A (if it exists) is denoted by $\max\limits_{x \in A} x$.

An analogous definition is stated for the minimum, which is denoted by min.

A set A in $\mathbb{R}$ is said to have supremum if the set of its majorants has minimum, and this minimum is called the supremum of the set A. So, the supremum is the least of the majorants.

A set need not have supremum, but if it has, then this supremum is unique. The supremum of a set A (if it exists) is denoted by $\sup\limits_{x\in A} x$.

If the supremum belongs to A, then it is the maximum, and vise versa. An analogous definition is stated for the infimum, which is denoted by inf. If the infimum belongs to A, then it is the minimum.

The following theorem holds, see, e.g., Bourbaki (1960).

Theorem 1.1.1 *An arbitrary, nonempty, majorized set in $\mathbb{R}$ has supremum, and an arbitrary, nonempty, minorized set has infimum. The supremum b of a majorized set A is characterized by the following statements:*

i) *$x \leq b$ for any $x \in A$;*

ii) *for an arbitrary b_1 such that $b_1 < b$, there exists at least one number $x \in A$ such that $b_1 < x \leq b$.*

The infimum c of a minorized set A is characterized by the following statements:

i) *$x \geq c$ for any $x \in A$;*

ii) *for an arbitrary c_1 such that $c_1 > c$, there exists at least one number $x \in A$ such that $c_1 > x \geq c$.*

For example,

$$\sup_{x\in(-\infty,0)} x = 0, \qquad \inf_{x\in(0,\infty)} x = 0, \qquad \sup_{x\in(-\infty,0]} x = \max_{x\in(-\infty,0]} x = 0,$$
$$\inf_{x\in[0,\infty)} x = \min_{x\in[0,\infty)} x = 0.$$

The set that is the union of the real numbers $\mathbb{R}$ and the set consisting of the two elements $-\infty$, $+\infty$ is called the completion of the set of real numbers $\mathbb{R}$, and is denoted by $\overline{\mathbb{R}}$. This set has minimum $-\infty$ and maximum $+\infty$. Every set in $\overline{\mathbb{R}}$ is bounded, and Theorem 1.1.1 holds true in $\overline{\mathbb{R}}$ without the assumption that A is majorized (or minorized), since this is always the case.

Let us introduce the notions of upper and lower limits. Let $\{x_n\}$ be a sequence of elements from $\mathbb{R}$. Denote by $\lim\mathrm{pt}\{x_n\}$ the collection of the limit points of the sequence $\{x_n\}$. Then, the upper limit L of the sequence $\{x_n\}$ is defined by

$$L = \sup\left\{\, x \mid x \in \lim\mathrm{pt}\{x_n\} \,\right\}.$$

It is denoted by $L = \limsup\{x_n\}$. Analogously, we can define a lower limit l:

$$l = \liminf\{x_n\} = \inf\left\{\, x \mid x \in \lim\mathrm{pt}\{x_n\} \,\right\}.$$

The upper limit is characterized by the following property. For arbitrary L_1 and L_2 such that $L_1 < L < L_2$, all x_n but a finite number are less than or equal to L_2, and there is an infinite set of n such that $x_n \geq L_1$.

We stress that every bounded sequence in $\mathbb{R}$ has supremum and infimum. A sequence in $\mathbb{R}$ converges if and only if the upper and lower limits of it are equal.

1.2 Topological, metric, and normed spaces

1.2.1 General notions

Topological spaces

A topological space X is a set X in which there has been chosen a collection τ of subsets, called open sets, that have the following properties:

a) the space X itself and the empty set $\varnothing$ are open;

b) the intersection of a finite number of open sets is open;

c) the union of either a finite or infinite number of open sets is open;

d) for arbitrary different points a and b of the space X, there exist two open disjoint sets from X that contain a and b, respectively.

Notice that open sets are usually supposed to satisfy only the conditions a)–c). If, additionally, the condition d) is satisfied, then the topological space is called a Hausdorff space.

The collection τ is called a topology in X. One set may be supplied with different topologies, so that we will indicate the topology with respect to which some property holds.

The collection of all topologies on the set X is naturally ordered: $\tau_1 \prec \tau_2$ if $\tau_1 \subset \tau_2$ (in the sense of the usual inclusion). If $\tau_1 \prec \tau_2$, then the topology τ_1 is said to be weaker than the topology τ_2. (The word "weaker" is used because more sequences converge in the topology τ_1 than in the topology τ_2, so that the τ_1-convergence is a weaker concept than the τ_2-convergence.)

Let A be a set in a topological space X. The set A is called closed if its complement is open. The closure $\overline{A}$ of the set A is the intersection of all the closed sets containing A.

The interior of the set A is the union of all the open sets contained in A. The interior of the set A is denoted by $\overset{\circ}{A}$; $\overset{\circ}{A}$ is the biggest open set contained in A; $\overset{\circ}{A}$ may be empty. The set $\overline{A} \setminus \overset{\circ}{A}$ is called the boundary of A. A neighborhood of a point a in the set X is an arbitrary subset of X containing an open set the point a belongs to.

A collection S of subsets of a topological space X is called a base of neighborhoods of a point a if each $N \in S$ is a neighborhood of a and, for an arbitrary neighborhood M of a, there exists $N \in S$ such that $N \subset M$.

A set M in a topological space X is called dense in X if its closure coincides with X, i.e., $\overline{M} = X$. A topological space is called separable if it has a countable dense set.

A sequence $\{x_n\}$ of points of a topological space X converges in X to some $x \in X$ if, for an arbitrary neighborhood U of x, there exists a natural number $N(U)$ such that all x_n, with $n > N(U)$, belong to U. If a sequence $\{x_n\}$ converges to x, then x is called its limit, and one writes $x_n \to x$ in X.

Induced topology. Product of topologies

Let E be a subset of a topological space X. On the set E, one can introduce a topology by taking as open sets the intersections of the open sets in X with E. Then, E is called the topological subspace of X, or the topological space with the topology induced (generated) by the topology of X.

Let X_1, X_2 be two topological spaces. On the product of these spaces, one can introduce a topology in the following way. A set E in $X_1 \times X_2$ is called open if, for each point $(x_1, x_2) \in E$, there are open sets A_1, A_2 in X_1, X_2, respectively, such that $x_1 \in A_1$, $x_2 \in A_2$, and $A_1 \times A_2 \subset E$. It can be verified that this definition is correct, i.e., the open sets defined satisfy all the axioms of a topological space. As easily seen, the product of two arbitrary open sets is again an open set, though there are also other open sets. Such a topology on $X_1 \times X_2$ is called the product of the topologies of X_1 and X_2, or the topological product of X_1 and X_2.

In the same way one defines the topological product of more than two spaces. For example, if $X_1 = X_2 = \cdots = X_n = \mathbb{R}$, then their product is just the space $\mathbb{R}^n$ equipped with the natural metric.

1.2.2 Metric spaces

A set E is called a metric space if the distance between its elements is defined, i.e., there is a mapping of $E \times E$ into the half-line $\mathbb{R}_+$ that places in correspondence to an arbitrary pair $(x, y) \in E \times E$ a number $d(x, y)$, which is called the distance between x and y.

The distance must possess the following three properties:

1. symmetry: $d(x, y) = d(y, x)$;
2. positivity: $d(x, y) > 0$ if $x \neq y$, and $d(x, x) = 0$;
3. the triangle inequality: $d(x, z) \leq d(x, y) + d(y, z)$.

In particular, the real line $\mathbb{R}$ with the distance $d(x, y) = |x - y|$ is a metric space. This metric is called the natural metric in $\mathbb{R}$. Everywhere below, the metric in $\mathbb{R}$ is supposed to be natural.

Let E be a metric space. Given an arbitrary point a from E, the set

$$\{\, x \mid x \in E,\ d(a, x) < r \,\}$$

is called the open ball centered at a with radius r, and the set

$$\{\, x \mid x \in E,\ d(a, x) \leq r \,\}$$

the closed ball centered at a with radius r, r being finite and nonzero. A set A in the metric space E is called open if, together with its every point, it contains a ball (open or closed) centered at this point.

Clearly, open sets in the space E possess the properties a), b), c), and d) stated above. Therefore, a metric space is a specific case of a topological space, and its topology is determined by the metric.

A topological space is said to be metrizable if there exists a metric generating its topology.

A sequence of points $\{x_n\}$ of a metric space E converges to a point $x \in E$ in the sense of the topology of this space if and only if $\lim_{n\to\infty} d(x_n, x) = 0$. Notice that, because of the triangle inequality, $d(x_m, x_n) \to 0$ as $m, n \to \infty$.

A sequence possessing this property is called a Cauchy sequence (or a fundamental sequence).

A metric space E is called complete if every fundamental sequence in it converges to an element of E.

1.2.3 Normed vector spaces

Basic definitions

A set E is called a vector space over the field of real numbers $\mathbb{R}$ if given are two operations on it, named multiplication by a scalar and addition. These operations are supposed to possess the following algebraic properties:

(i) to every pair of vectors (elements) $x, y \in E$, there corresponds a vector $x+y \in E$, and the equalities

$$x + y = y + x, \qquad x + (y + z) = (x + y) + z \tag{1.2.1}$$

hold; E contains a unique vector 0 (the zero vector) such that $x + 0 = x$ for any x from E; for every $x \in E$, there exists a unique vector $-x$ such that $x + (-x) = 0$;

(ii) to every pair (t, x), where $t \in \mathbb{R}$ and $x \in E$, there corresponds a vector $tx \in E$, and the equalities

$$1x = x, \qquad s(tx) = (st)x \qquad s \in \mathbb{R},$$
$$s(x + y) = sx + sy, \qquad (s + t)x = sx + tx.$$

hold.

A segment with ends x and y in the vector space E is the set of the elements of the form $tx + (1 - t)y$, $0 \leq t \leq 1$. It is denoted by $[x, y]$. A set $V \subset E$ is called convex if, for every pair (x, y) of its elements, the segment $[x, y]$ belongs to V.

A cone $K \subset E$ is defined to be a set containing, together with its arbitrary element x, the elements of the form tx, where $t > 0$. Note that 0 – the zero element of E – may either belong or not belong to K. Provided $0 \in K$, K is called a cone with vertex at 0. A cone that is a convex set is called a convex cone. Note that a cone need not be "sharpened." For example, an arbitrary subspace of E is a cone.

Here are two important examples of convex cones in the case $E = \mathbb{R}^n$.

The nonnegative orthant:

$$\{\, x = (x_1, \dots, x_n) \mid x_1 \geq 0, \dots, x_n \geq 0 \,\}.$$

The positive orthant:

$$\{ x = (x_1, \dots, x_n) \mid x_1 > 0, \dots, x_n > 0 \}.$$

Notice that the first example is a cone with vertex at 0.

Let f be a mapping of a convex set $V \subset E$ into $\mathbb{R}$. This mapping is called convex if, for arbitrary $x, y \in V$ and an arbitrary $\lambda \in [0,1]$, the following inequality holds:

$$f(\lambda x + (1-\lambda)y) \leq \lambda f(x) + (1-\lambda) f(y).$$

A norm in the vector space E is defined to be a function $x \to \|x\|$ that assigns a number $\|x\|$ to a vector x and possesses the following properties:

a) $\|x\| > 0$ if $x \neq 0$, and $\|0\| = 0$;
b) $\|x + y\| \leq \|x\| + \|y\|$ for any $x, y \in E$;
c) $\|tx\| = |t|\|x\|$ for any $x \in E$ and $t \in \mathbb{R}$,

In a normed vector space, one can define a distance between elements satisfying the properties 1), 2), 3) above by the formula $d(x,y) = \|x - y\|$. Therefore, an arbitrary normed vector space is a metric space.

In a normed vector space, one can introduce the notion of completeness (see Subsec. 1.2.2).

A complete normed vector space is called a Banach space.

Examples of Banach spaces are ℓ_∞ and $\ell_{\infty,0}$, which will be used in the sequel. ℓ_∞ is the space of bounded number sequences. The norm of an element $\xi = (\xi_1, \xi_2, \dots, \xi_n, \dots) \in \ell_\infty$ is defined by

$$\|\xi\|_{l_\infty} = \sup_i |\xi_i|,$$

$\ell_{\infty,0}$ is the subspace of ℓ_∞ consisting of the sequences converging to zero. The norm in $\ell_{\infty,0}$ is defined just as in ℓ_∞.

Two norms $\|\cdot\|_1$ and $\|\cdot\|_2$ defined in a vector space E are called equivalent if there exist constants m_1 and m_2 such that

$$\|x\|_1 \leq m_1 \|x\|_2, \qquad \|x\|_2 \leq m_2 \|x\|_1, \qquad x \in E.$$

The metrics $d_1(x,y) = \|x - y\|_1$ and $d_2(x,y) = \|x - y\|_2$ corresponding to equivalent norms define in E identical systems of open sets, i.e., they define identical topologies.

Theorem 1.2.1 *In a finite-dimensional vector space, arbitrary two-norms are equivalent. There exists one system of open sets for any norm introduced in this space.*

For the proof, see, e.g., Schwartz (1967).

For example, consider the finite-dimensional space $\mathbb{R}^n$. A norm on $\mathbb{R}^n$ can be defined in different ways. For example, if $x = (x_1, x_2, \dots, x_n)$, then

$$\|x\| = \left(\sum_{i=1}^n x_i^2 \right)^{\frac{1}{2}}.$$

This norm is called Euclidean. It is possible also to introduce a norm by the formulas $\|x\| = \max_i |x_i|$ or $\|x\| = \sum_{i=1}^n |x_i|$. All these norms are equivalent.

Let X and Y be vector normed spaces, and let A be a linear mapping of X into Y, i.e.,

$$A(\alpha x + \beta y) = \alpha Ax + \beta Ay, \qquad x, y \in X, \ \alpha, \beta \in \mathbb{R}.$$

The mapping A is called bounded if there exists a constant c such that

$$\|Ax\|_Y \leq c\|x\|_X, \qquad x \in X.$$

Denote by $\mathcal{L}(X, Y)$ the set of all bounded linear mappings of X into Y; in the case when $X = Y$, we use the notation $\mathcal{L}(X, Y) = \mathcal{L}(X)$. On the set $\mathcal{L}(X, Y)$, one can naturally define addition of elements and multiplication by a scalar. So, $\mathcal{L}(X, Y)$ is a vector space. In this space, we define a norm as follows:

$$\|A\| = \sup_{x \neq 0} \frac{\|Ax\|_Y}{\|x\|_X} = \sup_{\|x\|_X = 1} \|Ax\|_Y$$

for an arbitrary element $A \in \mathcal{L}(X, Y)$.

In the case $Y = \mathbb{R}$, the space $\mathcal{L}(X, \mathbb{R})$ is denoted by X^* and is called the dual of X. Elements of X^* are called bounded linear functionals in X, or bounded linear forms in X.

The value of a linear functional f from X^* at an element $x \in X$ is usually denoted by (f, x); the function $(\cdot, \cdot)$, defined on $X^* \times X$, being called the scalar product of X^* and X.

Theorem 1.2.2 *If E, F are normed vector spaces and F is a Banach space, then the space $\mathcal{L}(E, F)$ is also a Banach space. In particular, the dual E^* of the space E is a Banach space.*

For the proof, see, e.g., Schwartz (1967).

Let X be a Banach space, and let X^* be its dual, which is also a Banach space. The space X^*, in turn, has its dual, which is denoted by X^{**} and is called the second dual of X. For an arbitrary element $x \in X$, there exists a unique element $x^{**} \in X^{**}$ possessing the property

$$(x^{**}, x^*) = (x^*, x), \qquad x^* \in X^*.$$

Moreover, $\|x\|_X = \|x^{**}\|_{X^{**}}$. So, the space X may be considered as a subspace of the Banach space X^{**}.

A Banach space X is called reflexive if $X = X^{**}$.

Weak topology and weak convergence in a Banach space

In addition to the topology generated by the norm and called the strong topology, one introduces in a Banach space a weak topology.

The weak topology in a Banach space is the weakest topology on X with respect to which every functional $f \in X^*$ is continuous.

For the weak topology, a base of neighborhoods of zero is given by the following sets

$$N(f_1, f_2, \dots, f_n; \varepsilon) = \{ x \mid x \in X, \ |f_i(x)| < \varepsilon, \ i = 1, 2, \dots, n \}.$$

Here $f_1, f_2, \dots, f_n$ is a finite collection of arbitrary elements of X^*, ε is a positive number.

The convergence with respect to the strong topology in a Banach space X is called strong convergence, or the convergence in norm. Everywhere below, if we deal with a normed space and do not indicate which topology is under consideration, then the strong topology is understood.

The convergence with respect to the weak topology is called weak convergence; a sequence $\{x_n\}$ of elements of X converges weakly to $x_0 \in X$ if and only if, for an arbitrary $f \in X^*$, the number sequence $\{f(x_n)\}$ converges to $f(x_0)$.

The following theorem holds (see, e.g., Kantorovich and Akilov (1977)).

Theorem 1.2.3 *In a normed vector space X, the weak topology coincides with the strong topology if and only if the space X is finite-dimensional.*

We stress that, in an infinite-dimensional Banach space, the weak topology is not generated by any metric. Because of convexity of a norm, we have the following theorem, see Vainberg (1972), Gajewski et al. (1974).

Theorem 1.2.4 *If a sequence $\{x_n\}$ in a Banach space X weakly converges to x, then*

$$\|x\| \le \liminf_{n \to \infty} \|x_n\|.$$

A sequence $\{x_n\}$ is called weakly fundamental if, for any functional $f \in X^*$, there exists a finite limit $\lim_{n\to\infty} f(x_n)$. The space X is called sequentially weakly complete if every weakly fundamental sequence weakly converges to an element of X.

Theorem 1.2.5 *An arbitrary reflexive Banach space is sequentially weakly complete.*

For the proof, see, e.g., Yosida (1971).

In a normed vector space X, one can introduce the notions of a weakly closed set and a sequentially weakly closed set. A set $A \subset X$ is called weakly closed if it is closed in the weak topology of X. A set A is called sequentially weakly closed if the conditions $\{x_n\} \subset A$ and $x_n \to x$ weakly in X yield $x \in A$. A weakly closed set is sequentially weakly closed. However, in general, the inverse statement is not valid.

Note that an arbitrary closed (in the sense of the strong topology) convex set in a Banach space is sequentially weakly closed.

1.3 Continuous functions and compact spaces

1.3.1 Continuous and semicontinuous mappings

The continuity

Let E, F be topological spaces, let f be a mapping of E into F, and let A be a subset of E. By $f(A)$ we denote the subset of F consisting of all the elements $f(x)$, $x \in A$; $f(A)$ is called the image of the set A under the mapping f.

Let B be a subset of F. By $f^{-1}(B)$ we denote the set in E consisting of all elements x such that $f(x) \in B$. The set $f^{-1}(B)$ is called the prototype (or the inverse image) of the set B under the mapping f.

A mapping f of a topological space E into a topological space F is continuous at a point $a \in E$ if, for any neighborhood U of the point $f(a)$, there exists a neighborhood V of the point a such that $f(V) \subset U$. A mapping of E into F is called continuous if it is continuous at every point a of the space E.

Theorem 1.3.1 *A mapping f of a topological space E into a topological space F is continuous if and only if the prototype of an arbitrary open (closed) set in F is an open (closed) set in E.*

For the proof, see, e.g., Schwartz (1967), Kolmogorov and Fomin (1975).

In particular, a linear mapping A of a normed vector space X into a normed vector space Y is continuous if and only if it is bounded.

Let F, G be normed vector spaces, and let A be a linear continuous mapping of F into G, i.e., $A \in \mathcal{L}(F,G)$. This mapping is said to be invertible if it is a bijection and the inverse bijection A^{-1} is also continuous, i.e., $A^{-1} \in \mathcal{L}(G,F)$. Then, the mapping A is called an isomorphism.

A homeomorphism of a topological space E onto a topological space F is defined to be a bijection of E onto F such that both it and its inverse are continuous mappings.

Theorem 1.3.2 *Let F, G be Banach spaces and let $\mathcal{U}$ $(\mathcal{U}^{-1})$ be the set of the invertible elements of the space $\mathcal{L}(F,G)$ $(\mathcal{L}(G,F))$. Then, the bijection $f \to f^{-1}$ of the set $\mathcal{U}$ onto the set $\mathcal{U}^{-1}$ is a homeomorphism.*

For the proof, see, e.g., Schwartz (1967).

A mapping f of a topological space E into a topological space F is called sequentially continuous at a point $a \in E$ if the image of an arbitrary sequence of points from E converging to a is a sequence of points from F converging to $f(a)$. A mapping of E into F is called sequentially continuous if it is sequentially continuous at every point of the space E.

If topological spaces E and F are metrizable, then the notion of a continuous mapping is equivalent to the notion of a sequentially continuous mapping, that is, every continuous mapping of E into F is sequentially continuous, and vice versa.

If E and F are topological nonmetrizable spaces, or at least one of them is, then every continuous mapping of E into F is sequentially continuous, but the inverse is not true.

Everywhere below, permitting ourselves some familiarity, instead of saying "f is a sequentially continuous mapping of E into F," we will say "f is a continuous mapping of E into F."

Let f be a function defined on a topological space E and taking values in $\mathbb{R}$. This function is said to be upper semicontinuous at a point $a \in E$ if, for any given $b_1 > f(a)$, there exists in the space E a neighborhood U' of the point a such that $x \in U'$ yields $f(x) \leq b_1$. Analogously, f is lower semicontinuous at a point a if, for any given $b_2 < f(a)$, there exists a neighborhood U'' in the space E of the point a such that $x \in U''$ yields $f(x) \geq b_2$.

A function f is continuous at a point if and only if it is both upper and lower semicontinuous at this point.

A function f is called upper (lower) semicontinuous in E if it is upper (lower) semicontinuous at every point from E.

If E is a metrizable topological space, then a function $f\colon E \to \mathbb{R}$ is upper semicontinuous if and only if

$$\left.\begin{array}{l}\text{for an arbitrary sequence } \{u_n\} \text{ convergent in } E \text{ to } u, \text{ the} \\ \text{inequality } \limsup\limits_{n\to\infty} f(u_n) \leq f(u) \text{ holds.}\end{array}\right\} \tag{1.3.1}$$

Similarly, for a metrizable space E, a function $f\colon E \to \mathbb{R}$ is lower semicontinuous if and only if

$$\left.\begin{array}{l}\text{for an arbitrary sequence } \{u_n\} \text{ convergent in } E \text{ to } u, \text{ the} \\ \text{inequality } \liminf\limits_{n\to\infty} f(u_n) \geq f(u) \text{ holds.}\end{array}\right\} \tag{1.3.2}$$

In the sequel, every function $f\colon E \to \mathbb{R}$ satisfying the condition (1.3.1) (or (1.3.2)) will be called upper (lower) semicontinuous even though the space E is nonmetrizable.

Bilinear and k-linear continuous mappings

Let E, F, and G be vector spaces over $\mathbb{R}$. A mapping $x, y \to u(x, y)$ from the space $E \times F$ into the space G is called bilinear if, given one fixed variable, it is linear with respect to the other variable.

If we fix $x \in E$, then u determines the partial linear mapping $y \to u(x, y)$ of the space F into G, which is denoted by $u(x, \cdot)$. Analogously $y \in F$ being fixed, the partial mapping $x \to u(x, y)$, denoted by $u(\cdot, y)$, is a linear mapping from E into G.

In the case $G = \mathbb{R}$, the mapping u is called a bilinear form on $E \times F$.

The usual product of real numbers is a bilinear mapping of $\mathbb{R} \times \mathbb{R}$ into $\mathbb{R}$.

Theorem 1.3.3 *A bilinear mapping of a product $E \times F$ of normed vector spaces into a normed vector space G is continuous if and only if there exists a constant $M \geq 0$ such that*

$$\|u(x,y)\| \leq M\|x\|\,\|y\|, \qquad x \in E,\ y \in F.$$

For the proof, see, e.g., Schwartz (1967).

The set of the bilinear continuous mappings of $E \times F$ into G is a vector space over $\mathbb{R}$ in which the norm is defined by

$$\|u\| = \sup_{x \neq 0,\ y \neq 0} \frac{\|u(x,y)\|}{\|x\|\,\|y\|} = \sup_{\|x\| \leq 1,\ \|y\| \leq 1} \|u(x,y)\|.$$

This space is denoted by $\mathcal{L}_2(E,F;G)$.

Let E be a normed vector space, and let $x, y \to u(x,y)$ be a bilinear form on $E \times E$, i.e., a bilinear mapping of $E \times E$ into $\mathbb{R}$. This form is called symmetric if

$$u(x,y) = u(y,x), \qquad x, y \in E.$$

A form U is called coercive on $E \times E$ if

$$u(x,x) \geq c\|x\|^2, \qquad x \in E,\ c = \text{const} > 0.$$

Let now $E_1, E_2, \dots, E_k$ be vector spaces over $\mathbb{R}$, and let $E = E_1 \times E_2 \times \cdots \times E_k$. A mapping $u(x_1, x_2, \dots, x_k)$ of E into G, where G is a vector space over $\mathbb{R}$, is called k-linear if it is linear with respect to every $x_i \in E_i$. If E_i's and G are normed vector spaces, then a k-linear mapping of $E_1 \times E_2 \times \cdots \times E_k$ into G is continuous if and only if it is bounded, i.e., there exists a constant $M \geq 0$ such that

$$\|u(x_1, x_2, \dots, x_k)\| \leq M\|x_1\|\,\|x_2\| \cdots \|x_k\|, \qquad x = (x_1, x_2, \dots, x_k) \in E.$$

The set of k-linear continuous mappings of $E_1 \times E_2 \times \cdots \times E_k$ into G is a normed vector space. The least constant M for which the latter inequality holds is the norm of the mapping u.

1.3.2 Compact spaces

Let E be a topological space. A collection of sets in E such that every point of E belongs to at least one of these sets is called a covering of E. A subcovering of a covering is a covering formed by some sets of the original covering. A covering is called finite if it consists of a finite number of sets in E. A covering is called open if all the sets of this covering are open in E.

For example, the set of the intervals $(n-1, n+1)$, where n runs through the set of integers, forms an open covering of $\mathbb{R}$. Note that it has no subcoverings.

A topological space is called compact if, from every open covering of this space, one can choose at least one finite subcovering. A set A in a topological space E is called compact if, in the sense of the topology induced on A by the topology of E, A is a compact space. Notice that, in the latter case, the space E itself may be not compact.

Every set in E whose closure is compact is called relatively compact.

Theorem 1.3.4 *Let E be a topological space, and let F be a compact set in E. Then, F is a closed set in E.*

Theorem 1.3.5 *The topological product of two compact spaces is compact.*

A topological space E is called sequentially compact if, from an arbitrary sequence of elements of E, one can choose a subsequence converging to an element of E. We stress that, upon the Bolzano-Weierstrass theorem, for the metrizable topological spaces, the notions of compact and sequentially compact spaces are equivalent; that is, if E is a compact, metric space, it is also sequentially compact, and conversely a sequentially compact, metric space is a compact, metric space.

If E is a nonmetrizable compact topological space, then it is sequentially compact, too. However, the inverse is not true.

Theorem 1.3.6 *A set in a finite-dimensional normed vector space is compact if and only if it is closed and bounded.*

For the proofs of Theorems 1.3.4–1.3.6, see, e.g., Schwartz (1967).

The following theorem is a very important property of a reflexive space.

Theorem 1.3.7 *Every closed ball K in a reflexive Banach space E is sequentially compact in the weak topology, that is, from every sequence of elements of K one can choose a subsequence weakly converging to an element of K.*

For the proof, see Céa (1971).

1.3.3 Continuous functions on compact spaces

Theorem 1.3.8 *Let f be a continuous mapping of a topological compact space E into a topological space F. Then, the image $f(E)$ of the space E under the mapping f is compact in F.*

For the proof, see, e.g., Schwartz (1967).

The following theorem will be of great importance for us.

Theorem 1.3.9 (Weierstrass) *A continuous mapping of a nonempty compact space into $\mathbb{R}$ reaches its maximum and minimum.*

Proof. Let F be a continuous mapping of a compact space E into $\mathbb{R}$. Then, by Theorem 1.3.8, the image $f(E)$ is compact in $\mathbb{R}$, whence $f(E)$ is a bounded set in $\mathbb{R}$.

From Theorem 1.1.1, it follows that there exists a sequence $\{x_n\} \subset E$ such that

$$\lim_{n\to\infty} f(x_n) = b = \sup_{x\in E} f(x). \tag{1.3.3}$$

As E is a compact space, by the Bolzano-Weierstrass theorem, from the sequence $\{x_n\}$ one can choose a subsequence $\{x_m\}$ converging to an $a \in E$. Since f is a continuous function, by (1.3.3) we have

$$\lim_{m\to\infty} f(x_m) = f(a) = \sup_{x\in E} f(x).$$

The proof for the minimum of a function is analogous.

Let E and F be metric spaces. A mapping f of E into F is called uniformly continuous if, for an arbitrary $\varepsilon > 0$, there exists $\eta > 0$ such that the inequality $d(x', x'') \leq \eta$, $x', x'' \in E$, yields $d(f(x'), f(x'')) \leq \varepsilon$.

Theorem 1.3.10 *Every continuous mapping of a compact metric space E into a metric space F is uniformly continuous.*

For the proof, see, e.g., Schwartz (1967).

1.4 Maximum function and its properties

1.4.1 Discrete maximum function

Let U be a topological space and let $u \to f_i(u)$ be mappings of U into $\mathbb{R}$, where $i \in I = \{1, 2, \ldots, N\}$. Define a function $\varphi\colon U \to \mathbb{R}$ by

$$u \to \varphi(u) = \max_{i\in I} f_i(u). \tag{1.4.1}$$

The function φ is called a maximum function (discrete maximum function).

Theorem 1.4.1 *If for every $i \in I$, $u \to f_i(u)$ is a continuous mapping of U into $\mathbb{R}$, then so is $u \to \varphi(u)$.*

Proof. Let $\{u_n\}$ be a sequence converging to u in U. It is clear that

$$\begin{aligned} \max_{i\in I} f_i(u_n) &= \max_{i\in I} \left[f_i(u_0) + (f_i(u_n) - f_i(u_0))\right] \\ &\leq \max_{i\in I} f_i(u_0) + \max_{i\in I} |f_i(u_n) - f_i(u_0)|. \end{aligned} \tag{1.4.2}$$

By analogy, we get

$$\max_{i\in I} f_i(u_0) \leq \max_{i\in I} f_i(u_n) + \max_{i\in I} |f_i(u_n) - f_i(u_0)|. \tag{1.4.3}$$

Then, by (1.4.2) and (1.4.3), we have

$$|\varphi(u_n) - \varphi(u_0)| = \left|\max_{i\in I} f_i(u_n) - \max_{i\in I} f_i(u_0)\right| \leq \max_{i\in I} |f_i(u_n) - f_i(u_0)|.$$

Since $u_n \to u_0$ in U and f_i's are continuous mappings of U into $\mathbb{R}$, we get

$$\lim_{n\to\infty} \varphi(u_n) = \varphi(u_0),$$

and so the theorem is proved.

Theorem 1.4.2 *Let A be a convex set in a topological space U. If the functions $u \to f_i(u)$, $i \in I = \{1, 2, \ldots, N\}$, are convex on A, then the function φ is also convex on A.*

Proof. Let $u_1, u_2 \in A$ and $\alpha \in [0, 1]$. Then

$$\begin{aligned} f_i(\alpha u_1 + (1-\alpha)u_2) &\leq \alpha f_i(u_1) + (1-\alpha) f_i(u_2) \\ &\leq \alpha\varphi(u_1) + (1-\alpha)\varphi(u_2). \end{aligned}$$

Because these inequalities hold for all $i \in I$, we have

$$\varphi(\alpha u_1 + (1-\alpha)u_2) \leq \alpha\varphi(u_1) + (1-\alpha)\varphi(u_2),$$

which completes the proof.

Theorem 1.4.3 *Let U be a topological space and let $u \to f_i(u)$, $i \in I = \{1, 2, \ldots, N\}$, be mappings of U into $\mathbb{R}$ that are lower semicontinuous. Let the maximum function be of the form (1.4.1). Then $u \to \varphi(u)$ is a lower semicontinuous mapping of U into $\mathbb{R}$.*

Proof. Let $\{u_n\}$ be a sequence converging to u in U. Let us show that, for an arbitrary $\varepsilon > 0$, there exists n_ε such that

$$\varphi(u_n) \geq \varphi(u) - \varepsilon, \qquad n \geq n_\varepsilon. \tag{1.4.4}$$

Since $u \to f_i(u)$ is a lower semicontinuous mapping of U into $\mathbb{R}$, for an arbitrary $i \in I$, there exists $n_{i\varepsilon}$ such that

$$f_i(u_n) \geq f_i(u) - \varepsilon, \qquad n \geq n_{i\varepsilon}.$$

For some $k \in I$, we have

$$f_k(u) = \varphi(u),$$

whence

$$f_k(u_n) \geq \varphi(u) - \varepsilon, \qquad n \geq n_{k\varepsilon}.$$

Thus, setting $n_\varepsilon = \max_{i\in I} n_{i\varepsilon}$, we obtain the inequality (1.4.4), and the theorem is proved.

1.4.2 General maximum function

Let U and V be metric spaces, and assume that

$$\left.\begin{array}{l} u, v \to F(u,v) \text{ is a continuous mapping of } U \times V \\ \text{(equipped with the product of the topologies of } U \text{ and } V) \\ \text{into } \mathbb{R}. \end{array}\right\} \tag{1.4.5}$$

Further, suppose

$$G \text{ is a compact set in } V. \tag{1.4.6}$$

By (1.4.5), for an arbitrary fixed $u \in U$, the partial function $v \to F(u,v)$ is a continuous mapping of V into $\mathbb{R}$. As G is compact, by Theorem 1.3.9, for an arbitrary $u \in U$ there exists a function $w_u \in G$ such that

$$F(u, w_u) = \max_{v \in G} F(u,v).$$

Suppose now that

$$Q \text{ is a compact set in } U. \tag{1.4.7}$$

Define a maximum function $\varphi\colon Q \to \mathbb{R}$ by

$$u \to \varphi(u) = \max_{v \in G} F(u,v). \tag{1.4.8}$$

Theorem 1.4.4 *Let U, V be metric spaces and let the conditions* (1.4.5)–(1.4.7) *hold. Then, the function $u \to \varphi(u)$ defined by* (1.4.8) *is a continuous mapping of Q (equipped with the topology induced by the topology of U) into $\mathbb{R}$.*

Proof. Let us show that the following inequality holds:

$$\begin{aligned} |\varphi(w) - \varphi(u)| &= \left| \max_{v \in G} F(w,v) - \max_{v \in G} F(u,v) \right| \\ &\le \max_{v \in G} |F(w,v) - F(u,v)|, \qquad w, u \in Q. \end{aligned} \tag{1.4.9}$$

Indeed, we have

$$\begin{aligned} \max_{v \in G} F(w,v) &= \max_{v \in G} \left[F(u,v) + (F(w,v) - F(u,v)) \right] \\ &\le \max_{v \in G} F(u,v) + \max_{v \in G} |F(w,v) - F(u,v)|. \end{aligned} \tag{1.4.10}$$

By analogy, we have

$$\max_{v \in G} F(u,v) \le \max_{v \in G} F(w,v) + \max_{v \in G} |F(w,v) - F(u,v)|. \tag{1.4.11}$$

Then, (1.4.9) follows from (1.4.10) and (1.4.11).

Let $\{u_n\}$ be a sequence such that

$$\{u_n\} \subset Q, \qquad u_n \to u_0 \quad \text{in } Q. \tag{1.4.12}$$

(1.4.7) and (1.4.12) yield that $u_0 \in Q$. By Theorem 1.3.5, the set $Q \times G$ equipped with the product of the topologies of Q and G is compact. As Q and G are metric spaces, so is the space $Q \times G$. Let d_1 and d_2 be the metrics of the spaces U and V, respectively, and let $\delta = \max(d_1, d_2)$ be the metric on $U \times V$. Since $Q \times G$ is compact, from (1.4.5) and Theorem 1.3.10 it follows that

$$\text{the function } u, v \to F(u,v) \text{ is uniformly continuous on } Q \times G, \tag{1.4.13}$$

i.e., for an arbitrary $\varepsilon > 0$, there exists $\eta > 0$ such that the inequalities

$$d_1(u', u'') \leq \eta, \qquad d_2(v', v'') \leq \eta, \qquad u', u'' \in Q,\ v', v'' \in G,$$

yield

$$|F(u', v') - F(u'', v'')| \leq \varepsilon.$$

By (1.4.12), for an arbitrary $\eta > 0$, there exists N_η such that, provided $n \geq N_\eta$, the inequality $d_1(u_n, u_0) \leq \eta$ holds. Then,

$$|F(u_n, v) - F(u_0, v)| \leq \varepsilon, \qquad v \in G,\ n \geq N_\eta.$$

From here, by (1.4.9), we get

$$|\varphi(u_n) - \varphi(u_0)| \leq \max_{v \in G} |F(u_n, v) - F(u_0, v)| \leq \varepsilon, \qquad n \geq N_\eta.$$

The theorem is proved.

1.5 Hilbert space

1.5.1 Basic definitions and properties

Let H be a vector space over $\mathbb{R}$. A scalar product in H is a bilinear form $H \times H\colon x, y \to (x,y)_H \in \mathbb{R}$ that is symmetric, i.e., $(x,y)_H = (y,x)_H$, and positive definite, i.e., $(x,x)_H \geq 0$ and $(x,x)_H = 0$ if and only if $x = 0$. The function $x \to \|x\|_H = (x,x)_H^{\frac{1}{2}}$ defines a norm in H.

A vector space with a scalar product that is complete with respect to the topology generated by the scalar product is called a Hilbert space.

Every Hilbert space is clearly a Banach space.

For arbitrary elements x, y of a Hilbert space H, the following inequality – called the Schwartz inequality– holds

$$|(x,y)_H| \leq \|x\|_H \|y\|_H.$$

Theorem 1.5.1 (Riesz) *For every Hilbert space H, there exists a unique, one-to-one, linear mapping $\mathcal{R}$ of the dual space H^* onto H possessing the following properties:*

$$(\mathcal{R}f, x)_H = (f, x), \qquad \|\mathcal{R}f\|_H = \|f\|_{H^*}, \qquad f \in H^*,\ x \in H.$$

Notice that here (f, x) is the pairing of elements $f \in H^*$ and $x \in H$, and

$$\|f\|_{H^*} = \sup_{\|x\|_H \leq 1} |(f, x)|.$$

For the proof, see, e.g., Céa (1971).

The operator $\mathcal{R} \in \mathcal{L}(H^*, H)$ is called the Riesz operator for the space H. One can define a scalar product in H^* by setting $(f, g)_{H^*} = (\mathcal{R}f, \mathcal{R}g)_H$. Then, the dual of a Hilbert space becomes a Hilbert space, too. The Riesz theorem gives the possibility of identifying the spaces H^* and H (in that case, a linear functional $f \in H^*$ and the element $\mathcal{R}f \in H$ are identified).

Theorem 1.5.2 (Lax-Milgram) *Let H be a Hilbert space, and let $u, v \to a(u, v)$ be a bilinear form on $H \times H$ having the following properties:*

continuity: there exists a constant $c_1 > 0$ such that

$$|a(u, v)| \leq c_1 \|u\|_H \|v\|_H, \qquad u, v \in H,$$

coerciveness: there exists a constant $c_2 > 0$ such that

$$a(u, u) \geq c_2 \|u\|_H^2, \qquad u \in H.$$

Then, for every $f \in H^*$, there exists a unique element $w \in H$ such that

$$a(w, v) = (f, v), \qquad v \in H.$$

Moreover, the following estimates hold

$$\|w\|_H \leq c_2^{-1} \|f\|_{H^*}, \qquad \|f\|_{H^*} \leq c_1 \|w\|_H.$$

For the proof, see, e.g., Yosida (1971), Céa (1971).

Theorem 1.5.3 *Let H be a Hilbert space. If $u_n \to u$ weakly in H and $\|u_n\|_H \to \|u\|_H$, then $u_n \to u$ strongly in H.*

Proof. From the definition of a scalar product, it follows that

$$\|u_n - u\|_H^2 = (u_n - u, u_n - u)_H = \|u_n\|_H^2 - 2(u_n, u)_H + \|u\|_H^2. \tag{1.5.1}$$

Since $u_n \to u$ weakly in H, we have

$$\lim_{n \to \infty} (u_n, u)_H = \|u\|_H^2. \tag{1.5.2}$$

Taking into account that $\|u_n\|_H \to \|u\|_H$, by virtue of (1.5.1) and (1.5.2), we conclude that $u_n \to u$ strongly in H.

Two elements u, v of a Hilbert space H are called orthogonal if $(u, v)_H = 0$. Two subspaces F_1, F_2 of H are called orthogonal if every element of F_1 is orthogonal to every element of F_2.

Let $\{u_i\}$ be a sequence of nonzero orthogonal elements of H, i.e., $(u_i, u_j) = 0$ if $i \neq j$. If the orthogonal system $\{u_i\}$ is complete in H (that is, the least closed subspace containing $\{u_i\}$ is the whole H), then this system is called an (orthogonal) basis of H. If, additionally, the norm of each element of the system is equal to 1, then the system $\{u_i\}$ is called an orthonormal basis of H. Notice that, in the latter case, $(u_i, u_j)_H = \delta_{ij}$ where

$$\delta_{ij} = \begin{cases} 1, & \text{if } i = j, \\ 0, & \text{if } i \neq j \end{cases}$$

is the Kronecker delta.

Let F be a subspace of a Hilbert space H. The set of all the elements of H that are orthogonal to F is called the orthogonal complement of F and is denoted by $F^{\perp}$.

Lemma 1.5.1 *Let F be a closed subspace of a Hilbert space H. Then, every element $u \in H$ can be uniquely expanded in the sum $u = v + w$, where $v \in F$ and $w \in F^{\perp}$. The element v is characterized by the following equivalent properties*:

$$\|u - v\|_H \leq \|u - g\|_H, \qquad g \in F.$$
$$(u - v, g)_H = 0, \qquad g \in F.$$

For the proof, see, e.g., Céa (1971), Kantorovich and Akilov (1977).

Remark 1.5.1 If the hypothesis of Lemma 1.5.1 is satisfied, the element v is called the orthogonal projection of u onto F, and denoted by $v = Pu$. The operator P is called the orthogonal projection of H onto F;

$$Pu = u, \qquad u \in F, \qquad \|P\|_{\mathcal{L}(H,F)} = 1.$$

Remark 1.5.2 Let H be a Hilbert space with a scalar product $(f, g)_H$ and norm $\|f\|_H$. Let $(f, g)_1$ be another symmetric positive definite form on $H \times H$ and let $\|f\|_1 = (f, f)_1^{\frac{1}{2}}$. Suppose that there exist positive constants c_0 and c_1 such that

$$c_0 \|f\|_H \leq \|f\|_1 \leq c_1 \|f\|_H, \qquad f \in H,$$

i.e., the norms $\|\cdot\|_H$ and $\|\cdot\|_1$ are equivalent. Denote by H_1 the Hilbert space that coincides as a set with H and the scalar product which is given by the form $(\cdot, \cdot)_1$. Then, evidently, H and H_1 coincide as topological spaces, though they are different as Hilbert spaces. Despite this, in order to not introduce new notations, we will sometimes say that H is a Hilbert space with the scalar product $(\cdot, \cdot)_1$, i.e., we will use the same symbol for different spaces; in such a case, we will always indicate the form of the scalar product in H.

1.5.2 Compact and selfadjoint operators in a Hilbert space

Definitions

Let U, V be Hilbert spaces and let $J \in \mathcal{L}(U, V)$, i.e., J is a linear continuous operator acting from U into V.

The operator J is called compact if the image $J(B)$ of every bounded set $B \subset U$ is relatively compact, i.e., the closure $\overline{J(B)}$ is a compact set in V.

This definition is equivalent to the following: the image of every weakly convergent sequence from U under the action of the operator J is a sequence that strongly converges in V, i.e., $u_n \to u$ weakly in U implies $Ju_n \to Ju$ strongly in V.

Let H be a Hilbert space and let $A \in \mathcal{L}(H, H)$. The operator A is called selfadjoint if

$$(Au, v)_H = (u, Av)_H, \qquad u, v \in H,$$

A is called strictly positive if

$$(Au, u) \geq 0, \qquad u \in H,$$

and the equality to zero takes place only if $u = 0$.

A real number α for which the equation

$$Au = \alpha u$$

has a solution $u \neq 0$ is called an eigenvalue, and u is called an eigenfunction.

The vector subspace generated by all the eigenfunctions belonging to an eigenvalue α is a closed subspace of H, which is called the eigenspace of α. The dimension of this space is called the multiplicity of α.

Spectrum of a selfadjoint operator

Assume H to be an infinite-dimensional Hilbert space. For strictly positive operators, the following theorem holds.

Theorem 1.5.4 *Let A be a selfadjoint, strictly positive, linear, compact operator acting from H into H. Then, the set of all its eigenvalues α_i disposed in nonincreasing order, so that every eigenvalue is taken as many times as its multiplicity, forms an infinite sequence of positive numbers convergent to zero:*

$$\alpha_1 \geq \alpha_2 \geq \cdots \geq \alpha_n \geq \cdots, \qquad \lim_{n \to \infty} \alpha_n = 0. \tag{1.5.3}$$

There exists a countable orthonormal sequence of eigenfunctions $\{u_i\}$,

$$Au_i = \alpha_i u_i, \qquad (u_i, u_j)_H = \delta_{ij}, \qquad i, j \geq 1, \tag{1.5.4}$$

that forms a basis of H. Moreover,

$$\alpha_1 = \sup_{\substack{u \in H \\ u \neq 0}} \frac{(Au, u)_H}{(u, u)_H} = \frac{(Au_1, u_1)_H}{(u_1, u_1)_H}. \tag{1.5.5}$$

For the proof, see, e.g., Gould (1966).

We proceed to study the case when the operator A is not positive.

Theorem 1.5.5 *Let A be a selfadjoint, linear, compact operator acting from H into H. Then, the equation*

$$Au = \lambda u \tag{1.5.6}$$

has a countable set of eigenvalues $\{\lambda_i\}$ in which every eigenvalue is taken as many times as its multiplicity, and every nonzero eigenvalue has a finite multiplicity. Zero is the limit element of the sequence $\{\lambda_i\}$, i.e., $\lim_{i\to\infty} \lambda_i = 0$. All the eigenvalues of the equation (1.5.6) satisfy the condition

$$m = \inf_{\|v\|_H=1} (Av, v)_H \le \lambda_i \le \sup_{\|v\|_H=1} (Av, v)_H = M. \tag{1.5.7}$$

If $M \neq 0$, then M is an eigenvalue. Moreover, if u is an eigenfunction belonging to M such that $\|u\|_H = 1$, then $M = (Au, u)_H$. If $m \neq 0$, then m is an eigenvalue and if w is an eigenfunction belonging to m such that $\|w\|_H = 1$, then $(Aw, w)_H = m$.

For the proof, see, e.g., Riesz and Sz.-Nagy (1972).

Let us divide the positive and negative eigenvalues of the operator A into two sequences,

$$\lambda_1^+, \lambda_2^+, \ldots, \qquad \lambda_1^-, \lambda_2^-, \ldots,$$

the first sequence being nonincreasing and the second one nondecreasing, and every eigenvalue appearing in these sequences as many times as its multiplicity. Each of these sequences may be finite (and even empty) or infinite.

Let

$$u_1^+, u_2^+, \ldots, \qquad u_1^-, u_2^-, \ldots$$

be the corresponding sequences of the orthonormal eigenfunctions.

Lemma 1.5.2 *Under the assumptions of Theorem 1.5.5, for the n-th positive eigenvalue, the following representation holds:*

$$\lambda_n^+ = \max_{u\in Q_n^+} (Au, u)_H = (Au_n^+, u_n^+)_H, \tag{1.5.8}$$

where

$$Q_n^+ = \{\, u \mid u \in H,\ \|u\|_H = 1,\ (u, u_i^+)_H = 0,\ i = 1, 2, \ldots, n-1,$$

and for the n-th negative eigenvalue

$$\lambda_n^- = \min_{u\in Q_n^-} (Au, u)_H = (Au_n^-, u_n^-)_H,$$

where

$$Q_n^- = \{\, u \mid u \in H,\ \|u\|_H = 1,\ (u, u_i^-)_H = 0,\ i = 1, 2, \ldots, n-1 \,\}.$$

For the proof, see, e.g., Riesz and Sz.-Nagy (1972).

Lemma 1.5.2 allows one to calculate the n-th eigenvalue and eigenfunction, provided the eigenfunctions with less numbers are known. The following lemma is based on the minimax-method, which gives a direct rule of finding the n-th eigenvalue.

Lemma 1.5.3 *Let $h_1, h_2, \ldots, h_{n-1}$ be arbitrary $(n-1)$ elements of the space H and let $\nu = \nu(h_1, h_2, \ldots, h_{n-1})$ be the maximal value of $(Au, u)_H$ on the set of the elements $u \in H$ satisfying the condition*

$$\|u\|_H = 1, \qquad (u, h_i)_H = 0, \qquad i = 1, 2, \ldots, n-1.$$

Then, the minimal value of the function ν on all the possible h_i is equal to λ_n^+ and it is achieved for

$$h_i = u_i^+, \qquad i = 1, 2, \ldots, n-1.$$

An analogous statement on λ_n^- is obtained if we trade the places of "the minimal value" and "the maximal value."

For the proof, see Riesz and Sz.-Nagy (1972).

In the sequel, we will need to compare the eigenvalues of different operators, so that we will use the following statement.

Theorem 1.5.6 (Weyl) *Let A_1 and A be compact selfadjoint operators acting in H. Denote by λ_{1n}^+, λ_n^+ (λ_{1n}^-, λ_n^-) the n-th positive (negative) eigenvalues of the operators A_1, A, respectively. Then, the following estimates hold:*

$$\begin{aligned} |\lambda_n^+ - \lambda_{1n}^+| &\le \|A - A_1\| \qquad \forall n \\ |\lambda_n^- - \lambda_{1n}^-| &\le \|A - A_1\| \qquad \forall n. \end{aligned} \tag{1.5.9}$$

Proof. Introduce the set

$$Q_{1n}^+ = \{u \mid u \in H,\ \|u\|_H = 1,\ (u, u_{1i}^+)_H = 0,\ i = 1, 2, \ldots, n-1\},$$

where u_{1i}^+ is the eigenfunction of the operator A_1 belonging to the eigenvalue λ_{1i}^+.

By using Lemmas 1.5.2 and 1.5.3, we get

$$\begin{aligned} \lambda_n^+ &\le \sup_{u \in Q_{1n}^+} (Au, u)_H \\ &\le \sup_{u \in Q_{1n}^+} (A_1 u, u)_H + \sup_{u \in Q_{1n}^+} |(Au, u)_H - (A_1 u, u)_H| \le \lambda_{1n}^+ + \|A - A_1\|. \end{aligned}$$

Analogously, we get

$$\lambda_{1n}^+ \leq \lambda_n^+ + \|A - A_1\|.$$

Thus, the first estimate in (1.5.9) is proved.

Further, introduce the set

$$Q_{1n}^- = \big\{ u \mid u \in H,\ \|u\|_H = 1,\ (u, u_{1i}^-)_H = 0,\ i = 1, 2, \dots, n-1 \big\},$$

where u_{1i}^- is the eigenfunction of the operator A_1 belonging to the eigenvalue λ_{1i}^-. Again, by using Lemmas 1.5.2 and 1.5.3 , we get

$$\begin{aligned} \lambda_n^- &\geq \inf_{u \in Q_{1n}^-} (Au, u)_H \\ &\geq \inf_{u \in Q_{1n}^-} (A_1 u, u)_H - \sup_{u \in Q_{1n}^-} |(Au, u)_H - (A_1 u, u)_H| \geq \lambda_{1n}^- - \|A - A_1\| \end{aligned}$$

and

$$\lambda_{1n}^- \geq \lambda_n^- - \|A - A_1\|.$$

So, the second estimate in (1.5.9) is also proved.

Equation $Au = \lambda Bu$

Let U, V be two Hilbert spaces such that

$$U \subset V, \qquad \text{the embedding } U \to V \text{ is compact.} \tag{1.5.10}$$

This means that every bounded set in U is relatively compact in V.

As usual, by $(\cdot,\cdot)_U$ and $(\cdot,\cdot)_V$ we denote the scalar products in U and V; if H is a Hilbert space, $f \in H^*$, and $u \in H$, then (f, u) stands for the dual pairing of f and u.

Since, for a fixed $u \in U$, the form $v \to (u, v)_U$ is linear and continuous on U, we can write

$$(u, v)_U = (Au, v), \qquad Au \in U^*, \tag{1.5.11}$$

which defines an operator

$$A \in \mathcal{L}(U, U^*),$$

called the canonical isometry corresponding to the scalar product. A is evidently the inverse of the Riesz operator defined in Theorem 1.5.1 .

Let us consider the scalar product $(u, v)_V$ as a bilinear form defined on $U \times U$. Hence, for any $u, v \in U$, we write

$$(u, v)_V = (Bu, v), \qquad Bu \in U^*. \tag{1.5.12}$$

Thus, we have defined an operator

$$B \in \mathcal{L}(U, U^*). \tag{1.5.13}$$

Let us show that

$$B \text{ is a compact mapping of } U \text{ into } U^*. \tag{1.5.14}$$

Indeed, let $u_n \to u$ weakly in U, then $u_n \to u$ strongly in V, and so

$$|(Bu_n - Bu, v)| = |(u_n - u, v)_V| \leq c\|u_n - u\|_V \|v\|_U, \qquad c = \text{const} > 0.$$

Hence, $Bu_n \to Bu$ strongly in U^*.

Consider the problem of finding $(u_i, \lambda_i) \in U \times \mathbb{R}$, $u_i \neq 0$, such that

$$(u_i, v)_U = \lambda_i (u_i, v)_V, \qquad v \in U. \tag{1.5.15}$$

By virtue of (1.5.11) and (1.5.12), this problem is equivalent to the following one:

$$Au_i = \lambda_i B u_i. \tag{1.5.16}$$

Denote by A^{-1} the inverse operator of A, i.e., the Riesz operator. Then, (1.5.16) is equivalent to the relation

$$u_i = \lambda_i A^{-1} \circ B u_i. \tag{1.5.17}$$

Here, $\circ$ stands for the composition of mappings. Since $A^{-1} \in \mathcal{L}(U^*, U)$, it follows from (1.5.13) and (1.5.14) that

$$A^{-1} \circ B \text{ is a linear compact mapping of } U \text{ into } U. \tag{1.5.18}$$

Further, by using (1.5.11) and (1.5.12), we have

$$\begin{aligned} (A^{-1} \circ Bu, v)_U = (Bu, v) = (Bv, u) &= (A^{-1} \circ Bv, u)_U \\ &= (u, A^{-1} \circ Bv)_U, \qquad u, v \in U, \\ (A^{-1} \circ Bu, u)_U = (Bu, u) &= (u, u)_V, \qquad u \in U. \end{aligned} \tag{1.5.19}$$

Thus, $A^{-1} \circ B$ is a selfadjoint, strictly positive operator. Taking (1.5.18) into account, setting $\alpha_i = \frac{1}{\lambda_i}$, and applying Theorem 1.5.4 to the problem (1.5.17), we obtain the following result.

Theorem 1.5.7 *Let the condition* (1.5.10) *be satisfied and let operators A, B be defined by* (1.5.11), (1.5.12). *Then, the equation* (1.5.16) (*equivalently, the equation* (1.5.15)) *has a countable set of eigenvalues*

$$0 < \lambda_1 \leq \lambda_2 \leq \cdots \leq \lambda_i \leq \cdots, \qquad \lim_{i \to \infty} \lambda_i = \infty,$$

and a corresponding sequence of eigenfunctions $\{u_i\}$ that forms an orthogonal basis in U. If U is dense in V, then $\{u_i\}$ forms an orthogonal basis in V. Moreover,

$$\lambda_i = \inf_{\substack{u \in U \\ u \neq 0}} \frac{(u, u)_U}{(u, u)_V} = \frac{(u_1, u_1)_U}{(u_1, u_1)_V}.$$

1.5.3 Theorem on continuity of a spectrum

Suppose now that

$$\left.\begin{array}{l} U \text{ is a Hilbert space, } Y \text{ is a Banach space, } U \subset Y, \text{ and the} \\ \text{embedding } U \to Y \text{ is compact.} \end{array}\right\} \tag{1.5.20}$$

Suppose we are given a bilinear, symmetric, continuous form b on $Y \times Y$:

$$b(u,v) = b(v,u), \qquad u, v \in Y, \tag{1.5.21}$$

$$|b(u,v)| \le c\|u\|_Y \|v\|_Y, \qquad u, v \in Y, \;\; c = \text{const} > 0. \tag{1.5.22}$$

Consider the following eigenvalue problem:

$$(u_i, \lambda_i) \in U \times \mathbb{R}, \quad u_i \neq 0, \qquad \lambda_i (u_i, v)_U = b(u_i, v), \qquad v \in U. \tag{1.5.23}$$

For a fixed element $u \in U$, the form $v \to b(u,v)$ is linear and continuous on Y, and all the more on U. Therefore, there exists an operator $B \in \mathcal{L}(U, U^*)$ such that

$$b(u,v) = (Bu, v), \qquad Bu \in U^*. \tag{1.5.24}$$

By (1.5.21), the operator B is selfadjoint. By using (1.5.20), analogously to above, we establish that

$$B \text{ is a compact mapping of } U \text{ into } U^*. \tag{1.5.25}$$

The problem (1.5.23) is equivalent to the following one:

$$(u_i, \lambda_i) \in U \times \mathbb{R}, \qquad u_i \neq 0, \qquad \lambda_i A u_i = B u_i, \tag{1.5.26}$$

where A is the canonical isometry corresponding to the scalar product in U (see (1.5.11)). The problem (1.5.26) may be represented in the form

$$(u_i, \lambda_i) \in U \times \mathbb{R}, \qquad u_i \neq 0, \qquad A^{-1} \circ B u_i = \lambda_i u_i. \tag{1.5.27}$$

The operator $A^{-1} \circ B$ is selfadjoint (see (1.5.19)). Since $A^{-1} \in \mathcal{L}(U^*, U)$, it follows from (1.5.25) that $A^{-1} \circ B$ is a compact mapping of U into U. Now, by applying Theorem 1.5.5 to the problem (1.5.27), we get

Theorem 1.5.8 *Let the conditions* (1.5.20) *be satisfied and let b be a bilinear form on $Y \times Y$ for which the formulas* (1.5.21), (1.5.22) *hold. Then, the problem* (1.5.23) *has a countable set of eigenvalues $\{\lambda_i\}$, where every eigenvalue is taken as many times as its multiplicity; every eigenvalue that is not equal to zero is of finite multiplicity, and the following inequalities hold*

$$m = \inf_{\|u\|_U = 1} b(u,u) \le \lambda_i \le \sup_{\|u\|_U = 1} b(u,u) = M. \tag{1.5.28}$$

If $m \neq 0$, then m is an eigenvalue; if $M \neq 0$, then M is an eigenvalue. Moreover, $\lim_{i \to \infty} \lambda_i = 0$.

Note that for proving (1.5.28), one uses the equalities

$$(A^{-1} \circ Bu, u)_U = (Bu, u) = b(u, u), \qquad u \in U.$$

Remark 1.5.3 Let X be a Hilbert space and let φ be a bilinear, continuous, symmetric form on $X \times X$. Then, the following equality holds

$$\|\varphi\|_{\mathcal{L}_2(X,X;\mathbb{R})} = \sup_{\|u\|_X=1} |\varphi(u,u)|. \tag{1.5.29}$$

For let

$$Q = \sup_{\|u\|_X=1} |\varphi(u,u)|.$$

It follows from the definition of a norm of a bilinear form (see Subsec. 1.3.1) that

$$Q \leq \|\varphi\|_{\mathcal{L}_2(X,X;\mathbb{R})}.$$

Let us show that the inverse inequality holds. Indeed, we have

$$\begin{aligned} |\varphi(u,v)| &= \left|\frac{1}{4}\left(\varphi(u+v,u+v) - \varphi(u-v,u-v)\right)\right| \\ &\leq \frac{1}{4}Q\left(\|u+v\|_X^2 + \|u-v\|_X^2\right) = \frac{1}{2}Q\left(\|u\|_X^2 + \|v\|_X^2\right), \qquad u, v \in X. \end{aligned}$$

It follows from here that

$$\|\varphi\|_{\mathcal{L}_2(X,X;\mathbb{R})} = \sup_{\|u\|_X \leq 1,\, \|v\|_X \leq 1} |\varphi(u,v)| \leq Q,$$

and so (1.5.29) holds.

Denote by $\mathcal{U}_1$ the set of the bilinear, symmetric, continuous, coercive forms on $U \times U$:

$$\begin{aligned} \mathcal{U}_1 = \big\{ a \mid a \in \mathcal{L}_2(U,U;\mathbb{R}),\ a(u,v) = a(v,u),\ u, v \in U, \\ a(u,u) \geq c(a)\|u\|_U^2,\ u \in U,\ c(a) = \text{const} > 0 \big\}. \end{aligned} \tag{1.5.30}$$

Here, $c(a)$ denotes that the positive constant in the inequality of coerciveness depends on an element a.

Let further $\mathcal{U}_2$ be the set of the bilinear, symmetric, continuous forms on $Y \times Y$:

$$\mathcal{U}_2 = \big\{ b \mid b \in \mathcal{L}_2(Y,Y;\mathbb{R}),\ b(u,v) = b(v,u),\ u, v \in U \big\}. \tag{1.5.31}$$

Suppose now that

$$\{a_n\}_{n=1}^{\infty} \subset \mathcal{U}_1, \qquad a_0 \in \mathcal{U}_1, \quad a_n \to a_0 \quad \text{in } \mathcal{L}_2(U,U;\mathbb{R}), \tag{1.5.32}$$

$$\{b_n\}_{n=1}^{\infty} \subset \mathcal{U}_2, \qquad b_n \to b_0 \quad \text{in } \mathcal{L}_2(Y, Y; \mathbb{R}). \tag{1.5.33}$$

Consider the following eigenvalue problem:

$$\begin{aligned} &(u_{in}, \lambda_{in}) \in U \times \mathbb{R}, \qquad u_{in} \neq 0, \\ &\lambda_{in} a_n(u_{in}, v) = b_n(u_{in}, v), \qquad v \in U, \ n = 0, 1, 2, \ldots. \end{aligned} \tag{1.5.34}$$

Theorem 1.5.9 *Let the conditions* (1.5.20) *be satisfied and let sets* U_1, U_2 *be defined by* (1.5.30), (1.5.31). *Let* (1.5.32) *and* (1.5.33) *hold. Then, for any* $n = 0, 1, 2, \ldots,$, *there exists a countable set of eigenvalues of the problem* (1.5.34). *Let*

$$\lambda_{1n}^+, \lambda_{2n}^+, \ldots, \qquad \lambda_{1n}^-, \lambda_{2n}^-, \ldots \tag{1.5.35}$$

be the two sequences consisting of the positive and negative eigenvalues ordered so that the first sequence is nonincreasing and the second one is nondecreasing and every eigenvalue is taken as many times as its multiplicity. Then,

$$\lim_{n\to\infty} \lambda_{in}^+ = \lambda_{i0}^+ \qquad \forall i, \qquad \lim_{n\to\infty} \lambda_{jn}^- = \lambda_{j0}^- \qquad \forall j. \tag{1.5.36}$$

Moreover, the convergence is uniform, i.e., for an arbitrary $\varepsilon > 0$, *there exists* n_ε *such that, for all* $n \geq n_\varepsilon$,

$$\begin{aligned} |\lambda_{in}^+ - \lambda_{i0}^+| \leq \varepsilon \qquad \forall i, \\ |\lambda_{jn}^- - \lambda_{j0}^-| \leq \varepsilon \qquad \forall j. \end{aligned}$$

Remark 1.5.4 Each of the sequences in (1.5.35) may be either empty, or finite, or infinite.

To prove Theorem 1.5.9, we need the following lemma.

Lemma 1.5.4 *Let the conditions* (1.5.20) *be satisfied, let sets* $\mathcal{U}_1$, $\mathcal{U}_2$ *be defined by the expressions* (1.5.30), (1.5.31), *and let* (1.5.32), (1.5.33) *hold true. Let* U_1 *be a subspace of* U *(in particular, it may be that* $U_1 = U$*) and let*

$$f_n = \sup_{\substack{u \in U_1 \\ u \neq 0}} \left| \frac{b_n(u,u)}{a_n(u,u)} - \frac{b_0(u,u)}{a_0(u,u)} \right|, \qquad n = 1, 2, \ldots.$$

Then

$$\lim_{n\to\infty} f_n = 0. \tag{1.5.37}$$

Proof. Since the bilinear form $a_0(u, v)$ is symmetric, continuous, and coercive on $U \times U$, we get

$$f_n = \sup_{u \in U_0} \left| \frac{b_n(u,u)}{a_n(u,u)} - \frac{b_0(u,u)}{a_0(u,u)} \right|, \tag{1.5.38}$$

where

$$U_0 = \{ u \mid u \in U_1, \ a_0(u,u) = 1 \}.$$

By (1.5.32) and (1.5.33),

$$|a_n(u,u) - 1| = |a_n(u,u) - a_0(u,u)| \le \alpha_n, \qquad u \in U_0, \ \lim_{n\to\infty} \alpha_n = 0,$$
$$|b_n(u,u) - b_0(u,u)| \le \beta_n, \qquad u \in U_0, \ \lim_{n\to\infty} \beta_n = 0.$$

Therefore, for an arbitrary $\varepsilon > 0$, there exists k_ε such that

$$\left| \frac{b_n(u,u)}{a_n(u,u)} - \frac{b_0(u,u)}{a_0(u,u)} \right| \le \varepsilon, \qquad n \ge k_\varepsilon, \ u \in U_0. \tag{1.5.39}$$

Now, (1.5.37) follows from (1.5.38) and (1.5.39).

Proof of Theorem 1.5.9. Since $a_n \in \mathcal{U}_1$, $n = 0, 1, 2, \ldots$, for any n one can consider the Hilbert space V_n as the set U in which the scalar product and the norm are generated by the form a_n. Notice that all the norms generated by a_n's are equivalent to each other, and equivalent to the norm of the space U. Hence, (1.5.20) implies that the embedding of V_n into Y is compact.

Now, from Theorem 1.5.8 it follows that there exists a countable set of eigenvalues of the problem (1.5.34). Divide these eigenvalues into the two sequences (1.5.35), and let

$$\overset{+}{u}_{1n}, \overset{+}{u}_{2n}, \ldots, \qquad \overset{-}{u}_{1n}, \overset{-}{u}_{2n}, \ldots \tag{1.5.40}$$

be the corresponding sequences of eigenfunctions.

The problem (1.5.34) is equivalent to the following one

$$(u_{in}, \lambda_{in}) \in V_n \times \mathbb{R}, \qquad u_{in} \ne 0, \qquad \lambda_{in} u_{in} = A_n^{-1} \circ B_n u_{in}. \tag{1.5.41}$$

Here, A_n^{-1} is the inverse of the operator A_n generated by the scalar product in V_n, B_n the operator generated by the bilinear form b_n,

$$a_n(u,v) = (A_n u, v), \qquad u, v \in V_n, \tag{1.5.42}$$
$$b_n(u,v) = (B_n u, v), \qquad u, v \in V_n. \tag{1.5.43}$$

Analogously to above, we see that

$$A_n^{-1} \circ B_n \text{ is a linear, compact, selfadjoint operator in } V_n. \tag{1.5.44}$$

We have

$$a_n(A_n^{-1} \circ B_n u, u) = (B_n u, u) = b_n(u,u), \qquad u \in V_n, \ n = 0, 1, 2, \ldots. \tag{1.5.45}$$

From (1.5.41), (1.5.44), (1.5.45), and Lemma 1.5.2, we have that, for $n = 0, 1, 2, \ldots$,

$$\lambda_{in}^{+} = \sup_{u \in G_{in}^{+}} \frac{b_n(u,u)}{a_n(u,u)}, \tag{1.5.46}$$

where

$$G_{in}^{+} = \{\, u \mid u \in V_n,\ u \neq 0,\ a_n(u, u_{kn}^{+}) = 0,\ k = 1, 2, \ldots, i-1 \,\}. \tag{1.5.47}$$

Analogously

$$\lambda_{jn}^{-} = \inf_{u \in G_{jn}^{-}} \frac{b_n(u,u)}{a_n(u,u)}, \tag{1.5.48}$$

$$G_{jn}^{-} = \{\, u \mid u \in V_n,\ u \neq 0,\ a_n(u, u_{kn}^{-}) = 0,\ k = 1, 2, \ldots, j-1 \,\}. \tag{1.5.49}$$

Adding to the set G_{in}^{+} the zero of the space U, we get a subspace $\overline{G}_{in}^{+}$. If $\overline{G}_{in}^{+}$ is considered as a subspace of V_m, $m \neq n$, then the orthogonal complement of $\overline{G}_{in}^{+}$ to the space V_m has dimension $i-1$. Therefore, by using Lemma 1.5.3 and taking note of (1.5.45), we get

$$\lambda_{in}^{+} \leq \sup_{u \in G_{i0}^{+}} \frac{b_n(u,u)}{a_n(u,u)}, \qquad n = 1, 2, \ldots, \tag{1.5.50}$$

$$\lambda_{i0}^{+} \leq \sup_{u \in G_{in}^{+}} \frac{b_0(u,u)}{a_0(u,u)}, \qquad n = 1, 2, \ldots. \tag{1.5.51}$$

Now, from (1.5.46) and (1.5.51), we derive

$$\lambda_{i0}^{+} \leq \sup_{u \in G_{in}^{+}} \frac{b_0(u,u)}{a_0(u,u)} \leq \sup_{u \in G_{in}^{+}} \frac{b_n(u,u)}{a_n(u,u)} + \gamma_{in}^{+} = \lambda_{in}^{+} + \gamma_{in}^{+}, \tag{1.5.52}$$

where

$$\gamma_{in}^{+} = \sup_{u \in G_{in}^{+}} \left| \frac{b_0(u,u)}{a_0(u,u)} - \frac{b_n(u,u)}{a_n(u,u)} \right|. \tag{1.5.53}$$

From (1.5.46) and (1.5.50),

$$\lambda_{in}^{+} \leq \sup_{u \in G_{i0}^{+}} \frac{b_n(u,u)}{a_n(u,u)} \leq \sup_{u \in G_{i0}^{+}} \frac{b_0(u,u)}{a_0(u,u)} + \xi_{in}^{+} = \lambda_{i0}^{+} + \xi_{in}^{+}, \tag{1.5.54}$$

where

$$\xi_{in}^{+} = \sup_{u \in G_{i0}^{+}} \left| \frac{b_0(u,u)}{a_0(u,u)} - \frac{b_n(u,u)}{a_n(u,u)} \right|. \tag{1.5.55}$$

By (1.5.53), (1.5.55), and Lemma 1.5.4

$$\lim_{n\to\infty} \gamma_{in}^+ = 0, \quad \lim_{n\to\infty} \xi_{in}^+ = 0 \qquad \forall i.$$

Hence, (1.5.52) and (1.5.54) give

$$\lim_{n\to\infty} \lambda_{in}^+ = \lambda_{i0}^+ \qquad \forall i.$$

Next, taking into account Lemmas 1.5.2, 1.5.3 and (1.5.45), (1.5.48), we infer

$$\lambda_{j0}^- \geq \inf_{u\in G_{jn}^-} \frac{b_0(u,u)}{a_0(u,u)} \geq \inf_{u\in G_{jn}^-} \frac{b_n(u,u)}{a_n(u,u)} - \gamma_{jn}^- = \lambda_{jn}^- - \gamma_{jn}^-, \tag{1.5.56}$$

where

$$\gamma_{jn}^- = \sup_{u\in G_{jn}^-} \left| \frac{b_0(u,u)}{a_0(u,u)} - \frac{b_n(u,u)}{a_n(u,u)} \right|.$$

Analogously, we have

$$\lambda_{jn}^- \geq \inf_{u\in G_{j0}^-} \frac{b_n(u,u)}{a_n(u,u)} \geq \inf_{u\in G_{j0}^-} \frac{b_0(u,u)}{a_0(u,u)} - \xi_{jn}^- = \lambda_{j0}^- - \xi_{jn}^-,$$

$$\xi_{jn}^- = \sup_{u\in G_{j0}^-} \left| \frac{b_0(u,u)}{a_0(u,u)} - \frac{b_n(u,u)}{a_n(u,u)} \right|. \tag{1.5.57}$$

From Lemmma 1.5.4

$$\lim_{n\to\infty} \gamma_{jn}^- = 0, \qquad \lim_{n\to\infty} \xi_{jn}^- = 0 \qquad \forall j.$$

(1.5.56) and (1.5.57) yield

$$\lim_{n\to\infty} \lambda_{jn}^- = \lambda_{j0}^- \qquad \forall j.$$

Let

$$f_n = \sup_{\substack{u\in U\\ u\neq 0}} \left| \frac{b_n(u,u)}{a_n(u,u)} - \frac{b_0(u,u)}{a_0(u,u)} \right|.$$

Evidently

$$f_n \geq \gamma_{in}^+, \qquad f_n \geq \xi_{in}^+ \qquad \forall i,$$
$$f_n \geq \gamma_{jn}^-, \qquad f_n \geq \xi_{jn}^- \qquad \forall j.$$

Thus, taking note that $\lim_{n\to\infty} f_n = 0$ (by Lemma 1.5.4), we conclude that, for any $\varepsilon > 0$, there exists n_ε such that, for all $n \geq n_\varepsilon$,

$$|\lambda_{in}^+ - \lambda_{i0}^+| \leq \varepsilon \qquad \forall i, \qquad |\lambda_{jn}^- - \lambda_{j0}^-| \leq \varepsilon \qquad \forall j.$$

The theorem is proved.

Remark 1.5.5 Suppose the conditions of Theorem 1.5.9 hold and for $n = 0$ one of the sets in (1.5.35) is either empty, or finite, and zero is an eigenvalue of the problem (1.5.34) for $n = 0$. Let, for example, the set of the negative eigenvalues for $n = 0$ contain k elements, $k \geq 0$. In accordance with (1.5.35), we set

$$\lambda_{1n}^- \leq \lambda_{2n}^- \leq \cdots \leq \lambda_{kn}^- \leq \lambda_{(k+1)n}^-,$$

with $\lambda_{(k+1)0}^- = 0$. Then, by using an argument similar to that in the proof of Theorem 1.5.9, one can verify that $\lim_{n\to\infty} \lambda_{(k+1)n}^- = 0$.

1.5.4 Embedding of a Hilbert space in its dual

From the Riesz theorem it follows that a Hilbert space can be identified with its dual. However, such an identification is not always suitable. In study of very many problems, one deals with different Hilbert spaces that are ordered with respect to embedding. Then, one of the spaces is identified with its dual, and is called the zero space. Usually, as a zero space, one chooses a Hilbert space the scalar product in which is of a rather simple form (for instance, the space $L_2(\Omega)$). As for the other spaces, one supposes that the pairing of dual elements is determined by the extension of the scalar product of the zero space.

The following characterization of dense linear subsets holds.

Theorem 1.5.10 *Let V be a Hilbert space and let D be its linear subset. Then, D is dense in V if and only if any linear continuous functional on V vanishing on D vanishes on the whole space V.*

Proof. The "only if" part is obvious. Let us prove the "if" part. Let $\overline{D}$ be the closure of D, let P be the orthogonal projection on $\overline{D}$, and let P' be the transposition of P that is the mapping of V^* into V^* defined by

$$(P'f, u) = (f, Pu), \qquad f \in V^*, \ u \in V.$$

Here, as usual, (g, v) is the dual pairing of g and v. Evidently, $P'f - f \in V^*$ and

$$(P'f - f, u) = (f, Pu - u) = 0, \qquad u \in \overline{D}.$$

Hence, by the hypothesis of the theorem,

$$(P'f - f, v) = 0, \qquad v \in V.$$

Therefore, P' is the identical operator on V^*, and so P is the identity on V and $\overline{D} = V$, concluding the proof.

Further, let V, H be Hilbert spaces such that

$$V \subset H, \quad \text{the embedding is continuous and dense.} \tag{1.5.58}$$

Let us choose H as a zero space. Then, the scalar product $(u, v)_H$ is identified with the dual pairing on $H \times H$, i.e., for $f, v \in H$, we have $(f, v) = (f, v)_H$.

Theorem 1.5.11 *Let V, H be Hilbert spaces, let (1.5.58) hold, and let H be a zero space. Then, one can identify V and H with dense linear subsets of the dual space V^*,*

$$V \subset H \subset V^*, \tag{1.5.59}$$

where the embeddings are continuous and the dualization between V^* and V is identified with the unique extension of the scalar product in H.

Proof. To every element $u \in H$, one can place in correspondence the linear continuous functional f_u in the space V given by

$$(f_u, v) = (u, v)_H, \qquad v \in V. \tag{1.5.60}$$

Thus, one defines the linear continuous mapping $u \to f_u$ of the space H into V^*. This mapping is injective. Indeed, if $f_u = 0$, then by (1.5.60)

$$(u, v)_H = 0, \qquad v \in V,$$

and since V is dense in H, $u = 0$. So, one can identify the elements u and f_u, which gives

$$H \subset V^*.$$

Let us show that H is dense in V^*. To this end, we use Theorem 1.5.10, noticing that the linear form u defined on V^* is an element of V, because a Hilbert space is reflexive. So, $u \in V^{**} = V$ is a linear form on V^* vanishing on H. Then

$$(u, v) = (u, v)_H = 0, \qquad v \in H.$$

Hence, $u = 0$, which yields that H is dense in V^*.

Let now f be an arbitrary element of V^*. Since H is dense in V^*, there exists a sequence of elements of H, $\{u_n\}$, such that, for the linear forms $f_n \in V^*$ given by the relations

$$(f_n, v) = (u_n, v)_H, \qquad v \in V, \tag{1.5.61}$$

it holds that

$$\lim_{n\to\infty} \|f_n - f\|_{V^*} = 0.$$

From this and (1.5.61), we conclude

$$(f, v) = \lim_{n\to\infty} (u_n, v)_H, \qquad v \in V,$$

i.e., the dualization between V^* and V is identified with the unique extension of the scalar product $(u, v)_H$ defined on $H \times V \subset H \times H$, concluding the proof.

1.5.5 Scales of Hilbert spaces and compact embedding

Compact embedding of a Hilbert space

Let U, V be Hilbert spaces such that

$$\left.\begin{array}{l} U \subset V,\, U \text{ dense in } V, \text{ the embedding of } U \text{ into } V \text{ is} \\ \text{compact.} \end{array}\right\} \tag{1.5.62}$$

Choose V as a zero space, i.e., identify V with its dual V^*. Consider the following eigenvalue problem:

$$(u_j, v)_U = \lambda_j (u_j, v)_V, \qquad v \in U. \tag{1.5.63}$$

From (1.5.62) and Theorem 1.5.7 we deduce the existence of a countable set of eigenvalues $\{\lambda_j\}$ such that

$$0 < \lambda_1 \le \lambda_2 \le \cdots \le \lambda_j \le \cdots, \qquad \lim_{j\to\infty} \lambda_j = \infty, \tag{1.5.64}$$

and of a corresponding sequence of eigenfunctions $\{u_j\}$ which solve the problem (1.5.63) and form an orthonormal basis in V and orthogonal in U. Thus,

$$(u_i, u_j)_V = \delta_{ij}, \qquad (u_i, u_j)_U = \lambda_i \delta_{ij}. \tag{1.5.65}$$

Every element $u \in V$ has a representation

$$u = \sum_{i=1}^{\infty} (u, u_i)_V u_i. \tag{1.5.66}$$

By (1.5.65)

$$\|u\|_V^2 = \sum_{i=1}^{\infty} (u, u_i)_V^2 < \infty, \tag{1.5.67}$$

and if $u \in U$, then

$$\|u\|_U^2 = \sum_{i=1}^{\infty} \lambda_i (u, u_i)_V^2 < \infty. \tag{1.5.68}$$

From (1.5.62) and Theorem 1.5.11, it follows that V is dense in U^*, and since $\{u_i\}$ is a basis in V and U, every element $g \in U^*$ has a representation

$$g = \sum_{i=1}^{\infty} (g, u_i) u_i. \tag{1.5.69}$$

Let us show that

$$\|g\|_{U^*}^2 = \sum_{i=1}^{\infty} \lambda_i^{-1}(g, u_i)^2. \tag{1.5.70}$$

Indeed, in virtue of (1.5.66), (1.5.68), and (1.5.69), we get, for $g \in U^*$ and any $u \in U$,

$$\begin{aligned} |(g,u)| &= \left| \sum_{i=1}^{\infty} (g, u_i)(u, u_i)_V \right| \\ &= \left| \sum_{i=1}^{\infty} \left[\lambda_i^{-\frac{1}{2}} (g, u_i)\right] \left[\lambda_i^{\frac{1}{2}} (u, u_i)_V\right] \right| \\ &\le \left[\sum_{i=1}^{\infty} \lambda_i^{-1} (g, u_i)^2 \right]^{\frac{1}{2}} \|u\|_U. \end{aligned}$$

Hence,

$$\|g\|_{U^*} \le \left[\sum_{i=1}^{\infty} \lambda_i^{-1} (g, u_i)^2 \right]^{\frac{1}{2}}. \tag{1.5.71}$$

Let v be of the form

$$v = \sum_{i=1}^{n} \lambda_i^{-1} (g, u_i) u_i. \tag{1.5.72}$$

Evidently, $v \in U$ and by (1.5.66), (1.5.68) we have

$$\|v\|_U^2 = \sum_{i=1}^{n} \lambda_i^{-1} (g, u_i)^2. \tag{1.5.73}$$

From (1.5.69), (1.5.72), and (1.5.73), we infer

$$|(g,v)| = \sum_{i=1}^{n} \lambda_i^{-1} (g, u_i)^2 = \left[\sum_{i=1}^{n} \lambda_i^{-1} (g, u_i)^2 \right]^{\frac{1}{2}} \|v\|_U.$$

From here

$$\left[\sum_{i=1}^{n} \lambda_i^{-1} (g, u_i)^2 \right]^{\frac{1}{2}} \le \|g\|_{U^*}.$$

The latter inequality holds for an arbitrary n, and therefore

$$\left[\sum_{i=1}^{\infty} \lambda_i^{-1} (g, u_i)^2 \right]^{\frac{1}{2}} \le \|g\|_{U^*}.$$

This together with (1.5.71) yields (1.5.70).

Scales of Hilbert spaces

Let the conditions (1.5.62) be satisfied and let H_θ, $-1 \le \theta \le 1$, be the Hilbert spaces distinguished by

$$H_\theta = \Big\{ u \mid u \in U^*, \ \sum_{i=1}^{\infty} \lambda_i^\theta (u, u_i)^2 < \infty \Big\} \tag{1.5.74}$$

and endowed with the norm

$$\|u\|_{H_\theta} = \left[\sum_{i=1}^{\infty} \lambda_i^\theta (u, u_i)^2 \right]^{\frac{1}{2}}. \tag{1.5.75}$$

From (1.5.67), (1.5.68), and (1.5.70), it follows that

$$H_0 = V = V^*, \qquad H_1 = U, \qquad H_{-1} = U^*,$$

and

$$H_\theta^* = H_{-\theta}, \qquad \theta > 0.$$

Theorem 1.5.12 *Let U, V be two Hilbert spaces for which* (1.5.62) *holds and let spaces H_θ, $-1 \le \theta \le 1$, be defined by* (1.5.74), (1.5.75). *Then, the embedding of H_{θ_1} into H_{θ_2}, where $-1 \le \theta_2 < \theta_1 \le 1$, is compact.*

Proof. Let $\{v_n\} \subset H_{\theta_1}$ and

$$v_n \to 0 \qquad \text{weakly in } H_{\theta_1}. \tag{1.5.76}$$

Let us show that then

$$v_n \to 0 \qquad \text{strongly in } H_{\theta_2}. \tag{1.5.77}$$

From (1.5.75) and (1.5.76) we get

$$\|v_n\|^2_{H_{\theta_1}} = \sum_{i=1}^{\infty} \lambda_i^{\theta_1} (v_n, u_i)^2 \le \text{const} \qquad \forall n. \tag{1.5.78}$$

By (1.5.75),

$$\begin{aligned} \|v_n\|^2_{H_{\theta_2}} &= \sum_{i=1}^{\infty} \lambda_i^{\theta_2} (v_n, u_i)^2 \\ &\le \sum_{i=1}^{M} \lambda_i^{\theta_2} (v_n, u_i)^2 + \sum_{i=M+1}^{\infty} \lambda_{M+1}^{\theta_2 - \theta_1} \big[\lambda_i^{\theta_1} (v_n, u_i)^2 \big]. \end{aligned} \tag{1.5.79}$$

Let now ε be an arbitrary positive number. Taking note that $\lambda_i \to \infty$ as $i \to \infty$ and $\theta_1 > \theta_2$, we infer from (1.5.78) that there exists M such that

$$\sum_{i=M+1}^{\infty} \lambda_{M+1}^{\theta_2-\theta_1} \left[\lambda_i^{\theta_1} (v_n, u_i)^2\right] \leq \frac{\varepsilon}{2} \qquad \forall n. \tag{1.5.80}$$

By (1.5.76), there is N such that

$$\sum_{i=1}^{M} \lambda_i^{\theta_2} (v_n, u_i)^2 \leq \frac{\varepsilon}{2}, \qquad n \geq N. \tag{1.5.81}$$

Finally, (1.5.77) follows from (1.5.79)–(1.5.81), concluding the proof.

1.6 Functional spaces that are used in the investigation of boundary value and optimal control problems

1.6.1 Spaces of continuously differentiable functions

Let Ω be a domain in $\mathbb{R}^n$, i.e., an open connected set in $\mathbb{R}^n$, $x = (x_1, x_2, \ldots, x_n) \in \Omega$. Further, let $k = (k_1, k_2, \ldots, k_n)$ be an ordered row of nonnegative integers k_i, which will be referred to as a multi-index. To every multi-index k we place in correspondence the differential operator

$$D^k = \frac{\partial^{k_1+\cdots+k_n}}{\partial x_1^{k_1} \cdots \partial x_n^{k_n}}.$$

The nonnegative integer

$$|k| = k_1 + \cdots + k_n$$

is called the order of the operator D^k. If $|k| = 0$, then $D^k f = f$.

A real-valued function u determined in $\overline{\Omega}$ is said to belong to the space $C^l(\overline{\Omega})$ if it has continuous derivatives in $\overline{\Omega}$ up to and including order l. This means that every derivative of order $|k| \leq l$ exists at all points of Ω and coincides in these points with a continuous function on $\overline{\Omega}$.

The norm in $C^l(\overline{\Omega})$ is defined by the formula

$$\|u\|_{C^l(\overline{\Omega})} = \sum_{|k| \leq l} \sup_{x \in \Omega} \left|D^k u(x)\right|.$$

For $l = 0$, we get the space of continuous functions on $\overline{\Omega}$, which is denoted by $C(\overline{\Omega})$.

If Ω is a bounded open set in $\mathbb{R}^n$, then $C^l(\overline{\Omega})$ is a Banach space, and, in the definition of the norm, sup may be changed to max.

By $C^\infty(\overline{\Omega})$ we denote the space of infinitely differentiable functions in $\overline{\Omega}$.

1.6.2 Spaces of integrable functions

The space of (classes of) real-valued functions f which are measurable on Ω and such that

$$\|f\|_{L_p(\Omega)} = \left(\int_\Omega |f(x)|^p dx \right)^{\frac{1}{p}} < \infty, \qquad 1 \leq p < \infty,$$

will be referred to as $L_p(\Omega)$.

The space $L_p(\Omega)$ is a separable Banach space. As long as $1 < p < \infty$, the spaces $L_p(\Omega)$ are reflexive. The space $L_2(\Omega)$ is a Hilbert space equipped with the scalar product

$$(u, v) = \int_\Omega u(x)\, v(x)\, dx.$$

A measurable function $u\colon \Omega \to \mathbb{R}$ is called essentially bounded if it is equivalent to some bounded function, i.e., if there exists a number M such that $|u(x)| \leq M$ for almost all $x \in \Omega$. The precise lower bound of such constants is denoted by $\operatorname*{vrai\,max}_{x\in\Omega} |u(x)|$.

The space of (classes of) measurable, essentially bounded functions will be referred to as $L_\infty(\Omega)$, which is a Banach space with respect to the norm

$$\|u\|_{L_\infty(\Omega)} = \operatorname*{vrai\,max}_{x\in\Omega} |u(x)|.$$

$L_1(\Omega)$ is called the space of integrable functions. If Ω is a bounded domain in $\mathbb{R}^n$, then $L_p(\Omega) \subset L_1(\Omega)$ for all $p \in (1, \infty]$.

Theorem 1.6.1 (Lebesgue) *Let $\{u_k\}$ be a sequence of integrable functions on Ω which converges almost everywhere (a.e.) on Ω to a function u, i.e., for almost all $x \in \Omega$ the numerical sequence $\{u_k(x)\}$ tends to $u(x)$ as $k \to \infty$. Assume that there exists an integrable function v on Ω such that*

$$|u_k(x)| \leq v(x) \qquad \forall k \ \text{ a.e. on } \Omega.$$

Then, the function u is integrable on Ω and

$$\lim_{k\to\infty} \int_\Omega u_k\, dx = \int_\Omega u\, dx.$$

1.6.3 Test and generalized functions

Let $x \to f(x)$ be an arbitrary continuous function on $\overline{\Omega}$. The support of f (denote by $\operatorname{supp} f$) is the least closed set in $\overline{\Omega}$ outside of which $f(x)$ vanishes, i.e.,

$$\operatorname{supp} f = \overline{\{\, x \mid f(x) \neq 0 \,\}} \cap \overline{\Omega}.$$

By $\mathcal{D}(\Omega)$ we denote the space of real-valued functions u which are infinitely differentiable in the domain Ω and have compact supports in Ω. The space $\mathcal{D}(\Omega)$

is endowed with the inductive limit topology (see, e.g., Schwartz (1966), Rudin (1973)). A sequence $\{u_i\} \subset \mathcal{D}(\Omega)$ converges to a function $u \in \mathcal{D}(\Omega)$ as $i \to \infty$ if the following conditions hold:

1. there exists a compact set $B \subset \Omega$ such that the supports of all the functions u_i are contained in B;
2. the derivatives of an arbitrary order $m \geq 1$ of the functions u_i uniformly converge to the corresponding derivatives of the function u as $i \to \infty$.

By $\mathcal{D}^*(\Omega)$ we denote the dual of $\mathcal{D}(\Omega)$ (the space of distributions or generalized functions), i.e., the space of linear functionals on $\mathcal{D}(\Omega)$ that are continuous with respect to the topology of $\mathcal{D}(\Omega)$. If $T \in \mathcal{D}^*(\Omega)$ and $u \in \mathcal{D}(\Omega)$, then the value of T at u is denoted by (T, u) and called the dual pairing between T and u.

The space $\mathcal{D}^*(\Omega)$ is equipped with the strong dual topology, that is, with the topology of uniform convergence on every bounded set in $\mathcal{D}(\Omega)$. Notice that the convergence in the weak topology of the space $\mathcal{D}^*(\Omega)$ is equivalent to the convergence in the strong topology. Thus, a sequence of distributions $\{T_i\}$ is said to converge to a distribution T if, for every function $u \in \mathcal{D}(\Omega)$,

$$(T_i, u) \to (T, u).$$

For $T \in \mathcal{D}^*(\Omega)$, the derivative $\frac{\partial T}{\partial x_i}$ is defined by the equality

$$\left(\frac{\partial T}{\partial x_i}, u\right) = -\left(T, \frac{\partial u}{\partial x_i}\right), \qquad u \in \mathcal{D}(\Omega),$$

which determines the linear continuous mapping

$$T \to \frac{\partial T}{\partial x_i}$$

of $\mathcal{D}^*(\Omega)$ into $\mathcal{D}^*(\Omega)$.

Placing in correspondence to every function $f \in L_p(\Omega)$ the distribution (generalized function) $\tilde{f}$ given by the formula

$$u \to (\tilde{f}, u) = \int_\Omega f(x)\, u(x)\, dx, \qquad u \in \mathcal{D}(\Omega),$$

we obtain the linear, continuous, injective mapping $f \to \tilde{f}$ of the space $L_p(\Omega)$ into $\mathcal{D}^*(\Omega)$. Then, f and $\tilde{f}$ can be identified. Thus, we have

$$\mathcal{D}(\Omega) \subset L_p(\Omega) \subset \mathcal{D}^*(\Omega). \tag{1.6.1}$$

Taking (1.6.1) into account, one derives the distribution derivatives $\frac{\partial f}{\partial x_i}$ of any function $f \in L_p(\Omega)$. In general, for $f \in \mathcal{D}^*(\Omega)$, we set

$$\left(D^k f, u\right) = (-1)^{|k|}\left(f, D^k u\right), \qquad u \in \mathcal{D}(\Omega). \tag{1.6.2}$$

According to (1.6.2), a function $f \in L_p(\Omega)$ is said to have the distribution derivative $D^k f$ which belongs to $L_p(\Omega)$ if there exists a function $w_k \in L_p(\Omega)$ such that

$$(w_k, u) = (-1)^{|k|}\left(f, D^k u\right), \qquad u \in \mathcal{D}(\Omega).$$

The function w_k is called the distribution derivative of $f \in L_p(\Omega)$, and the notation $w_k = D^k f$ is used just as for the ordinary derivative.

1.6.4 Sobolev spaces

Definitions. The embedding theorem

For $p \geq 1$ and $m \in \mathbb{N}$ ($\mathbb{N}$ denoting the set of natural numbers), the Sobolev space $W_p^m(\Omega)$ is defined as the set all $v \in L_p(\Omega)$ whose distribution derivatives $D^k v$, $|k| \leq m$, belong to $L_p(\Omega)$, and this set is endowed with the norm

$$\|v\|_{W_p^m(\Omega)} = \left(\sum_{|k| \leq m} \left\|D^k v\right\|^p_{L_p(\Omega)} \right)^{\frac{1}{p}}. \tag{1.6.3}$$

As the differential operator D^k maps continuously $\mathcal{D}^*(\Omega)$ into itself, $W_p^m(\Omega)$ is a Banach space, and it is reflexive for $1 < p < \infty$.

When $m = 0$, the space $W_p^m(\Omega)$ is identified with the space $L_p(\Omega)$, and, for $p = 2$, $W_2^m(\Omega)$ is a Hilbert space.

For domains with Lipschitz continuous boundaries, the set of functions $C^\infty(\overline{\Omega})$ is dense in $W_p^m(\Omega)$.

By $\overset{\circ}{W}{}_p^m(\Omega)$ we denote the closure of $\mathcal{D}(\Omega)$ in the norm of the space $W_p^m(\Omega)$.

The embedding of a space $W_p^m(\Omega)$ into a space $W_{p_1}^{m_1}(\Omega_1)$, $\Omega_1 \subset \Omega$, is the operator which maps every function from $W_p^m(\Omega)$ into itself, being considered as an element of $W_{p_1}^{m_1}(\Omega_1)$. There are a number of results known as embedding theorems. Let us present one of them; see, Sobolev (1963), Adams (1975), Brezis (1983).

Theorem 1.6.2 (embedding theorem) *Let Ω be a bounded domain in $\mathbb{R}^n$ with a Lipschitz continuous boundary. Then, the embedding of $W_p^m(\Omega)$ into $L_q(S_r)$, where S_r is the intersection of Ω with an r-dimensional plane, in particular, $S_n = \Omega$, is a bounded operator if $n > mp$, $r > n - mp$, $q \leq \frac{pr}{n-mp}$, and is a compact operator if $q < \frac{pr}{n-mp}$. For $n = mp$, it is a compact operator for every finite q. For $n < mp$, any function $u \in W_p^m(\Omega)$ is continuous in $\overline{\Omega}$, and the operator of embedding of $W_p^m(\Omega)$ into $C(\overline{\Omega})$ is bounded and compact. If $n < mp$ and $0 < l < m - \frac{n}{p}$, the operator of embedding of $W_p^m(\Omega)$ into $C^l(\overline{\Omega})$ is bounded and compact.*

Theorem 1.6.3 (Calderon) *Let Ω be a bounded domain in $\mathbb{R}^n$ with a Lipschitz continuous boundary. Then, there exists a continuation operator P such that*

$$P \in \mathcal{L}\left(W_p^m(\Omega), W_p^m(\mathbb{R}^n)\right),$$
$$Pu = u \qquad \text{a.e. in } \Omega.$$

For the proof see, Calderon (1961) and Fikhtengolts (1966) for the case $n = 2$.

Averaging (regularization) of functions

Let $\omega(\xi)$ be a nonnegative, infinitely differentiable function on

$$\mathbb{R}_+ = \{ t \,|\, t \in \mathbb{R},\ t \geq 0 \},$$

which vanishes for $\xi \geq 1$. For example,

$$\omega(\xi) = \begin{cases} \exp \frac{1}{\xi^2 - 1}, & \text{if } 0 \leq \xi < 1, \\ 0, & \text{if } 1 \leq \xi < \infty. \end{cases}$$

If $x \in \mathbb{R}^n$, then the function $x \to \omega\big(\frac{|x|}{\rho}\big)$ vanishes for $|x| \geq \rho$,

$$|x| = \Big(\sum_{i=1}^n x_i^2 \Big)^{\frac{1}{2}}$$

and

$$\int_{\mathbb{R}^n} \omega\Big(\frac{|x|}{\rho}\Big)\, dx = \chi \rho^n,$$

where χ is a constant. As an averaging kernel we take

$$\omega_\rho(x) = \frac{1}{\chi \rho^n}\, \omega\Big(\frac{|x|}{\rho}\Big). \tag{1.6.4}$$

If f is a locally integrable function in $\mathbb{R}^n$, then the averaging of this function has the form

$$f_\rho(x) = \int_{\mathbb{R}^n} \omega_\rho(x - y)\, f(y)\, dy \tag{1.6.5}$$

(actually, the integral is taken over the ball $|x - y| \leq \rho$, not over the whole $\mathbb{R}^n$). If f is defined only in the domain Ω, then f_ρ is defined in the domain $\Omega_\rho \subset \Omega$ the boundary of which is remote from the boundary of Ω on distance ρ, and $f_\rho \in C^\infty(\overline{\Omega}_\rho)$. The averaging has the following properties (see Sobolev (1963)).

1. Let $f \in L_p(\Omega)$, $p \geq 1$. Set $f(x) = 0$ for $x \notin \Omega$. Then, the function f_ρ is well defined in $\mathbb{R}^n$, infinitely differentiable, and $f_\rho \to f$ in $L_p(\Omega)$ as $\rho \to 0$.
2. Let $f \in L_p(\Omega)$, $p \geq 1$. Set $f(x) = 0$ for $x \notin \Omega$. Then, for all $\rho > 0$, the function $f \mapsto f_\rho$ is a linear continuous mapping of $L_p(\Omega)$ into $C^m(\overline{\Omega})$ for an arbitrary integer $m > 0$.
3. Let K be an arbitrary compact set in Ω and let $f \in W_p^m(\Omega)$. Then, $D^k f_\rho = (D^k f)_\rho$ in $\overset{\circ}{K}$ as long as $|k| \leq m$ and ρ is sufficiently small. Moreover, $D^k f_\rho \to D^k f$ in $L_p(\overset{\circ}{K})$ as $\rho \to 0$.

Remark 1.6.1 Property 3 and Theorem 1.6.3 imply that, if Ω is a bounded domain in $\mathbb{R}^n$ with a Lipschitz continuous boundary, then every function $u \in W_p^m(\Omega)$ can be extended to a function $\overline{u} = Pu \in W_p^m(\mathbb{R}^n)$, and moreover $\overline{u}_\rho \to u$ in $W_p^m(\Omega)$ as $\rho \to 0$.

Theorem on equivalent norms

We will need the following statement.

Theorem 1.6.4 *Let l_j, $j = 1, 2, \dots, N$, be linear continuous functionals in $W_p^m(\Omega)$, where Ω is a bounded domain in $\mathbb{R}^n$ and $1 < p < \infty$. Assume that, for each polynomial*

$$g(x) = \sum_{|k| \le m-1} c_k x_1^{k_1} \cdots x_n^{k_n}, \qquad |k| = k_1 + \cdots + k_n, \ c_k = \text{const}, \tag{1.6.6}$$

of degree $\le m-1$, the condition $l_j(g) = 0$, $j = 1, 2, \dots, N$, yields that $g = 0$. Then, the norm in $W_p^m(\Omega)$ defined by the formula

$$\|v\| = \Big(\sum_{|k|=m} \|D^k v\|^p_{L_p(\Omega)} + \sum_{j=1}^N |l_j(v)|^p \Big)^{\frac{1}{p}}, \tag{1.6.7}$$

is equivalent to the norm (1.6.3).

Proof. Let us establish the inequality

$$\|v\|_{W_p^m(\Omega)} \le c\|v\|, \qquad v \in W_p^m(\Omega), \tag{1.6.8}$$

where c is a constant and the norms on the left- and right-hand sides are defined by the relations (1.6.3) and (1.6.7). Indeed, if (1.6.8) is not valid, then there exists a sequence $\{v_i\} \subset W_p^m(\Omega)$ such that

$$\|v_i\|_{W_p^m(\Omega)} = 1 \qquad \forall i, \tag{1.6.9}$$

$$\sum_{|k|=m} \|D^k v_i\|^p_{L_p(\Omega)} \to 0 \qquad \text{as } i \to \infty, \tag{1.6.10}$$

$$l_j(v_i) \to 0 \qquad \text{as } i \to \infty, \quad j = 1, 2, \dots, N. \tag{1.6.11}$$

By virtue of (1.6.9) and Theorem 1.6.2, from the sequence $\{v_i\}$ one can choose a subsequence $\{v_\mu\}$ such that

$$v_\mu \to v_0 \qquad \text{weakly in } W_p^m(\Omega), \tag{1.6.12}$$

$$D^k v_\mu \to D^k v_0 \qquad \text{weakly in } L_p(\Omega), \ |k| = m, \tag{1.6.13}$$

$$D^k v_\mu \to D^k v_0 \qquad \text{strongly in } L_p(\Omega), \ |k| \le m-1. \tag{1.6.14}$$

(1.6.10) and (1.6.13) imply that

$$D^k v_0 = 0 \qquad \text{a.e. in } \Omega, \ |k| = m. \tag{1.6.15}$$

Let K be an arbitrary compact set in Ω. For sufficiently small $\rho > 0$, in $\overset{\circ}{K}$ we can define the function $v_{0\rho}$ which is the averaging of the function v_0. (1.6.15) and Property 3 of the averaging imply that

$$D^k v_{0\rho} = \left(D^k v_0\right)_\rho = 0 \qquad \text{in } \overset{\circ}{K}, \ |k| = m.$$

From here, taking into account the infinite differentiability of the function $v_{0\rho}$ in $\overset{\circ}{K}$, we get

$$v_{0\rho} = \sum_{|k| \le m-1} a_k^\rho x_1^{k_1} \cdots x_n^{k_n} \qquad \text{in } \overset{\circ}{K}, \ |k| = k_1 + \cdots + k_n, \tag{1.6.16}$$

a_k^ρ being constants. By Property 3 of the averaging, we have $v_{0\rho} \to v_0$ in $L_p(\overset{\circ}{K})$ as $\rho \to 0$. Thus, (1.6.16) and the closedness of a finite-dimensional subspace of $L_p(\overset{\circ}{K})$ yield

$$v_0 = \sum_{|k| \le m-1} a_k x_1^{k_1} \cdots x_n^{k_n} \qquad \text{in } \overset{\circ}{K}, \ |k| = k_1 + \cdots + k_n. \tag{1.6.17}$$

Let K_1 be a compact neighborhood of K, and let $K_1 \subset \Omega$. Similarly to the above argument, we conclude that the function v_0 is defined by the relation (1.6.17) in $\overset{\circ}{K}_1$ and, hence, in K. There exists a sequence of compact sets $\{K_\nu\}$ such that $K_\nu \subset \Omega$, $K_\nu \subset K_{\nu+1}$ for all ν, and $\Omega = \bigcup_{\nu=1}^\infty K_\nu$. Since in every K_ν the function v_0 has the form (1.6.17), it is determined by this formula in Ω.

(1.6.11) and (1.6.12) imply

$$l_j(v_0) = 0, \qquad j = 1, 2, \ldots, N.$$

Since v_0 is a polynomial in Ω of degree $\le m-1$, the hypothesis of the theorem yields

$$v_0 = 0. \tag{1.6.18}$$

By (1.6.10), (1.6.14), and (1.6.15), we have

$$v_\mu \to v_0 \qquad \text{strongly in } W_p^m(\Omega).$$

This makes a contradiction with (1.6.9) and (1.6.18), so that (1.6.8) holds. The inequality inverse of (1.6.8) is implied by the boundedness of the functionals l_j. The theorem is proved.

The following statement is a consequence of the proof of Theorem 1.6.4.

Corollary 1.6.1 *Let Ω be a bounded domain in $\mathbb{R}^n$ and let $1 < p < \infty$. Assume that E is a subspace of $W_p^m(\Omega)$ such that the condition that $g \in E$ is a polynomial of degree $\leq m-1$ yields $g = 0$. Then, in the space E, a norm defined by the relation*

$$\|v\| = \Big(\sum_{|k|=m} \|D^k v\|^p_{L_p(\Omega)} \Big)^{\frac{1}{p}},$$

is equivalent to the norm (1.6.3).

Spaces $H^s(\Omega)$

In the case $p = 2$, one uses the notation

$$H^m(\Omega) = W_2^m(\Omega).$$

If the boundary S of the domain Ω is regular, more precisely, if S is a manifold of dimension $n-1$ and of the C^∞ class, and Ω is placed at one side of S, then one can define the spaces $H^s(\Omega)$ for any real $s \geq 0$ as intermediate spaces between $H^m(\Omega)$, $m \in \mathbb{N}$, and $H^0(\Omega) = L_2(\Omega)$ (see Lions and Magenes (1972) and Subsec. 1.5.5).

In case of a smooth boundary, this definition of the space $H^s(\Omega)$ is independent of a choice of m; $H^s(\Omega)$ is a Hilbert space.

In particular, if Ω is a bounded regular domain,

$$H^s(\Omega) = \Big\{ u \,|\, u \in H^0(\Omega),\ \sum_{i=1}^{\infty} \lambda_i^\theta (u, u_i)^2_{H^0(\Omega)} < \infty,\ \theta m = s \Big\},$$

where $\theta \in [0,1]$, λ_i and u_i are the eigenvalues and eigenfunctions of the problem

$$(u_i, v)_{H^m(\Omega)} = \lambda_i (u_i, v)_{H^0(\Omega)}, \qquad v \in H^m(\Omega),$$

and $(u_i, u_j)_{H^0(\Omega)} = \delta_{ij}$. The norm in $H^s(\Omega)$ can be defined by the formula

$$\|u\|_{H^s(\Omega)} = \Big(\sum_{i=1}^{\infty} \lambda_i^\theta (u, u_i)^2_{H^0(\Omega)} \Big)^{\frac{1}{2}}, \qquad \theta m = s.$$

By $H_0^s(\Omega)$ we denote the closure of $\mathcal{D}(\Omega)$ in $H^s(\Omega)$, and by $H^{-s}(\Omega)$ the dual space of $H_0^s(\Omega)$, $s > 0$, i.e., $(H_0^s(\Omega))^* = H^{-s}(\Omega)$.

By using a system of local maps, one introduces the notion of a trace of a function as an element of the space $H^t(S)$, $t > 0$, $(H^t(S))^* = H^{-t}(S)$ for $t \geq 0$. The following theorem on the trace space holds (see Lions and Magenes (1972)).

Theorem 1.6.5 *The mapping*

$$C^\infty(\overline{\Omega}) \ni u \to \Big\{ \frac{\partial^i u}{\partial \nu^i}\Big|_S,\ i = 0, 1, 2, \ldots, m-1 \Big\} \in \big(C^\infty(S)\big)^m, \tag{1.6.19}$$

where $\frac{\partial^i u}{\partial \nu^i}\Big|_S$ *is the normal derivative of order* i *on* S, *can be extended by continuity to a linear continuous mapping of* $H^m(\Omega)$ *into* $\prod_{i=0}^{m-1} H^{m-i-\frac{1}{2}}(S)$.

The mapping (1.6.19) *is surjective and its kernel coincides with* $H_0^m(\Omega) = \overset{\circ}{W}{}_2^m(\Omega)$.

Remark 1.6.2 If Ω is a bounded domain with a Lipschitz continuous boundary, then the trace $u|_S$ of a function $f \in H^m(\Omega)$, $m \geq 1$, on S belongs to the space $L_2(S)$ and the corresponding embedding operator $u \to u|_S$ is compact, see, e.g., Ladyzhenskaya and Uraltseva (1973).

Remark 1.6.3 If the boundary S of a domain Ω is not smooth, but consists of a finite number of smooth, open (in S), and disjoint subsets S_i such that

$$S = \bigcup_{i=1}^{\nu} \overline{S}_i,$$

then to any set S_i one can apply Theorem 1.6.5. More precisely, if S_i is an $(n-1)$-dimensional manifold of the C^∞ class, Ω is placed at one side of S, and S_i' is an open set in S_i such that $\overline{S_i'} \subset S_i$, then

$$u \to \left\{ \frac{\partial^i u}{\partial \nu^i}\Big|_{S_i'},\ i = 0, 1, 2, \ldots, m-1 \right\}$$

is a continuous, surjective mapping of $H^m(\Omega)$ onto $\prod_{i=0}^{m-1} H^{m-i-\frac{1}{2}}(S_i')$.

1.7 Inequalities of coerciveness

1.7.1 Coercive systems of operators

Let $W = \prod_{r=1}^{m} W_2^{l_r}(\Omega)$ be a topological product of spaces $W_2^{l_r}(\Omega)$, $l_r \geq 1$, $r = 1, 2, \ldots, m$, i.e., W is the space of vector functions $v = (v_1, v_2, \ldots, v_m)$ defined on $\Omega \subset \mathbb{R}^n$ and taking values in $\mathbb{R}^m$, where $v_r \in W_2^{l_r}(\Omega)$. The norm in W is defined through the formula

$$\|v\|_W^2 = \sum_{r=1}^{m} \|v_r\|_{W_2^{l_r}(\Omega)}^2. \tag{1.7.1}$$

Further, let N_i, $i = 1, 2, \ldots, \nu$, be the linear continuous mappings of W into $L_2(\Omega)$ defined by the formula

$$v \to N_i v = \sum_{r=1}^{m} \sum_{|k| \leq l_r} g_{irk} D^k v_r, \tag{1.7.2}$$

where

$$g_{irk} \in L_\infty(\Omega). \tag{1.7.3}$$

The system of operators $\{N_i\}_{i=1}^{\nu}$ is called W-coercive with respect to $(L_2(\Omega))^m$ if there exists a constant $c > 0$ such that

$$\sum_{i=1}^{\nu} \|N_i v\|^2_{L_2(\Omega)} + \sum_{r=1}^{m} \|v_r\|^2_{L_2(\Omega)} \geq c\|v\|^2_W, \qquad v \in W. \tag{1.7.4}$$

Let V be a closed subspace of W (in particular, we can take $V = W$). The system of operators $\{N_i\}_{i=1}^{\nu}$ is called coercive in V if there exists a constant $c > 0$ such that

$$\sum_{i=1}^{\nu} \|N_i v\|^2_{L_2(\Omega)} \geq c\|v\|^2_W, \qquad v \in V. \tag{1.7.5}$$

For an arbitrary $\xi = (\xi_1, \ldots, \xi_n) \in \mathbb{C}^n$ ($\mathbb{C}$ being the field of complex numbers), we set

$$N_{ir}(x, \xi) = \sum_{|k|=l_r} g_{irk}(x)\xi_1^{k_1} \cdots \xi_n^{k_n}. \tag{1.7.6}$$

An open set $\Omega \subset \mathbb{R}^n$ is said to satisfy the cone condition if

$$x + U(e(x), H) \subset \Omega, \qquad x \in \Omega,$$

where $U(e(x), H)$ is a straight circular cone with vertex at the origin, of a fixed angle, and of a fixed height H, $0 < H < \infty$, whose axis has direction $e(x)$ depending on x. In particular, a Lipschitz domain satisfies the cone condition.

The following statement is valid (Besov et al. (1975), Hlavaček and Nečas (1970)).

Theorem 1.7.1 *Let Ω be a bounded domain in $\mathbb{R}^n$ satisfying the cone condition. The system of differential operators*

$$v \to N_i v = \sum_{r=1}^{m} \sum_{|k| \leq l_r} g_{irk} D^k v_r, \qquad i = 1, 2, \ldots, \nu,$$

where $g_{irk} \in L_\infty(\Omega)$ if $|k| < l_r$ and $g_{irk} \in C(\overline{\Omega})$ if $|k| = l_r$, is W-coercive with respect to $(L_2(\Omega))^m$ if (and, in the case when g_{irk} are constants for $|k| = l_r$, only if) the rank of the matrix $\{N_{ir}(x, \xi)\}$ is equal to m for an arbitrary complex $\xi \in \mathbb{C}^n$, $\xi \neq 0$, and any $x \in \overline{\Omega}$.

In the sequel, we will need also the following theorem.

Theorem 1.7.2 *Let $\{N_i\}_{i=1}^{\nu}$ be a system of operators that is W-coercive with respect to $(L_2(\Omega))^m$ and let $u, v \to a(u, v)$ be a bilinear, symmetric, continuous form on $W \times W$ such that*

$$a(v, v) \geq 0, \qquad v \in W, \tag{1.7.7}$$

and the conditions

$$w \in W, \qquad \sum_{i=1}^{\nu} \|N_i w\|^2_{L_2(\Omega)} + a(w,w) = 0 \tag{1.7.8}$$

imply $w = 0$. *Then, there exists a positive constant* c_0 *such that*

$$\sum_{i=1}^{\nu} \|N_i v\|^2_{L_2(\Omega)} + a(v,v) \geq c_0 \|v\|^2_W, \qquad v \in W. \tag{1.7.9}$$

Proof. Let us show first that there exists a positive constant c_1 such that

$$\sum_{i=1}^{\nu} \|N_i v\|^2_{L_2(\Omega)} + a(v,v) \geq c_1 \|v\|^2_{(L_2(\Omega))^m}, \qquad v \in W. \tag{1.7.10}$$

Here

$$\|v\|^2_{(L_2(\Omega))^m} = \sum_{r=1}^{m} \|v_r\|^2_{L_2(\Omega)}.$$

Assume that the inequality (1.7.10) is not valid. Then, there exists a sequence $\{v^{(n)}\}$ of elements of W such that

$$\|v^{(n)}\|_{(L_2(\Omega))^m} = 1, \tag{1.7.11}$$

$$\sum_{i=1}^{\nu} \|N_i v^{(n)}\|^2_{L_2(\Omega)} \to 0, \qquad a\left(v^{(n)}, v^{(n)}\right) \to 0. \tag{1.7.12}$$

From (1.7.4), (1.7.11), and (1.7.12), we conclude that the sequence $\{v^{(n)}\}$ is bounded in W. Let us choose from it a subsequence $\{v^{(k)}\}$ such that

$$v^{(k)} \to w \qquad \text{weakly in } W, \tag{1.7.13}$$

$$v^{(k)} \to w \qquad \text{strongly in } (L_2(\Omega))^m. \tag{1.7.14}$$

The formulas (1.7.11) and (1.7.14) yield

$$\|w\|_{(L_2(\Omega))^m} = 1. \tag{1.7.15}$$

(1.7.13) implies

$$N_i v^{(k)} \to N_i w \qquad \text{weakly in } L_2(\Omega) \text{ as } k \to \infty, \quad i = 1, 2, \ldots, \nu.$$

Combining this with (1.7.12) and taking into consideration Theorem 1.2.4, we obtain

$$\lim_{k\to\infty} \sum_{i=1}^{\nu} \|N_i v^{(k)}\|^2_{L_2(\Omega)} = \liminf_{k\to\infty} \sum_{i=1}^{\nu} \|N_i v^{(k)}\|^2_{L_2(\Omega)} \geq \sum_{i=1}^{\nu} \|N_i w\|^2_{L_2(\Omega)} = 0. \tag{1.7.16}$$

One can easily derive the equality

$$a(v^{(k)}, v^{(k)}) = a(w, w) + 2a(w, v^{(k)} - w) + a(v^{(k)} - w, v^{(k)} - w).$$

Then, (1.7.7), (1.7.12), and (1.7.13) yield

$$\lim_{k\to\infty} a(v^{(k)}, v^{(k)}) = \liminf_{k\to\infty} a(v^{(k)}, v^{(k)}) \geq a(w, w) = 0. \tag{1.7.17}$$

By virtue of (1.7.16) and (1.7.17), the element w satisfies the conditions (1.7.8), so that $w = 0$. But this makes a contradiction to the equality (1.7.15), hence (1.7.10) is valid. The inequality (1.7.10) implies that

$$\sum_{i=1}^{\nu} \|N_i v\|^2_{L_2(\Omega)} + a(v, v) \geq c_2 \left(\sum_{i=1}^{\nu} \|N_i v\|^2_{L_2(\Omega)} + \|v\|^2_{(L_2(\Omega))^m} \right), \qquad v \in W, \tag{1.7.18}$$

c_2 being a positive constant.

Now, from (1.7.4) and (1.7.18) we deduce the inequality (1.7.9) with the positive constant $c_0 = cc_2$.

Remark 1.7.1 Assume that the conditions of Theorem 1.7.2 are satisfied and that V is a closed subspace of W defined through the relation

$$V = \{ u \,|\, u \in W,\ a(u, u) = 0 \}.$$

Then, the relation (1.7.9) implies the inequality

$$\sum_{i=1}^{\nu} \|N_i v\|^2_{L_2(\Omega)} \geq c_0 \|v\|^2_W, \qquad v \in V,$$

i.e., the system of operators $\{N_i\}$ is coercive in V.

Theorem 1.7.3 *Let $\{N_i\}_{i=1}^{\nu}$ be a system of operators that is W-coercive with respect to $(L_2(\Omega))^m$ and let V be a closed subspace of W such that the condition*

$$w \in V, \qquad \sum_{i=1}^{\nu} \|N_i w\|^2_{L_2(\Omega)} = 0$$

yields $w = 0$. Then, there exists a positive number c such that

$$\sum_{i=1}^{\nu} \|N_i v\|^2_{L_2(\Omega)} \geq c \|v\|^2_W, \qquad v \in V$$

i.e., the system of operators $\{N_i\}_{i=1}^{\nu}$ is coercive in V.

Proof. We will use Theorem 1.7.2. Since $V \subset W$, $\{N_i\}_{i=1}^{\nu}$ is V-coercive with respect to $(L_2(\Omega))^m$. Choose the zero form for the bilinear, symmetric, continuous, nonnegative form a on $W \times W$, i.e.,

$$u, v \to a(u, v) = 0, \qquad u, v \in W.$$

Now, setting $W = V$ in Theorem 1.7.2, we get Theorem 1.7.3 as a consequence of Theorem 1.7.2.

1.7.2 Korn's inequality

We will now obtain Korn's inequality for a two-dimensional space as an example of application of Theorem 1.7.1; we will need this inequality below. Let Ω be a bounded domain in $\mathbb{R}^2$ with a Lipschitz continuous boundary, and let $W = \big(W_2^1(\Omega)\big)^2$.

Introduce a system of linear continuous operators N_i mapping W into $L_2(\Omega)$ through the following formulas:

$$v = (v_1, v_2) \in W,$$
$$N_1 v = 2\,\frac{\partial v_1}{\partial x_1}, \qquad N_2 v = \sqrt{2}\left(\frac{\partial v_1}{\partial x_2} + \frac{\partial v_2}{\partial x_1}\right), \qquad N_3 v = 2\,\frac{\partial v_2}{\partial x_2}. \tag{1.7.19}$$

Comparing (1.7.2) and (1.7.19) we see that g_{irk} are constants in the present setting, so that the matrix $(N_{ir}(x,\xi))$ does not depend on x, and due to (1.7.6) it has the form

$$(N_{ir}(x,\xi)) = \begin{bmatrix} 2\xi_1 & 0 \\ \sqrt{2}\xi_2 & \sqrt{2}\xi_1 \\ 0 & 2\xi_2 \end{bmatrix}. \tag{1.7.20}$$

It is easy to see that, for any $\xi \in \mathbb{C}^2, \xi \neq 0$, the columns of the matrix (1.7.20) are linearly independent, hence the rank of this matrix is equal to 2. Applying Theorem 1.7.1 and taking into consideration (1.7.19), we get

$$\int_\Omega \sum_{i,j=1}^{2} \left(\frac{\partial v_i}{\partial x_j} + \frac{\partial v_j}{\partial x_i}\right)^2 dx + \sum_{i=1}^{2} \|v_i\|^2_{L_2(\Omega)} \geq c \sum_{i=1}^{2} \|v_i\|^2_{W_2^1(\Omega)}, \qquad v \in W, \tag{1.7.21}$$

c being a positive constant.

(1.7.21) is called Korn's inequality. From the physical point of view, the expressions $N_i v$ defined by the formulas (1.7.19) determine up to constant multipliers the components of the strain tensor of a continuum,

$$\varepsilon_{ij}(v) = \frac{1}{2}\left(\frac{\partial v_i}{\partial x_j} + \frac{\partial v_j}{\partial x_i}\right), \tag{1.7.22}$$

caused by a vector function of displacements $v = (v_1, v_2)$.

To apply Theorem 1.7.2, we must find the intersection of the kernel spaces of the operators N_i defined through the formulas (1.7.19), i.e., we must find the subspace of functions $v \in W$ such that

$$\varepsilon_{ij}(v) = \frac{1}{2}\left(\frac{\partial v_i}{\partial x_j} + \frac{\partial v_j}{\partial x_i}\right) = 0, \qquad i,j = 1,2. \tag{1.7.23}$$

From the physical point of view, these functions define "infinitesimal rigid displacements of continuum." We denote the subspace of these functions by Q,

$$Q = \left\{ v \,|\, v \in W,\ \frac{\partial v_i}{\partial x_j} + \frac{\partial v_j}{\partial x_i} = 0,\ i,j = 1,2 \right\}. \tag{1.7.24}$$

The space Q is known to be of the form (see, e.g., Hlavaček and Nečas (1970))

$$Q = \{ v \,|\, v = (v_1, v_2),\ v_1 = a_1 + a_3 x_2,\ v_2 = a_2 - a_3 x_1,\ a_i \in \mathbb{R} \}. \tag{1.7.25}$$

Let S be the boundary of Ω and let S_1 be an open set in S. Define a bilinear symmetric form on $W \times W$ by

$$a(u,v) = \int_{S_1} \sum_{i=1}^{2} u_i v_i \, ds.$$

Taking note of Remark 1.6.2, we obtain

$$|a(u,v)| \leq \sum_{i=1}^{2} \|u_i\|_{L_2(S_1)} \|v_i\|_{L_2(S_1)} \leq c\|u\|_W \|v\|_W.$$

Hence, the form $u, v \to a(u,v)$ is continuous on $W \times W$.

Further, let us show that, if $v \in Q$ and

$$\int_{S_1} (v_1^2 + v_2^2) \, ds = 0, \tag{1.7.26}$$

then $v = 0$. Indeed, suppose that $v \in Q$. In view of (1.7.25), the function $x \to v(x)$ has the following representation

$$v(x) = a + Ax, \tag{1.7.27}$$

where

$$v = \begin{Bmatrix} v_1 \\ v_2 \end{Bmatrix}, \qquad a = \begin{Bmatrix} a_1 \\ a_2 \end{Bmatrix}, \qquad A = \begin{bmatrix} 0 & a_3 \\ -a_3 & 0 \end{bmatrix}. \tag{1.7.28}$$

Since S_1 is an open set in S, there exist two different points $x^{(1)}$, $x^{(2)}$ belonging to S_1. By virtue of (1.7.26)–(1.7.28), we get

$$A(x^{(1)} - x^{(2)}) = 0,$$

i.e., the determinant of the matrix A should be equal to zero. So, (1.7.28) implies $a_3 = 0$, and from (1.7.26), (1.7.27) we deduce that $a_1 = a_2 = 0$.

Now, from (1.7.21) and Theorem 1.7.2 we derive the inequality

$$\int_\Omega \sum_{i,j=1}^{2} \left(\frac{\partial v_i}{\partial x_j} + \frac{\partial v_j}{\partial x_i} \right)^2 dx + \int_{S_1} (v_1^2 + v_2^2) \, ds \geq c\|v\|_W^2, \tag{1.7.29}$$
$$v \in W, \ c = \text{const} > 0.$$

Introduce the notation

$$V = \{ u \,|\, u \in W, \ \int_{S_1} (u_1^2 + u_2^2) \, ds = 0 \}. \tag{1.7.30}$$

V is obviously a subspace of W. The set V is closed in W since it is the prototype of the closed set $\{0\}$ under the continuous mapping

$$u \to \int_{S_1} (u_1^2 + u_2^2)\, ds$$

of W into $\mathbb{R}$. (1.7.29) and (1.7.30) yield

$$\int_\Omega \sum_{i,j=1}^{2} \left(\frac{\partial v_i}{\partial x_j} + \frac{\partial v_j}{\partial x_i} \right)^2 dx \geq c\|v\|_W^2, \qquad v \in V. \tag{1.7.31}$$

1.8 Theorem on the continuity of solutions of functional equations

Let F, G be normed vector spaces and let A be a linear continuous operator acting from F into G, i.e., $A \in \mathcal{L}(F, G)$. Recall that a mapping is called invertible if it is a bijection and the inverse bijection A^{-1} is also continuous, i.e., $A^{-1} \in \mathcal{L}(G, F)$.

Theorem 1.8.1 *Let Y be a topological space, let F, G be normed vector spaces, and let $\mathcal{U}$ be the set of the invertible elements of the space $\mathcal{L}(F, G)$ equipped with the topology induced by the topology of $\mathcal{L}(F, G)$. Assume that $B\colon h \to B(h)$ is a continuous mapping of Y into $\mathcal{U}$. Define a function $Q\colon (h, f) \to u_{h,f}$, where $(h, f) \in Y \times G$ and $u_{h,f}$ is the solution to the problem*

$$u_{h,f} \in F, \qquad B(h)u_{h,f} = f. \tag{1.8.1}$$

Then, Q is a continuous mapping of $Y \times G$ into F.

Proof. Let $\{h_n, f_n\}$ be a sequence of elements of $Y \times G$ such that

$$h_n \to h_0 \ \text{ in } Y, \qquad f_n \to f_0 \ \text{ in } G. \tag{1.8.2}$$

The first relation in (1.8.2) implies that

$$B(h_n) \to B(h_0) \qquad \text{in } \mathcal{U}. \tag{1.8.3}$$

Denote the set of the invertible elements of the space $\mathcal{L}(G, F)$ by $\mathcal{U}^{-1}$. Then the bijection $A \to A^{-1}$ of $\mathcal{U}$ onto $\mathcal{U}^{-1}$ is a homeomorphism (cf. Schwartz (1967)). Hence, (1.8.3) yields

$$(B(h_n))^{-1} \to (B(h_0))^{-1} \qquad \text{in } \mathcal{U}^{-1}. \tag{1.8.4}$$

Here, $(B(h_n))^{-1}, (B(h_0))^{-1}$ are the bijective inverses of $B(h_n)$ and $B(h_0)$, and the set $\mathcal{U}^{-1}$ is equipped with the topology induced by the topology of the space $\mathcal{L}(G, F)$.

Let

$$u_n = u_{h_n, f_n}, \quad B_n = B(h_n), \quad B_n^{-1} = (B(h_n))^{-1}, \qquad n = 0, 1, 2, \dots. \tag{1.8.5}$$

Then

$$u_n = B_n^{-1} f_n, \qquad u_0 = B_0^{-1} f_0.$$

Therefore

$$\begin{aligned} \|u_n - u_0\|_F &= \|B_n^{-1}(f_n - f_0) + (B_n^{-1} - B_0^{-1}) f_0\|_F \\ &\le \|B_n^{-1}\|_{\mathcal{L}(G,F)} \|f_n - f_0\|_G + \|B_n^{-1} - B_0^{-1}\|_{\mathcal{L}(G,F)} \|f_0\|_G. \end{aligned}$$

This inequality together with (1.8.2), (1.8.4), (1.8.5) implies $u_n \to u_0$ in F, completing the proof.

1.9 Differentiation in Banach spaces and the implicit function theorem

1.9.1 Fréchet derivative and its properties

Let f be a mapping of an open set U in a Banach space E into a Banach space F. The mapping f is said to be Fréchet differentiable at a point $a \in U$ if there exists a linear continuous mapping $L \in \mathcal{L}(E, F)$ such that, for any $a + h \in U$, the equality

$$f(a + h) = f(a) + Lh + \varphi(h)\|h\|_E,$$

holds, where $\varphi(h) \in F$ and $\|\varphi(h)\|_F \to 0$ as $h \to 0$ in E and $h \neq 0$.

The mapping L is called the Fréchet derivative of the mapping f at the point a, and it is denoted by $L = f'(a)$, or $L = f'_a$.

A mapping f is said to be continuously Fréchet differentiable in U if, at every point $x \in U$, the mapping f is Fréchet differentiable and the function $x \to f'(x)$ is a continuous mapping of U endowed with the topology induced by the topology of E into the space $\mathcal{L}(E, F)$.

Let us list some properties of the Fréchet derivative.

1. If a mapping f has Fréchet derivative at a point a, then this derivative is unique.
2. If $f \in \mathcal{L}(E, F)$, then $f'(a) = f$ at every point $a \in E$.
3. If f is a mapping of $U \subset E$ into F, g is a mapping of U into F, and both f and g have Fréchet derivatives at a point $a \in U$, then the function $f + g \colon x \to f(x) + g(x)$ has Fréchet derivative at the point a, which is equal to the sum of the derivatives, $(f + g)'(a) = f'(a) + g'(a)$.
4. Let E, F, and G be Banach spaces, let U be an open set in E, and let U_1 be an open set in F. Let f be a mapping of U into U_1 and g a mapping of U_1 into G. If f has Fréchet derivative $f'(a) \in \mathcal{L}(E, F)$ at a point $a \in U$

and g has Fréchet derivative $g'(b) \in \mathcal{L}(F, G)$ at the point $b = f(a)$, then the composition of the mappings $h = g \circ f$ has Fréchet derivative at the point a, which is the composition of the derivatives, i.e.,

$$h'(a) = g'(b) \circ f'(a) = g'(f(a)) \circ f'(a).$$

The latter formula is called the chain rule, or the theorem on a composite function.

Properties 2 and 3 are obvious. For the proof of Properties 1 and 4, see Schwartz (1967).

If E is a product of Banach spaces E_1 and E_2, then a mapping f of an open set $U \subset E_1 \times E_2$ into F is a function of two variables, $f(x_1, x_2)$, $x_1 \in E_1$, $x_2 \in E_2$. As x_1 is fixed at a point a_1, one can consider the partial mapping $f_{a_1} : x_2 \to f(a_1, x_2)$ and look for the Fréchet derivative at a point a_2. If the Fréchet derivative of the mapping f_{a_1} at a point a_2 exists, it is called the partial Fréchet derivative of the mapping f with respect to x_2 at the point (a_1, a_2) and denoted by $\frac{\partial f}{\partial x_2}(a_1, a_2)$.

Notice that a mapping f is continuously Fréchet differentiable in U if and only if it has partial Fréchet derivatives $\frac{\partial f}{\partial x_1}$ and $\frac{\partial f}{\partial x_2}$ which are continuous in U. Then, the Fréchet derivative of the mapping f at a point $(a_1, a_2) \in U$ is given by the formula

$$f'(a_1, a_2)(X_1, X_2) = \frac{\partial f}{\partial x_1}(a_1, a_2)X_1 + \frac{\partial f}{\partial x_2}(a_1, a_2)X_2,$$

where $X_1 \in E_1$ and $X_2 \in E_2$ (see Schwartz (1967)).

1.9.2 Implicit function

Let E, F, and G be Banach spaces, let f be a mapping of $E \times F$ into G, and let c be a point of the space G. Consider the equation

$$f(x, y) = c. \tag{1.9.1}$$

Assume that there exists a partial solution of this equation $x = a$, $y = b$. Suppose that, for some neighborhood of the point a in E, the equation $f(x, y) = c$ with respect to y has a unique solution in a neighborhood of the point b in F. Then, this equation defines y as a function $g(x)$ of the variable x. $g(x)$ is called the implicit function determined by the equation (1.9.1), and it is characterized by the following property

$$f(x, g(x)) = c. \tag{1.9.2}$$

The following theorems on the existence and differentiability of an implicit function hold (see, e.g., Schwartz (1967)).

Theorem 1.9.1 *Let E, F, and G be Banach spaces, let U be an open set in $E \times F$, and let $(a, b) \in U$. Let f be a continuous mapping of U into G and let $f(a, b) = c$.*

Assume that, for every fixed x, the function f has the partial Fréchet derivative $\frac{\partial f}{\partial y}(x,y) \in \mathcal{L}(F,G)$ and $x, y \to \frac{\partial f}{\partial y}(x,y)$ is a continuous mapping of U into $\mathcal{L}(F,G)$. Moreover, suppose that $Q = \frac{\partial f}{\partial y}(a,b)$ is an invertible mapping of F onto G, i.e., Q is a bijection and $Q^{-1} \in \mathcal{L}(G,F)$. Then, there exist open sets A and B in the spaces E and F containing a and b, respectively, such that, for any $x \in A$, the equation (1.9.1) with respect to y has a unique solution in B. The function $y = g(x)$ determined by this solution is a continuous mapping of A into B.

Theorem 1.9.2 *Let E, F, and G be Banach spaces, let U be an open set in $E \times F$, and let f be a mapping of U into G. Let A and B be open sets in the spaces E and F, respectively, $A \times B \subset U$, and let g be a mapping of A into B which meets (1.9.2). Further, let the mapping f be Fréchet differentiable at a point (a,b), $b = g(a)$, let its partial Fréchet derivatives*

$$\frac{\partial f}{\partial x}(a,b), \quad \frac{\partial f}{\partial y}(a,b)$$

be linear continuous mappings of E and F into G, and let $\frac{\partial f}{\partial y}(a,b)$ be invertible. Then, if the mapping g is continuous at the point a, then it is Fréchet differentiable at this point and the Fréchet derivative is given by the relation

$$g'(a) = -\left(\frac{\partial f}{\partial y}(a,b)\right)^{-1} \circ \left(\frac{\partial f}{\partial x}(a,b)\right). \tag{1.9.3}$$

1.9.3 The Gâteaux derivative and its connection with the Fréchet derivative

Let f be a mapping of an open set U in a Banach space E into a Banach space F, and let a be a point of U. Assume that, for an arbitrary fixed h from E, there exists the derivative of the real-valued function $t \to f(a+th)$ at $t = 0$. This derivative is denoted by $\delta f(a,h)$ and is called the Gâteaux differential, or variation, or derivative of the function f in direction h at the point a. Thus, we have

$$\delta f(a,h) = \frac{d}{dt} f(a+th)\big|_{t=0} = \lim_{\substack{t \neq 0,\, t \to 0 \\ a+th \in U}} \frac{f(a+th) - f(a)}{t}. \tag{1.9.4}$$

The convergence on the right-hand side of (1.9.4) is understood in the norm of the space F. The function $h \to \delta f(a,h)$ is homogeneous, that is, $\delta f(a,ch) = c\delta f(a,h)$, $c \in \mathbb{R}$, however, it is not always linear.

Suppose that $U = E$, i.e., f is a mapping of E into F and there exists an operator $L \in \mathcal{L}(E,F)$ such that $\delta f(a,h) = Lh$ for all $h \in E$. The operator L is called the Gâteaux derivative of the mapping f at the point a and is denoted by $f'_{\mathrm{G}}(a)$. Thus

$$\frac{d}{dt} f(a+th)\big|_{t=0} = f'_{\mathrm{G}}(a)h, \qquad h \in E, \quad f'_{\mathrm{G}}(a) \in \mathcal{L}(E,F).$$

Theorem 1.9.3 *Let f be a mapping of a Banach space E into a Banach space F. Let V be an open set in E. Suppose that, at each point $x \in V$, there exists the Gâteaux derivative $f'_{\mathrm{G}}(x) \in \mathcal{L}(E, F)$. Suppose that $x \to f'_{\mathrm{G}}(x)$ is a continuous mapping of V (in the topology generated by the topology of E) into the space $\mathcal{L}(E, F)$. Then, f is a continuously Fréchet differentiable mapping of V into F and its Gâteaux derivative is the Fréchet derivative, i.e., $f'_{\mathrm{G}}(x) = f'(x)$, $x \in V$.*

For the proof see, e.g., Vainberg (1972), Kolmogorov and Fomin (1975).

1.10 Differentiation of the norm in the space $W_p^m(\Omega)$

Apparently, Mazur (1933) was the first to establish that the norm in the space $L_p(\Omega)$ is Fréchet differentiable at any nonzero point as long as $p > 1$. Some items connected with the Gâteaux and Fréchet differentiability of the norm are dealt with in Vainberg (1972).

Let us show that, if Ω is a bounded domain in $\mathbb{R}^n$, then the norm in $W_p^m(\Omega)$, $p > 1$, is a continuously Fréchet differentiable functional everywhere but zero.

1.10.1 Auxiliary statement

Lemma 1.10.1 *For any $t > 1$, there exists a constant $c > 0$, depending on t, such that*

$$|z - y|^t \le c\big||z|^t \operatorname{sign} z - |y|^t \operatorname{sign} y\big|, \qquad z, y \in \mathbb{R}, \tag{1.10.1}$$

where

$$\operatorname{sign} x = \begin{cases} -1, & \text{if } x < 0, \\ 0, & \text{if } x = 0, \\ 1, & \text{if } x > 0. \end{cases}$$

Proof. Introduce the function

$$f(z, y) = \frac{|z - y|^t}{\big||z|^t \operatorname{sign} z - |y|^t \operatorname{sign} y\big|}, \tag{1.10.2}$$

defined on the complement to the set of the points of the line $z = y$ in $\mathbb{R}^2$. Dividing the numerator and denominator in (1.10.2) by $|z|^t$, $z \neq 0$, we obtain

$$f(z, y) = \frac{\big|\operatorname{sign} z - \frac{y}{|z|}\big|^t}{\big|\operatorname{sign} z - \big|\frac{y}{z}\big|^t \operatorname{sign} y\big|}, \tag{1.10.3}$$

that is, the function f depends only on the sign of z and on the ratio $\frac{y}{|z|}$. One can easily see that

$$\lim_{(y/|z|) \to 0} f(z, y) = 1, \qquad \lim_{(y/|z|) \to \infty} f(z, y) = 1, \qquad \lim_{(y/|z|) \to -\infty} f(z, y) = 1, \tag{1.10.4}$$

To investigate the behaviour of the function f in the neighborhood of the line $z = y$ we use the L'Hospital rule of removing indeterminacy. Let z_0 be an arbitrary fixed point in $\mathbb{R}$, $z_0 \neq 0$. Calculating the partial derivative in y of the numerator and denominator in (1.10.3), we have

$$\lim_{y \to z_0} f(z_0, y) = 0, \qquad z_0 \in \mathbb{R}, \; z_0 \neq 0. \tag{1.10.5}$$

Now, taking into account (1.10.2)–(1.10.5), we get the inequality (1.10.1) with a constant c dependent on t.

1.10.2 Theorem on differentiability

Theorem 1.10.1 *Let Ω be a bounded domain in $\mathbb{R}^n$ and let f be the functional in the space $W_p^m(\Omega)$ defined by the expression*

$$f(u) = \Big(\sum_{|k| \leq m} \int_\Omega |D^k u|^p dx \Big)^{\frac{1}{p}} = \|u\|_{W_p^m(\Omega)}. \tag{1.10.6}$$

If $p > 1$, the functional f is continuously Fréchet differentiable in the complement to the zero element of $W_p^m(\Omega)$ and its Fréchet derivative is given by

$$f'(u)v = \|u\|^{1-p}_{W_p^m(\Omega)} \sum_{|k| \leq m} \int_\Omega |D^k u|^{p-2} D^k u D^k v \, dx. \tag{1.10.7}$$

Proof. 1. Denote

$$\varphi(w) = \|w\|^p_{W_p^m(\Omega)}. \tag{1.10.8}$$

Let u, v be arbitrary fixed elements of $W_p^m(\Omega)$.

Introduce the function

$$g(t) = \varphi(u + tv) = \sum_{|k| \leq m} \int_\Omega |D^k (u + tv)|^p \, dx, \qquad t \in (-1, 1). \tag{1.10.9}$$

For almost every $x \in \Omega$, the function under the integral sign in (1.10.9) is differentiable in t, and the absolute value of its derivative is majorized by an integrable function in Ω that is independent of t. Hence, applying the theorem on the differentiability of a function represented as an integral (e.g., Schwartz (1967)), we obtain

$$\frac{dg}{dt}(0) = \frac{d}{dt} \varphi(u + tv)\big|_{t=0} = \sum_{|k| \leq m} p \int_\Omega D^k v |D^k u|^{p-1} \operatorname{sign} D^k u \, dx. \tag{1.10.10}$$

Applying the Hölder inequality with indices p and $q = \frac{p}{p-1}$ ($\frac{1}{p} + \frac{1}{q} = 1$) to the right-hand side of (1.10.10), we get

$$\Big| \frac{d}{dt} \varphi(u + tv)\big|_{t=0} \Big| \leq p \sum_{|k| \leq m} \|D^k v\|_{L_p(\Omega)} \|D^k u\|^{p-1}_{L_p(\Omega)}.$$

Hence, for a fixed u, the right-hand side of (1.10.10) defines a linear continuous functional of $v \in W_p^m(\Omega)$. Consequently, for all $u \in W_p^m(\Omega)$, the functional φ defined by (1.10.8) is Gâteaux differentiable and its Gâteaux derivative is given by

$$\varphi'_{\mathrm{G}}(u)v = \sum_{|k|\leq m} p \int_\Omega D^k v |D^k u|^{p-1} \operatorname{sign} D^k u \, dx. \tag{1.10.11}$$

2. Now let us show that

$$\left.\begin{array}{l}\text{the function } u \to \varphi'_{\mathrm{G}}(u) \text{ is a continuous mapping of} \\ W_p^m(\Omega) \text{ into } \mathcal{L}(W_p^m(\Omega);\mathbb{R}) = (W_p^m(\Omega))^*.\end{array}\right\} \tag{1.10.12}$$

Let $\{u_n\}$ be a sequence such that

$$u_n \to u \qquad \text{in } W_p^m(\Omega). \tag{1.10.13}$$

Taking into account (1.10.11) and applying the Hölder inequality with indices p and $q = \frac{p}{p-1}$, we have

$$|(\varphi'_{\mathrm{G}}(u_n) - \varphi'_{\mathrm{G}}(u))\, v| \leq \alpha_n \|v\|_{W_p^m(\Omega)}, \tag{1.10.14}$$

where

$$\alpha_n = \sum_{|k|\leq m} p \Big(\int_\Omega \big||D^k u_n|^{p-1} \operatorname{sign} D^k u_n - |D^k u|^{p-1} \operatorname{sign} D^k u\big|^{\frac{p}{p-1}} \, dx \Big)^{\frac{p-1}{p}}. \tag{1.10.15}$$

Using the inequality (1.10.1) at every point $x \in \Omega$ for which the function under the integral sign in (1.10.15) is well defined, we conclude that, for $t = \frac{p}{p-1}$,

$$\begin{aligned}\alpha_n &\leq \sum_{|k|\leq m} c_1 \Big(\int_\Omega \big||D^k u_n|^p \operatorname{sign} D^k u_n - |D^k u|^p \operatorname{sign} D^k u\big| \, dx \Big)^{\frac{p-1}{p}} \\ &\leq c_1 \sum_{|k|\leq m} (\beta_{kn} + \gamma_{kn})^{\frac{p-1}{p}},\end{aligned} \tag{1.10.16}$$

where

$$\begin{aligned}\beta_{kn} &= \int_\Omega \big|(|D^k u_n|^p - |D^k u|^p) \operatorname{sign} D^k u_n\big| \, dx \\ &= \int_\Omega \big||D^k u_n|^p - |D^k u|^p\big| \, dx,\end{aligned} \tag{1.10.17}$$

$$\gamma_{kn} = \int_\Omega |D^k u|^p | \operatorname{sign} D^k u_n - \operatorname{sign} D^k u| \, dx. \tag{1.10.18}$$

Introduce the notation

$$f_{kn} = |D^k u_n|^p - |D^k u|^p. \tag{1.10.19}$$

The continuity of the norm together with (1.10.13) implies that

$$\lim_{n\to\infty} \int_{\Omega'} f_{kn}\,dx = 0, \qquad |k| = 0, 1, 2, \dots, m, \tag{1.10.20}$$

Ω' being an arbitrary measurable subset of the set Ω. Known results (Natanson (1974)) and (1.10.20) yield the functions f_{kn} to have uniformly absolutely continuous integrals, that is, for any $\varepsilon > 0$ there exists $\delta > 0$ such that, for any measurable set $\Omega' \subset \Omega$ with $\operatorname{mes}\Omega' < \delta$ (mes standing for the Lebesgue measure), the following estimate holds:

$$\left| \int_{\Omega'} f_{kn}\,dx \right| < \varepsilon \qquad \forall n.$$

Further, let

$$\Omega'_+ = \{\, x \,|\, x \in \Omega',\ f_{kn}(x) \geq 0 \,\},$$
$$\Omega'_- = \{\, x \,|\, x \in \Omega', f_{kn}(x) < 0 \,\}.$$

Then

$$\int_{\Omega'} |f_{kn}|\,dx = \int_{\Omega'_+} f_{kn}\,dx - \int_{\Omega'_-} f_{kn}\,dx \leq 2\varepsilon \qquad \forall n,$$

i.e., the functions $|f_{kn}|$ have uniformly absolutely continuous integrals.

Due to (1.10.13) and (1.10.19), we can extract from the sequence $\{f_{kn}\}_{n=1}^\infty$ a subsequence $\{f_{kn_i}\}_{i=1}^\infty$ such that $f_{kn_i} \to 0$ a.e. in Ω, and so $f_{kn_i} \to 0$ in measure. Now, because of the Vitali theorem (see Natanson (1974)), we obtain

$$\lim_{i\to\infty} \int_{\Omega} |f_{kn_i}|\,dx = 0. \tag{1.10.21}$$

We may, in turn, extract from any subsequence $\{f_{kn_s}\}_{s=1}^\infty$ a subsubsequence $\{f_{kn_i}\}_{i=1}^\infty$ that makes (1.10.21) true, so that

$$\lim_{n\to\infty} \int_{\Omega} |f_{kn}|\,dx = 0.$$

This equality together with (1.10.17), (1.10.19) implies that

$$\lim_{n\to\infty} \beta_{kn} = 0, \qquad |k| = 0, 1, 2, \dots, m. \tag{1.10.22}$$

Using the Lebesgue theorem, we easily see that the numbers γ_{kn} determined by the formulas (1.10.18) satisfy the following relation:

$$\lim_{n\to\infty} \gamma_{kn} = 0, \qquad |k| = 0, 1, 2, \dots, m. \tag{1.10.23}$$

Now, (1.10.12) is a consequence of (1.10.14), (1.10.16), (1.10.22), (1.10.23).

3. On account of (1.10.12) and Theorem 1.9.3, we deduce that the function φ from (1.10.8) is a continuously Fréchet differentiable mapping of $W_p^m(\Omega)$ into $\mathbb{R}$ and that

$$\varphi'_{\mathrm{G}}(u) = \varphi'(u), \qquad u \in W_p^m(\Omega). \tag{1.10.24}$$

Obviously the functional f from (1.10.6) is the composition of the mapping $\varphi\colon W_p^m(\Omega) \to \mathbb{R}_+$ and the function $\psi\colon \mathbb{R}_+ \to \mathbb{R}_+$, $y \to \psi(y) = y^{1/p}$.

The function ψ is continuously differentiable in $\mathbb{R}_+ \setminus \{0\}$. Now, by using the rule of the differentiation of a composite function (see Subsec. 1.9.1, Property 4), we conclude that the functional f is continuously Fréchet differentiable in the complement to zero in $W_p^m(\Omega)$, and its Fréchet derivative is determined by the expression (1.10.7). The theorem is proved.

1.11 Differentiation of eigenvalues

"'The owl was a very respectable old bird, terribly well educated,' the mouse said, 'she knew more than the night watchman and almost as much as I ... She proved to me that the night watchman could not hoot unless he used the horn that hung from his shoulder.'"

– *H. Ch. Andersen*
"How to Cook Soup
upon a Sausage Pin"

1.11.1 The eigenvalue problem

Suppose that we are given spaces $\mathcal{U}$, V such that

$$\left.\begin{array}{l}\mathcal{U} \text{ is a Banach space, } V \text{ is a Hilbert space, } V \subset \mathcal{U}, \text{ the} \\ \text{embedding } V \to \mathcal{U} \text{ is compact.}\end{array}\right\} \tag{1.11.1}$$

Let also X be a Banach space, let G be an open set in X, and let $h \to a_h$, $h \to b_h$ be functions such that

$$\left.\begin{array}{l}h \to a_h \text{ is a continuously Fréchet differentiable mapping of} \\ G \text{ into } \mathcal{L}_2(V, V; \mathbb{R}),\end{array}\right\} \tag{1.11.2}$$

$$\left.\begin{array}{l}h \to b_h \text{ is a continuously Fréchet differentiable mapping of} \\ G \text{ into } \mathcal{L}_2(\mathcal{U}, \mathcal{U}; \mathbb{R}).\end{array}\right\} \tag{1.11.3}$$

We suppose also that, for all $h \in G$, the bilinear form a_h is symmetric and coercive and the bilinear form b_h is symmetric, i.e.,

$$a_h(u, v) = a_h(v, u), \qquad u, v \in V, \tag{1.11.4}$$

$$a_h(u,u) \geq c(h)\|u\|_V^2, \qquad u \in V,\ c(h) = \text{const} > 0, \tag{1.11.5}$$
$$b_h(u,v) = b_h(v,u), \qquad u,v \in \mathcal{U}. \tag{1.11.6}$$

Consider the following eigenvalue problem:

$$(u_i(h), \mu_i(h)) \in V \times \mathbb{R}, \qquad u_i(h) \neq 0,$$
$$\mu_i(h) a_h(u_i(h), v) = b_h(u_i(h), v), \qquad v \in V. \tag{1.11.7}$$

Theorem 1.11.1 *Let the conditions* (1.11.1)–(1.11.6) *be fulfilled. Then, for any* $h \in G$, *there exists a countable set* $\{\mu_i(h)\}_{i=1}^{\infty}$ *of eigenvalues of the problem* (1.11.7) *in which every eigenvalue appears as many times as its multiplicity and every nonzero eigenvalue is of finite multiplicity. The corresponding countable set of eigenfunctions* $\{u_i(h)\}_{i=1}^{\infty}$ *forms a basis in the space* V, *and these functions can be chosen so that the condition*

$$a_h(u_i(h), u_j(h)) = \delta_{ij} \qquad \forall i,j. \tag{1.11.8}$$

is satisfied.

Proof. Let h be an arbitrary element of G. By virtue of (1.11.2), (1.11.4), (1.11.5), the bilinear form a_h defines a scalar product in V (see Remark 1.5.2). Now, (1.11.1), (1.11.3), (1.11.6), and Theorem 1.5.8 imply the problem (1.11.7) to have a countable set of eigenvalues $\{\mu_i(h)\}_{i=1}^{\infty}$ in which every eigenvalue appears as many times as its multiplicity and every nonzero eigenvalue is of finite multiplicity.

Introduce the operators $A(h)$, $B(h)$ generated by the bilinear forms a_h and b_h

$$(A(h)u, v) = a_h(u,v), \qquad u,v \in V, \tag{1.11.9}$$
$$(B(h)u, v) = b_h(u,v), \qquad u,v \in V. \tag{1.11.10}$$

Now, instead of (1.11.7), we have the following problem

$$(u_i(h), \mu_i(h)) \in V \times \mathbb{R}, \qquad u_i(h) \neq 0,$$
$$A(h)^{-1} \circ B(h) u_i(h) = \mu_i(h) u_i(h). \tag{1.11.11}$$

Consider the bilinear form a_h as a scalar product in V. Then, by (1.11.6), (1.11.9), and (1.11.10), we get

$$a_h(A(h)^{-1} \circ B(h)u, v) = (B(h)u, v) = (B(h)v, u)$$
$$= a_h(A(h)^{-1} \circ B(h)v, u) = a_h(u, A(h)^{-1} \circ B(h)v), \qquad u,v \in V. \tag{1.11.12}$$

Hence, $A(h)^{-1} \circ B(h)$ is a selfadjoint operator in V with respect to the scalar product $a_h(\cdot,\cdot)$.

Using (1.11.1) and the fact that $a_h \in \mathcal{L}_2(V,V;\mathbb{R})$, $b_h \in \mathcal{L}_2(\mathcal{U},\mathcal{U};\mathbb{R})$, we deduce (see Subsec. 1.5.2) that $A(h)^{-1} \circ B(h)$ is a compact operator in V. Known results (see, e.g., Kantorovich and Akilov (1977)) yield that the set of eigenfunctions $\{u_i(h)\}_{i=1}^{\infty}$ forms a basis in the space V and the condition (1.11.8) is satisfied provided an appropriate choice of these functions is made.

1.11.2 Differentiation of an operator-valued function

In the proof of Theorem 1.11.1, we defined the scalar product in V via the bilinear form a_h. However, since h runs through the set G, in what follows we will assume that the space V is endowed with a scalar product $(\cdot,\cdot)_V$ that is independent of h. In this case, the relations (1.11.9), (1.11.10) still define operators $A(h), B(h) \in \mathcal{L}(V, V^*)$, and since the operator $A(h)$ is invertible (see Theorem 1.5.2), the problem (1.11.7) reduces to the problem (1.11.11). Notice that the operator $A(h)^{-1} \circ B(h)$ is no longer selfadjoint in V.

Introduce the notation

$$L(h) = A(h)^{-1} \circ B(h). \tag{1.11.13}$$

Lemma 1.11.1 *Let the conditions* (1.11.1)–(1.11.6) *be satisfied and let operators* $A(h)$, $B(h)$, $L(h)$ *be determined by the relations* (1.11.9), (1.11.10), (1.11.13). *Then, the function* $h \to L(h)$ *is a continuously Fréchet differentiable mapping of* G *into* $\mathcal{L}(V, V)$, *and at an arbitrary point* $h_0 \in G$ *the Fréchet derivative* $L'(h_0)$ *of the function* $h \to L(h)$ *is given by*

$$L'(h_0)q = A(h_0)^{-1} \circ B'(h_0)q - A(h_0)^{-1} \circ A'(h_0)q \circ A(h_0)^{-1} \circ B(h_0), \qquad q \in X. \tag{1.11.14}$$

Here, $A'(h_0)$, $B'(h_0)$ *are the Fréchet derivatives of the functions* $h \to A(h)$, $h \to B(h)$ *at the point* h_0.

Proof. Using (1.11.2) and (1.11.9), we easily see that

$$\left.\begin{array}{l}\text{the function } h \to A(h) \text{ is a continuously Fréchet}\\ \text{differentiable mapping of } G \text{ into } \mathcal{L}(V, V^*),\end{array}\right\} \tag{1.11.15}$$

and

$$((A'(h_0)q)u, v) = a'_{h_0}q(u, v), \qquad h_0 \in G,\ q \in X,\ u, v \in V. \tag{1.11.16}$$

Here, a'_{h_0} is the Fréchet derivative of the function $h \to a_h$ at the point h_0 and $a'_{h_0}q \in \mathcal{L}_2(V, V; \mathbb{R})$. By (1.11.1), (1.11.3), (1.11.10), we obtain that

$$\left.\begin{array}{l}\text{the function } h \to B(h) \text{ is a continuously Fréchet}\\ \text{differentiable mapping of } G \text{ into } \mathcal{L}(V, V^*),\end{array}\right\} \tag{1.11.17}$$

and

$$((B'(h_0)q)u, v) = b'_{h_0}q(u, v), \qquad h_0 \in G,\ q \in X,\ u, v \in V. \tag{1.11.18}$$

Denote

$$f(h) = A(h)^{-1}, \quad h \in G, \tag{1.11.19}$$

and let K (K^{-1}) be the set of the invertible elements of the space $\mathcal{L}(V, V^*)$ (respectively, $\mathcal{L}(V^*, V)$). The function f is the composition of the mapping $f_1 \colon h \to f_1(h) = A(h)$ acting from G into K and the mapping $f_2 \colon A \to f_2(A) = A^{-1}$ acting from K to K^{-1}.

By virtue of (1.11.15), the function f_1 is continuously Fréchet differentiable. The function f_2 is a continuously Fréchet differentiable mapping of K into K^{-1} (see Schwartz (1967)) and

$$f_2'(A)T = -A^{-1} \circ T \circ A^{-1}, \qquad A \in K,\ T \in \mathcal{L}(V, V^*). \tag{1.11.20}$$

By using the theorem on the differentiation of a composite function (e.g., Schwartz (1967)) we conclude that $h \to f(h) = A(h)^{-1}$ is a continuously Fréchet differentiable mapping of G into $\mathcal{L}(V^*, V)$ (more exactly, into K^{-1}) and

$$\begin{aligned} f'(h_0)q &= f_2'(f_1(h_0)) \circ f_1'(h_0)q \\ &= -A(h_0)^{-1} \circ A'(h_0)q \circ A(h_0)^{-1}, \qquad q \in X. \end{aligned} \tag{1.11.21}$$

The function $h \to L(h)$ defined by the relation (1.11.13) is the composition of the mapping $\varphi \colon h \to \{A(h)^{-1}, B(h)\}$ acting from G into $\mathcal{L}(V^*, V) \times \mathcal{L}(V, V^*)$ and the bilinear continuous mapping $\varphi_2 \colon L_1, L_2 \to L_1 \circ L_2$ acting from $\mathcal{L}(V^*, V) \times \mathcal{L}(V, V^*)$ into $\mathcal{L}(V, V)$.

Applying the theorem on the differentiation of a composite function and noticing (1.11.13), (1.11.17), (1.11.19), we obtain that the function $h \to L(h)$ is a continuously Fréchet differentiable mapping G into $\mathcal{L}(V, V)$ and

$$L'(h_0)q = f(h_0) \circ B'(h_0)q + f'(h_0)q \circ B(h_0), \qquad h_0 \in G.$$

Therefore, making use of (1.11.19) and (1.11.21), we get (1.11.14), concluding the proof.

1.11.3 Eigenspaces and projections

Projections

As before, we suppose that the space V is equipped with a scalar product $(\cdot, \cdot)_V$ that is independent of h. Consider the following eigenvalue problem for the operator $L(h)$ defined by the relations (1.11.9), (1.11.10), (1.11.13):

$$(u_i(h), \mu_i(h)) \in V \times \mathbb{R}, \qquad u_i(h) \neq 0, \qquad L(h)u_i(h) = \mu_i(h)u_i(h). \tag{1.11.22}$$

Obviously, the problems (1.11.7) and (1.11.22) have the same eigenvalues and eigenfunctions. By virtue of Theorem 1.11.1, the set of the eigenfunctions of the problem (1.11.22) forms a basis in the space V. So, given $h \in G$, an arbitrary element u of V may be uniquely represented as

$$u = \sum_{i=1}^{\infty} c_i(h, u)u_i(h), \tag{1.11.23}$$

where

$$c_i(h,u) = a_h(u,u_i). \tag{1.11.24}$$

Here, we suppose the eigenfunctions to satisfy the condition (1.11.8).

Define operators $P_i(h) \in \mathcal{L}(V,V)$, $i = 1,2,\ldots$, by

$$P_i(h)u = c_i(h,u)u_i(h). \tag{1.11.25}$$

The operator $P_i(h)$ is the projection onto the one-dimensional subspace generated by the eigenfunction $u_i(h)$. However, since the scalar product in the space V is not defined by the bilinear form a_h, $P_i(h)$ is not an orthogonal projection.

The formulas (1.11.8), (1.11.23)–(1.11.25) yield that the operators $P_i(h)$ satisfy the following relations:

$$P_i(h) \circ P_j(h) = \delta_{ij}P_j(h), \qquad \sum_{i=1}^{\infty} P_i(h) = I, \tag{1.11.26}$$

I being the identity operator in V.

We will refer to the complexification of the space V as $\check{V}$ (see, e.g., Kantorovich and Akilov (1977)). An arbitrary element $w \in \check{V}$ is represented in the form $w = u + iv$, where $u,v \in V$ and i is the imaginary unit. If $A \in \mathcal{L}(V,V)$, setting

$$\check{A}w = \check{A}(u+iv) = Au + iAv, \qquad u,v \in V, \tag{1.11.27}$$

we obtain the operator $\check{A} \in \mathcal{L}(\check{V},\check{V})$ which is the complex extension of the operator A.

By analogy with (1.11.27), we get the complex extensions of the operators $P_i(h)$, $L(h)$, I, still denoted by the same letters. So,

$$P_i(h),\ L(h),\ I \text{ belong to } \mathcal{L}(\check{V},\check{V}) \text{ and } \mathcal{L}(V,V). \tag{1.11.28}$$

Let $\xi \in \mathbb{C}$. Theorem 1.11.1 and the relations (1.11.23), (1.11.25), (1.11.26) imply

$$L(h) - \xi I = \sum_{i=1}^{\infty} (\mu_i(h) - \xi)P_i(h). \tag{1.11.29}$$

Thus, if ξ belongs to the complement of the spectrum of the operator $L(h)$ in $\mathbb{C}$, then there exists the operator $(L(h) - \xi I)^{-1}$, the inverse of $L(h) - \xi I$, and the following formula holds true:

$$(L(h) - \xi I)^{-1} = \sum_{i=1}^{\infty} (\mu_i(h) - \xi)^{-1}P_i(h). \tag{1.11.30}$$

This relation is the consequence of (1.11.26) and (1.11.29).

Let (a,b) be an interval in $\mathbb{R}$ such that $\mu_i(h) \in (a,b)$ for $i = k, k+1, \ldots, k+l$, $\mu_i(h) < a$ for $i < k$, and $\mu_i(h) > b$ for $i > k+l$. Let also Γ be the positive oriented circle in the complex plane with radius $\frac{b-a}{2}$, centered at the point $\frac{a+b}{2}$. Then, the following relation holds

$$\sum_{i=k}^{k+l} P_i(h) = -\frac{1}{2\pi i} \int_\Gamma (L(h) - \xi I)^{-1} \, d\xi, \tag{1.11.31}$$

which is easily verified by using (1.11.29) and the theorem on the residues of a meromorphic function (e.g., Lavrentiev and Shabat (1973), Schwartz (1967)).

Lemma 1.11.2 *Let the condition* (1.11.1)–(1.11.6) *be satisfied and let an operator* $L(h)$ *be determined by the relations* (1.11.9), (1.11.10), (1.11.13). *Assume that* $h_0 \in G$ *and that* μ *is a nonzero eigenvalue of the operator* $L(h_0)$ *of multiplicity* $m \geq 1$, *i.e.,*

$$\mu_i(h_0) = \mu \quad \text{for } i \in J, \qquad \mu_i(h_0) \neq \mu \quad \text{for } i \notin J, \tag{1.11.32}$$

$$J = \{\, j, j+1, \ldots, j+m-1 \,\}, \tag{1.11.33}$$

j *being an index.*

Let also Γ_0 *be a positive oriented circle in the complex plane centered at the point* μ *and of so small radius that all the eigenvalues of the operator* $L(h_0)$ *which are not equal to* μ *lie outside of* Γ_0. *Then, there exists an open neighborhood* ω *of the point* h_0 *in* G *such that, for any* $h \in \omega$, *the interior of* Γ_0 *contains exactly* m *eigenvalues (with regard for their multiplicity) of the operator* $L(h)$, *and the other eigenvalues of the operator* $L(h)$ *lie outside of* Γ_0, *i.e., if* Ω_0 *is the open circle in* $\mathbb{C}$ *with the boundary* Γ_0, *then*

$$\mu_i(h) \in \Omega_0, \qquad i \in J, \ h \in \omega,$$
$$\mu_i(h) \in \mathbb{C} \setminus (\Omega_0 \cup \Gamma_0), \qquad i \notin J, \ h \in \omega.$$

The proof of Lemma 1.11.2 is almost identical to the proof of Theorem 1.5.9, so we omit it.

Now define a function $P \colon \omega \to \mathcal{L}(\check{V}, \check{V})$ by

$$P(h) = \int_{\Gamma_0} (L(h) - \xi I)^{-1} \, d\xi. \tag{1.11.34}$$

The above argument implies that $P(h) = \sum_{i \in J} P_i(h)$ and $P(h_0)$ is the projection onto the eigenspace of the operator $L(h_0)$ corresponding to the eigenvalue μ.

Differentiability of the function P

Lemma 1.11.3 *Let the conditions* (1.11.1)–(1.11.6) *be fulfilled and let operators* $A(h)$, $B(h)$, $L(h)$ *be defined by the relations* (1.11.9), (1.11.10), (1.11.13). *Then, the function* $h \to P(h)$ *determined by* (1.11.34) *is a continuously Fréchet differentiable mapping of* ω *into* $\mathcal{L}(\check{V}, \check{V})$ *and of* ω *into* $\mathcal{L}(V, V)$.

Proof. Denote

$$R(h,\xi) = (L(h) - \xi I)^{-1}. \tag{1.11.35}$$

By virtue of the theorem on the differentiation of a composite function, we establish, just as in the proof of Lemma 1.11.1 (see (1.11.20), (1.11.21)), that for each $\xi \in \Gamma_0$ the partial function $h \to R(h,\xi)$ is a continuously differentiable mapping of ω into $\mathcal{L}(\check{V},\check{V})$, and at a point $y \in \omega$ its Fréchet derivative $\frac{\partial R}{\partial h}(y,\xi)$ is given by

$$\frac{\partial R}{\partial h}(y,\xi)q = -R(y,\xi) \circ L'(y)q \circ R(y,\xi), \qquad q \in X. \tag{1.11.36}$$

Here, $L'(y)$ stands for the Fréchet derivative of the function $h \to L(h)$ at a point y, $L'(y) \in \mathcal{L}(X, \mathcal{L}(\check{V},\check{V}))$.

The function $h,\xi \to R(h,\xi)$ is the composition of the continuous mapping $f_1\colon h,\xi \to f_1(h,\xi) = L(h) - \xi I$ acting from $\omega \times \Gamma_0$ into $\mathcal{L}(\check{V},\check{V})$ and the continuous mapping $f_2\colon A \to f_2(A) = A^{-1}$ acting in K_1, the set of the invertible elements of the space $\mathcal{L}(\check{V},\check{V})$. Hence, the function $h,\xi \to R(h,\xi)$ is a continuous mapping of $\omega \times \Gamma_0$ into $\mathcal{L}(\check{V},\check{V})$.

Thus, by Lemma 1.11.1 and (1.11.36), we get that

$$\left.\begin{array}{l}\text{the function } h,\xi \to \dfrac{\partial R}{\partial h}(h,\xi) \text{ is a continuous mapping of} \\ \omega \times \Gamma_0 \text{ into } \mathcal{L}(X, \mathcal{L}(\check{V},\check{V})).\end{array}\right\} \tag{1.11.37}$$

So, if y is an arbitrary element of ω, there exists a neighborhood ω_0 of the point y such that

$$\omega_0 \subset \omega, \qquad \left\| \frac{\partial R}{\partial h}(h,\xi) \right\|_{\mathcal{L}(X,\mathcal{L}(\check{V},\check{V}))} \le \text{const}, \qquad h \in \omega_0,\ \xi \in \Gamma_0.$$

Now, applying the theorem on the differentiability of a function represented as an integral (e.g., Schwartz (1967)) to the expression (1.11.34), we conclude that $h \to P(h)$ a continuously Fréchet differentiable mapping of ω into $\mathcal{L}(\check{V},\check{V})$. Since $P(h) \in \mathcal{L}(V,V)$, the function $h \to P(h)$ is also a continuously Fréchet differentiable mapping of ω into $\mathcal{L}(V,V)$, concluding the proof.

1.11.4 Differentiation of eigenvalues

Gâteaux differential of the function $h \to \mu_i(h)$

Let h_0 be an arbitrary fixed element of G, and let μ be a nonzero eigenvalue of the operator $L(h_0)$ of multiplicity $m \ge 1$, i.e., (1.11.32), (1.11.33) hold. Let also q be an arbitrary fixed element of X.

There exist numbers $a < 0$, $b > 0$ such that $h_0 + \chi q \in \omega$ for all $\chi \in (a,b)$, where ω is the open neighborhood of the point h_0 defined in Lemma 1.11.2. Introduce the notations:

$$\tilde{L}(\chi) = L(h_0 + \chi q), \qquad \tilde{\mu}_i(\chi) = \mu_i(h_0 + \chi q), \qquad i \in J,\ \chi \in (a,b). \tag{1.11.38}$$

We suppose that the eigenvalues $\tilde{\mu}_i(\chi), i \in J$, are enumerated in such a way that, for arbitrary $i, i+1$ from J, the following inequalities hold:

$$\begin{aligned} \tilde{\mu}_i(\chi) &\geq \tilde{\mu}_{i+1}(\chi) \qquad \text{for } b > \chi \geq 0, \\ \tilde{\mu}_i(\chi) &\leq \tilde{\mu}_{i+1}(\chi) \qquad \text{for } a < \chi \leq 0. \end{aligned} \tag{1.11.39}$$

Then, by (1.11.32) and (1.11.38), we get

$$\tilde{\mu}_i(0) = \mu_i(h_0) = \mu, \qquad i \in J. \tag{1.11.40}$$

Theorem 1.11.2 *Let the conditions* (1.11.1)–(1.11.6) *be fulfilled and let operators* $A(h)$, $B(h)$, $L(h)$ *be defined by the relations* (1.11.9), (1.11.10), (1.11.13). *Assume that* $h_0 \in G$ *and* μ *is an eigenvalue of the operator* $L(h_0)$ *of multiplicity* $m \geq 1$, *that is,* (1.11.32), (1.11.33) *hold true. Let also* q *be an arbitrary element of* X *and let the eigenvalues of the operator* $\tilde{L}(\chi)$, $\tilde{\mu}_i(\chi), i \in J$, *(see* (1.11.38)*) be enumerated so that the inequalities* (1.11.39) *are valid. Then, the functions* $\chi \to \tilde{\mu}_i(\chi)$ *are differentiable at zero, i.e.,*

$$\tilde{\mu}_i(\chi) = \mu + \chi\mu_i^{(1)} + o(\chi), \qquad i \in J.$$

Moreover, $\mu_i^{(1)}$ *are the eigenvalues of the* $m \times m$*-dimensional matrix* $\tau = [\tau_{lk}]$ *whose elements are defined by the formula*

$$\tau_{lk} = \tau_{kl} = (((B'(h_0) - \mu A'(h_0))q)u_l(h_0), u_k(h_0)), \qquad l, k \in J. \tag{1.11.41}$$

Here, $A'(h_0)$, $B'(h_0)$ *are the Fréchet derivatives of the functions* $h \to A(h)$ *and* $h \to B(h)$ *at the point* h_0, *and* $u_i(h_0)$ *are eigenfunctions of the operator* $L(h_0)$ *belonging to the eigenvalue* μ *such that*

$$(A(h_0)u_l(h_0), u_k(h_0)) = \delta_{lk}, \qquad l, k \in J. \tag{1.11.42}$$

In particular, if $m = 1$, *then* $J = \{j\}$ *and*

$$\dot{\mu}_j^{(1)} = (((B'(h_0) - \mu A'(h_0))q)u_j(h_0), u_j(h_0)). \tag{1.11.43}$$

Proof. 1. Denote

$$\begin{aligned} \tilde{P}(\chi) &= P(h_0 + \chi q), \qquad & \tilde{P}_i(\chi) &= P_i(h_0 + \chi q), \\ M(\chi) &= \tilde{P}(\chi)V, \qquad & M_i(\chi) &= \tilde{P}_i(\chi)V, \qquad i \in J, \ \chi \in (a, b). \end{aligned} \tag{1.11.44}$$

Here, the function P is determined by the expression (1.11.34), $P_i(h_0 + \chi q)$ is the projection onto the one-dimensional eigenspace of the operator $L(h_0 + \chi q)$ belonging to the eigenvalue $\tilde{\mu}_i(\chi)$ (see (1.11.25)).

With the notations accepted, the following equality holds true:

$$\tilde{P}(\chi) = \sum_{i \in J} \tilde{P}_i(\chi), \qquad \chi \in (a, b). \tag{1.11.45}$$

Known results (see, e.g., Kato (1976)) and Lemma 1.11.3 imply that, for any $\chi \in (a,b)$, there exists an operator $U(\chi) \in \mathcal{L}(V,V)$ such that

$$\begin{aligned} \tilde{P}(\chi) &= U(\chi) \circ \tilde{P}(0) \circ U(\chi)^{-1}, \\ \tilde{P}_i(\chi) &= U(\chi) \circ \tilde{P}_i(0) \circ U(\chi)^{-1}. \end{aligned} \tag{1.11.46}$$

Moreover, $\chi \to U(\chi)$ and $\chi \to U(\chi)^{-1}$ are continuously differentiable mappings of (a,b) into $\mathcal{L}(V,V)$ and

$$\begin{aligned} \tilde{P}(0) \circ \frac{dU^{-1}}{d\chi}(0) \circ \tilde{L}(0) \circ \tilde{P}(0) = 0, \\ \tilde{P}(0) \circ \tilde{L}(0) \circ \frac{dU}{d\chi}(0) \circ \tilde{P}(0) = 0, \qquad i \in J. \end{aligned} \tag{1.11.47}$$

For any $\chi \in (a,b)$, $U(\chi)$ is a one-to-one mapping of $M(0)$ into $M(\chi)$, and of $M_i(0)$ into $M_i(\chi)$ for all $i \in J$. This operator is called a transforming function.

With regard to the notations (1.11.38) and (1.11.44), we get

$$\tilde{L}(\chi) \circ \tilde{P}_i(\chi) = \tilde{\mu}_i(\chi)\tilde{P}_i(\chi), \qquad i \in J,\ \chi \in (a,b). \tag{1.11.48}$$

Multiplying both-hand sides of the equality (1.11.48) by $U(\chi)^{-1}$ from the left and by $U(\chi)$ from the right, and noticing (1.11.46), we obtain

$$\hat{L}(\chi) \circ \tilde{P}_i(0) = \tilde{\mu}_i(\chi)\tilde{P}_i(0), \qquad i \in J,\ \chi \in (a,b), \tag{1.11.49}$$

where

$$\hat{L}(\chi) = U(\chi)^{-1} \circ \tilde{L}(\chi) \circ U(\chi). \tag{1.11.50}$$

By virtue of (1.11.45), the equality (1.11.49) means that $\tilde{\mu}_i(\chi)$, $i \in J$, are the eigenvalues of the operator $\hat{L}(\chi)$ in the m-dimensional subspace $M(0) = \tilde{P}(0)V$.

2. By (1.11.38), (1.11.50), taking into account Lemma 1.11.1 and noticing that $\chi \to U(\chi)$ and $\chi \to U(\chi)^{-1}$ are continuously differentiable mappings of (a,b) into $\mathcal{L}(V,V)$, we have

$$\left.\begin{array}{l} \chi \to \hat{L}(\chi) \text{ is a continuously differentiable mapping of } (a,b) \\ \text{into } \mathcal{L}(V,V). \end{array}\right\} \tag{1.11.51}$$

By known results (e.g., Kato (1976)) and (1.11.51), the eigenvalues $\tilde{\mu}_i(\chi)$, $i \in J$, are differentiable at zero and their derivatives $\frac{d\tilde{\mu}_i}{d\chi}(0) = \mu_i^{(1)}$ are the eigenvalues of the operator

$$T = \tilde{P}(0) \circ \frac{d\hat{L}}{d\chi}(0) \circ \tilde{P}(0)$$

in the subspace $M(0)$. By (1.11.50) and (1.11.47), we deduce that

$$T = \tilde{P}(0) \circ \frac{d\tilde{L}}{d\chi}(0) \circ \tilde{P}(0). \tag{1.11.52}$$

The eigenfunctions $\{u_i(h_0)\}$, $i \in J$, of the operator $\tilde{L}(0)$ form a basis in the subspace $M(0)$. We choose these functions so that they satisfy (1.11.42).

Now, we will be occupied with investigation of the representation of the operator T in the matrix form with respect to the basis $\{u_i(h_0)\}$, $i \in J$. For convenience of calculation, we assume that the scalar product in V is defined by the expression

$$(u,v)_V = (A(h_0)u, v) = a_{h_0}(u,v). \tag{1.11.53}$$

From (1.11.38) and Lemma 1.11.1, we deduce that

$$\frac{d\tilde{L}}{d\chi}(0) = L'(h_0)q. \tag{1.11.54}$$

Taking into account (1.11.33), (1.11.42), (1.11.52)–(1.11.54), we obtain that the operator T in the basis $\{u_i(h_0)\}$, $i \in J$, is represented as the $m \times m$-dimensional matrix $[\tau_{lk}]$ whose elements are given by

$$\tau_{lk} = ((A(h_0) \circ L'(h_0)q)u_l(h_0), u_k(h_0)), \qquad l,k \in J. \tag{1.11.55}$$

From here, noticing (1.11.11), (1.11.14) and (1.11.40), we get

$$\begin{aligned}\tau_{lk} &= ((B'(h_0)q - A'(h_0)q \circ A(h_0)^{-1} \circ B(h_0))u_l(h_0), u_k(h_0)) \\ &= (((B'(h_0) - \mu A'(h_0))q)u_l(h_0), u_k(h_0)), \qquad l,k \in J.\end{aligned} \tag{1.11.56}$$

Since the bilinear forms a_h and b_h are symmetric, we easily see that

$$\begin{aligned}((A'(h_0)q)u, v) &= ((A'(h_0)q)v, u), \\ ((B'(h_0)q)u, v) &= ((B'(h_0)q)v, u), \qquad u, v \in V.\end{aligned}$$

These equalities together with (1.11.56) yield that $\tau_{lk} = \tau_{kl}$.

If μ is a simple eigenvalue of the operator $L(h_0)$, then $m = 1$, $J = \{j\}$, and the operator T from (1.11.52) in the one-dimensional subspace $M(0) = \tilde{P}(0)V$ is just the multiplication by the number equal to the right-hand side of the equality (1.11.43). The theorem is proved.

Remark 1.11.1 Since the problems (1.11.7) and (1.11.22) are equivalent, Theorem 1.11.2 states that, if the conditions (1.11.1)–(1.11.6) are fulfilled, then every eigenvalue $\mu_i(h)$ of the problem (1.11.7) is differentiable in any direction, i.e., if h_0 is an arbitrary element of G, then for any i and $q \in X$ there exists the derivative

$$\frac{d}{d\chi}\mu_i(h_0 + \chi q)\big|_{\chi=0} = \delta\mu_i(h_0, q) = \mu_i^{(1)}, \tag{1.11.57}$$

which is the Gâteaux differential of the function $h \to \mu_i(h)$ at the point h_0 (see Subsec. 1.9.3). If the multiplicity of an eigenvalue $\mu = \mu_i(h_0)$ equals $m > 1$, then (1.11.57) has the following sense: there exists an enumerating of the eigenvalues (see (1.11.38), (1.11.39)) such that (1.11.57) holds.

Remark 1.11.2 Due to (1.11.16), (1.11.18), (1.11.41), the elements τ_{lk} of the matrix τ can be represented as

$$\tau_{lk} = \left(b'_{h_0} - \mu a'_{h_0}\right) q(u_l(h_0), u_k(h_0)), \qquad l, k \in J. \tag{1.11.58}$$

Remark 1.11.3 Let the conditions of Theorem 1.11.2 be fulfilled, let the multiplicity m of an eigenvalue μ be 2, and let $J = \{1, 2\}$. (It should be noted that the case of a double eigenvalue has been considered by Bratus and Seiranian (1983b) in the study of the problem of maximization of the minimal eigenvalue of a selfadjoint operator.) Then the eigenvalues of the matrix τ, $\mu_1^{(1)}$ and $\mu_2^{(1)}$, are the roots of the following equation

$$\det \begin{bmatrix} \tau_{11} - \lambda & \tau_{12} \\ \tau_{21} & \tau_{22} - \lambda \end{bmatrix} = 0,$$

or, equivalently,

$$\lambda^2 - (\tau_{11} + \tau_{22})\lambda + (\tau_{11}\tau_{22} - \tau_{12}^2) = 0.$$

The roots of the latter equation are given by the formula

$$\mu_{1,2}^{(1)} = \frac{\tau_{11} + \tau_{22}}{2} \pm \left(\frac{(\tau_{11} + \tau_{22})^2}{4} - \tau_{11}\tau_{22} + \tau_{12}^2 \right)^{\frac{1}{2}}. \tag{1.11.59}$$

Thus, from (1.11.41) it follows that, in the case when the multiplicity of an eigenvalue μ is greater than 1, the Gâteaux differential $\delta\mu_i(h_0, q) = \mu_i^{(1)}$ is not, generally speaking, a linear function of q, so that the Gâteaux derivative of the function $h \to \mu_i(h)$, $i \in J$, at the point h_0 does not exist (see Subsec. 1.9.3).

Fréchet differentiability

Theorem 1.11.3 *Let the conditions* (1.11.1)–(1.11.6) *be satisfied, let* $h_0 \in G$, *and let* μ *be a nonzero simple eigenvalue of the problem* (1.11.7) *for* $h = h_0$, *i.e., there exists* j *such that*

$$\mu_j(h_0) = \mu, \qquad \mu_i(h_0) \neq \mu, \qquad i \neq j. \tag{1.11.60}$$

Then, there exists an open neighborhood ω *of the point* h_o *in* G *such that the function* $h \to \mu_j(h)$ *is a continuously Fréchet differentiable mapping of* ω *into* $\mathbb{R}$, *and its Fréchet derivative at a point* $h \in \omega$ *is given by*

$$\mu'_j(h)q = (((B'(h) - \mu_j(h)A'(h))q)u_j(h), u_j(h)), \qquad q \in X. \tag{1.11.61}$$

Here, $A'(h)$, $B'(h)$ *are the Fréchet derivatives at the point* h *of the functions* $h \to A(h)$, $h \to B(h)$ *defined by the relations* (1.11.9), (1.11.10), *and the eigenfunction* $u_j(h)$ *satisfies the equality*

$$a_h(u_j(h), u_j(h)) = 1. \tag{1.11.62}$$

Proof. By virtue of (1.11.60), there exists an open circle Ω_0 in the complex plane $\mathbb{C}$ with center at μ and boundary Γ_0 of so small radius that $\mu_i(h_0) \in \mathbb{C} \setminus (\Omega_0 \cup \Gamma_0)$ for all $i \neq j$. Lemma 1.11.2 yields the existence of an open neighborhood ω of the point h_0 in G such that

$$\begin{aligned} \mu_j(h) \in \Omega_0, &\qquad h \in \omega, \\ \mu_i(h) \in \mathbb{C} \setminus (\Omega_0 \cup \Gamma_0), &\qquad i \neq j,\ h \in \omega. \end{aligned} \tag{1.11.63}$$

Theorem 1.11.2 implies that the function $h \to \mu_j(h)$ considered as a mapping of ω into $\mathbb{R}$ is Gâteaux differentiable at any point $h \in \omega$ and its Gâteaux derivative $\mu'_{j\,\mathrm{G}}(h)$ at a point h is given by

$$\begin{aligned} \mu'_{j\,\mathrm{G}}(h)q &= \frac{d}{d\chi}\mu_j(h + \chi q)\big|_{\chi=0} \\ &= (((B'(h) - \mu_j(h)A'(h))q)u_j(h), u_j(h)), \qquad q \in X. \end{aligned} \tag{1.11.64}$$

The eigenfunction $u_j(h)$ meets the condition (1.11.62). From (1.11.2), (1.11.3), (1.11.9), and (1.11.10), we infer that

$$\left. \begin{array}{l} h \to A'(h) \text{ and } h \to B'(h) \text{ are continuous mappings of } \omega \\ \text{into } \mathcal{L}(X, \mathcal{L}(V, V^*)). \end{array} \right\} \tag{1.11.65}$$

Denote by S the set of the eigenfunctions of index j corresponding to all h from ω satisfying the condition (1.11.62), i.e.,

$$S = \{\, u \mid u = u_j(h),\ h \in \omega,\ a_h(u_j(h),\ u_j(h)) = 1 \,\}. \tag{1.11.66}$$

We endow the set S with the topology generated by the topology of V. Let us show that

$$h \to u_j(h) \text{ is a continuous mapping of } \omega \text{ into } S. \tag{1.11.67}$$

Let $t_0 \in \omega$, $\{t_n\}_{n=1}^{\infty} \subset \omega$ and

$$t_n \to t_0 \quad \text{in } X. \tag{1.11.68}$$

Lemma 1.11.3 implies that

$$P(t_n)u_j(t_0) \to P(t_0)u_j(t_0) = u_j(t_0) \quad \text{in } V, \tag{1.11.69}$$

$P(t_n), P(t_0)$ being the projections onto the one-dimensional subspaces generated by the eigenfunctions $u_j(t_n)$, $u_j(t_0)$. We assume that

$$a_{t_i}(u_j(t_i), u_j(t_i)) = 1, \qquad i = 0, 1, 2, \ldots, \tag{1.11.70}$$

that is, $u_j(t_i) \in S$. By (1.11.69), we deduce that $P(t_n)u_j(t_0) \neq 0$ for n sufficiently large, so

$$c_n P(t_n)u_j(t_0) = u_j(t_n), \tag{1.11.71}$$

where c_n is a constant. Due to (1.11.2), (1.11.68)–(1.11.71), we get $\lim_{n\to\infty} c_n = 1$. From this and (1.11.69), (1.11.71), we conclude that $u_j(t_n) \to u_j(t_0)$ in V as $n \to \infty$, i.e., (1.11.67) holds. The relations (1.11.1)–(1.11.6) and Theorem 1.5.9 imply

$$h \to \mu_j(h) \text{ is a continuous mapping of } \omega \text{ into } \mathbb{R}. \tag{1.11.72}$$

Using (1.11.64), (1.11.65), (1.11.67), and (1.11.72), we easily see that

$$h \to \mu'_{j\,\mathrm{G}}(h) \text{ is a continuous mapping of } \omega \text{ into } X^*. \tag{1.11.73}$$

Now, (1.11.64), (1.11.73), and Theorem 1.9.3 yield that $h \to \mu_j(h)$ is a continuously Fréchet differentiable mapping of ω into $\mathbb{R}$ and its Fréchet derivative is determined by the relation (1.11.61).

1.12 The Lagrange principle in smooth extremum problems

Let X, Y be Banach spaces and let U be an open set in X. Suppose we are given functionals g_i mapping U into $\mathbb{R}$, $i = 0, 1, 2, \dots, m$, and a function F mapping U into Y.

Define a set U_{ad} in the following way:

$$U_{\mathrm{ad}} = \{\, u \mid u \in U,\ F(u) = 0,\ g_i(u) \le 0,\ i = 1, 2, \dots, m \,\}. \tag{1.12.1}$$

Supposing U_{ad} is not empty, consider the following extremum problem: Find a function $\hat{u}$ such that

$$\hat{u} \in U_{\mathrm{ad}}, \qquad g_0(\hat{u}) = \inf_{u \in U_{\mathrm{ad}}} g_0(u). \tag{1.12.2}$$

The function $\hat{u}$ is called a point of minimum, or simply a minimum of the problem (1.12.2). A function $\hat{u} \in U_{\mathrm{ad}}$ is called a point of local minimum, or simply a local minimum of the problem (1.12.2) if there exists a neighborhood U_0 of the point $\hat{u}$ in U such that

$$g_0(\hat{u}) = \inf_{u \in U_{\mathrm{ad}} \cap U_0} g_0(u). \tag{1.12.3}$$

We place in correspondence to the problem (1.12.2) the following so-called Lagrange functional:

$$\mathcal{L}(u, w, \lambda) = \sum_{i=0}^{m} \lambda_i g_i(u) + (w, F(u)), \tag{1.12.4}$$

where $\lambda = (\lambda_0, \lambda_1, \dots, \lambda_m) \in \mathbb{R}^{m+1}$, $w \in Y^*$ are the Lagrange multipliers and $u \in U$.

The following theorem states necessary conditions for a function to be a local minimum of the problem (1.12.2).

Theorem 1.12.1 *Let X, Y be Banach spaces, let U be an open set in X, let F be a continuous mapping of U into Y, and let $g_0, g_1, \ldots, g_m$ be continuous mappings of U into $\mathbb{R}$. Suppose that a nonempty set U_{ad} is defined by the relation* (1.12.1) *and that a function $\hat{u}$ from U_{ad} is a local minimum of the problem* (1.12.2), *i.e., $\hat{u}$ meets the condition* (1.12.3), *where U_0 is a neighborhood of the point $\hat{u}$ in U. Assume also that the functionals $g_0, g_1, \ldots, g_m$ are Fréchet differentiable in U_0, and the mapping F is continuously Fréchet differentiable in U_0. At last, suppose that $\mathfrak{R}F'(\hat{u})$ is a closed subspace of Y, $\mathfrak{R}F'(\hat{u})$ being the range of the mapping $F'(\hat{u})$. Then, there exist Lagrange multipliers $\hat{\lambda} = (\hat{\lambda}_0, \hat{\lambda}_1, \ldots, \hat{\lambda}_m) \in \mathbb{R}^{m+1}$, $\hat{w} \in Y^*$ not all equal to zero and such that*

$$\hat{\lambda}_0 \geq 0, \ldots, \hat{\lambda}_m \geq 0, \tag{1.12.5}$$

$$\frac{\partial \mathfrak{L}}{\partial u}(\hat{u}, \hat{w}, \hat{\lambda}) = \sum_{i=0}^{m} \hat{\lambda}_i g_i'(\hat{u}) + (F'(\hat{u}))^* \hat{w} = 0, \tag{1.12.6}$$

$$\hat{\lambda}_i g_i(\hat{u}) = 0, \qquad i = 1, 2, \ldots, m, \tag{1.12.7}$$

$(F'(\hat{u}))^*$ being the adjoint operator of $F'(\hat{u})$. Moreover, if $\mathfrak{R}F'(u) = Y$ and there exists an element $v \in X$ such that $F'(\hat{u})v = 0$, $g_i'(\hat{u})v < 0$, $i = 1, 2, \ldots, m$, then $\hat{\lambda}_0 \neq 0$ and without loss of generality one may take $\hat{\lambda}_0 = 1$.

For the proof, see, e.g., Ioffe and Tikhomirov (1979).

We adduce also the following theorem, see Pshenichny (1980).

Theorem 1.12.2 *Let X be a Banach space and let M be a convex set in X. Let also U be an open set in X and $M \subset U$. Assume that $\hat{u}$ is a solution of the problem*

$$\hat{u} \in U_{\text{ad}}, \qquad g_0(\hat{u}) = \inf_{u \in U_{\text{ad}}} g_0(u), \tag{1.12.8}$$

where

$$U_{\text{ad}} = \big\{ u \mid u \in M,\ g_i(u) \leq 0,\ i = 1, \ldots, r,\ g_i(u) = 0,\ i = r+1, r+2, \ldots, r+m \big\},$$

and g_i's are given functionals that are continuously Fréchet differentiable in U. Then, there exist constants λ_i not all equal to zero such that

$$\sum_{i=0}^{r+m} \lambda_i g_i'(\hat{u})(u - \hat{u}) \geq 0, \qquad u \in M, \tag{1.12.9}$$

$$\lambda_i \geq 0, \qquad i = 0, 1, \ldots, r, \tag{1.12.10}$$

$$\lambda_i g_i(\hat{u}) = 0, \qquad i = 1, \ldots, r. \tag{1.12.11}$$

1.13 *G*-convergence and *G*-closedness of linear operators

Let V be a separable Hilbert space over $\mathbb{R}$. A linear continuous operator $A\colon V \to V^*$ is called coercive if there exists a constant $c > 0$ such that

$$(Au, u) \geq c\|u\|_V^2, \qquad u \in V. \tag{1.13.1}$$

Denote by $\mathcal{L}_c(V, V^*)$ the set of linear, continuous, coercive operators acting from V into V^*. For an operator $A \in \mathcal{L}_c(V, V^*)$, define a bilinear form a on $V \times V$ in the following way:

$$a(u, v) = (Au, v), \qquad u, v \in V. \tag{1.13.2}$$

(1.13.1), (1.13.2), and Theorem 1.5.2 imply the existence of the operator $A^{-1} \in \mathcal{L}_c(V^*, V)$.

A sequence $\{A_n\} \subset \mathcal{L}_c(V, V^*)$ is said to G-converge to an operator $A \in \mathcal{L}_c(V, V^*)$ (notation: $A_n \overset{\mathrm{G}}{\to} A$) if

$$\lim_{n\to\infty} \left(g, A_n^{-1} f\right) = \left(g, A^{-1} f\right), \quad f, g \in V^*. \tag{1.13.3}$$

Theorem 1.13.1 *Let V be a Hilbert space over $\mathbb{R}$ and let a set $Q(c_1, c_2)$ be defined by*

$$\begin{aligned} Q(c_1, c_2) = \{\, A \,|\, A \in \mathcal{L}(V, V^*),\ (Au, v) = (Av, u),\ u, v \in V, \\ c_1\|u\|_V^2 \leq (Au, u) \leq c_2\|u\|_V^2,\ u \in V \,\}, \end{aligned} \tag{1.13.4}$$

c_1, c_2 being positive constants, Then, from any sequence $\{A_n\} \subset Q(c_1, c_2)$ one can extract a subsequence $\{A_m\}$ which G*-converges to an operator $A \in Q(c_1, c_2)$.*

For the proof, see Zhikov et al. (1993).

It should be noted that the G-convergence of elliptic operators is not connected with the convergence of their coefficients, and any kind of convergence of coefficients making the solutions converge in the distribution sense is stronger than the G-converges (see Marcellini (1979), Zhikov et al. (1993)).

Further, let $Q_1(c_1, c_2)$ be a subset of $Q(c_1, c_2)$. In particular, $Q_1(c_1, c_2)$ may coincide with $Q(c_1, c_2)$. The set $Q_1(c_1, c_2)$ is said to be G-closed if it contains all limit operators in the sense of the G-convergence, i.e., for any sequence $\{A_n\} \subset Q_1(c_1, c_2)$ the condition $A_n \overset{\mathrm{G}}{\to} A$ implies $A \in Q_1(c_1, c_2)$.

Theorem 1.13.1 yields the set $Q(c_1, c_2)$ to be G-closed. Problems of the G-closedness for second-order elliptic operators have been studied by Tartar (1975), Marcellini (1979), Raitum (1989).

1.14 Diffeomorphisms and invariance of Sobolev spaces with respect to diffeomorphisms

1.14.1 Diffeomorphisms and the relations between the derivatives

Let Q be an open or closed set in $\mathbb{R}^n$ and let f be a function defined on Q and taking values in $\mathbb{R}^k$. We say that f is of the C^m class if it is m times continuously differentiable in Q. In the case when Q is a closed set, it means that, in the interior of Q, f has all derivatives of order $\leq m$ and these derivatives coincide with some continuous functions in Q. Let also $n = k$ and let f be a bijection of Q onto $f(Q)$. The bijection f is called a C^m-diffeomorphism if both f and the inverse bijection f^{-1} are of the C^m class.

We will use the following two theorems, see, e.g., Schwartz (1967).

Theorem 1.14.1 *Let Q be an open set in $\mathbb{R}^n$ and let f be an injection of Q into $\mathbb{R}^n$ of the C^m class. Suppose that, at an arbitrary point $x \in Q$, the Fréchet derivative $f'(x)$ is an invertible element of the space $\mathcal{L}(\mathbb{R}^n, \mathbb{R}^n)$. Then, f is a C^m-diffeomorphism of Q onto $f(Q)$.*

Theorem 1.14.2 *Let Q and Ω be open sets in $\mathbb{R}^n$ and let f be a bijection of Q onto Ω. Suppose that f and the inverse bijection f^{-1} are Fréchet differentiable at every point of Q and Ω, respectively. Then, at each point $a \in Q$, the Fréchet derivative $f'(a)$ is a bijection of $\mathbb{R}^n$ onto $\mathbb{R}^n$, and the inverse bijection $(f'(a))^{-1}$ is the Fréchet derivative of f^{-1} at the point $f(a)$, i.e.,*

$$(f'(a))^{-1} = (f^{-1})'(b), \qquad b = f(a). \tag{1.14.1}$$

Notice that, passing to the inverse mappings in (1.14.1), we get

$$f'(a) = \left((f^{-1})'(f(a))\right)^{-1}, \qquad a \in Q, \tag{1.14.2}$$

or

$$f'(f^{-1}(b)) = \left((f^{-1})'(b)\right)^{-1}, \qquad b \in \Omega. \tag{1.14.3}$$

The formula (1.14.1) allows one to express the partial derivatives of the function f^{-1} by the derivatives of the function f, while by (1.14.2), (1.14.3) the partial derivatives of f may be expressed by those of f^{-1}.

Let us derive the latter relations. Denote points of Q by $y = (y_1, \ldots, y_n)$ and points of Ω by $x = (x_1, \ldots, x_n)$. By f_i and t_i, $i = 1, \ldots, n$, we denote the components of the mappings f and f^{-1}, see Fig. 1.14.1, i.e.,

$$f = (f_1, \ldots, f_n), \qquad f^{-1} = (t_1, \ldots, t_n), \tag{1.14.4}$$

$$(f^{-1})'(x) = \begin{bmatrix} \frac{\partial t_1}{\partial x_1}(x) & \frac{\partial t_1}{\partial x_2}(x) & \cdots & \frac{\partial t_1}{\partial x_n}(x) \\ \frac{\partial t_2}{\partial x_1}(x) & \frac{\partial t_2}{\partial x_2}(x) & \cdots & \frac{\partial t_2}{\partial x_n}(x) \\ \vdots & \vdots & \vdots & \vdots \\ \frac{\partial t_n}{\partial x_1}(x) & \frac{\partial t_n}{\partial x_2}(x) & \cdots & \frac{\partial t_n}{\partial x_n}(x) \end{bmatrix}. \tag{1.14.5}$$

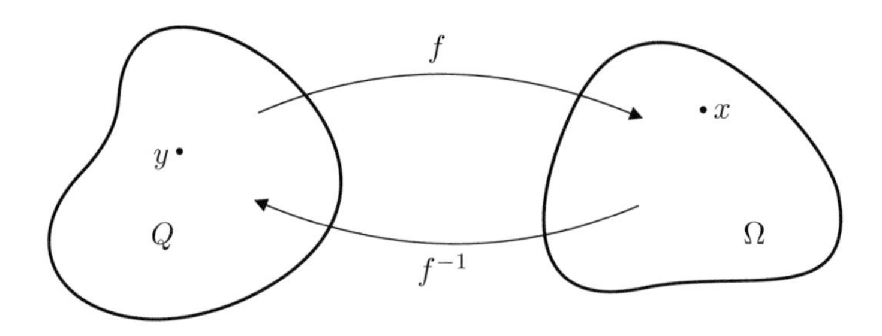

Figure 1.14.1: Sets Q, Ω and bijections f, f^{-1}

By Cramer's rule, we obtain

$$\left((f^{-1})'(x)\right)^{-1} = z(x) \begin{bmatrix} a_{11}(x) & a_{21}(x) & \dots & a_{n1}(x) \\ a_{12}(x) & a_{22}(x) & \dots & a_{n2}(x) \\ \vdots & \vdots & \vdots & \vdots \\ a_{1n}(x) & a_{2n}(x) & \dots & a_{nn}(x) \end{bmatrix}. \tag{1.14.6}$$

Here, $z(x) = \left\{ \det((f^{-1})'(x))\right\}^{-1}$ and $a_{ij}(x)$ is the cofactor of the element $\frac{\partial t_i}{\partial x_j}(x)$ of the matrix $(f^{-1})'(x)$. We have

$$f'(y) = \begin{bmatrix} \frac{\partial f_1}{\partial y_1}(y) & \frac{\partial f_1}{\partial y_2}(y) & \dots & \frac{\partial f_1}{\partial y_n}(y) \\ \frac{\partial f_2}{\partial y_1}(y) & \frac{\partial f_2}{\partial y_2}(y) & \dots & \frac{\partial f_2}{\partial y_n}(y) \\ \vdots & \vdots & \vdots & \vdots \\ \frac{\partial f_n}{\partial y_1}(y) & \frac{\partial f_n}{\partial y_2}(y) & \dots & \frac{\partial f_n}{\partial y_n}(y) \end{bmatrix}. \tag{1.14.7}$$

Put $y = f^{-1}(x)$, then (1.14.3), (1.14.6), and (1.14.7) yield

$$\frac{\partial f_k}{\partial y_i}(f^{-1}(x)) = z(x)a_{ik}(x), \qquad i, k = 1, \dots, n. \tag{1.14.8}$$

Particularly, for $n = 2$, we get

$$z(x) = \left(\frac{\partial t_1}{\partial x_1}(x)\,\frac{\partial t_2}{\partial x_2}(x) - \frac{\partial t_1}{\partial x_2}(x)\,\frac{\partial t_2}{\partial x_1}(x)\right)^{-1}, \tag{1.14.9}$$

$$a_{11}(x) = \frac{\partial t_2}{\partial x_2}(x), \qquad a_{12}(x) = -\frac{\partial t_2}{\partial x_1}(x),$$
$$a_{21}(x) = -\frac{\partial t_1}{\partial x_2}(x), \qquad a_{22}(x) = \frac{\partial t_1}{\partial x_1}(x), \tag{1.14.10}$$

and the relations (1.14.8) take the form

$$\frac{\partial f_1}{\partial y_1}(f^{-1}(x)) = z(x)\,\frac{\partial t_2}{\partial x_2}(x), \qquad \frac{\partial f_1}{\partial y_2}(f^{-1}(x)) = -z(x)\,\frac{\partial t_1}{\partial x_2}(x),$$

$$\frac{\partial f_2}{\partial y_1}(f^{-1}(x)) = -z(x)\frac{\partial t_2}{\partial x_1}(x), \qquad \frac{\partial f_2}{\partial y_2}(f^{-1}(x)) = z(x)\frac{\partial t_1}{\partial x_1}(x). \tag{1.14.11}$$

The formula (1.14.8) expresses the partial derivatives of the functions f_k by those of the functions t_j, $j,k = 1,2,\dots,n$. If the conditions of Theorem 1.14.1 are fulfilled, then by using the formula (1.14.8) and applying sequentially the theorem on a composite function, we can obtain formulas expressing the partial derivatives of f_k of order $l \leq m$ by those of t_j. For example,

$$\begin{aligned}\frac{\partial^2 f_k}{\partial y_j \partial y_i}(f^{-1}(x)) &= \frac{\partial}{\partial y_j}\left(\frac{\partial f_k}{\partial y_i}(y)\right)\Big|_{y=f^{-1}(x)} \\ &= \sum_{l=1}^{n}\left[\frac{\partial}{\partial x_l}\left(\frac{\partial f_k}{\partial y_i}(f^{-1}(x))\right)\right]\frac{\partial f_l}{\partial y_j}(f^{-1}(x)) \\ &= \sum_{l=1}^{n}\left[\frac{\partial}{\partial x_l}(z(x)a_{ik}(x))\right]z(x)a_{jl}(x).\end{aligned} \tag{1.14.12}$$

1.14.2 Sequential Fréchet derivatives and partial derivatives of a composite function

Let Q and Ω be two bounded domains in $\mathbb{R}^n$ and let $f = (f_1,\dots,f_n)$ be a C^m-diffeomorphism of Q onto Ω (see Fig. 1.14.1). Suppose u is a function of the C^m class defined on Ω and taking values in $\mathbb{R}$. Then, $\hat{u} = u \circ f$ is an m times continuously differentiable function in Q. Let us consider the question on computation of the derivatives of the function $\hat{u}$. By the chain rule (see Subsec. 1.9.1), the Fréchet derivative of $\hat{u}$ at a point $y \in Q$ is given by

$$\hat{u}'(y)X = u'(f(y)) \circ f'(y)X, \qquad X \in \mathbb{R}^n. \tag{1.14.13}$$

Sequentially applying the chain rule, we can compute the Fréchet derivatives of higher orders. Specifically, the second order Fréchet derivative is given by

$$\begin{gathered}\hat{u}''(y)(X,Y) = u''(f(y))(f'(y)X, f'(y)Y) + u'(f(y)) \circ f''(y)(X,Y), \\ X,Y \in \mathbb{R}^n.\end{gathered} \tag{1.14.14}$$

The Fréchet derivative of order $s \leq m$ of the function $\hat{u} = u \circ f$ at a point $y \in Q$ is given by (see, e.g., Schwartz (1967))

$$\begin{aligned}\hat{u}^{(s)}(y) = \sum_{k_1+2k_2+\cdots+sk_s=s} &\frac{s!}{k_1!\,k_2!\dots k_s!\,(1!)^{k_1}(2!)^{k_2}\dots(s!)^{k_s}} \\ &\times u^{(k_1+k_2+\cdots+k_s)}(f')^{k_1}(f'')^{k_2}\dots(f^{(s)})^{k_s},\end{aligned} \tag{1.14.15}$$

where $f^{(p)} = f^{(p)}(y)$ and $u^{(q)} = u^{(q)}(f(y))$ are the Fréchet derivatives of orders p and q at points y and $f(y)$, respectively.

Let $\{e_i\}_{i=1}^n$ be a basis in $\mathbb{R}^n$ such that e_i is the unit vector directed along the y_i coordinate axis. We have

$$\frac{\partial \hat{u}}{\partial y_i}(y) = \hat{u}'(y)e_i, \qquad i = 1, 2, \dots, n, \tag{1.14.16}$$

$$\frac{\partial^{|l|}\hat{u}}{\partial y_1^{l_1} \dots \partial y_n^{l_n}}(y) = \hat{u}^{(|l|)}(y)(e_{\alpha_1}, \dots, e_{\alpha_n}), \qquad |l| = l_1 + \dots + l_n,\ |l| \le m, \tag{1.14.17}$$

where e_{α_i} denotes that the vector e_i appears l_i times in the parentheses. The relations (1.14.13), (1.14.16), and the equality

$$f'(y)e_i = \left(\frac{\partial f_1}{\partial y_i}(y), \dots, \frac{\partial f_n}{\partial y_i}(y)\right), \tag{1.14.18}$$

yield

$$\frac{\partial \hat{u}}{\partial y_i}(y) = \sum_{k=1}^n \frac{\partial u}{\partial x_k}(f(y))\,\frac{\partial f_k}{\partial y_i}(y). \tag{1.14.19}$$

Putting $X = e_j$, $Y = e_i$ in (1.14.14) and taking into account (1.14.17), (1.14.18), we get

$$\begin{aligned} \frac{\partial^2 \hat{u}}{\partial y_j\, \partial y_i}(y) &= \sum_{k,l=1}^n \frac{\partial^2 u}{\partial x_k\, \partial x_l}(f(y))\,\frac{\partial f_k}{\partial y_i}(y)\,\frac{\partial f_l}{\partial y_j}(y) \\ &\quad + \sum_{k=1}^n \frac{\partial u}{\partial x_k}(f(y))\,\frac{\partial^2 f_k}{\partial y_j\, \partial y_i}(y). \end{aligned} \tag{1.14.20}$$

Analogously, the partial derivatives of $\hat{u}$ of order $|l| > 2$ (under the condition $m \ge |l|$) may be calculated by applying (1.14.15), (1.14.17), (1.14.18). These may also be directly computed by sequential differentiation of (1.14.20), by applying the chain rule.

1.14.3 Theorem on the invariance of Sobolev spaces

Theorem 1.14.3 *Let Q and Ω be bounded domains in $\mathbb{R}^n$ with Lipschitz continuous boundaries. Let f be a C^m-diffeomorphism of $\overline{Q}$ onto $\overline{\Omega}$, $m \ge 1$. Suppose $u \in W_p^m(\Omega)$, with $p \in [1, \infty)$. Then, $\hat{u} = u \circ f \in W_p^m(Q)$ and the mapping $u \to u \circ f$ is an isomorphism of $W_p^m(\Omega)$ onto $W_p^m(Q)$.*

Proof. 1. Suppose, first, that $u \in C^m(\overline{\Omega})$ and let us prove that

$$\|\hat{u}\|_{W_p^m(Q)} \le c\|u\|_{W_p^m(\Omega)}, \tag{1.14.21}$$

where $\hat{u} = u \circ f$ and c is independent of u. Denote by x and y points of Ω and Q. By the rule of change of variables in a multiple integral (see, e.g., Schwartz (1967)), we have

$$\int_Q |\hat{u}(y)|^p\, dy = \int_\Omega |u(x)|^p |\det(f^{-1})'(x)|\, dx \le c_1 \int_\Omega |u(x)|^p\, dx, \tag{1.14.22}$$

where

$$c_1 = \max_{x\in\overline{\Omega}} |\det(f^{-1})'(x)|. \tag{1.14.23}$$

Since $u \in C^m(\overline{\Omega})$, we have $\hat{u} \in C^m(\overline{Q})$, and due to results of Subsec. 1.14.2, we get

$$\begin{gathered}\frac{\partial^{|l|}\hat{u}}{\partial y_1^{l_1}\cdots\partial y_n^{l_n}}(y) = \sum_{0<|s|\le|l|} a_s(y)\,\frac{\partial^{|s|}u}{\partial x_1^{s_1}\cdots\partial x_n^{s_n}}(f(y)),\\ |l| = l_1+\cdots+l_n,\ |s| = s_1+\cdots+s_n.\end{gathered} \tag{1.14.24}$$

Here, $0 < |l| \le m$ and a_s are functions continuous in Q and depending on the derivatives of f of orders $\le |l|$, and independent of u. The equality (1.14.24) and the triangle inequality for the norm of $L_p(\Omega)$ give

$$\begin{aligned}&\left(\int_Q \left|\frac{\partial^{|l|}\hat{u}}{\partial y_1^{l_1}\cdots\partial y_n^{l_n}}(y)\right|^p dy\right)^{\frac{1}{p}}\\ &\quad = \left(\int_\Omega \left|\sum_{0<|s|\le|l|} a_s(f^{-1}(x))\,\frac{\partial^{|s|}u}{\partial x_1^{s_1}\cdots\partial x_n^{s_n}}(x)\right|^p |\det(f^{-1})'(x)|\, dx\right)^{\frac{1}{p}}\\ &\quad \le c_2 \sum_{0<|s|\le|l|}\left(\int_\Omega \left|\frac{\partial^{|s|}u}{\partial x_1^{s_1}\cdots\partial x_n^{s_n}}(x)\right|^p dx\right)^{\frac{1}{p}}, \qquad |l|\le m,\end{aligned} \tag{1.14.25}$$

where c_2 is independent of u. Now, (1.14.22) and (1.14.25) yield (1.14.21).

2. Next, let u be an arbitrary function from $W_p^m(\Omega)$ and let $\hat{u} = u \circ f$. Since the boundary of Ω is Lipschitz continuous, there exists a sequence $\{u_\mu\}$ such that

$$u_\mu \in C^m(\overline{\Omega}), \qquad u_\mu \to u \quad \text{in } W_p^m(\Omega), \tag{1.14.26}$$

and due to the above proved

$$\|\hat{u}_\mu\|_{W_p^m(Q)} \le \|u_\mu\|_{W_p^m(\Omega)} \qquad \forall \mu,\ \hat{u}_\mu = u_\mu \circ f. \tag{1.14.27}$$

We have

$$\begin{aligned}&\int_Q \left|\frac{\partial^{|s|}u_\mu}{\partial x_1^{s_1}\cdots\partial x_n^{s_n}}(f(y)) - \frac{\partial^{|s|}u}{\partial x_1^{s_1}\cdots\partial x_n^{s_n}}(f(y))\right|^p dy\\ &\qquad \le c_1\|u_\mu - u\|^p_{W_p^m(\Omega)}, \qquad 0\le|s|\le m,\end{aligned} \tag{1.14.28}$$

where c_1 is the constant from (1.14.23). By (1.14.24), for an arbitrary $w \in \mathcal{D}(Q)$,

$$(-1)^{|l|} \int_Q \hat{u}_\mu(y) \frac{\partial^{|l|} w}{\partial y_1^{l_1} \cdots \partial y_n^{l_n}}(y)\, dy$$
$$= \int_Q \sum_{0<|s|\le|l|} a_s(y) \frac{\partial^{|s|} u_\mu}{\partial x_1^{s_1} \cdots \partial x_n^{s_n}}(f(y)) w(y)\, dy, \qquad 1 \le |l| \le m. \quad (1.14.29)$$

By (1.14.26) and (1.14.28), we pass to the limit in (1.14.29), which gives

$$(-1)^{|l|} \int_Q \hat{u}(y) \frac{\partial^{|l|} w}{\partial y_1^{l_1} \cdots \partial y_n^{l_n}}(y)\, dy = \int_Q \sum_{0<|s|\le|l|} a_s(y) \frac{\partial^{|s|} u}{\partial x_1^{s_1} \cdots \partial x_n^{s_n}}(f(y)) w(y)\, dy,$$
$$w \in \mathcal{D}(Q), \qquad (1.14.30)$$

Therefore, the partial distribution derivatives of the function $\hat{u}$ are still determined by (1.14.24) and $\hat{u} \in W_p^m(Q)$. It follows from (1.14.24), (1.14.26), and (1.14.28) that

$$\hat{u}_\mu \to \hat{u} \qquad \text{in } W_p^m(Q). \qquad (1.14.31)$$

Now, (1.14.26), (1.14.27), and (1.14.31) give (1.14.21). Analogously, we prove that

$$\|u \circ f^{-1}\|_{W_p^m(\Omega)} \le c_2 \|u\|_{W_p^m(Q)}, \qquad u \in W_p^m(Q),$$

and so the mapping $u \to u \circ f$ is an isomorphism of $W_p^m(\Omega)$ onto $W_p^m(Q)$.

1.14.4 Transformation of derivatives under the change of variables

Let, as above, Q and Ω be two bounded domains in $\mathbb{R}^n$ with Lipschitz continuous boundaries, and let f be a C^m-diffeomorphism of $\overline{Q}$ onto $\overline{\Omega}$. By Theorem 1.14.3, the function $u \to \hat{u} = u \circ f$ is an isomorphism of $W_p^m(\Omega)$ onto $W_p^m(Q)$, and the partial derivatives of the function $\hat{u} = u \circ f$ are defined by the formula (1.14.24) (in particular, for $|l| = 1$ and $|l| = 2$, by (1.14.19) and (1.14.20)).

For solution of several problems, we will use the change of variables, and so we have to derive formulas that express the function

$$x \to \frac{\partial^{|l|} \hat{u}}{\partial y_1^{l_1} \cdots \partial y_n^{l_n}}(f^{-1}(x))$$

by the derivatives of u and f^{-1}.

For $|l| = 1$, setting $y = f^{-1}(x)$ in (1.14.19) and taking into account (1.14.8), we get

$$\frac{\partial \hat{u}}{\partial y_i}(f^{-1}(x)) = \sum_{k=1}^{n} \frac{\partial u}{\partial x_k}(x) z(x) a_{ik}(x). \qquad (1.14.32)$$

For $|l| = 2$, setting $y = f^{-1}(x)$ in (1.14.20) and taking into account (1.14.8), (1.14.12), we obtain

$$\begin{aligned} \frac{\partial^2 \hat{u}}{\partial y_j\, \partial y_i}(f^{-1}(x)) &= \sum_{k,l=1}^{n} \frac{\partial^2 u}{\partial x_k\, \partial x_l}(x)(z(x))^2 a_{ik}(x) a_{jl}(x) \\ &+ \sum_{k=1}^{n} \frac{\partial u}{\partial x_k}(x) \Big\{ \sum_{l=1}^{n} \Big[\frac{\partial}{\partial x_l}\big(z(x) a_{ik}(x)\big) \Big] z(x) a_{jl}(x) \Big\}. \end{aligned} \tag{1.14.33}$$

Notice that the formula (1.14.33) and the formulas for the partial derivatives of higher orders may also be obtained in the following way. Let the function

$$\varphi(x) = \frac{\partial^{|l|} \hat{u}}{\partial y_1^{l_1} \cdots \partial y_j^{l_j} \cdots \partial y_n^{l_n}}(f^{-1}(x)) \tag{1.14.34}$$

be known. By (1.14.8) and the chain rule, we get

$$\begin{aligned} \frac{\partial^{|l|+1} \hat{u}}{\partial y_1^{l_1} \cdots \partial y_j^{l_j+1} \cdots \partial y_n^{l_n}}(f^{-1}(x)) &= \sum_{k=1}^{n} \frac{\partial \varphi}{\partial x_k}(x) \frac{\partial f_k}{\partial y_j}(f^{-1}(x)) \\ &= \sum_{k=1}^{n} \frac{\partial \varphi}{\partial x_k}(x) z(x) a_{jk}(x). \end{aligned} \tag{1.14.35}$$

Chapter 2

Optimal Control by Coefficients in Elliptic Systems

> "'The time has come,' the Walrus said,
> 'To talk of many things:
> Of shoes – and ships – and sealing-wax –
> Of cabbages – and kings –
> And why the sea is boiling hot –
> And whether pigs have wings.'"
>
> – *Lewis Carrol*
> "Through the Looking-Glass"

Almost all this chapter is concerned with problems of control by coefficients in elliptic systems defined on a bounded domain $\Omega \subset \mathbb{R}^2$. Nevertheless, the technique developed is easily transferred to the case of $\Omega \subset \mathbb{R}^n$ where $n > 2$. One has just to raise either the smoothness index or the degree of integrability of elements of a Sobolev space which contains the set of admissible controls.

2.1 Direct problem

2.1.1 Coercive forms and operators

Let Ω be a bounded Lipschitz domain in $\mathbb{R}^2$, let $(x, y) \in \Omega$, $W = \prod_{s=1}^{\nu} W_2^{l_s}(\Omega)$, where $l_s \geq 1$, and let V be a closed infinite-dimensional subspace of W with the norm of the space W:

$$P_j \in \mathcal{L}(W, L_2(\Omega)), \qquad j = 1, 2, \ldots, k, \tag{2.1.1}$$

i.e., P_j is a linear, continuous mapping of W into $L_2(\Omega)$.

Define a set

$$Y_p = \{\, h \mid h \in W_p^1(\Omega),\, e_1 \leq h \leq e_2 \,\}, \tag{2.1.2}$$

where

$$e_1, e_2 \text{ are positive constants.} \tag{2.1.3}$$

Set a family of bilinear, continuous forms a_h on the space $V \times V$, which depend on a parameter h from Y_p, by the expression

$$a_h(u,v) = \iint_\Omega \sum_{i,j=1}^{k} a_{ij}(h)(P_i u)(P_j v)\, dx\, dy. \tag{2.1.4}$$

Here

$$a_{ij} \in C([e_1, e_2]), \qquad i,j = 1,2,\dots,k, \tag{2.1.5}$$

$$a_{ij} = a_{ji}, \qquad i,j = 1,2,\dots,k, \tag{2.1.6}$$

$$\sum_{i,j=1}^{k} a_{ij}(t)\xi_i\xi_j \geq c\sum_{i=1}^{k}\xi_i^2, \qquad \xi \in \mathbb{R}^k,\, t \in [e_1,e_2],\, c = \text{const} > 0. \tag{2.1.7}$$

The formulas (2.1.1)–(2.1.5) imply continuity of the form a_h in the following sense:

$$|a_h(u,v)| \leq c_1\|u\|_V\|v\|_V, \qquad u,v \in V,\ h \in Y_p,\ c_1 = \text{const} > 0, \tag{2.1.8}$$

and due to (2.1.6), the form a_h is symmetric, i.e.,

$$a_h(u,v) = a_h(v,u), \qquad u,v \in V. \tag{2.1.9}$$

Further, we assume the system of operators $\{P_j\}_{j=1}^k$ to be coercive in V, i.e.,

$$\iint_\Omega \sum_{j=1}^{k}(P_j u)^2\, dx\, dy \geq c_2\|u\|_V^2, \qquad u \in V,\, c_2 = \text{const} > 0. \tag{2.1.10}$$

The relations (2.1.2)–(2.1.4), (2.1.7), (2.1.10) imply coercivity of the form a_h in the following sense:

$$a_h(u,u) \geq c_3\|u\|_V^2, \qquad u \in V,\, h \in Y_p,\, c_3 = \text{const} > 0. \tag{2.1.11}$$

2.1.2 Boundary value problem

Consider the problem: For given elements $f \in V^*$ and $h \in Y_p$ (V^* stands for the dual space of V), find a function u_h such that

$$u_h \in V, \qquad a_h(u_h, v) = (f,v), \qquad v \in V. \tag{2.1.12}$$

Here, (f,v) denotes the value of the functional $f \in V^*$ at the element $v \in V$.

By using the Riesz theorem, or the Lax-Milgram one (see Theorems 1.5.1 and 1.5.2) we get the following theorem:

Theorem 2.1.1 *Let a_h be a bilinear, symmetric form on $V \times V$ defined by* (2.1.1), (2.1.4)–(2.1.6), *and let the inequalities* (2.1.7), (2.1.10) *hold. A set Y_p is supposed to be defined by* (2.1.2), (2.1.3). *Then, for any $h \in Y_p$, $f \in V^*$, the problem* (2.1.12) *has a unique solution.*

Remark 2.1.1 In fact, the solution u_h to problem (2.1.12) depends not only on h, but also on f. However, in what follows, an element f is supposed to be fixed, so dependence of the solution on f will not be indicated in the notations, i.e., we will write u_h instead of $u(h, f)$.

We will need the following statement.

Lemma 2.1.1 *Let the conditions of Theorem* 2.1.1 *be satisfied, let $\{h_n\}_{n=1}^{\infty} \subset Y_p$, and let $u_n = u_{h_n}$ be a solution to the problem* (2.1.12) *for $h = h_n$. Then, for $p = 2$, the condition $h_n \to h_0$ weakly in $W_2^1(\Omega)$ yields $u_n \to u_0$ weakly in V, where u_0 is the solution to the problem* (2.1.12) *for $h = h_0$, and for $p > 2$ the condition $h_n \to h_0$ weakly in $W_p^1(\Omega)$ implies $u_n \to u_0$ strongly in V.*

Proof. 1. Let $\{h_n\}_{n=1}^{\infty} \subset Y_2$ and

$$h_n \to h_0 \qquad \text{weakly in } W_2^1(\Omega). \tag{2.1.13}$$

Introducing the notations

$$a_n(u, v) = a_{h_n}(u, v), \qquad n = 0, 1, 2, \ldots, \tag{2.1.14}$$

and taking into account (2.1.11), (2.1.12), we get

$$c_3 \|u_n\|_V^2 \le a_n(u_n, u_n) = (f, u_n) \le \|f\|_{V^*} \|u_n\|_V.$$

Hence, for all n,

$$\|u_n\|_V \le \text{const}. \tag{2.1.15}$$

Due to (2.1.13), (2.1.15) we may extract a subsequence $\{h_m, u_m\}_{m=1}^{\infty}$ such that

$$h_m \to h_0 \qquad \text{strongly in } L_2(\Omega) \text{ and a.e. in } \Omega, \tag{2.1.16}$$

$$u_m \to u_0 \qquad \text{weakly in } V, \tag{2.1.17}$$

u_0 being an element of V. Further, let us show that u_0 is a solution of the problem (2.1.12) for $h = h_0$, i.e., by using the notations (2.1.14), we have to show that

$$a_0(u_0, v) = (f, v), \qquad v \in V. \tag{2.1.18}$$

From (2.1.2), (2.1.16), we obtain

$$e_1 \le h_n \le e_2, \qquad n = 0, 1, 2, \ldots. \tag{2.1.19}$$

Therefore, by virtue of (2.1.1), (2.1.5), (2.1.16) and of the Lebesgue theorem, we get, for an arbitrary fixed $v \in V$ and for fixed i, j,

$$a_{ij}(h_m)P_j v \to a_{ij}(h_0)P_j v \qquad \text{strongly in } L_2(\Omega) \text{ as } m \to \infty. \tag{2.1.20}$$

By (2.1.1), (2.1.17), we have

$$P_i u_m \to P_i u_0 \qquad \text{weakly in } L_2(\Omega) \text{ as } m \to \infty. \tag{2.1.21}$$

With the notations (2.1.14), we obtain from (2.1.4), (2.1.20), and (2.1.21) that

$$\lim_{m\to\infty} a_m(u_m, v) = a_0(u_0, v), \qquad v \in V. \tag{2.1.22}$$

Since

$$a_m(u_m, v) = (f, v), \qquad v \in V,\ m \geq 1, \tag{2.1.23}$$

the relation (2.1.22) implies (2.1.18).

So, supposing (2.1.13) is true, we have established the existence of a subsequence $\{u_m\}_{m=1}^{\infty}$ of the sequence $\{u_n = u_{h_n}\}_{n=1}^{\infty}$ such that (2.1.17) holds with $u_0 = u_{h_0}$. Let us prove that the relation (2.1.17) remains valid for the entire sequence $\{u_n\}$, i.e.,

$$u_n \to u_0 \qquad \text{weakly in } V.$$

Assume the contrary. Then, there exist a functional $g \in V^*$, a number $\varepsilon > 0$, and a subsequence $\{u_k, h_k\}_{k=1}^{\infty}$ such that

$$|(g, u_k) - (g, u_0)| \geq \varepsilon \qquad \forall k, \tag{2.1.24}$$

$$h_k \to h_0 \ \text{ strongly in } L_2(\Omega) \text{ and a.e. in } \Omega, \tag{2.1.25}$$

$$u_k \to \tilde{u} \ \text{ weakly in } V, \tag{2.1.26}$$

$\tilde{u}$ being an element from V. Here, just as before, we denote $u_k = u_{h_k}$. Letting k tend to infinity, we get, by virtue of (2.1.25), (2.1.26),

$$a_0(\tilde{u}, v) = (f, v), \qquad v \in V. \tag{2.1.27}$$

However, by Theorem 2.1.1, there exists a unique element u_0 in V satisfying (2.1.18). Combining this with (2.1.27), we obtain $\tilde{u} = u_0$, hence the relations (2.1.24) and (2.1.26) contradict each other.

2. Let now $\{h_n\}_{n=1}^{\infty} \subset Y_p$ and

$$h_n \to h_0 \ \text{ weakly in } W_p^1(\Omega), \qquad p > 2. \tag{2.1.28}$$

By virtue of the embedding theorem (Theorem 1.6.2), we get

$$\lim_{n\to\infty} \|h_n - h_0\|_{C(\overline{\Omega})} = 0, \tag{2.1.29}$$

$C(\overline{\Omega})$ being the space of continuous functions on $\overline{\Omega}$.

It follows from the above argument that (2.1.28) implies

$$u_n \to u_0 \qquad \text{weakly in } V. \tag{2.1.30}$$

By the condition of the lemma, (2.1.8), (2.1.9), (2.1.11) hold , i.e., the form $a_0 = a_{h_0}$ defines a scalar product in V and a norm which is equivalent to the original one in this space. Hence, if we prove that

$$\lim_{n\to\infty} a_0(u_n, u_n) = a_0(u_0, u_0), \tag{2.1.31}$$

then (2.1.30) will imply that $u_n \to u_0$ strongly in V (see Theorem 1.5.3). So, let us establish the equality (2.1.31).

We have

$$a_n(u_n, u_n) = (f, u_n). \tag{2.1.32}$$

Setting $v = u_n$ in (2.1.18), we get, because of (2.1.32),

$$a_0(u_0, u_n) = a_n(u_n, u_n). \tag{2.1.33}$$

By (2.1.30),

$$\lim_{n\to\infty} a_0(u_0, u_n) = a_0(u_0, u_0).$$

This equality together with (2.1.33) yields

$$\lim_{n\to\infty} a_n(u_n, u_n) = a_0(u_0, u_0). \tag{2.1.34}$$

By using the notation (2.1.14) and the relations (2.1.1), (2.1.4), we conclude

$$\begin{aligned} |a_n(u_n, u_n) - a_0(u_n, u_n)| &\le \alpha_n \iint_\Omega \sum_{i,j=1}^{k} |(P_i u_n)(P_j u_n)| \, dx\, dy \\ &\le \alpha_n c \|u_n\|_V^2, \qquad c = \text{const} > 0, \end{aligned} \tag{2.1.35}$$

where

$$\alpha_n = \max_{i,j} \max_{(x,y)\in\overline{\Omega}} |a_{ij}(h_n) - a_{ij}(h_0)| \,. \tag{2.1.36}$$

By (2.1.5), (2.1.29), (2.1.36), and Theorem 1.3.10 we, get

$$\lim_{n\to\infty} \alpha_n = 0. \tag{2.1.37}$$

By virtue of (2.1.30), $\|u_n\|_V \le$ const for all n, and therefore (2.1.35) and (2.1.37) yield

$$\lim_{n\to\infty} |a_n(u_n, u_n) - a_0(u_n, u_n)| = 0. \tag{2.1.38}$$

Combining (2.1.34) and (2.1.38), we obtain (2.1.31), and so the lemma is proved.

Remark 2.1.2 Theorem 2.1.1 and Lemma 2.1.1 remain valid without the assumption that the form $a_h(u,v)$ is symmetric, i.e., when the condition (2.1.6) does not hold. Indeed, in this case, Theorem 2.1.1 is true by virtue of the Lax-Milgram theorem. Evidently, the relations (2.1.11), (2.1.30), (2.1.31) are satisfied without the assumption (2.1.6). By (2.1.11), we have

$$\begin{aligned} a_0(u_n - u_0, u_n - u_0) &= a_0(u_n, u_n) - a_0(u_0, u_n) - a_0(u_n, u_0) + a_0(u_0, u_0) \\ &\geq c_3 \|u_n - u_0\|_V^2 . \end{aligned}$$

Therefore, taking into account (2.1.30) and (2.1.31), we get $u_n \to u_0$ strongly in V.

2.2 Optimal control problem

2.2.1 Nonregular control

Basic assumptions

Let us introduce a set of admissible controls by

$$\begin{aligned} Q_{\text{ad}} = \{\, h \,|\, h \in W_2^1(\Omega), \|h\|_{W_2^1(\Omega)} \leq c, \check{h} \leq h \leq \hat{h}, \\ \Psi_k(h, u_h) \leq 0, k = 1, 2, \dots, l \,\}. \end{aligned} \tag{2.2.1}$$

Here

c, $\check{h}$, $\hat{h}$ are positive numbers such that $e_1 < \check{h} < \hat{h} < e_2$, e_1 and e_2 being the numbers from (2.1.2); (2.2.2)

$h, u \to \Psi_k(h,u)$ are lower semicontinuous functionals given on $Y_2 \times V$ (endowed with the topology generated by the product of the weak topology of $W_2^1(\Omega)$ and of the weak topology of V), $k = 1, 2, \dots, l$. (2.2.3)

The assumption (2.2.3) means that, provided $\{h_n\} \subset Y_2$, $h_n \to h$ weakly in $W_2^1(\Omega)$ and $\{u_n\} \subset V$, $u_n \to u$ weakly in V, we have

$$\liminf_{n\to\infty} \Psi_k(h_n, u_n) \geq \Psi_k(h,u), \qquad k = 1, 2, \dots, l.$$

We assume the set Q_{ad} to be nonempty. It should be stressed that the function u_h in the expression of $\Psi_k(h, u_h)$ from (2.2.1) is a solution to the problem (2.1.12) for given h.

Let us introduce a goal functional $h \to f_1(h)$ satisfying the following condition

$h \to f_1(h)$ is a continuous mapping of Y_2 (equipped with the topology induced by the $W_2^1(\Omega)$-weak topology) into $\mathbb{R}$. (2.2.4)

Existence theorem

The optimal control problem consists in finding a function h_0 such that

$$h_0 \in Q_{\text{ad}}, \qquad f_1(h_0) = \inf_{h \in Q_{\text{ad}}} f_1(h). \tag{2.2.5}$$

Theorem 2.2.1 *Let a_h be a bilinear, symmetric form on $V \times V$ defined by the relations* (2.1.1), (2.1.4)–(2.1.6) *and let the inequalities* (2.1.7), (2.1.10) *hold. Also, let a nonempty set Q_{ad} be defined by* (2.2.1), (2.2.2), (2.2.3) *and let the goal functional satisfy the condition* (2.2.4). *Then, the problem* (2.2.5) *has a solution.*

Proof. Since the set Q_{ad} is not empty by the hypothesis, by virtue of Theorem 1.1.1 and the argument below it about the completed real line $\overline{\mathbb{R}}$, there exists a sequence $\{h_n\}$ such that

$$h_n \in Q_{\text{ad}} \qquad \forall n, \tag{2.2.6}$$

$$\lim_{n\to\infty} f_1(h_n) = \inf_{h \in Q_{\text{ad}}} f_1(h). \tag{2.2.7}$$

By Theorem 2.1.1, to every element h_n there corresponds an element $u_n = u_{h_n}$ giving a solution to the problem (2.1.12) for $h = h_n$. By using (2.1.11), (2.1.12), and the notations (2.1.14), we get

$$c_3 \|u_n\|_V^2 \leq a_n(u_n, u_n) = (f, u_n) \leq \|f\|_{V^*} \|u_n\|_V. \tag{2.2.8}$$

Hence,

$$\|u_n\|_V \leq \text{const} \qquad \forall n. \tag{2.2.9}$$

By the definition of the set Q_{ad}, the sequence $\{h_n\}$ is bounded in $W_2^1(\Omega)$. This fact and (2.2.9) yield the existence of a subsequence $\{h_m, u_m\}_{m=1}^{\infty}$ such that

$$h_m \to h_0 \qquad \text{weakly in } W_2^1(\Omega), \tag{2.2.10}$$

$$h_m \to h_0 \qquad \text{strongly in } L_2(\Omega) \text{ and a.e. in } \Omega, \tag{2.2.11}$$

$$u_m \to u_0 \qquad \text{weakly in } V. \tag{2.2.12}$$

Passing to the limit as $m \to \infty$, we get, just as in the proof of Lemma 2.1.1,

$$a_0(u_0, v) = (f, v), \qquad v \in V, \tag{2.2.13}$$

where $a_0 = a_{h_0}$. So, we have established that $u_0 = u_{h_0}$.

Let us show that

$$h_0 \in Q_{\text{ad}}. \tag{2.2.14}$$

By (2.2.1), (2.2.6), (2.2.11), we obtain

$$\check{h} \leq h_0 \leq \hat{h}, \tag{2.2.15}$$

and by (2.2.1), (2.2.6), (2.2.10), and Theorem 1.2.4,

$$\|h_0\|_{W_2^1(\Omega)} \leq \liminf_{m\to\infty} \|h_m\|_{W_2^1(\Omega)} \leq c, \tag{2.2.16}$$

c being the number from (2.2.1).

At last, by virtue of (2.2.1), (2.2.3), (2.2.6), (2.2.10), and (2.2.12), we get

$$0 \geq \liminf_{m\to\infty} \Psi_k(h_m, u_m) \geq \Psi_k(h_0, u_0), \qquad k = 1, 2, \ldots, l, \tag{2.2.17}$$

where $u_0 = u_{h_0}$.

Now, (2.2.14) follows from (2.2.15), (2.2.16), and (2.2.17). Taking note of (2.2.4), (2.2.7), and (2.2.10), we get

$$f_1(h_0) = \lim_{m\to\infty} f_1(h_m) = \inf_{h\in Q_{\mathrm{ad}}} f_1(h).$$

The theorem is proved.

2.2.2 Regular control

Basic assumptions

Let us study the optimal control problem when the set of admissible controls is "more regular" than in Subsec. 2.2.1.

Let a set of admissible controls U_{ad} be given by

$$U_{\mathrm{ad}} = \big\{ h \,|\, h \in W_p^1(\Omega),\ \|h\|_{W_p^1(\Omega)} \leq c,\ \check{h} \leq h \leq \hat{h}, \\ \Psi_k(h, u_h) \leq 0,\ k = 1, 2, \ldots, l \big\}. \tag{2.2.18}$$

Here,

$$\left.\begin{array}{l} c,\ \check{h},\ \hat{h} \text{ are positive numbers such that } e_1 < \check{h} < \hat{h} < e_2 \\ (e_1,\ e_2 \text{ being the positive numbers from (2.1.2)) and } p > 2, \end{array}\right\} \tag{2.2.19}$$

$$\left.\begin{array}{l} h, u \to \Psi_k(h, u) \text{ is a continuous mapping of } Y_p \times V \\ \text{(endowed with the topology generated by the product of} \\ \text{the } W_p^1(\Omega)\text{-weak topology and the } V\text{-strong topology)} \\ \text{into } \mathbb{R},\ k = 1, 2, \ldots, l. \end{array}\right\} \tag{2.2.20}$$

The set U_{ad} is equipped with the topology induced by the topology of $W_p^1(\Omega)$ on U_{ad} and is supposed to be nonempty.

Let us introduce a goal functional $h \to f_1(h)$ satisfying the following condition

$$\left.\begin{array}{l} h \to f_1(h) \text{ is a continuous mapping of } Y_p \text{ (endowed with} \\ \text{the topology induced by the } W_p^1(\Omega)\text{-weak topology) into} \\ \mathbb{R}. \end{array}\right\} \tag{2.2.21}$$

Existence theorem

Let us consider the problem of finding a function h_0 such that

$$h_0 \in U_{\text{ad}}, \qquad f_1(h_0) = \inf_{h \in U_{\text{ad}}} f_1(h). \tag{2.2.22}$$

Theorem 2.2.2 *Let a_h be a bilinear symmetric form on $V \times V$ defined by the relations* (2.1.1), (2.1.4)–(2.1.6) *and let the inequalities* (2.1.7), (2.1.10) *hold true. Let a nonempty set U_{ad} be defined by* (2.2.18)–(2.2.20) *and let the goal functional satisfy the condition* (2.2.21). *Then, the problem* (2.2.22) *has a solution.*

Proof. Let $\{h_n\}_{n=1}^{\infty}$ be a minimizing sequence, that is,

$$h_n \in U_{\text{ad}} \qquad \forall n, \tag{2.2.23}$$

$$f_1(h_n) \to \inf_{h \in U_{\text{ad}}} f_1(h). \tag{2.2.24}$$

By virtue of (2.2.18), the sequence $\{h_n\}$ is bounded in $W_p^1(\Omega)$. Choose a subsequence $\{h_m\}_{m=1}^{\infty}$ such that

$$h_m \to h_0 \qquad \text{weakly in } W_p^1(\Omega), \tag{2.2.25}$$

$$h_m \to h_0 \qquad \text{strongly in } C(\overline{\Omega}). \tag{2.2.26}$$

Let $u_m = u_{h_m}$ be a solution to the problem (2.1.12) when $h = h_m$, i.e.,

$$u_m \in V, \qquad a_m(u_m, v) = (f, v), \qquad v \in V,\ m = 0, 1, 2, \dots, \tag{2.2.27}$$

where

$$a_m(u, v) = a_{h_m}(u, v). \tag{2.2.28}$$

By (2.2.25) and Lemma 2.1.1, we have

$$u_m \to u_0 \qquad \text{strongly in } V. \tag{2.2.29}$$

Moreover, u_0 satisfies (2.2.27), (2.2.28) when $m = 0$. By virtue of (2.2.23) and (2.2.26), we get

$$\check{h} \leq h_0 \leq \hat{h}. \tag{2.2.30}$$

By (2.2.23), (2.2.25), we obtain

$$c \geq \liminf_{m \to \infty} \|h_m\|_{W_p^1(\Omega)} \geq \|h_0\|_{W_p^1(\Omega)}, \tag{2.2.31}$$

c being the constant from (2.2.18).

By using (2.2.18), (2.2.20), (2.2.23), (2.2.25), and (2.2.29), we get

$$0 \geq \lim_{m \to \infty} \Psi_k(h_m, u_m) = \Psi_k(h_0, u_0), \qquad k = 1, 2, \dots, l. \tag{2.2.32}$$

From (2.2.30)–(2.2.32), it follows that $h_0 \in U_{\text{ad}}$. At last, the relations (2.2.21), (2.2.24), and (2.2.25) imply

$$f_1(h_0) = \lim_{m \to \infty} f_1(h_m) = \inf_{h \in U_{\text{ad}}} f_1(h),$$

which gives the theorem.

2.2.3 Regular problem and necessary conditions of optimality

Modification of the restrictions

In the space $W_p^1(\Omega)$, $p > 2$, define the following functionals

$$\varphi_1(h) = \max_{(x,y)\in\overline{\Omega}} (\check{h} - h(x,y)), \qquad \varphi_2(h) = \max_{(x,y)\in\overline{\Omega}} (h(x,y) - \hat{h}), \tag{2.2.33}$$

$\check{h}$ and $\hat{h}$ being the numbers from (2.2.18). Then, the expression (2.2.18) may be rewritten as

$$U_{\text{ad}} = \{ h \,|\, h \in W_p^1(\Omega),\ \|h\|_{W_p^1(\Omega)} \le c,\ \varphi_1(h) \le 0, \\ \varphi_2(h) \le 0,\ \Psi_k(h, u_h) \le 0,\ k = 1, 2, \ldots, l \}. \tag{2.2.34}$$

The functionals φ_1 and φ_2 from (2.2.33) are not Fréchet differentiable. Therefore, to regularize the control problem we change the conditions

$$\varphi_1(h) \le 0, \qquad \varphi_2(h) \le 0 \tag{2.2.35}$$

by the following ones

$$Q_i(h) \le 0, \qquad i = 1, 2, \ldots, 2r, \tag{2.2.36}$$

r being a natural number. The functionals Q_i are chosen so that they are Fréchet differentiable and the conditions (2.2.36) approximate the conditions (2.2.35) sufficiently well.

Let an ε-net be chosen on the set $\overline{\Omega}$, i.e., a set of points $\Omega_r = \{x_i, y_i\}_{i=1}^r \subset \overline{\Omega}$ is chosen so that, for any point $(x, y) \in \overline{\Omega}$, there exists at least one point $(x_k, y_k) \in \Omega_r$ satisfying

$$((x - x_k)^2 + (y - y_k)^2)^{1/2} \le \varepsilon.$$

Define functionals Q_i by

$$Q_i(h) = \check{h} - h(x_i, y_i), \qquad i = 1, 2, \ldots, r, \\ Q_i(h) = h(x_{i-r}, y_{i-r}) - \hat{h}, \qquad i = r + 1, r + 2, \ldots, 2r. \tag{2.2.37}$$

Since $W_p^1(\Omega)$, $p > 2$, is continuously embedded into the Hölder space (see Triebel (1978)), for any fixed $p > 2$, we can find an ε-net such that the conditions (2.2.36) approximate the conditions (2.2.35) sufficiently well. More precisely, for any $\gamma > 0$, there exists $\varepsilon > 0$ such that, for the corresponding ε-net, the conditions

$$\|h\|_{W_p^1(\Omega)} \le c, \qquad Q_i(h) \le 0, \qquad i = 1, 2, \ldots, 2r(\varepsilon), \tag{2.2.38}$$

will imply

$$\varphi_1(h) \le \gamma, \qquad \varphi_2(h) \le \gamma. \tag{2.2.39}$$

In what follows, we assume the net to be chosen in such a way that the condition (2.2.38) yields $h \in Y_p$, that is, $\gamma \le \max(\check{h} - e_1, e_2 - \hat{h})$ (see (2.1.2) and (2.2.19)). Since, for $p > 2$, the embedding of $W_p^1(\Omega)$ into $C(\overline{\Omega})$ is continuous and the mappings Q_i from (2.2.37) are affine functionals, these mappings are Fréchet differentiable in $W_p^1(\Omega)$ and their Fréchet derivatives are determined by the following formulas:

$$\begin{aligned} &Q_i'(h)q = -q(x_i, y_i), && i = 1, 2, \dots, r,\ q \in W_p^1(\Omega), \\ &Q_i'(h)q = q(x_{i-r}, y_{i-r}), && i = r+1, r+2, \dots, 2r. \end{aligned} \tag{2.2.40}$$

Now, let us replace the set U_{ad} from (2.2.34) with the following one

$$\begin{aligned} \tilde{U}_{\text{ad}} = \{\, h \mid h \in W_p^1(\Omega),\ \|h\|_{W_p^1(\Omega)} \le c,\ Q_i(h) \le 0, \\ i = 1, 2, \dots, 2r,\ \Psi_k(h, u_h) \le 0,\ k = 1, 2, \dots, l \,\}. \end{aligned} \tag{2.2.41}$$

Further, let a functional Ψ_0 on the set $Y_p \times V$ be defined such that

$$\left.\begin{array}{l} h, u \to \Psi_0(h, u) \text{ is a continuous mapping of } Y_p \times V \\ \text{(endowed with the topology generated by the product of} \\ \text{the } W_p^1(\Omega)\text{-weak topology and the } V\text{-strong one) into } \mathbb{R}. \end{array}\right\} \tag{2.2.42}$$

Consider the problem of finding a function h_0 such that

$$h_0 \in \tilde{U}_{\text{ad}}, \qquad \Psi_0(h_0, u_{h_0}) = \inf_{h \in \tilde{U}_{\text{ad}}} \Psi_0(h, u_h). \tag{2.2.43}$$

By using the argument from the proof of Theorem 2.2.2, we establish the following fact.

Theorem 2.2.3 *Let a_h be a bilinear, symmetric form on $V \times V$ determined by the relations* (2.1.1), (2.1.4)–(2.1.6) *and let the inequalities* (2.1.7), (2.1.10) *hold; let a nonempty set $\tilde{U}_{\text{ad}}$ be defined by* (2.2.41), (2.2.37), (2.2.19), (2.2.20) *and let* (2.2.42) *hold true. Then, the problem* (2.2.43) *has a solution.*

Reformulation of the problem

Now, by using Theorem 1.11.1, we will establish necessary conditions of optimality in the problem (2.2.43).

To this end, we have to reformulate the problem (2.2.43).

Let us introduce the following sets:

$$\begin{aligned} &X = W_p^1(\Omega) \times V, \qquad p > 2, \\ &G = \{\, h \mid h \in W_p^1(\Omega),\ e_1 < h < e_2 \,\}, \\ &U = G \times V, \end{aligned} \tag{2.2.44}$$

e_1 and e_2 being the positive numbers from (2.1.2).

Since the embedding of $W_p^1(\Omega)$ into $C(\overline{\Omega})$ is continuous, G is an open set in $W_p^1(\Omega)$, hence, U is an open set in X.

Define a function $F\colon U \to V^*$ through the formula

$$(h,u) \in U, \qquad (F(h,u),v) = a_h(u,v) - (f,v), \qquad v \in V, \tag{2.2.45}$$

where a_h is the bilinear form defined in Subsec. 2.1.1, and f a fixed element from V^* (see (2.1.12)) for which the problem (2.2.43) is being solved.

Let us define on the set U functionals

$$g_0(h,u) = \Psi_0(h,u), \tag{2.2.46}$$
$$g_1(h,u) = \|h\|_{W_p^1(\Omega)} - c, \tag{2.2.47}$$
$$g_i(h,u) = Q_{i-1}(h), \qquad i = 2,3,\dots,2r+1, \tag{2.2.48}$$
$$g_i(h,u) = \Psi_{i-2r-1}(h,u), \qquad i = 2r+2, 2r+3, \dots, 2r+l+1. \tag{2.2.49}$$

Further, let

$$\check{U}_{\text{ad}} = \{\, (h,u) \mid (h,u) \in U,\ F(h,u) = 0,\ g_i(h,u) \le 0,$$
$$i = 1,2,\dots,m \ (m = 2r+l+1)\}. \tag{2.2.50}$$

Now, the problem (2.2.43) reduces to the following one: Find a pair h_0, u_0 such that

$$(h_0,u_0) \in \check{U}_{\text{ad}}, \qquad g_0(h_0,u_0) = \inf_{(h,u)\in\check{U}_{\text{ad}}} g_0(h,u). \tag{2.2.51}$$

Notice that the relations (2.1.12), (2.2.45), (2.2.50), (2.2.51) yield $u_0 = u_{h_0}$.

Auxiliary statements

Lemma 2.2.1 *Let a_h be a bilinear form on $V \times V$ determined by the relations* (2.1.1), (2.1.4)–(2.1.6) *and let the inequalities* (2.1.7), (2.1.10) *hold. Assume that the functions $t \to a_{ij}(t)$ are continuously differentiable on $[e_1,e_2]$, that is,*

$$a_{ij} \in C^1([e_1,e_2]), \qquad i,j = 1,2,\dots,k. \tag{2.2.52}$$

Then, the function F determined by (2.2.45) *is a continuously Fréchet differentiable mapping of U into V^* and its Fréchet derivative is given by*

$$F'(h,u)(q,v) = \frac{\partial F}{\partial h}(h,u)q + F(h,v) + f, \qquad q \in W_p^1(\Omega),\ v \in V, \tag{2.2.53}$$

where $\frac{\partial F}{\partial h}(h,u)$ belongs to the space $\mathcal{L}(W_p^1(\Omega),V^*)$ and is determined through the relation

$$\left(\frac{\partial F}{\partial h}(h,u)q, w\right) = \iint_\Omega \sum_{i,j=1}^k \frac{da_{ij}}{dt}(h)\, q(P_iu)(P_jw)\, dx\, dy, \tag{2.2.54}$$
$$q \in W_p^1(\Omega),\ w \in V.$$

Proof. Given a pair $(h, u) \in U$, define an operator $\Lambda_{hu} \in \mathcal{L}(W_p^1(\Omega), V^*)$ by the relation

$$(\Lambda_{hu} q, w) = \iint\limits_{\Omega} \sum_{i,j=1}^{k} \frac{da_{ij}}{dt}(h)\, q(P_i u)(P_j w)\, dx\, dy, \tag{2.2.55}$$

$$q \in W_p^1(\Omega),\ w \in V.$$

Since $h \in G$ and G is an open set in $W_p^1(\Omega)$ (see (2.2.44)), there exists $r > 0$ such that

$$h + z \in G, \qquad z \in d(r, 0),$$

where

$$d(r, 0) = \left\{ z \mid z \in W_p^1(\Omega),\ \|z\|_{W_p^1(\Omega)} \le r \right\}.$$

Let $q \in d(r, 0)$. Then, (2.1.4) and (2.2.45) obviously imply the inequality

$$\begin{aligned} &\|F(h+q, u) - F(h, u) - \Lambda_{hu} q\|_{V^*} \\ &\quad \le c \sum_{i,j=1}^{k} \left\{ \int_{\Omega}\!\!\int \left[a_{ij}(h+q) - a_{ij} h - \frac{da_{ij}}{dt}(h) q \right]^2 (P_i u)^2\, dx\, dy \right\}^{1/2}. \end{aligned} \tag{2.2.56}$$

Let (x, y) be an arbitrary point from $\overline{\Omega}$. By the mean value theorem, we obtain

$$\begin{aligned} &\left| a_{ij}(h(x,y) + q(x,y)) - a_{ij}(h(x,y)) - \frac{da_{ij}}{dt}(h(x,y)) q(x,y) \right| \\ &\quad = \left| \left(\frac{da_{ij}}{dt}(h(x,y) + \xi q(x,y)) - \frac{da_{ij}}{dt}(h(x,y)) \right) q(x,y) \right| \\ &\quad \le \omega\left(\frac{da_{ij}}{dt}; \|q\|_{C(\overline{\Omega})} \right) \|q\|_{C(\overline{\Omega})}, \qquad \xi \in (0, 1). \end{aligned} \tag{2.2.57}$$

Here $\omega(\cdot, \cdot)$ stands for the continuity modulus:

$$\omega\left(\frac{da_{ij}}{dt}; \varepsilon \right) = \sup \left| \frac{da_{ij}}{dt}(t') - \frac{da_{ij}}{dt}(t'') \right|, \qquad \varepsilon > 0, \tag{2.2.58}$$

where the supremum is taken over $t', t'' \in [e_1, e_2]$, $|t' - t''| \le \varepsilon$. By (2.2.52) and Theorem 1.3.10, we get

$$\omega\left(\frac{da_{ij}}{dt}; \|q\|_{C(\overline{\Omega})} \right) \to 0 \qquad \text{as } \|q\|_{C(\overline{\Omega})} \to 0.$$

From (2.2.56), (2.2.57), and the continuity of the embedding of $W_p^1(\Omega)$ into $C(\overline{\Omega})$, it follows that Λ_{hu} is the partial Fréchet derivative of the function F with respect to h, i.e.,

$$\frac{\partial F}{\partial h}(h, u) = \Lambda_{h,u}\,. \tag{2.2.59}$$

Assume that

$$\begin{aligned} &\{h_n\}_{n=1}^{\infty} \subset G, \qquad h \in G, \qquad \{u_n\}_{n=1}^{\infty} \subset V, \\ &h_n \to h \;\text{ in } W_p^1(\Omega), \qquad u_n \to u \;\text{ in } V. \end{aligned} \tag{2.2.60}$$

Using (2.2.55), one can easily get the estimate

$$\begin{aligned} &\|(\Lambda_{h_n u_n} - \Lambda_{hu})\, q\|_{V^*} \\ &\le c \sum_{i,j=1}^{k} \biggl(\iint_{\Omega} \Bigl(\frac{da_{ij}}{dt}(h_n) P_i u_n - \frac{da_{ij}}{dt}(h) P_i u \Bigr)^2 dx\, dy \biggr)^{1/2} \|q\|_{C(\overline{\Omega})}. \end{aligned} \tag{2.2.61}$$

By (2.1.1) and (2.2.52) we obtain

$$\begin{aligned} &\biggl(\iint_{\Omega} \Bigl(\frac{da_{ij}}{dt}(h_n) P_i u_n - \frac{da_{ij}}{dt}(h) P_i u \Bigr)^2 dx\, dy \biggr)^{1/2} \\ &\quad \le \biggl(\iint_{\Omega} \Bigl(\frac{da_{ij}}{dt}(h_n)(P_i u_n - P_i u) \Bigr)^2 dx\, dy \biggr)^{1/2} \\ &\qquad + \biggl(\iint_{\Omega} \Bigl[\Bigl(\frac{da_{ij}}{dt}(h_n) - \frac{da_{ij}}{dt}(h) \Bigr) P_i u \Bigr]^2 dx\, dy \biggr)^{1/2} \\ &\quad \le c\|u_n - u\|_V + c_1 \omega\Bigl(\frac{da_{ij}}{dt}; \|h_n - h\|_{C(\overline{\Omega})} \Bigr). \end{aligned} \tag{2.2.62}$$

Due to (2.2.60), the right-hand side of the inequality (2.2.62) tends to zero. So, (2.2.59)–(2.2.61) yield

$$h, u \to \frac{\partial F}{\partial h}(h,u) \text{ is a continuous mapping of } U \text{ into } \mathcal{L}(W_p^1(\Omega), V^*). \tag{2.2.63}$$

It follows from (2.2.45) that, for any fixed $h \in G$, the partial function $u \to F(h,u)$ is a continuous affine mapping of V into V^*. Therefore, it is Fréchet differentiable and its Fréchet derivative $\frac{\partial F}{\partial u}(h,u)$ is given by the formula

$$\frac{\partial F}{\partial u}(h,u)v = F(h,v) + f, \qquad v \in V. \tag{2.2.64}$$

Let the conditions (2.2.60) hold true again. Taking into account (2.1.4), (2.2.45), (2.2.52), and (2.2.64), we get

$$\begin{aligned} &\Bigl\| \Bigl(\frac{\partial F}{\partial u}(h_n, u_n) - \frac{\partial F}{\partial u}(h,u) \Bigr) v \Bigr\|_{V^*} \\ &\quad \le c \sum_{i,j=1}^{k} \biggl(\iint_{\Omega} ((a_{ij}(h_n) - a_{ij}(h)) P_i v)^2 \, dx\, dy \biggr)^{1/2} \le c_1 \varphi_n \|v\|_V, \end{aligned}$$

where $\varphi_n = \max\limits_{i,j} \omega(a_{ij}; \|h_n - h\|_{C(\overline{\Omega})})$.

However, due to (2.2.60), $\varphi_n \to 0$, so

$$h, u \to \frac{\partial F}{\partial u}(h,u) \text{ is a continuous mapping of } U \text{ into } \mathcal{L}(V, V^*). \tag{2.2.65}$$

Now, (2.2.55), (2.2.59), (2.2.63)–(2.2.65) imply (see Subsec. 1.9.1) that the function F is a Fréchet continuously differentiable mapping of U into V^* and its Fréchet derivative is given through the relations (2.2.53), (2.2.54).

Lemma 2.2.2 *Let the conditions of Lemma* 2.2.1 *be satisfied and let h, u be an arbitrary pair from U. Then, the operator $F'(h,u)$ maps $W_p^1(\Omega) \times V$ onto the whole space V^*, that is,*

$$\mathfrak{R}F'(h,u) = V^*, \qquad (h,u) \in U. \tag{2.2.66}$$

Proof. Given a pair h, u from U, let us show that, for an arbitrary $z \in V^*$, there exists a pair q, v such that

$$(q,v) \in W_p^1(\Omega) \times V, \qquad F'(h,u)(q,v) = z. \tag{2.2.67}$$

By (2.2.45) and Theorem 2.1.1, we conclude the existence of a function $\hat{v}$ satisfying the conditions

$$\hat{v} \in V, \qquad (F(h,\hat{v}), w) + (f, w) = a_h(\hat{v}, w) = (z, w), \qquad w \in V. \tag{2.2.68}$$

From here and (2.2.53), we deduce that the pair $q = 0, v = \hat{v}$ is a solution to the problem (2.2.67), which completes the proof.

Necessary conditions of optimality

Let us return to the problem (2.2.50), (2.2.51). Introduce the Lagrange functional connected with it:

$$\mathfrak{L}(h,u,w,\lambda) = \sum_{i=0}^{m} \lambda_i g_i(h,u) + (F(h,u), w), \tag{2.2.69}$$

where $\lambda = (\lambda_0, \lambda_1, \ldots, \lambda_m) \in \mathbb{R}^{m+1}$, $u, w \in V$, $h \in G$ (see (2.2.44)).

Theorem 2.2.4 *Let a_h be a bilinear, symmetric form on $V \times V$ determined by the relations* (2.1.1), (2.1.4), (2.1.6), (2.2.52) *and let the inequalities* (2.1.7), (2.1.10) *hold. A nonempty set $\breve{U}_{\text{ad}}$ is defined by* (2.2.50), (2.2.19), (2.2.20), (2.2.37), (2.2.45), (2.2.47)–(2.2.49) *and a goal functional g_0 by* (2.2.46), (2.2.42). *Then, there exists a pair h_0, u_0 solving the problem* (2.2.50), (2.2.51). *If the functionals $h, u \to \Psi_i(h,u)$ are Fréchet differentiable in U, $i = 0, 1, 2, \ldots, l$, then there exist Lagrange multipliers $\hat{\lambda} = (\hat{\lambda}_0, \hat{\lambda}_1, \ldots, \hat{\lambda}_m) \in \mathbb{R}^{m+1}$ and $\hat{w} \in V$ which do not vanish simultaneously and satisfy the following conditions:*

$$\hat{\lambda}_i \geq 0, \qquad i = 0, 1, 2, \ldots, m, \tag{2.2.70}$$

$$\hat{\lambda}_i g_i(h_0, u_0) = 0, \qquad i = 1, 2, \ldots, m, \tag{2.2.71}$$

$$\begin{aligned} \frac{\partial \mathfrak{L}}{\partial h}(h_0, u_0, \hat{w}, \hat{\lambda})q = \sum_{i=0}^{m} \hat{\lambda}_i \frac{\partial g_i}{\partial h}(h_0, u_0)q \\ + \left(\frac{\partial F}{\partial h}(h_0, u_0)q, \hat{w}\right) = 0, \qquad q \in W_p^1(\Omega), \end{aligned} \tag{2.2.72}$$

$$\frac{\partial \mathfrak{L}}{\partial u}(h_0, u_0, \hat{w}, \hat{\lambda})v = a_{h_0}(\hat{w}, v) + \sum_{i=0}^{m} \hat{\lambda}_i \frac{\partial g_i}{\partial u}(h_0, u_0)v = 0, \qquad v \in V. \tag{2.2.73}$$

If a pair $\tilde{h}$, $\tilde{u}$ exists such that

$$\begin{aligned} (\tilde{h}, \tilde{u}) \in W_p^1(\Omega) \times V, \qquad F'(h_0, u_0)(\tilde{h}, \tilde{u}) = 0, \\ g_i'(h_0, u_0)(\tilde{h}, \tilde{u}) < 0, \qquad i = 1, 2, \ldots, m, \end{aligned} \tag{2.2.74}$$

then $\hat{\lambda}_0 \neq 0$ and we may take $\hat{\lambda}_0 = 1$.

Proof. The problem (2.2.50), (2.2.51) being a reformulation of the problem (2.2.41), (2.2.43), the existence of a solution to the problem (2.2.50), (2.2.51) follows from Theorem 2.2.3.

Let us verify that, in the present setting, the conditions of Theorem 1.12.1 hold true. The sets X and U are defined by (2.2.44), and $Y = V^*$. Due to Lemma 2.2.1, the function F is a Fréchet continuously differentiable mapping of U into Y. The equalities (2.2.46), (2.2.49) and the assumption of the theorem imply the functionals $g_0, g_{2r+2}, g_{2r+3}, \ldots, g_{2r+l+1}$ to be Fréchet differentiable in U.

Theorem 1.10.1 and (2.2.47) yield g_1 to be a Fréchet continuously differentiable functional in U. Further, from (2.2.37) and (2.2.48), we deduce the functionals $g_2, g_3, \ldots, g_{2r+1}$ to be Fréchet continuously differentiable in U, too.

At last, by virtue of Lemma 2.2.2, we get

$$\mathfrak{R}F'(h, u) = V^* = Y, \qquad (h, u) \in U. \tag{2.2.75}$$

Now, Theorem 1.12.1 implies the existence of Lagrange multipliers

$$\hat{\lambda} = (\hat{\lambda}_0, \hat{\lambda}_1, \ldots, \hat{\lambda}_m) \in \mathbb{R}^{m+1}, \qquad \hat{w} \in Y^* = V,$$

which do not vanish simultaneously and satisfy the conditions (2.2.70), (2.2.71), and the following one (cf. Subsec. 1.9.1)

$$\begin{gathered} \frac{\partial \mathfrak{L}}{\partial h}(h_0, u_0, \hat{w}, \hat{\lambda})q + \frac{\partial \mathfrak{L}}{\partial u}(h_0, u_0, \hat{w}, \hat{\lambda})v = 0 \\ q \in W_p^1(\Omega), \ v \in V. \end{gathered} \tag{2.2.76}$$

Setting $v = 0$ in this equality and taking into account (2.2.69), we obtain (2.2.72). Now, let us take in (2.2.76) $q = 0$. By virtue of (2.2.69) and (2.2.64), we get

$$\frac{\partial \mathfrak{L}}{\partial u}(h_0, u_0, \hat{w}, \hat{\lambda})v = \sum_{i=0}^{m} \hat{\lambda}_i \frac{\partial g_i}{\partial u}(h_0, u_0)v + (F(h_0, v) + f, \hat{w}) = 0, \qquad v \in V. \tag{2.2.77}$$

Since the form a_h is symmetric and (2.2.45) holds, we have

$$(F(h_0, v) + f, \hat{w}) = a_{h_0}(v, \hat{w}) = a_{h_0}(\hat{w}, v),$$

which, together with (2.2.77), yields (2.2.73). Finally, due to Theorem 1.12.1, if there exists a pair $\tilde{h}, \tilde{u}$ satisfying the conditions (2.2.74), then $\hat{\lambda}_0 \neq 0$, so we can take $\hat{\lambda}_0 = 1$. The theorem is proved.

2.2.4 Nonsmooth (discontinuous) control

On the incorrectness of the initial problem

Let us consider the optimal control problem when the controls belong to the space $L_\infty(\Omega)$. More precisely, define set a U_{ad} as

$$U_{\text{ad}} = \{h \mid h \in L_\infty(\Omega),\, \check{h} \le h \le \hat{h} \ \text{ a.e. in } \Omega\}, \tag{2.2.78}$$

$\check{h}, \hat{h}$ being the positive numbers introduced above. Again, we assume the conditions (2.1.1), (2.1.4)–(2.1.7), (2.1.10) are satisfied. By virtue of Theorem 1.5.2, for a given fixed element $f \in V^*$ and for any $h \in U_{\text{ad}}$, there exists a unique function u_h such that

$$u_h \in V, \qquad a_h(u_h, v) = (f, v), \qquad v \in V, \tag{2.2.79}$$

and the inequality

$$\|u_h\|_V \le \text{const}, \qquad h \in U_{\text{ad}}, \tag{2.2.80}$$

holds. We suppose the goal functional is of the form

$$f_1(h) = \|u_h - z\|_V^2, \tag{2.2.81}$$

z being a fixed element from V.

The optimal control problem consists in finding a function h_0 such that

$$h_0 \in U_{\text{ad}}, \qquad f_1(h_0) = \inf_{h \in U_{\text{ad}}} f_1(h). \tag{2.2.82}$$

Let $\{h_n\}_{n=1}^\infty$ be a minimizing sequence, i.e.,

$$h_n \in U_{\text{ad}} \quad \forall n, \qquad \lim_{n\to\infty} f_1(h_n) = \inf_{h \in U_{\text{ad}}} f_1(h). \tag{2.2.83}$$

Because of (2.2.78), the sequence $\{h_n\}$ is bounded in $L_\infty(\Omega)$, and (2.2.80) implies the sequence $\{u_{h_n}\}$ to be bounded in V. This is why we can find a subsequence $\{h_m\}$ such that

$$h_m \to \tilde{h} \qquad *\text{-weakly in } L_\infty(\Omega), \tag{2.2.84}$$

$$u_{h_m} \to \tilde{u} \qquad \text{weakly in } V. \tag{2.2.85}$$

The relation (2.2.84) means that

$$\lim_{m\to\infty} \iint_\Omega h_m g \, dx \, dy = \iint_\Omega \tilde{h} g \, dx \, dy, \qquad g \in L_1(\Omega).$$

Concerning the $*$-weak convergence, see, e.g., Yosida (1971).

However, the relations (2.2.84) and (2.2.85) are insufficient to pass to the limit in the expression of $a_{h_m}(u_{h_m}, v)$, so we cannot state that the function $\tilde{u}$ coincides with $u_{\tilde{h}}$. (In subsecs. 2.2.1 and 2.2.2, we passed to the limit in the expression $a_{h_m}(u_{h_m}, v)$ using Lemma 2.1.1, which cannot be applicable now.)

For an arbitrary function $h \in U_{\mathrm{ad}}$, denote by A_h the operator generated by the bilinear form a_h in the following way:

$$(A_h u, v) = a_h(u, v), \qquad u, v \in V. \tag{2.2.86}$$

Provided (2.1.1), (2.1.4)–(2.1.7), and (2.1.10) hold, A_h is a linear, continuous, selfadjoint, coercive operator acting from V into V^*, and there exist positive numbers c_1, c_2 such that

$$c_1\|u\|_V^2 \le (A_h u, u) \le c_2\|u\|_V^2, \qquad u \in V,\ h \in U_{\mathrm{ad}}. \tag{2.2.87}$$

For $h \in U_{\mathrm{ad}}$ and $g \in V^*$, there exists a unique function $u(h, g)$ such that

$$u(h, g) \in V, \qquad A_h u(h, g) = g. \tag{2.2.88}$$

By Theorem 1.13.1 there exists a subsequence $\{A_{h_k}\}_{k=1}^\infty$ of the sequence $\{A_{h_m}\}_{m=1}^\infty$ such that A_{h_k} G-converges to a linear, continuous, selfadjoint, coercive operator A as $k \to \infty$, i.e.,

$$\lim_{k\to\infty} \left(A_{h_k}^{-1} g, q\right) = \left(A^{-1} g, q\right), \qquad g, q \in V^*. \tag{2.2.89}$$

The inequalities (2.2.87) hold for the operator A with the same numbers c_1 and c_2. However, the operator A is not, in general, generated by a function h from U_{ad}, that is, there does not exist $h \in U_{\mathrm{ad}}$ such that $A = A_h$, and the set of the operators $\{A_h\}, h \in U_{\mathrm{ad}}$, is not G-closed (see Section 1.13 and the example below).

We denote by a the bilinear form generated by the operator A

$$a(u, v) = (Au, v), \qquad u, v \in V. \tag{2.2.90}$$

The function $\tilde{u}$ from (2.2.85) is a solution to the following problem

$$\tilde{u} \in V, \qquad a(\tilde{u}, v) = (f, v), \qquad v \in V,$$

where $a \ne a_{\tilde{h}}$.

It follows from the preceding argument that the problem (2.2.78), (2.2.79), (2.2.81), (2.2.82) is ill-posed, so we need a new setting of it.

One of the possible ways is the use of smoother controls. For example, we may assume that $h \in W_p^1(\Omega)$ with $p \geq 2$. A problem of this kind was considered above, in subsecs. 2.2.1 and 2.2.2.

Another way requires an essential change of the problem. The function h is not a control any more. As a set of admissible controls we take either a set of operators or a set of elements of some functional space such that, to every element of this set, there corresponds an operator and the set of such operators is G-closed.

Let us investigate the latter way by considering an example.

Optimization problem for a second-order elliptic equation

Let $V = \overset{\circ}{W}{}_2^1(\Omega)$ and let a bilinear form a_h be given by

$$a_h(u,v) = \iint\limits_{\Omega} h\left(\frac{\partial u}{\partial x}\frac{\partial v}{\partial x} + \frac{\partial u}{\partial y}\frac{\partial v}{\partial y}\right) dx\,dy, \qquad u,v \in \overset{\circ}{W}{}_2^1(\Omega). \tag{2.2.91}$$

We suppose the set U_{ad} and the goal functional f_1 to be defined by (2.2.78), (2.2.81).

The set of the operators $\{A_h\}, h \in U_{\text{ad}}$, determined by the relations (2.2.86), (2.2.91) is not G-closed since a sequence of operators $\{A_{h_k}\}, h_k \in U_{\text{ad}}$ can G-converge to an operator $A_b \in \mathcal{L}(V, V^*)$ of the form (see Marino and Spagnolo (1969)):

$$\begin{aligned}(A_b u, v)& \\ = \iint\limits_{\Omega} \Big(b_{11}\frac{\partial u}{\partial x}\frac{\partial v}{\partial x} &+ b_{12}\frac{\partial u}{\partial x}\frac{\partial v}{\partial y} + b_{21}\frac{\partial u}{\partial y}\frac{\partial v}{\partial x} + b_{22}\frac{\partial u}{\partial y}\frac{\partial v}{\partial y}\Big)\, dx\,dy,\end{aligned} \tag{2.2.92}$$

where $b_{ij} \in L_\infty(\Omega)$, $i,j = 1,2$, $b_{12} = b_{21} \neq 0$.

Now, let us define a set P_{ad} by the formula

$$\begin{aligned}P_{\text{ad}} = \Big\{ b \,|\, b = \{b_{ij}\},\ i,j = 1,2,\ b_{ij} \in L_\infty(\Omega),\ b_{12} = b_{21},& \\ \check{h}\left(\xi_1^2 + \xi_2^2\right) \leq \sum_{i,j=1}^{2} b_{ij}(x,y)\xi_i\xi_j \leq \hat{h}\left(\xi_1^2 + \xi_2^2\right),\ \xi_1,\xi_2 \in \mathbb{R} \text{ a.e. in } \Omega \Big\}.&\end{aligned} \tag{2.2.93}$$

To each element $b \in P_{\text{ad}}$, there corresponds an operator

$$A_b \in \mathcal{L}\big(\overset{\circ}{W}{}_2^1(\Omega), \big(\overset{\circ}{W}{}_2^1(\Omega)\big)^*\big)$$

determined by (2.2.92).

Given a fixed element $f \in \big(\overset{\circ}{W}{}_2^1(\Omega)\big)^*$, there exists a unique function u_b such that

$$u_b \in \overset{\circ}{W}{}_2^1(\Omega), \qquad A_b u_b = f. \tag{2.2.94}$$

Let the goal functional look like

$$f_1(b) = \|u_b - z\|^2_{\overset{\circ}{W}{}^1_2(\Omega)}, \tag{2.2.95}$$

z being a given element from $\overset{\circ}{W}{}^1_2(\Omega)$. The optimal control problem consists in finding an element $b^{(0)} = \{b^{(0)}_{ij}\}$ such that

$$b^{(0)} \in P_{\text{ad}}, \qquad f_1\left(b^{(0)}\right) = \inf_{b \in P_{\text{ad}}} f_1(b). \tag{2.2.96}$$

Theorem 2.2.5 *Let an operator* $A_b \in \mathcal{L}\big(\overset{\circ}{W}{}^1_2(\Omega),\ \big(\overset{\circ}{W}{}^1_2(\Omega)\big)^*\big)$ *be determined by* (2.2.92), *and let a set* P_{ad} *and a goal functional be defined by* (2.2.93), (2.2.95). *Then, the problem* (2.2.96) *has a solution.*

Proof. Let $\{b^{(n)}\}$ be a minimizing sequence:

$$b^{(n)} \in P_{\text{ad}} \qquad \forall n, \tag{2.2.97}$$

$$\lim_{n\to\infty} f_1\left(b^{(n)}\right) = \inf_{b \in P_{\text{ad}}} f_1(b). \tag{2.2.98}$$

By virtue of the compactness theorem (see Marcellini (1979)), we can find a subsequence $\{A_{b^{(m)}}\}_{m=1}^{\infty}$ of the sequence $\{A_{b^{(n)}}\}_{n=1}^{\infty}$ such that

$$u_{b^{(m)}} \to u_{b^{(0)}} \qquad \text{weakly in } \overset{\circ}{W}{}^1_2(\Omega), \tag{2.2.99}$$

where

$$b^{(0)} \in P_{\text{ad}}. \tag{2.2.100}$$

By (2.2.99), we get $\liminf\limits_{m\to\infty} \|u_{b^{(m)}}\|_{\overset{\circ}{W}{}^1_2(\Omega)} \geq \|u_{b^{(0)}}\|_{\overset{\circ}{W}{}^1_2(\Omega)}$. Therefore, taking into account (2.2.95) and (2.2.99), we obtain

$$\liminf_{m\to\infty} f_1\left(b^{(m)}\right) \geq f_1\left(b^{(0)}\right). \tag{2.2.101}$$

Now, (2.2.98), (2.2.100), (2.2.101) yield the vector function $b^{(0)}$ to satisfy the condition (2.2.96). The theorem is proved.

An example of the existence on a G-nonclosed set

Here we expound an example from Céa and Malanowski (1970).

Let $V = \overset{\circ}{W}{}^1_2(\Omega)$, let a bilinear form a_h be defined by the relation (2.2.91) and let the set U_{ad} look like

$$U_{\text{ad}} = \Big\{ h \mid h \in L_\infty(\Omega),\ \check{h} \leq h \leq \hat{h} \ \text{ a.e. in } \Omega,\ \iint\limits_{\Omega} h\, dx\, dy = c \Big\}. \tag{2.2.102}$$

We assume that $\check{h} \operatorname{mes} \Omega < c < \hat{h} \operatorname{mes} \Omega$, which implies that the set U_{ad} is not empty.

Let a goal functional be of the form

$$f_1(h) = \iint\limits_{\Omega} h \left[\left(\frac{\partial u_h}{\partial x} \right)^2 + \left(\frac{\partial u_h}{\partial y} \right)^2 \right] dx\, dy, \tag{2.2.103}$$

u_h being a solution to the problem (2.2.79) when f is a fixed function from $\big(\overset{\circ}{W}{}_2^1(\Omega)\big)^*$.

Theorem 2.2.6 *Let $V = \overset{\circ}{W}{}_2^1(\Omega)$, let a bilinear form a_h be defined by the relation* (2.2.91), *and let a set U_{ad} and a goal functional f_1 be determined by* (2.2.102), (2.2.103). *Then, there exists a function $\tilde{h}$ such that*

$$\tilde{h} \in U_{\mathrm{ad}}, \qquad f_1(\tilde{h}) = \inf_{h \in U_{\mathrm{ad}}} f_1(h). \tag{2.2.104}$$

Proof. Let us show that

$$\left.\begin{array}{l} h \to f_1(h) \text{ is a lower semicontinuous function that maps} \\ U_{\mathrm{ad}} \text{ (endowed with the topology generated by the } *\text{-weak} \\ \text{topology of } L_\infty(\Omega)) \text{ into } \mathbb{R}. \end{array}\right\} \tag{2.2.105}$$

Suppose that

$$h_n \in U_{\mathrm{ad}}, \qquad h_n \to h_0 \quad *\text{-weakly in } L_\infty(\Omega). \tag{2.2.106}$$

Taking into account (2.2.102) and (2.2.106), we easily deduce that $h_0 \in U_{\mathrm{ad}}$. Introduce the notation

$$u_n = u_{h_n}, \qquad n = 0, 1, 2, \dots . \tag{2.2.107}$$

By (2.2.79), (2.2.91), and (2.2.107), we get

$$\begin{aligned} (f, u_0) &= \iint\limits_{\Omega} h_n \left(\frac{\partial u_n}{\partial x} \frac{\partial u_0}{\partial x} + \frac{\partial u_n}{\partial y} \frac{\partial u_0}{\partial y} \right) dx\, dy \\ &= \iint\limits_{\Omega} h_0 \left[\left(\frac{\partial u_0}{\partial x} \right)^2 + \left(\frac{\partial u_0}{\partial y} \right)^2 \right] dx\, dy. \end{aligned} \tag{2.2.108}$$

In view of (2.2.103) and (2.2.108), we obtain

$$\begin{aligned} f_1(h_n) - f_1(h_0) = &\iint\limits_{\Omega} h_n \left[\left(\frac{\partial u_n}{\partial x} - \frac{\partial u_0}{\partial x} \right)^2 + \left(\frac{\partial u_n}{\partial y} - \frac{\partial u_0}{\partial y} \right)^2 \right] dx\, dy \\ &- \iint\limits_{\Omega} (h_n - h_0) \left[\left(\frac{\partial u_0}{\partial x} \right)^2 + \left(\frac{\partial u_0}{\partial y} \right)^2 \right] dx\, dy. \end{aligned} \tag{2.2.109}$$

Now, (2.2.106) and (2.2.109) yield

$$\liminf_{n\to\infty} f_1(h_n) \geq f_1(h_0).$$

Thus, (2.2.105) holds true. Let $\{q_n\}$ be a minimizing sequence:

$$q_n \in U_{\text{ad}}, \qquad \lim_{n\to\infty} f_1(q_n) = \inf_{q\in U_{\text{ad}}} f_1(q). \tag{2.2.110}$$

From (2.2.102) it follows that the sequence $\{q_n\}$ is bounded in $L_\infty(\Omega)$. Hence, we can extract a subsequence $\{q_m\}$ such that

$$q_m \to q_0 \quad *\text{-weakly in } L_\infty(\Omega). \tag{2.2.111}$$

From (2.2.102) and (2.2.111), we get $q_0 \in U_{\text{ad}}$.

Now, the relations (2.2.105), (2.2.110), (2.2.111) make the function $\tilde{h} = q_0$ satisfy the condition (2.2.104).

Remark 2.2.1 The above argument makes it clear that, if the set U_{ad} is sequentially $*$-weakly closed in $L_\infty(\Omega)$ and the goal functional is lower semicontinuous, then the optimal control problem has a solution even if the set of the operators corresponding to the set U_{ad} is not G-closed.

2.2.5 Some remarks on the use of regular and discontinuous controls

"The natural is rounded, the artificial is made up of angles ... Beauty is Nature in perfection; circularity is its chief attribute. Behold the full moon, the domes of splendid temples, the huckleberry pie ... On the other hand, straight lines show that Nature has been deflected."

– *O. Henry*
"Squaring the Circle"

The results of Subsec. 2.2.4 show that, if the state of a system is described by the bilinear form a_h from Subsec. 2.1.1, then the optimal control problem is ill-posed when one uses controls from $L_\infty(\Omega)$, so one has to enlarge the set of bilinear forms to make the set of operators generated by these forms G-closed.

Such enlarging is not always justified from the point of view of physics. In fact, a G-limit operator might have no physical meaning. Besides, the class of admissible functionals determining the set of admissible controls and the goal functional is much larger in the case of admissible controls from $W_p^1(\Omega)$ than that in the case of controls belonging to $L_\infty(\Omega)$.

We point out that often, in optimization problems of physics and technics, one has to use regular controls. For instance, in problems of control by a function of either the thickness of a plate or a shell, or by the form of the surface of a shell, discontinuous controls cause concentration of stresses, leading to destruction of the structure. Moreover, mathematical models of plates and shells do not describe the concentration of stresses caused by discontinuity of the function of thickness and nonsmoothness of the function of the midsurface of the shell. Besides, these models are applicable only in cases when the function of the thickness of a plate or a shell is smooth and its derivatives are comparatively small.

On the other hand, the use of nonsmooth controls when they are considered as admissible enables one to expand significantly the set of admissible controls and to get, generally speaking, much benefit in the values of the goal functional in comparison with the case of smooth controls.

We note also that G-convergence is widely used for averaging of differential operators; see, e.g., Duvaut (1976), Bensoussan et al. (1978), Sanchez-Palencia (1980), Bakhvalov and Panasenko (1984), Zhikov et al. (1993). In turn, the averaged differential operators are applied to the optimization of nonhomogeneous media, in particular, composites and structures, see Rozvany (1989), Lurie (1993), Bendsøe (1994), and references therein.

2.3 The finite-dimensional problem

Let us consider the problem of approximation of the solution to the regular control problem formulated in Subsec. 2.2.2 by a solution to a finite-dimensional problem.

We suppose $\{H_n\}_{n=1}^{\infty}$ to be a sequence of finite-dimensional subspaces of $W_p^1(\Omega)$ satisfying the limit density condition, i.e.,

$$\lim_{n\to\infty}\inf_{h\in H_n}\|h-w\|_{W_p^1(\Omega)}=0, \qquad w\in W_p^1(\Omega). \tag{2.3.1}$$

The finite-dimensional problem consists in finding a function h_n such that

$$h_n\in H_n\cap U_{\mathrm{ad}}, \qquad f_1(h_n)=\inf_{h\in H_n\cap U_{\mathrm{ad}}} f_1(h). \tag{2.3.2}$$

Theorem 2.3.1 *Let the assumptions of Theorem* 2.2.2 *hold true and let* $\{H_n\}$ *be a sequence of finite-dimensional subspaces of* $W_p^1(\Omega)$ *satisfying the condition* (2.3.1). *Let also a sequence* $\{q_n\}_{n=1}^{\infty}$ *exist such that*

$$q_n\in \overset{\circ}{U}_{\mathrm{ad}} \qquad \forall n, \tag{2.3.3}$$

$$q_n\to h_0 \;\text{ in } W_p^1(\Omega), \tag{2.3.4}$$

where $\overset{\circ}{U}_{\mathrm{ad}}$ *is the set of interior points of* U_{ad}, *and* h_0 *a solution to the problem* (2.2.22). *Then, for each* n *sufficiently large (say,* $n\ge k$*), there exists a solution* h_n *to the problem* (2.3.2) *and*

$$\lim_{n\to\infty} f_1(h_n)=f_1(h_0)=\inf_{h\in U_{\mathrm{ad}}} f_1(h). \tag{2.3.5}$$

Moreover, there exists a subsequence $\{h_m\}_{m=1}^{\infty}$ *of the sequence* $\{h_n\}_{n=k}^{\infty}$ *such that* $h_m \to h_0$ *weakly in* $W_p^1(\Omega)$.

Remark 2.3.1 The set U_{ad} defined by the relations (2.2.18)–(2.2.20) is equipped with the topology induced by the topology of $W_p^1(\Omega)$ on U_{ad}. This is why the condition $q_n \in \overset{\circ}{U}_{\text{ad}}$ is equivalent to the existence of a number $\varepsilon_n > 0$ such that $d(\varepsilon_n, q_n) \subset U_{\text{ad}}$, where

$$d(\varepsilon_n, q_n) = \{\, h \,|\, h \in W_p^1(\Omega),\ \|h - q_n\|_{W_p^1(\Omega)} \le \varepsilon_n \,\}. \tag{2.3.6}$$

Remark 2.3.2 It follows from the proof of Theorem 2.3.1 that the problem (2.3.2) has a solution for any n if the set $H_n \cap U_{\text{ad}}$ is not empty for any n.

Proof of Theorem 2.3.1. For any $\varepsilon > 0$, there exists an element $h_\varepsilon \in \overset{\circ}{U}_{\text{ad}}$ such that

$$\|h_\varepsilon - h_0\|_{W_p^1(\Omega)} \le \varepsilon. \tag{2.3.7}$$

Indeed, by virtue of (2.3.3), (2.3.4), we can take $h_\varepsilon = q_n$ for n sufficiently large. Since $h_\varepsilon \in \overset{\circ}{U}_{\text{ad}}$, there exists a number γ such that

$$0 < \gamma \le \varepsilon, \qquad d(\gamma, h_\varepsilon) \subset U_{\text{ad}}, \tag{2.3.8}$$

where

$$d(\gamma, h_\varepsilon) = \{\, h \mid h \in W_p^1(\Omega),\ \|h - h_\varepsilon\|_{W_p^1(\Omega)} \le \gamma \,\}. \tag{2.3.9}$$

From (2.3.1), we have that, for n sufficiently large, an element g from H_n exists such that $g \in d(\gamma, h_\varepsilon)$, and so, because of (2.3.8), $g \in U_{\text{ad}}$. By (2.3.7)–(2.3.9), we get

$$\|g - h_0\|_{W_p^1(\Omega)} \le \|g - h_\varepsilon\|_{W_p^1(\Omega)} + \|h_\varepsilon - h_0\|_{W_p^1(\Omega)} \le 2\varepsilon. \tag{2.3.10}$$

The above argument implies the existence of a sequence $\{g_n\}_{n=k}^{\infty}$ such that

$$g_n \in H_n \cap U_{\text{ad}}, \tag{2.3.11}$$

$$g_n \to h_0 \ \text{ strongly in } W_p^1(\Omega). \tag{2.3.12}$$

By (2.2.21), (2.3.12), we obtain

$$\lim_{n\to\infty} f_1(g_n) = f_1(h_0). \tag{2.3.13}$$

Next, the set $H_n \cap U_{\text{ad}}$ is not empty if $n \ge k$, since it contains the function g_n. Due to (2.2.18), this set is bounded in $W_p^1(\Omega)$ and in H_n (H_n being endowed with the norm of the space $W_p^1(\Omega)$). Obviously, the set

$$X_n = \{\, h \,|\, h \in H_n,\ \|h\|_{W_p^1(\Omega)} \le c,\ \check{h} \le h \le \hat{h} \,\}$$

is closed in H_n.

By virtue of (2.2.20) and Lemma 2.1.1, the function $h \to \Psi_k(h, u_h)$ is a continuous mapping of Y_p (endowed with the topology generated by the $W_p^1(\Omega)$-weak one) into $\mathbb{R}$ for $p > 2$. Therefore, the set

$$Z_n = \{ h \,|\, h \in H_n \cap Y_p,\ \Psi_k(h, u_h) \leq 0,\ k = 1, 2, \ldots, l \}$$

is closed in H_n. Taking into account that $H_n \cap U_{\text{ad}} = X_n \cap Z_n$, we deduce $H_n \cap U_{\text{ad}}$ to be closed in H_n. Since the space H_n is finite-dimensional and $H_n \cap U_{\text{ad}}$ is bounded and closed, the set $H_n \cap U_{\text{ad}}$ is compact.

By (2.2.21), $h \to f_1(h)$ is a continuous mapping of $H_n \cap U_{\text{ad}}$ into $\mathbb{R}$. Now, Theorem 1.3.9 implies that the problem (2.3.2) is solvable for any $n \geq k$. Moreover, if for any n the set $H_n \cap U_{\text{ad}}$ is not empty, then the problem (2.3.2) is solvable for all n.

By (2.2.22), (2.3.2), (2.3.11),

$$f_1(g_n) \geq f_1(h_n) \geq f_1(h_0).$$

From here and (2.3.13), we get

$$\lim_{n \to \infty} f_1(h_n) = f_1(h_0),$$

i.e., (2.3.5) holds. By virtue of (2.3.2) and (2.3.5), the elements h_n satisfy the relations (2.2.23), (2.2.24). Now, the argument used in the proof of Theorem 2.2.2 shows that we can find a subsequence $\{h_m\}$ of the sequence $\{h_n\}$ such that $h_m \to h_0$ weakly in $W_p^1(\Omega)$. We note that the problems (2.2.22) and (2.3.2) might have, in general, a set of solutions, and from the sequence $\{h_n\}$ of the solutions to the problem (2.3.2) one can choose subsequences $\{h_m\}$ convergent to the solutions of the problem (2.2.22).

2.4 The finite-dimensional problem (another approach)

2.4.1 The set $U^{(t)}$

In Section 2.3, in studying the approximation of a solution to the problem (2.2.22) by a solution to the finite-dimensional problem, we supposed the existence of a sequence of interior points of U_{ad} convergent to the solution to the problem (2.2.22). Below, we propose a technique for finding an approximate solution of the problem (2.2.22) that makes no use of this assumption.

Let $t \in \mathbb{R}$, $1 \leq t \leq \min\left(2 - \frac{e_1}{\check{h}}, \frac{e_2}{\hat{h}}\right)$, see (2.1.2) and (2.2.19). Define a set

$$\begin{aligned} U^{(t)} = \{ h \,|& h \in W_p^1(\Omega),\ \|h\|_{W_p^1(\Omega)} \leq tc, \\ & \check{h}(2-t) \leq h \leq t\hat{h},\ \Psi_k(h, u_h) \leq t - 1,\ k = 1, 2, \ldots, l \}. \end{aligned} \tag{2.4.1}$$

Here, Ψ_k are the functionals defined in Subsec. 2.2.2 (see (2.2.20)).

We equip the set $U^{(t)}$ with the topology induced by the $W_p^1(\Omega)$-one. Comparing (2.2.18) and (2.4.1), we get

$$U_{\text{ad}} \subset U^{(t)}, \qquad U_{\text{ad}} = U^{(1)}. \tag{2.4.2}$$

Directly from the proof of Theorem 2.2.2 we get the following fact.

Theorem 2.4.1 *Let a_h be a bilinear, symmetric form on $V \times V$ defined by the relations* (2.1.1), (2.1.4)–(2.1.6) *and let the inequalities (*2.1.7), (2.1.10) *hold; let a set $U^{(t)}$ be defined by the relations* (2.4.1), (2.2.19), (2.2.20) *and*

$$1 < t \leq \min\left(2 - \frac{e_1}{\check{h}}, \frac{e_2}{\hat{h}}\right). \tag{2.4.3}$$

Also, let the set $U^{(t)}$ be nonempty and let the goal functional satisfy the condition (2.2.21). *Then, there exists a function h_t such that*

$$h_t \in U^{(t)}, \qquad f_1(h_t) = \inf_{h \in U^{(t)}} f_1(h). \tag{2.4.4}$$

Remark 2.4.1 The right-hand side of (2.4.3) is introduced in order to make the inequality $e_1 \leq h \leq e_2$ hold for all $h \in U^{(t)}$ (see (2.1.2)). We need this because the functionals Ψ_k and f_1 are defined on Y_p.

We will refer to the set of interior points of $U^{(t)}$ as $\overset{\circ}{U}{}^{(t)}$. Recall that the topology on $U^{(t)}$ is generated by the topology of $W_p^1(\Omega)$.

Lemma 2.4.1 *Let a set $U^{(t)}$ be defined by the relations* (2.4.1), (2.4.3), (2.2.19), (2.2.20), *and let h_0 be a solution to the problem* (2.2.22). *Then, $h_0 \in \overset{\circ}{U}{}^{(t)}$.*

Proof. For all $h \in W_p^1(\Omega)$, we obtain

$$\|h\|_{W_p^1(\Omega)} \leq \|h_0\|_{W_p^1(\Omega)} + \|h_0 - h\|_{W_p^1(\Omega)}. \tag{2.4.5}$$

By virtue of (2.2.18), (2.2.22), we have $\|h_0\|_{W_p^1(\Omega)} \leq c$. From here and (2.4.5), we deduce the existence of a constant $\gamma_1 > 0$ such that, for any $h \in d(\gamma_1, h_0)$, the inequality

$$\|h\|_{W_p^1(\Omega)} \leq tc, \tag{2.4.6}$$

is satisfied, $d(\gamma_1, h_0)$ being the ball in $W_p^1(\Omega)$ centered at h_0, with radius γ_1, defined by (2.3.9).

The embedding of $W_p^1(\Omega)$ into $C(\overline{\Omega})$ being continuous for $p > 2$ and $\check{h} \leq h_0 \leq \hat{h}$, there exists a number $\gamma_2 > 0$ such that, for any $h \in d(\gamma_2, h_0)$,

$$(2 - t)\check{h} \leq h \leq t\hat{h}. \tag{2.4.7}$$

Lemma 2.1.1 and (2.2.20) yield the function $h \to \Psi_k(h, u_h)$ to be a continuous mapping of Y_p (equipped with the topology generated by the $W_p^1(\Omega)$-weak one) into $\mathbb{R}$ for $p > 2$. Hence, noticing that $\Psi_k(h_0, u_{h_0}) \leq 0$, $k = 1, 2, \ldots, l$, we conclude the existence of a number $\gamma_3 > 0$ such that, for any $h \in d(\gamma_3, h_0)$,

$$\Psi_k(h, u_h) \leq t - 1, \qquad k = 1, 2, \ldots, l. \tag{2.4.8}$$

Setting $\gamma = \min(\gamma_1, \gamma_2, \gamma_3)$, by (2.4.6)–(2.4.8) we get that $d(\gamma, h_0) \subset U^{(t)}$, consequently, $h_0 \in \overset{\circ}{U}{}^{(t)}$.

Remark 2.4.2 When proving Lemma 2.4.1, we used only the fact that $h_0 \in U_{\text{ad}}$. So, under the assumption of Lemma 2.4.1, not only $h_0 \in \overset{\circ}{U}{}^{(t)}$, but also $U_{\text{ad}} \subset \overset{\circ}{U}{}^{(t)}$.

2.4.2 Approximate solution of the problem (2.2.22)

Let $\{H_n\}_{n=1}^{\infty}$ be a sequence of finite-dimensional subspaces of $W_p^1(\Omega)$ satisfying the conditions (2.3.1), and

$$H_n \subset H_{n+1}. \tag{2.4.9}$$

Consider the problem of finding a function h_n such that

$$h_n \in H_n \cap U^{(t)}, \tag{2.4.10}$$

$$f_1(h_n) = \inf_{h \in H_n \cap U^{(t)}} f_1(h). \tag{2.4.11}$$

Theorem 2.4.2 *Let the conditions of Theorems 2.2.2 and 2.4.1 be fulfilled, let $\{H_n\}_{n=1}^{\infty}$ be a sequence of finite-dimensional subspaces of $W_p^1(\Omega)$ satisfying the conditions (2.3.1), (2.4.9), and let h_0, h_t be solutions to the problems (2.2.22), (2.4.4), respectively. Then, for any n sufficiently large (say, $n \geq k$), the problem (2.4.10), (2.4.11) has a solution h_n such that*

$$f_1(h_t) \leq \lim_{n \to \infty} f_1(h_n) \leq f_1(h_0). \tag{2.4.12}$$

Also, one can choose a subsequence $\{h_m\}_{m=1}^{\infty}$ from the sequence $\{h_n\}_{n=k}^{\infty}$ such that $h_m \to \tilde{h}$ weakly in $W_p^1(\Omega)$, $\tilde{h} \in U^{(t)}$, and

$$\lim_{m \to \infty} f_1(h_m) = \lim_{n \to \infty} f_1(h_n) = f_1(\tilde{h}). \tag{2.4.13}$$

Proof. Due to Lemma 2.4.1, $h_0 \in \overset{\circ}{U}{}^{(t)}$. Consequently, for some $\gamma > 0$, $d(\gamma, h_0) \subset U^{(t)}$, $d(\gamma, h_0)$ being defined by (2.3.9). From here and (2.3.1), we deduce that the set $H_n \cap U^{(t)}$ is nonempty for n sufficiently large (say, $n \geq k$) and that a sequence $\{g_n\}_{n=k}^{\infty}$ exists such that

$$g_n \in H_n \cap U^{(t)}, \tag{2.4.14}$$

$$g_n \to h_0 \quad \text{strongly in } W_p^1(\Omega). \tag{2.4.15}$$

By virtue of (2.2.21), (2.4.15), we get

$$\lim_{n\to\infty} f_1(g_n) = f_1(h_0). \tag{2.4.16}$$

The argument similar to the one used in the proof of Theorem 2.3.1 shows that the set $H_n \cap U^{(t)}$ is a compact topological space. Hence, by (2.2.21), we conclude the problem (2.4.10), (2.4.11) to be solvable for any $n \geq k$, and for any n if the set $H_n \cap U^{(t)}$ is not empty for all n.

The relations (2.4.4), (2.4.11) imply

$$f_1(h_t) \leq f_1(h_n) \qquad \forall n. \tag{2.4.17}$$

Due to (2.4.9), (2.4.11),

$$f_1(h_{n+1}) \leq f_1(h_n).$$

Consequently, using (2.4.17), we deduce the sequence of the numbers $\{f_1(h_n)\}_{n=k}^{\infty}$ to be convergent and

$$\lim_{n\to\infty} f_1(h_n) \geq f_1(h_t). \tag{2.4.18}$$

By (2.4.11), (2.4.14), we get

$$f_1(g_n) \geq f_1(h_n). \tag{2.4.19}$$

Now, (2.4.16), (2.4.18), (2.4.19) yield (2.4.12). By (2.4.10) and (2.4.1), we conclude the sequence $\{h_n\}_{n=k}^{\infty}$ to be bounded in $W_p^1(\Omega)$. Let us extract from it a subsequence $\{h_m\}_{m=1}^{\infty}$ such that

$$h_m \to \tilde{h} \qquad \text{weakly in } W_p^1(\Omega), \tag{2.4.20}$$

$$h_m \to \tilde{h} \qquad \text{strongly in } C(\overline{\Omega}), \tag{2.4.21}$$

Let us show that $\tilde{h} \in U^{(t)}$. Lemma 2.1.1 and the relations (2.2.20), (2.4.20) imply

$$\lim_{m\to\infty} \Psi_k(h_m, u_m) = \Psi_k(\tilde{h}, u_{\tilde{h}}), \qquad k = 1, 2, \ldots, l. \tag{2.4.22}$$

Here, $u_m = u_{h_m}$ is a solution to the problem (2.1.12) when $h = h_m$. By (2.4.1), (2.4.10), (2.4.22), we have

$$\Psi_k(\tilde{h}, u_{\tilde{h}}) \leq t - 1, \qquad k = 1, 2, \ldots, l. \tag{2.4.23}$$

Combining (2.4.1), (2.4.10), (2.4.21), we obtain

$$\check{h}(2 - t) \leq \tilde{h} \leq t\hat{h}. \tag{2.4.24}$$

Because of (2.4.1), (2.4.10), (2.4.20), we get

$$tc \geq \liminf_{m\to\infty} \|h_m\|_{W_p^1(\Omega)} \geq \|\tilde{h}\|_{W_p^1(\Omega)}. \tag{2.4.25}$$

Now, the relations (2.4.23)–(2.4.25) yield $\tilde{h} \in U^{(t)}$. At last, (2.2.21) and (2.4.20) imply (2.4.13). Thus, the theorem is proved.

We note that the above technique of the construction of approximate solutions to the problem (2.2.22) is also applicable to the problem (2.2.41), (2.2.43).

2.4.3 Approximate solution of the optimal control problem when the set $\overset{\circ}{U}_{\text{ad}}$ is empty

So far we investigated the approximation of a solution to the optimal control problem by a solution to the finite-dimensional problem when the set of admissible controls is in $W_p^1(\Omega)$, $p > 2$. In both approaches considered in Sections 2.3 and 2.4, we used the existence of a sequence of interior points of the set U_{ad} or $U^{(t)}$ converging to the solution h_0 of the optimal control problem.

In the case when the set of admissible controls Q_{ad} is defined by (2.2.1), i.e., when $Q_{\text{ad}} \subset W_2^1(\Omega)$, and the topology on Q_{ad} is generated by the $W_2^1(\Omega)$-strong one, the set of the interior points of Q_{ad} is empty, so both methods considered above are not applicable.

This reason makes us study the general optimal control problem in the case when the set of admissible controls has no interior points.

Suppose $\mathcal{U}$ is a Banach space and

$$\mathcal{U}_{\text{ad}} \text{ is a closed, bounded set in } \mathcal{U} \tag{2.4.26}$$

The set $\mathcal{U}_{\text{ad}}$ is endowed with the topology induced by that of $\mathcal{U}$, and in this topology $\mathcal{U}_{\text{ad}}$ may contain no interior points. Suppose also that

$$h \to f(h) \text{ is a continuous mapping of } \mathcal{U}_{\text{ad}} \text{ into } \mathbb{R}. \tag{2.4.27}$$

Theorem 2.4.3 *Let the conditions* (2.4.26), (2.4.27) *be satisfied and let there exist a function h_0 such that*

$$h_0 \in \mathcal{U}_{\text{ad}}, \qquad f(h_0) = \inf_{h \in \mathcal{U}_{\text{ad}}} f(h).$$

Suppose $\{H_n\}$ is a sequence of finite-dimensional subspaces of $\mathcal{U}$ such that $H_n \subset H_{n+1}$ for all n and the set $H_1 \cap \mathcal{U}_{\text{ad}}$ is not empty. Assume there exists a sequence $\{g_n\}$ such that

$$g_n \in H_n \cap \mathcal{U}_{\text{ad}}, \qquad g_n \to h_0 \text{ strongly in } \mathcal{U}. \tag{2.4.28}$$

Then, for any n, there exists a solution to the finite-dimensional problem

$$h_n \in H_n \cap \mathcal{U}_{\text{ad}}, \qquad f(h_n) = \inf_{h \in H_n \cap \mathcal{U}_{\text{ad}}} f(h), \tag{2.4.29}$$

and moreover

$$\lim_{n \to \infty} f(h_n) = f(h_0) = \inf_{h \in \mathcal{U}_{\text{ad}}} f(h). \tag{2.4.30}$$

Proof. By the hypothesis of the theorem, the set $H_n \cap \mathcal{U}_{\text{ad}}$ is not empty for each n, and since the space H_n is finite-dimensional, we conclude $H_n \cap \mathcal{U}_{\text{ad}}$ to be a compact set. From this and (2.4.27), we deduce that the problem (2.4.29) has a solution for any n.

By (2.4.27), (2.4.28), we get

$$\lim_{n\to\infty} f(g_n) = f(h_0).$$

This equality together with (2.4.29) implies (2.4.30).

Apply Theorem 2.4.3 to the problem (2.2.5). Set

$$\mathcal{U} = W_2^1(\Omega), \qquad \mathcal{U}_{\text{ad}} = Q_{\text{ad}}, \qquad f(h) = f_1(h),$$

Q_{ad} being defined by (2.2.1)–(2.2.3) and the function f_1 satisfying the condition (2.2.4). By virtue of (2.2.1), the set Q_{ad} is bounded in $W_2^1(\Omega)$. It follows from the proof of Theorem 2.2.1 that the set Q_{ad} is sequentially weakly closed, and so it is closed in the topology induced by the $W_2^1(\Omega)$-strong one.

Let $\{\mathcal{H}_n\}$ be a sequence of finite-dimensional subspaces of $W_2^1(\Omega)$ satisfying the conditions

$$\lim_{n\to\infty} \inf_{u\in\mathcal{H}_n} \|u - v\|_{W_2^1(\Omega)} = 0, \qquad v \in W_2^1(\Omega),\ H_n \subset H_{n+1} \quad \forall n,$$

and suppose that the set $\mathcal{H}_1 \cap Q_{\text{ad}}$ is not empty. Then, Theorem 2.4.3 implies the existence of a function h_n for every n such that

$$h_n \in Q_{\text{ad}} \cap \mathcal{H}_n, \qquad f_1(h_n) = \inf_{h\in Q_{\text{ad}}\cap\mathcal{H}_n} f_1(h),$$

and if a sequence $\{q_n\}$ exists such that

$$q_n \in Q_{\text{ad}} \cap \mathcal{H}_n, \qquad q_n \to h_0 \ \text{ strongly in } W_2^1(\Omega),$$

h_0 meeting the relation (2.2.5), then

$$\lim_{n\to\infty} f_1(h_n) = \inf_{h\in Q_{\text{ad}}} f_1(h).$$

Remark 2.4.3 All the results established above in Sections 2.1–2.4 remain valid without assuming the form $a_h(u, v)$ to be symmetric (see Remark 2.1.2).

2.4.4 On the computation of the functional $h \to \Psi_k(h, u_h)$

Passage to the limit

Solution of the finite-dimensional optimization problems (2.3.2) and (2.4.10), (2.4.11) is connected with calculation of the functional $h \to \Psi_k(h, u_h)$, where $h \in Y_p$ (see (2.1.2)) and the function u_h is a solution to the problem

$$a_h(u_h, v) = (f, v), \qquad v \in V. \tag{2.4.31}$$

The problem (2.4.31) is infinite-dimensional, since so is the space V. For every $h \in Y_p$, let there be given a sequence $\{u_{hm}\}_{m=1}^{\infty}$ of approximate solutions to the problem (2.4.31) such that

$$u_{hm} \to u_h \ \text{ strongly in } V \text{ as } m \to \infty. \tag{2.4.32}$$

Then, by (2.2.20), we have

$$\Psi_k(h, u_h) = \lim_{m\to\infty} \Psi_k(h, u_{hm}), \qquad h \in Y_p. \tag{2.4.33}$$

The latter relation can be used when one solves the problems (2.3.2) and (2.4.10), (2.4.11).

There exist a number of methods to find an approximate solution to the problem (2.4.31) providing one with a sequence $\{u_{hm}\}_{m=1}^{\infty}$ satisfying the condition (2.4.32). Below, we will consider the Riesz method.

Reduction of the problem (2.4.31) to a variational one

For every $h \in Y_p$, a_h is a bilinear, symmetric, continuous, coercive form on $V \times V$, and so it determines a scalar product in V and a norm

$$\|v\|_1 = [a_h(v, v)]^{1/2}, \tag{2.4.34}$$

which is equivalent to the original norm of the space V.

Define on the space V a functional

$$\Psi_h(v) = a_h(v, v) - 2(f, v). \tag{2.4.35}$$

Let u_h be a solution to the problem (2.4.31). For any $w \in V$ and any $\alpha \in \mathbb{R}$, we have

$$\Psi_h(u_h + \alpha w) = \Psi_h(u_h) + 2\alpha[a_h(u_h, w) - (f, w)] + \alpha^2 a_h(w, w).$$

Hence, taking $\alpha = 1$ and using (2.4.31), we obtain

$$\Psi_h(u_h + w) \geq \Psi_h(u_h), \qquad w \in V.$$

The form a_h being coercive, the equality holds only when $w = 0$.

So, if u_h is a solution to the problem (2.4.31), the functional (2.4.35) has its minimum in V at $v = u_h$.

Conversely, if $\Psi_h(v)$ takes its minimal value in V on some element u_h then $\Psi_h(u_h + \alpha w) \geq \Psi_h(u_h)$ for all $w \in V$ and all $\alpha \in \mathbb{R}$, that is, for every fixed w, the function $\alpha \to \Psi_h(u_h + \alpha w)$ reaches its minimal value at $\alpha = 0$. But then the derivative of this function vanishes at $\alpha = 0$, i.e.,

$$\frac{d}{d\alpha}\Psi_h(u_h + \alpha w)\Big|_{\alpha=0} = 2\big[a_h(u_h, w) - (f, w)\big] = 0, \qquad w \in V.$$

Hence, if the functional (2.4.35) reaches its minimum on an element u_h, then u_h is a solution to the problem (2.4.31).

The Riesz method

Define a Riesz operator Q_h to be the canonical isometry of V^* onto V defined, for any $f \in V^*$, by the relation

$$a_h(Q_h f, v) = (f, v), \qquad v \in V. \tag{2.4.36}$$

By (2.4.31), (2.4.36), taking into account the existence of a unique solution to the problem (2.4.31), we get

$$u_h = Q_h f. \tag{2.4.37}$$

Due to (2.4.35)–(2.4.37), the functional (2.4.35) may be represented as

$$\Psi_h(v) = a_h(v, v) - 2a_h(u_h, v) = a_h(v - u_h, v - u_h) - a_h(u_h, u_h). \tag{2.4.38}$$

Further, let $\{V_m\}_{m=1}^{\infty}$ be a sequence of finite-dimensional subspaces of V satisfying the limit density condition in V, i.e.,

$$\lim_{m\to\infty} \inf_{u\in V_m} \|u - w\|_V = 0, \qquad w \in V. \tag{2.4.39}$$

A Riesz approximate solution to the problem (2.4.31) is defined as a unique element $u_{hm} \in V_m$ such that

$$\Psi_h(u_{hm}) = \inf_{v\in V_m} \Psi_h(v). \tag{2.4.40}$$

The element u_{hm} is characterized by the relation

$$a_h(u_{hm}, v) = (f, v), \qquad v \in V_m. \tag{2.4.41}$$

From (2.4.38) and (2.4.40),

$$\begin{aligned}\Psi_h(u_{hm}) &= a_h(u_{hm} - u_h, u_{hm} - u_h) - a_h(u_h, u_h) \\ &= \inf_{v\in V_m} a_h(v - u_h, v - u_h) - a_h(u_h, u_h).\end{aligned}$$

Hence, the function u_{hm} is an element of the best approximation to the explicit solution u_h of the problem (2.4.31) in the space V_m in the norm $\|\cdot\|_1$ determined by the relation (2.4.34).

Now, taking into account (2.4.39) and the equivalence of the norms $\|\cdot\|_1$ and the original one of the space V, we establish the relation (2.4.32) to hold true for the Riesz approximate solutions.

We formulate the results obtained as

Theorem 2.4.4 *Let a_h be a bilinear form on $V \times V$ determined by the relations* (2.1.1), (2.1.4)–(2.1.6) *and let the inequalities* (2.1.7), (2.1.10) *hold. Let also the set Y_p be determined by the relations* (2.1.2), (2.1.3) *and $f \in V^*$.*

Assume that $\{V_m\}_{m=1}^{\infty}$ is a sequence of finite-dimensional subspaces of V satisfying the limit density condition (2.4.39). For any $h \in Y_p$ and any m, define an approximate solution to the problem (2.4.31) as the unique element $u_{hm} \in V_m$ satisfying (2.4.41). Then

$$\lim_{m\to\infty} \|u_{hm} - u_h\|_V = 0,$$

where u_h is a solution to the problem (2.4.31).

2.4.5 Calculation and use of the Fréchet derivative of the functional $h \to \Psi_m(h, u_h)$

Finite-dimensional regular problem and Fréchet derivatives

Consider the regular control problem (2.2.41), (2.2.43) supposing the hypothesis of Theorem 2.2.4 to be fulfilled. We may use the techniques given in Sections 2.3 and 2.4 in order to get an approximate solution to this problem. By using the technique of Section 2.3, an approximate solution h_n to the problem (2.2.41), (2.2.43) is defined as

$$h_n \in H_n \cap \tilde{U}_{\mathrm{ad}}, \qquad \Psi_0(h_n, u_{h_n}) = \inf_{h\in H_n\cap\tilde{U}_{\mathrm{ad}}} \Psi_0(h, u_h), \tag{2.4.42}$$

H_n being a finite-dimensional subspace of $W_p^1(\Omega)$. When applying the technique of Section 2.4, we replace the set $\tilde{U}_{\mathrm{ad}}$ in (2.4.42) by a bigger one, namely $U^{(t)}$.

Effective methods of solution of the regular finite-dimensional problem (2.4.42) are the linearization ones (see Céa (1971), Pshenichny (1983), and Haug et al. (1986)). To use them, one needs to calculate the Fréchet derivatives of the functionals $h \to \Psi_m(h, u_h)$, $m = 0, 1, 2, \dots, l$, $h \to \|h\|_{W_p^1(\Omega)}$, and $h \to Q_i(h)$, $i = 1, 2, \dots, 2r$ (see (2.2.41)). These functionals are thought of as those defined on the set G determined by (2.2.44).

The functional $h \to \|h\|_{W_p^1(\Omega)}$ is Fréchet continuously differentiable in G and its Fréchet derivative is given by the formula (1.10.7). The functionals Q_i, $i = 1, 2, \dots, 2r$, are also Fréchet continuously differentiable and their Fréchet derivatives are given by (2.2.40).

Now, let us study the question of differentiability of the functionals $h \to \Psi_m(h, u_h)$.

The function $h \to u_h$

According to the relations (2.1.12), (2.2.45) and Theorem 2.1.1, to every element $h \in G$ there corresponds a unique function u_h being the solution of the following problem:

$$u_h \in V, \qquad F(h, u_h) = 0. \tag{2.4.43}$$

So, a function $\varphi\colon G \to V$ is defined as

$$h \to \varphi(h) = u_h. \tag{2.4.44}$$

Theorem 2.4.5 *Let a_h be a bilinear, symmetric form on $V \times V$ determined by the relations* (2.1.1), (2.1.4), (2.1.6), (2.2.52) *and let the inequalities* (2.1.7), (2.1.10) *hold true. Then the function $\varphi\colon G \to V$ defined by the relations* (2.4.44), (2.1.12) *(f assumed to be fixed in* (2.1.12)*) is Fréchet continuously differentiable and its Fréchet derivative at an arbitrary point $q \in G$ is given by*

$$\varphi'(q)h = -\left(\frac{\partial F}{\partial u}(q, \varphi(q))\right)^{-1} \circ \frac{\partial F}{\partial h}(q, \varphi(q))h, \qquad h \in W_p^1(\Omega), \tag{2.4.45}$$

$\left(\frac{\partial F}{\partial u}(q,\varphi(q))\right)^{-1}$ *being the inverse mapping of $\frac{\partial F}{\partial u}(q,\varphi(q))$, and $\frac{\partial F}{\partial u}, \frac{\partial F}{\partial h}$ being defined by the formulas* (2.2.64), (2.2.54).

Proof. From (2.4.43) and (2.4.44), it follows that $h \to \varphi(h)$ is the implicit function defined by the equation $F(h,u) = 0$.

To prove the Fréchet differentiability of the function φ, we will use Theorem 1.9.2. The function F is thought of as a mapping of the open set $U = G \times V$ (see (2.2.44)) into V^*.

Let q be an arbitrary element from G. The argument of the proof of Lemma 2.2.1 implies that the mapping F is Fréchet differentiable at the point $(q, \varphi(q)) \in U$ and the partial Fréchet derivatives are:

$$\frac{\partial F}{\partial u}(q, \varphi(q)) \in \mathcal{L}(V, V^*), \qquad \frac{\partial F}{\partial h} \in \mathcal{L}(W_p^1(\Omega), V^*).$$

The relations (2.2.45), (2.2.64) imply

$$\left(\frac{\partial F}{\partial u}(q, \varphi(q))v, w\right) = (F(q,v) + f, w) = a_q(v,w), \qquad v, w \in V. \tag{2.4.46}$$

Since for every q from G, the bilinear form a_q is symmetric, continuous, and coercive (see Subsec. 2.1.1), from (2.4.46) and the Riesz theorem we conclude that the mapping $\frac{\partial F}{\partial u}(q, \varphi(q))$ is invertible and the inverse mapping

$$\left(\frac{\partial F}{\partial u}(q, \varphi(q))\right)^{-1} \in \mathcal{L}(V^*, V)$$

is given through the relation

$$\left(\frac{\partial F}{\partial u}(q, \varphi(q))\right)^{-1} g = u, \tag{2.4.47}$$

where

$$u \in V, \qquad a_q(u,w) = (g,w), \qquad w \in V. \tag{2.4.48}$$

Due to Lemma 2.1.1, $h \to \varphi(h) = u_h$ is a continuous mapping of G into V. Now, applying Theorem 1.9.2, we deduce that the mapping φ is Fréchet differentiable at an arbitrary point $q \in G$ and its Fréchet derivative is given by (2.4.45).

It remains to show that

$$q \to \varphi'(q) \text{ is a continuous mapping of } G \text{ into } \mathcal{L}(W_p^1(\Omega), V). \tag{2.4.49}$$

Let

$$\{q_n\}_{n=1}^{\infty} \subset G, \qquad q_0 \in G, \qquad q_n \to q_0 \text{ in } W_p^1(\Omega). \tag{2.4.50}$$

By virtue of Lemma 2.1.1, we get

$$\varphi(q_n) \to \varphi(q_0) \text{ in } V. \tag{2.4.51}$$

Introduce the notations

$$A_n = \frac{\partial F}{\partial u}(q_n, \varphi(q_n)), \qquad B_n = \frac{\partial F}{\partial h}(q_n, \varphi(q_n)), \qquad n = 0, 1, 2, \dots. \tag{2.4.52}$$

By (2.2.63), (2.2.65), and (2.4.50)–(2.4.52), we obtain that

$$A_n \to A_0 \text{ in } \mathcal{L}(V, V^*), \tag{2.4.53}$$

$$B_n \to B_0 \text{ in } \mathcal{L}(W_p^1(\Omega), V^*). \tag{2.4.54}$$

By using the well-known result (see, e.g., Schwartz (1967)), from (2.4.53) we have

$$A_n^{-1} \to A_0^{-1} \text{ in } \mathcal{L}(V^*, V). \tag{2.4.55}$$

Now, we get

$$\begin{aligned} &\|A_n^{-1} \circ B_n - A_0^{-1} \circ B_0\|_{\mathcal{L}(W_p^1(\Omega), V)} \\ &\quad \le \|A_n^{-1} \circ (B_n - B_0)\|_{\mathcal{L}(W_p^1(\Omega), V)} + \|(A_n^{-1} - A_0^{-1}) \circ B_0\|_{\mathcal{L}(W_p^1(\Omega), V)} \\ &\quad \le c\big(\|B_n - B_0\|_{\mathcal{L}(W_p^1(\Omega), V^*)} + \|A_n^{-1} - A_0^{-1}\|_{\mathcal{L}(V^*, V)} \big), \end{aligned}$$

and by virtue of (2.4.54), (2.4.55), the right-hand side of this inequality tends to zero as $n \to \infty$. Hence, (2.4.45), (2.4.50), and (2.4.52) imply (2.4.49). The theorem is proved.

Remark 2.4.4 Combining (2.4.45) and (2.4.46), we easily conclude that

$$a_q(\varphi'(q)h, w) = -\Big(\frac{\partial F}{\partial h}(q, \varphi(q))h, w\Big), \qquad h \in W_p^1(\Omega),\ w \in V. \tag{2.4.56}$$

The relation (2.4.46) allows us to rewrite (2.4.56) as

$$F(q, \varphi'(q)h) + f = -\frac{\partial F}{\partial h}(q, \varphi(q))h, \qquad h \in W_p^1(\Omega). \tag{2.4.57}$$

The Fréchet derivative of the functional $h \to \Psi_m(h, u_h)$

Define functionals $\Phi_m : G \to \mathbb{R}$ in the following way:

$$\Phi_m(h) = \Psi_m(h, u_h), \qquad m = 0, 1, 2, \ldots, l. \tag{2.4.58}$$

Theorem 2.4.6 *Let a_h be a bilinear, symmetric form on $V \times V$ determined by the relations* (2.1.1), (2.1.4), (2.1.6), (2.2.52) *and let the inequalities* (2.1.7), (2.1.10) *hold true. Assume $h, v \to \Psi_m(h, v)$ are Fréchet continuously differentiable mappings of $G \times V$ into $\mathbb{R}$, $m = 0, 1, 2, \ldots, l$. Then, the functionals Φ_m defined by* (2.4.58), (2.1.12) *(in* (2.1.12) *f is supposed to be fixed) are Fréchet continuously differentiable on G, and at an arbitrary point $q \in G$ the Fréchet derivative of the functional Φ_m is determined by the relation*

$$\Phi'_m(q)h = \frac{\partial \Psi_m}{\partial h}(q, u_q)h + \iint\limits_{\Omega} \sum_{i,j=1}^{k} \frac{da_{ij}}{dt}(q)h(P_i u_q)\left(P_j v^{(m)}\right) dx\, dy, \tag{2.4.59}$$

the function $v^{(m)}$ being the solution to the following problem

$$v^{(m)} \in V, \qquad a_q\left(v^{(m)}, w\right) = -\left(\frac{\partial \Psi_m}{\partial u}(q, u_q), w\right), \tag{2.4.60}$$
$$w \in V, \ m = 0, 1, 2, \ldots, l.$$

Proof. The functional Φ_m is the composition of the mappings $\Phi_m^{(1)} : G \to G \times V$, $h \to \Phi_m^{(1)}(h) = \{h, u_h\}$ and $\Phi_m^{(2)} : G \times V \to \mathbb{R}$, $h, v \to \Phi_m^{(2)}(h, v) = \Psi_m(h, v)$.

By virtue of Theorem 2.4.5, $\Phi_m^{(1)}$ is a Fréchet continuously differentiable mapping of G into $G \times V$. By the hypothesis, $\Phi_m^{(2)}$ is a Fréchet continuously differentiable mapping of $G \times V$ into $\mathbb{R}$.

Hence, the functional Φ_m is Fréchet continuously differentiable in G (see Schwartz (1967)), and the Fréchet derivative of the mapping Φ_m at an arbitrary point $q \in G$ is given by the relation

$$\Phi'_m(q)h = \frac{\partial \Psi_m}{\partial h}(q, u_q)h + \frac{\partial \Psi_m}{\partial u}(q, u_q) \circ \varphi'(q)h, \tag{2.4.61}$$
$$h \in W_p^1(\Omega), \quad m = 0, 1, 2, \ldots, l.$$

Since

$$\frac{\partial \Psi_m}{\partial u}(q, u_q) \in V^*, \qquad \varphi'(q)h \in V,$$

the following formula is true

$$\frac{\partial \Psi_m}{\partial u}(q, u_q) \circ \varphi'(q)h = \left(\frac{\partial \Psi_m}{\partial u}(q, u_q), \varphi'(q)h\right). \tag{2.4.62}$$

Further, let the function $v^{(m)}$ be a solution to the problem (2.4.60). We take in (2.4.60) $w = \varphi'(q)h$. Then, bearing in mind that the form a_q is symmetric and making use of (2.4.62), we get

$$\begin{aligned} \frac{\partial \Psi_m}{\partial u}(q, u_q) \circ \varphi'(q)h &= -a_q\left(v^{(m)}, \varphi'(q)h\right) \\ &= -a_q\left(\varphi'(q)h, v^{(m)}\right). \end{aligned} \tag{2.4.63}$$

Setting in (2.4.56) $w = v^{(m)}$ and taking into account that $\varphi(q) = u_q$ (see (2.4.44)), we have

$$-a_q\left(\varphi'(q)h, v^{(m)}\right) = \left(\frac{\partial F}{\partial h}(q, u_q)h, v^{(m)}\right). \tag{2.4.64}$$

By (2.4.63), (2.4.64), and (2.2.54), we obtain

$$\begin{aligned} \frac{\partial \Psi_m}{\partial u}(q, u_q) \circ \varphi'(q)h &= \left(\frac{\partial F}{\partial h}(q, u_q)h, v^{(m)}\right) \\ &= \iint\limits_{\Omega} \sum_{i,j=1}^{k} \frac{da_{ij}}{dt}(q)h(P_i u_q)\left(P_j v^{(m)}\right)\, dx\, dy, \end{aligned} \tag{2.4.65}$$

Now, (2.4.59) follows from the relations (2.4.61) and (2.4.65), which completes the proof of the theorem.

2.5 Spectral problem

2.5.1 Eigenvalue problem

Assume that

$$\left.\begin{array}{l} \mathcal{U} \text{ is a Banach space, } V \subset \mathcal{U} \subset (L_2(\Omega))^\nu, \text{ and the} \\ \text{embedding } V \to \mathcal{U} \text{ is compact.} \end{array}\right\} \tag{2.5.1}$$

Let there be defined a family of bilinear, symmetric, continuous forms b_h on $\mathcal{U} \times \mathcal{U}$ depending on parameter h running over the set Y_p.

Consider the following eigenvalue problem:

$$\begin{aligned} &(u_i, \mu_i) \in V \times \mathbb{R}, \qquad u_i \neq 0, \\ &\mu_i a_h(u_i, v) = b_h(u_i, v), \qquad v \in V. \end{aligned} \tag{2.5.2}$$

Thinking of the bilinear form a_h as a scalar product in V (see Remark 1.5.2) and applying Theorem 1.5.8, we get the following

Theorem 2.5.1 *Let a_h be a bilinear, symmetric form on $V \times V$ determined by the relations* (2.1.1), (2.1.4)–(2.1.6) *and let the inequalities* (2.1.7), (2.1.10) *hold. Assume b_h to be a bilinear, symmetric, continuous form on $\mathcal{U} \times \mathcal{U}$ depending on a parameter $h \in Y_p$ and* (2.5.1) *to be valid. Then, the problem* (2.5.2) *has a countable set of eigenvalues $\{\mu_i\}$. Each eigenvalue in the sequence $\{\mu_i\}$ is repeated as many times as its multiplicity; each nonzero eigenvalue has a finite multiplicity and the following estimate holds:*

$$m = \inf_{a_h(u,u)=1} b_h(u,u) \leq \mu_i \leq \sup_{a_h(u,u)=1} b_h(u,u) = M. \tag{2.5.3}$$

If $m \neq 0$, then m is an eigenvalue, and if $M \neq 0$, then M is an eigenvalue. Also, $\lim\limits_{i\to\infty} \mu_i = 0$.

Remark 2.5.1 If the hypothesis of Theorem 2.5.1 is valid, the form b_h is strictly positive on V, i.e.,

$$b_h(u,u) \geq 0, \qquad u \in V, \tag{2.5.4}$$

and the equality takes place only for $u = 0$, then $\mu = 0$ cannot be an eigenvalue of the problem (2.5.2). In this case, all the eigenvalues of the problem (2.5.2) are positive. They form a sequence converging to zero and every eigenvalue is of finite multiplicity.

2.5.2 On the continuity of the spectrum

Let us denote

$$\mathcal{N}_1 = \{\, a \mid a \in \mathcal{L}_2(V,V;\mathbb{R}), a(u,v) = a(v,u) \quad u,v \in V,$$
$$a(u,u) \geq c(a)\|u\|_V^2 \quad u \in V,\, c(a) = \text{const} > 0 \,\}. \tag{2.5.5}$$
$$\mathcal{N}_2 = \{\, b \,|\, b \in \mathcal{L}_2(\mathcal{U},\mathcal{U};\mathbb{R}),\; b(u,v) = b(v,u) \quad u,v \in \mathcal{U} \,\}. \tag{2.5.6}$$

The set $\mathcal{N}_1$ is endowed with the topology induced by the $\mathcal{L}_2(V,V;\mathbb{R})$ one and in this topology $\mathcal{N}_1$ is a metric space. If $a_1, a_2 \in \mathcal{N}_1$, then $d(a_1,a_2)$ – the distance between a_1 and a_2 – is defined, because of Remark 1.5.3, by the relation

$$d(a_1,a_2) = \sup_{\|u\|_V=1} |a_1(u,u) - a_2(u,u)|; \tag{2.5.7}$$

$\mathcal{N}_2$ is a normed space equipped with the norm

$$\|b\|_{\mathcal{N}_2} = \sup_{\|u\|_{\mathcal{U}}=1,\ \|v\|_{\mathcal{U}}=1} |b(u,v)|. \tag{2.5.8}$$

Lemma 2.5.1 *Let a_h be a bilinear, symmetric form on $V \times V$ determined by the relations* (2.1.1), (2.1.4)–(2.1.6) *and let the inequalities* (2.1.7), (2.1.10) *hold. Then $h \to a_h$ is a continuous mapping of Y_p (endowed with the topology induced by the $W_p^1(\Omega)$-weak one) into $\mathcal{N}_1$ when $p > 2$.*

Proof. Let $\{h_n\}_{n=1}^{\infty} \subset Y_p$, $p > 2$ and $h_n \to h_0$ weakly in $W_p^1(\Omega)$. Then, by the embedding theorem (see Theorem 1.6.2)

$$h_n \to h_0 \qquad \text{strongly in } C(\overline{\Omega}). \tag{2.5.9}$$

By (2.1.1), (2.1.4), and (2.1.5), we have

$$\begin{aligned} &|a_{h_n}(u,u) - a_{h_0}(u,u)| \\ &\quad = \left| \iint_{\Omega} \sum_{i,j=1}^{k} (a_{ij}(h_n) - a_{ij}(h_0))\,(P_i u)(P_j u))\,dx\,dy \right| \\ &\quad \le \alpha_n c \|u\|_V^2, \qquad u \in V. \end{aligned} \tag{2.5.10}$$

Here, $c = \text{const} > 0$ and

$$\alpha_n = \sup_{i,j} \sup_{x \in \overline{\Omega}} |a_{ij}(h_n) - a_{ij}(h_0)|. \tag{2.5.11}$$

The relation (2.1.5) and Theorem 1.3.10 imply $t \to a_{ij}(t)$ to be a uniformly continuous mapping of the interval $[e_1, e_2]$ into $\mathbb{R}$. Hence, by (2.5.9), (2.5.11), we obtain $\lim_{n\to\infty} \alpha_n = 0$. Now, (2.5.7) and (2.5.10) yield

$$\lim_{n\to\infty} d\,(a_{h_n}, a_{h_0}) = 0,$$

concluding the proof of the lemma.

The dependence of the bilinear form b_h on a parameter h running over the set Y_p defines the mapping $h \to b_h$. We suppose this mapping to satisfy the following condition:

$$\left. \begin{array}{l} h \to b_h \text{ is a continuous mapping of } Y_p \text{ (endowed with the} \\ \text{topology generated by the } W_p^1(\Omega)\text{-weak one) into } \mathcal{N}_2. \end{array} \right\} \tag{2.5.12}$$

The relation (2.5.12) means that, provided $\{h_n\} \subset Y_p$ and $h_n \to h$ weakly in $W_p^1(\Omega)$, we get $\|b_{h_n} - b_h\|_{\mathcal{N}_2} \to 0$ as $n \to \infty$.

Recall that the vector normed space of bounded number sequences converging to zero is referred to as $\ell_{\infty,0}$. The norm of an element $\xi = \{\xi_1, \xi_2, \dots\} \in \ell_{\infty,0}$ is defined as

$$\|\xi\|_{\ell_{\infty,0}} = \sup_i |\xi_i|. \tag{2.5.13}$$

Since the bilinear forms a_h and b_h depend on a parameter $h \in Y_p$, the eigenvalues μ_i of the problem (2.5.2) depend on h, too, i.e., $\mu_i = \mu_i(h)$. By virtue of Theorem 2.5.1, to every $h \in Y_p$ there corresponds an element

$$\mu(h) = \{\mu_1(h), \mu_2(h), \dots\} \in \ell_{\infty,0},$$

$\mu_i(h)$ being the eigenvalues of the problem (2.5.2) ordered so that

$$|\mu_1(h)| \ge |\mu_2(h)| \ge \cdots. \tag{2.5.14}$$

Under such ordering, if the set of the nonzero eigenvalues is countable (i.e., the number of the nonzero eigenvalues is not finite) and zero is an eigenvalue, then the set $\{\mu_i(h)\}_{i=1}^{\infty}$ does not contain zero.

Thus, we defined the mapping $h \to \mu(h)$ of the set Y_p into $\ell_{\infty,0}$. The following statement is valid.

Theorem 2.5.2 *Let the assumptions of Theorem* 2.5.1 *be satisfied and* (2.5.12) *hold. Then,* $h \to \mu(h)$ *is a continuous mapping of* Y_p *(endowed with the topology induced by the* $W_p^1(\Omega)$*-weak one) into* $\ell_{\infty,0}$ *when* $p > 2$.

Proof. Let $\{h_n\} \subset Y_p$, $p > 2$, and $h_n \to h$ weakly in $W_p^1(\Omega)$. Then, Lemma 2.5.1 and (2.5.12) imply that

$$a_{h_n} \to a_h \text{ in } \mathcal{N}_1, \qquad b_{h_n} \to b_h \text{ in } \mathcal{N}_2.$$

Now, Theorem 2.5.2 is a consequence of Theorem 1.5.9 and Remark 1.5.5.

2.6 Optimization of the spectrum

2.6.1 Formulation of the problem and the existence theorem

Define a set of admissible controls as

$$U_{\text{ad}} = \Big\{ h \mid h \in W_p^1(\Omega),\ \|h\|_{W_p^1(\Omega)} \le C,\ \iint_{\Omega} hg\,dx\,dy \le C_1,\ \check{h} \le h \le \hat{h} \Big\}. \quad (2.6.1)$$

Here,

$$\left.\begin{array}{l} p > 2,\ C,\ C_1,\ \check{h}, \text{ and } \hat{h} \text{ are positive numbers and} \\ e_1 \le \check{h} < \hat{h} \le e_2, \text{ where } e_1,\ e_2 \text{ are the constants from} \\ (2.1.2); \end{array}\right\} \quad (2.6.2)$$

$$g \in L_1(\Omega), \qquad g \ge c = \text{const} > 0 \text{ a.e. in } \Omega. \quad (2.6.3)$$

We suppose that

$$\text{the set } U_{\text{ad}} \text{ is nonempty.} \quad (2.6.4)$$

Further, assume that we are given a mapping A such that

$$\mu \to A\mu \text{ is a continuous mapping of } \ell_{\infty,0} \text{ into } \mathbb{R}. \quad (2.6.5)$$

Define a goal functional as

$$f(h) = A\mu(h). \quad (2.6.6)$$

Here, $\mu(h) = \{\mu_1(h), \mu_2(h), \dots\}$ and $\mu_i(h) = \mu_i$ are the eigenvalues of the problem (2.5.2).

The optimal control problem consists in finding a function h_0 such that

$$h_0 \in U_{\mathrm{ad}}, \qquad f(h_0) = \inf_{h \in U_{\mathrm{ad}}} f(h). \tag{2.6.7}$$

Let us consider a few realizations of the mapping A.

I. Let $A\mu = \|\mu\|_{\ell_{\infty,0}} = \sup_i |\mu_i|$. In this case, the condition (2.6.5) is satisfied and the problem (2.6.7) means minimization on U_{ad} of the absolute value of the first eigenvalue of the problem (2.5.2).

II. Let $\xi_i \in \mathbb{R}$, $\xi_i \neq 0$, $i = 1, 2, \ldots, m$. Also, let $t \to \varphi_i(t)$ be continuous mappings of $\mathbb{R}$ into $\mathbb{R}$ such that $\varphi_i(t) \geq 0$, $\varphi_i(\xi_i) \neq 0$, the support of each φ_i belongs to a sufficiently small neighborhood of the point ξ_i and does not contain zero, $\int_{\mathbb{R}} \varphi_i(t)\, dt = 1$, and the supports of the functions φ_i and φ_j are disjoint when $i \neq j$. Define a mapping A by

$$\mu \in \ell_{\infty,0}, \qquad A\mu = \sum_{i=1}^{m} \sum_{j=1}^{\infty} c_i \varphi_i(\mu_j), \qquad c_i = \mathrm{const} > 0. \tag{2.6.8}$$

Since the support of φ_i does not contain zero, the summation in (2.6.8) is taken over a finite number of indices and the mapping A meets the condition (2.6.5). The problem (2.6.7) means that we search the h_0 for which the number of eigenvalues contained in given neighborhoods of the points ξ_i, which are defined by the supports of the functions φ_i, is minimized. In other words, the solution to the problem (2.6.7) defines an element h_0 for which the eigenvalues of the problem (2.5.2) are "maximally shifted" from the given numbers ξ_i.

Theorem 2.6.1 *Let a_h be a bilinear form on $V \times V$ determined by the relations* (2.1.1), (2.1.4)–(2.1.7), (2.1.10). *Also, let a mapping $h \to b_h$ satisfying the condition* (2.5.12) *be defined. Suppose that a set U_{ad} is determined by the relations* (2.6.1)–(2.6.3) *and satisfies* (2.6.4), *and a goal functional $f(h)$ is defined by* (2.6.5), (2.6.6). *Then, the problem* (2.6.7) *has a solution.*

Proof. Let $\{h_n\}_{n=1}^{\infty}$ be a sequence such that

$$h_n \in U_{\mathrm{ad}}, \tag{2.6.9}$$

$$\lim_{n \to \infty} f(h_n) = \inf_{h \in U_{\mathrm{ad}}} f(h). \tag{2.6.10}$$

By virtue of (2.6.1), (2.6.2), the sequence $\{h_n\}_{n=1}^{\infty}$ is bounded in $W_p^1(\Omega)$ when $p > 2$. Hence, we may find a subsequence $\{h_m\}_{m=1}^{\infty}$ of it such that

$$h_m \to z \qquad \text{weakly in } W_p^1(\Omega), \tag{2.6.11}$$

$$h_m \to z \qquad \text{strongly in } C(\overline{\Omega}). \tag{2.6.12}$$

By (2.6.11) and by Theorem 2.5.2, we get

$$\mu(h_m) \to \mu(z) \qquad \text{in } \ell_{\infty,0}. \tag{2.6.13}$$

Taking into account (2.6.5), (2.6.6), (2.6.10), we obtain

$$\lim_{m\to\infty} f(h_m) = f(z) = \inf_{h\in U_{\rm ad}} f(h). \tag{2.6.14}$$

By (2.6.9) and (2.6.12), we have

$$\check{h} \le z \le \hat{h}. \tag{2.6.15}$$

Using (2.6.3), (2.6.9), (2.6.12), and the Lebesgue theorem (Theorem 1.6.1), we get

$$\lim_{m\to\infty} \iint\limits_{\Omega} h_m g\, dx\, dy = \iint\limits_{\Omega} zg\, dx\, dy \le C_1, \tag{2.6.16}$$

where C_1 is the number from (2.6.1). By (2.6.9), (2.6.11), we obtain

$$C \ge \liminf_{m\to\infty} \|h_m\|_{W_p^1(\Omega)} \ge \|z\|_{W_p^1(\Omega)}, \tag{2.6.17}$$

where C is the number from (2.6.1). The relations (2.6.1), (2.6.15)–(2.6.17) imply $z \in U_{\rm ad}$. Finally, from (2.6.14) we deduce that the function $h_0 = z$ is a solution to the problem (2.6.7).

2.6.2 Finite-dimensional approximation of the optimal control problem

Lemma on the approximation from the interior of $U_{\rm ad}$

We equip the set $U_{\rm ad}$ defined by the relations (2.6.1)–(2.6.3) with the topology generated by the $W_p^1(\Omega)$-one. We will refer to the set of the interior points of $U_{\rm ad}$ as $\overset{\circ}{U}_{\rm ad}$.

Lemma 2.6.1 *Let a set $U_{\rm ad}$ be defined by the relations* (2.6.1)–(2.6.3) *and meet the condition* (2.6.4). *Also, let h be an arbitrary function from $U_{\rm ad}$ that is not identically equal to $\check{h}$ in Ω. Then, there exists a sequence $\{h_n\} \subset \overset{\circ}{U}_{\rm ad}$ such that*

$$\lim_{n\to\infty} \|h_n - h\|_{W_p^1(\Omega)} = 0. \tag{2.6.18}$$

Proof. 1. By virtue of the embedding theorem for a two-dimensional domain Ω, the functions from the space $W_p^1(\Omega)$ are equivalent to continuous functions in $\overline{\Omega}$ when $p > 2$. By (2.6.1), we get $h \ge \check{h}$, and since h is not identically equal to $\check{h}$ by the hypothesis, we infer from (2.6.3) that

$$\iint\limits_{\Omega} hg\, dx\, dy > \iint\limits_{\Omega} \check{h}g\, dx\, dy, \qquad \iint\limits_{\Omega} h\, dx\, dy > \iint\limits_{\Omega} \check{h}\, dx\, dy.$$

Then, there exists a constant t such that

$$\hat{h} > t > \check{h}, \tag{2.6.19}$$

$$\iint_\Omega hg\,dx\,dy > \iint_\Omega tg\,dx\,dy > \iint_\Omega \check{h}g\,dx\,dy, \tag{2.6.20}$$

$$\iint_\Omega h\,dx\,dy > \iint_\Omega t\,dx\,dy > \iint_\Omega \check{h}\,dx\,dy. \tag{2.6.21}$$

Represent the element h as

$$h = t + q. \tag{2.6.22}$$

Then, on account of (2.6.20) and (2.6.21), we obtain

$$\iint_\Omega qg\,dx\,dy > 0, \tag{2.6.23}$$

$$\iint_\Omega q\,dx\,dy > 0. \tag{2.6.24}$$

Define functions h_n as follows:

$$h_n = t + l_n q, \qquad 0 < l_n < 1, \qquad \lim_{n\to\infty} l_n = 1. \tag{2.6.25}$$

The relations (2.6.22), (2.6.23), and (2.6.25) yield

$$\iint_\Omega hg\,dx\,dy - \iint_\Omega h_n g\,dx\,dy = (1 - l_n)\iint_\Omega qg\,dx\,dy > 0.$$

Hence,

$$\iint_\Omega h_n g\,dx\,dy < C_1 \qquad \forall n, \tag{2.6.26}$$

where C_1 is the number from (2.6.1).

Taking into account that $\check{h} \le h \le \hat{h}$, the relations (2.6.19), (2.6.22), and (2.6.25) imply

$$\check{h} < h_n < \hat{h} \qquad \forall n. \tag{2.6.27}$$

Let us suppose that the inequality

$$\|h_n\|_{W_p^1(\Omega)} < C \qquad \forall n, \tag{2.6.28}$$

holds, where C is the number from (2.6.1). Since the embedding of $W_p^1(\Omega)$ into $C(\overline{\Omega})$ is continuous when $p > 2$ (see Theorem 1.6.2), the relations (2.6.26)–(2.6.28) imply that, for any n, a number $\varepsilon_n > 0$ exists such that

$$d(\varepsilon_n, h_n) \subset U_{\text{ad}}, \tag{2.6.29}$$

where

$$d(\varepsilon_n, h_n) = \{\, z \mid z \in W_p^1(\Omega),\ \|z - h_n\|_{W_p^1(\Omega)} \le \varepsilon_n \,\}. \tag{2.6.30}$$

Hence, the functions h_n defined in (2.6.25) belong to $\overset{\circ}{U}_{\mathrm{ad}}$. From (2.6.22), (2.6.25), we easily deduce that $h_n \to h$ in $W_p^1(\Omega)$. So, if the estimate (2.6.28) holds, then the lemma will be proved.

2. Let us establish the inequality (2.6.28). By (2.6.22), (2.6.25), we get

$$\begin{aligned} \iint\limits_{\Omega} \left|\frac{\partial h}{\partial x}\right|^p dx\,dy &= \iint\limits_{\Omega} \left|\frac{\partial q}{\partial x}\right|^p dx\,dy \\ &\ge \iint\limits_{\Omega} l_n^p \left|\frac{\partial q}{\partial x}\right|^p dx\,dy = \iint\limits_{\Omega} \left|\frac{\partial h_n}{\partial x}\right|^p dx\,dy. \end{aligned} \tag{2.6.31}$$

Similarly, we obtain

$$\iint\limits_{\Omega} \left|\frac{\partial h}{\partial y}\right|^p dx\,dy \ge \iint\limits_{\Omega} \left|\frac{\partial h_n}{\partial y}\right|^p dx\,dy. \tag{2.6.32}$$

Introduce the function

$$[0,1] \ni s \to f(s) = \iint\limits_{\Omega} \left(|t+q|^p - |t+sq|^p\right) dx\,dy. \tag{2.6.33}$$

(Notice that the function $(x,y) \to \alpha_s(x,y) = t + sq(x,y)$ is positive on Ω for every $s \in [0,1]$, so we could write below $(t+sq)^p$ instead of $|t+sq|^p$.)

From (2.6.22), (2.6.25), and (2.6.33), it follows that

$$f(l_n) = \iint\limits_{\Omega} |h|^p\, dx\,dy - \iint\limits_{\Omega} |h_n|^p\, dx\,dy. \tag{2.6.34}$$

The function $s \to f(s)$ is differentiable on the interval (0,1). Applying to it the mean value theorem, we get

$$\begin{aligned} f(s) &= f(1) + \frac{\partial f}{\partial s}(\theta)(s-1) \\ &= -(s-1)p \iint\limits_{\Omega} |t+\theta q|^{p-1} q\, dx\,dy, \qquad s \in (0,1),\ \theta \in (s,1). \end{aligned} \tag{2.6.35}$$

Let us denote

$$\Omega_1 = \{\, (x,y) \mid (x,y) \in \Omega,\ q(x,y) > 0 \,\}, \tag{2.6.36}$$

$$\Omega_2 = \{\, (x,y) \mid (x,y) \in \Omega,\ q(x,y) < 0 \,\}. \tag{2.6.37}$$

By (2.6.24), we have

$$a_1 = \iint\limits_{\Omega_1} q\,dx\,dy > 0, \qquad a_2 = \iint\limits_{\Omega_2} q\,dx\,dy \le 0, \tag{2.6.38}$$

and

$$\iint\limits_{\Omega} q\,dx\,dy = a_1 + a_2 > 0. \tag{2.6.39}$$

Obviously, mes $\Omega_1 > 0$ and

$$|t+\theta q|^{p-1} > t^{p-1} \qquad \text{on } \Omega_1, \tag{2.6.40}$$

$$0 < |t+\theta q|^{p-1} < t^{p-1} \qquad \text{on } \Omega_2. \tag{2.6.41}$$

Due to (2.6.36), (2.6.38), and (2.6.40), we obtain

$$\iint\limits_{\Omega_1} |t+\theta q|^{p-1} q\,dx\,dy > \iint\limits_{\Omega_1} t^{p-1} q\,dx\,dy = t^{p-1} a_1. \tag{2.6.42}$$

By virtue of (2.6.37), (2.6.38), and (2.6.41), we get

$$\iint\limits_{\Omega_2} |t+\theta q|^{p-1} q\,dx\,dy \ge t^{p-1} a_2. \tag{2.6.43}$$

Further, by (2.6.36)–(2.6.39), (2.6.42), and (2.6.43), we have

$$\iint\limits_{\Omega} |t+\theta q|^{p-1} q\,dx\,dy = \iint\limits_{\Omega_1} |t+\theta q|^{p-1} q\,dx\,dy$$
$$+ \iint\limits_{\Omega_2} |t+\theta q|^{p-1} q\,dx\,dy > t^{p-1}(a_1+a_2) > 0, \qquad \theta \in (0,1).$$

From here and from the relation (2.6.35)

$$f(s) > 0, \qquad s \in (0,1). \tag{2.6.44}$$

Now, using (2.6.31), (2.6.32), (2.6.34), and (2.6.44) and noticing that $0 < l_n < 1$, we obtain

$$\|h\|_{W_p^1(\Omega)} > \|h_n\|_{W_p^1(\Omega)}.$$

This inequality yields (2.6.28). The lemma is proved.

Finite-dimensional problem

Let $\{H_n\}_{n=1}^{\infty}$ be a sequence of finite-dimensional subspaces of $W_p^1(\Omega)$ satisfying the limit density condition, i.e.,

$$\lim_{n\to\infty} \inf_{z\in H_n} \|h-z\|_{W_p^1(\Omega)} = 0, \qquad h \in W_p^1(\Omega). \tag{2.6.45}$$

Consider the following problem: Find a function h_n such that

$$h_n \in H_n \cap U_{\text{ad}}, \tag{2.6.46}$$

$$f(h_n) = \inf_{h\in H_n\cap U_{\text{ad}}} f(h). \tag{2.6.47}$$

Theorem 2.6.2 *Let the conditions of Theorem 2.6.1 be fulfilled and let a function h_0 solving the problem (2.6.7) be not identically equal to $\check{h}$ in Ω. Also, let $\{H_n\}_{n=1}^{\infty}$ be a sequence of finite-dimensional subspaces of $W_p^1(\Omega)$ satisfying the condition (2.6.45). Then, for any n sufficiently large (say, $n \geq k$), the problem (2.6.46), (2.6.47) has a solution. This problem has a solution for each n if the set $H_n \cap U_{\text{ad}}$ is not empty for all n. Moreover,*

$$\lim_{n\to\infty} f(h_n) = f(h_0) = \inf_{h\in U_{\text{ad}}} f(h). \tag{2.6.48}$$

One may choose a subsequence $\{h_m\}_{m=1}^{\infty}$ of the sequence $\{h_n\}_{n=k}^{\infty}$ such that $h_m \to h_0$ weakly in $W_p^1(\Omega)$.

Proof. 1. By virtue of Lemma 2.6.1, for any $\varepsilon > 0$, there exists an element $h_\varepsilon \in \overset{\circ}{U}_{\text{ad}}$ such that

$$\|h_0 - h_\varepsilon\|_{W_p^1(\Omega)} \leq \varepsilon. \tag{2.6.49}$$

Since $h_\varepsilon \in \overset{\circ}{U}_{\text{ad}}$, there exists $\gamma \in \mathbb{R}$ such that

$$0 < \gamma \leq \varepsilon, \qquad d(\gamma, h_\varepsilon) \subset U_{\text{ad}}, \tag{2.6.50}$$

where

$$d(\gamma, h_\varepsilon) = \{\, h \mid h \in W_p^1(\Omega),\ \|h - h_\varepsilon\|_{W_p^1(\Omega)} \leq \gamma \,\}. \tag{2.6.51}$$

From (2.6.45), it follows that, for any n sufficiently large (say, $n \geq k$), we can find an element g_n from H_n such that $g_n \in d(\gamma, h_\varepsilon)$; moreover, due to (2.6.50), $g_n \in U_{\text{ad}}$. By (2.6.49)–(2.6.51), we have

$$\|g_n - h_0\|_{W_p^1(\Omega)} \leq \|g_n - h_\varepsilon\|_{W_p^1(\Omega)} + \|h_\varepsilon - h_0\|_{W_p^1(\Omega)} \leq 2\varepsilon.$$

The above argument implies the existence of a sequence $\{g_n\}_{n=k}^{\infty}$ such that

$$g_n \in H_n \cap U_{\text{ad}}, \tag{2.6.52}$$

$$g_n \to h_0 \ \text{ in } W_p^1(\Omega). \tag{2.6.53}$$

2. So, the set $H_n \cap U_{\text{ad}}$ is not empty when $n \geq k$. It is easy to see that $H_n \cap U_{\text{ad}}$ is a bounded and closed subset of H_n (H_n being endowed with the topology of $W_p^1(\Omega)$). Since the set H_n is finite-dimensional, we conclude that $H_n \cap U_{\text{ad}}$ is a compact topological space.

Theorem 2.5.2 and the relations (2.6.5), (2.6.6) yield that

$$\left.\begin{array}{l} h \to f(h) \text{ is a continuous mapping of } Y_p \text{ (endowed with} \\ \text{the topology generated by the } W_p^1(\Omega)\text{-weak one) into } \mathbb{R} \\ \text{when } p > 2. \end{array}\right\} \tag{2.6.54}$$

Now, Theorem 1.3.9 implies the problem (2.6.46), (2.6.47) to have a solution for $n \geq k$, and for any n provided the set $H_n \cap U_{\text{ad}}$ is not empty for every n.

By (2.6.53), (2.6.54), we have

$$\lim_{n \to \infty} f(g_n) = f(h_0). \tag{2.6.55}$$

Due to (2.6.7), (2.6.47), (2.6.52), we get

$$f(g_n) \geq f(h_n) \geq f(h_0) \qquad \forall n.$$

From here, taking into account (2.6.55), we obtain (2.6.48). By virtue of (2.6.46), (2.6.48), the elements h_n satisfy the conditions (2.6.9), (2.6.10). The argument from the proof of Theorem 2.6.1 implies the existence of a subsequence $\{h_m\}$ of the sequence $\{h_n\}$ such that $h_m \to h_0$ weakly in $W_p^1(\Omega)$.

Note that the problems (2.6.7) and (2.6.46), (2.6.47) may, in general, possess a set of solutions, so we may choose subsequences of the sequence $\{h_n\}$ convergent to the solutions to the problem (2.6.7).

2.6.3 Computation of eigenvalues

Solution of the optimal control problem (2.6.7) and its finite-dimensional analogs – the problems (2.6.46), (2.6.47) –is connected with computation of the eigenvalues of the problem (2.5.2). Let us show that these can be found by means of a passage to the limit, on the basis of the Galerkin approximations.

Let $\{V_m\}_{m=1}^{\infty}$ be a sequence of finite-dimensional subspaces of V satisfying the limit density condition (2.4.39). We consider the following eigenvalue problem in the space V_m:

$$\begin{aligned} &(u_{im}, \mu_{im}) \in V_m \times \mathbb{R}, \qquad u_{im} \neq 0 \\ &\mu_{im} a_h(u_{im}, v) = b_h(u_{im}, v), \qquad v \in V_m. \end{aligned} \tag{2.6.56}$$

Here, u_{im}, μ_{im} are approximate eigenfunctions and eigenvalues of the problem (2.5.2). Since, for any $h \in Y_p$, the bilinear form $a_h(u, v)$ is symmetric, continuous, and coercive on $V \times V$, we may consider it as a scalar product on V (see Remark

1.5.2). Because $b_h(u,v)$ is a bilinear, symmetric, continuous form on $\mathcal{U} \times \mathcal{U}$ and (2.5.1) is valid, we deduce from the Riesz theorem that

$$b_h(u,v) = a_h(B_h u, v), \qquad B_h u \in V, \ u,v \in V. \tag{2.6.57}$$

In this case,

$$B_h \text{ is a linear, compact, selfadjoint operator acting in } V. \ \} \tag{2.6.58}$$

Theorem 2.6.3 *Let the conditions of Theorem* 2.5.1 *be satisfied and let* $\{V_m\}$ *be a sequence of finite-dimensional subspaces of* V *satisfying the condition* (2.4.39). *Also, let* u_{im}, μ_{im}, $i = 1,2,\ldots,k_m$, *be determined by a solution to the problem* (2.6.56) (k_m *being the dimension of the space* V_m). *Then, the following estimate for the eigenvalue error holds*

$$|\mu_{im} - \mu_i| \leq 2^{1/2} \|(I - P_m) \circ B_h\|_{\mathcal{L}(V,V)}, \qquad i = 1,2,\ldots,k_m, \tag{2.6.59}$$

where μ_i *are the eigenvalues of the problem* (2.5.2), I *is the identity operator in* V, P_m *the orthogonal projection of* V *onto* V_m. *Moreover,*

$$\lim_{m\to\infty} \|(I - P_m) \circ B_h\|_{\mathcal{L}(V,V)} = 0. \tag{2.6.60}$$

Proof. Considering the form $a_h(u,v)$ as a scalar product in V and noticing (2.6.57), we may rewrite the problem (2.5.2) as

$$(u_i, \mu_i) \in V \times \mathbb{R}, \qquad u_i \neq 0, \qquad B_h u_i = \mu_i u_i. \tag{2.6.61}$$

Let P_m be the orthogonal projection of V onto V_m in the scalar product generated by a_h. Then (see Lemma 1.5.1 and Remark 1.5.1), we may present the problem (2.6.56) in the form

$$(u_{im}, \mu_{im}) \in V_m \times \mathbb{R}, \qquad u_{im} \neq 0, \qquad P_m(\mu_{im} u_{im} - B_h u_{im}) = 0.$$

Taking into account that $P_m u = u$ for any $u \in V_m$, the latter problem can be rewritten as

$$(u_{im}, \mu_{im}) \in V \times \mathbb{R}, \qquad u_{im} \neq 0, \qquad \mu_{im} u_{im} = (P_m \circ B_h \circ P_m) u_{im}. \tag{2.6.62}$$

The operator P_m being selfadjoint, we deduce from (2.6.58) that $P_m \circ B_h \circ P_m$ is a linear, compact, selfadjoint operator acting in V. Now, applying Theorem 1.5.6 to the operators B_h and $P_m \circ B_h \circ P_m$ (to the problems (2.6.61) and (2.6.62)), we get

$$|\mu_{im} - \mu_i| \leq \|B_h - P_m \circ B_h \circ P_m\|_{\mathcal{L}(V,V)}, \qquad i = 1,2,\ldots,k_m. \tag{2.6.63}$$

For an arbitrary $u \in V$, we set

$$w = (B_h - P_m \circ B_h \circ P_m)\, u. \tag{2.6.64}$$

We have

$$P_m w = P_m \circ B_h \circ P^{(m)} u, \qquad P^{(m)} w = P^{(m)} \circ B_h u,$$

where $P^{(m)} = I - P_m$. Therefore, taking note that $B_h \circ P^{(m)}$ is the adjoint operator of $P^{(m)} \circ B_h$, we get the estimates

$$\begin{gathered} \|P_m w\|_V \leq \|P^{(m)} \circ B_h\|_{\mathcal{L}(V,V)} \|u\|_V, \\ \|P^{(m)} w\|_V \leq \|P^{(m)} \circ B_h\|_{\mathcal{L}(V,V)} \|u\|_V, \\ \|w\|_V^2 = \|P_m w\|_V^2 + \|P^{(m)} w\|_V^2 \leq 2\|P^{(m)} \circ B_h\|^2_{\mathcal{L}(V,V)} \|u\|_V^2. \end{gathered} \tag{2.6.65}$$

By (2.6.64) and (2.6.65), we have

$$\|B_h - P_m \circ B_h \circ P_m\|_{\mathcal{L}(V,V)} \leq 2^{1/2} \|P^{(m)} \circ B_h\|_{\mathcal{L}(V,V)}.$$

This inequality together with (2.6.63) implies the estimate (2.6.59).

Denote the closed unit ball in V by Q. The operator B_h being compact, the set $\overline{B_h(Q)}$ is compact. Due to (2.4.39), the sequence of operators $\{P_m\}$ converges simply to the operator I, that is,

$$\lim_{m \to \infty} \|P_m u - u\|_V = 0, \qquad u \in V.$$

Hence, taking into account that the set $\overline{B_h(Q)}$ is compact, by the Banach-Steinhaus theorem (see, e.g., Schwartz (1967)), we obtain that $P_m \to I$ uniformly on $\overline{B_h(Q)}$, i.e., (2.6.60) holds. The theorem is proved.

Note that the eigenvalue error estimates mentioned above, as well as more precise ones, may be found in Krasnoselsky et al. (1969), Michlin (1970), Gould (1966).

2.7 Control under restrictions on the spectrum

2.7.1 Optimal control problem

Suppose we are given mappings A_i such that

$$\left. \begin{array}{l} \mu \to A_i \mu \text{ is a continuous mapping of } \ell_{\infty,0} \text{ into } \mathbb{R}, \\ i = 1, 2, \ldots, k. \end{array} \right\} \tag{2.7.1}$$

Define a set of admissible controls as

$$\begin{aligned} U_{\text{ad}} = \big\{ h \,|\, h \in W_p^1(\Omega),\ \|h\|_{W_p^1(\Omega)} \leq C,\ \check{h} \leq h \leq \hat{h}, \\ \Psi(h) \leq 0,\ A_i \mu(h) \leq 0\ (i = 1, 2, \ldots, k) \big\}. \end{aligned} \tag{2.7.2}$$

Here,

$$\left. \begin{array}{l} p > 2,\ C,\ \check{h},\ \hat{h} \text{ are positive numbers such that} \\ e_1 < \check{h} < \hat{h} < e_2, \text{ where } e_1,\ e_2 \text{ are the positive numbers} \\ \text{from (2.1.2);} \end{array} \right\} \tag{2.7.3}$$

$$\left.\begin{array}{l} h \to \Psi(h) \text{ is a continuous mapping of } Y_p \text{ (endowed with} \\ \text{the topology generated by the } W_p^1(\Omega)\text{-weak one) into } \mathbb{R}; \end{array}\right\} \quad (2.7.4)$$

$$\mu(h) = \{\mu_1(h), \mu_2(h), \dots\}, \quad (2.7.5)$$

$\mu_i(h) = \mu_i$ being the eigenvalues of the problem (2.5.2).

We suppose the goal functional to satisfy the following condition:

$$\left.\begin{array}{l} h \to f(h) \text{ is a continuous mapping of } Y_p \text{ (equipped with} \\ \text{the topology generated by the } W_p^1(\Omega)\text{-weak one) into } \mathbb{R}. \end{array}\right\} \quad (2.7.6)$$

The optimal control problem consists in finding a function h_0 such that

$$h_0 \in U_{\text{ad}}, \qquad f(h_0) = \inf_{h \in U_{\text{ad}}} f(h). \quad (2.7.7)$$

Theorem 2.7.1 *Let a_h be a bilinear form on $V \times V$ determined by the relations* (2.1.1), (2.1.4)–(2.1.7), *and* (2.1.10). *Let also a mapping $h \to b_h$ satisfying the condition* (2.5.12) *be given. Assume that the nonempty set U_{ad} is determined by the relations* (2.7.1)–(2.7.5), *and the goal functional satisfies the condition* (2.7.6). *Then, the problem* (2.7.7) *has a solution.*

Proof. Let $\{h_n\}_{n=1}^{\infty}$ be a sequence such that

$$h_n \in U_{\text{ad}} \qquad \forall n, \quad (2.7.8)$$

$$\lim_{n \to \infty} f(h_n) = \inf_{h \in U_{\text{ad}}} f(h). \quad (2.7.9)$$

Due to (2.7.2), (2.7.3), the sequence $\{h_n\}_{n=1}^{\infty}$ is bounded in $W_p^1(\Omega)$ with $p > 2$. Hence, we may choose a subsequence $\{h_m\}_{m=1}^{\infty}$ such that

$$h_m \to z \text{ weakly in } W_p^1(\Omega), \qquad h_m \to z \text{ strongly in } C(\overline{\Omega}). \quad (2.7.10)$$

From (2.7.8), (2.7.10), we deduce that

$$C \geq \liminf_{m \to \infty} \|h_m\|_{W_p^1(\Omega)} \geq \|z\|_{W_p^1(\Omega)}, \quad (2.7.11)$$

$$\check{h} \leq z \leq \hat{h}, \quad (2.7.12)$$

C being the number from (2.7.2).

Taking into account (2.7.4), we obtain from (2.7.8) and (2.7.10) that

$$\Psi(z) \leq 0. \quad (2.7.13)$$

By virtue of Theorem 2.5.2, (2.7.1), (2.7.8), and (2.7.10), we have

$$A_i \mu(z) \leq 0, \qquad i = 1, 2, \dots, k.$$

Therefore, from (2.7.11)–(2.7.13) we conclude that $z \in U_{\text{ad}}$. Finally, the relations (2.7.6), (2.7.9), (2.7.10) imply

$$\lim_{m \to \infty} f(h_m) = f(z) = \inf_{h \in U_{\text{ad}}} f(h).$$

Consequently, the function $h_0 = z$ is a solution to the problem (2.7.7).

2.7.2 Approximate solution of the problem (2.7.7)

Let $\{H_n\}_{n=1}^{\infty}$ be a sequence of finite-dimensional subspaces of $W_p^1(\Omega)$ satisfying the condition (2.6.45). Consider the problem: Find a function h_n such that

$$h_n \in H_n \cap U_{\text{ad}}, \qquad f(h_n) = \inf_{h \in H_n \cap U_{\text{ad}}} f(h). \tag{2.7.14}$$

We endow the set U_{ad} defined by the relations (2.7.1)–(2.7.5) with the topology generated by the $W_p^1(\Omega)$-one.

Theorem 2.7.2 *Let the conditions of Theorem* 2.7.1 *be satisfied and let* $\{H_n\}_{n=1}^{\infty}$ *be a sequence of finite-dimensional subspaces of* $W_p^1(\Omega)$ *meeting the condition* (2.6.45). *Also assume the existence of a sequence* $\{q_n\}_{n=1}^{\infty}$ *such that* $q_n \in \overset{\circ}{U}_{\text{ad}}$ *for each* n *and* $q_n \to h_0$ *in* $W_p^1(\Omega)$, h_0 *being a solution to the problem* (2.7.7). *Then, for every* n *sufficiently large (say, for* $n \geq l$*), the problem* (2.7.14) *has a solution. This problem has a solution for every* n *if the set* $H_n \cap U_{\text{ad}}$ *is not empty for each* n. *Moreover,*

$$\lim_{n \to \infty} f(h_n) = f(h_0) = \inf_{h \in U_{\text{ad}}} f(h). \tag{2.7.15}$$

One may choose a subsequence $\{h_m\}$ *of the sequence* $\{h_n\}$ *such that* $h_m \to h_0$ *weakly in* $W_p^1(\Omega)$.

Proof. Making use of the assumption about the existence of a sequence $\{q_n\} \subset \overset{\circ}{U}_{\text{ad}}$ convergent to h_0 in $W_p^1(\Omega)$, we deduce, just as in the proof of Theorem 2.6.2, that there exists a sequence $\{g_n\}_{n=l}^{\infty}$ such that

$$g_n \in H_n \cap U_{\text{ad}}, \tag{2.7.16}$$

$$g_n \to h_0 \quad \text{in } W_p^1(\Omega). \tag{2.7.17}$$

The set $H_n \cap U_{\text{ad}}$ equipped with the topology induced by the $W_p^1(\Omega)$-one is a compactum. Hence, (2.7.6) implies the problem (2.7.14) to be solvable for $n \geq l$, and for each n if the set $H_n \cap U_{\text{ad}}$ is not empty for all n.

By (2.7.6) and (2.7.17), we obtain

$$\lim_{n \to \infty} f(g_n) = f(h_0). \tag{2.7.18}$$

By (2.7.7), (2.7.14), (2.7.16), we have

$$f(g_n) \geq f(h_n) \geq f(h_0),$$

whence, noticing (2.7.18), we obtain (2.7.15). Thus, the sequence $\{h_n\}$ satisfies the conditions (2.7.8), (2.7.9). Now, the argument of Theorem 2.7.1 yields the existence of a subsequence $\{h_m\}$ of the sequence $\{h_n\}$ such that $h_m \to h_0$ weakly in $W_p^1(\Omega)$.

2.7.3 Second method of approximate solution of the problem (2.7.7)

The sequence $\{q_n\}$

In Subsec. 2.7.2, we assumed the existence of a sequence of interior points of the set U_{ad} convergent to the solution of the problem (2.7.7). However, the verification of this assumption is rather difficult. This is why we consider another approach that makes no use of that condition.

Let $t \in \mathbb{R}$, $1 \le t \le \frac{e_2}{\hat{h}}$ (e_2 being the number from (2.1.2)). Define a set

$$U^{(t)} = \{\, h \,|\, h \in W_p^1(\Omega),\ \|h\|_{W_p^1(\Omega)} \le tC,\ \check{h} \le h \le t\hat{h},$$
$$\Psi(h) \le t-1,\ A_i\mu(h) \le t-1\ (i = 1, 2, \dots, k)\,\}. \quad (2.7.19)$$

Here,

$$C,\ \check{h},\ \hat{h} \text{ are the positive numbers from (2.7.2)},\ p > 2. \quad (2.7.20)$$

Combining (2.7.2) and (2.7.19), we see that $U_{\text{ad}} \subset U^{(t)}$, $U_{\text{ad}} = U^{(1)}$.

Supposing the conditions of Theorem 2.7.1 to be fulfilled, we consider the problem: For a fixed t such that $t > 1$, $t\hat{h} \le e_2$ (e_2 being the positive number from (2.1.2)), find a function h_t satisfying

$$h_t \in U^{(t)}, \qquad f(h_t) = \inf_{h \in U^{(t)}} f(h). \quad (2.7.21)$$

Due to Theorem 2.7.1, the problem (2.7.21) has a solution, and the embedding $U_{\text{ad}} \subset U^{(t)}$ implies

$$f(h_t) \le f(h_0), \quad (2.7.22)$$

h_0 being the solution to the problem (2.7.7).

Define functions q_n by

$$q_n = t_n h_0, \quad (2.7.23)$$

where

$$1 < t_n < t, \qquad \lim_{n\to\infty} t_n = 1. \quad (2.7.24)$$

By (2.7.23) and (2.7.24),

$$q_n \to h_0 \text{ in } W_p^1(\Omega). \quad (2.7.25)$$

We endow the set $U^{(t)}$ with the topology induced by the $W_p^1(\Omega)$-one. Let us show that

$$q_n \in \overset{\circ}{U}{}^{(t)}, \qquad n \ge l, \quad (2.7.26)$$

where l is a positive number.

Since $h_0 \in U_{\text{ad}}$, from (2.7.2), (2.7.23), and (2.7.24), we get

$$\|q_n\|_{W_p^1(\Omega)} = t_n\|h_o\|_{W_p^1(\Omega)} < t\|h_o\|_{W_p^1(\Omega)} \le tC \qquad \forall n, \quad (2.7.27)$$

$$\check{h} \le h_0 < t_n h_0 = q_n < t\hat{h} \qquad \forall n, \quad (2.7.28)$$

C, $\check{h}$, and $\hat{h}$ being the numbers from (2.7.2). By using (2.7.2), (2.7.4), and (2.7.25), we infer the following estimate for n sufficiently large (say, $n \geq l$)

$$\Psi(q_n) < t - 1, \qquad n \geq l. \tag{2.7.29}$$

By (2.7.1) and Theorem 2.5.2, we obtain that

$$\left.\begin{array}{l} h \to A_i\mu(h) \text{ is a continuous mapping of } Y_p \text{ (endowed with} \\ \text{the topology generated by the } W_p^1(\Omega)\text{-weak one, } p > 2) \\ \text{into } \mathbb{R}\ (i = 1, 2, \dots, k). \end{array}\right\} \tag{2.7.30}$$

Since $A_i\mu(h_0) \leq 0$ and $t > 1$, the formulas (2.7.25) and (2.7.30) yield the following inequality to hold for l sufficiently large:

$$A_i\mu(q_n) < t - 1, \qquad n \geq l,\ i = 1, 2, \dots, k. \tag{2.7.31}$$

Due to (2.7.4), (2.7.29)–(2.7.31), for any $n \geq l$, there exists $\varepsilon_n > 0$ such that

$$\Psi(h) \leq t - 1, \qquad h \in d(\varepsilon_n, q_n),\ n \geq l, \tag{2.7.32}$$

$$A_i\mu(h) \leq t - 1, \qquad h \in d(\varepsilon_n, q_n),\ n \geq l,\ i = 1, 2, \dots, k, \tag{2.7.33}$$

$$d(\varepsilon_n, q_n) = \left\{ h \,|\, h \in W_p^1(\Omega),\ \|h - q_n\|_{W_p^1(\Omega)} \leq \varepsilon_n \right\}. \tag{2.7.34}$$

The embedding $W_p^1(\Omega) \to C(\overline{\Omega})$ being continuous for $p > 2$, from (2.7.27), (2.7.28), we deduce the existence of a ball $d(\lambda_n, q_n)$, $\lambda_n \leq \varepsilon_n$, defined by the relation (2.7.34) such that the following estimate holds:

$$\|h\|_{W_p^1(\Omega)} \leq tC, \qquad \check{h} \leq h \leq t\hat{h}, \qquad h \in d(\lambda_n, q_n),\ n \geq l. \tag{2.7.35}$$

Now, (2.7.19), (2.7.27), (2.7.32), (2.7.33), and (2.7.35) imply (2.7.26).

Approximate solution of the problem (2.7.7)

Let $\{H_n\}_{n=1}^{\infty}$ be a sequence of finite-dimensional subspaces of $W_p^1(\Omega)$ satisfying the condition (2.6.45) and

$$H_n \subset H_{n+1} \qquad \forall n. \tag{2.7.36}$$

Consider the problem: Find a function h_n such that

$$h_n \in H_n \cap U^{(t)}, \tag{2.7.37}$$

$$f(h_n) = \inf_{h \in H_n \cap U^{(t)}} f(h). \tag{2.7.38}$$

Theorem 2.7.3 *Let the conditions of Theorem 2.7.1 be fulfilled and let a set $U^{(t)}$ be determined by the relations (2.7.19), (2.7.20) when $t > 1$, $t\hat{h} \leq e_2$ (e_2 being the positive number from (2.1.2)). Further, let $\{H_n\}$ be a sequence of finite-dimensional subspaces of $W_p^1(\Omega)$ meeting the conditions (2.6.45), (2.7.36) and let h_0, h_t be solutions to the problems (2.7.7), (2.7.21), respectively. Assume that the set $H_1 \cap U^{(t)}$ is not empty.*

Then, for each n, the problem (2.7.37), (2.7.38) *has a solution such that*

$$f(h_t) \leq \lim_{n\to\infty} f(h_n) \leq f(h_0). \tag{2.7.39}$$

A subsequence $\{h_m\}$ of the sequence $\{h_n\}$ may be chosen such that $h_m \to g$ weakly in $W_p^1(\Omega)$, $g \in U^{(t)}$, and

$$\lim_{m\to\infty} f(h_m) = \lim_{n\to\infty} (h_n) = f(g). \tag{2.7.40}$$

Proof. By the hypothesis, the set $H_1 \cap U^{(t)}$ is not empty. Hence, (2.7.36) implies the set $H_n \cap U^{(t)}$ to be nonempty for each n. Taking into account (2.7.4) and (2.7.30), we easily see that the set $H_n \cap U^{(t)}$ is a compact topological space with the topology generated by the $W_p^1(\Omega)$-one. Now, (2.7.6) implies the solvability of the problem (2.7.37), (2.7.38) for each n.

Combining (2.7.21), (2.7.38), we get

$$f(h_t) \leq f(h_n) \qquad \forall n. \tag{2.7.41}$$

From (2.7.36) and (2.7.38)

$$f(h_{n+1}) \leq f(h_n) \qquad \forall n.$$

Therefore, by (2.7.41), we conclude the number sequence $\{f(h_n)\}_{n=1}^{\infty}$ to converge and

$$\lim_{n\to\infty} f(h_n) \geq f(h_t). \tag{2.7.42}$$

Using (2.6.45), (2.7.25), and (2.7.26), analogously to the proof of Theorem 2.6.2, one shows the existence of a sequence $\{w_n\}_{n=1}^{\infty}$ such that

$$w_n \in H_n \cap U^{(t)}, \tag{2.7.43}$$

$$w_n \to h_0 \quad \text{in } W_p^1(\Omega). \tag{2.7.44}$$

By (2.7.6) and (2.7.44), we have

$$\lim_{n\to\infty} f(w^n) = f(h_0). \tag{2.7.45}$$

Due to (2.7.38) and (2.7.43), we get

$$f(w_n) \geq f(h_n). \tag{2.7.46}$$

Now, (2.7.42), (2.7.45), and (2.7.46) yield (2.7.39). (2.7.37) implies the sequence $\{h_n\}_{n=1}^{\infty}$ to be bounded in $W_p^1(\Omega)$. Let us choose a subsequence $\{h_m\}_{m=1}^{\infty}$ such that

$$h_m \to g \qquad \text{weakly in } W_p^1(\Omega), \tag{2.7.47}$$

$$h_m \to g \qquad \text{strongly in } C(\overline{\Omega}). \tag{2.7.48}$$

By (2.7.4), (2.7.19), (2.7.30), (2.7.37), and (2.7.47), we get

$$\Psi(g) = \lim_{m\to\infty} \Psi(h_m) \leq t - 1, \tag{2.7.49}$$

$$A_i\mu(g) = \lim_{m\to\infty} A_i\mu(h_m) \leq t - 1, \qquad i = 1, 2, \ldots, k. \tag{2.7.50}$$

The relations (2.7.19), (2.7.37), and (2.7.48) imply

$$\check{h} \leq g \leq t\hat{h}. \tag{2.7.51}$$

By virtue of (2.7.19), (2.7.37), and (2.7.47), we obtain

$$tC \geq \liminf_{m\to\infty} \|h_m\|_{W_p^1(\Omega)} \geq \|g\|_{W_p^1(\Omega)}. \tag{2.7.52}$$

(2.7.49)–(2.7.52) yield $g \in U^{(t)}$. Finally, by (2.7.6), (2.7.47), we get (2.7.40), concluding the proof.

Remark 2.7.1 The function g and the functions h_n solving the problem (2.7.37), (2.7.38) belong to $U^{(t)}$ and, generally speaking, they do not belong to U_{ad}. But since the parameter t can be chosen as near to 1 as desired, "the getting of the functions h_n and g out of the region U_{ad}" can be made arbitrarily small.

2.7.4 Differentiation of the functionals $h \to A_i\mu(h)$ and necessary conditions of optimality

Theorem on the differentiation of functionals

Let A be a continuous functional defined on the space $\ell_{\infty,0}$. The functional A is said to satisfy at a point $\mu \in \ell_{\infty,0}$ the condition α with integer indices $j_1, j_2, \ldots, j_m$, if there exists a neighborhood $d(\mu)$ of the point μ in $\ell_{\infty,0}$ such that

$$A\eta = f_\mu(\eta_{j_1}, \eta_{j_2}, \ldots, \eta_{j_m}), \qquad \eta = \{\eta_j\}_{j=1}^{\infty} \in d(\mu),$$

f_μ being a continuously differentiable function defined on an open set Q_μ in $\mathbb{R}^m$ such that

$$(\eta_{j_1}, \eta_{j_2}, \ldots, \eta_{j_m}) \in Q_\mu, \qquad \eta \in d(\mu).$$

We endow the set G defined in (2.2.44) with the topology generated by the $W_p^1(\Omega)$-one. Provided the hypothesis of Theorem 2.5.2 is fulfilled, define a functional Ψ on G by

$$\Psi(h) = A\mu(h). \tag{2.7.53}$$

From the continuity of the functional A and Theorem 2.5.2, we conclude Ψ to be a continuous functional on G. The following theorem on the differentiability of this functional holds.

Theorem 2.7.4 *Let the conditions of Theorem* 2.5.1 *be satisfied and let* (2.2.52), (2.5.12) *hold. Also, let $h \to b_h$ be a Fréchet continuously differentiable mapping of G into $\mathcal{N}_2$.*

Assume that $\tilde{h} \in G$ and a continuous functional A meets at the point $\mu(\tilde{h})$ the condition α with integer indices $j_1, j_2, \dots, j_m$, and $\mu_{j_1}(\tilde{h}), \mu_{j_2}(\tilde{h}), \dots, \mu_{j_m}(\tilde{h})$ are simple nonzero eigenvalues. Then, there exists a neighborhood $d(\tilde{h})$ of the point $\tilde{h}$ in G such that the functional Ψ defined by (2.7.53) *is Fréchet continuously differentiable in $d(\tilde{h})$ and its Fréchet derivative $\Psi'(h)$ at a point $h \in d(\tilde{h})$ is given by the formula*

$$\Psi'(h)q = \sum_{s=1}^{m} \frac{\partial f_{\mu(\tilde{h})}}{\partial \eta_{j_s}} (\mu_{j_1}(h), \mu_{j_2}(h), \dots, \mu_{j_m}(h)) \\ \times (b'_h q - \mu_{j_s}(h) a'_h q) (u_{j_s}(h), u_{j_s}(h)), \qquad q \in W_p^1(\Omega), \quad (2.7.54)$$

b'_h, a'_h being the Fréchet derivatives of the functions $h \to a_h$ and $h \to b_h$ at point h, $u_{j_s}(h)$ the eigenfunctions of the problem (2.5.2) *corresponding to the eigenvalues $\mu_{j_s}(h)$ and satisfying the condition $a_h(u_{j_s}(h), u_{j_s}(h)) = 1$.*

Proof. The inequality (2.2.57) and the argument below it yield $h \to a_h$ to be a Fréchet continuously differentiable mapping of G into $\mathcal{L}_2(V, V; \mathbb{R})$, and its Fréchet derivative a'_h at point $h \in G$ to be given by the relation

$$a'_h q(u, v) = \iint_{\Omega} \sum_{i,j=1}^{k} \frac{da_{ij}}{dt}(h) q (P_i u)(P_j v)\, dx\, dy, \\ q \in W_p^1(\Omega), \ u, v \in V.$$

Now, Theorem 1.11.3 implies the existence of a neighborhood $d_1(\tilde{h})$ of the point $\tilde{h}$ in G such that the functionals $h \to \mu_{j_s}(h)$, $s = 1, 2, \dots, m$, are Fréchet continuously differentiable in $d_1(\tilde{h})$ and their Fréchet derivatives $\mu'_{j_s}(h)$ at a point $h \in d_1(\tilde{h})$ are given by

$$\mu'_{j_s}(h)q = (b'_h q - \mu_{j_s}(h) a'_h q) (u_{j_s}(h), u_{j_s}(h)), \qquad q \in W_p^1(\Omega), \ s = 1, 2, \dots, m.$$

Further, the well-known result on the differentiation of a composite function (see Subsec. 1.9.1, Property 4) yields the existence of a neighborhood $d(\tilde{h})$ of the point $\tilde{h}$ in G such that the functional Ψ from (2.7.53) is Fréchet continuously differentiable in $d(\tilde{h})$, and its Fréchet derivatives $\Psi'(h)$ at a point $h \in d(\tilde{h})$ is given by the formula (2.7.54).

By using Theorem 2.7.4, one can find the Fréchet derivatives of the functionals $h \to A_i\mu(h)$ when this theorem is applicable. These derivatives may be used when one solves the finite-dimensional problem (2.7.14).

Restatement of the problem and necessary conditions of optimality

The restrictions $\check{h} \leq h \leq \hat{h}$ lead to the functionals φ_1 and φ_2 defined by (2.2.33) which are not Fréchet differentiable. As noted in Subsec. 2.2.3, by introducing a corresponding ε-net, one can approximate the restrictions $\check{h} \leq h \leq \hat{h}$ by those of the form $Q_i(h) \leq 0$, as precisely as desired (Q_i being the Fréchet differentiable functionals determined by the formulas (2.2.37)). Now, replace the set U_{ad} from (2.7.2) by the following one

$$\tilde{U}_{\mathrm{ad}} = \{\, h \,|\, h \in W_p^1(\Omega),\ \|h\|_{W_p^1(\Omega)} \leq C,\ Q_i(h) \leq 0,\ i = 1, 2, \ldots, 2r, \\ \Psi(h) \leq 0,\ A_j\mu(h) \leq 0,\ j = 1, 2, \ldots, k \,\}. \qquad (2.7.55)$$

Given a goal functional f, consider the problem of finding a function h_0 such that

$$h_0 \in \tilde{U}_{\mathrm{ad}}, \qquad f(h_0) = \inf_{h \in \tilde{U}_{\mathrm{ad}}} f(h). \qquad (2.7.56)$$

Using the argument from the proof of Theorem 2.7.1, we establish the following fact.

Theorem 2.7.5 *Let a_h be a bilinear form on $V \times V$ defined by the relations* (2.1.1), (2.1.4), (2.1.6), (2.1.7), (2.1.10), (2.2.52). *Assume that $h \to b_h$ is a continuously Fréchet differentiable mapping of G into $\mathcal{N}_2$, and* (2.5.12) *holds. Suppose that a nonempty set $\tilde{U}_{\mathrm{ad}}$ is determined by the relations* (2.7.55), (2.2.37), (2.7.1), (2.7.3), (2.7.4), *and the goal functional f meets the condition* (2.7.6). *Then, there exists a solution h_0 to the problem* (2.7.56).

Let $J = (1, 2, \ldots, k)$ and let $J_0 = (i_1, i_2, \ldots, i_l)$ be a subset of J such that

$$A_j\mu(h_0) = 0, \qquad j \in J_0,$$
$$A_j\mu(h_0) < 0, \qquad j \in J \setminus J_0.$$

Introduce the following notations:

$$g_0(h) = f(h), \qquad g_1(h) = \|h\|_{W_p^1(\Omega)} - C,$$
$$g_i(h) = Q_{i-1}(h), \quad i = 2, 3, \ldots, 2r+1, \qquad g_{2r+2}(h) = \Psi(h), \qquad (2.7.57)$$
$$g_{2r+2+s}(h) = A_{i_s}\mu(h), \qquad s = 1, 2, \ldots, l,\ i_s \in J_0,$$

where $h \in G$.

Theorem 2.7.6 *Let the conditions of Theorem 2.7.5 be fulfilled and let $d(h_0)$ be some neighborhood of the function h_0 in G. Suppose that the functionals f and Ψ are Fréchet differentiable on $d(h_0)$. Also assume the functionals A_j, $j \in J_0$, meet at a point $\mu(h_0)$ the condition α with indices $i_1(j), i_2(j), \ldots, i_{m(j)}(j)$, and the corresponding eigenvalues $\mu_{i_1(j)}(h_0), \mu_{i_2(j)}(h_0), \ldots, \mu_{i_{m(j)}(j)}(h_0)$ are simple ones* *and do not equal zero. Then, there exist real numbers* $\lambda_0, \lambda_1, \dots, \lambda_{2r+l+2}$ *that do not vanish simultaneously and such that*

$$\lambda_0 \geq 0,\ \lambda_1 \geq 0, \dots, \lambda_{2r+l+2} \geq 0, \qquad \sum_{i=0}^{2r+l+2} \lambda_i g_i'(h_0) = 0,$$
$$\lambda_i g_i(h_0) = 0, \qquad i = 1, 2, \dots, 2r + l + 2,$$

$g_i'(h_0)$ *being the Fréchet derivatives of the functionals* g_i *at a point* h_0 *(see* (2.7.57)*).*

If there exists an element $v \in W_p^1(\Omega)$ *such that* $g_i'(h_0)v < 0$ *for* $i = 1, 2, \dots, 2r + l + 2$*, then* $\lambda_0 \neq 0$ *and we can take* $\lambda_0 = 1$*.*

Proof. By virtue of Theorem 2.7.4, the functionals g_{2r+2+s}, $s = 1, 2, \dots, l$, defined in (2.7.57) are continuously Fréchet differentiable in some neighborhood of the point h_0 in G.

The functional g_1 is also continuously differentiable in this neighborhood (see Theorem 1.10.1). From (2.2.37), (2.7.57), and the embedding theorem we have $g_2, g_3, \dots, g_{2r+1}$ to be affine continuous functionals in $W_p^1(\Omega)$, so they are continuously differentiable in G.

Since by the hypothesis of the theorem the functionals f and Ψ are Fréchet differentiable in $d(h_0)$, due to (2.7.57) the functionals g_0 and g_{2r+2} are Fréchet differentiable in $d(h_0)$.

Next, use Theorem 1.12.1. Set $X = W_p^1(\Omega)$, $U = G$, $Y = \{0\} \subset \mathbb{R}$, and let $Fh = 0$ for all $h \in G$. Then, Theorem 2.7.6 is a consequence of Theorem 1.12.1.

2.8 The basic optimal control problem

2.8.1 Setting of the problem. Existence theorem

Let the set of admissible controls be of the form

$$U_{\text{ad}} = \left\{ h \,\middle|\, h \in W_p^1(\Omega),\ \|h\|_{W_p^1(\Omega)} \leq C,\ \check{h} \leq h \leq \hat{h},\ \Phi_k(h) \leq 0\ (k = 1, 2, \dots, q) \right\}. \tag{2.8.1}$$

Here

$p > 2$, C, $\check{h}$, $\hat{h}$ are positive numbers such that $e_1 < \check{h} < \hat{h} < e_2$, where e_1, e_2 are the positive numbers from (2.1.2); (2.8.2)

$h \to \Phi_k(h)$ is a continuous mapping of Y_p (equipped with the topology induced by the $W_p^1(\Omega)$-weak one) into $\mathbb{R}$ when $p > 2$ $(k = 1, 2, \dots, q)$. (2.8.3)

We suppose that

$$\text{the set } U_{\text{ad}} \text{ is not empty.} \tag{2.8.4}$$

Let f be a goal functional satisfying the following condition:

$$\left.\begin{array}{l} h \to f(h) \text{ is a continuous mapping of } Y_p \text{ (equipped with} \\ \text{the topology induced by the } W_p^1(\Omega)\text{-weak one) into } \mathbb{R}. \end{array}\right\} \tag{2.8.5}$$

The optimal control problem consists in finding a function h_0 such that

$$h_0 \in U_{\text{ad}}, \qquad f(h_0) = \inf_{h \in U_{\text{ad}}} f(h). \tag{2.8.6}$$

Theorem 2.8.1 *Let a set* U_{ad} *be determined by the relations* (2.8.1)–(2.8.4), *and let a goal functional satisfy the condition* (2.8.5). *Then, there exists a solution to the problem* (2.8.6).

Proof. By virtue of (2.8.4), there exists a sequence $\{h_n\}$ such that

$$h_n \in U_{\text{ad}} \qquad \forall n, \tag{2.8.7}$$

$$\lim_{n \to \infty} f(h_n) = \inf_{h \in U_{\text{ad}}} f(h). \tag{2.8.8}$$

By the definition of the set U_{ad}, the sequence $\{h_n\}$ is bounded in $W_p^1(\Omega)$. So, we can choose a subsequence $\{h_m\}$ such that

$$h_m \to g \qquad \text{weakly in } W_p^1(\Omega), \tag{2.8.9}$$

g being an element of $W_p^1(\Omega)$. Then

$$C \geq \liminf_{m \to \infty} \|h_m\|_{W_p^1(\Omega)} \geq \|g\|_{W_p^1(\Omega)}, \tag{2.8.10}$$

C being the constant from (2.8.1).

By using (2.8.9) and the compactness of the embedding $W_p^1(\Omega) \to C(\overline{\Omega})$ as $p > 2$, we conclude that

$$\check{h} \leq g \leq \hat{h}. \tag{2.8.11}$$

Further, the relations (2.8.1), (2.8.3), (2.8.7), and (2.8.9) yield

$$\lim_{m \to \infty} \Phi_k(h_m) = \Phi_k(g) \leq 0, \qquad k = 1, 2, \ldots, q. \tag{2.8.12}$$

The formulas (2.8.10)–(2.8.12) imply that $g \in U_{\text{ad}}$, and from (2.8.5), (2.8.8), and (2.8.9), we deduce

$$f(g) = \lim_{m \to \infty} f(h_m) = \inf_{h \in U_{\text{ad}}} f(h).$$

So, the function $h_0 = g$ is a solution to the problem (2.8.6).

2.8.2 Approximate solution of the problem (2.8.6)

Expansion of the set U_{ad}

Let $t \in \mathbb{R}$ and $1 \leq t \leq \min(2 - \frac{e_1}{\check{h}}, \frac{e_2}{\hat{h}})$, where e_1 and e_2 are the numbers from (2.1.2).

Define a set

$$U^{(t)} = \big\{ h \,|\, h \in W_p^1(\Omega),\ \|h\|_{W_p^1(\Omega)} \leq tC,\ \check{h}(2-t) \leq h \leq t\hat{h},\\ \Phi_k(h) \leq t-1 \ \ (k = 1, 2, \dots, q) \big\}. \quad (2.8.13)$$

Here, $\Phi_k(h)$ are the functionals defined in Subsec. 2.8.1.

Comparing (2.8.1) and (2.8.13) gives

$$U_{\mathrm{ad}} \subset U^{(t)}, \qquad U_{\mathrm{ad}} = U^{(1)},$$

and the argument from the proof of Theorem 2.8.1 implies the following statement.

Theorem 2.8.2 *Let a set $U^{(t)}$ be determined by the formulas* (2.8.13), (2.8.2), (2.8.3), *and*

$$1 < t \leq \min\left(2 - \frac{e_1}{\check{h}}, \frac{e_2}{\hat{h}}\right). \quad (2.8.14)$$

and let the set $U^{(t)}$ be nonempty. Let a goal functional f satisfy the condition (2.8.5). *Then, there exists a function h_t such that*

$$h_t \in U^{(t)}, \qquad f(h_t) = \inf_{h \in U^{(t)}} f(h). \quad (2.8.15)$$

Remark 2.8.1 We introduce the right-hand side of the estimate (2.8.14) to make the inequality $e_1 \leq h \leq e_2$ hold for all $h \in U^{(t)}$, e_1 and e_2 being the positive numbers from (2.1.2), i.e., in order to get the inclusion $U^{(t)} \subset Y_p$, which is needed because the functionals Φ_k and f are defined on Y_p.

Endow the set $U^{(t)}$ with the topology induced by the strong topology of the space $W_p^1(\Omega)$. The following statement is valid.

Lemma 2.8.1 *Let a set U_{ad} be determined through the relations* (2.8.1)–(2.8.4), *and a set $U^{(t)}$ through the relations* (2.8.13), (2.8.14), (2.8.2), *and* (2.8.3). *Then,* $U_{\mathrm{ad}} \subset \overset{\circ}{U}{}^{(t)}$.

Proof. Let us show that $g \in \overset{\circ}{U}{}^{(t)}$ for an arbitrary element g from U_{ad}. For any $h \in W_p^1(\Omega)$, we have

$$\|h\|_{W_p^1(\Omega)} \leq \|g\|_{W_p^1(\Omega)} + \|h - g\|_{W_p^1(\Omega)}. \quad (2.8.16)$$

Since $g \in U_{\text{ad}}$, we have $\|g\|_{W_p^1(\Omega)} \leq C$, where C is the constant from (2.8.1). Hence, (2.8.16) implies the existence of a number $\gamma_1 > 0$ such that

$$\|h\|_{W_p^1(\Omega)} \leq tC, \qquad h \in d(\gamma_1, g), \tag{2.8.17}$$

where

$$d(\gamma_1, g) = \{\, h \,|\, h \in W_p^1(\Omega),\ \|h - g\|_{W_p^1(\Omega)} \leq \gamma_1 \,\}.$$

As the embedding $W_p^1(\Omega) \to C(\overline{\Omega})$ is continuous for $p > 2$ and $\check{h} \leq g \leq \hat{h}$, there exists a number $\gamma_2 > 0$ such that

$$\check{h}(2-t) \leq h \leq t\hat{h}, \qquad h \in d(\gamma_2, g). \tag{2.8.18}$$

The function $h \to \Phi_k(h)$ is continuous in the topology generated by the $W_p^1(\Omega)$-weak one, so that it is continuous in the topology generated by the strong one of $W_p^1(\Omega)$. Thus, from the estimate $\Phi_k(g) \leq 0$, we deduce the existence of a number $\gamma_3 > 0$ such that

$$\Phi_k(h) \leq t - 1, \qquad h \in d(\gamma_3, g). \tag{2.8.19}$$

Setting $\gamma = \min(\gamma_1, \gamma_2, \gamma_3)$, taking into account (2.8.17)–(2.8.19), we get $d(\gamma, g) \subset U^{(t)}$. Hence, $g \in \overset{\circ}{U}{}^{(t)}$ and the lemma is proved.

Approximate solution of the basic problem

Let $\{H_n\}$ be a sequence of finite-dimensional subspaces of $W_p^1(\Omega)$ satisfying the conditions (2.6.45) and (2.7.36). Let us consider the problem of finding a function h_n such that

$$h_n \in H_n \cap U^{(t)}, \tag{2.8.20}$$

$$f(h_n) = \inf_{h \in H_n \cap U^{(t)}} f(h). \tag{2.8.21}$$

Theorem 2.8.3 *Let the assumptions of Theorems* 2.8.1 *and* 2.8.2 *be fulfilled, let* $\{H_n\}$ *be a sequence of finite-dimensional subspaces of* $W_p^1(\Omega)$ *satisfying the conditions* (2.6.45) *and* (2.7.36), *and let* h_0, h_t *be solutions to the problems* (2.8.6), (2.8.15), *respectively. Suppose that the set* $H_1 \cap U^{(t)}$ *is not empty. Then, for any* n, *the problem* (2.8.20), (2.8.21) *has a solution* h_n *such that*

$$f(h_t) \leq \lim_{n\to\infty} f(h_n) \leq f(h_0). \tag{2.8.22}$$

One can choose from the sequence $\{h_n\}$ *a subsequence* $\{h_m\}$ *such that* $h_m \to \tilde{h}$ *weakly in* $W_p^1(\Omega)$, $\tilde{h} \in U^{(t)}$, *and*

$$\lim_{m\to\infty} f(h_m) = \lim_{n\to\infty} f(h_n) = f(\tilde{h}). \tag{2.8.23}$$

Proof. The set $H_1 \cap U^{(t)}$ is not empty, by the hypothesis. From here and from (2.7.36), we deduce $H_n \cap U^{(t)}$ to be nonempty for all n. The set $H_n \cap U^{(t)}$ is a compact topological space endowed with the topology generated by the topology of $W_p^1(\Omega)$. This conclusion together with (2.8.5) yields the problem (2.8.20), (2.8.21) to be solvable for any n. The relations (2.8.15) and (2.8.21) imply

$$f(h_t) \leq f(h_n) \qquad \forall n. \tag{2.8.24}$$

From (2.7.36) and (2.8.21), we get

$$f(h_{n+1}) \leq f(h_n).$$

Combining this inequality with (2.8.24), we deduce the sequence of numbers $\{f(h_n)\}_{n=1}^{\infty}$ to be convergent and

$$\lim_{n\to\infty} f(h_n) \geq f(h_t). \tag{2.8.25}$$

Due to Lemma 2.8.1, $h_0 \in \overset{\circ}{U}{}^{(t)}$. From this and from (2.6.45), we derive the existence of a sequence $\{g_n\}$ such that

$$g_n \in H_n \cap U^{(t)}, \qquad \lim_{n\to\infty} \|g_n - h_0\|_{W_p^1(\Omega)} = 0. \tag{2.8.26}$$

Hence,

$$\lim_{n\to\infty} f(g_n) = f(h_0). \tag{2.8.27}$$

The relations (2.8.20), (2.8.21), and (2.8.26) yield

$$f(g_n) \geq f(h_n).$$

Therefore, by (2.8.25) and (2.8.27), we get (2.8.22). By (2.8.13) and (2.8.20), we see that the sequence $\{h_n\}$ is bounded in $W_p^1(\Omega)$. Choose a subsequence $\{h_m\}$ from it such that

$$h_m \to \tilde{h} \qquad \text{weakly in } W_p^1(\Omega), \tag{2.8.28}$$

$$h_m \to \tilde{h} \qquad \text{strongly in } C(\overline{\Omega}). \tag{2.8.29}$$

Repeating the above argument and making use of (2.8.3), (2.8.13), (2.8.20), (2.8.28), and (2.8.29), we easily see that $\tilde{h} \in U^{(t)}$. Now, (2.8.5) and (2.8.28) yield (2.8.23), concluding the proof.

2.9 The combined problem

Let us consider a combined problem as an application of the basic problem. Suppose that

$$\left.\begin{array}{l} u, v \to a_h^{(i)}(u,v) \text{ is a bilinear form on } V \times V \text{ depending on} \\ \text{parameter } h \in Y_p \text{ that is determined by the relations} \\ \text{(2.1.1), (2.1.4)–(2.1.7), and (2.1.10) } (i = 1, 2, 3); \end{array}\right\} \tag{2.9.1}$$

$$\left.\begin{array}{l} u, v \to b_h(u,v) \text{ is a bilinear, symmetric, continuous form} \\ \text{on } X_1 \times X_1 \text{ depending on parameter } h \in Y_p; \end{array}\right\} \tag{2.9.2}$$

$$\left.\begin{array}{l} u, v \to G_{h,w}(u,v) \text{ is a bilinear, symmetric, continuous} \\ \text{form on } X_2 \times X_2 \text{ depending on parameter } (h,w) \in Y_p \times V. \end{array}\right\} \tag{2.9.3}$$

Here

$$\left.\begin{array}{l} X_i \text{ is a Banach space } (i = 1,2),\ V \subset X_i, \text{ the embedding} \\ V \to X_i \text{ is compact.} \end{array}\right\} \tag{2.9.4}$$

We assume also that

$$\left.\begin{array}{l} h \to b_h \text{ is a continuous mapping of } Y_p \text{ (endowed with the} \\ \text{topology induced by the } W_p^1(\Omega)\text{-weak one) into} \\ \mathcal{L}_2(X_1, X_1, \mathbb{R}); \end{array}\right\} \tag{2.9.5}$$

$$\left.\begin{array}{l} h, w \to G_{h,w} \text{ is a continuous mapping of } Y_p \times V \text{ (equipped} \\ \text{with the topology generated by the product of the} \\ W_p^1(\Omega)\text{-weak topology and of the } V\text{-strong one) into} \\ \mathcal{L}_2(X_2, X_2, \mathbb{R}). \end{array}\right\} \tag{2.9.6}$$

Fix an element f_1 from V^*. For a given element $h \in Y_p$, consider the following three problems.

First problem: Find a function u_h such that

$$u_h \in V, \qquad a_h^{(1)}(u_h, v) = (f_1, v), \qquad v \in V.$$

Second problem (*the eigenvalue problem*): Find $(\lambda_i, u_i) \in \mathbb{R} \times V$ such that

$$\lambda_i a_h^{(2)}(u_i, v) = b_h(u_i, v), \qquad v \in V.$$

Third problem: Find $(\mu_i, u_i) \in \mathbb{R} \times V$ such that

$$\mu_i a_h^{(3)}(u_i, v) = G_{h,u_h}(u_i, v), \qquad v \in V,$$

u_h being a solution to the first problem.

We are going to set control by the function h. So, let us define a set of admissible controls by

$$U_{\text{ad}} = \big\{ h \,\big|\, h \in W_p^1(\Omega),\ \|h\|_{W_p^1(\Omega)} \le C,\ \check{h} \le h \le \hat{h},$$
$$\Psi_1(h, u_h) \le 0,\ \Psi_2(\lambda) \le 0,\ \Psi_3(\mu) \le 0 \big\}. \tag{2.9.7}$$

Here,

$$\left.\begin{array}{l} p > 2,\ C,\ \check{h},\ \hat{h} \text{ are positive numbers such that} \\ e_1 < \check{h} < \hat{h} < e_2, \text{ where } e_1 \text{ and } e_2 \text{ are the positive numbers} \\ \text{from (2.1.2);} \end{array}\right\} \tag{2.9.8}$$

$$\left.\begin{array}{l} h, u \to \Psi_1(h,u) \text{ is a continuous mapping of } Y_p \times V \\ \text{(equipped with the topology generated by the product of} \\ \text{the } W_p^1(\Omega)\text{-weak topology and of the } V\text{-strong one) into } \mathbb{R}; \end{array}\right\} \tag{2.9.9}$$

$$\Psi_2,\ \Psi_3 \text{ are continuous mappings of the space } \ell_{\infty,0} \text{ into } \mathbb{R}, \} \tag{2.9.10}$$

$\lambda = \{\lambda_1, \lambda_2, \lambda_3, \dots\}$ and $\mu = \{\mu_1, \mu_2, \mu_3, \dots\}$ are the eigenvalues of the second and third problems, respectively. The function u_h appearing as an argument of functional Ψ_1 (see (2.9.7)) is a solution to the first problem for a fixed $f_1 \in V^*$.

About the goal functional f we suppose that

$$\left.\begin{array}{l} h \to f(h) \text{ is a continuous mapping of } Y_p \text{ (endowed with} \\ \text{the topology induced by the } W_p^1(\Omega)\text{-weak one) into } \mathbb{R}. \end{array}\right\} \tag{2.9.11}$$

Consider the following optimal control problem: Find a function h_0 such that

$$h_0 \in U_{\text{ad}}, \qquad f(h_0) = \inf_{h \in U_{\text{ad}}} f(h). \tag{2.9.12}$$

Notice that the problem (2.9.12) is an abstract analog of the problem of optimization of a plate (shell) under restrictions on the strength, stability, and free oscillation frequencies (cf. Subsec. 5.3.4).

Theorem 2.9.1 *Let the relations* (2.9.1)–(2.9.6) *be fulfilled and let a nonempty set* U_{ad} *be defined by the relations* (2.9.7)–(2.9.10). *Assume that the goal functional satisfies the condition* (2.9.11). *Then, the problem* (2.9.12) *has a solution.*

Proof. Define a functional $\Phi_1(h)$ as follows

$$h \to \Phi_1(h) = \Psi_1(h, u_h),$$

where u_h is a solution to the first problem. By (2.9.9) and Lemma 2.1.1, we conclude the mapping $h \to \Phi_1(h)$ to satisfy the condition (2.8.3). Further, introduce a functional $\Phi_2(h)$

$$h \to \Phi_2(h) = \Psi_2(\lambda(h)),$$

where $\lambda(h)$ is the set of the eigenvalues of the second problem, depending on parameter $h \in Y_p$. Combining (2.9.2), (2.9.5), (2.9.10), and Theorem 2.5.2, we obtain that the mapping $h \to \Phi_2(h)$ meets the condition (2.8.3).

Define a functional

$$h \to \Phi_3(h) = \Psi_3(\mu(h)),$$

where $\mu(h)$ is the set of the eigenvalues of the third problem.

Using Lemma 2.1.1, Theorem 2.5.2, and the relations (2.9.3), (2.9.6), (2.9.10), it is easy to see that the mapping $h \to \Phi_3(h)$ satisfies the condition (2.8.3). With the notations introduced, the problem (2.9.12) is reduced to the problem (2.8.6). Now, Theorem 2.8.1 implies that (2.9.12) has a solution.

It should be noted that, provided the set U_{ad} is determined by the relations (2.9.7)–(2.9.10), an approximate solution to the problem (2.9.12) may be constructed by using the technique of Subsec. 2.8.2.

2.10 Optimal control problem for the case when the state of the system is characterized by a set of functions

2.10.1 Setting of the problem

Until now we studied optimal control problems when the control h belonged to the set Y_p determined by the relations (2.1.2) and (2.1.3), and the state of the system u_h was a solution to the problem (2.1.12). The function f in (2.1.12) was supposed to be fixed and given. In this case, to every control $h \in Y_p$ there corresponds a function $u_h \in V$ that determines the state of the system.

We will now study control problems when the function $f \in V^*$ is not fixed, and instead of f we will be given some set $T \subset V^*$. This problem is an abstract analog of the problem of optimization of a construction for a class of loads (T corresponding to this class).

Provided a form a_h is determined by the relations (2.1.1), (2.1.4)–(2.1.6) and the inequalities (2.1.7) and (2.1.10) hold, for any element $h \in Y_p$ and for any function $g \in V^*$, there exists a unique function $u(h, g)$ such that

$$u(h,g) \in V, \qquad a_h(u(h,g), v) = (g, v), \qquad v \in V. \tag{2.10.1}$$

Now, for every control $h \in Y_p$, the state of the system Q_h coincides with the set of functions $u(h, g)$, where g runs through the set T, i.e.,

$$Q_h = \{u(h,g)\}_{g \in T}.$$

Here, one can reason in another way. Since, for a given element $h \in Y_p$, the bilinear form a_h is symmetric, continuous, and coercive on $V \times V$, the Riesz theorem implies that problem (2.10.1) has a unique solution $u(h, g)$ and well defined is an operator $N_h \in \mathcal{L}(V^*, V)$ such that

$$N_h g = u(h, g). \tag{2.10.2}$$

Then, for a given element $h \in Y_p$, the state of the system is the image of the set T under the mapping N_h, i.e.,

$$Q_h = N_h\{T\} = \{u(h,g)\}_{g \in T}. \tag{2.10.3}$$

Introduce the notation

$$B = \{\, h \mid h \in C(\overline{\Omega}),\ e_1 \le h \le e_2 \,\}, \tag{2.10.4}$$

e_1, e_2 being the positive numbers from (2.1.2). By virtue of the embedding theorem, $Y_p \subset B$ when $p > 2$ (the set Y_p defined by (2.1.2)).

Let mappings $A_i \colon B \times V \times V^* \to \mathbb{R}$ be defined such that

$$\left.\begin{array}{l} h, u, g \to A_i(h,u,g) \text{ is a continuous mapping of} \\ B \times V \times V^* \text{ (equipped with the topology generated by the} \\ \text{product of the topologies of } C(\overline{\Omega}),\ V,\ V^*) \text{ into } \mathbb{R} \\ (i = 1, 2, \dots, q). \end{array}\right\} \tag{2.10.5}$$

Define mappings $\Gamma_i\colon B \times V^* \to \mathbb{R}$ as follows:

$$h, g \to \Gamma_i(h, g) = A_i(h, u(h, g), g). \tag{2.10.6}$$

Here, $i = 1, 2, \ldots, q$, and $u(h, g)$ is a solution to the problem (2.10.1). Further, assume that

$$T \text{ is a compact set in } V^*, \tag{2.10.7}$$

and the functions $\Psi_i\colon B \to \mathbb{R}$ have the form

$$h \to \Psi_i(h) = \sup_{g \in T} \Gamma_i(h, g), \qquad i = 1, 2, \ldots, q. \tag{2.10.8}$$

Define a set of admissible controls U_{ad} as

$$U_{\mathrm{ad}} = \big\{\, h \,|\, h \in W_p^1(\Omega),\ \|h\|_{W_p^1(\Omega)} \le C,\ \check{h} \le h \le \hat{h},\ \Psi_i(h) \le 0\ (i = 1, 2, \ldots, q) \,\big\}. \tag{2.10.9}$$

Here,

$$\left.\begin{array}{l} p > 2,\ C,\ \check{h},\ \hat{h} \text{ are positive numbers such that} \\ e_1 < \check{h} < \hat{h} < e_2, \text{ where } e_1,\ e_2 \text{ are the positive numbers} \\ \text{from (2.1.2).} \end{array}\right\} \tag{2.10.10}$$

Introduce a goal functional $h \to f(h)$ such that

$$\left.\begin{array}{l} h \to f(h) \text{ is a continuous mapping of } Y_p \text{ (endowed with} \\ \text{the topology generated by the } W_p^1(\Omega)\text{-weak one) into } \mathbb{R}. \end{array}\right\} \tag{2.10.11}$$

The optimal control problem consists in finding a function h_0 such that

$$h_0 \in U_{\mathrm{ad}}, \qquad f(h_0) = \inf_{h \in U_{\mathrm{ad}}} f(h). \tag{2.10.12}$$

Notice that the basic difference between the optimal control problem (2.10.12) and the problems investigated in Section 2.2 is that now the state of the system is determined not by the single function u_h, but by the set Q_h (see (2.10.3)), which is not, in general, a countable one, and that the set of admissible controls U_{ad} is determined by Q_h.

2.10.2 The existence theorem

Theorem 2.10.1 *Let a_h be a bilinear form on $V \times V$ determined by the relations* (2.1.1), (2.1.4)–(2.1.7), *and* (2.1.10). *Suppose that the state of the system Q_h is defined by the formulas* (2.10.1)–(2.10.3), *and the set of admissible controls U_{ad} is defined by the relations* (2.10.4)–(2.10.10) *and is nonempty. Let also the goal functional $f(h)$ meet the condition* (2.10.11). *Then, the problem* (2.10.12) *has a solution.*

To prove Theorem 2.10.1 we will need two lemmas.

Lemma 2.10.1 *Let a_h be a bilinear form on $V \times V$ determined by the relations (2.1.1), (2.1.4)–(2.1.7), and (2.1.10). Then, the function $h, g \to u(h, g)$ defined by the solution to the problem (2.10.1) is a continuous mapping of $B \times V^*$ (endowed with the topology generated by the product of the topologies of the spaces $C(\overline{\Omega})$ and V^*) into V.*

Proof. Let $\{h_n, g_n\}_{n=1}^{\infty} \subset B \times V^*$ and

$$h_n \to h_0 \qquad \text{in } C(\overline{\Omega}), \tag{2.10.13}$$

$$g_n \to g_0 \qquad \text{in } V^*. \tag{2.10.14}$$

Then, $(h_0, g_0) \in B \times V^*$, and we need to prove that

$$u(h_n, g_n) \to u(h_0, g_0) \qquad \text{in } V \text{ as } n \to \infty. \tag{2.10.15}$$

Taking into account (2.1.1), (2.1.4), and (2.1.5), we obtain

$$|a_{h_n}(u, u) - a_{h_0}(u, u)| \le \alpha_n c \|u\|_V^2, \qquad u \in V. \tag{2.10.16}$$

Here, c is a positive number and

$$\alpha_n = \max_{i,j} \max_{(x,y) \in \overline{\Omega}} |a_{ij}(h_n) - a_{ij}(h_0)|. \tag{2.10.17}$$

Combining (2.1.5), (2.10.13), and Theorem 1.3.10, we deduce that

$$\lim_{n \to \infty} \alpha_n = 0. \tag{2.10.18}$$

For all $h \in B$, the bilinear form a_h satisfies the following condition (cf. Subsec. 1.5.2)

$$a_h(u, v) = (A_h u, v), \qquad u, v \in V, \tag{2.10.19}$$

where $A_h \in \mathcal{L}(V, V^*)$, and the problem (2.10.1) is equivalent to the following one:

$$u(h, g) \in V, \qquad A_h u(h, g) = g. \tag{2.10.20}$$

Due to (2.1.9), the operator A_h is selfadjoint, i.e.,

$$(A_h u, v) = (A_h v, u), \qquad u, v \in V,$$

hence,

$$\|A_h\|_{\mathcal{L}(V,V^*)} = \sup_{\|u\|_V = 1} |(A_h u, u)|.$$

Thus, by (2.10.16), (2.10.18), and (2.10.19), we conclude that

$$\lim_{n \to \infty} \|A_{h_n} - A_{h_0}\|_{\mathcal{L}(V,V^*)} = 0. \tag{2.10.21}$$

We will denote by $\mathcal{U}$ the set of the invertible elements of the space $\mathcal{L}(V, V^*)$ endowed with the topology generated by that of the space $\mathcal{L}(V, V^*)$. The form a_h being symmetric, continuous, and coercive on $V \times V$ for all $h \in B$, the Riesz theorem and (2.10.19) imply that

$$A_h \in \mathcal{U}, \qquad h \in B.$$

By (2.10.13) and (2.10.21), we have

$$h \to A_h \text{ is a continuous mapping of } B \text{ into } \mathcal{U}. \tag{2.10.22}$$

Combining (2.10.13), (2.10.14), (2.10.20), and (2.10.22) and applying Theorem 1.8.1, we get (2.10.15). The lemma is proved.

Introduce a set

$$Y_{pC} = \{\, h \mid h \in W_p^1(\Omega),\ \|h\|_{W_p^1(\Omega)} \le C,\ e_1 \le h \le e_2 \,\}, \tag{2.10.23}$$

e_1, e_2, and C being the positive numbers from (2.1.2) and (2.10.9).

Lemma 2.10.2 *Let a_h be a bilinear form on $V \times V$ determined by the relations* (2.1.1), (2.1.4)–(2.1.7), *and* (2.1.10). *Then, the function $h \to \Psi_i(h)$, $i = 1, 2, \dots, q$, defined by the formulas* (2.10.5)–(2.10.8), *where $u(h, g)$ is a solution to the problem* (2.10.1), *is a continuous mapping of Y_{pC} (endowed with the topology generated by the $W_p^1(\Omega)$-weak one) into $\mathbb{R}$ when $p > 2$.*

Proof. By Lemma 2.10.1 and the relations (2.10.5), (2.10.6), we obtain

$$\left.\begin{array}{l} h, g \to \Gamma_i(h, g) \text{ is a continuous mapping of } B \times V^* \\ \text{(equipped with the topology induced by the product of} \\ \text{the topologies of } C(\overline{\Omega}) \text{ and } V^*) \text{ into } \mathbb{R}. \end{array}\right\} \tag{2.10.24}$$

Denote the closure of Y_{pC} in $C(\overline{\Omega})$ as $\overline{Y}_{pC}$. The embedding $W_p^1(\Omega) \to C(\overline{\Omega})$ being compact when $p > 2$, by (2.10.4) and (2.10.23), we deduce that

$$\left.\begin{array}{l} \overline{Y}_{pC} \text{ is a compact set in } B \text{ (with respect to the topology} \\ \text{generated by the } C(\overline{\Omega})\text{-one).} \end{array}\right\} \tag{2.10.25}$$

Now, the theorem on the continuity of the maximum function (Theorem 1.4.4) together with (2.10.7), (2.10.8), (2.10.24), and (2.10.25) yields that

$$\left.\begin{array}{l} h \to \Psi_i(h) = \sup\limits_{g \in T} \Gamma_i(h, g) \text{ is a continuous mapping of } \overline{Y}_{pC} \\ \text{(equipped with the topology generated by the } C(\overline{\Omega})\text{-one)} \\ \text{into } \mathbb{R}. \end{array}\right\} \tag{2.10.26}$$

Let $\{h_n\}$ be a sequence such that

$$\{h_n\} \subset Y_{pC}, \qquad h_n \to h \text{ weakly in } W_p^1(\Omega), \qquad p > 2. \tag{2.10.27}$$

Then, $h \in Y_{pC}$ and

$$h_n \to h \qquad \text{strongly in } C(\overline{\Omega}).$$

Therefore, from (2.10.26), we deduce that

$$\lim_{n\to\infty} \Psi_i(h_n) = \Psi_i(h), \qquad i = 1, 2, \dots, q,$$

concluding the proof of the lemma.

Proof of Theorem 2.10.1. Let $\{h_n\}_{n=1}^{\infty}$ be a minimizing sequence, i.e.,

$$\{h_n\}_{n=1}^{\infty} \subset U_{\mathrm{ad}}, \tag{2.10.28}$$

$$f(h_n) \to \inf_{h\in U_{\mathrm{ad}}} f(h). \tag{2.10.29}$$

By virtue of (2.10.9) and (2.10.10), the sequence $\{h_n\}$ is bounded in $W_p^1(\Omega)$, $p > 2$; let us chose a subsequence $\{h_m\}$ from it such that

$$h_m \to \tilde{h} \qquad \text{weakly in } W_p^1(\Omega), \tag{2.10.30}$$

$$h_m \to \tilde{h} \qquad \text{strongly in } C(\overline{\Omega}). \tag{2.10.31}$$

The relations (2.10.28) and (2.10.30) yield

$$C \geq \liminf_{m\to\infty} \|h_m\|_{W_p^1(\Omega)} \geq \|\tilde{h}\|_{W_p^1(\Omega)}, \tag{2.10.32}$$

C being the number from (2.10.9). Taking note of (2.10.9), (2.10.28), and (2.10.31), we get

$$\check{h} \leq \tilde{h} \leq \hat{h}. \tag{2.10.33}$$

Combining (2.10.9), (2.10.28), (2.10.30), and Lemma 2.10.2 we obtain

$$\Psi_i(\tilde{h}) = \lim_{m\to\infty} \Psi_i(h_m) \leq 0, \qquad i = 1, 2, \dots, q.$$

This equality and (2.10.32), (2.10.33) imply $\tilde{h} \in U_{\mathrm{ad}}$. Further, using (2.10.11), (2.10.29), and (2.10.30), we deduce

$$\lim_{m\to\infty} f(h_m) = f(\tilde{h}) = \inf_{h\in U_{\mathrm{ad}}} f(h).$$

Hence, the function $h_0 = \tilde{h}$ is a solution to the problem (2.10.12).

2.11 The general control problem

So far we have studied optimal control problems connected with the bilinear form a_h (see (2.1.4)) whose coefficients a_{ij} were functions of control $h \in W_p^1(\Omega)$. Now, we proceed to study problems in which not only coefficients but also operators of the bilinear form depend on the control; moreover, we suppose that the right-hand sides of the equations depend on the control. We consider optimal control problems under restrictions on solutions to the equations and on eigenvalues. We do not specify the space of controls in these problems, which allows us to apply the results obtained to general situations.

2.11.1 Bilinear form a_q and the corresponding equation

Let, as before, Ω be a bounded, Lipschitz domain in $\mathbb{R}^2$, $(x, y) \in \Omega$,

$$W = \prod_{s=1}^{\nu} W_2^{l_s}(\Omega),$$

where $l_s \geq 1$, $s = 1, 2, \ldots, \nu$, and let V be a closed subspace of W with the norm of the space W. Let also B_1 be a Banach space, let Q_1 be a closed set in B_1 endowed with the topology induced by the B_1-strong one, and let mappings $r \to P_i^{(r)}$ be defined such that

$$\left. \begin{array}{l} r \to P_i^{(r)} \text{ is a continuous mapping of } Q_1 \text{ into } \mathcal{L}(W, L_2(\Omega)), \\ i = 1, 2, \ldots, k, \end{array} \right\} \tag{2.11.1}$$

$\mathcal{L}(W, L_2(\Omega))$ being the space of linear continuous mappings of W into $L_2(\Omega)$.

We are given the family of the bilinear, continuous forms a_r on the space $V \times V$ depending on a parameter r running through the set Q_1 defined by

$$a_r(u, v) = \iint_{\Omega} \sum_{i,j=1}^{k} a_{ij}^{(r)} \big(P_i^{(r)} u\big)\big(P_j^{(r)} v\big)\, dx\, dy \quad u, v \in V. \tag{2.11.2}$$

Here, $P_i^{(r)} u$ is the image of an element $u \in V$ under the mapping

$$P_i^{(r)} \in \mathcal{L}(W, L_2(\Omega)),$$

and

$$a_{ij}^{(r)} \in L_\infty(\Omega), \qquad a_{ij}^{(r)} = a_{ji}^{(r)}, \qquad r \in Q_1, \; i, j = 1, 2, \ldots, k. \tag{2.11.3}$$

Assume that

$$\left. \begin{array}{l} r \to a_{ij}^{(r)} \text{ is a continuous mapping of } Q_1 \text{ into } L_\infty(\Omega) \\ (i, j = 1, 2, \ldots, k), \end{array} \right\} \tag{2.11.4}$$

$$\left. \begin{array}{l} \displaystyle\sum_{i,j=1}^{k} a_{ij}^{(r)}(x, y)\xi_i\xi_j \geq c_r \sum_{i=1}^{k} \xi_i^2, \\ (x, y) \in \Omega, \; \xi \in \mathbb{R}^k, \; c_r = \text{const} > 0, \\ \text{where } c_r \text{ depends on } r \in Q_1. \end{array} \right\} \tag{2.11.5}$$

Due to (2.11.3), the form a_r is symmetric:

$$a_r(u, v) = a_r(v, u), \qquad u, v \in V, \; r \in Q_1. \tag{2.11.6}$$

(2.11.1)–(2.11.3) imply the form a_r to be continuous on $V \times V$ for any $r \in Q_1$, i.e.,

$$|a_r(u, v)| \leq \tilde{c}_r \|u\|_V \|v\|_V, \qquad u, v \in V, \tag{2.11.7}$$

$\tilde{c}_r$ being a positive number depending on $r \in Q_1$.

Further, suppose that, for all $r \in Q_1$, the system of operators $\{P_i^{(r)}\}$ is coercive in V, i.e.,

$$\iint\limits_{\Omega} \sum_{i=1}^{k} \left(P_i^{(r)} u\right)^2 dx\, dy \geq \mu_r \|u\|_V^2, \qquad u \in V,\ \mu_r = \text{const} > 0. \tag{2.11.8}$$

The relations (2.11.2), (2.11.5), and (2.11.8) yield the form a_r to be coercive in V for all $r \in Q_1$, i.e.,

$$a_r(u,u) \geq \lambda_r \|u\|_V^2, \qquad u \in V,\ \lambda_r = \text{const} > 0. \tag{2.11.9}$$

We stress that the number λ_r depends on $r \in Q_1$. Now, consider the following problem: Given $r \in Q_1$ and $f \in V^*$, find a function $u_{r,f}$ such that

$$u_{r,f} \in V, \qquad a_r(u_{r,f}, v) = (f, v), \qquad v \in V. \tag{2.11.10}$$

Theorem 2.11.1 *Let a_r be a bilinear form on $V \times V$ defined by the relations* (2.11.1)–(2.11.5), *and let the inequality* (2.11.8) *hold. Then, for any $r \in Q_1$ and $f \in V^*$, the problem* (2.11.10) *has a unique solution and the function $r, f \to u_{r,f}$ determined by this solution is a continuous mapping of $Q_1 \times V^*$ into V.*

Proof. The existence and uniqueness of a solution to the problem (2.11.10) is a consequence of the relations (2.11.6), (2.11.7), (2.11.9), and of the Riesz theorem. Let $\{r_n, f_n\}_{n=1}^{\infty} \subset Q_1 \times V^*$ and

$$r_n \to r_0 \qquad \text{in } Q_1, \tag{2.11.11}$$

$$f_n \to f_0 \qquad \text{in } V^*. \tag{2.11.12}$$

To prove the theorem, we need to establish that

$$u_{r_n,f_n} \to u_{r_0,f_0} \ \text{ in } V, \tag{2.11.13}$$

where the functions u_{r_n,f_n} satisfy the conditions

$$u_{r_n,f_n} \in V, \qquad a_{r_n}(u_{r_n,f_n}, v) = (f_n, v), \qquad v \in V,\ n = 0,1,2,\dots. \tag{2.11.14}$$

Let us show that

$$\begin{gathered} |a_{r_n}(u,u) - a_{r_0}(u,u)| \leq \alpha_n \|u\|_V^2, \qquad u \in V. \\ \lim_{n\to\infty} \alpha_n = 0. \end{gathered} \tag{2.11.15}$$

Indeed, for every addendum on the left-hand side of (2.11.15), due to (2.11.2), we have for any $u \in V$

$$\begin{aligned}
&\left| \iint_\Omega \left[a_{ij}^{(r_n)} \left(P_i^{(r_n)} u\right)\left(P_j^{(r_n)} u\right) - a_{ij}^{(r_0)} \left(P_i^{(r_0)} u\right)\left(P_j^{(r_0)} u\right)\right] dx\, dy \right| \\
&\quad = \left| \iint_\Omega \left\{ a_{ij}^{(r_n)} \left[\left(P_i^{(r_n)} u\right)\left(P_j^{(r_n)} u - P_j^{(r_0)} u\right) + \left(P_j^{(r_0)} u\right)\left(P_i^{(r_n)} u - P_i^{(r_0)} u\right)\right] \right.\right. \\
&\qquad \left.\left. + \left(a_{ij}^{(r_n)} - a_{ij}^{(r_0)}\right)\left(P_i^{(r_0)} u\right)\left(P_j^{(r_0)} u\right)\right\} dx\, dy \right| \\
&\quad \leq c_1 \left\| a_{ij}^{(r_n)} \right\|_{L_\infty(\Omega)} \|u\|_V^2 \left(\left\| P_i^{(r_n)} \right\|_{\mathcal{L}(W, L_2(\Omega))} \left\| P_j^{(r_n)} - P_j^{(r_0)} \right\|_{\mathcal{L}(W, L_2(\Omega))} \right. \\
&\qquad \left. + \left\| P_i^{(r_n)} - P_i^{(r_0)} \right\|_{\mathcal{L}(W, L_2(\Omega))} \right) + c_2 \|u\|_V^2 \left\| a_{ij}^{(r_n)} - a_{ij}^{(r_0)} \right\|_{L_\infty(\Omega)}, \qquad u \in V,
\end{aligned} \tag{2.11.16}$$

c_1 and c_2 being positive numbers. Now, the relations (2.11.1), (2.11.4), (2.11.11), and (2.11.16) imply (2.11.15).

For all $r \in Q_1$, the bilinear form a_r satisfies the formula

$$a_r(u, v) = (A_r u, v), \qquad u, v \in V, \tag{2.11.17}$$

where $A_r \in \mathcal{L}(V, V^*)$, so that the problem (2.11.10) is equivalent to the following one

$$u_{r,f} \in V, \qquad A_r u_{r,f} = f. \tag{2.11.18}$$

By virtue of (2.11.6), the operator A_r is selfadjoint , i.e.,

$$(A_r u, v) = (A_r v, u), \qquad u, v \in V.$$

Hence, for all $r \in Q_1$, the following relation holds

$$\|A_r\|_{\mathcal{L}(V,V^*)} = \sup_{\|u\|_V = 1} |(A_r u, u)|.$$

Thus, from (2.11.15), (2.11.17), we conclude

$$\lim_{n \to \infty} \|A_{r_n} - A_{r_0}\|_{\mathcal{L}(V,V^*)} = 0. \tag{2.11.19}$$

Denote by $\mathcal{U}$ the set of the invertible elements of the space $\mathcal{L}(V, V^*)$ equipped with the topology induced by that of $\mathcal{L}(V, V^*)$. Since for all $r \in Q_1$ the bilinear form a_r is symmetric, continuous, and coercive on $V \times V$, the Riesz theorem together with (2.11.17) yields that $A_r \in \mathcal{U}$ for all $r \in Q_1$.

From (2.11.11) and (2.11.19), we deduce $r \to A_r$ to be a continuous mapping of Q_1 into $\mathcal{U}$. Now, taking into account (2.11.11), (2.11.12), (2.11.18) and applying Theorem 1.8.1, we get (2.11.13). The theorem is proved.

2.11.2 Bilinear form b_r and the spectral problem

Suppose that

$$\left.\begin{array}{l} Y \text{ is a Banach space, } V \subset Y \subset (L_2(\Omega))^\nu, \text{ the embedding} \\ V \to Y \text{ is compact.} \end{array}\right\} \tag{2.11.20}$$

Suppose we are given a family of bilinear, symmetric, continuous forms b_r on $Y \times Y$ depending on a parameter r running through Q_1, i.e.,

$$b_r(u,v) = b_r(v,u), \qquad u, v \in Y, \tag{2.11.21}$$

$$|b_r(u,v)| \le c_r \|u\|_Y \|v\|_Y, \qquad u, v \in Y, \tag{2.11.22}$$

c_r being a positive number depending on $r \in Q_1$.

The dependence of the bilinear form b_r on a parameter $r \in Q_1$ defines the mapping $r \to b_r$ which is supposed to satisfy the following condition

$$r \to b_r \text{ is a continuous mapping of } Q_1 \text{ into } \mathcal{L}_2(Y,Y;\mathbb{R}), \quad \} \tag{2.11.23}$$

$\mathcal{L}_2(Y,Y;\mathbb{R})$ being the vector normed space of bilinear continuous forms on $Y \times Y$ equipped with the norm

$$\|b_r\|_{\mathcal{L}_2(Y,Y;\mathbb{R})} = \sup_{\|u\|_Y \le 1,\, \|v\|_Y \le 1} |b_r(u,v)|.$$

Given an element $r \in Q_1$, consider the following eigenvalue problem

$$\begin{aligned} &\left(u_i^{(r)}, \mu_i^{(r)}\right) \in V \times \mathbb{R}, \qquad u_i^{(r)} \ne 0, \\ &\mu_i^{(r)} a_r\left(u_i^{(r)}, v\right) = b_r\left(u_i^{(r)}, v\right), \qquad v \in V. \end{aligned} \tag{2.11.24}$$

By virtue of Theorem 1.5.8, the problem (2.11.24) has a countable set of eigenvalues $\{\mu_i^{(r)}\}_{i=1}^\infty$, which are ordered so that (cf. Subsec. 2.5.2)

$$|\mu_1^{(r)}| \ge |\mu_2^{(r)}| \ge \cdots, \qquad \lim_{i\to\infty} \mu_i^{(r)} = 0.$$

So, we have defined the mapping $r \to \mu(r) = \{\mu_i^{(r)}\}_{i=1}^\infty$ of the topological space Q_1 into $\ell_{\infty,0}$.

Theorem 2.11.2 *Let a_r be a bilinear form on $V \times V$ determined by the relations (2.11.1)–(2.11.5) and (2.11.8). Let also (2.11.20) hold and let the bilinear form b_r satisfy the conditions (2.11.21)–(2.11.23). Then, the function $r \to \mu(r) = \{\mu_i^{(r)}\}_{i=1}^\infty$, where $\mu_i^{(r)}$ are the eigenvalues of the problem (2.11.24), is a continuous mapping of Q_1 into $\ell_{\infty,0}$.*

Proof. Let $\{r_n\}_{n=1}^\infty$ be a sequence such that

$$\{r_n\}_{n=1}^\infty \subset Q_1, \qquad r_n \to r_0 \text{ in } Q_1. \tag{2.11.25}$$

Then, the argument from the proof of Theorem 2.11.1 (see (2.11.15)), together with Remark 1.5.3 implies that

$$a_{r_n} \to a_{r_0} \text{ in } \mathcal{L}_2(V, V; \mathbb{R}). \tag{2.11.26}$$

By (2.11.23) and (2.11.25), we get

$$b_{r_n} \to b_{r_0} \text{ in } \mathcal{L}_2(Y, Y; \mathbb{R}). \tag{2.11.27}$$

Now, taking into account (2.11.26), (2.11.27) and applying Theorem 1.5.9, we obtain

$$\lim_{n\to\infty} \sup_i \left|\mu_i^{(r_n)} - \mu_i^{(r_0)}\right| = 0,$$

which concludes the proof of the theorem.

2.11.3 Basic control problem

Setting of the problem. Existence theorem

In subsection 2.11.1, we introduced the Banach space B_1 and the closed set Q_1 in B_1 equipped with the topology induced by the B_1-strong one. Additionally, assume that

B_2 is a reflexive Banach space, $B_2 \subset B_1$, and the embedding $B_2 \to B_1$ is compact, (2.11.28)

Q_2 is a convex, closed, bounded set in B_2, $Q_2 \subset Q_1$, (2.11.29)

$$Q_3 \text{ is an open set in } B_2, \ \ Q_2 \subset Q_3 \subset Q_1. \tag{2.11.30}$$

Define a set of admissible controls by

$$U_{\text{ad}} = \left\{ r \,|\, r \in Q_2, \ \Psi_k(r) \le 0, \ k = 1, 2, \ldots, l \right\}. \tag{2.11.31}$$

Here

$r \to \Psi_k(r)$ is a continuous mapping of Q_3 (endowed with the topology generated by the B_2-weak one) into $\mathbb{R}$, $k = 1, 2, \ldots, l$. (2.11.32)

We suppose that a goal functional φ satisfies the condition

$r \to \varphi(r)$ is a continuous mapping of Q_3 (equipped with the topology generated by the B_2-weak one) into $\mathbb{R}$. (2.11.33)

The optimal control problem consists in finding an element r_0 such that

$$r_0 \in U_{\text{ad}}, \qquad \varphi(r_0) = \inf_{r \in U_{\text{ad}}} \varphi(r). \tag{2.11.34}$$

Theorem 2.11.3 *Let a nonempty set $U_{\rm ad}$ be determined by the relations* (2.11.28)–(2.11.32), *and let* (2.11.33) *hold. Then, the problem* (2.11.34) *has a solution.*

Proof. Let $\{r_n\}_{n=1}^\infty$ be a minimizing sequence, i.e.,

$$\{r_n\}_{n=1}^\infty \subset U_{\rm ad}, \qquad \lim_{n\to\infty} \varphi(r_n) = \inf_{r\in U_{\rm ad}} \varphi(r). \tag{2.11.35}$$

By virtue of (2.11.28) and (2.11.29) we can choose from the sequence $\{r_n\}$ a subsequence $\{r_m\}$ such that

$$r_m \to y \ \text{ weakly in } B_2, \tag{2.11.36}$$

$$y \in Q_2. \tag{2.11.37}$$

Since $r_m \in U_{\rm ad}$, (2.11.32) and (2.11.36) yield that $\Psi_k(y) \le 0$, $k = 1, 2, \dots, l$. Therefore, from (2.11.37) we get $y \in U_{\rm ad}$. Finally, the relations (2.11.33), (2.11.35), and (2.11.36) imply that

$$\lim_{m\to\infty} \varphi(r_m) = \varphi(y) = \inf_{r\in U_{\rm ad}} \varphi(r).$$

So, the function $r_0 = y$ is a solution to the problem (2.11.34).

Approximate solution of the basic problem

In the same way as it was done before (see, for example, Section 2.3) we can search approximate solutions to the problem (2.11.34) on the set $H_n \cap U_{\rm ad}$ ($\{H_n\}$ being a sequence of finite-dimensional subspaces of B_2 satisfying the limit density condition). If there exists a sequence of interior points of the set $U_{\rm ad}$ convergent to the solution r_0 of the problem (2.11.34) (in this case, the set $U_{\rm ad}$ is equipped with the topology generated by that of the space B_2), then the corresponding sequence of approximate solutions is a minimizing one and a subsequence that converges to r_0 weakly in B_2 may be extracted.

Below we will consider another approach to approximate solution of the problem (2.11.34) that makes no use of the existence of a sequence of interior points from $U_{\rm ad}$ convergent to r_0.

Let a family of sets $Q_2^{(t)}$ depending on parameter t running through the set $[1, \alpha]$, $\alpha > 1$, be given such that

$$\left.\begin{array}{l} Q_2^{(t)} \text{ is a convex, closed, bounded set in } B_2 \text{ for any} \\ t \in [1,\alpha], \end{array}\right\} \tag{2.11.38}$$

$$\left.\begin{array}{l} Q_2 \subset \overset{\circ}{Q}_2{}^{(t)} \subset Q_2^{(t)} \subset Q_3 \text{ for any } t \in (1,\alpha],\ Q_2 = Q_2^{(1)}, \\ Q_2^{(t_1)} \subset \overset{\circ}{Q}_2{}^{(t_2)} \text{ if } t_2 > t_1. \end{array}\right\} \tag{2.11.39}$$

Here $\overset{\circ}{Q}_2{}^{(t)}$ is the interior of the set $Q_2^{(t)}$, the latter being equipped with the topology generated by the B_2-one.

Additionally, we suppose that

$$\lim_{t\to 1}\Big(\sup_{q\in Q_2^{(t)}}\inf_{r\in Q_2}\|q-r\|_{B_2}\Big)=0. \tag{2.11.40}$$

The relation (2.11.40) means that the distance between an element $q\in Q_2^{(t)}$ and the set Q_2 tends to zero uniformly in $q\in Q_2^{(t)}$ as $t\to 1$.

For $t\in(1,\alpha]$, define a set

$$U^{(t)}=\{\,r\,|\,r\in Q_2^{(t)},\ \Psi_k(r)\le t-1,\ k=1,2,\dots,l\,\}. \tag{2.11.41}$$

Here, Ψ_k are the functionals from (2.11.31). The following statement is a direct consequence of the proof of Theorem 2.11.3.

Theorem 2.11.4 *Let a set $U^{(t)}$ be defined by the relations* (2.11.38), (2.11.39), (2.11.41), *and* (2.11.32), *and let the set $U^{(1)}=U_{\rm ad}$ be not empty. Assume that a function φ satisfies the condition* (2.11.33). *Then, for all $t\in[1,\alpha]$, there exists an element r_t such that*

$$r_t\in U^{(t)},\qquad \varphi(r_t)=\inf_{r\in U^{(t)}}\varphi(r). \tag{2.11.42}$$

Let $\{H_n\}_{n=1}^{\infty}$ be a sequence of finite-dimensional subspaces of B_2 meeting the conditions

$$\lim_{n\to\infty}\inf_{r\in H_n}\|q-r\|_{B_2}=0,\qquad q\in B_2, \tag{2.11.43}$$

$$H_n\subset H_{n+1}\qquad \forall n. \tag{2.11.44}$$

Consider the problem of finding a function r_n such that

$$r_n\in H_n\cap U^{(t)}, \tag{2.11.45}$$

$$\varphi(r_n)=\inf_{H_n\cap U^{(t)}}\varphi(r). \tag{2.11.46}$$

Theorem 2.11.5 *Let the assumptions of Theorems* 2.11.3 *and* 2.11.4 *be fulfilled, let $\{H_n\}_{n=1}^{\infty}$ be a sequence of finite-dimensional subspaces of B_2 satisfying the conditions* (2.11.43) *and* (2.11.44), *and let r_0, r_t be solutions to the problems* (2.11.34) *and* (2.11.42), *respectively, where $t\in(1,\alpha]$. Suppose that the set $H_1\cap U^{(t)}$ is not empty. Then, for any n, the problem* (2.11.45), (2.11.46) *has a solution r_n, and*

$$\varphi(r_t)\le\lim_{n\to\infty}\varphi(r_n)\le\varphi(r_0). \tag{2.11.47}$$

From the sequence $\{r_n\}$ one can choose a subsequence $\{r_m\}$ such that

$$r_m\to\tilde r\ \text{ weakly in } B_2,\qquad \tilde r\in U^{(t)},$$
$$\lim_{m\to\infty}\varphi(r_m)=\lim_{n\to\infty}\varphi(r_n)=\varphi(\tilde r). \tag{2.11.48}$$

Proof. By hypothesis, the set $H_1 \cap U^{(t)}$ is not empty. So, (2.11.44) implies $H_n \cap U^{(t)}$ to be not empty for all n. Since $H_n \cap U^{(t)}$ is a compactum, (2.11.33) yields that problem (2.11.45), (2.11.46) has a solution for all n.

By (2.11.42) and (2.11.46),

$$\varphi(r_t) \leq \varphi(r_n) \qquad \forall n. \tag{2.11.49}$$

The relations (2.11.44) and (2.11.46) imply $\varphi(r_{n+1}) \leq \varphi(r_n)$. Therefore, from (2.11.49) we deduce that the number sequence $\{\varphi(r_n)\}_{n=1}^{\infty}$ converges and

$$\lim_{n\to\infty} \varphi(r_n) \geq \varphi(r_t). \tag{2.11.50}$$

By using (2.11.31), (2.11.32), (2.11.39), and (2.11.41), it is easy to see that $r_0 \in \overset{\circ}{U}{}^{(t)}$ ($\overset{\circ}{U}{}^{(t)}$ being the interior of $U^{(t)}$ which is equipped with the topology generated by the B_2-one). This together with (2.11.43) implies the existence of a sequence $\{z_n\}$ such that

$$z_n \in H_n \cap U^{(t)}, \qquad z_n \to r_0 \ \text{ in } B_2. \tag{2.11.51}$$

Hence,

$$\lim_{n\to\infty} \varphi(z_n) = \varphi(r_0). \tag{2.11.52}$$

From (2.11.45), (2.11.46), and (2.11.51), we have $\varphi(z_n) \geq \varphi(r_n)$. Thus, (2.11.50) and (2.11.52) yield (2.11.47).

By (2.11.38), we get that the sequence $\{r_n\}$ is bounded in B_2. Choose a subsequence $\{r_m\}$ from it such that

$$r_m \to \tilde{r} \qquad \text{weakly in } B_2. \tag{2.11.53}$$

Using (2.11.32), (2.11.38), and (2.11.53), we easily obtain $\tilde{r} \in U^{(t)}$. Finally, (2.11.33) and (2.11.53) imply (2.11.48).

Remark 2.11.1 We did not use the supposition (2.11.40) in Theorem 2.11.5. However, if (2.11.40) holds, then for t sufficiently close to 1, "the getting of the functions r_n and $\tilde{r}$ out of the domain U_{ad}" can be made as small as desired, and in this case (2.11.47) and (2.11.48) hold.

2.11.4 Application of the basic control problem (combined problem)

Suppose that

$$\left.\begin{array}{l} u, v \to a_r(u,v) \text{ is a bilinear form on } V \times V \text{ depending on} \\ \text{parameter } r \in Q_1 \text{ and determined by the relations} \\ \text{(2.11.1)–(2.11.5), (2.11.8),} \end{array}\right\} \tag{2.11.54}$$

$$\left.\begin{array}{l} u,v \to b_r(u,v) \text{ is a bilinear, symmetric, continuous form} \\ \text{on } Y \times Y \text{ depending on parameter } r \text{ running through } Q_1 \\ \text{and (2.11.20), (2.11.23) hold,} \end{array}\right\} \tag{2.11.55}$$

$$r \to f_r \quad \text{is a continuous mapping of } Q_1 \text{ into } V^*. \tag{2.11.56}$$

Provided $r \in Q_1$ is given, consider the following problems.

First problem: Find a function u_r such that

$$u_r \in V, \qquad a_r(u_r, v) = (f_r, v), \qquad v \in V. \tag{2.11.57}$$

Second problem: Find eigenvalues $\mu_i^{(r)}$ of the equation

$$\mu_i^{(r)} a_r\big(u_i^{(r)}, v\big) = b_r\big(u_i^{(r)}, v\big), \qquad v \in V, \tag{2.11.58}$$

$u_i^{(r)}$ being the eigenfunctions, $u_i^{(r)} \in V$.

A solution to the second problem determines a continuous mapping $r \to \mu(r) = \{\mu_i^{(r)}\}_{i=1}^{\infty}$ of the set Q_1 (endowed with the topology generated by the B_1-one) into $\ell_{\infty,0}$ (see Theorem 2.11.2).

We will control by the function r. In this connection, define a set of admissible controls as

$$\begin{aligned} U_{\text{ad}} = \big\{\, r \,|\, r \in Q_2,\ &\Psi_k^{(1)}(r, u_r) \le 0 \ (k = 1, 2, \dots, n_1), \\ &\Psi_i^{(2)}(r, \mu(r)) \le 0 \ (i = 1, 2, \dots, n_2) \,\big\}. \end{aligned} \tag{2.11.59}$$

Here, the set Q_2 is determined by the relations (2.11.28), (2.11.29), and u_r, $\mu(r)$ are solutions to the first and second problems,

$$\left.\begin{array}{l} r, w \to \Psi_k^{(1)}(r, w) \text{ is a continuous mapping of } Q_1 \times V \text{ into} \\ \mathbb{R}\ (k = 1, 2, \dots, n_1), \end{array}\right\} \tag{2.11.60}$$

$$\left.\begin{array}{l} r, z \to \Psi_i^{(2)}(r, z) \text{ is a continuous mapping of } Q_1 \times \ell_{\infty,0} \text{ into} \\ \mathbb{R}\ (i = 1, 2, \dots, n_2). \end{array}\right\} \tag{2.11.61}$$

Provided the goal functional $\varphi(r)$ satisfies the condition (2.11.33), consider the optimal control problem: Find a function r_0 such that

$$r_0 \in U_{\text{ad}}, \qquad \varphi(r_0) = \inf_{r \in U_{\text{ad}}} \varphi(r). \tag{2.11.62}$$

Theorem 2.11.6 *Let the relations* (2.11.54)–(2.11.56) *hold and let a nonempty set* U_{ad} *be determined by the formulas* (2.11.28)–(2.11.30), (2.11.59)–(2.11.61). *Let also the goal functional meet the condition* (2.11.33). *Then, the problem* (2.11.62) *has a solution.*

Proof. To apply Theorem 2.11.3, define functionals $\Psi_k(r)$ in the following way:

$$\Psi_k(r) = \Psi_k^{(1)}(r, u_r), \qquad k = 1, 2, \dots, n_1, \tag{2.11.63}$$

$$\Psi_{n_1+i}(r) = \Psi_i^{(2)}(r, \mu(r)), \qquad i = 1, 2, \dots, n_2. \tag{2.11.64}$$

The functionals $\Psi_k(r)$, $k = 1, 2, \ldots, n_1 + n_2$, are considered on the set Q_3 (see (2.11.30)). Further, let $\{r_n\}_{n=1}^{\infty} \subset Q_3$, $r_0 \in Q_3$, and $r_n \to r_0$ weakly in B_2. Then, by virtue of (2.11.28), $r_n \to r_0$ strongly in B_1, i.e.,

$$r_n \to r_0 \quad \text{in } Q_1, \tag{2.11.65}$$

and taking into account (2.11.56), we deduce $f_{r_n} \to f_{r_0}$ in V^*. Now, Theorem 2.11.1 yields

$$u_{r_n} \to u_{r_0} \quad \text{in } V. \tag{2.11.66}$$

By using (2.11.60), (2.11.65), and (2.11.66), we get

$$\lim_{n\to\infty} \Psi_k^{(1)}(r_n, u_{r_n}) = \Psi_k^{(1)}(r_0, u_{r_0}), \qquad k = 1, 2, \ldots, n_1.$$

So, the functionals $\Psi_k(r)$ determined by the formula (2.11.63) satisfy the condition (2.11.32). By (2.11.65) and Theorem 2.11.2, we obtain $\mu(r_n) \to \mu(r_0)$ in $\ell_{\infty,0}$. Therefore, from (2.11.61), (2.11.65) we deduce that

$$\lim_{n\to\infty} \Psi_i^{(2)}(r_n, \mu(r_n)) = \Psi_i^{(2)}(r_0, \mu(r_0)), \qquad i = 1, 2, \ldots, n_2,$$

i.e., the functionals $\Psi_k(r)$ determined by the formula (2.11.64) meet the condition (2.11.32). Now, Theorem 2.11.3 yields that the problem (2.11.62) has a solution.

Notice that in order to solve the problem (2.11.62) approximately, one can use the technique from Subsec. 2.11.3.

2.12 Optimization by the shape of domain and by operators

Hitherto, the domain Ω was assumed to be fixed. However, in applications, there arise many problems connected with the search of an optimal domain. So, we proceed now to study problems in which controls are the shape of domain and operators.

2.12.1 Domains and bilinear forms

Let Ω be a bounded region in $\mathbb{R}^2$, let M_l be a topological space (l being a natural number), and let there correspond to any $q \in M_l$ a domain Ω_q in $\mathbb{R}^2$ and a diffeomorphism P_q of the set $\overline{\Omega}_q$ onto $\overline{\Omega}$ such that

$$P_q \in C^l\left(\overline{\Omega}_q, \overline{\Omega}\right), \qquad P_q^{-1} \in C^l\left(\overline{\Omega}, \overline{\Omega}_q\right), \tag{2.12.1}$$

where P_q^{-1} is the inverse mapping of P_q. The relations (2.12.1) mean that $P_q = (P_{q1}, P_{q2})$, $P_q^{-1} = \left(P_{q1}^{-1}, P_{q2}^{-1}\right)$, where P_{qi} and P_{qi}^{-1} are scalar functions and $P_{qi} \in C^l\left(\overline{\Omega}_q\right)$, $P_{qi}^{-1} \in C^l\left(\overline{\Omega}\right)$, $i = 1, 2$.

Further, let $W(\Omega) = \prod_{i=1}^{\nu} W_2^{l_i}(\Omega)$, $W(\Omega_q) = \prod_{i=1}^{\nu} W_2^{l_i}(\Omega_q)$, and let $V(\Omega)$, $V(\Omega_q)$ be closed subspaces of $W(\Omega)$ and $W(\Omega_q)$, respectively. Suppose that $l = \max_i l_i$. Then, from (2.12.1) and Theorem 1.14.3, it follows that the mapping $u \to u \circ P_q$ is an isomorphism of $W(\Omega)$ onto $W(\Omega_q)$.

Assume the following condition to be fulfilled:

$$\left.\begin{array}{l}\text{for all } q \in M_l \text{ the mapping } u \to u \circ P_q \text{ is an isomorphism}\\ \text{of } V(\Omega) \text{ onto } V(\Omega_q).\end{array}\right\} \tag{2.12.2}$$

Provided Y is a Hilbert space, we will refer to the set of the bilinear, symmetric, continuous, coercive forms on $Y \times Y$ as $\mathcal{L}_{2,\,\mathrm{sc}}(Y, Y; \mathbb{R})$.

Suppose we are given a form

$$a_q \in \mathcal{L}_{2,\,\mathrm{sc}}(V(\Omega_q), V(\Omega_q); \mathbb{R}).$$

Denote by $P_q a_q$ the bilinear form on $V(\Omega) \times V(\Omega)$ that is the image of the form a_q under the mapping P_q and is given by

$$P_q a_q(u, v) = a_q(u \circ P_q, v \circ P_q), \qquad u, v \in V(\Omega). \tag{2.12.3}$$

To get the explicit form of $P_q a_q$, we have to change the variable $y \in \overline{\Omega}_q$ to $x = P_q(y) \in \overline{\Omega}$.

We stress that $P_q a_q \in \mathcal{L}_{2,\,\mathrm{sc}}(V(\Omega), V(\Omega); \mathbb{R})$. Indeed, the form $P_q a_q$ is symmetric and continuous because of (2.12.2), (2.12.3), and of the symmetry and continuity of the form a_q. Since a_q is coercive, by (2.12.2) and (2.12.3), we have

$$\begin{gathered} P_q a_q(u, u) = a_q(u \circ P_q, u \circ P_q) \geq c\|u \circ P_q\|^2_{V(\Omega_q)} \geq c_1\|u\|^2_{V(\Omega)}, \\ u \in V(\Omega), \ c_1 = \mathrm{const} > 0. \end{gathered}$$

2.12.2 Optimization problem connected with solution of an operator equation

Let G be a topological space such that

$$\left.\begin{array}{l}\text{from every sequence } \{h_n\} \subset G \text{ one can choose a}\\ \text{subsequence } \{h_m\} \text{ convergent to some } h \in G.\end{array}\right\} \tag{2.12.4}$$

Also to every pair $(h, q) \in G \times M_l$ let there correspond a form a_{hq} such that

$$a_{hq} \in \mathcal{L}_{2,\,\mathrm{sc}}(V(\Omega_q), V(\Omega_q); \mathbb{R}). \tag{2.12.5}$$

Define a form $P_q a_{hq}$ by the equality (2.12.3) in which a_q is replaced by a_{hq}. By the above argument, we have

$$P_q a_{hq} \in \mathcal{L}_{2,\,\mathrm{sc}}(V(\Omega), V(\Omega); \mathbb{R}).$$

Suppose that

$$\left.\begin{array}{l} h,q \to P_q a_{hq} \text{ is a continuous mapping of } G \times M_l \text{ into} \\ \mathcal{L}_{2,\,\mathrm{sc}}(V(\Omega),V(\Omega);\mathbb{R}). \end{array}\right\} \tag{2.12.6}$$

Assume we are given a mapping B satisfying the following condition

$$\left.\begin{array}{l} h,q \to B(h,q) \text{ is a continuous mapping of } G \times M_l \text{ into} \\ (V(\Omega))^*. \end{array}\right\} \tag{2.12.7}$$

Now, consider the problem: Given an element $(h,q) \in G \times M_l$, find a function u_{hq} such that

$$u_{hq} \in V(\Omega_q),\ a_{hq}(u_{hq},v) = \left(B(h,q), v \circ P_q^{-1}\right), \qquad v \in V(\Omega_q). \tag{2.12.8}$$

Since $B(h,q) \in (V(\Omega))^*$, the relation (2.12.2) yields that

$$v \to \left(B(h,q), v \circ P_q^{-1}\right)$$

is a linear continuous mapping of $V(\Omega_q)$ into $\mathbb{R}$. This together with (2.12.5) implies that the problem (2.12.8) has a unique solution.

By virtue of (2.12.2) and (2.12.3), the problem (2.12.8) is equivalent to the following one: Find a function $\tilde{u}_{hq}$ such that

$$\tilde{u}_{hq} \in V(\Omega), \qquad P_q a_{hq}(\tilde{u}_{hq},w) = (B(h,q),w), \qquad w \in V(\Omega), \tag{2.12.9}$$

and $\tilde{u}_{hq} = u_{hq} \circ P_q^{-1}$.

Let a topological space N_l be defined such that

$$\left.\begin{array}{l} \text{from every sequence } \{q_n\} \subset N_l \text{ one can choose a} \\ \text{subsequence } \{q_m\} \text{ convergent to some } q \in N_l,\ N_l \subset M_l, \\ \text{the embedding } N_l \text{ into } M_l \text{ being continuous.} \end{array}\right\} \tag{2.12.10}$$

Assume that functionals Ψ_i are given such that

$$\left.\begin{array}{l} h,q,u \to \Psi_i(h,q,u) \text{ is a continuous mapping of} \\ G \times N_l \times V(\Omega) \text{ into } \mathbb{R}\ (i = 0,1,2,\dots,m). \end{array}\right\} \tag{2.12.11}$$

Define a set of admissible controls as

$$U_{\mathrm{ad}} = \left\{ (h,q) \,|\, (h,q) \in G \times N_l,\ \Psi_i(h,q,\tilde{u}_{hq}) \le 0,\ i = 1,2,\dots,m \right\}, \tag{2.12.12}$$

and let a goal functional be of the form

$$J(h,q) = \Psi_0(h,q,\tilde{u}_{hq}), \tag{2.12.13}$$

$\tilde{u}_{hq}$ being a solution to the problem (2.12.9).

The optimal control problem consists in finding an element (h_0,q_0) such that

$$(h_0,q_0) \in U_{\mathrm{ad}}, \qquad J(h_0,q_0) = \inf_{(h,q)\in U_{\mathrm{ad}}} J(h,q). \tag{2.12.14}$$

Theorem 2.12.1 *Let the conditions* (2.12.2)–(2.12.7) *and* (2.12.9)–(2.12.13) *hold, and let the set* U_{ad} *be nonempty. Then, the problem* (2.12.14) *has a solution.*

Proof. Let $\{h_n, q_n\}$ be a minimizing sequence, i.e.,

$$\{h_n, q_n\} \subset U_{\mathrm{ad}}, \qquad \lim_{n\to\infty} J(h_n, q_n) = \inf_{(h,q)\in U_{\mathrm{ad}}} J(h,q). \tag{2.12.15}$$

(2.12.4) and (2.12.10) imply that we can choose a subsequence $\{h_m, g_m\}$ from the sequence $\{h_n, q_n\}$ such that

$$h_m \to h' \text{ in } G, \qquad q_m \to q' \text{ in } N_l, \qquad q_m \to q' \text{ in } M_l. \tag{2.12.16}$$

From here, taking into account (2.12.6) and (2.12.7), we get

$$\begin{aligned} P_{q_m} a_{h_m q_m} &\to P_{q'} a_{h'q'} \text{ in } \mathcal{L}_{2,\,\mathrm{sc}}(V(\Omega), V(\Omega); \mathbb{R}), \\ B(h_m, q_m) &\to B(h', q') \text{ in } (V(\Omega))^*. \end{aligned} \tag{2.12.17}$$

Using Theorem 1.8.1 and (2.12.17), we obtain

$$\tilde{u}_{h_m q_m} \to \tilde{u}_{h'q'} \text{ in } V(\Omega), \tag{2.12.18}$$

where $\tilde{u}_{h_m q_m}$ and $\tilde{u}_{h'q'}$ are solutions to the problem (2.12.9) for $h = h_m$, $q = q_m$ and $h = h'$, $q = q'$, respectively. Now, taking into consideration (2.12.11), (2.12.12), (2.12.16), and (2.12.18) we deduce that $(h', q') \in U_{\mathrm{ad}}$. Further, due to (2.12.11), (2.12.13), (2.12.15), (2.12.16), and (2.12.18), we conclude

$$J(h', q') = \inf_{(h,q)\in U_{\mathrm{ad}}} J(h,q).$$

Thus, the pair $h_0 = h'$, $q_0 = q'$ is a solution to the problem (2.12.14).

2.12.3 Eigenvalue optimization problem

Under the conditions imposed above, we suppose also that $Y(\Omega)$ and $Y(\Omega_q)$ are Banach spaces of functions defined on Ω and Ω_q, respectively, such that

$$\left.\begin{array}{l} V(\Omega) \subset Y(\Omega), \text{ the embedding of } V(\Omega) \text{ into } Y(\Omega) \text{ is} \\ \text{compact,} \end{array}\right\} \tag{2.12.19}$$

$$\left.\begin{array}{l} \text{for all } q \in M_l, \text{ the mapping } u \to u \circ P_q \text{ is an} \\ \text{isomorphism of } Y(\Omega) \text{ into } Y(\Omega_q). \end{array}\right\} \tag{2.12.20}$$

By (2.12.2), (2.12.19), and (2.12.20), we get that

$$\left.\begin{array}{l} V(\Omega_q) \subset Y(\Omega_q), \text{ the embedding of } V(\Omega_q) \text{ into } Y(\Omega_q) \text{ is} \\ \text{compact for all } q \in M_l. \end{array}\right\} \tag{2.12.21}$$

Further, to every pair $(h,q) \in G \times M_l$ let there correspond a bilinear, symmetric, continuous form b_{hq} on $Y(\Omega_q) \times Y(\Omega_q)$. We denote by $P_q b_{hq}$ the bilinear, symmetric, continuous form on $Y(\Omega) \times Y(\Omega)$ that is determined by

$$P_q b_{hq}(u,v) = b_{hq}(u \circ P_q, v \circ P_q), \qquad u,v \in Y(\Omega). \tag{2.12.22}$$

Assume that

$$\left.\begin{array}{l} h,q \to P_q b_{hq} \text{ is a continuous mapping of } G \times M_l \text{ into} \\ \mathcal{L}_2(Y(\Omega), Y(\Omega); \mathbb{R}). \end{array}\right\} \tag{2.12.23}$$

Now, consider the problem: Given a pair $(h,q) \in G \times M_l$, find the eigenvalues $\{\mu_{hq}^{(i)}\}$ of the following problem:

$$\begin{aligned} &u_{hq}^{(i)} \in V(\Omega_q), \qquad u_{hq}^{(i)} \neq 0, \qquad \mu_{hq}^{(i)} \in \mathbb{R}, \\ &\mu_{hq}^{(i)} a_{hq}\big(u_{hq}^{(i)}, v\big) = b_{hq}\big(u_{hq}^{(i)}, v\big), \qquad v \in V(\Omega_q). \end{aligned} \tag{2.12.24}$$

Making use of (2.12.5), (2.12.21), and of Theorem 1.5.8, we deduce that there exists a sequence of the eigenfunctions of the problem (2.12.24) convergent to zero, i.e., $\{\mu_{hq}^{(i)}\}_{i=1}^{\infty} = \mu_{hq} \in \ell_{\infty,0}$ and $|\mu_{hq}^{(1)}| \geq |\mu_{hq}^{(2)}| \geq |\mu_{hq}^{(3)}| \geq \cdots$. By virtue of (2.12.2), (2.12.3), and (2.12.22), $\mu_{hq}^{(i)}$ are the eigenvalues of the following problem

$$\begin{aligned} &\tilde{u}_{hq}^{(i)} \in V(\Omega), \qquad \tilde{u}_{hq}^{(i)} \neq 0, \qquad \mu_{hq}^{(i)} \in \mathbb{R}, \\ &\mu_{hq}^{(i)} P_q a_{hq}\big(\tilde{u}_{hq}^{(i)}, w\big) = P_q b_{hq}\big(\tilde{u}_{hq}^{(i)}, w\big), \qquad w \in V(\Omega), \end{aligned} \tag{2.12.25}$$

where $\tilde{u}_{hq}^{(i)} = u_{hq}^{(i)} \circ P_q^{-1}$, $u_{hq}^{(i)}$ being the eigenfunctions of the problem (2.12.24).

Further, suppose we are given mappings A_i and Ψ_j such that

$$\left.\begin{array}{l} \mu \to A_i \mu \text{ is a continuous mapping of } \ell_{\infty,0} \text{ into } \mathbb{R} \\ (i = 1,2,\ldots,k), \end{array}\right\} \tag{2.12.26}$$

$$\left.\begin{array}{l} h,q \to \Psi_j(h,q) \text{ is a continuous mapping of } G \times N_l \\ \text{into } \mathbb{R} \ (j = 0,1,2,\ldots,m). \end{array}\right\} \tag{2.12.27}$$

Define a set of admissible controls as

$$\begin{aligned} U_{\text{ad}} = \{\, (h,q) \mid (h,q) \in G \times N_l,\ \Psi_j(h,q) \leq 0 \ (j = 1,2,\ldots,m), \\ A_i \mu_{hq} \leq 0 \ (i = 1,2,\ldots,k) \,\}. \end{aligned} \tag{2.12.28}$$

Consider the problem: Find a pair (h_0, q_0) such that

$$(h_0, q_0) \in U_{\text{ad}}, \qquad \Psi_0(h_0, q_0) = \inf_{(h,q) \in U_{\text{ad}}} \Psi_0(h,q). \tag{2.12.29}$$

Theorem 2.12.2 *Let the conditions* (2.12.2)–(2.12.6), (2.12.10), (2.12.19), (2.12.20), (2.12.22), (2.12.23), *and* (2.12.26)–(2.12.28) *hold, and let the set* U_{ad} *be nonempty. Then, the problem* (2.12.29) *has a solution.*

Proof. By using (2.12.6), (2.12.23), (2.12.25), and Theorem 1.5.9, we conclude that $h, q \to \mu_{hq}$ is a continuous mapping of $G \times M_l$ into $\ell_{\infty,0}$. Now, choosing from the minimizing sequence a convergent subsequence and passing to the limit, we see that the limit element is a solution of the problem (2.12.29).

Remark 2.12.1 In some important problems, the spaces $V(\Omega_q)$ are defined so that the condition (2.12.2) is not satisfied, while the mapping $u \to u \circ P_q$ is an isomorphism of $W(\Omega) = \prod_{i=1}^{\nu} W_2^{l_i}(\Omega)$ onto $W(\Omega_q) = \prod_{i=1}^{\nu} W_2^{l_i}(\Omega_q)$ for an arbitrary $q \in M_l$, $l = \max_i l_i$. Particularly, (2.12.2) does not hold when $V(\Omega_q)$ consists of solenoidal vector-valued functions. In this case, one can reformulate the direct problem. The general approach to such optimization problems is considered in Litvinov (1989).

2.12.4 Some realizations of the spaces M_l and N_l

Double-connected domain

Suppose Q is a double-connected domain in $\mathbb{R}^2$ that is star-shaped with respect to the origin (Fig. 2.12.1), i.e., the boundary S of the domain Q consists of two connected components S_1 and S_2. Suppose also that S_1 is given by the equations

$$y_1 = q^{(1)}(\alpha) \cos \alpha, \qquad y_2 = q^{(1)}(\alpha) \sin \alpha \tag{2.12.30}$$

and the equations for S_2 are of the form

$$y_1 = q^{(2)}(\alpha) \cos \alpha, \qquad y_2 = q^{(2)}(\alpha) \sin \alpha. \tag{2.12.31}$$

Here, $0 \leq \alpha \leq 2\pi$ and the functions $q^{(i)}$ characterizing the domain Q are periodic, i.e., $q^{(i)}(0) = q^{(i)}(2\pi)$, $i = 1, 2$.

Assume that

$$r_1, \ r_2, \ \varepsilon \ \text{ are positive numbers}, \quad r_1 + \varepsilon < r_2. \tag{2.12.32}$$

Define a set M_l in the following way:

$$M_l = \big\{ q = (q^{(1)}, q^{(2)}) \,|\, q \in (\tilde{C}^l([0, 2\pi]))^2, \ r_1 \leq q^{(1)}(\alpha), \\ q^{(1)}(\alpha) + \varepsilon \leq q^{(2)}(\alpha) \leq r_2 \ \ \alpha \in [0, 2\pi] \big\}. \tag{2.12.33}$$

Here, $\tilde{C}^l([0, 2\pi])$ is the space of functions defined on the interval $[0, 2\pi]$ having continuous derivatives up to the l-th order and satisfying the periodicity condition, i.e.,

$$\tilde{C}^l([0, 2\pi]) = \Big\{ f \,|\, f \in C^l([0, 2\pi]), \ \frac{d^k f}{d\alpha^k}(0) = \frac{d^k f}{d\alpha^k}(2\pi), \ k = 0, 1, 2, \ldots, l \Big\}. \tag{2.12.34}$$

The space $\tilde{C}^l([0, 2\pi])$ can be considered as the restriction on the interval $[0, 2\pi]$ of the set of 2π-periodical functions defined on $\mathbb{R}$ and having continuous derivatives in $\mathbb{R}$ up to the l-th order.

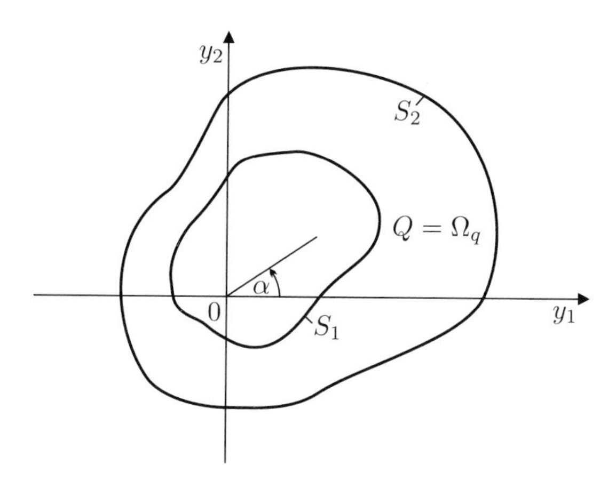

Figure 2.12.1: Double-connected star-shaped domain Q

Define a set N_l by

$$N_l = \{ q = (q^{(1)}, q^{(2)}) \mid q \in (\tilde{W}_2^{l+1}(0, 2\pi))^2, \ \|q\|_{(\tilde{W}_2^{l+1}(0,2\pi))^2} \leq r_3,$$
$$r_1 \leq q^{(1)}(\alpha), \ q^{(1)}(\alpha) + \varepsilon \leq q^{(2)}(\alpha) \leq r_2 \ \alpha \in [0, 2\pi] \}. \quad (2.12.35)$$

Here, $\tilde{W}_2^{l+1}(0, 2\pi)$ is the closure of $\tilde{C}^{l+1}([0, 2\pi])$ in $W_2^{l+1}(0, 2\pi)$, and r_3 is a positive number.

Now, if we endow M_l and N_l with the strong topology of the space $C^l([0, 2\pi])^2$ and the weak topology of $(\tilde{W}_2^{l+1}(0, 2\pi))^2$, respectively, then, due to the embedding theorem, the condition (2.12.10) is satisfied. Moreover, to every $q \in M_l$ there corresponds the double-connected domain $\Omega_q \subset \mathbb{R}^2$, the connected components of the boundary of which are defined by the formulas (2.12.30) and (2.12.31).

Define a set Ω by

$$\Omega = \{ x = (x_1, x_2) \mid x \in \mathbb{R}^2, \ 1 < x_1^2 + x_2^2 < 4 \}. \quad (2.12.36)$$

Let F stand for the mapping that transfers polar coordinates of a point into Cartesian ones:

$$F : (r, \varphi) \to F(r, \varphi) = (x_1, x_2), \qquad x_1 = r \cos \varphi, \ x_2 = r \sin \varphi; \quad (2.12.37)$$

F is a bijection of the set

$$Q_1 = \{ (r, \varphi) \mid (r, \varphi) \in \mathbb{R}^2, \ 0 < r < \infty, \ 0 \leq \varphi < 2\pi \}$$

onto the complement of the origin in $\mathbb{R}^2$, and the restriction of F on

$$Q_2 = \{ (r, \varphi) \mid 0 < r < \infty, \ 0 < \varphi < 2\pi \}$$

is a diffeomorphism of the C^∞ class of the set Q_2 onto $F(Q_2)$.

Denote by F^{-1} the mapping inverse of F. Then, for $q \in M_l$, define a mapping $P_q : \overline{\Omega}_q \to \overline{\Omega}$ as

$$P_q = F \circ \mathcal{D} \circ F^{-1}, \tag{2.12.38}$$

where $\mathcal{D}\colon F^{-1}(\overline{\Omega}_q) \to F^{-1}(\overline{\Omega})$,

$$\begin{gathered}(\rho, \alpha) \to \mathcal{D}(\rho, \alpha) = (r, \varphi),\\ r = \frac{\rho - 2q^{(1)}(\alpha) + q^{(2)}(\alpha)}{q^{(2)}(\alpha) - q^{(1)}(\alpha)}, \qquad \varphi = \alpha;\end{gathered} \tag{2.12.39}$$

$\mathcal{D}$ is a bijection, and the inverse bijection $\mathcal{D}^{-1}\colon F^{-1}(\overline{\Omega}) \to F^{-1}(\overline{\Omega}_q)$ is of the form

$$\begin{gathered}(r, \varphi) \to \mathcal{D}^{-1}(r, \varphi) = (\rho, \alpha),\\ \rho = 2q^{(1)}(\varphi) - q^{(2)}(\varphi) + \big[q^{(2)}(\varphi) - q^{(1)}(\varphi)\big] r, \qquad \alpha = \varphi.\end{gathered} \tag{2.12.40}$$

Taking note that $q \in M_l$, we easily conclude the mapping P_q determined by the relations (2.12.37)–(2.12.39) to satisfy the condition (2.12.1).

Single-connected domain

Now, define sets M_l and N_l by the following relations

$$\begin{aligned} M_l &= \big\{\, q \,|\, q \in \tilde{C}^l([0, 2\pi]),\ r_1 \le q(\alpha) \le r_2 \ \ \alpha \in [0, 2\pi] \,\big\},\\ N_l &= \big\{\, q \,|\, q \in \tilde{W}_2^{l+1}(0, 2\pi),\ \|q\|_{\tilde{W}_2^{l+1}(0,2\pi)} \le r_3,\ r_1 \le q(\alpha) \le r_2 \ \ \alpha \in [0, 2\pi] \,\big\},\end{aligned} \tag{2.12.41}$$

where $r_3 = \text{const} > 0$ and r_1, r_2 meet (2.12.32). We endow M_l with the strong topology of the space $C^l([0, 2\pi])$ and N_l with the weak topology of $\tilde{W}_2^{l+1}(0, 2\pi)$. Then, due to the embedding theorem, the condition (2.12.10) holds. To every $q \in M_l$ there corresponds the single-connected, star-shaped domain $\Omega_q \subset \mathbb{R}^2$ whose boundary is defined by the function $\alpha \to q(\alpha)$, where α is the polar angle.

Denote by $B(0, c)$ the open ball in $\mathbb{R}^2$ centered at 0, of radius $c > 0$, i.e., $B(0, c) = \{(x_1, x_2) \,|\, 0 \le x_1^2 + x_2^2 < c^2\}$, and let

$$\Omega = B(0, r_1), \qquad r_4 \in \mathbb{R}, \quad 0 < r_4 < r_1. \tag{2.12.42}$$

Define a mapping $P_q^{-1} : \overline{\Omega} \to \overline{\Omega}_q$ by

$$P_q^{-1}(x_1, x_2) = \begin{cases} (x_1, x_2), & \text{if } (x_1, x_2) \in B(0, r_4),\\ F \circ \mathcal{D}_1 \circ F^{-1}(x_1, x_2), & \text{if } (x_1, x_2) \in \overline{\Omega} \setminus B(0, r_4). \end{cases} \tag{2.12.43}$$

Here,

$$\mathcal{D}_1\colon F^{-1}(\overline{\Omega} \setminus B(0, r_4)) \to \mathcal{D}_1(F^{-1}(\overline{\Omega} \setminus B(0, r_4))), \tag{2.12.44}$$

$$r, \varphi \to \mathcal{D}_1(r, \varphi) = (\rho, \alpha), \qquad \rho(r, \varphi) = \sum_{i=0}^{l+1} a_i(\varphi) r^i, \qquad \alpha = \varphi, \tag{2.12.45}$$

and the functions $\varphi \to a_i(\varphi)$ are determined by the relations

$$\begin{aligned} &\rho(r_4,\varphi) = r_4, \qquad \rho(r_1,\varphi) = q(\varphi), \\ &\frac{\partial \rho}{\partial r}(r_4,\varphi) = 1, \qquad \frac{\partial^k \rho}{\partial r^k}(r_4,\varphi) = 0, \qquad k = 2,3,\dots,l. \end{aligned} \tag{2.12.46}$$

Let us consider the case when $l = 1$. Then, the solution to the system (2.12.45), (2.12.46) is given by the formulas

$$a_2 = \frac{q - r_1}{(r_1 - r_4)^2}, \qquad a_1 = 1 - 2r_4 a_2, \qquad a_0 = q - r_1 a_1 - r_1^2 a_2. \tag{2.12.47}$$

We have $\frac{\partial \rho}{\partial r} = 1 + 2(r - r_4)\, a_2$, and since $q \in M_1$, it follows that $a_2(\varphi) \geq 0$ for all $\varphi \in [0, 2\pi]$. Hence, $\frac{\partial \rho}{\partial r} \geq 1$ as $r \geq r_4$, which implies that the mapping $\mathcal{D}_1$ is invertible. Therefore, there exists a mapping P_q inverse of P_q^{-1}. It is easy to see that $P_q^{-1} \in C^1(\overline{\Omega}, \overline{\Omega}_q)$, and the Jacobian of the mapping P_q^{-1} at every point $(x_1, x_2) \in \overline{\Omega}$ is not equal to zero. Now, the results on the differentiation of an implicit function (see Schwartz (1967)) imply that

$$P_q \in C^1(\overline{\Omega}_q, \overline{\Omega}).$$

Consider the case $l = 2$. Then, the solution to the system (2.12.45), (2.12.46) is given by the formulas

$$\begin{aligned} &a_3 = \frac{q - r_1}{(r_1 - r_4)^3}, \qquad a_2 = -3r_4 a_3, \qquad a_1 = 1 - 2r_4 a_2 - 3r_4^2 a_3, \\ &a_0 = q - r_1 a_1 - r_1^2 a_2 - r_1^3 a_3. \end{aligned} \tag{2.12.48}$$

From (2.12.45) and (2.12.48), we conclude

$$\frac{\partial \rho}{\partial r} = 1 + 3a_3 (r - r_4)^2.$$

Since $q \in M_2$, we get $a_3(\varphi) \geq 0$ for all $\varphi \in [0, 2\pi]$. Hence, $\frac{\partial \rho}{\partial r} \geq 1$ as $r \geq r_4$. Now, an argument similar to the above one shows that there exists a mapping P_q inverse of P_q^{-1} and

$$P_q^{-1} \in C^2(\overline{\Omega}, \overline{\Omega}_q), \qquad P_q \in C^2(\overline{\Omega}_q, \overline{\Omega}).$$

Notice that the above stated approach to the domain shape optimization is based on the construction of diffeomorphisms P_q of sets $\overline{\Omega}_q$, $q \in M_l$, onto a fixed set $\overline{\Omega}$ and on the transition to the optimization problem in the fixed domain $\overline{\Omega}$. In somewhat different form, this approach was suggested and analyzed by Murat and Simon (1976).

2.13 Optimization problems with smooth solutions of state equations

We have already studied optimization problems in which the weak (generalized) solutions of the state equations in the form of systems of elliptic equations were used. In some applications, though, only smooth solutions of the state equations are allowed. Particularly, such situations occur in the optimization problems in which restrictions are imposed on the values of the solution of the state equation and on the derivatives of this solution at points of the domain under consideration. For example, in some formulations of optimization problems for elastic solids, plates, and shells with restrictions on stiffness and strength, the displacements and stresses must be bounded at each point of the body, the stresses being defined via the derivatives of the function of displacements.

Below, we consider domain shape optimization problems for objects described by systems of equations that are elliptic in the sense of Douglis-Nirenberg, see Agmon et al. (1964).

2.13.1 Systems of elliptic equations

Let the state of some object be described by the following system of differential equations:

$$A(x,D)u(x) = f(x), \qquad x \in \Omega, \tag{2.13.1}$$

$$B(x,D)u(x) = g(x), \qquad x \in S. \tag{2.13.2}$$

Here, Ω is a bounded domain in $\mathbb{R}^n$ with a boundary S, $u = (u_1, \dots, u_m)$ and $f = (f_1, \dots, f_m)$ are m-dimensional vector-valued functions defined in Ω, $x = (x_1, \dots, x_n)$ are points of Ω, $A(x,D)$ is a square $m \times m$ matrix with elements $A_{ij}(x,D)$, $D = (D_1, \dots, D_n)$, $D_i = \frac{\partial}{\partial x_i}$, A_{ij} are polynomials in D with coefficients depending on $x \in \Omega$.

The matrix $A(x,D)$ is supposed to satisfy the following conditions. The orders of the operators $A_{ij}(x,D)$ depend on two systems of integers $s_1, \dots, s_m$, $\max_i s_i = 0$, and $t_1, \dots, t_m$, s_i corresponding to the i-th equation and t_j to the j-th dependent variable u_j, the order of the operator $A_{ij}(x,D)$ does not exceed $s_i + t_j$, and if $s_i + t_j < 0$, then $A_{ij}(x,D) = 0$.

One supposes also that the following conditions hold:

Condition of Ellipticity. At each point $x \in \Omega \cup S$, for an arbitrary $\xi = (\xi_1, \dots, \xi_n) \in \mathbb{R}^n$, $\xi \neq 0$,

$$\mathcal{H}(x,\xi) = \det\left(A'_{ij}(x,\xi)\right) \neq 0,$$

where $A'_{ij}(x,\xi)$ consists of the terms of $A_{ij}(x,\xi)$ that are exactly of the order $s_i + t_j$. The matrix with the elements $A'_{ij}(x,\xi)$ is denoted by $A^0(x,\xi)$.

Supplementary Condition. Let $x \in S$, $\xi = \zeta + \tau\nu$, where ζ and ν are arbitrary tangent and normal vectors to S at x. Then, the polynomial $\mathcal{H}(x, \zeta + \tau\nu)$ in the complex variable τ has exactly r roots with positive imaginary part and r roots with negative imaginary part. Thus, the degree of the polynomial $\mathcal{H}(x, \xi)$ (with respect to ξ) is $2r$, and

$$\sum_{i=1}^{m}(s_i + t_i) = 2r.$$

In (2.13.2), $B(x, D)$ is a rectangular matrix that has r rows and m columns with elements $B_{qj}(x, D)$, $g = (g_1, \dots, g_r)$ is an r-dimensional vector-valued function defined on S, $B_{qj}(x, D)$ are polynomials in D with coefficients depending on $x \in S$. Let β_{qj} be the order of $B_{qj}(x, \xi)$, and let $\sigma_q = \max_{j=1,\dots,m}(\beta_{qj} - t_j)$. Then, $\sigma_q + t_j \geq \beta_{qj}$. The sum of all the terms of the polynomial $B_{qj}(x, \xi)$ which are of degree $\sigma_q + t_j$ is denoted by $B^0_{qj}(x, \xi)$, and the matrix with the elements $B^0_{qj}(x, \xi)$ is designated by $B^0(x, \xi)$.

At any point $x \in S$, for arbitrary tangent $\zeta \neq 0$ and normal ν to S, we denote by $\tau_j^+(x, \zeta)$, $j = 1, \dots, r$, the roots (in τ) of the equation $\mathcal{H}(x, \zeta + \tau\nu) = 0$ with positive imaginary part. The existence of these roots is assured by the supplementary condition. Set

$$M^+(x, \zeta, \tau) = \prod_{j=1}^{r}(\tau - \tau_j^+(x, \zeta)).$$

Let $\hat{A}(x, \xi)$ denote the matrix adjoint of $A^0(x, \xi)$:

$$A^0(x, \xi)\hat{A}(x, \xi) = \mathcal{H}(x, \xi)I,$$

where I is the unit matrix.

Complementing Boundary Condition. For any $x \in S$ and any nonzero vector ζ tangent to S at x, the rows of the matrix

$$B^0(x, \zeta + \tau\nu)\hat{A}(x, \zeta + \tau\nu)$$

(which are polynomials in τ) are linearly independent modulo $M^+(x, \zeta, \tau)$ (for a detailed description, see Agmon et al. (1964), Solonnikov (1964)).

The problem (2.13.1), (2.13.2) is considered in the Hölder spaces $C^l(\overline{\Omega})$, where $l > 0$ is not an integer. The $C^l(\overline{\Omega})$ is provided with the norm

$$\|u\|_{C^l(\overline{\Omega})} = \|u\|_{C^{[l]}(\overline{\Omega})} + \sum_{|k|=[l]} \sup_{x,x' \in \Omega} \frac{|D^k u(x) - D^k u(x')|}{|x - x'|^{l-[l]}}, \tag{2.13.3}$$

where $[l]$ is an integer such that $l - [l] \in (0, 1)$, $|x| = \left(\sum_{i=1}^{n} x_i^2\right)^{\frac{1}{2}}$, and

$$\|u\|_{C^{[l]}(\overline{\Omega})} = \sum_{|k| \leq [l]} \max_{x \in \overline{\Omega}} |D^k u(x)|. \tag{2.13.4}$$

Define spaces V_l and H_l as follows:

$$\begin{aligned} V_l &= \prod_{j=1}^{m} C^{l+t_j}(\overline{\Omega}), \\ H_l &= \prod_{i=1}^{m} C^{l-s_i}(\overline{\Omega}) \times \prod_{q=1}^{r} C^{l-\sigma_q}(S). \end{aligned} \tag{2.13.5}$$

We also define an operator $L \in \mathcal{L}(V_l, H_l)$ by

$$L\colon u \to Lu = (A(x,D)u, B(x,D)u),$$

and let $\hat{V}_l = \ker L$, $\check{H}_l = L(V_l)$. Then, the following representations are valid:

$$V_l = \hat{V}_l \oplus \check{V}_l, \qquad H_l = \hat{H}_l \oplus \check{H}_l, \tag{2.13.6}$$

where $\oplus$ stands for direct sum of subspaces. The dimensions of $\hat{V}_l$ and $\hat{H}_l$ are finite, and $\hat{V}_l$ and $\hat{H}_l$ do not depend on l provided the coefficients of the operators $A(x,D)$, $B(x,D)$ and the boundary S are of the C^∞ class.

We say that (2.13.1), (2.13.2) is an elliptic problem if the condition of ellipticity, the supplementary condition, and the complementing boundary condition hold.

Theorem 2.13.1 *Let* (2.13.1), (2.13.2) *be an elliptic problem, and let l be not an integer, $l > \max(0, \sigma_1, \dots, \sigma_r)$. Let also the boundary S be of the $C^{l+t_{\max}}$ class, where $t_{\max} = \max(t_1, \dots, t_m)$, and let the coefficients of the operators $A_{ij}(x,D)$ and $B_{qj}(x,D)$ belong to $C^{l-s_i}(\overline{\Omega})$ and $C^{l-\sigma_q}(S)$, respectively. Then, the operator L is an isomorphism of $\check{V}_l$ onto $\check{H}_l$.*

A proof of this theorem can be found in Solonnikov (1966).

By using Theorem 2.13.1, it is easy to obtain the following

Theorem 2.13.2 *Assume the conditions of Theorem* 2.13.1 *hold and the dimensions of $\hat{V}_l$ and $\hat{H}_l$ are equal. Let $\{\varphi_i\}_{i=1}^k$ be a basis in $\hat{V}_l$ and let $\{\psi_i\}_{i=1}^k$ be a basis in $\hat{H}_l$. Define an operator $T \in \mathcal{L}(V_l, H_l)$ by*

$$T\varphi_i = \psi_i, \qquad i = 1, \dots, k, \qquad Tu = 0, \qquad u \in \check{V}_l.$$

Then, the operator $L_1\colon u \to L_1 u = Lu + Tu$ is an isomorphism of V_l onto H_l.

2.13.2 Elliptic problems in domains and in a fixed domain

Let M be a space of controls. We assume that M is an open set in an affine normed space X and M is provided with the topology generated by that of X. Suppose a domain Ω_q in $\mathbb{R}^n$ with a boundary S_q of the $C^{l+t_{\max}}$ and a diffeomorphism P_q of $\overline{\Omega}_q$ onto $\overline{\Omega}$ of the $C^{[l]+1+t_{\max}}$ class are given for every $q \in M$, that is,

$$P_q \in C^{[l]+1+t_{\max}}(\overline{\Omega}_q, \overline{\Omega}), \qquad P_q^{-1} \in C^{[l]+1+t_{\max}}(\overline{\Omega}, \overline{\Omega}_q). \tag{2.13.7}$$

Then, the mapping $u \to u \circ P_q$ is an isomorphism of $C^p(\overline{\Omega})$ onto $C^p(\overline{\Omega}_q)$ and of $C^p(S)$ onto $C^p(S_q)$ for any $p \in [0, l + t_{\max}]$. An elliptic problem in $\overline{\Omega}_q$ of the above type is also given for every $q \in M$

$$\begin{aligned} A_q(y, D)u(y) &= f_q(y), \qquad y \in \Omega_q, \\ B_q(y, D)u(y) &= g_q(y), \qquad y \in S_q. \end{aligned} \tag{2.13.8}$$

We denote by V_{lq} and H_{lq} the spaces V_l and H_l (see (2.13.5)) in which Ω and S are substituted by Ω_q and S_q, respectively. We suppose also that

$$\left.\begin{array}{c} \text{for any } q \in M, \text{ the operator} \\ L_q\colon u \to L_q u = (A_q(y, D)u, B_q(y, D)u) \\ \text{is an isomorphism of } V_{lq} \text{ onto } H_{lq}. \end{array}\right\} \tag{2.13.9}$$

For every $q \in M$, define an operator $\widetilde{L}_q = (\widetilde{A}_q, \widetilde{B}_q) \in \mathcal{L}(V_l, H_l)$ by

$$\begin{aligned} \widetilde{L}_q u &= (\widetilde{A}_q u, \widetilde{B}_q u), \\ \widetilde{A}_q u &= (A_q(u \circ P_q)) \circ P_q^{-1}, \\ \widetilde{B}_q u &= (B_q(u \circ P_q)) \circ P_q^{-1}. \end{aligned} \tag{2.13.10}$$

Here, A_q and B_q are the operators from (2.13.8), and we write A_q, B_q instead of $A_q(y, D)$, $B_q(y, D)$. The operator $\widetilde{L}_q$ is obtained from the operator L_q under the change of variables corresponding to the mapping P_q. Owing to (2.13.7) and (2.13.9), we have

$$\text{the operator } \widetilde{L}_q \text{ is an isomorphism of } V_l \text{ onto } H_l, \tag{2.13.11}$$

and if a function $\tilde{u} \in V_l$ is a solution of the problem

$$\begin{aligned} \widetilde{A}_q \tilde{u} &= f_q \circ P_q^{-1}, \qquad \text{in } \Omega, \\ \widetilde{B}_q \tilde{u} &= g_q \circ P_q^{-1}, \qquad \text{on } S, \end{aligned} \tag{2.13.12}$$

where $(f_q, g_q) \in H_{lq}$, then the function $u = \tilde{u} \circ P_q$ is a solution of the problem (2.13.8). On the contrary, if u is a solution of the problem (2.13.8), then $\tilde{u} = u \circ P_q^{-1}$ is a solution of (2.13.12).

Suppose that

$$\left.\begin{array}{l} q \to \widetilde{L}_q = (\widetilde{A}_q, \widetilde{B}_q) \text{ is a continuous mapping of } M \text{ into} \\ \mathcal{L}(V_l, H_l), \end{array}\right\} \tag{2.13.13}$$

$$\left.\begin{array}{l} q \to (f_q \circ P_q^{-1}, g_q \circ P_q^{-1}) \text{ is a continuous mapping of } M \\ \text{into } H_l. \end{array}\right\} \tag{2.13.14}$$

Theorem 2.13.3 *Let the conditions* (2.13.7), (2.13.9), (2.13.13), *and* (2.13.14) *hold. Then, for every $q \in M$, there exists a unique solution $\tilde{u}$ of the problem* (2.13.12), *and the function $\lambda\colon q \to \lambda(q) = \tilde{u}$ determined by this solution is a continuous mapping of M into V_l.*

Proof. The existence and uniqueness of the solution to the problem (2.13.12) follow from (2.13.7) and (2.13.9). Let $q \in M$, $\{q_k\}_{k=1}^{\infty} \subset M$, and $q_k \to q$ in M. By (2.13.13) and (2.13.14), we have

$$\widetilde{L}_{q_k} \to \widetilde{L}_q \qquad \text{in } \mathcal{L}(V_l, H_l), \tag{2.13.15}$$

$$(f_{q_k} \circ P_{q_k}^{-1}, g_{q_k} \circ P_{q_k}^{-1}) \to (f_q \circ P_q^{-1}, g_q \circ P_q^{-1}) \qquad \text{in } H_l. \tag{2.13.16}$$

From (2.13.15) and from the invertibility of the operators $\widetilde{L}_{q_k}$ and $\widetilde{L}_q$, we conclude the convergence of the inverse operators (e.g., Schwartz (1967)), that is,

$$\widetilde{L}_{q_k}^{-1} \to \widetilde{L}_q^{-1} \qquad \text{in } \mathcal{L}(H_l, V_l). \tag{2.13.17}$$

Now, from (2.13.16) and (2.13.17), we obtain that $\lambda(q_k) \to \lambda(q)$ in V_l, concluding the proof.

Determine a mapping $\mathcal{T}\colon M \times V_l \to H_l$ by

$$\begin{gathered} q \in M, \quad u \in V_l \\ \mathcal{T}(q,u) = (\widetilde{A}_q u - f_q \circ P_q^{-1}, \widetilde{B}_q u - g_q \circ P_q^{-1}). \end{gathered} \tag{2.13.18}$$

Obviously, the function $\lambda\colon M \to V_l$ introduced in the formulation of Theorem 2.13.3 is the implicit function determined by the mapping $\mathcal{T}$, that is,

$$\mathcal{T}(q, \lambda(q)) = 0, \tag{2.13.19}$$

and $\lambda(q) = \tilde{u}$, where $\tilde{u}$ is the solution of the problem (2.13.12). The existence and continuity of the implicit function λ follow from Theorem 2.13.3. Consider now the question of the differentiability of λ.

Theorem 2.13.4 *Let the conditions* (2.13.7), (2.13.9) *hold and let $q \to \widetilde{L}_q = (\widetilde{A}_q, \widetilde{B}_q)$, $q \to (f_q \circ P_q^{-1}, g_q \circ P_q^{-1})$ be Fréchet continuously differentiable mappings of M into $\mathcal{L}(V_l, H_l)$ and into H_l, respectively. Then, the function λ determined by the equation* (2.13.19) *is a Fréchet continuously differentiable mapping of M into V_l, and the Fréchet derivative λ' of λ at point $q \in M$ is given by*

$$\lambda'(q)h = -\widetilde{L}_q^{-1} \circ \frac{\partial \mathcal{T}}{\partial q}(q, \lambda(q))h, \qquad h \in X, \tag{2.13.20}$$

where the operator $\frac{\partial \mathcal{T}}{\partial q}(q,u)$ is given by the formula

$$\begin{gathered} \frac{\partial \mathcal{T}}{\partial q}(q,u)h = \{(\widetilde{A}_q' h)u - (f_q \circ P_q^{-1})'h, (\widetilde{B}_q' h)u - (g_q \circ P_q^{-1})'h\}, \\ h \in X. \end{gathered} \tag{2.13.21}$$

Here, $\widetilde{A}'_q$, $\widetilde{B}'_q$, $(f_q \circ P_q^{-1})'$, $(g_q \circ P_q^{-1})'$ *are the Fréchet derivatives at a point* q *of the mappings* $q \to \widetilde{A}_q$, $q \to \widetilde{B}_q$, $q \to (f_q \circ P_q^{-1})$, $q \to (g_q \circ P_q^{-1})$.

Proof. It is obvious that $\mathcal{T}$ is a Fréchet continuously differentiable mapping of $M \times V_l$ into H_l. Then, $\frac{\partial \mathcal{T}}{\partial q}(q,u)$ is determined by the formula (2.13.21), and $\frac{\partial \mathcal{T}}{\partial u}(q,u) = \widetilde{L}_q = (\widetilde{A}_q, \widetilde{B}_q)$. From here, taking into account (2.13.9), we obtain that the operator $\frac{\partial \mathcal{T}}{\partial u}(q,u)$ is an isomorphism of V_l onto H_l. Now, the theorem follows from Theorem 1.9.2.

Remark 2.13.1 We assumed above the condition (2.13.9) to hold. Let it be not satisfied. Then, upon (2.13.6), the following representations are valid

$$V_{lq} = \hat{V}_{lq} \oplus \check{V}_{lq}, \qquad H_{lq} = \hat{H}_{lq} \oplus \check{H}_{lq}, \tag{2.13.22}$$

where $\hat{V}_{lq} = \ker L_q$, $\check{H}_{lq} = L_q(V_{lq})$. Suppose that

$$\dim \hat{V}_{lq} = \dim \hat{H}_{lq} = k_q, \qquad q \in M, \tag{2.13.23}$$

where k_q is a natural number. Analogously to the operator T from Theorem 2.13.2, we define an operator T_q by

$$T_q \varphi_{qi} = \psi_{qi}, \qquad i = 1, \ldots, k_q, \qquad T_q u = 0, \qquad u \in \check{V}_{lq}, \tag{2.13.24}$$

where φ_{qi} and ψ_{qi} are basis functions in $\hat{V}_{lq}$ and $\hat{H}_{lq}$. Then, the operator $L_{q1} = L_q + T_q$ is an isomorphism of V_{lq} onto H_{lq}. Therefore, under suitable conditions, the results stated above remain true if we substitute the operator L_q for L_{q1}.

It should be noted that

$$\varkappa(L_q) = \dim \hat{V}_{lq} - \dim \hat{H}_{lq}$$

is called the index of the elliptic operator $L_q = (A_q, B_q)$. So, (2.13.23) means that

$$\varkappa(L_q) = 0, \qquad q \in M. \tag{2.13.25}$$

It is known that the index is an unstable characteristic of an operator (see, e.g., Rempel and Schulze (1982)), and for an arbitrary elliptic operator L_q the condition (2.13.25) may be unsatisfiable. But usually, in physical problems, the spaces $\hat{V}_{lq}$ and $\hat{H}_{lq}$ have some physical meaning that does not depend on a control q, i.e., on the shape of the domain Ω_q, and because of this meaning the dimensions of the spaces $\hat{V}_{lq}$ and $\hat{H}_{lq}$ are equal (see, e.g., Section 5.9). So, (2.13.23) is a natural assumption for applied problems.

2.13.3 The problem of domain shape optimization

Let functionals Ψ_i over $M \times V_l$ be given such that

$$\left. \begin{array}{l} (q,u) \to \Psi_i(q,u) \text{ is a continuous mapping of } M \times V_l \text{ into} \\ \mathbb{R},\ i = 0, 1, \ldots, k. \end{array} \right\} \tag{2.13.26}$$

Define functionals Φ_i over M by

$$\Phi_i(q) = \Psi_i(q, \lambda(q)), \qquad i = 0, 1, \dots, k, \tag{2.13.27}$$

where $\lambda(q)$ is determined by the expression (2.13.19). Let M_1 be a compact set in M. We take a set of admissible controls U_{ad} in the form

$$U_{\text{ad}} = \big\{\, q \mid q \in M_1,\ \Phi_i(q) \le 0,\ i = 1, 2, \dots, k \,\big\}. \tag{2.13.28}$$

The optimization problem consists in finding q_0 such that

$$q_0 \in U_{\text{ad}}, \qquad \Phi_0(q_0) = \inf_{q \in U_{\text{ad}}} \Phi_0(q). \tag{2.13.29}$$

Theorem 2.13.5 *Let the conditions* (2.13.7), (2.13.9), (2.13.13), (2.13.14), *and* (2.13.26) *hold, let M_1 be a compact set in M, and let a nonempty set U_{ad} be determined by* (2.13.28). *Then, there exists a solution of the problem* (2.13.29).

Proof. As M_1 is a nonempty set, there exists a minimizing sequence $\{q_n\}$ such that

$$\{q_n\} \subset U_{\text{ad}}, \qquad \lim \Phi_0(q_n) = \inf_{q \in U_{\text{ad}}} \Phi_0(q). \tag{2.13.30}$$

As M_1 is a compact set in M, we can choose a subsequence $\{q_m\}$ such that $q_m \to z$ in M, $z \in M_1$. By Theorem 2.13.3, $\lambda(q_m) \to \lambda(z)$ in V_l, where $q(z)$ is a solution of the problem (2.13.12) with $q = z$. It is easily seen that $q_0 = z$ is a solution of the problem (2.13.29), concluding the proof.

In connection with finding a solution of the problem (2.13.29), there arises the question of differentiability of the functionals Φ_i. By using Theorem 2.13.4 and the chain rule, we obtain the following

Theorem 2.13.6 *Let the conditions of Theorem* 2.13.4 *hold, and let $\Psi_i\colon (q,u) \to \Psi_i(q,u)$ be a Fréchet continuously differentiable mapping of $M \times V_l$ into $\mathbb{R}$. Then, the functional Φ_i defined by* (2.13.27) *is a Fréchet continuously differentiable mapping of M into $\mathbb{R}$, and the Fréchet derivative Φ_i' of Φ_i at a point $q \in M$ is given by*

$$\Phi_i'(q)h = \frac{\partial \Psi_i}{\partial q}(q, \lambda(q))h + \left(\frac{\partial \Psi_i}{\partial u}(q, \lambda(q)) \circ \lambda'(q)\right)h, \qquad h \in X.$$

2.13.4 Approximate solution of the direct problem ensuring convergence in the norm of a space of smooth functions

Approximate solution of the optimization problem (2.13.29) is connected with obtaining approximations of the direct problem (2.13.1), (2.13.2) (or (2.13.12)) which converge to the exact solution in the norm of the space V_l. In many applications there also arises the problem of construction of approximate solutions converging to the exact one in a strong norm. Galerkin and Riesz methods provide convergence in weak norms corresponding to generalized solutions, and so these cannot be used in that case. We consider now a method of minimization of residual for the problem (2.13.1), (2.13.2).

Let us examine first the case when $\hat{V}_l$ contains only zero element and $(f, g) \in \check{H}_l$. Then, by Theorem 2.13.1, there exists a unique solution u of the problem (2.13.1), (2.13.2) and the following estimate holds:

$$\|u\|_{V_l} \leq c\|(f, g)\|_{H_l}, \tag{2.13.31}$$

where $\|\cdot\|_{V_l}$ and $\|\cdot\|_{H_l}$ are the norms of the spaces V_l and H_l and c is a constant independent of f and g.

Let $\{w_i\}$ be a sequence such that

1. $w_i \in V_l$ for all i;
2. the elements $w_1, \ldots, w_\nu$ are linearly independent for an arbitrary ν;
3. finite linear combinations $\sum c_i w_i$ $(c_i \in \mathbb{R})$ are dense in V_l.

Let $V_{l\nu}$ be the linear span of the elements $w_1, \ldots, w_\nu$, i.e.,

$$V_{l\nu} = \Big\{ v \mid v = \sum_{i=1}^{\nu} c_i w_i, \ c_i \in \mathbb{R} \Big\}.$$

Define an approximate solution u_ν of the problem (2.13.1), (2.13.2) as follows:

$$u_\nu \in V_{l\nu}, \qquad \|Lu_\nu - (f, g)\|_{H_l} = \min_{v \in V_{l\nu}} \|Lv - (f, g)\|_{H_l}. \tag{2.13.32}$$

One can see that, under the above mentioned assumptions, for each ν there exists a solution of the problem (2.13.32) and

$$\lim_{\nu \to \infty} \|u_\nu - u\|_{V_l} = 0. \tag{2.13.33}$$

However, (2.13.32) is a difficult problem of minimization of a maximum function. So, from the viewpoint of computation, it is convenient to replace the space H_l with a Hilbert space $\mathcal{H}_l$ given by

$$\mathcal{H}_l = \prod_{i=1}^{m} H^{\beta_i}(\Omega) \times \prod_{q=1}^{r} H^{\gamma_q}(S),$$

$$C^{l-s_i}(\overline{\Omega}) \supset H^{\beta_i}(\Omega), \qquad C^{l-\sigma_q}(S) \supset H^{\gamma_q}(S), \qquad i = 1, \ldots, m,$$

$q = 1, \ldots, r$ (see (2.13.5)), β_i and γ_q being integers. We suppose that $(f, g) \in \mathcal{H}_l$ and the functions w_i satisfy Conditions 1, 2, 3 with V_l replaced by V_{l_1} such that $\mathcal{H}_l \supset V_{l_1}$. The space V_{l_1} may be defined by the embedding theorem. Then, we can replace the problem (2.13.32) with the following one: Find $\tilde{u}_\nu$ satisfying

$$\begin{aligned} &\tilde{u}_\nu \in V_{l_1\nu}, \\ &\|L\tilde{u}_\nu - (f, g)\|^2_{\mathcal{H}_l} = \min_{v \in V_{l_1\nu}} \|Lv - (f, g)\|^2_{\mathcal{H}_l}. \end{aligned} \tag{2.13.34}$$

This is the problem of minimization of a quadratic functional and it is significantly easier than the problem (2.13.32). For an arbitrary ν there exists a solution of the problem (2.13.34) and it may be shown that the above mentioned conditions imply $\tilde{u}_\nu \to u$ in V_l.

Suppose now the dimension of the space $\hat{V}_l$ is not equal to zero and $(f,g) \in \check{H}_l$. By Theorem 2.13.1, there exists a unique solution u of the problem (2.13.1), (2.13.2) such that $u \in \check{V}_l$. Let $\{\varphi_1, \dots, \varphi_k\}$ be a basis in $\hat{V}_l$. For $\nu > k$, define subspaces $\widetilde{V}_{l_1\nu} \subset V_{l_1}$ as follows:

$$\widetilde{V}_{l_1\nu} = \big\{ v \mid v \in V_{l_1\nu},\ (v, \varphi_i)_{L_2(\Omega)^m} = 0,\ i = 1, \dots, k \big\}. \tag{2.13.35}$$

Define an approximate solution $\check{u}_\nu$ of the problem (2.13.1), (2.13.2) by

$$\begin{aligned} &\check{u}_\nu \in \widetilde{V}_{l_1\nu}, \\ \|L\check{u}_\nu - (f,g)\|^2_{\mathcal{H}_l} &= \min_{v \in \widetilde{V}_{l_1\nu}} \|Lu - (f,g)\|^2_{\mathcal{H}_l}. \end{aligned} \tag{2.13.36}$$

The problem (2.13.36) may be solved by, e.g., Lagrange multipliers, and it may be shown that $\check{u}_\nu \to u$ in $\check{V}_l$.

Chapter 3

Control by the Right-hand Sides in Elliptic Problems

"'Let us sit on this log at the road side,' says I, 'and forget the inhumanity and ribaldry of the poets. It is in the glorious columns of ascertained facts and legalised measures that beauty is to be found.'"

– *O. Henry*
"The Handbook of Hymen"

3.1 On the minimum of nonlinear functionals

3.1.1 Setting of the problem. Auxiliary statements

Let U be a reflexive Banach space, and suppose

$$\left.\begin{array}{l} u, v \to \pi(u,v) \text{ is a bilinear, symmetric, continuous,} \\ \text{positive form on } U \times U, \end{array}\right\} \tag{3.1.1}$$

i.e.,

$$\begin{gathered} \pi(u,v) = \pi(v,u), \qquad u, v \in U, \\ |\pi(u,v)| \le c\|u\|_U \|v\|_U, \qquad u, v \in U,\ c = \text{const} > 0, \\ \pi(u,u) \ge 0, \qquad u \in U. \end{gathered}$$

Suppose also that

$$v \to Q(v) \ \text{ is a linear continuous form on } U, \tag{3.1.2}$$

$$U_{\text{ad}} \text{ is a convex, closed, bounded set in } U. \tag{3.1.3}$$

Consider the quadratic functional

$$J(v) = \pi(v, v) - 2Q(v), \tag{3.1.4}$$

which is to be minimized on the set U_{ad}.

Before considering this problem, we state two auxiliary statements.

Lemma 3.1.1 *Let $u, v \to \pi(u, v)$ be a bilinear, symmetric, continuous, positive form on $U \times U$, where U is a Banach space. Then, the function $v \to \pi(v, v)$ is lower semicontinuous in the weak topology of the space U.*

Proof. It is easy to verify the following equality

$$\pi(v, v) = \pi(u, u) + 2\pi(u, v - u) + \pi(v - u, v - u), \qquad u, v \in U. \tag{3.1.5}$$

Let $v_n \to v_0$ weakly in U. Setting in (3.1.5) $v = v_n$, $u = v_0$, we get

$$\pi(v_n, v_n) = \pi(v_0, v_0) + 2\pi(v_0, v_n - v_0) + \pi(v_n - v_0, v_n - v_0). \tag{3.1.6}$$

Since the form $\pi(u, v)$ is continuous on $U \times U$, the function $v \to \pi(v_0, v)$ is continuous in the weak topology of the space U. Hence,

$$\lim_{n \to \infty} \pi(v_0, v_n - v_0) = 0.$$

Making use of this equality, (3.1.6), and of the fact that $\pi(u, u) \geq 0$ for all $u \in U$, we obtain

$$\liminf_{n \to \infty} \pi(v_n, v_n) \geq \pi(v_0, v_0).$$

Lemma 3.1.2 *Let $u, v \to \pi(u, v)$ be a bilinear, symmetric, continuous form on $U \times U$, where U is a Banach space, and let*

$$\pi(u, u) \geq 0, \qquad u \in U. \tag{3.1.7}$$

Then, the function $u \to \pi(u, u)$ is convex, i.e., for all $u_1, u_2 \in U$ and for all $t \in (0, 1)$, the following estimate is true:

$$\pi((1 - t)u_1 + tu_2, (1 - t)u_1 + tu_2) \leq (1 - t)\pi(u_1, u_1) + t\pi(u_2, u_2). \tag{3.1.8}$$

If the equality in (3.1.7) occurs only for $u = 0$, then the function $u \to \pi(u, u)$ is strictly convex, i.e., the equality in (3.1.8) takes place only when $u_1 = u_2$.

Proof. Introduce the notation

$$u_0 = (1 - t)u_1 + tu_2. \tag{3.1.9}$$

By setting in (3.1.5) $v = u_1$, $u = u_0$, and taking into consideration (3.1.7), we have

$$\pi(u_1, u_1) \geq \pi(u_0, u_0) + 2\pi(u_0, u_1 - u_0). \tag{3.1.10}$$

In the same way, we get

$$\pi(u_2, u_2) \geq \pi(u_0, u_0) + 2\pi(u_0, u_2 - u_0). \tag{3.1.11}$$

If the function $u \to \pi(u,u)$ vanishes only at $u = 0$, then the equality in (3.1.10) holds only for $u_1 = u_0$, and in (3.1.11) it holds only for $u_2 = u_0$, i.e., if and only if $u_1 = u_2$.

By multiplying the inequalities (3.1.10) and (3.1.11) by $1-t$ and t, respectively, where $t \in (0,1)$, and summing them up, we obtain in view of (3.1.9) that

$$\begin{aligned}(1-t)\pi(u_1, u_1) &+ t\pi(u_2, u_2)\\ &\geq \pi(u_0, u_0) + 2\pi(u_0, (1-t)u_1 + tu_2 - u_0)\\ &= \pi(u_0, u_0) + 2\pi(u_0, 0) = \pi(u_0, u_0).\end{aligned} \tag{3.1.12}$$

If the condition $\pi(u,u) = 0$ yields $u = 0$, then in (3.1.12) the equality holds only when $u_1 = u_2$. Now, (3.1.8) is implied by (3.1.9) and (3.1.12), concluding the proof.

3.1.2 The existence theorem

Theorem 3.1.1 *Let the conditions* (3.1.1)–(3.1.3) *hold, and let a functional* $J(v)$ *be determined by the relation* (3.1.4). *Then, the subset* X *defined by the formula*

$$X = \{\, u \,|\, u \in U_{\text{ad}},\ J(u) = \inf_{v \in U_{\text{ad}}} J(v) \,\}$$

is nonempty, closed in U_{ad}, *and convex. If the function* $v \to \pi(v,v)$ *vanishes only at* $v = 0$, *then the set* X *contains only one element.*

Proof. Let $\{v_n\} \subset U_{\text{ad}}$ be a minimizing sequence, i.e.,

$$J(v_n) \to \inf_{v \in U_{\text{ad}}} J(v). \tag{3.1.13}$$

By (3.1.3), we conclude

$$\|v_n\|_U \leq \text{const} \qquad \forall n.$$

Therefore, we can choose from the sequence $\{v_n\}$ a subsequence $\{v_m\}$ such that

$$v_m \to w \ \text{ weakly in } U. \tag{3.1.14}$$

Since U_{ad} is a closed, convex set in the Banach space U, by the Mazur theorem (see, e.g., Yosida (1971)), U_{ad} is sequentially weakly closed. Hence, (3.1.14) yields

$$w \in U_{\text{ad}}. \tag{3.1.15}$$

By virtue of (3.1.2), the function $v \to Q(v)$ is continuous in the weak topology of the space U. This and Lemma 3.1.1 imply that the function $v \to J(v)$ is lower semicontinuous in the weak topology of U. Now, due to (3.1.14), we have

$$\liminf_{m \to \infty} J(v_m) \geq J(w).$$

Then, by (3.1.13) and (3.1.15), we conclude

$$w \in U_{\mathrm{ad}}, \qquad J(w) \leq \inf_{v \in U_{\mathrm{ad}}} J(v).$$

Thus,

$$J(w) = \inf_{v \in U_{\mathrm{ad}}} J(v),$$

and the set X is nonempty. Let $\{u_n\}$ be a sequence of elements from X, and let $u_n \to u$ in U. Then

$$u \in U_{\mathrm{ad}}, \qquad J(u) = \lim_{n \to \infty} J(u_n) = \inf_{v \in U_{\mathrm{ad}}} J(v).$$

Hence, $u \in X$ and the set X is closed in U_{ad}.

To prove that X is convex, take $u_1, u_2 \in X$; then

$$J(u_1) = J(u_2) = \inf_{v \in U_{\mathrm{ad}}} J(v). \tag{3.1.16}$$

By virtue of Lemma 3.1.2, the function $v \to \pi(v, v)$ is convex. Hence, the function $v \to J(v)$ is also convex, and if $t \in (0, 1)$, then

$$J((1-t)u_1 + tu_2) \leq (1-t)J(u_1) + tJ(u_2) = \inf_{v \in U_{\mathrm{ad}}} J(v).$$

From (3.1.3)

$$(1-t)u_1 + tu_2 \in U_{\mathrm{ad}},$$

consequently,

$$(1-t)u_1 + tu_2 \in X.$$

Now, let the function $v \to \pi(v, v)$ vanish only at $v = 0$. Let us show that, in this case, the set X contains only one element.

Suppose that there exist two elements u_1, u_2 of the set U_{ad} such that $u_1 \neq u_2$ and (3.1.16) holds. Since the function $v \to \pi(v, v)$ vanishes only at $v = 0$, by virtue of Lemma 3.1.2 this function is strictly convex. Therefore, the function $v \to J(v)$ is also strictly convex, and

$$J\Big(\frac{1}{2}(u_1 + u_2)\Big) < \frac{1}{2}(J(u_1) + J(u_2)) = \inf_{v \in U_{\mathrm{ad}}} J(v). \tag{3.1.17}$$

Since the set U_{ad} is convex, we have $\frac{1}{2}(u_1 + u_2) \in U_{\mathrm{ad}}$. So, the relation (3.1.17) contains a contradiction. Hence, $u_1 = u_2$, and the theorem is proved.

Remark 3.1.1 Let the above assumptions hold and let the function $v \to \pi(v, v)$ meet the condition

$$\pi(v, v) \geq c\|v\|_U^2, \qquad v \in U, \ c = \text{const} > 0. \tag{3.1.18}$$

Then, due to the following inequality

$$J(v) = \pi(v,v) - 2Q(v) \geq c\|v\|_U^2 - c_1\|v\|_U,$$

a sequence $\{v_n\}$ satisfying the condition (3.1.13) is bounded in U. Therefore, if U_{ad} is a closed, convex set in U (which is not necessarily bounded), then there exists a function $u \in U_{\text{ad}}$ such that

$$J(u) = \inf_{v \in U_{\text{ad}}} J(v). \tag{3.1.19}$$

Moreover, due to (3.1.18), the function $v \to \pi(v,v)$ vanishes only at $v = 0$. So, there exists a unique element $u \in U_{\text{ad}}$ satisfying (3.1.19).

Remark 3.1.2 Let U be endowed with the strong topology, and let U_{ad} be a compactum in U (which is not necessarily convex). By (3.1.1), (3.1.2), and (3.1.4), the function $v \to J(v)$ is continuous in the strong topology of the space U, so that there exists an element $u \in U_{\text{ad}}$ that satisfies the condition (3.1.19).

Remark 3.1.3 In the proof of Theorem 3.1.1, we needed the closedness and convexity of the set U_{ad} only to derive (3.1.15) from (3.1.14). Therefore, the statement of Theorem 3.1.1 remains valid if we assume that the set U_{ad} is bounded in the strong topology of U and sequentially weakly closed, instead of the condition (3.1.3).The sequential weak closedness means that, if $\{v_n\} \subset U_{\text{ad}}$, $v_n \to v$ weakly in U, then $v \in U_{\text{ad}}$ (see Subsec. 1.2.3).

3.1.3 Characterization of a minimizing element

Theorem 3.1.2 *If the assumptions of Theorem* 3.1.1 *hold, then an element u of U_{ad} belongs to the set X if and only if the following inequality holds*

$$\pi(u, v-u) \geq Q(v-u), \qquad v \in U_{\text{ad}}. \tag{3.1.20}$$

Proof. Let $u \in X$. Since the set U_{ad} is convex, for all $v \in U_{\text{ad}}$ and $t \in (0,1)$ we have

$$J(u) \leq J((1-t)u + tv),$$

whence

$$\frac{J(u + t(v-u)) - J(u)}{t} \geq 0, \qquad t \in (0,1). \tag{3.1.21}$$

The functional $J(v)$ being defined by the equality (3.1.4), we easily conclude that, for all $w, h \in U$, the following relation is valid

$$\lim_{t \to 0} \frac{J(w+th) - J(w)}{t} = J'(w)h = 2[\pi(w,h) - Q(h)].$$

Here, $J'(w)$ is the linear continuous form on U given by

$$y \to J'(w)y = 2[\pi(w,y) - Q(y)]. \tag{3.1.22}$$

So, in (3.1.21), we can pass to the limit as t tends to zero from the right. Thus, we get

$$J'(u)(v-u) \geq 0, \qquad v \in U_{\text{ad}}. \tag{3.1.23}$$

Therefore, from (3.1.22) we deduce (3.1.20).

Conversely, let the relation (3.1.20) hold (so that (3.1.23) holds, too). By Lemma 3.1.2, the function $v \to J(v)$ is convex, and so for any $t \in (0,1)$

$$J(v) - J(w) \geq \frac{J((1-t)w + tv) - J(w)}{t}, \qquad v, w \in U.$$

Passing to the limit as $t \to 0$ on the right-hand side of this inequality, we obtain

$$J(v) - J(w) \geq J'(w)(v-w).$$

Setting here $w = u$ and taking into account (3.1.23), we deduce that

$$J(v) - J(u) \geq 0, \qquad v \in U_{\text{ad}},$$

concluding the proof.

Remark 3.1.4 Suppose that $\pi(u,v)$ is a bilinear, symmetric, continuous, coercive form on $U \times U$, where U is a Hilbert space over $\mathbb{R}$. Further, let $v \to Q(v)$ be a linear continuous form on U, and let U_{ad} be a closed, convex cone with peak at the origin of coordinates. Then, there exists a unique function $u \in U_{\text{ad}}$ satisfying the conditions (3.1.19) and (3.1.20) (see Remark 3.1.1). Moreover, (3.1.20) is equivalent to the following relations

$$\pi(u,v) \geq Q(v), \qquad v \in U_{\text{ad}}, \qquad \pi(u,u) = Q(u).$$

Indeed, since U_{ad} is a cone, $u + v \in U_{\text{ad}}$ for all $v \in U_{\text{ad}}$. By substituting in (3.1.20) $u+v$ for v, we get the first inequality. By setting $v = 0$ in (3.1.20), we obtain $\pi(u,u) \leq Q(u)$, hence, $\pi(u,u) = Q(u)$. Conversely, adding the equality $\pi(u,-u) = Q(-u)$ to the inequality $\pi(u,v) \geq Q(v)$, we have (3.1.20).

3.1.4 Functionals continuous in the weak topology

Hitherto, $v \to J(v)$ was assumed to be a quadratic functional. Below, we will present a statement in which the functional $J(v)$ is not supposed to be quadratic.

Assume, as before, that U is a reflexive Banach space, and

$$\left.\begin{array}{l} U_{\text{ad}} \subset U,\ U_{\text{ad}} \text{ is bounded in the strong topology of } U \text{ and} \\ \text{sequentially weakly closed.} \end{array}\right\} \tag{3.1.24}$$

Suppose also that

$$\left.\begin{array}{l} v \to J(v) \text{ is a lower semicontinuous mapping of } U_{\text{ad}} \text{ into } \mathbb{R} \\ \text{with respect to the topology generated by the weak one of} \\ \text{the space } U. \end{array}\right\} \tag{3.1.25}$$

Theorem 3.1.3 *If the conditions* (3.1.24) *and* (3.1.25) *are fulfilled, then there exists a function* u *such that*

$$u \in U_{\mathrm{ad}}, \qquad J(u) = \inf_{v \in U_{\mathrm{ad}}} J(v). \tag{3.1.26}$$

Proof. Let $\{v_n\}$ be a minimizing sequence, i.e.,

$$v_n \in U_{\mathrm{ad}} \quad \forall n, \qquad \lim_{n\to\infty} J(v_n) = \inf_{v \in U_{\mathrm{ad}}} J(v). \tag{3.1.27}$$

By (3.1.24), we conclude

$$\|v_n\| \le \mathrm{const} \qquad \forall n.$$

Hence, we can choose from the sequence $\{v_n\}$ a subsequence $\{v_m\}$ such that

$$v_m \to w \qquad \text{weakly in } U. \tag{3.1.28}$$

Since U_{ad} is sequentially weakly closed by the hypothesis, (3.1.28) yields

$$w \in U_{\mathrm{ad}}. \tag{3.1.29}$$

By combining (3.1.25), (3.1.27), and (3.1.28), we conclude

$$\inf_{v \in U_{\mathrm{ad}}} J(v) = \liminf_{m\to\infty} J(v_m) \ge J(w).$$

Therefore, from (3.1.29) we deduce the function $u = w$ to satisfy the relation (3.1.26).

3.2 Approximate solution of the minimization problem

3.2.1 Inner point lemma

Below, we will need the following

Lemma 3.2.1 *Let* A *be a convex set in the Banach space* U *which contains at least one inner point* y_0, *and let* $\overline{A}$ *be the closure of* A. *If* $y \in \overline{A}$, *then every point of the open interval with the ends* y_0 *and* y *is an inner point of the set* A.

Proof. Let z be an arbitrary point of the open interval with the ends y and y_0, i.e.,

$$z = (1-t)y + ty_0 = y + t(y_0 - y), \qquad t \in (0,1). \tag{3.2.1}$$

Further, let $g \to f(g)$ be a mapping of U into U defined by the relation

$$f(g) = z + \lambda(g - z), \tag{3.2.2}$$

where

$$\lambda = -\frac{t}{1-t}. \tag{3.2.3}$$

Obviously, $g \to f(g)$ is continuous; the inverse mapping is given by

$$y \to f^{-1}(y) = \frac{1}{\lambda}\,(y + (\lambda - 1)z),$$

and maps U into U continuously, too. Hence,

$$g \to f(g) \ \text{ is a homeomorphism of } U \text{ onto } U. \tag{3.2.4}$$

Taking into consideration (3.2.1), one can easily see that

$$f(y_0) = y. \tag{3.2.5}$$

If Q is an open set in U containing the point y_0 and belonging to A, then, by virtue of (3.2.4), $f(Q)$ is an open set in U, $y \in f(Q)$ (because of (3.2.5)), and there exists a point h such that

$$h \in Q, \qquad f(h) \in A.$$

Indeed, if $\gamma > 0$ and sufficiently small, then the following inclusion holds

$$q = y + \gamma(y_0 - y) \in f(Q), \qquad q \in A.$$

So, we can take $h = f^{-1}(q)$.

By (3.2.2), we get

$$f(h) - z = \lambda(h - z) = \lambda(h - f(h)) + \lambda(f(h) - z).$$

Hence,

$$z = f(h) + \frac{\lambda}{\lambda - 1}\,(h - f(h)), \tag{3.2.6}$$

i.e., z is the image of the function h under the mapping

$$g \to \varphi(g) = f(h) + \frac{\lambda}{\lambda - 1}\,(g - f(h)). \tag{3.2.7}$$

Since $\lambda < 0$ in view of (3.2.3), we get

$$0 < \frac{\lambda}{\lambda - 1} < 1. \tag{3.2.8}$$

From this we conclude that the function $g \to \varphi(g)$ is a homeomorphism of U onto U, and so

$$\varphi(Q) \ \text{ is an open set in } U. \tag{3.2.9}$$

Since $h \in Q$, from (3.2.6) we deduce that

$$z \in \varphi(Q). \tag{3.2.10}$$

Let g be an arbitrary point from Q. Since $f(h) \in A$ and the set A is convex, we conclude that $\varphi(g) \in A$ from (3.2.7) and (3.2.8). Thus,

$$\varphi(Q) \subset A. \tag{3.2.11}$$

Finally, by (3.2.9)–(3.2.11), we conclude z to be an inner point of the set A. The lemma is proved.

3.2.2 Finite-dimensional problem

Below we suppose U to be a reflexive, separable Banach space. Then, there exists a sequence of finite-dimensional subspaces $\{U_n\} \subset U$ which satisfies the limit density condition:

$$\lim_{n\to\infty} \inf_{v\in U_n} \|v - w\|_U = 0, \qquad w \in U. \tag{3.2.12}$$

The problem of minimization of the functional $J(v)$ on U_{ad} is replaced by the following one: Find a function u_n such that

$$u_n \in U_n \cap U_{\text{ad}}, \qquad J(u_n) = \inf_{v\in U_n\cap U_{\text{ad}}} J(v). \tag{3.2.13}$$

Finite-dimensional problem for the quadratic functional

Theorem 3.2.1 *Let U be a reflexive, separable Banach space, let the assumptions of Theorem* 3.1.1 *be satisfied, and let $\{U_n\}$ be a sequence of finite-dimensional subspaces of U meeting the relation* (3.2.12). *Suppose that the set U_{ad} is endowed with the topology generated by the strong one of the space U, and in this topology the set U_{ad} has at least one inner point. Then, for all n sufficiently large, the problem* (3.2.13) *has a solution u_n and*

$$\lim_{n\to\infty} J(u_n) = \inf_{v\in U_{\text{ad}}} J(v). \tag{3.2.14}$$

One can choose a subsequence $\{u_m\}$ of the sequence $\{u_n\}$ such that $u_m \to u$ weakly in U, and $u \in X$. If the set X consists of one element u only, then the sequence $\{u_n\}$ weakly converges to u in U.

Proof. Let $u \in X$, i.e.,

$$u \in U_{\text{ad}}, \qquad J(u) = \inf_{v\in U_{\text{ad}}} J(v). \tag{3.2.15}$$

Denote by $\overset{\circ}{U}_{\text{ad}}$ the interior of U_{ad}. By the hypothesis, there exists a function g belonging to $\overset{\circ}{U}_{\text{ad}}$.

Let $\{t_i\}_{i=1}^{\infty}$ be a sequence such that $t_i \in (0,1)$ for all i, and $\lim_{i\to\infty} t_i = 0$. From Lemma 3.2.1, we deduce that

$$z_i = u + t_i(g - u) \in \overset{\circ}{U}_{\text{ad}},$$

and

$$\lim_{i\to\infty} \|z_i - u\|_U = 0. \tag{3.2.16}$$

Let ε be an arbitrary positive number. In view of (3.2.16), we can choose $i = i_\varepsilon$ such that

$$\|z_{i_\varepsilon} - u\|_U \le \frac{\varepsilon}{2}. \tag{3.2.17}$$

Since $z_{i_\varepsilon} \in \overset{\circ}{U}_{\mathrm{ad}}$, there exists a constant γ such that

$$0 < \gamma \leq \frac{\varepsilon}{2}, \qquad d(\gamma, z_{i_\varepsilon}) \subset U_{\mathrm{ad}}, \tag{3.2.18}$$

where

$$d(\gamma, z_{i_\varepsilon}) = \{ v \,|\, v \in U, \ \|v - z_{i_\varepsilon}\|_U \leq \gamma \}. \tag{3.2.19}$$

By (3.2.12), there is $l > 0$ such that, for each $n \geq l$, there exists $g_n \in U_n$ satisfying

$$\|g_n - z_{i_\varepsilon}\|_U \leq \gamma. \tag{3.2.20}$$

From (3.2.17), (3.2.18), and (3.2.20), we deduce that

$$\|g_n - u\|_U \leq \|g_n - z_{i_\varepsilon}\|_U + \|z_{i_\varepsilon} - u\|_U \leq \varepsilon, \qquad n \geq l,$$

and the relations (3.2.18)–(3.2.20) yield that $g_n \in U_{\mathrm{ad}}$.

From the above argument, we conclude the existence of a sequence $\{g_n\}_{n=l}^{\infty}$ such that

$$g_n \in U_n \cap U_{\mathrm{ad}}, \qquad \lim_{n\to\infty} \|g_n - u\|_U = 0. \tag{3.2.21}$$

Hence,

$$\lim_{n\to\infty} J(g_n) = J(u) = \inf_{v \in U_{\mathrm{ad}}} J(v). \tag{3.2.22}$$

When $n \geq l$, the set $U_n \cap U_{\mathrm{ad}}$ is nonempty. By (3.1.3), $U_n \cap U_{\mathrm{ad}}$ is a compactum, and since $v \to J(v)$ is a continuous mapping of U into $\mathbb{R}$, the problem (3.2.13) has a solution u_n. Further, (3.2.13), (3.2.15), and (3.2.21) imply that

$$J(g_n) \geq J(u_n) \geq J(u).$$

Combining this with (3.2.22), we get (3.2.14).

By virtue of (3.1.3), the sequence $\{u_n\}$ is bounded in U. Let us choose a subsequence $\{u_m\}$ from it such that

$$u_m \to w \qquad \text{weakly in } U.$$

Then, $w \in U_{\mathrm{ad}}$, because U_{ad} is a convex set that is closed in the strong topology of U. Taking into account that the mapping $v \to J(v)$ is lower semicontinuous in the weak topology of U, and using (3.2.14), we have

$$\liminf_{m\to\infty} J(u_m) = \lim_{m\to\infty} J(u_m) = \inf_{v \in U_{\mathrm{ad}}} J(v) \geq J(w).$$

Thus,

$$J(w) = \inf_{v \in U_{\mathrm{ad}}} J(v),$$

i.e., $w \in X$, and we can set $u = w$.

Suppose that the set X contains only one element u. Let us show that then

$$u_n \to u \qquad \text{weakly in } U. \tag{3.2.23}$$

Assume that (3.2.23) is not true. Then, there exist an element $q \in U^*$, a constant $\varepsilon > 0$, and a subsequence $\{u_k\}$ of the sequence $\{u_n\}$ such that

$$|(q, u_k - u)| \geq \varepsilon \qquad \forall k. \tag{3.2.24}$$

Since $\{u_k\} \subset U_{\text{ad}}$, the subsequence $\{u_k\}$ is bounded. Let us choose a subsequence $\{u_i\}$ from it such that

$$u_i \to y \qquad \text{weakly in } U. \tag{3.2.25}$$

Therefore, from (3.2.14) we get

$$\liminf_{i\to\infty} J(u_i) = \lim_{i\to\infty} J(u_i) = \inf_{v\in U_{\text{ad}}} J(v) \geq J(y).$$

The set U_{ad} being sequentially weakly closed, from (3.2.25) we deduce $y \in U_{\text{ad}}$. Consequently, $y \in X$ and $y = u$ because the set X contains only one element u. Thus, the formulas (3.2.24) and (3.2.25) contain a contradiction. The theorem is proved.

Remark 3.2.1 Let, under the assumptions of Theorem 3.2.1, U be a Hilbert space, and let a form $\pi(v, w)$ be coercive, i.e.,

$$\pi(v, v) \geq c\|v\|_U^2, \qquad v \in U, \ c = \text{const} > 0.$$

Then the form $\pi(v, w)$ generates in U a scalar product and a norm equivalent to the original one of the space U. Moreover, there exists a unique element u satisfying (3.2.15) (see Remark 3.1.1), and (3.2.14) implies that

$$\pi(u_n, u_n) \to \pi(u, u).$$

Therefore, $u_n \to u$ weakly in U yields that $u_n \to u$ strongly in U.

Example

Let Ω be a bounded open set in $\mathbb{R}^m$, $x = \{x_1, \ldots, x_m\} \in \Omega$, $dx = dx_1 \ldots dx_m$, $U = L_2(\Omega)$.

For $v, w \in L_2(\Omega)$, we set

$$\pi(v, w) = \int_\Omega vw\, dx, \qquad v \to Q(v) = \int_\Omega gv\, dx, \qquad g \in L_2(\Omega).$$

Obviously, the form π is coercive. Let the set U_{ad} be of the form

$$U_{\text{ad}} = \{\, v \,|\, v \in L_2(\Omega), \ \|v\|_{L_2(\Omega)} \leq c \,\},$$

c being a positive number.

On the set $U_{\rm ad}$, we introduce the topology generated by that of the space $L_2(\Omega)$. Then, $U_{\rm ad}$ is a bounded, closed, convex set in $U = L_2(\Omega)$ with a nonempty interior $\overset{\circ}{U}_{\rm ad}$. So, Theorems 3.1.1 and 3.2.1 can be applied.

Suppose now that the set $U_{\rm ad}$ is determined by the following relation

$$U_{\rm ad} = \big\{ v \,|\, v \in L_2(\Omega),\ c_0 \le v(x) \le c_1 \ \text{ a.e. in } \Omega; \\ c_0 < c_1;\ c_0 \text{ and } c_1 \text{ are constants} \big\}. \tag{3.2.26}$$

It is easy to see that $U_{\rm ad}$ is a closed, bounded, convex subset of the space $U = L_2(\Omega)$.

In this case, we can use Theorem 3.1.1. From this theorem and Remark 3.1.1 we deduce the existence of a unique function u such that

$$u \in U_{\rm ad}, \qquad J(u) = \inf_{v \in U_{\rm ad}} J(v), \tag{3.2.27}$$

where

$$J(v) = \int_\Omega v^2\,dx - 2\int_\Omega gv\,dx. \tag{3.2.28}$$

However, Theorem 3.2.1 cannot be applied, because the set $U_{\rm ad}$ defined by (3.2.26) is “thin,” that is, the interior $\overset{\circ}{U}_{\rm ad}$ of $U_{\rm ad}$ in the topology generated by the strong one of the space $L_2(\Omega)$ is empty.

We proceed as follows. Let $\{H_n\}$ be a sequence of finite-dimensional subspaces of $L_2(\Omega)$ such that the set $H_n \cap U_{\rm ad}$ is nonempty for all n. Then, by the coercivity of the form π, for each n there exists a unique function u_n such that

$$u_n \in H_n \cap U_{\rm ad}, \qquad J(u_n) = \inf_{H_n \cap U_{\rm ad}} J(v).$$

If we prove the existence of a sequence $\{g_n\}$ such that

$$g_n \in H_n \cap U_{\rm ad}, \qquad g_n \to u \ \text{ strongly in } L_2(\Omega), \tag{3.2.29}$$

where u satisfies the conditions (3.2.27), then by Theorem 2.4.3 we will conclude that

$$\lim_{n\to\infty} J(u_n) = J(u) = \inf_{v \in U_{\rm ad}} J(v). \tag{3.2.30}$$

Moreover, the boundedness of the set $U_{\rm ad}$ will imply that $u_n \to u$ weakly in $L_2(\Omega)$, and since (3.2.30) yields that

$$\|u_n\|_{L_2(\Omega)} \to \|u\|_{L_2(\Omega)},$$

we get that $u_n \to u$ strongly in $L_2(\Omega)$.

Let us construct the subspaces H_n for the problem under consideration. To every index n we set a corresponding partition of Ω in a finite number of disjoint subsets $\Omega_{1n}, \Omega_{2n}, \dots, \Omega_{N_n n}$ which are Lebesgue measurable. Denote such a partition by Δ_n.

For every set Ω_{in}, $i = 1, 2, \ldots, N_n$, define its diameter

$$\delta_{in} = \sup_{x,y \in \Omega_{in}} \|x - y\|_{\mathbb{R}^n},$$

and let $\delta_n = \max_{1 \le i \le N_n} \delta_{in}$. We suppose that

$$\lim_{n \to \infty} \delta_n = 0. \tag{3.2.31}$$

Denote by χ_{in} the characteristic function of the set Ω_{in}:

$$\chi_{in}(x) = \begin{cases} 1 & \text{if } x \in \Omega_{in}, \\ 0, & \text{if } x \notin \Omega_{in}. \end{cases}$$

Now define subspaces H_n by

$$H_n = \Big\{ v \,|\, v = \sum_{i=1}^{N_n} c_i \chi_{in},\ c_i \text{ are constants} \Big\}. \tag{3.2.32}$$

The subspaces H_n satisfy the limit density condition in $L_2(\Omega)$ (see, e.g., Schwartz (1967)), moreover, for every function $w \in U_{\rm ad}$, where $U_{\rm ad}$ is defined by (3.2.26), there exists a sequence $\{w_n\}$ such that

$$w_n \in H_n \cap U_{\rm ad}, \qquad w_n \to w \ \text{ strongly in } L_2(\Omega).$$

Thus, if one solves the problem (3.2.13) with the functional (3.2.28), with the set $U_{\rm ad}$ determined by the formula (3.2.26), and with H_n defined by (3.2.32), (3.2.31), then (3.2.30) holds and $u_n \to u$ strongly in $L_2(\Omega)$.

Nonquadratic functional

Now, we are going to examine the case when the set $U_{\rm ad}$ and the functional $J(v)$ meet the conditions (3.1.24) and (3.1.25). By virtue of Theorem 3.1.3, there exists a function u such that

$$u \in U_{\rm ad}, \qquad J(u) = \inf_{v \in U_{\rm ad}} J(v). \tag{3.2.33}$$

As before, we will denote by $\overset{\circ}{U}_{\rm ad}$ the interior of the set $U_{\rm ad}$, provided $U_{\rm ad}$ is equipped with the topology generated by the U-strong one.

Theorem 3.2.2 *Let U be a reflexive, separable Banach space, let $\{U_n\}$ be a sequence of finite-dimensional subspaces of U which satisfy the relation* (3.2.12), *and let $U_{\rm ad}$ meet the condition* (3.1.24).

Assume that the functional $J(v)$ satisfies the condition (3.1.25), *and, moreover,*

$$\left.\begin{array}{l} v \to J(v) \text{ is a continuous mapping of } U_{\mathrm{ad}} \text{ into } \mathbb{R} \text{ with} \\ \text{respect to the topology generated by the } U\text{-strong one.} \end{array}\right\} \tag{3.2.34}$$

Also let there exist a sequence $\{q_n\}$ such that

$$q_n \in \overset{\circ}{U}_{\mathrm{ad}} \quad \forall n, \qquad q_n \to u \text{ strongly in } U, \tag{3.2.35}$$

where u is the function satisfying (3.2.33).

Then, for sufficiently large n, the problem (3.2.13) *has a solution u_n, and*

$$\lim_{n\to\infty} J(u_n) = J(u) = \inf_{v\in U_{\mathrm{ad}}} J(v). \tag{3.2.36}$$

From the sequence $\{u_n\}$ one can choose a subsequence $\{u_m\}$ such that $u_m \to u$ weakly in U. If there exists a unique function u for which (3.2.33) *is valid, then $u_n \to u$ weakly in U.*

Proof. Analogously to the proof of Theorem 3.2.1, by using (3.2.12) and (3.2.35), one can show that there exists a sequence $\{g_n\}_{n=k}^{\infty}$, k being a positive number, such that

$$g_n \in U_n \cap U_{\mathrm{ad}}, \qquad g_n \to u \text{ strongly in } U. \tag{3.2.37}$$

Therefore, from (3.2.34) we have

$$\lim_{n\to\infty} J(g_n) = J(u) = \inf_{v\in U_{\mathrm{ad}}} J(v). \tag{3.2.38}$$

When $n \geq k$, the set $U_n \cap U_{\mathrm{ad}}$ is not empty, and in view of (3.1.24) it is a compactum. So, (3.2.34) implies that the problem (3.2.13) has a solution u_n for any $n \geq k$. By (3.2.13), (3.2.33), and (3.2.37), we deduce that

$$J(g_n) \geq J(u_n) \geq J(u).$$

This inequality together with (3.2.38) yields (3.2.36).

By virtue of (3.1.24), the sequence $\{u_n\}$ is bounded. Let us choose a subsequence $\{u_m\}$ from it such that

$$u_m \to w \qquad \text{weakly in } U. \tag{3.2.39}$$

Hence, taking into account (3.1.25) and (3.2.36), we have

$$\liminf_{m\to\infty} J(u_m) = \lim_{m\to\infty} J(u_m) = \inf_{v\in U_{\mathrm{ad}}} J(v) \geq J(w). \tag{3.2.40}$$

From (3.1.24) and (3.2.39) it follows that $w \in U_{\mathrm{ad}}$. Now, by (3.2.40), we conclude

$$J(w) = \inf_{v\in U_{\mathrm{ad}}} J(v),$$

and the function $u = w$ satisfies the relations (3.2.33). If there exists a unique function u that satisfies (3.2.33), then one can conclude the convergence $u_n \to u$ weakly in U in the same way as in the proof of Theorem 3.2.1.

Remark 3.2.2 In Theorems 3.2.1 and 3.2.2, we supposed that $\overset{\circ}{U}_{\text{ad}}$ was not empty. If $\overset{\circ}{U}_{\text{ad}}$ is an empty set, then to investigate the solvability of the problem (3.2.13) and the convergence of its solutions, one can apply Theorem 2.4.3. In this case, the proof of the existence of a sequence $\{g_n\}$ which meets (2.4.28) is the most serious obstacle when this theorem is used. However, as has been seen in the example of the present subsection, when the set U_{ad} has a comparatively simple structure, one does manage to apply Theorem 2.4.3.

3.3 Control by the right-hand side in elliptic problems provided the goal functional is quadratic

3.3.1 Setting of the problem

Let H be a Hilbert space over $\mathbb{R}$ and let

$$\left.\begin{array}{l} a(w,v) \text{ be a bilinear form on } H \times H \text{ that is symmetric,} \\ \text{continuous, and coercive.} \end{array}\right\} \tag{3.3.1}$$

Then, for a given element $f \in H^*$, there exists a unique $u \in H$ such that

$$a(u,v) = (f,v), \qquad v \in H,$$

and $f \to u$ is a linear continuous mapping of H^* into H.

Suppose we are given a set of controls U such that

$$U \text{ is a reflexive Banach space} \tag{3.3.2}$$

and

$$B \in \mathcal{L}(U, H^*). \tag{3.3.3}$$

Suppose that some system is described by the form $a(w,v)$, i.e., for every control $g \in U$, the state of the system $u \in H$ is determined as the solution of the problem

$$a(u,v) = (f + Bg, v), \qquad v \in H. \tag{3.3.4}$$

It is obvious that u depends on g (the element f is assumed to be fixed), so that we will write $u(g)$. Then

$$a(u(g),v) = (f + Bg, v), \qquad v \in H. \tag{3.3.5}$$

In view of (3.3.1), the equation (3.3.5) determines uniquely the state $u(g)$ Moreover, suppose we are also given an observation

$$z(g) = Lu(g),$$

where $L \in \mathcal{L}(H, \mathcal{H})$, $\mathcal{H}$ is a Hilbert space.

To every control $g \in U$ there corresponds the value of the goal functional

$$J(g) = \|Lu(g) - z_0\|^2_{\mathcal{H}}, \tag{3.3.6}$$

z_0 being a given element of the space $\mathcal{H}$.

About the set of admissible controls we suppose that

$$U_{\text{ad}} \text{ is a convex, closed, bounded set in the space } U. \tag{3.3.7}$$

It should be stressed that here the closedness and boundedness is considered in the strong topology of U; as we agreed in Subsec. 1.2.3, if the precise reference to a topology of a normed space is omitted, then we mean the strong one.

The optimal control problem consists in finding y such that

$$y \in U_{\text{ad}}, \qquad J(y) = \inf_{g \in U_{\text{ad}}} J(g).$$

3.3.2 Existence of a solution. Optimality conditions

Due to (3.3.5), $g \to u(g)$ is an affine mapping of U into H, and

$$a((u(g) - u(0)) + u(0), v) = (f + Bg, v), \qquad v \in H.$$

Therefore,

$$a(u(g) - u(0), v) = (Bg, v), \qquad v \in H.$$

Hence,

$$\left.\begin{array}{l} g \to (u(g) - u(0)) \text{ is a linear continuous mapping of } U \text{ into} \\ H. \end{array}\right\} \tag{3.3.8}$$

Rewrite $J(g)$ as

$$J(g) = \|L(u(g) - u(0)) + Lu(0) - z_0\|^2_{\mathcal{H}}.$$

If, for arbitrary elements y, g from U, we set

$$\pi(y, g) = (L(u(y) - u(0)), L(u(g) - u(0)))_{\mathcal{H}}, \tag{3.3.9}$$

$$(F, g) = (z_0 - Lu(0), L(u(g) - u(0)))_{\mathcal{H}}, \tag{3.3.10}$$

then

$$J(g) = \pi(g, g) - 2(F, g) + \|z_0 - Lu(0)\|^2_{\mathcal{H}}. \tag{3.3.11}$$

Since $L \in \mathcal{L}(H, \mathcal{H})$, (3.3.8)–(3.3.10) imply that $F \in U^*$ and that $y, g \to \pi(y, g)$ is a bilinear, symmetric, continuous, positive form on $U \times U$.

By applying Theorem 3.1.1, we obtain the following

Theorem 3.3.1 *Let the conditions* (3.3.1)–(3.3.3) *be fulfilled, and let the state of the system be determined as the solution of the problem* (3.3.4), *where f is a fixed element of H^*, $g \in U$. Let also a goal functional J be defined by the relation* (3.3.6), *where $z_0 \in \mathcal{H}$, $L \in \mathcal{L}(H, \mathcal{H})$, and let the set $U_{\rm ad}$ meet the condition* (3.3.7). *Then, the subset X defined by the relation*

$$X = \{\, y \,|\, y \in U_{\rm ad},\ J(y) = \inf_{g \in U_{\rm ad}} J(g) \,\},$$

is nonempty, closed in $U_{\rm ad}$, and convex. If the function $g \to \pi(g,g)$ determined by the formula (3.3.9) *vanishes only at $g = 0$, then the subset X contains only one element.*

Remark 3.3.1 Assume that the above suppositions hold, U is a finite-dimensional subspace of H^*, B is an embedding of U into H^*, and the function $g \to \pi(g,g)$ vanishes only at $g = 0$. Then, the form $\pi(y,g)$ determines a scalar product and a norm in U. Since in a finite-dimensional vector space every two norms are equivalent, the bilinear form $\pi(y,g)$ is coercive in U. Hence, Remark 3.1.1 yields that if $U_{\rm ad}$ is a closed, convex set in U (not necessarily bounded), then the subset X contains only one element.

Now, let us apply Theorem 3.1.2 to the functional (3.3.6) or, equivalently, to (3.3.11). If $y \in X$, then this theorem together with (3.3.9)–(3.3.11) yields

$$\begin{aligned}\frac{1}{2} J'(y)(g-y) &= \pi(y, g-y) - (F, g-y)\\ &= (L(u(y)-u(0)), L(u(g-y)-u(0)))_{\mathcal{H}}\\ &\qquad - (z_0 - Lu(0), L(u(g-y)-u(0)))_{\mathcal{H}}\\ &= (Lu(y) - z_0, L(u(g-y)-u(0)))_{\mathcal{H}} \geq 0, \qquad g \in U_{\rm ad}. \end{aligned} \tag{3.3.12}$$

By (3.3.5), we conclude

$$\begin{aligned} a(u(g), v) &= (f + Bg, v), \qquad v \in H,\\ a(u(y), v) &= (f + By, v), \qquad v \in H. \end{aligned}$$

Therefore,

$$a(u(g) - u(y), v) = (B(g-y), v), \qquad v \in H. \tag{3.3.13}$$

Further, we have

$$a(u(g-y), v) = (f + B(g-y), v), \qquad v \in H.$$

From this equality, taking into account that

$$a(u(0), v) = (f, v), \qquad v \in H,$$

we obtain

$$a(u(g-y) - u(0), v) = (B(g-y), v), \qquad v \in H. \tag{3.3.14}$$

In view of (3.3.1), the bilinear form $a(w,v)$ generates a scalar product and a norm in the space H, which is equivalent to the original norm of the space H. So, by (3.3.13) and (3.3.14), we get

$$u(g-y) - u(0) = u(g) - u(y).$$

Hence, we can represent the optimality condition (3.3.12) in the form

$$(Lu(y) - z_0, L(u(g) - u(y)))_{\mathcal{H}} \geq 0, \qquad g \in U_{\text{ad}}. \tag{3.3.15}$$

Thus, we have proved the following

Theorem 3.3.2 *Let the conditions of Theorem* 3.3.1 *hold. Then, the function* y *belongs to the set* X *if and only if the inequality* (3.3.15) *holds.*

Notice that in order to get an approximate solution of the problem of minimization of the functional (3.3.6) on the set U_{ad} that satisfies the condition (3.3.7), one can apply Theorem 3.2.1 in the case when the set U_{ad} has at least one interior point. If it is not the case, one can use Theorem 2.4.3 and the reasoning from the example in Subsec. 3.2.2.

Nevertheless, it is better to choose a set of controls U in such a way that the set U_{ad} contains interior points (see also Remark 3.2.2 and the example below).

3.3.3 An example of a system described by the Dirichlet problem

Direct problem

Let $H = H_0^1(\Omega)$, is Ω be a bounded domain in $\mathbb{R}^n$, and let

$$a(w,v) = \sum_{i,j=1}^{n} \int_\Omega a_{ij} \frac{\partial w}{\partial x_i} \frac{\partial v}{\partial x_j}\, dx + \int_\Omega a_0 wv\, dx, \tag{3.3.16}$$

where

$$a_{ij}, a_0 \in L_\infty(\Omega), \qquad a_{ij} = a_{ji}, \tag{3.3.17}$$

$$\left.\begin{array}{l} \displaystyle\sum_{i,j=1}^{n} a_{ij}(x)\xi_i\xi_j \geq \alpha \sum_{i=1}^{n} \xi_i^2 \text{ a.e. in } \Omega \text{ for all } \xi \in \mathbb{R}^n, \\ \alpha = \text{const} > 0,\ a_0(x) \geq \alpha > 0 \text{ a.e.in } \Omega. \end{array}\right\} \tag{3.3.18}$$

From (3.3.16)–(3.3.18)

$$a(v,v) \geq \alpha \|v\|^2_{H^1(\Omega)}.$$

Thus, from (3.3.17) we get

$$\left.\begin{array}{l} \text{the bilinear form } a(w,v) \text{ determined by (3.3.16) is} \\ \text{continuous, symmetric, and coercive on } H_0^1(\Omega) \times H_0^1(\Omega). \end{array}\right\} \tag{3.3.19}$$

Hence, if $q \in \left(H_0^1(\Omega)\right)^* = H^{-1}(\Omega)$, then there exists a unique element $u \in H_0^1(\Omega)$ such that

$$a(u,v) = (q,v), \qquad v \in H_0^1(\Omega). \tag{3.3.20}$$

Define the elliptic operator A of the second order:

$$Aw = -\sum_{i,j=1}^{n} \frac{\partial}{\partial x_j}\left(a_{ij}\frac{\partial}{\partial x_i}w\right) + a_0 w.$$

Then, the equation (3.3.20) is equivalent to the following one

$$Au = q, \qquad u \in H_0^1(\Omega). \tag{3.3.21}$$

Indeed, since $\mathcal{D}(\Omega)$ is dense in $H_0^1(\Omega)$, (3.3.20) is equivalent to the equation

$$a(u,\varphi) = (q,\varphi), \qquad \varphi \in \mathcal{D}(\Omega),$$

which, in turn, is equivalent to the equation (3.3.21) by the definition of a distribution derivative.

The problem (3.3.21) is called the Dirichlet problem.

Optimization problem

Suppose

$$\left.\begin{array}{l} U \subset H^{-1}(\Omega), \text{ the embedding } U \to H^{-1}(\Omega) \text{ is continuous,} \\ B \text{ is the operator of the embedding of the space } U \text{ into} \\ H^{-1}(\Omega),\ \mathcal{H} = L_2(\Omega),\ L \text{ is the embedding of the space} \\ H_0^1(\Omega) \text{ into } L_2(\Omega). \end{array}\right\} \tag{3.3.22}$$

Thus, the state of the system $u(g)$ is determined as the solution of the problem

$$a(u(g),v) = (f+g,v), \qquad v \in H_0^1(\Omega),\ u(g) \in H_0^1(\Omega), \tag{3.3.23}$$

f being a fixed function from $H^{-1}(\Omega)$. The solution $u(g)$ of the problem (3.3.23) is a function defined on Ω:

$$x \to u(x;g).$$

Under the conditions (3.3.22), the goal functional is of the form

$$J(g) = \int_\Omega (u(x;g) - z_0(x))^2\,dx. \tag{3.3.24}$$

By (3.3.9), to this functional there corresponds the bilinear form

$$\pi(y,g) = \int_\Omega (u(x;y) - u(x;0))(u(x;g) - u(x;0))\,dx. \tag{3.3.25}$$

Let us show that the function $g \to \pi(g,g)$ considered as a mapping of U into $\mathbb{R}$ vanishes if and only if $g = 0$.

By (3.3.25), we conclude that $\pi(0,0) = 0$. Suppose $\pi(g,g) = 0$. Then

$$u(x;g) = u(x;0), \tag{3.3.26}$$

and from the relations (3.3.19) and (3.3.26), by using the Riesz theorem, we deduce $g = 0$.

Now, Theorem 3.3.1 implies that, if the set U_{ad} meets the condition (3.3.7), then there exists a unique function y such that

$$y \in U_{\text{ad}}, \qquad \int_\Omega (u(y) - z_0)^2\, dx = \inf_{g \in U_{\text{ad}}} \int_\Omega (u(x;g) - z_0)^2\, dx. \tag{3.3.27}$$

Let us examine some examples of the spaces U and the set U_{ad} for our problem.

Example I. Let

$$\begin{aligned} U &= L_2(\Omega),\\ U_{\text{ad}} &= \{\, g \mid g \in L_2(\Omega),\ \xi_0(x) \le g \le \xi_1(x) \text{ a.e. in } \Omega \,\}, \end{aligned} \tag{3.3.28}$$

where ξ_0 and ξ_1 are fixed functions from $L_\infty(\Omega)$.

It is easy to see that the condition (3.3.7) holds and the embedding $U \to H^{-1}(\Omega)$ is continuous. Hence, there exists a unique function y satisfying (3.3.27).

However, since functions from $L_2(\Omega)$ are not, in general, bounded almost everywhere in Ω, even in the case when

$$\xi_1(x) - \xi_0(x) \ge c \qquad \text{a.e. in } \Omega,\ c = \text{const} > 0, \tag{3.3.29}$$

the set U_{ad} has no interior points.

Example II. Let

$$U = W_p^l(\Omega), \qquad p > 1,\ pl > n, \tag{3.3.30}$$

$$\begin{aligned} U_{\text{ad}} = \{\, g \mid g \in W_p^l(\Omega),\ \|g\|_{W_p^l(\Omega)} \le c_1,\\ \xi_0(x) \le g(x) \le \xi_1(x)\ x \in \Omega,\ c_1 = \text{const} > 0 \,\}, \end{aligned} \tag{3.3.31}$$

where, just as above, ξ_0, ξ_1 are fixed functions from $L_\infty(\Omega)$.

The condition (3.3.7) holds and the embedding $U \to H^{-1}(\Omega)$ is continuous. So, for the space U_{ad} determined by the relations (3.3.30) and (3.3.31), there exists a unique function y satisfying (3.3.27).

Since $W_p^l(\Omega) \subset C(\overline{\Omega})$ (by (3.3.30) and the embedding theorem), under the conditions (3.3.29)–(3.3.31), the set U_{ad} has a nonempty interior $\overset{\circ}{U}_{\text{ad}}$, so that one can apply Theorem 3.2.1.

Notice that, if the condition (3.3.29) is not satisfied, then, by setting

$$\xi_1' = \xi_1 + \varepsilon,$$

ε being an arbitrarily small positive number, we have

$$\xi_1'(x) - \xi_0(x) \geq \varepsilon, \qquad \text{a.e. in } \Omega.$$

Thus, the set $\tilde{U}_{\text{ad}}$ determined by the right-hand side of (3.3.31) with ξ_1' substituted for ξ_1 has a nonempty interior.

It should be noted that we have enlarged the set U_{ad}, by passing from the "thin" set U_{ad} to its extension $\tilde{U}_{\text{ad}}$, which does have a nonempty interior. Under such an extension, in the case when the topology on U_{ad} is good enough, the set U_{ad} changes slightly (in the example considered, the deviation of U_{ad} from $\tilde{U}_{\text{ad}}$ is determined by the parameter ε, which can be taken as small as desired).

Now let us enlarge the set U_{ad} defined by the relations (3.3.28) and (3.3.29), for example, up to a set $\check{U}_{\text{ad}}$ which has the interior point $p = \xi_0 + \frac{c}{2}$:

$$\check{U}_{\text{ad}} = \left\{ g \,|\, g = \xi_0 + \tfrac{1}{2}\, c + q,\ \|q\|_{L_2(\Omega)} \leq \varepsilon,\ \varepsilon > 0 \right\}.$$

In this case, since functions from $L_2(\Omega)$ are unbounded in Ω, for every positive ε functions g from $\check{U}_{\text{ad}}$ may take arbitrarily large values, so that they cannot be bounded by functions $\xi_0, \xi_1 \in L_\infty(\Omega)$.

Another goal functional

Let U be a closed subspace of $H^{-1}(\Omega)$ with the norm of $H^{-1}(\Omega)$, let B be the operator of the embedding of the space U into $H^{-1}(\Omega)$, $\mathcal{H} = (L_2(\Omega))^{n+1}$,

$$L \in \mathcal{L}\left(H_0^1(\Omega), (L_2(\Omega))^{n+1}\right), \qquad L\colon v \to Lv = \left\{ v, \frac{\partial v}{\partial x_1}, \dots, \frac{\partial v}{\partial x_n} \right\}.$$

Define a norm in the space $\mathcal{H}$ by

$$v = \{v_0, v_1, \dots, v_n\}, \qquad \|v\|_{\mathcal{H}}^2 = \sum_{i=0}^{n} b_i \|v_i\|_{L_2(\Omega)}^2, \tag{3.3.32}$$

$$b_i = \text{const} > 0.$$

The state of the system $u(g)$ is determined as the solution of the problem (3.3.23), and according to (3.3.6) the goal functional has the form

$$J(g) = \sum_{i=1}^{n} b_i \int_\Omega \left(\frac{\partial u(g)}{\partial x_i} - z_{0i} \right)^2 dx + b_0 \int_\Omega (u(g) - z_{00})^2\, dx. \tag{3.3.33}$$

Here z_{0i} are given elements of $L_2(\Omega)$, $i = 0, 1, \dots, n$.

By (3.3.9), to the goal functional (3.3.33) there corresponds the bilinear form

$$\pi(y,g) = \sum_{i=1}^{n} b_i \int_\Omega \left(\frac{\partial u(y)}{\partial x_i} - \frac{\partial u(0)}{\partial x_i} \right) \left(\frac{\partial u(g)}{\partial x_i} - \frac{\partial u(0)}{\partial x_i} \right) dx$$
$$+ b_0 \int_\Omega (u(y) - u(0))(u(g) - u(0))\, dx. \quad (3.3.34)$$

The bilinear form $\pi(y,g)$ is obviously symmetric and continuous on $U \times U$. By (3.3.32) and (3.3.34), we obtain

$$\pi(g,g) \geq c\|u(g) - u(0)\|^2_{H^1_0(\Omega)}, \qquad g \in U,\ c = \text{const} > 0. \quad (3.3.35)$$

Since

$$a(u(g) - u(0), v) = (g, v), \qquad v \in H^1_0(\Omega),$$

from (3.3.19) and the Riesz theorem we conclude

$$\|u(g) - u(0)\|_{H^1_0(\Omega)} \geq c_1 \|g\|_{H^{-1}(\Omega)}, \qquad g \in U,\ c_1 = \text{const} > 0.$$

Combining this with (3.3.35), we get the form $\pi(y,g)$ to be coercive, i.e., there exists a positive number c_2 such that

$$\pi(g,g) \geq c_2 \|g\|^2_{H^{-1}(\Omega)}, \qquad g \in U.$$

Now, by Remark 3.1.1, we get that, if U_{ad} is a closed, convex set in U, which is not necessarily bounded, then there exists a unique function $y \in U_{\text{ad}}$ such that

$$J(y) = \inf_{g \in U_{\text{ad}}} J(g).$$

3.4 Minimax control problems

Let H be a Hilbert space over $\mathbb{R}$, let the conditions (3.3.1)–(3.3.3) hold, and let the state of the system $u(g)$ be defined as a solution of the problem (3.3.5), where f is a fixed element of H^* and $g \in U$.

Suppose we are given mappings $P_k \colon H \times U \to \mathbb{R}$ such that

$$\left.\begin{array}{l} v, q \to P_k(v,q) \text{ is a lower semicontinuous mapping of} \\ H \times U \text{ into } \mathbb{R} \text{ with respect to the topology generated by} \\ \text{the product of the weak topologies of } H \text{ and } U,\ k \in I, \\ I = \{1, 2, \ldots, l\}. \end{array}\right\} \quad (3.4.1)$$

Define mappings $Q_k \colon U \to \mathbb{R}$ by

$$g \to Q_k(g) = P_k(u(g), g), \qquad k \in I, \quad (3.4.2)$$

and let a goal functional be of the form

$$J(g) = \max_{k \in I} Q_k(g). \tag{3.4.3}$$

Suppose that

$$\left. \begin{array}{l} U_{\text{ad}} \subset U,\ U_{\text{ad}} \text{ is bounded in the strong topology of } U \text{ and} \\ \text{sequentially weakly closed.} \end{array} \right\} \tag{3.4.4}$$

Theorem 3.4.1 *Let the conditions* (3.3.1)–(3.3.3) *hold, and let the state of the system* $u(g)$ *be determined as the solution of the problem* (3.3.5), *where* f *is a fixed element of* H^* *and* $g \in U$. *Let also a goal functional* $J(g)$ *be defined by the formulas* (3.4.1)–(3.4.3), *and let the set* U_{ad} *meet the condition* (3.4.4). *Then, there exists a function* y *such that*

$$y \in U_{\text{ad}}, \qquad J(y) = \inf_{g \in U_{\text{ad}}} J(g). \tag{3.4.5}$$

To prove this statement, we need the following lemma.

Lemma 3.4.1 *Let the conditions* (3.3.1)–(3.3.3) *hold, and let the state of the system* $u(g)$ *be determined as the solution of the problem* (3.3.5), *where* f *is a fixed element of* H^* *and* $g \in U$. *Then,* $g \to u(g)$ *is a continuous mapping of the space* U *endowed with the weak topology into the space* H *equipped with the weak topology.*

Proof. Let $g \in U$ and let $\{g_n\}$ be a sequence of elements of U such that

$$g_n \to g \qquad \text{weakly in } U. \tag{3.4.6}$$

From (3.3.5)

$$a(u(g_n) - u(g), v) = (B(g_n - g), v), \qquad v \in H. \tag{3.4.7}$$

For a fixed $v \in H$, the function $q \to (Bq, v)$ is a linear continuous mapping of U into $\mathbb{R}$. So, (3.4.6) yields

$$(B(g_n - g), v) \to 0.$$

Combining this with (3.4.7), we get

$$\lim_{n \to \infty} a(u(g_n) - u(g), v) = 0, \qquad v \in H.$$

Taking note of the Riesz theorem, from the latter relation and (3.3.1), we deduce that

$$u(g_n) \to u(g) \qquad \text{weakly in } H.$$

Thus, the lemma is proved.

Proof of Theorem 3.4.1. Let $\{q_n\}$ be a minimizing sequence, i.e.,

$$\{q_n\} \subset U_{\text{ad}}, \qquad J(q_n) \to \inf_{g \in U_{\text{ad}}} J(g). \tag{3.4.8}$$

(3.4.4) implies

$$\|q_n\|_U \leq \text{const} \qquad \forall n.$$

Therefore, from the sequence $\{q_n\}$ we can choose a subsequence $\{q_m\}$ such that

$$q_m \to q \qquad \text{weakly in } U, \tag{3.4.9}$$

moreover, (3.4.4) yields that

$$q \in U_{\text{ad}}. \tag{3.4.10}$$

By using Lemma 3.4.1, (3.4.1), and (3.4.2), we deduce that $g \to Q_k(g)$ is a lower semicontinuous mapping of U equipped with the weak topology into $\mathbb{R}$. Theorem 1.4.3 implies

$$\left.\begin{array}{l} g \to J(g) = \max\limits_{k \in I} Q_k(g) \text{ is a lower semicontinuous mapping} \\ \text{of the space } U \text{ endowed with the weak topology into } \mathbb{R}. \end{array}\right\} \tag{3.4.11}$$

By (3.4.9) and (3.4.11), we get

$$\liminf_{m \to \infty} J(q_m) \geq J(q).$$

Hence, from (3.4.8) we get

$$\liminf_{m \to \infty} J(q_m) = \inf_{g \in U_{\text{ad}}} J(g) \geq J(q). \tag{3.4.12}$$

By (3.4.12), taking into account (3.4.10), we obtain

$$J(q) = \inf_{g \in U_{\text{ad}}} J(g).$$

Hence, the function $y = q$ is a solution of the problem (3.4.5), proving the theorem.

Example. Assume that $H = H_0^1(\Omega)$, Ω is a bounded domain in $\mathbb{R}^n$, the form $a(w, v)$ is determined by the relations (3.3.16)–(3.3.18), $U \subset H^{-1}(\Omega)$, the embedding $U \to H^{-1}(\Omega)$ is continuous, and B is the embedding of U into $H^{-1}(\Omega)$.

The state of the system $u(g)$ is defined as the solution of the problem: Find $u(g) \in H_0^1(\Omega)$ such that

$$a(u(g), v) = (f + g, v), \qquad v \in H_0^1(\Omega),$$

where f is a fixed element of $H^{-1}(\Omega)$, $g \in U$.

Let Ω be partitioned into a finite number of measurable subsets Ω_i, i.e., $\Omega_i \subset \Omega$ and $\Omega = \bigcup_{i=1}^m \Omega_i$. Define functions

$$\begin{aligned} g \to Q_{ki}(g) &= \int_{\Omega_i} \left(\frac{\partial u(g)}{\partial x_k} - \chi_{ki} \right)^2 dx \qquad k = 1, \dots, n; \ i = 1, \dots, m, \\ g \to Q_{0i}(g) &= \int_{\Omega_i} (u(g) - \chi_{0i})^2 \, dx. \end{aligned} \tag{3.4.13}$$

Here, χ_{ki}, χ_{0i} are given functions from $L_2(\Omega_i)$.

Assign a goal functional through the relation

$$J(g) = \max_{(k,i) \in I \times Y} Q_{ki}(g),$$

where $I = \{0, 1, 2, \dots, n\}$ and $Y = \{1, 2, \dots, m\}$.

Lemma 3.4.1 implies that $g \to u(g)$ is a continuous mapping of the space U equipped with the weak topology into $H_0^1(\Omega)$ endowed with the weak topology.

Making use of Lemma 3.1.1, it is easy to see that

$$\left. \begin{array}{l} \text{the function } u \to \displaystyle\int_{\Omega_i} \left(\frac{\partial u}{\partial x_k} - \chi_{ki} \right)^2 dx \text{ is lower} \\ \text{semicontinuous in the weak topology of } H_0^1(\Omega). \end{array} \right\} \tag{3.4.14}$$

The embedding theorem yields that

$$\left. \begin{array}{l} u \to \displaystyle\int_{\Omega_i} (u - \chi_{0i})^2 \, dx \text{ is a continuous mapping of the space} \\ H_0^1(\Omega) \text{ endowed with the weak topology into } \mathbb{R}. \end{array} \right\} \tag{3.4.15}$$

Now, Theorem 3.4.1 and the relations (3.4.13)–(3.4.15) imply that, if the set U_{ad} meets the condition (3.4.4), then there exists a function y such that

$$y \in U_{\text{ad}}, \qquad J(y) = \inf_{g \in U_{\text{ad}}} J(g).$$

3.5 Control of systems whose state is described by variational inequalities

3.5.1 Setting of the problem

So far we have considered optimal control problems when the state of the system is defined as the solution of an elliptic problem, and this solution minimizes the quadratic functional in a corresponding Hilbert space.

Now, we will investigate the optimal control problems in which the state of the system (the solution of the direct problem) is the solution of a variational inequality, i.e., the state of the system is the function minimizing a quadratic functional on some set from the initial space.

Let H be a Hilbert space over $\mathbb{R}$, and suppose that

$$\left.\begin{array}{l}a(w,v) \text{ is a symmetric, continuous, coercive bilinear form}\\ \text{on } H\times H,\end{array}\right\} \tag{3.5.1}$$

$$\widetilde{H} \text{ is a convex, closed set in } H. \tag{3.5.2}$$

We assume also that

$$\left.\begin{array}{l}U \text{ is a reflexive Banach space, } U\subset H^*, \text{ the embedding}\\ U\to H^* \text{ is compact.}\end{array}\right\} \tag{3.5.3}$$

For a given element $g\in U$, the state of the system $u(g)$ is defined as the solution of the problem

$$\begin{gathered}u(g)\in\widetilde{H},\\ a(u(g),u(g))-2(f+g,u(g))=\inf_{v\in\widetilde{H}}[a(v,v)-2(f+g,v)],\end{gathered} \tag{3.5.4}$$

f being a fixed element of H^*.

By Theorem 3.1.1 and Remark 3.1.1, for every $g\in U$ there exists a unique function $u(g)$ satisfying (3.5.4). Theorem 3.1.2 implies the function $u(g)$ to be characterized by the formula

$$u(g)\in\widetilde{H},\qquad a(u(g),v-u(g))\geq(f+g,v-u(g)),\qquad v\in\widetilde{H}.$$

Further, suppose we are given a mapping $P\colon H\times U\to\mathbb{R}$ such that

$$\left.\begin{array}{l}v,g\to P(v,g) \text{ is a lower semicontinuous mapping of}\\ H\times U \text{ into } \mathbb{R} \text{ with respect to the topology of the product}\\ \text{of weak topologies of } H \text{ and } U,\end{array}\right\} \tag{3.5.5}$$

and let a goal functional be of the form

$$g\to J(g)=P(u(g),g). \tag{3.5.6}$$

Here, the function $u(g)$ is a solution of the problem (3.5.4).

Suppose that

$$\left.\begin{array}{l}U_{\mathrm{ad}}\subset U,\ U_{\mathrm{ad}} \text{ is bounded in the strong topology of } U \text{ and}\\ \text{sequentially weakly closed.}\end{array}\right\} \tag{3.5.7}$$

The optimal control problem consists in finding a function y such that

$$y\in U_{\mathrm{ad}},\qquad J(y)=\inf_{g\in U_{\mathrm{ad}}}J(g). \tag{3.5.8}$$

3.5.2 The existence theorem

Theorem 3.5.1 *Let H be a Hilbert space over $\mathbb{R}$, let the conditions* (3.5.1)–(3.5.3) *hold, and let the state of the system $u(g)$ be defined as the solution of the problem* (3.5.4), *f being a fixed element of H^*, $g \in U$. Assume that a goal functional $J(g)$ is determined by the relations* (3.5.5) *and* (3.5.6), *and the set U_{ad} meets the condition* (3.5.7). *Then, the problem* (3.5.8) *has a solution.*

To prove Theorem 3.5.1, we need the following lemma.

Lemma 3.5.1 *Let H be a Hilbert space over $\mathbb{R}$, let the conditions* (3.5.1)–(3.5.3) *be satisfied, and let a function $f \in H^*$ be given. Then, the function $g \to u(g)$ determined by the solution of the problem* (3.5.4) *is a continuous mapping of the space U equipped with the weak topology into the space H endowed with the weak topology.*

Proof. 1) Let $\{g_n\}$ be a sequence of elements of U such that

$$g_n \to g \qquad \text{weakly in } U, \tag{3.5.9}$$

and let $\{u(g_n)\}$ be the corresponding sequence of solutions of the problem (3.5.4). Let us show that there exists a positive number c such that

$$\|u(g_n)\|_H \le c \qquad \forall n. \tag{3.5.10}$$

Introduce the notations

$$u_n = u(g_n), \tag{3.5.11}$$

$$\Psi_n(u_n) = a(u_n, u_n) - 2(f + g_n, u_n), \tag{3.5.12}$$

and assume the condition (3.5.10) is not valid. Then, from the sequence $\{u_n\}$ one can choose a subsequence $\{u_m\}$ such that

$$\|u_m\|_H \to \infty. \tag{3.5.13}$$

By (3.5.1) and (3.5.12), we conclude that

$$\Psi_m(u_m) \ge c_1 \|u_m\|_H^2 - 2\|f + g_m\|_{H^*} \|u_m\|_H \qquad \forall m, \tag{3.5.14}$$

where c_1 is a positive number.

By (3.5.3) and (3.5.9), there exists a positive number c_2 such that

$$\|f + g_m\|_{H^*} \le c_2 \qquad \forall m.$$

Combining this with (3.5.13) and (3.5.14), we get

$$\Psi_m(u_m) \to \infty. \tag{3.5.15}$$

Now let v be an arbitrary element of $\widetilde{H}$. (3.5.1) and (3.5.9) yield

$$|a(v,v) - 2(f + g_m, v)| \le c_3\|v\|_H^2 + c_4\|v\|_H = c_5 \qquad \forall m$$

where c_3, c_4, and c_5 are positive numbers.

Hence, from (3.5.4) and (3.5.12), we obtain

$$\Psi_m(u_m) \le c_5 \qquad \forall m.$$

The latter inequality makes a contradiction with (3.5.15), so that (3.5.10) is true.

2) (3.5.3), (3.5.9), and (3.5.10) imply that from the sequence $\{g_n, u_n\}$ one can choose a subsequence $\{g_m, u_m\}$ such that

$$g_m \to g \qquad \text{strongly in } H^*, \tag{3.5.16}$$

$$u_m \to w \qquad \text{weakly in } H. \tag{3.5.17}$$

Since $u_m \in \widetilde{H}$, from (3.5.2) and (3.5.17) we deduce

$$w \in \widetilde{H}. \tag{3.5.18}$$

(3.5.4) and (3.5.11) yield

$$a(u_m, u_m) - 2(f + g_m, u_m) \le a(v,v) - 2(f + g_m, v), \qquad v \in \widetilde{H}. \tag{3.5.19}$$

Taking into account (3.5.16) and (3.5.17), we have

$$\begin{aligned} \liminf_{m\to\infty} a(u_m, u_m) &\ge a(w,w), \\ \lim_{m\to\infty} (f + g_m, u_m) &= (f + g, w), \\ \lim_{m\to\infty} (f + g_m, v) &= (f + g, v). \end{aligned} \tag{3.5.20}$$

By using (3.5.20), we pass in (3.5.19) to the limit in m for an arbitrary fixed element v from $\widetilde{H}$. Then, we get

$$a(w,w) - 2(f + g, w) \le a(v,v) - 2(f + g, v), \qquad v \in \widetilde{H}. \tag{3.5.21}$$

From this and (3.5.18), we see that $w = u(g)$. We have only to prove that (3.5.17) holds true not only for the subsequence $\{u_m\}$, but for the whole sequence $\{u_n\}$, i.e.,

$$u_n \to w \qquad \text{weakly in } H. \tag{3.5.22}$$

Indeed, assume that (3.5.22) is not valid. Then, there exist $\varepsilon > 0$, $z \in H^*$, and a subsequence $\{u_k\}$ of $\{u_n\}$ such that

$$|(z, u_k - w)| \ge \varepsilon \qquad \forall k. \tag{3.5.23}$$

Since the subsequence $\{u_k\}$ is bounded, we can choose a subsequence $\{u_l\}$ from it such that

$$u_l \to u_0 \ \text{ weakly in } H, \qquad u_0 \in \widetilde{H}. \tag{3.5.24}$$

Then, by passing to the limit just as it was done above, we infer that the function u_0 satisfies the relation (3.5.21) in which w is replaced by u_0. For a given element $g \in U$, there exists a unique function $w \in \widetilde{H}$ satisfying (3.5.21) (see Remark 3.1.1). Hence,

$$u(g) = w = u_0,$$

and the relations (3.5.23) and (3.5.24) make a contradiction. So, (3.5.22) holds, and the lemma is proved.

Proof of Theorem 3.5.1. Let $\{g_n\}$ be a minimizing sequence, i.e.,

$$\{g_n\} \in U_{\text{ad}}, \qquad J(g_n) \to \inf_{g \in U_{\text{ad}}} J(g). \tag{3.5.25}$$

By (3.5.7), we conclude that from the sequence $\{g_n\}$ one can choose a subsequence $\{g_m\}$ such that

$$g_m \to z \ \text{ weakly in } U, \tag{3.5.26}$$

$$z \in U_{\text{ad}}. \tag{3.5.27}$$

By (3.5.26), using Lemma 3.5.1, we get

$$u(g_m) \to u(z) \qquad \text{weakly in } H. \tag{3.5.28}$$

Further, the relations (3.5.5), (3.5.6), (3.5.26), and (3.5.28) yield

$$\liminf_{m \to \infty} J(g_m) \geq J(z).$$

Combining this with (3.5.25) and (3.5.27), we have

$$J(z) = \inf_{g \in U_{\text{ad}}} J(g).$$

Thus, the function $y = z$ is a solution to problem (3.5.8), concluding the proof.

3.5.3 An example of control of a system described by a variational inequality

Direct problem

Let $H = H^1(\Omega)$, where Ω is a bounded open set in $\mathbb{R}^n$ with a smooth boundary S, and let a form $a(w, v)$ be determined by the relations (3.3.16)–(3.3.18). Suppose that the set $\widetilde{H}$ is of the form

$$\widetilde{H} = \big\{ v \,|\, v \in H^1(\Omega),\ v \geq 0 \text{ a.e. on } S \big\}. \tag{3.5.29}$$

In this case, $\widetilde{H}$ is a convex, closed cone in H with vertex at the origin. Indeed, the convexity of the set $\widetilde{H}$ is obvious, and the closedness of $\widetilde{H}$ in H is implied by the continuity of the mapping $v \to v|_S$ of the space $H^1(\Omega)$ onto $H^{\frac{1}{2}}(\Omega)$ (see Theorem 1.6.5).

Since the bilinear form $a(w,v)$ determined by the formulas (3.3.16)–(3.3.18) is symmetric, continuous, and coercive on $H^1(\Omega) \times H^1(\Omega)$, by virtue of Remark 3.1.4 for any element $g \in (H^1(\Omega))^*$ there exists a unique function $u(g)$ such that

$$u(g) \in \widetilde{H}, \qquad a(u(g),u(g)) - 2(g,u(g)) \leq a(v,v) - 2(g,v), \qquad v \in \widetilde{H}, \quad (3.5.30)$$

the function $u(g)$ being characterized by the relations

$$u(g) \in \widetilde{H}, \qquad a(u(g),v) \geq (g,v), \qquad v \in \widetilde{H}, \quad (3.5.31)$$

$$a(u(g),u(g)) = (g,u(g)). \quad (3.5.32)$$

Let an element $g \in (H^1(\Omega))^*$ be determined by the relation

$$(g,v) = \int_\Omega g^{(1)} v\,dx + \int_S g^{(2)} v\,ds, \qquad g^{(1)} \in L_2(\Omega),\ g^{(2)} \in L_2(S). \quad (3.5.33)$$

Notice that the element g in (3.5.33) indeed determines a linear continuous form on $H^1(\Omega)$ (i.e., it belongs to $(H^1(\Omega))^*$) because $v \to v|_S$ is a continuous mapping of $H^1(\Omega)$ onto $H^{\frac{1}{2}}(S)$.

The formulas (3.5.31) and (3.5.32) are interpreted in the following way. Let $v = \pm\varphi$, where $\varphi \in \mathcal{D}(\Omega)$, so that $v \in \widetilde{H}$ due to (3.5.29). Then, (3.5.31) implies

$$a(u(g),\varphi) = (g,\varphi), \qquad \varphi \in \mathcal{D}(\Omega).$$

From this, by using (3.3.16), and (3.5.33), we get

$$Au(g) = -\sum_{i,j=1}^{n} \frac{\partial}{\partial x_j}\left(a_{ij}\frac{\partial u(g)}{\partial x_i}\right) + a_0 u(g) = g^{(1)} \qquad \text{in } \Omega, \quad (3.5.34)$$

the derivatives being understood in the sense of distributions on Ω.

Multiplying (3.5.34) by $v \in \widetilde{H}$ and applying Green's formula, we have

$$-\int_S \frac{\partial u(g)}{\partial \nu_A} v\,ds + a(u(g),v) = \int_\Omega g^{(1)} v\,dx; \quad (3.5.35)$$

here

$$\frac{\partial u(g)}{\partial \nu_A} = \sum_{i,j=1}^{n} a_{ij}\frac{\partial u(g)}{\partial x_i}\cos(\nu,x_j) \text{ on } S, \qquad \frac{\partial u(g)}{\partial \nu_A} \in H^{-\frac{1}{2}}(S),$$

and $\cos(\nu,x_j)$ is the j-th component of the external unit normal ν to the boundary S of the domain Ω.

Note that, for a function $u \in H^1(\Omega)$ such that $Au = f \in L_2(\Omega)$, one can determine uniquely $\frac{\partial u}{\partial \nu_A}$ on S, in this case $\frac{\partial u}{\partial \nu_A} \in H^{-\frac{1}{2}}(S)$, and Green's formula is valid:

$$-\int_S \frac{\partial u}{\partial \nu_A} v\,ds + a(u,v) = \int_\Omega fv\,dx \qquad v \in H^1(\Omega),$$

see Lions and Magenes (1972), Aubin (1972).

From (3.5.31), (3.5.33), and (3.5.35)

$$\int_S \left(\frac{\partial u(g)}{\partial \nu_A} - g^{(2)}\right) v\,ds \geq 0. \tag{3.5.36}$$

Since here $v \in \widetilde{H}$, we have $v \geq 0$ on S, so that the inequality (3.5.36) is equivalent to the following one

$$\frac{\partial u(g)}{\partial \nu_A} - g^{(2)} \geq 0 \quad \text{on } S \qquad \text{(in the sense of } H^{-\frac{1}{2}}(S)). \tag{3.5.37}$$

Moreover, by (3.5.32), (3.5.33), and (3.5.35)

$$\int_S \left(\frac{\partial u(g)}{\partial \nu_A} - g^{(2)}\right) u(g)\,ds = 0.$$

Taking into account this equality, (3.5.37), and the relation $u(g) \geq 0$ on S, we get

$$u(g)\left(\frac{\partial u(g)}{\partial \nu_A} - g^{(2)}\right) = 0 \qquad \text{on } S.$$

Thus, in the space $H^1(\Omega)$, there exists a unique function $u(g)$ that satisfies the equation (3.5.34) and the following boundary conditions

$$u(g) \geq 0 \ \text{ on } S, \quad \frac{\partial u(g)}{\partial \nu_A} - g^{(2)} \geq 0 \ \text{ on } S, \quad u(g)\left(\frac{\partial u(g)}{\partial \nu_A} - g^{(2)}\right) = 0 \ \text{ on } S. \tag{3.5.38}$$

According to the last condition in (3.5.38), there exists a subset S_0 of the boundary S on which $u(g) = 0$; then $\frac{\partial u(g)}{\partial \nu_A} - g^{(2)} = 0$ on $S \setminus S_0$. It is obvious that S_0 is not known beforehand, in particular, S_0 may be an empty set.

The optimal control problem

Let us consider the function $g = \left(g^{(1)}, g^{(2)}\right) \in L_2(\Omega) \times L_2(S)$ as a control. So,

$$U = L_2(\Omega) \times L_2(S). \tag{3.5.39}$$

Through the formula (3.5.33), to every element $g \in U$ there corresponds a linear continuous functional in the space $H^1(\Omega)$, which will also be denoted g. Thus, we defined the embedding $U \to (H^1(\Omega))^*$.

By Theorem 1.5.12, since the embeddings $H^1(\Omega) \to L_2(\Omega)$ and $H^{\frac{1}{2}}(S) \to L_2(S)$ are compact, so are the embeddings $L_2(\Omega) \to (H^1(\Omega))^*$, $L_2(S) \to H^{-\frac{1}{2}}(S)$. Hence,

$$\text{the embedding } U \to (H^1(\Omega))^* \text{ is compact.} \tag{3.5.40}$$

Define a mapping $P\colon H^1(\Omega) \to \mathbb{R}$ by

$$w \to P(w) = \sum_{i=1}^{n} b_i \int_\Omega \left(\frac{\partial w}{\partial x_i} - z_i \right)^2 dx + b_0 \int_\Omega (w - z_0)^2 \, dx, \tag{3.5.41}$$

z_i and z_0 being given functions from $L_2(\Omega)$, b_i being positive numbers.

Making use of Lemma 3.1.1, we see that

$$\left. \begin{array}{l} \text{the function } w \to P(w) \text{ is lower semicontinuous in the} \\ \text{weak topology of the space } H^1(\Omega). \end{array} \right\} \tag{3.5.42}$$

Let a goal functional $J\colon U \to \mathbb{R}$ have the form

$$g \to J(g) = P(u(g)), \tag{3.5.43}$$

$u(g)$ being the solution of the problem (3.5.30) (or, equivalently, of the problem (3.5.31), (3.5.32)).

Now, if the set U_{ad} satisfies the condition (3.5.7), then by (3.5.40), (3.5.42), and Theorem 3.5.1, we obtain that, for the goal functional $J(g)$ determined by the relations (3.5.41) and (3.5.43), there exists an element y such that

$$y \in U_{\text{ad}}, \qquad J(y) = \inf_{g \in U_{\text{ad}}} J(g).$$

Chapter 4

Direct Problems for Plates and Shells

"You must next be told why a strong man came to fall a victim to a Beauty Hint"

– *O. Henry*
"The Indian Summer of Dry Valley Johnson"

4.1 Bending and free oscillations of thin plates

4.1.1 Basic relations of the theory of bending of thin plates

Consider a homogeneous thin plate of variable thickness (see Fig. 4.1.1). We suppose that the plate has a midplane such that the plate is symmetric with respect to it. Take the midplane of the plate to be the (x, y) plane, and let the z axis be directed downwards. Denote displacements of points of the midplane along the z axis by w and assume that the so-called Kirchhoff hypotheses hold:

1. Normals to the midplane before the bending go over into the normals to the midsurface after the bending, and their length does not change during the bending.
2. Inside of the plate, the stresses normal to the midsurface are small as compared to other stress components, so that they can be neglected in relations between the stresses and strains (the plane stress state hypothesis).
3. Under bending, elements of the midsurface of the plate are not subject to tension and pressure.

Under these assumptions, the components $u(x, y, z)$ and $v(x, y, z)$ of the vector of displacements of points of the plate in the directions of the x and y axes

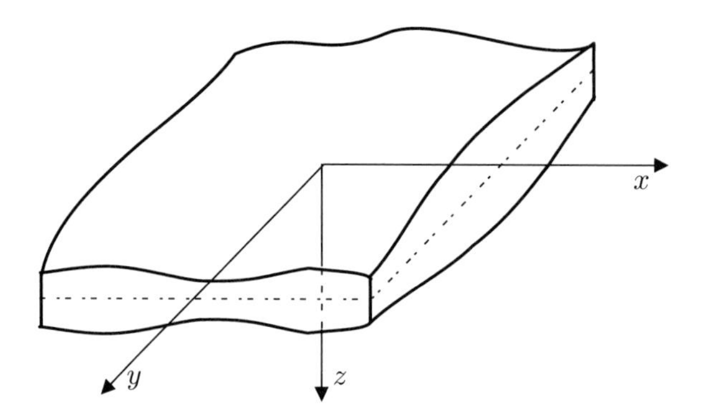

Figure 4.1.1: Thin plate of variable thickness

have the form

$$u(x,y,z) = -z\,\frac{\partial w}{\partial x}, \qquad v(x,y,z) = -z\,\frac{\partial w}{\partial y},$$

where $-\frac{h}{2} \le z \le \frac{h}{2}$ and h is the thickness of the plate, depending on x and y.

The strain components are of the form

$$\varepsilon_1 = \frac{\partial u}{\partial x} = -z\,\frac{\partial^2 w}{\partial x^2}, \qquad \varepsilon_2 = \frac{\partial v}{\partial y} = -z\,\frac{\partial^2 w}{\partial y^2},$$
$$\varepsilon_3 = \frac{\partial u}{\partial y} + \frac{\partial v}{\partial x} = -2z\frac{\partial^2 w}{\partial x \partial y}. \tag{4.1.1}$$

For an isotropic plate, the stress components are determined by the relations

$$\sigma_1 = \frac{E}{1-\mu^2}(\varepsilon_1 + \mu\varepsilon_2) = -\frac{Ez}{1-\mu^2}\left(\frac{\partial^2 w}{\partial x^2} + \mu\frac{\partial^2 w}{\partial y^2}\right),$$
$$\sigma_2 = \frac{E}{1-\mu^2}(\varepsilon_2 + \mu\varepsilon_1) = -\frac{Ez}{1-\mu^2}\left(\frac{\partial^2 w}{\partial y^2} + \mu\,\frac{\partial^2 w}{\partial x^2}\right), \tag{4.1.2}$$
$$\sigma_3 = G\varepsilon_3 = \frac{E}{2(1+\mu)}\,\varepsilon_3 = -\frac{Ez}{1+\mu}\,\frac{\partial^2 w}{\partial x \partial y},$$

where E is the elasticity modulus, G is the shearing modulus, and μ is the Poisson ratio.

Introduce the notations

$$\chi_1 = -\frac{\partial^2 w}{\partial x^2}, \qquad \chi_2 = -\frac{\partial^2 w}{\partial y^2}, \qquad \chi_3 = -\frac{\partial^2 w}{\partial x \partial y}; \tag{4.1.3}$$

χ_1 and χ_2 are called the components of the bending strains of the midplane, χ_3 is the component of the torsion strains.

The bending moments M_1, M_2 and the torque M_3 are given by

$$\begin{aligned}
M_1 &= \int_{-\frac{h}{2}}^{\frac{h}{2}} \sigma_1(z)\, z\, dz = D(\chi_1 + \mu\chi_2), \\
M_2 &= \int_{-\frac{h}{2}}^{\frac{h}{2}} \sigma_2(z)\, z\, dz = D(\chi_2 + \mu\chi_1), \\
M_3 &= \int_{-\frac{h}{2}}^{\frac{h}{2}} \sigma_3(z)\, z\, dz = D(1-\mu)\chi_3,
\end{aligned} \tag{4.1.4}$$

where D is the cylindrical stiffness of the plate,

$$D = \frac{Eh^3}{12(1-\mu^2)}. \tag{4.1.5}$$

The energy of elastic deformation accumulated in the plate, i.e., the strain energy, is defined by the following formula

$$\Phi(w) = \frac{1}{2} \iint\limits_{\Omega} dx\, dy \int_{-\frac{h}{2}}^{\frac{h}{2}} (\sigma_1\varepsilon_1 + \sigma_2\varepsilon_2 + \sigma_3\varepsilon_3)\, dz, \tag{4.1.6}$$

where Ω is the domain occupied by the midplane of the plate, which is supposed to be bounded in what follows.

From this, using (4.1.1)–(4.1.3), we obtain

$$\begin{aligned}
\Phi(w) &= \iint\limits_{\Omega} \frac{D}{2} \left[(\chi_1 + \chi_2)^2 - 2(1-\mu)(\chi_1\chi_2 - \chi_3^2) \right] dx\, dy \\
&= \iint\limits_{\Omega} \frac{D}{2} \left[\left(\frac{\partial^2 w}{\partial x^2} + \frac{\partial^2 w}{\partial y^2} \right)^2 - 2(1-\mu) \left(\frac{\partial^2 w}{\partial x^2} \frac{\partial^2 w}{\partial y^2} - \left(\frac{\partial^2 w}{\partial x \partial y} \right)^2 \right) \right] dx\, dy.
\end{aligned} \tag{4.1.7}$$

4.1.2 Orthotropic plates

The strain components ε_1, ε_2, and ε_3 for an orthotropic plate are determined by the formulas (4.1.1), and the stress components are given by

$$\begin{aligned}
\sigma_1 &= E_{11}\varepsilon_1 + E_{12}\varepsilon_2 = -E_{11} z \frac{\partial^2 w}{\partial x^2} - E_{12} z \frac{\partial^2 w}{\partial y^2}, \\
\sigma_2 &= E_{21}\varepsilon_1 + E_{22}\varepsilon_2 = -E_{21} z \frac{\partial^2 w}{\partial x^2} - E_{22} z \frac{\partial^2 w}{\partial y^2}, \\
\sigma_3 &= G\varepsilon_3 = -2Gz \frac{\partial^2 w}{\partial x \partial y}.
\end{aligned} \tag{4.1.8}$$

Here

$$\begin{aligned} E_{ii} &= \frac{E_i}{1-\mu_1\mu_2}, \qquad i=1,2, \\ E_{12} &= E_{21} = \mu_2 E_{11} = \mu_1 E_{22}, \end{aligned} \tag{4.1.9}$$

E_1, E_2, G, μ_1, and μ_2 being the elasticity characteristics of the material. By (4.1.2), (4.1.8), and (4.1.9), we conclude the isotropic material to be a partial case of the orthotropic one, and for it the following relations hold

$$E_1 = E_2 = E, \qquad \mu_1 = \mu_2 = \mu, \qquad G = \frac{E}{2(1+\mu)}.$$

For the bending moments and torque, we have in view of (4.1.8) and (4.1.9),

$$\begin{aligned} M_1 &= \int_{-\frac{h}{2}}^{\frac{h}{2}} \sigma_1(z)\, z\, dz = D_1\chi_1 + D_{12}\chi_2, \\ M_2 &= \int_{-\frac{h}{2}}^{\frac{h}{2}} \sigma_2(z)\, z\, dz = D_{21}\chi_1 + D_2\chi_2, \\ M_3 &= \int_{-\frac{h}{2}}^{\frac{h}{2}} \sigma_3(z)\, z\, dz = D_3\chi_3. \end{aligned} \tag{4.1.10}$$

Here, χ_1, χ_2, and χ_3 are determined by the formulas (4.1.3) and

$$\begin{aligned} D_i &= \frac{h^3 E_i}{12(1-\mu_1\mu_2)}, \qquad i=1,2, \\ D_{12} &= D_{21} = \mu_2 D_1 = \mu_1 D_2, \qquad D_3 = \frac{h^3 G}{6}. \end{aligned} \tag{4.1.11}$$

Taking into consideration the relations (4.1.1), (4.1.6), (4.1.8), and (4.1.10), we get the following expression for the strain energy of the orthotropic plate

$$\begin{aligned} \Phi(w) &= \frac{1}{2}\iint\limits_{\Omega} \left(D_1\chi_1^2 + 2D_{12}\chi_1\chi_2 + D_2\chi_2^2 + 2D_3\chi_3^2\right) dx\, dy \\ &= \frac{1}{2}\iint\limits_{\Omega} \left(M_1\chi_1 + M_2\chi_2 + 2M_3\chi_3\right) dx\, dy. \end{aligned} \tag{4.1.12}$$

4.1.3 Bilinear form corresponding to the strain energy of the plate

Isotropic plate

Let V be a closed subspace of $W_2^2(\Omega)$ with the norm of the space $W_2^2(\Omega)$. According to (4.1.7), to the strain energy of an isotropic plate there corresponds the following symmetric bilinear form on $V \times V$

$$a_h(u,v) = \iint\limits_{\Omega} D\Bigg[\left(\frac{\partial^2 u}{\partial x^2} + \frac{\partial^2 u}{\partial y^2}\right)\left(\frac{\partial^2 v}{\partial x^2} + \frac{\partial^2 v}{\partial y^2}\right) - (1-\mu)\left(\frac{\partial^2 u}{\partial x^2}\frac{\partial^2 v}{\partial y^2} + \frac{\partial^2 u}{\partial y^2}\frac{\partial^2 v}{\partial x^2} - 2\frac{\partial^2 u}{\partial x \partial y}\frac{\partial^2 v}{\partial x \partial y}\right)\Bigg]\,dx\,dy, \qquad u,v \in V. \tag{4.1.13}$$

Obviously,

$$a_h(u,u) = 2\Phi(u), \tag{4.1.14}$$

where $\Phi(u)$ is defined by (4.1.7).

Suppose that

$$E = \text{const} > 0, \tag{4.1.15}$$

$$h \in Y_p \tag{4.1.16}$$

$$\mu = \text{const}, \qquad 0 \le \mu < 1, \tag{4.1.17}$$

where the set Y_p is defined by the formulas (2.1.2) and (2.1.3). Then, (4.1.5) and (4.1.13) yield the continuity of the form a_h in the following sense

$$|a_h(u,v)| \le c\|u\|_V\|v\|_V, \qquad u,v \in V,\ h \in Y_p,\ c = \text{const} > 0. \tag{4.1.18}$$

Taking into account the inequality

$$2\left|\frac{\partial^2 u}{\partial x^2}\frac{\partial^2 u}{\partial y^2}\right| \le \left(\frac{\partial^2 u}{\partial x^2}\right)^2 + \left(\frac{\partial^2 u}{\partial y^2}\right)^2, \tag{4.1.19}$$

by (4.1.5), (4.1.13), and (4.1.17), we get

$$\begin{aligned} a_h(u,u) &= \iint\limits_{\Omega} D\Bigg[\left(\frac{\partial^2 u}{\partial x^2}\right)^2 + \left(\frac{\partial^2 u}{\partial y^2}\right)^2 + 2\mu\frac{\partial^2 u}{\partial x^2}\frac{\partial^2 u}{\partial y^2} + 2(1-\mu)\left(\frac{\partial^2 u}{\partial x \partial y}\right)^2\Bigg]\,dx\,dy \\ &\ge c_1 \iint\limits_{\Omega}\Bigg[\left(\frac{\partial^2 u}{\partial x^2}\right)^2 + \left(\frac{\partial^2 u}{\partial y^2}\right)^2 + \left(\frac{\partial^2 u}{\partial x \partial y}\right)^2\Bigg]\,dx\,dy \end{aligned} \tag{4.1.20}$$

$$u \in V,\ h \in Y_p,\ c_1 = \text{const} > 0.$$

Now suppose that the following condition holds:

$$\left.\begin{array}{l}\text{the relations } u \in V,\ u = b_0 + b_1 x + b_2 y, \text{ where } b_0,\ b_1, \text{ and}\\ b_2 \text{ are constants, imply } b_0 = b_1 = b_2 = 0.\end{array}\right\} \tag{4.1.21}$$

Then, by Corollary 1.6.1, the formula

$$\|u\|_1 = \left\{ \iint_\Omega \left[\left(\frac{\partial^2 u}{\partial x^2}\right)^2 + \left(\frac{\partial^2 u}{\partial y^2}\right)^2 + \left(\frac{\partial^2 u}{\partial x \partial y}\right)^2 \right] dx\, dy \right\}^{\frac{1}{2}} \tag{4.1.22}$$

defines a norm in V, which is equivalent to the original one, i.e., to the norm of $W_2^2(\Omega)$. Now, by (4.1.20) and (4.1.22), we have

$$a_h(u,u) \geq c_2 \|u\|_V^2, \qquad u \in V,\ h \in Y_p,\ c_2 = \text{const} > 0. \tag{4.1.23}$$

Thus, the following theorem is proven.

Theorem 4.1.1 *Let the conditions* (4.1.15)–(4.1.17) *hold and let a set* Y_p *be defined by the relations* (2.1.2) *and* (2.1.3). *Also, let* V *be a closed subspace of* $W_2^2(\Omega)$ *meeting* (4.1.21). *Then, the bilinear form* $a_h(u,v)$ *determined by the formulas* (4.1.5) *and* (4.1.13) *is symmetric, continuous, and coercive on* $V \times V$ *in the sense of the inequalities* (4.1.18) *and* (4.1.23).

Orthotropic plate

According to (4.1.3) and (4.1.12), to the strain energy of an orthotropic plate there corresponds the following bilinear form on $V \times V$

$$\begin{aligned} a_h(u,v) = \iint_\Omega \Big[& D_1 \frac{\partial^2 u}{\partial x^2}\frac{\partial^2 v}{\partial x^2} + D_2 \frac{\partial^2 u}{\partial y^2}\frac{\partial^2 v}{\partial y^2} \\ & + D_{12}\left(\frac{\partial^2 u}{\partial x^2}\frac{\partial^2 v}{\partial y^2} + \frac{\partial^2 u}{\partial y^2}\frac{\partial^2 v}{\partial x^2}\right) + 2D_3 \frac{\partial^2 u}{\partial x \partial y}\frac{\partial^2 v}{\partial x \partial y}\Big]\, dx\, dy, \end{aligned} \tag{4.1.24}$$

where D_i, $i = 1,2,3$, and D_{12} are defined by (4.1.11). Obviously, $a_h(u,u) = 2\Phi(u)$, where $\Phi(u)$ is determined by the formulas (4.1.3) and (4.1.12).

Assume that

$$E_1, E_2, G \ \text{ are positive numbers,} \tag{4.1.25}$$

$$h \in Y_p, \tag{4.1.26}$$

$$\mu_1 \text{ and } \mu_2 \text{ are constants,} \quad 0 \leq \mu_i < 1, i = 1,2. \tag{4.1.27}$$

Making use of (4.1.11) and (4.1.25)–(4.1.27), one can easily see that, for the bilinear form a_h determined by (4.1.24), the inequality (4.1.18) is valid. From (4.1.24), by using (4.1.11), (4.1.19), and (4.1.25)–(4.1.27), we obtain for arbitrary $u \in V$, $h \in Y_p$,

$$a_h(u,u) \geq \iint_\Omega \left\{ D_1 \left(\frac{\partial^2 u}{\partial x^2}\right)^2 + D_2 \left(\frac{\partial^2 u}{\partial y^2}\right)^2 + 2D_3 \left(\frac{\partial^2 u}{\partial x \partial y}\right)^2 \right.$$

$$- D_{12}\left[\left(\frac{\partial^2 u}{\partial x^2}\right)^2 + \left(\frac{\partial^2 u}{\partial y^2}\right)^2\right]\Bigg\} dx\,dy$$

$$\geq c_1 \iint_\Omega \left[\left(\frac{\partial^2 u}{\partial x^2}\right)^2 + \left(\frac{\partial^2 u}{\partial y^2}\right)^2 + \left(\frac{\partial^2 u}{\partial x \partial y}\right)^2\right] dx\,dy,$$

c_1 being a positive number.

Thus, we have proved the following

Theorem 4.1.2 *Let the conditions* (4.1.25)–(4.1.27) *hold and let a set* Y_p *be defined by the relations* (2.1.2) *and* (2.1.3). *Assume* V *to be a closed subspace in* $W_2^2(\Omega)$ *meeting* (4.1.21). *Then, the bilinear form* $a_h(u,v)$ *determined by the formulas* (4.1.11) *and* (4.1.24) *is symmetric, continuous, and coercive on* $V \times V$ *in the sense of the inequalities* (4.1.18) *and* (4.1.23).

Remark 4.1.1 Theorem 4.1.2 remains valid if the conditions (4.1.27) are replaced by the following ones

$$\mu_1 \text{ and } \mu_2 \text{ are constants}, \qquad \mu_1\mu_2 < 1, \qquad E_1 - \mu_1^2 E_2 > 0. \tag{4.1.28}$$

Indeed, by (4.1.25) and the Sylvester criterion, we deduce the quadratic form

$$\frac{E_1}{1-\mu_1\mu_2}\,\xi_1^2 + \frac{2\mu_1 E_2}{1-\mu_1\mu_2}\,\xi_1\xi_2 + \frac{E_2}{1-\mu_1\mu_2}\,\xi_2^2, \qquad \xi_1, \xi_2 \in \mathbb{R}$$

to be positive definite. Hence, we have

$$a_h(u,u) \geq c \iint_\Omega \left[\left(\frac{\partial^2 u}{\partial x^2}\right)^2 + \left(\frac{\partial^2 u}{\partial y^2}\right)^2 + \left(\frac{\partial^2 u}{\partial x \partial y}\right)^2\right] dx\,dy,$$
$$u \in V,\ h \in Y_p,\ c = \text{const} > 0.$$

In the sequel, in considering various orthotropic plates and shells, we suppose the conditions (4.1.27) to be fulfilled, since the majority of actual orthotropic materials are of that kind.

However, in what follows, the relations (4.1.27) could be replaced by less restrictive assumptions (4.1.28).

4.1.4 Problem of bending of a plate

Setting of the problem. Examples of boundary conditions and loads

Let f be a load acting on a plate, which is identified with some element of the space V^*. Then, the problem of the bending of the plate reduces to the problem of finding a function u such that

$$u \in V, \qquad a_h(u,v) = (f,v), \qquad v \in V, \tag{4.1.29}$$

the form a_h being determined by the relation (4.1.13) for an isotropic plate and by the formula (4.1.24) for an orthotropic one.

By Theorems 4.1.1 and 4.1.2, for any $f \in V^*$ and $h \in Y_p$, the problem (4.1.29) has a unique solution for the isotropic and orthotropic plates. From the physical point of view, the problem (4.1.29) means that one searches for a function u which minimizes the stored energy of the system on a set of functions meeting smoothness conditions and main boundary conditions which correspond to the fastening of the plate. The stored energy of the system $\Psi(w)$ is determined by

$$\Psi(w) = \frac{1}{2} a_h(w, w) - (f, w), \qquad w \in V, \tag{4.1.30}$$

where $-(f, w)$ is the potential energy of the load f.

Let us consider some examples of the space V and the load f. Let S be the boundary of the domain Ω occupied by the midplane of the plate, and $S_1 \subset S$ (in particular, it may happen that $S_1 = S$). We suppose that S_1 contains three points which do not belong to a single straight line, and the domain Ω is bounded. Define the space V as follows:

$$V = \{\, u \,|\, u \in W_2^2(\Omega),\; u|_{S_1} = 0 \,\}. \tag{4.1.31}$$

In this case, the space V corresponds to the fastening (supporting) of the plate at the part S_1 of the boundary. Let $u = b_0 + b_1 x + b_2 y$, where b_0, b_1, b_2 are constants and $u \in V$. Since u vanishes in three points which do not belong to a single straight line, we easily deduce that $b_0 = b_1 = b_2 = 0$, i.e., (4.1.21) holds.

Let us consider the clamp of the plate at S_1. In this case, the space V looks as follows

$$V = \Big\{\, u \,|\, u \in W_2^2(\Omega),\; u|_{S_1} = 0,\; \frac{\partial u}{\partial \nu}\Big|_{S_1} = 0 \,\Big\}, \tag{4.1.32}$$

$\frac{\partial}{\partial \nu}$ being the derivative with respect to the normal to the boundary S. Here, by the theorem on the trace space (see Theorem 1.6.5 and Remarks 1.6.1, 1.6.2), the conditions $u|_{S_1} = 0$ and $\frac{\partial u}{\partial \nu}\big|_{S_1} = 0$ hold, and (4.1.21) is valid.

In solving problems of plate bending, one uses also the condition of a simple support. In this case,

$$u\big|_{S_1} = 0, \qquad M_\nu = M_1 \nu_x^2 + M_2 \nu_y^2 + 2M_3 \nu_x \nu_y = 0 \;\text{ on } S_1,$$

ν_x and ν_y being the components of the exterior unit normal ν to S. This condition makes no sense for an arbitrary function u from $W_2^2(\Omega)$; it is called the condition of transversality, or the natural condition (see "Interpretation of the problem (4.1.29)" below).

Let now the space V be of the form

$$V = \{\, u \,|\, u \in W_2^2(\Omega),\; u(x_1, y_1) = u(x_2, y_2) = u(x_3, y_3) = 0 \,\}. \tag{4.1.33}$$

Here, (x_1, y_1), (x_2, y_2), and (x_3, y_3) are three points from $\overline{\Omega}$ which do not belong to a single straight line. Since the embedding of $W_2^2(\Omega)$ into $C(\overline{\Omega})$ is continuous, the conditions $u(x_i, y_i) = 0$, $i = 1, 2, 3$, hold, and the space V determined by (4.1.33) meets the condition (4.1.21). Notice that the space V defined by (4.1.33) and similar ones appear in the calculation of elements of aircraft constructions when the plate is fastened at the center and at two parts of the edge surface by hinges (Fig. 4.1.2).

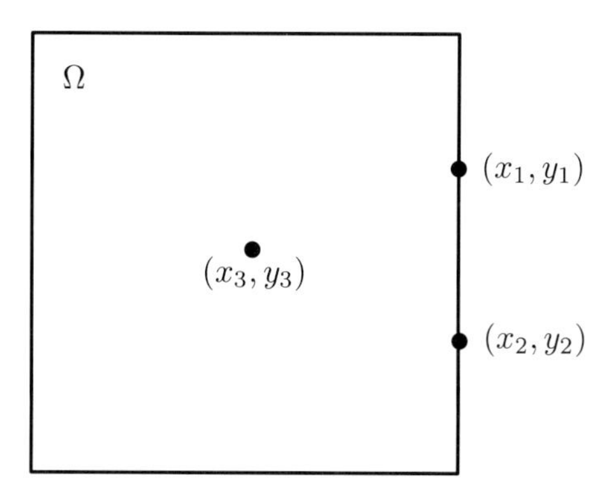

Figure 4.1.2: Plate fastened at three points (x_1, y_1), (x_2, y_2), and (x_3, y_3)

Now, instead of (4.1.33), one can use the following spaces

$$V = \left\{ u \,|\, u \in W_2^2(\Omega),\ u = 0 \ \text{ on } \Omega_1,\ u|_{S_2} = u|_{S_3} = 0 \right\}, \qquad (4.1.34)$$

$$V = \left\{ u \,|\, u \in W_2^2(\Omega),\ u(x_3, y_3) = 0,\ u|_{S_2} = u|_{S_3} = 0 \right\} \qquad (4.1.35)$$

(see Fig. 4.1.3), where Ω_1 is a subset of Ω, and S_2, S_3 are segments. In the case when the space V is determined by the relation (4.1.35), an approximation of a solution of the problem (4.1.29) was studied by Litvinov (1981a).

All the spaces V defined by the formulas (4.1.31)–(4.1.35) are closed subspaces of $W_2^2(\Omega)$.

Let us consider an example of a load which is represented by an element $f \in V^*$. Let the space V be determined by (4.1.31), let S_1 be an open set in S, and let S_2 be the interior of $S \setminus S_1$. Define an element $f \in V^*$ by

$$v \to (f, v) = \iint\limits_{\Omega} f_1 v \, dx \, dy + \int_{S_2} f_2 v \, ds + \int_S f_3 \frac{\partial v}{\partial \nu} \, ds. \qquad (4.1.36)$$

Here, $\frac{\partial v}{\partial \nu}$ is the derivative with respect to the normal of S, f_1 is the distributed load, f_2 is the cutting force, and f_3 is the bending moment, which act on Ω, S_2, and S, respectively. If we use the space V from (4.1.32), then the latter integral in (4.1.36) is to be taken over S_2, instead of S, because now $\frac{\partial v}{\partial \nu} = 0$ on S_1.

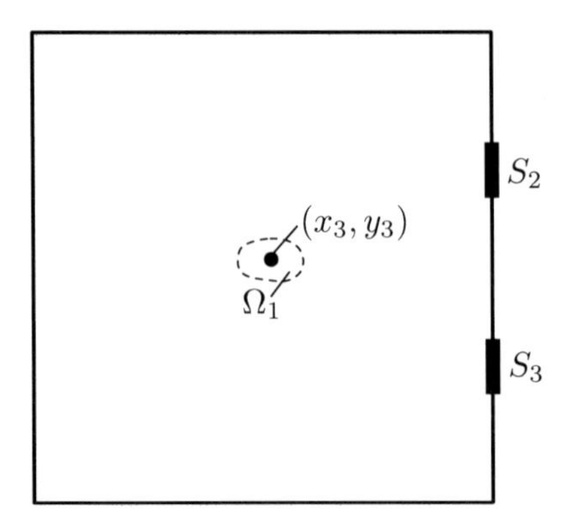

Figure 4.1.3: Plate fastened at a subset Ω_1 or at a point (x_3, y_3) and at segments S_2, S_3

We suppose also that

$$f_1 \in L_2(\Omega), \qquad f_2 \in H^{-\frac{3}{2}}(S_2), \qquad f_3 \in H^{-\frac{1}{2}}(S), \tag{4.1.37}$$

and the boundary S is regular enough.

The results of Subsec. 1.6.4 imply that, if (4.1.37) holds true, then the functional f defined by (4.1.36) is continuous on V, i.e., $f \in V^*$.

Interpretation of the problem (4.1.29)

Let us interpret the problem (4.1.29), supposing that the bilinear form a_h is defined by the relation (4.1.13), f by (4.1.36), and V by (4.1.31). Take in (4.1.29) $v = \varphi \in \mathcal{D}(\Omega)$. Then

$$Au = f_1 \qquad \text{in } \Omega, \tag{4.1.38}$$

the operator A being defined by the relation

$$\begin{aligned} Au = {} & \frac{\partial^2}{\partial x^2}\left[D\left(\frac{\partial^2 u}{\partial x^2} + \mu\frac{\partial^2 u}{\partial y^2}\right)\right] + \frac{\partial^2}{\partial y^2}\left[D\left(\frac{\partial^2 u}{\partial y^2} + \mu\frac{\partial^2 u}{\partial x^2}\right)\right] \\ & + 2(1-\mu)\frac{\partial^2}{\partial x \partial y}\left(D\frac{\partial^2 u}{\partial x \partial y}\right), \end{aligned} \tag{4.1.39}$$

where the derivatives are taken in the sense of distributions on Ω.

Suppose that the function u is smooth. Then, by multiplying both-hand sides of the equality (4.1.38) by $v \in V$ and making use of Green's formula, we get

$$\iint_\Omega (Au)v\,dx\,dy = -\int_S (B_1 u)\frac{\partial v}{\partial \nu}\,ds - \int_{S_2} (B_2 u)v\,ds + a_h(u,v) = \iint_\Omega f_1 v\,dx\,dy. \tag{4.1.40}$$

Here,

$$B_1 u = D \left[\mu \left(\frac{\partial^2 u}{\partial x^2} + \frac{\partial^2 u}{\partial y^2} \right) + \right.$$
$$\left. (1-\mu) \left(\frac{\partial^2 u}{\partial x^2} \nu_x^2 + \frac{\partial^2 u}{\partial y^2} \nu_y^2 + 2 \frac{\partial^2 u}{\partial x \partial y} \nu_x \nu_y \right) \right] \quad \text{on } S, \tag{4.1.41}$$

$$B_2 u = -\frac{\partial}{\partial \nu} \left[D \left(\frac{\partial^2 u}{\partial x^2} + \frac{\partial^2 u}{\partial y^2} \right) \right] - (1-\mu) \frac{\partial}{\partial \tau} \left\{ D \left[\frac{\partial^2 u}{\partial x \partial y} (\nu_y^2 - \nu_x^2) \right. \right.$$
$$\left. \left. + \left(\frac{\partial^2 u}{\partial x^2} - \frac{\partial^2 u}{\partial y^2} \right) \nu_x \nu_y \right] \right\} - \left(\frac{\partial D}{\partial x} \frac{\partial^2 u}{\partial x \partial y} - \frac{\partial D}{\partial y} \frac{\partial^2 u}{\partial x^2} \right) \nu_y$$
$$+ \left(\frac{\partial D}{\partial x} \frac{\partial^2 u}{\partial y^2} - \frac{\partial D}{\partial y} \frac{\partial^2 u}{\partial x \partial y} \right) \nu_x \quad \text{on } S_2. \tag{4.1.42}$$

Here, ν_x and ν_y are the components of the exterior unit normal ν to S, and $\frac{\partial}{\partial \tau}$ is the tangent derivative, i.e., the derivative along the unit tangent vector τ (see Fig. 4.1.4).

Notice that, if $u \in V$ is not smooth, but $Au = f_1 \in L_2(\Omega)$, then the relation (4.1.40) can be substantiated by using the abstract Green's formula (see Aubin (1972)). In this case,

$$B_1 u \in H^{-\frac{1}{2}}(S), \qquad B_2 u \in H^{-\frac{3}{2}}(S_2),$$

and the formula (4.1.40) takes the form

$$\iint_\Omega (Au)\, v\, dx\, dy = -\left(B_1 u, \left.\frac{\partial v}{\partial \nu}\right|_S \right) - \left(B_2 u, v|_{S_2} \right) + a_h(u,v)$$
$$= \iint_\Omega f_1 v\, dx\, dy, \qquad v \in V. \tag{4.1.43}$$

Due to (4.1.29) and (4.1.36), we have

$$a_h(u,v) = \iint_\Omega f_1 v\, dx\, dy + \left(f_2, v|_{S_2} \right) + \left(f_3, \left.\frac{\partial v}{\partial \nu}\right|_S \right), \qquad v \in V. \tag{4.1.44}$$

This equality together with (4.1.43) implies

$$\left(f_2 - B_2 u, v|_{S_2} \right) + \left(f_3 - B_1 u, \left.\frac{\partial v}{\partial \nu}\right|_S \right) = 0.$$

Now, Theorem 1.6.5 yields

$$f_2 = B_2 u \ \text{ on } S_2, \qquad f_3 = B_1 u \ \text{ on } S. \tag{4.1.45}$$

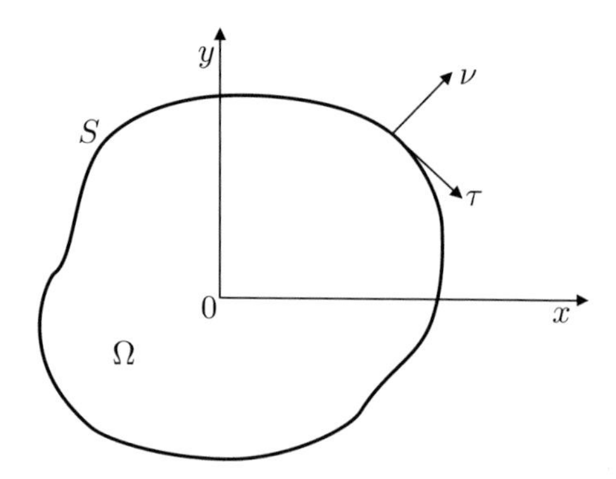

Figure 4.1.4: Domain Ω and unit normal ν and tangent τ vectors

Thus, the function u solving the problem (4.1.44) in which the form $a_h(u,v)$ is defined by (4.1.13) and f_i meets (4.1.37) is shown to satisfy the conditions (4.1.38) and (4.1.45). Conversely, using Green's formula, we get

$$\iint_\Omega (Aw)\, v\, dx\, dy = -\left(B_1 w, \left.\frac{\partial v}{\partial \nu}\right|_S\right) - (B_2 w, v|_{S_2}) + a_h(w,v),$$

which is valid for any function $w \in V$ such that $Aw \in L_2(\Omega)$ and for any $v \in V$ (see Aubin (1972)), and one can easily see that, if the function u satisfies (4.1.38) and (4.1.45), then it is a solution of the problem (4.1.44). By virtue of Theorem 4.1.1, the latter problem has a unique solution.

We stress that the boundary conditions (4.1.45) are called the natural conditions, or the transversality conditions, and the boundary conditions defining the space V (in this case $u\big|_{S_1} = 0$, see (4.1.31)) are called main or stable conditions. If in (4.1.44) $f_3 = 0$, i.e., no bending moments act on the plate along the boundary, then one says that the plate has a simple support. By (4.1.45), we have the following natural boundary condition

$$B_1 u = 0 \qquad \text{on } S. \tag{4.1.46}$$

If no bending moments and cutting forces act on the boundary S, i.e., $f_2 = 0$ and $f_3 = 0$ in (4.1.44), then we obtain the following natural boundary conditions

$$B_1 u = 0 \ \text{ on } S, \qquad B_2 u = 0 \ \text{ on } S_2,$$

which are referred to as the conditions of the free edge.

In the same way, the problem (4.1.29) can be interpreted for the orthotropic plate.

4.1.5 Problem of free oscillations of a plate

Let, as before, V be a closed subspace of $W_2^2(\Omega)$ meeting (4.1.21). Denote by $C^2((0,\infty);V)$ the set of twice continuously differentiable functions which are defined on the interval $(0,\infty)$ and take values in V. Let the plate make free transversal oscillations. A function $U(x,y,t)$ determining the deflection of the plate, which depends on the coordinates x, y and time t, is defined to be the solution of the problem

$$U \in C^2((0,\infty);V), \tag{4.1.47}$$

$$\iint\limits_{\Omega} \left[\rho h \frac{\partial^2 U}{\partial t^2} v + \frac{\rho h^3}{12} \left(\frac{\partial^3 U}{\partial t^2 \partial x} \frac{\partial v}{\partial x} + \frac{\partial^3 U}{\partial t^2 \partial y} \frac{\partial v}{\partial y} \right) \right] dx\, dy + a_h(U,v) = 0,$$

$$v \in V, \ t \in (0,\infty). \tag{4.1.48}$$

Here, the bilinear form a_h is defined by the relations (4.1.13) and (4.1.24) for isotropic and orthotropic plates, respectively, ρ being the density of the material of the plate,

$$\rho = \text{const} > 0. \tag{4.1.49}$$

The first term under the integral sign in (4.1.48) determines the work of the inertia forces on virtual transversal displacements, and the second one determines the work of these forces on virtual longitudinal displacements. Indeed, by the Kirchhoff hypothesis (see Subsec. 4.1.1), the components P_x and P_y of the inertia force distributed over the volume have the form

$$P_x = -\rho z \frac{\partial^3 U}{\partial t^2 \partial x}, \qquad P_y = -\rho z \frac{\partial^3 U}{\partial t^2 \partial y}.$$

If v is a virtual transversal displacement, i.e., a virtual deflection of the plate, then, by using again the Kirchhoff hypothesis, we conclude that the corresponding longitudinal displacements v_x, v_y in the x and y directions are

$$v_x = -z \frac{\partial v}{\partial x}, \qquad v_y = -z \frac{\partial v}{\partial y}.$$

The work of the inertia forces on virtual longitudinal displacements is given by

$$\iint\limits_{\Omega} dx\, dy \int_{-\frac{h}{2}}^{\frac{h}{2}} (P_x v_x + P_y v_y)\, dz = \iint\limits_{\Omega} dx\, dy \int_{-\frac{h}{2}}^{\frac{h}{2}} \rho z^2 \left(\frac{\partial^3 U}{\partial t^2 \partial x} \frac{\partial v}{\partial x} + \frac{\partial^3 U}{\partial t^2 \partial y} \frac{\partial v}{\partial y} \right) dz.$$

Implementing the integration in z on the right-hand side of this equality, we obtain the second term under the integral sign in (4.1.48).

We will search for the solution of the problem (4.1.47), (4.1.48) in the form

$$U(x,y,t) = (C_1 \cos \omega t + C_2 \sin \omega t) u(x,y), \qquad u(x,y) \in V, \tag{4.1.50}$$

C_1 and C_2 being constants.

The functions from (4.1.50) solving the problem (4.1.47), (4.1.48) are called natural oscillations of the plate. By substituting (4.1.50) into (4.1.48), we get the following equation for determining $u(x, y)$

$$u \in V, \qquad a_h(u, v) = \lambda b_h(u, v), \qquad v \in V, \tag{4.1.51}$$

$$\lambda = \omega^2, \tag{4.1.52}$$

$$b_h(u, v) = \iint\limits_{\Omega} \left[\rho h u v + \frac{\rho h^3}{12} \left(\frac{\partial u}{\partial x} \frac{\partial v}{\partial x} + \frac{\partial u}{\partial y} \frac{\partial v}{\partial y} \right) \right] dx\, dy. \tag{4.1.53}$$

The second term in (4.1.53) is called the work of forces caused by the inertia of rotation. For thin plates, $h \gg h^3$, and so in many cases one can neglect the second term. Then, the form b_h takes the form

$$b_h(u, v) = \iint\limits_{\Omega} \rho h u v \, dx\, dy. \tag{4.1.54}$$

Theorem 4.1.3 *Let the assumptions of Theorem* 4.1.1 (*for an isotropic plate*) *or of Theorem* 4.1.2 (*for an orthotropic plate*) *hold true and let* (4.1.49) *be fulfilled. Let also the form b_h be determined by the relation* (4.1.53) *or* (4.1.54). *Then, for any $h \in Y_p$, the spectral problem* (4.1.51) *for both the orthotropic and isotropic plates possesses a sequence of nonzero solutions $\{u_i\} \in V$ corresponding to a sequence of eigenvalues $\{\lambda_i\}$ such that*

$$a_h(u_i, v) = \lambda_i b_h(u_i, v), \qquad v \in V, \tag{4.1.55}$$

$$0 < \lambda_1 \le \lambda_2 \le \lambda_3 \le \cdots, \qquad \lim_{i \to \infty} \lambda_i = \infty,$$

and

$$\lambda_1 = \frac{a_h(u_1, u_1)}{b_h(u_1, u_1)} = \inf_{\substack{u \in V \\ u \neq 0}} \frac{a_h(u, u)}{b_h(u, u)}.$$

Each eigenvalue of the sequence $\{\lambda_i\}$ appears as many times as its multiplicity, and the multiplicity of each eigenvalue is finite.

Proof. By virtue of Theorems 4.1.1 and 4.1.2 (for isotropic and orthotropic plates, respectively), the bilinear form a_h defines a scalar product in V. Denote by $\tilde{V}$ the space coinciding with V as a set and with the scalar product defined by the form a_h. Making use of Theorems 4.1.1 and 4.1.2, it is easy to see that $\tilde{V}$ is a Hilbert space.

Being considered as a topological space, $\tilde{V}$ coincides with V and the norm generated by the form a_h is equivalent to the one of the space V, i.e., to the norm of $W_2^2(\Omega)$.

The bilinear forms b_h from (4.1.53) and (4.1.54) are obviously symmetric and continuous on $W_2^1(\Omega)$. Moreover, for these forms, the following estimate holds

$$b_h(v, v) \ge c\|v\|^2_{L_2(\Omega)}, \qquad v \in W_2^1(\Omega), \tag{4.1.56}$$

c being a positive constant. From (4.1.55) and (4.1.56) we conclude that

$$\lambda_i = \frac{a_h(u_i, u_i)}{b_h(u_i, u_i)} > 0 \qquad \forall i.$$

To finish the proof, we use Theorem 1.5.8, taking into account that the embedding of $\tilde{V}$ into $W_2^1(\Omega)$ is compact, and pass from the problem (4.1.55) to the problem

$$\mu_i a_h(u_i, v) = b_h(u_i, v), \qquad v \in V,$$

where $\mu_i = \frac{1}{\lambda_i}$.

4.2 Problem of stability of a thin plate

"The little hut was so wretched that it knew not on which side to fall, and therefore remained standing"

– *H. Ch. Andersen*
"The Ugly Duckling"

4.2.1 Stored energy of a plate

Let, as before, the (x, y) plane coincide with the midplane of the plate, let the z axis be directed downwards (see Fig. 4.1.1), let Ω be the domain occupied by the midplane of the plate, and let S be the boundary of Ω. Suppose that the plate is fastened at a part S_1 of the boundary, and at another part $S_2 = S \setminus S_1$ the plate is exposed to longitudinal forces $\lambda Q = (\lambda Q_1, \lambda Q_2)$, which are proportional to a number parameter λ. Here, Q_1 and Q_2 are the components of the vector Q in the x and y axes, which are functions of $s \in S_2$.

Denote by u, v, w the components of the vector ω of displacements of points of the midplane in the x, y, z axes, i.e., $\omega = (u, v, w)$. The strain components of the midplane of the plate are determined by the following relations

$$\varepsilon_{11}(\omega) = \frac{\partial u}{\partial x} + \frac{1}{2}\left(\frac{\partial w}{\partial x}\right)^2, \qquad \varepsilon_{22}(\omega) = \frac{\partial v}{\partial y} + \frac{1}{2}\left(\frac{\partial w}{\partial y}\right)^2,$$
$$\varepsilon_{12}(\omega) = \frac{\partial u}{\partial y} + \frac{\partial v}{\partial x} + \frac{\partial w}{\partial x}\frac{\partial w}{\partial y}, \tag{4.2.1}$$

and the bending and torsion strains are described by the formulas (4.1.3). So, the strain components at a point (x, y, z) of the plate, denoted by $\varepsilon_{11}^z(\omega)$, $\varepsilon_{22}^z(\omega)$, $\varepsilon_{12}^z(\omega)$, are given by

$$\varepsilon_{11}^z(\omega) = \varepsilon_{11}(\omega) + z\chi_1(\omega),$$
$$\varepsilon_{22}^z(\omega) = \varepsilon_{22}(\omega) + z\chi_2(\omega),$$
$$\varepsilon_{12}^z(\omega) = \varepsilon_{12}(\omega) + 2z\chi_3(\omega),$$

where

$$\chi_1(\omega) = -\frac{\partial^2 w}{\partial x^2}, \qquad \chi_2(\omega) = -\frac{\partial^2 w}{\partial y^2}, \qquad \chi_3(\omega) = -\frac{\partial^2 w}{\partial x\, \partial y}.$$

We assume that the material of the plate is orthotropic, and the principal directions of the elasticity of the material coincide with the directions of the x and y coordinate lines. Then, the stress components at a point (x, y, z) of the plate are determined by

$$\begin{aligned}
\sigma_{11}^z(\omega) &= E_{11}\varepsilon_{11}^z(\omega) + E_{12}\varepsilon_{22}^z(\omega),\\
\sigma_{22}^z(\omega) &= E_{21}\varepsilon_{11}^z(\omega) + E_{22}\varepsilon_{22}^z(\omega),\\
\sigma_{12}^z(\omega) &= G\varepsilon_{12}^z(\omega).
\end{aligned}$$

Using the foregoing formulas, we get the following relation for the strain energy of the plate

$$\begin{aligned}
&\Phi_1(\omega)\\
&= \frac{1}{2}\iint\limits_{\Omega} dx\, dy \int_{-\frac{h}{2}}^{\frac{h}{2}} \left[\sigma_{11}^z(\omega)\varepsilon_{11}^z(\omega) + \sigma_{22}^z(\omega)\varepsilon_{22}^z(\omega) + \sigma_{12}^z(\omega)\varepsilon_{12}^z(\omega)\right] dz\\
&= \frac{1}{2}\iint\limits_{\Omega} \left[c_{11}(\varepsilon_{11}(\omega))^2 + 2c_{12}\varepsilon_{11}(\omega)\varepsilon_{22}(\omega) + c_{22}(\varepsilon_{22}(\omega))^2 + c_{33}(\varepsilon_{12}(\omega))^2\right] dx\, dy\\
&+ \frac{1}{2}\iint\limits_{\Omega} \left[D_1(\chi_1(\omega))^2 + 2D_{12}\chi_1(\omega)\chi_2(\omega) + D_2(\chi_2(\omega))^2 + 2D_3(\chi_3(\omega))^2\right] dx\, dy.
\end{aligned} \tag{4.2.2}$$

The first term on the right-hand side of (4.2.2) is the energy caused by the deformations of the midplane of the plate, while the second one coincides with the expression (4.1.12) and determines the bending and torsion energies. The coefficients in (4.2.2) are given by the following formulas

$$\begin{gathered}
c_{ik} = hE_{ik}, \qquad i, k = 1, 2, \qquad c_{33} = hG,\\
D_i = \frac{h^3}{12} E_{ii} \qquad i = 1, 2, \qquad D_{12} = \mu_2 D_1 = \mu_1 D_2, \qquad D_3 = \frac{h^3 G}{6},\\
E_{ii} = \frac{E_i}{1 - \mu_1\mu_2}, \qquad i = 1, 2, \qquad E_{12} = E_{21} = \mu_2 E_{11} = \mu_1 E_{22}.
\end{gathered} \tag{4.2.3}$$

In particular, if

$$E_1 = E_2 = E, \qquad \mu_1 = \mu_2 = \mu, \qquad G = \frac{E}{2(1+\mu)}, \tag{4.2.4}$$

then the relations (4.2.2) and (4.2.3) determine the energy of elastic deformation accumulated in the isotropic plate.

The stored energy of the orthotropic plate is

$$\Phi_\lambda(\omega) = \Phi_1(\omega) - \lambda \int_{S_2} (Q_1 u + Q_2 v)\, ds. \tag{4.2.5}$$

The functions Q_1 and Q_2 can vanish at a part of S_2. At the part $S_1 = S \setminus S_2$, the plate is fastened. Thus, we will consider the functional (4.2.5) in the space V given by

$$V = V_1 \times V_2, \tag{4.2.6}$$

and if $\omega = (u, v, w) \in V$, then $(u, v) \in V_1$ and $w \in V_2$. The space V_1 is defined by

$$V_1 = \{\, g \,|\, g = (g_1, g_2),\ g \in (W_2^1(\Omega))^2,\ g\,|_{S_1} = 0 \,\} \tag{4.2.7}$$

and equipped with the norm of the space $(W_2^1(\Omega))^2$, i.e.,

$$\|g\|_{V_1} = \|g\|_{(W_2^1(\Omega))^2}, \qquad g \in V_1. \tag{4.2.8}$$

In this section, the set S_1 is assumed to be of positive one-dimensional measure.

From the results of Subsec. 1.7.2, we conclude that the formula

$$\|g\|_1 = \left\{ \iint\limits_\Omega \left[\left(\frac{\partial g_1}{\partial x}\right)^2 + \left(\frac{\partial g_2}{\partial y}\right)^2 + \left(\frac{\partial g_1}{\partial y} + \frac{\partial g_2}{\partial x}\right)^2 \right] dx\, dy \right\}^{1/2} \tag{4.2.9}$$

defines a norm in the space V_1 which is equivalent to the original one, i.e., to the norm of $(W_2^1(\Omega))^2$. In other words, there exist positive numbers c_1 and c_2 such that

$$c_1\|g\|_1 \le \|g\|_{V_1} \le c_2\|g\|_1, \qquad g \in V_1. \tag{4.2.10}$$

We assume the space V_2 to meet the following condition

$$\left.\begin{array}{l} V_2 \text{ is a closed subspace of } W_2^2(\Omega), \text{ the conditions } q \in V_2, \\ q = b_0 + b_1 x + b_2 y, \text{ where } b_0, b_1, \text{ and } b_2 \text{ are constants,} \\ \text{imply that } b_0 = b_1 = b_2 = 0. \end{array}\right\} \tag{4.2.11}$$

The space V_2 is endowed with the norm of $W_2^2(\Omega)$, so that we denote

$$\|q\|_{V_2} = \|q\|_{W_2^2(\Omega)}, \qquad q \in V_2. \tag{4.2.12}$$

In particular, if the plate is supported on S_1, then

$$V_2 = \{\, q \,|\, q \in W_2^2(\Omega), q|_{S_1} = 0 \,\},$$

and if it is clamped on S_1, then

$$V_2 = \left\{\, q \,|\, q \in W_2^2(\Omega),\ q|_{S_1} = 0,\ \left.\frac{\partial q}{\partial \nu}\right|_{S_1} = 0 \,\right\},$$

$\frac{\partial}{\partial \nu}$ being the derivative with respect to the normal to S.

Suppose that

$$Q = (Q_1, Q_2) \in (L_2(S_2))^2 \tag{4.2.13}$$

and the relations (4.1.25)–(4.1.27) hold. Then, the functional (4.2.5) is well defined on the whole space V. We will search for the stationary points of the functional (4.2.5) in the space V, which characterize the equilibrium states of the plate being under the longitudinal load λQ. In other words, we look for functions ω_λ such that

$$\omega_\lambda \in V, \qquad \frac{d}{dt}\,\Phi_\lambda(\omega_\lambda + th)\big|_{t=0} = 0, \qquad h \in V. \tag{4.2.14}$$

4.2.2 Conditions of stationarity

Define bilinear continuous mappings

$$T_1,\ T_2,\ T_{12} \in \mathcal{L}_2(V_2, V_2; L_2(\Omega)) \tag{4.2.15}$$

as follows

$$\begin{gathered} w',\ w'' \in V_2, \\ T_1(w', w'') = \frac{1}{2}E_{11}h\left(\frac{\partial w'}{\partial x}\frac{\partial w''}{\partial x} + \mu_2\frac{\partial w'}{\partial y}\frac{\partial w''}{\partial y}\right), \\ T_2(w', w'') = \frac{1}{2}E_{22}h\left(\frac{\partial w'}{\partial y}\frac{\partial w''}{\partial y} + \mu_1\frac{\partial w'}{\partial x}\frac{\partial w''}{\partial x}\right), \\ T_{12}(w', w'') = \frac{1}{2}Gh\left(\frac{\partial w'}{\partial x}\frac{\partial w''}{\partial y} + \frac{\partial w'}{\partial y}\frac{\partial w''}{\partial x}\right). \end{gathered} \tag{4.2.16}$$

Further, introduce linear continuous mappings

$$N_1,\ N_2,\ N_{12} \in \mathcal{L}(V_1, L_2(\Omega)) \tag{4.2.17}$$

by the relations

$$\begin{gathered} g = (g_1, g_2) \in V_1, \qquad N_1(g) = E_{11}h\left(\frac{\partial g_1}{\partial x} + \mu_2\frac{\partial g_2}{\partial y}\right), \\ N_2(g) = E_{22}h\left(\frac{\partial g_2}{\partial y} + \mu_1\frac{\partial g_1}{\partial x}\right), \qquad N_{12}(g) = Gh\left(\frac{\partial g_1}{\partial y} + \frac{\partial g_2}{\partial x}\right). \end{gathered} \tag{4.2.18}$$

We point out that $T_1(w, w)$, $T_2(w, w)$, and $T_{12}(w, w)$ determine inner forces acting in the midplane of the plate which are caused by the vector of displacements $\omega = (0, 0, w)$, $w \in V_2$, and the relations (4.2.18) define the forces in the midplane which are caused by the vector of displacements $\omega = (g_1, g_2, 0)$, $g = (g_1, g_2) \in V_1$.

In the general case, when $\omega = (g_1, g_2, w) \in V_1 \times V_2$, the forces acting in the midplane are determined by formulas

$$\begin{aligned}\tilde{N}_1(g,w) &= N_1(g) + T_1(w,w),\\ \tilde{N}_2(g,w) &= N_2(g) + T_2(w,w),\\ \tilde{N}_{12}(g,w) &= N_{12}(g) + T_{12}(w,w).\end{aligned} \tag{4.2.19}$$

From (4.2.15) and (4.2.17), we conclude

$$\begin{aligned}\tilde{N}_i\colon V_1 \times V_2 &\to L_2(\Omega), \qquad i = 1,2,\\ \tilde{N}_{12}\colon V_1 \times V_2 &\to L_2(\Omega).\end{aligned} \tag{4.2.20}$$

Also, introduce a bilinear continuous mapping

$$N \in \mathcal{L}_2(V_2, V_2; V_1^*) \tag{4.2.21}$$

through the following relations

$$w',\ w'' \in V_2, \quad g = (g_1, g_2) \in V_1,$$

$$\begin{aligned}(N(w',w''),g) = -\iint\limits_{\Omega} \bigg[T_1(w',w'')\frac{\partial g_1}{\partial x} &+ T_2(w',w'')\frac{\partial g_2}{\partial y}\\ &+ T_{12}(w',w'')\left(\frac{\partial g_1}{\partial y} + \frac{\partial g_2}{\partial x}\right)\bigg]\, dx\, dy.\end{aligned} \tag{4.2.22}$$

(4.2.16) and (4.2.22) yield N to be a symmetric mapping, i.e.,

$$N(w',w'') = N(w'',w'), \qquad w', w'' \in V_2. \tag{4.2.23}$$

Further, define a mapping

$$A\colon V_1 \times V_2 \to V_2^*$$

by the formulas

$$g = (g_1, g_2) \in V_1, \quad w', w'' \in V_2, \quad A(g,w') \in V_2^*,$$

$$\begin{aligned}(A(g,w'),w'') = \iint\limits_{\Omega} \bigg[\tilde{N}_1(g,w')\frac{\partial w'}{\partial x}\frac{\partial w''}{\partial x} &+ \tilde{N}_2(g,w')\frac{\partial w'}{\partial y}\frac{\partial w''}{\partial y}\\ &+ \tilde{N}_{12}(g,w')\left(\frac{\partial w'}{\partial x}\frac{\partial w''}{\partial y} + \frac{\partial w'}{\partial y}\frac{\partial w''}{\partial x}\right)\bigg]\, dx\, dy.\end{aligned} \tag{4.2.24}$$

Let ω_λ be a function satisfying the condition (4.2.14), i.e., it is a stationary point of the functional Φ_λ. Introduce the notations

$$\omega_\lambda = (u_\lambda, v_\lambda, w_\lambda), \tag{4.2.25}$$

$$g_\lambda = (u_\lambda, v_\lambda). \tag{4.2.26}$$

It is easy to see that (4.2.14) is equivalent to the following system

$$a^{(1)}(g_\lambda, g) = (N(w_\lambda, w_\lambda), g) + \lambda(Q, g), \qquad g \in V_1, \tag{4.2.27}$$

$$a^{(2)}(w_\lambda, w) + (A(g_\lambda, w_\lambda), w) = 0, \qquad w \in V_2. \tag{4.2.28}$$

Here, we used the notations

$$a^{(1)}(p, g) = \iint_\Omega h \left[E_{11} \frac{\partial p_1}{\partial x} \frac{\partial g_1}{\partial x} + E_{22} \frac{\partial p_2}{\partial y} \frac{\partial g_2}{\partial y} + E_{12} \left(\frac{\partial p_1}{\partial x} \frac{\partial g_2}{\partial y} + \frac{\partial p_2}{\partial y} \frac{\partial g_1}{\partial x} \right) \right.$$
$$\left. + G \left(\frac{\partial p_1}{\partial y} + \frac{\partial p_2}{\partial x} \right) \left(\frac{\partial g_1}{\partial y} + \frac{\partial g_2}{\partial x} \right) \right] dx\, dy$$
$$= \iint_\Omega \left[N_1(p) \frac{\partial g_1}{\partial x} + N_2(p) \frac{\partial g_2}{\partial y} + N_{12}(p) \left(\frac{\partial g_1}{\partial y} + \frac{\partial g_2}{\partial x} \right) \right] dx\, dy \tag{4.2.29}$$

$$p = (p_1, p_2) \in V_1, \quad g = (g_1, g_2) \in V_1,$$

$$(Q, g) = \int_{S_2} (Q_1 g_1 + Q_2 g_2)\, ds, \tag{4.2.30}$$

$$a^{(2)}(w', w'') = \iint_\Omega \frac{h^3}{12} \left[E_{11} \frac{\partial^2 w'}{\partial x^2} \frac{\partial^2 w''}{\partial x^2} + E_{22} \frac{\partial^2 w'}{\partial y^2} \frac{\partial^2 w''}{\partial y^2} \right.$$
$$\left. + E_{12} \left(\frac{\partial^2 w'}{\partial x^2} \frac{\partial^2 w''}{\partial y^2} + \frac{\partial^2 w'}{\partial y^2} \frac{\partial^2 w''}{\partial x^2} \right) + 4G \frac{\partial^2 w'}{\partial x \partial y} \frac{\partial^2 w''}{\partial x \partial y} \right] dx\, dy, \tag{4.2.31}$$
$$w', w'' \in V_2.$$

Notice that the equality (4.2.29) follows from (4.2.3) and (4.2.18).

4.2.3 Auxiliary statements

Lemma 4.2.1 *Let the conditions (4.1.25)–(4.1.27) be fulfilled and let a set Y_p be defined by the formulas (2.1.2) and (2.1.3). Further, let a space V_1 be determined by (4.2.7) and endowed with the norm of $(W_2^1(\Omega))^2$ (see (4.2.8)). Then, the bilinear form $a^{(1)}$ defined by (4.2.3) and (4.2.29), is symmetric, continuous, and coercive on $V_1 \times V_1$ for every $h \in Y_p$, i.e.,*

$$a^{(1)}(p, g) = a^{(1)}(g, p), \qquad p, g \in V_1, \tag{4.2.32}$$

$$|a^{(1)}(p, g)| \le \tilde{c}_1 \|p\|_{V_1} \|g\|_{V_1}, \qquad p, g \in V_1, \tag{4.2.33}$$

$$a^{(1)}(g, g) \ge \tilde{c}_2 \|g\|^2_{V_1}, \qquad g \in V_1, \tag{4.2.34}$$

$\tilde{c}_1$ and $\tilde{c}_2$ being positive numbers that are independent of $h \in Y_p$.

Proof. The relations (4.2.32) and (4.2.33) are obvious. Let us prove the inequality (4.2.34). From the formulas (2.1.2), (2.1.3), (4.2.3), and (4.2.29), we deduce that

$$a^{(1)}(g,g) \geq \iint_\Omega h\left[E_{11}\left(\frac{\partial g_1}{\partial x}\right)^2 + E_{22}\left(\frac{\partial g_2}{\partial y}\right)^2 - 2\mu_1 E_{22}\left|\frac{\partial g_1}{\partial x}\frac{\partial g_2}{\partial y}\right| \right.$$
$$\left. + G\left(\frac{\partial g_1}{\partial y} + \frac{\partial g_2}{\partial x}\right)^2\right] dx\,dy, \qquad g \in V_1. \quad (4.2.35)$$

From this, taking into account the estimate

$$2\left|\frac{\partial g_1}{\partial x}\frac{\partial g_2}{\partial y}\right| \leq \left(\frac{\partial g_1}{\partial x}\right)^2 + \left(\frac{\partial g_2}{\partial y}\right)^2,$$

(4.1.27), and the relation $\mu_1 E_{22} = \mu_2 E_{11}$ (see (4.2.3)), we get

$$a^{(1)}(g,g) \geq k\|g\|_1^2, \qquad g \in V_1. \quad (4.2.36)$$

Here, k is a positive number independent of $h \in Y_p$, and $\|g\|_1$ is determined by (4.2.9). Now, (4.2.34) is a consequence of (4.2.10) and (4.2.36), concluding the proof.

By virtue of (4.2.21), Lemma 4.2.1, and the Riesz theorem, we have that, for any $w \in V_2$, there exists a unique function $P(w)$ such that

$$P(w) \in V_1, \qquad a^{(1)}(P(w),g) = (N(w,w),g), \qquad g \in V_1. \quad (4.2.37)$$

Thus, we have defined the mapping $V_2 \ni w \to P(w) \in V_1$.

Lemma 4.2.2 *Let the hypotheses of Lemma* 4.2.1 *be fulfilled, let a space* V_2 *be defined by* (4.2.11) *and equipped with the norm of* $W_2^2(\Omega)$ *(see* (4.2.12)*). Let also a mapping* N *be determined by the relations* (4.2.16) *and* (4.2.22)*. Then,* $P\colon w \to P(w)$*,* $P(w)$ *being defined by* (4.2.37)*, is a continuously Fréchet differentiable mapping of* V_2 *into* V_1*, i.e., at every point* $w \in V_2$ *there exists the Fréchet derivative* $P'_w \in \mathcal{L}(V_2, V_1)$ *of the mapping* P *and, moreover,* $w \to P'_w$ *is a continuous mapping of* V_2 *into* $\mathcal{L}(V_2, V_1)$.

Proof. Let w and h be arbitrary elements from V_2. Taking into consideration (4.2.21) and (4.2.23), from (4.2.37) we get that

$$\begin{aligned} a^{(1)}(P(w+h),g) &= (N(w+h,w+h),g) \\ &= (N(w,w),g) + 2(N(w,h),g) + (N(h,h),g), \qquad g \in V_1. \end{aligned} \quad (4.2.38)$$

Thus, (4.2.37) implies that

$$a^{(1)}(P(w+h),g) = a^{(1)}(P(w),g) + 2(N(w,h),g) + (N(h,h),g), \qquad g \in V_1. \quad (4.2.39)$$

By virtue of Lemma 4.2.1, the bilinear form $a^{(1)}$ generates a scalar product and a norm on V_1, which is equivalent to the original norm of the space V_1, i.e., the norm of $\left(W_2^1(\Omega)\right)^2$. Hence, one can consider the Hilbert space $\tilde{V}_1$ as the set of elements of V_1 equipped with the scalar product and norm generated by the bilinear form $a^{(1)}$. (Obviously, V_1 and $\tilde{V}_1$ coincide as topological spaces, since they contain the same elements and their norms are equivalent.) Then, by (4.2.39) we have

$$P(w+h) = P(w) + 2\mathcal{R}\circ N(w,h) + \mathcal{R}\circ N(h,h). \tag{4.2.40}$$

Here, $\mathcal{R}$ is the Riesz operator (see Subsec. 1.5.1)

$$\mathcal{R} \in \mathcal{L}(\tilde{V}_1^*, \tilde{V}_1), \qquad \|\mathcal{R}\|_{\mathcal{L}(\tilde{V}_1^*,\tilde{V}_1)} = 1. \tag{4.2.41}$$

The operator $\mathcal{R}$ is defined by the relation

$$a^{(1)}(\mathcal{R}f, q) = (f, q), \qquad f \in \tilde{V}_1^*,\ q \in \tilde{V}_1. \tag{4.2.42}$$

By (4.2.15) and (4.2.22), we conclude

$$|(N(h,h), g)| \le c\|h\|_{V_2}^2 \|g\|_{\tilde{V}_1}, \qquad h \in V_2,\ g \in \tilde{V}_1,\ c = \text{const} > 0. \tag{4.2.43}$$

This yields

$$\|N(h,h)\|_{\tilde{V}_1^*} \le c\|h\|_{V_2}^2, \qquad h \in V_2. \tag{4.2.44}$$

Further, taking into account (4.2.34), (4.2.41), and (4.2.44), we obtain

$$\|\mathcal{R}\circ N(h,h)\|_{V_1} \le c_1\|\mathcal{R}\circ N(h,h)\|_{\tilde{V}_1} \le c_2\|h\|_{V_2}^2, \qquad h \in V_2,$$

c_1 and c_2 being positive numbers.

From this inequality we deduce that

$$\frac{\|\mathcal{R}\circ N(h,h)\|_{V_1}}{\|h\|_{V_2}} \to 0, \qquad \text{as } \|h\|_{V_2} \to 0. \tag{4.2.45}$$

(4.2.21) and (4.2.41) imply that, for a given $w \in V_2$, the function

$$h \to 2\mathcal{R}\circ N(w,h)$$

is a linear continuous mapping of V_2 into V_1.

Now, (4.2.40) and (4.2.45) yield that, at every point $w \in V_2$, the mapping $P\colon V_2 \to V_1$ is Fréchet differentiable (see Subsec. 1.9.1) and the Fréchet derivative $P'_w \in \mathcal{L}(V_2, V_1)$ is given by the formula

$$P'_w h = 2\mathcal{R}\circ N(w,h). \tag{4.2.46}$$

Let $\{w_n\}$ be a sequence of elements of V_2 such that

$$w_n \to w_0 \qquad \text{in } V_2. \tag{4.2.47}$$

By using (4.2.21), (4.2.41), and (4.2.46), we have

$$\begin{aligned} \left\|\left(P'_{w_n} - P'_{w_0}\right) h\right\|_{V_1} &= 2\|\mathcal{R} \circ N(w_n, h) - \mathcal{R} \circ N(w_0, h)\|_{V_1} \\ &= 2\|\mathcal{R} \circ N((w_n - w_0), h)\|_{V_1} \\ &\leq c\|w_n - w_0\|_{V_2}\|h\|_{V_2} \\ h &\in V_2, \; c = \text{const} > 0. \end{aligned}$$

From this, we get

$$\left\|P'_{w_n} - P'_{w_0}\right\|_{\mathcal{L}(V_2, V_1)} \leq c\|w_n - w_0\|_{V_2}.$$

At last, (4.2.47) implies

$$P'_{w_n} \to P'_{w_0} \;\text{ in } \mathcal{L}(V_2, V_1),$$

concluding the proof.

Theorem 4.1.2 yields that the bilinear form $a^{(2)}(w', w'')$ determined by the relation (4.2.31), is symmetric, continuous, and coercive on $V_2 \times V_2$, and one can define a Hilbert space $\tilde{V}_2$ as the set of functions from V_2 endowed with the norm and scalar product which are generated by the form $a^{(2)}$, i.e.,

$$(w', w'')_{\tilde{V}_2} = a^{(2)}(w', w''), \qquad \|w\|^2_{\tilde{V}_2} = a^{(2)}(w, w). \tag{4.2.48}$$

(Notice that, in this section, the form a_h defined by the formulas (4.1.11) and (4.1.24) is denoted by $a^{(2)}$ and the space V_2 corresponds to the space V introduced in Section 4.1.)

The spaces V_2 and $\tilde{V}_2$ obviously coincide provided they are considered as topological ones, since they contain the same elements and are equipped with equivalent norms. Now, the function $w \to P(w)$, $P(w)$ determined by (4.2.37), can be treated as a mapping of $\tilde{V}_2$ into V_1, and Lemma 4.2.2 implies the following

Corollary 4.2.1 *Let the hypotheses of Lemma* 4.2.2 *be fulfilled and let* $\tilde{V}_2$ *be the Hilbert space containing the elements of* V_2, *with the norm and scalar product defined by the formulas* (4.2.31) *and* (4.2.48). *Then, the function* $w \to P(w)$, $P(w)$ *determined by* (4.2.37), *is a continuously Fréchet differentiable mapping of* $\tilde{V}_2$ *into* V_1.

4.2.4 Transformation of the problem (4.2.27), (4.2.28)

From (4.2.13), (4.2.30), Lemma 4.2.1, and the Riesz theorem, we conclude the existence of a unique function $\tilde{g}$ such that

$$\tilde{g} \in V_1, \qquad a^{(1)}(\tilde{g}, g) = (Q, g), \qquad g \in V_1. \tag{4.2.49}$$

Introduce the bilinear form

$$b_{\tilde{g}} \in \mathcal{L}_2(\tilde{V}_2, \tilde{V}_2; \mathbb{R}), \tag{4.2.50}$$

given by

$$\begin{aligned} b_{\tilde{g}}(w', w'') = \iint_{\Omega} \Big[& N_1(\tilde{g}) \frac{\partial w'}{\partial x} \frac{\partial w''}{\partial x} + N_2(\tilde{g}) \frac{\partial w'}{\partial y} \frac{\partial w''}{\partial y} \\ & + N_{12}(\tilde{g}) \left(\frac{\partial w'}{\partial x} \frac{\partial w''}{\partial y} + \frac{\partial w'}{\partial y} \frac{\partial w''}{\partial x} \right) \Big] \, dx \, dy, \qquad w', w'' \in \tilde{V}_2, \end{aligned} \tag{4.2.51}$$

where $N_1(\tilde{g})$, $N_2(\tilde{g})$, and $N_{12}(\tilde{g})$ are given by the formulas (4.2.18).

By (4.2.27), (4.2.37), and (4.2.49), we get

$$g_\lambda = \lambda \tilde{g} + P(w_\lambda). \tag{4.2.52}$$

Substituting this equality into (4.2.28) and taking into consideration (4.2.19), (4.2.24), and (4.2.51), we reduce the problem (4.2.27), (4.2.28) to that of finding a function w_λ such that

$$\begin{gathered} w_\lambda \in \tilde{V}_2, \\ a^{(2)}(w_\lambda, w) + \lambda b_{\tilde{g}}(w_\lambda, w) + (G w_\lambda, w) = 0, \qquad w \in \tilde{V}_2. \end{gathered} \tag{4.2.53}$$

Here, G is the mapping of $\tilde{V}_2$ into $\tilde{V}_2^*$ determined by

$$\begin{aligned} (Gw', w'') = \iint_{\Omega} \Big\{ & [N_1(P(w')) + T_1(w', w')] \frac{\partial w'}{\partial x} \frac{\partial w''}{\partial x} \\ & + [N_2(P(w')) + T_2(w', w')] \frac{\partial w'}{\partial y} \frac{\partial w''}{\partial y} \\ & + [N_{12}(P(w')) + T_{12}(w', w')] \left(\frac{\partial w'}{\partial x} \frac{\partial w''}{\partial y} + \frac{\partial w'}{\partial y} \frac{\partial w''}{\partial x} \right) \Big\} \, dx \, dy, \end{aligned} \tag{4.2.54}$$

$$w', w'' \in \tilde{V}_2.$$

If a function w_λ solving the problem (4.2.53) is found, then one can determine the function g_λ from the equation (4.2.52). We will need now some properties of the mapping G.

Lemma 4.2.3 *Let the conditions* (4.1.25)–(4.1.27) *be fulfilled and let a set* Y_p *be determined by the relations* (2.1.2), (2.1.3). *Assume that spaces* V_1 *and* V_2 *are defined by the expressions* (4.2.7), (4.2.8), (4.2.11), *and* (4.2.12), *and* $\tilde{V}_2$ *is the set of functions from* V_2 *equipped with the norm* (4.2.48). *Then, the mapping* $G\colon \tilde{V}_2 \to \tilde{V}_2^*$

given by the relations (4.2.16), (4.2.18), (4.2.22), (4.2.37), *and* (4.2.54) *is continuously Fréchet differentiable, i.e., at every point* $w \in \tilde{V}_2$ *there exists the Fréchet derivative* G'_w *of the mapping* G *and, moreover,* $w \to G'_w$ *is a continuous mapping of* $\tilde{V}_2$ *into* $\mathcal{L}(\tilde{V}_2, \tilde{V}_2^*)$ *and the equality* $G'_\theta = \Theta$ *holds, where* θ *is the zero element of the space* $\tilde{V}_2$*, and* Θ *is the zero element of the space* $\mathcal{L}(\tilde{V}_2, \tilde{V}_2^*)$.

Proof. Let us represent the mapping G in the form

$$(Gw', w'') = \sum_{i=1}^{4} (G_i w', w''), \qquad w', w'' \in \tilde{V}_2, \tag{4.2.55}$$

where

$$\begin{aligned}
(G_1 w', w'') &= \iint_\Omega N_1(P(w')) \frac{\partial w'}{\partial x} \frac{\partial w''}{\partial x}\, dx\, dy \\
(G_2 w', w'') &= \iint_\Omega N_2(P(w')) \frac{\partial w'}{\partial y} \frac{\partial w''}{\partial y}\, dx\, dy \\
(G_3 w', w'') &= \iint_\Omega N_{12}(P(w')) \left(\frac{\partial w'}{\partial x} \frac{\partial w''}{\partial y} + \frac{\partial w'}{\partial y} \frac{\partial w''}{\partial x} \right) dx\, dy \\
(G_4 w', w'') &= \iint_\Omega \left[T_1(w', w') \frac{\partial w'}{\partial x} \frac{\partial w''}{\partial x} + T_2(w', w') \frac{\partial w'}{\partial y} \frac{\partial w''}{\partial y} \right. \\
&\qquad \left. + T_{12}(w', w') \left(\frac{\partial w'}{\partial x} \frac{\partial w''}{\partial y} + \frac{\partial w'}{\partial y} \frac{\partial w''}{\partial x} \right) \right] dx\, dy.
\end{aligned} \tag{4.2.56}$$

It is sufficient to verify that Lemma 4.2.3 is valid for every mapping $G_i \colon \tilde{V}_2 \to \tilde{V}_2^*$, $i = 1, 2, 3, 4$. The mapping G_1 is the composition of the mappings A and B, i.e.,

$$G_1 = B \circ A.$$

Here

$$A \colon \tilde{V}_2 \to L_2(\Omega) \times W_2^1(\Omega), \qquad w \to Aw = \left(N_1(P(w)), \frac{\partial w}{\partial x} \right),$$

and B is a bilinear, continuous mapping of the space $L_2(\Omega) \times W_2^1(\Omega)$ into $\tilde{V}_2^*$; if $(p, g) \in L_2(\Omega) \times W_2^1(\Omega)$, then the value of the functional $B(p, g)$ at an element $w \in \tilde{V}_2$ is given by the formula

$$(B(p, g), w) = \iint_\Omega pg \frac{\partial w}{\partial x}\, dx\, dy.$$

From Lemma 4.2.2 and Property 4 of the Fréchet derivative (see Subsec. 1.9.1), we deduce that the mapping A is Fréchet differentiable at every point

$w \in \tilde{V}_2$ and the derivative A'_w has the form

$$A'_w w' = \left(N_1 \circ P'_w w', \frac{\partial w'}{\partial x} \right), \qquad w' \in \tilde{V}_2.$$

Further, the mapping B is also Fréchet differentiable, and the derivative $B'_{(p,g)}$ is given by

$$(B'_{(p,g)}(p_1, g_1), w) = \iint_\Omega (p_1 g + p g_1) \frac{\partial w}{\partial x} \, dx \, dy,$$

$$(p, g), \, (p_1, g_1) \in L_2(\Omega) \times W_2^1(\Omega), \; w \in \tilde{V}_2.$$

From Property 4 of the Fréchet derivative, we conclude the mapping G_1 to be Fréchet differentiable at any point $w \in \tilde{V}_2$ and the derivative G'_{1w} is given by

$$(G'_{1w} w', w'') = \iint_\Omega \left[(N_1 \circ P'_w w') \frac{\partial w}{\partial x} + (N_1 \circ P(w)) \frac{\partial w'}{\partial x} \right] \frac{\partial w''}{\partial x} \, dx \, dy,$$

$$w', w'' \in \tilde{V}_2. \tag{4.2.57}$$

Taking into account that $w \to P(w)$ is a continuously differentiable mapping of $\tilde{V}_2$ into V_1 (see Lemma 4.2.2), and using (4.2.17) and (4.2.57), it is easy to see that $w \to G'_{1w}$ is a continuous mapping of $\tilde{V}_2$ into $\mathcal{L}(\tilde{V}_2, \tilde{V}_2^*)$.

(4.2.37) implies that $P(\theta) = \tilde{\theta}$, where θ and $\tilde{\theta}$ are the zero elements of the spaces $\tilde{V}_2$ and V_1, respectively. Thus, (4.2.57) yields

$$G'_{1\theta} = \Theta,$$

Θ being the zero element of the space $\mathcal{L}(\tilde{V}_2, \tilde{V}_2^*)$.

So, Lemma 4.2.3 is proved to be valid for the mapping G_1. In the same way, we derive it to hold true for the mappings G_2 and G_3. At last, G_4 is a trilinear continuous mapping of $\tilde{V}_2$ into $\tilde{V}_2^*$, and the validity of Lemma 4.2.3 for it is also easily verified.

A nonlinear operator A acting from a Hilbert space H into its dual space H^* is said to be compact if the image $A(Q)$ of every bounded set $Q \subset H$ is relatively compact, i.e., the closure $\overline{A(Q)}$ of the set $A(Q)$ is a compactum in H^*.

Obviously, if the condition

$$\{u_n\} \subset H, \quad u_n \to u_0 \text{ weakly in } H$$

implies $Au_n \to Au_0$ strongly in H^*, then the operator A is compact.

Lemma 4.2.4 *Under the suppositions of Lemma* 4.2.3, G *is a compact mapping of* $\tilde{V}_2$ *into* $\tilde{V}_2^*$.

Proof. Let $\{w_n\}$ be a sequence of elements of $\tilde{V}_2$ such that

$$w_n \to w_0 \quad \text{weakly in } \tilde{V}_2.$$

Hence, due to the compactness of the embedding $W_2^2(\Omega)$ into $W_4^1(\Omega)$ (see Theorem 1.6.2), we have

$$w_n \to w_0 \quad \text{strongly in } W_4^1(\Omega). \tag{4.2.58}$$

By (4.2.16), (4.2.22), and (4.2.58), we get

$$N(w_n, w_n) \to N(w_0, w_0) \quad \text{strongly in } V_1^*. \tag{4.2.59}$$

Now, taking into account Lemma 4.2.1, (4.2.37), and (4.2.59), we obtain

$$P(w_n) \to P(w_0) \quad \text{strongly in } V_1. \tag{4.2.60}$$

Further, by virtue of (4.2.16), (4.2.17), (4.2.54), (4.2.58), and (4.2.60), we deduce

$$Gw_n \to Gw_0 \quad \text{strongly in } \tilde{V}_2^*,$$

concluding the proof.

Consider the following eigenvalue problem: Find $(\lambda, w) \in \mathbb{R} \times \tilde{V}_2$, $w \neq \theta$ such that

$$a^{(2)}(w, w') + \lambda b_{\tilde{g}}(w, w') = 0, \qquad w' \in \tilde{V}_2. \tag{4.2.61}$$

Since the bilinear form $b_{\tilde{g}}$ is symmetric (see (4.2.51)), the embedding of $W_2^2(\Omega)$ into $W_4^1(\Omega)$ is compact and the bilinear form $a^{(2)}$ defines a scalar product in $\tilde{V}_2$, from Theorem 1.5.8 we obtain the following

Lemma 4.2.5 *Let the conditions* (4.1.25)–(4.1.27) *be fulfilled and let a set* Y_p *be determined by the relations* (2.1.2), (2.1.3). *Let* $\tilde{V}_2$ *be the Hilbert space containing the elements of the space* V_2 *(see* (4.2.11)*) with the scalar product generated by the form* $a^{(2)}$, *and let* $b_{\tilde{g}}$ *be the bilinear form defined by the formulas* (4.2.30), (4.2.49), (4.2.51), *and let* (4.2.13) *be valid. Then, the problem* (4.2.61) *has a countable set of eigenvalues* $\{\lambda_i\}$

$$0 < |\lambda_1| \leq |\lambda_2| \leq \cdots, \qquad \lim_{n\to\infty} |\lambda_n| = \infty.$$

Each eigenvalue appears as many times as its multiplicity, and the multiplicity of each eigenvalue is finite.

4.2.5 Stability of a plate and bifurcation

We will need now the notion of a bifurcation. Let H be a Hilbert space over $\mathbb{R}$, let A be a continuous mapping of H into its dual space H^*, and let B be a compact mapping of H into H^*. Let also

$$A\theta = B\theta = \Theta,$$

where θ and Θ are the zero elements of the spaces H and H^*, respectively. A real number λ_0 is said to be a bifurcation point of the equation

$$Au + \lambda Bu = \Theta \tag{4.2.62}$$

if for any $\varepsilon, \delta > 0$ there exist u_ε and λ_δ such that

$$Au_\varepsilon + \lambda_\delta Bu_\varepsilon = \Theta, \qquad |\lambda_0 - \lambda_\delta| < \delta, \qquad 0 < \|u_\varepsilon\|_H < \varepsilon.$$

Let us again consider the problem (4.2.53). The solution of this problem determines the transversal displacements of the plate (the function w_λ) on which the longitudinal forces λQ act, the forces being distributed on the part S_2 of the boundary S (see (4.2.30)).

By virtue of (4.2.54), $G\theta = \theta_1$, where θ and θ_1 are the zero elements of the spaces $\tilde{V}_2$ and $\tilde{V}_2^*$, respectively. Thus, for each $\lambda \in \mathbb{R}$, there exists the trivial solution $w_\lambda = \theta$ of the problem (4.2.53). We are interested in the bifurcation points of the problem (4.2.53), i.e., the values of λ in neighborhoods of which nontrivial solutions of the problem (4.2.53) appear. It is natural to connect with the smallest positive bifurcation point the loss of the stability of the plate.

We note that the problem (4.2.53) reduces to the problem (4.2.62) if one defines the operators $A\colon \tilde{V}_2 \to \tilde{V}_2^*$ and $B\colon \tilde{V}_2 \to \tilde{V}_2^*$ by

$$(Aw', w'') = a^{(2)}(w', w'') + (Gw', w''),$$
$$(Bw', w'') = b_{\tilde{g}}(w', w''),$$

where w', $w'' \in \tilde{V}_2$.

Theorem 4.2.1 *Let the conditions* (4.1.25)–(4.1.27) *be fulfilled and let a set* Y_p *be defined by the relations* (2.1.2) *and* (2.1.3). *Let also* $\tilde{V}_2$ *be the Hilbert space containing the elements of the space* V_2 *(see* (4.2.11)*) equipped with the scalar product generated by the form* $a^{(2)}$, *let* $b_{\tilde{g}}$ *be the bilinear form determined by the formulas* (4.2.30), (4.2.49), *and* (4.2.51), *let* (4.2.13) *hold, and let a mapping* $G\colon \tilde{V}_2 \to \tilde{V}_2^*$ *be defined by* (4.2.54). *If* λ_0 *is the bifurcation point of the problem* (4.2.53), *then* λ_0 *is an eigenvalue of the problem* (4.2.61), *i.e., there exists a function* w_0 *such that*

$$w_0 \in \tilde{V}_2, \qquad w_0 \neq \theta, \qquad a^{(2)}(w_0, w) + \lambda_0 b_{\tilde{g}}(w_0, w) = 0, \qquad w \in \tilde{V}_2. \tag{4.2.63}$$

Proof. By the Riesz theorem, the bilinear form $b_{\tilde{g}}$ generates an operator $B_{\tilde{g}} \in \mathcal{L}(\tilde{V}_2, \tilde{V}_2)$ such that

$$b_{\tilde{g}}(w', w'') = a^{(2)}\left(B_{\tilde{g}}w', w''\right), \qquad w', w'' \in \tilde{V}_2. \tag{4.2.64}$$

Since the embedding of $W_2^2(\Omega)$ into $W_4^1(\Omega)$ is compact,

$$B_{\tilde{g}} \text{ is a linear compact mapping of } \tilde{V}_2 \text{ into } \tilde{V}_2. \tag{4.2.65}$$

Further, let $\mathcal{R}_1$ be the Riesz operator in the space $\tilde{V}_2$ which is defined by the relation

$$(f, w) = a^{(2)}(\mathcal{R}_1 f, w), \qquad f \in \tilde{V}_2^*, \ w \in \tilde{V}_2. \tag{4.2.66}$$

Taking into account (4.2.64) and (4.2.66), the problem (4.2.53) can be represented in the form

$$w_\lambda \in \tilde{V}_2, \qquad w_\lambda + \lambda B_{\tilde{g}} w_\lambda + \mathcal{R}_1 \circ G w_\lambda = \theta. \tag{4.2.67}$$

The problem (4.2.63) now takes the form

$$w_0 \in \tilde{V}_2, \qquad w_0 \neq \theta, \qquad w_0 + \lambda_0 B_{\tilde{g}} w_0 = \theta. \tag{4.2.68}$$

Define a mapping $P_2 \colon \mathbb{R} \times \tilde{V}_2 \to \tilde{V}_2$ by

$$P_2(\lambda, w) = w + \lambda B_{\tilde{g}} w + \mathcal{R}_1 \circ G w, \qquad (\lambda, w) \in \mathbb{R} \times \tilde{V}_2. \tag{4.2.69}$$

Then, the problem (4.2.67) reduces to the following one

$$w_\lambda \in \tilde{V}_2, \qquad P_2(\lambda, w_\lambda) = \theta. \tag{4.2.70}$$

Making use of Lemma 4.2.3 and of the properties of Fréchet derivative (see Subsec. 1.9.1), one easily sees that, for every fixed $\lambda \in \mathbb{R}$, the function $w \to P_2(\lambda, w)$ has the Fréchet derivative

$$\frac{\partial P_2}{\partial w}(\lambda, w) \in \mathcal{L}(\tilde{V}_2, \tilde{V}_2) \tag{4.2.71}$$

which is given by the formula

$$\frac{\partial P_2}{\partial w}(\lambda, w) = I + \lambda B_{\tilde{g}} + \mathcal{R}_1 \circ G'_w. \tag{4.2.72}$$

Here, I is the identity operator in $\tilde{V}_2$. Since $G'_\theta = \Theta$ (see Lemma 4.2.3), from (4.2.72) we conclude

$$\frac{\partial P_2}{\partial w}(\lambda, \theta) = I + \lambda B_{\tilde{g}}. \tag{4.2.73}$$

Since $w \to G'_w$ is a continuous mapping of $\tilde{V}_2$ into $\mathcal{L}(\tilde{V}_2, \tilde{V}_2^*)$ (see Lemma 4.2.3) and $\mathcal{R}_1 \in \mathcal{L}(\tilde{V}_2^*, \tilde{V}_2)$, from (4.2.72) we deduce that

$$\left.\begin{array}{l}\lambda, w \to \dfrac{\partial P_2}{\partial w}(\lambda, w) \text{ is a continuous mapping of } \mathbb{R} \times \tilde{V}_2 \text{ into} \\ \mathcal{L}(\tilde{V}_2, \tilde{V}_2).\end{array}\right\} \tag{4.2.74}$$

Let λ_0 be a bifurcation point of the problem (4.2.53). Then, there exists a sequence $\{\lambda_n, w_n\}$ such that

$$|\lambda_n - \lambda_0| < \frac{1}{n}, \qquad 0 < \|w_n\|_{\tilde{V}_2} < \frac{1}{n}, \tag{4.2.75}$$

$$P_2(\lambda_n, w_n) = \theta. \tag{4.2.76}$$

For any $\lambda \in \mathbb{R}$, the function $w = \theta$ is a solution to the problem (4.2.53), so that

$$P_2(\lambda, \theta) = \theta, \qquad \lambda \in \mathbb{R}. \tag{4.2.77}$$

By virtue of (4.2.73), we have

$$\frac{\partial P_2}{\partial w}(\lambda_0, \theta) = I + \lambda_0 B_{\tilde{g}}. \tag{4.2.78}$$

If λ_0 is not an eigenvalue of the problem (4.2.61), or, equivalently, of the problem (4.2.68), then (4.2.65) together with the Fredholm alternative implies that

$$\frac{\partial P_2}{\partial w}(\lambda_0, \theta) \text{ is an invertible mapping of } \tilde{V}_2 \text{ into } \tilde{V}_2. \tag{4.2.79}$$

Then, by virtue of (4.2.74) and (4.2.79), from the implicit function theorem (Theorem 1.9.1) we deduce the existence of an open set U_1 in $\mathbb{R}$ containing the point λ_0, and of an open set U_2 in $\tilde{V}_2$ containing the point $w = \theta$ such that, for all $\lambda \in U_1$, the equation $P_2(\lambda, w) = \theta$ has a unique solution with respect to w. Hence, by (4.2.77), this solution is $w = \theta$ for all $\lambda \in U_1$. Thus, the relations (4.2.75) and (4.2.76) make a contradiction, concluding the proof.

So, only the eigenvalues of the problem (4.2.61) can be bifurcation points of the problem (4.2.53). The statement below gives a sufficient condition for the existence of bifurcation points.

Theorem 4.2.2 *Under the suppositions of Theorem* 4.2.1, *each eigenvalue of the problem* (4.2.61) *having odd multiplicity is a bifurcation point of the problem* (4.2.53).

Proof. Introduce a mapping $\Phi_\lambda \colon \tilde{V}_2 \to \tilde{V}_2$ by

$$\Phi_\lambda = I + \lambda B_{\tilde{g}} + \mathcal{R}_1 \circ G. \tag{4.2.80}$$

Since the problem (4.2.53) is equivalent to (4.2.67), the problem (4.2.53) is equivalent to that of finding a function w_λ satisfying the relations

$$w_\lambda \in \tilde{V}_2, \qquad \Phi_\lambda w_\lambda = \theta. \tag{4.2.81}$$

Now let λ_0 be an eigenvalue of the problem (4.2.61) of odd multiplicity. By virtue of Lemma 4.2.5, the problem (4.2.61) has a discrete spectrum. Since only points of the spectrum of the problem (4.2.61) can be bifurcation points of the problem (4.2.53) (see Theorem 4.2.1), for any fixed, sufficiently small $\varepsilon > 0$, there exists a ball d_ε in $\tilde{V}_2$ centered at $w = \theta$ in which the problem

$$w \in d_\varepsilon, \qquad \Phi_{\lambda_0 - \varepsilon} w = \theta$$

has a unique solution $w = \theta$.

Similarly, the problem

$$w \in d_\varepsilon, \qquad \Phi_{\lambda_0+\varepsilon} w = \theta$$

has a unique solution $w = \theta$.

Lemmas 4.2.3, 4.2.4, and (4.2.65) imply that

$$F\colon w \to Fw = B_{\tilde{g}} w + \mathcal{R}_1 \circ Gw$$

is a compact, continuously Fréchet differentiable mapping of $\tilde{V}_2$ into $\tilde{V}_2$, and the following equality holds

$$F'_\theta = B_{\tilde{g}}.$$

Now, the Leray-Schauder theorem (see Krasnoselskii (1956)) implies that, if L is the boundary of an arbitrary open subset of d_ε which contains θ, then the rotations of the mappings $\Phi_{\lambda_0-\varepsilon}$ and $\Phi_{\lambda_0+\varepsilon}$ on L have different signs, i.e., the mappings are not homotopic on L. Let us show that there exists a pair $(\tilde{w}, \tilde{\lambda})$ such that

$$\tilde{w} \in L, \qquad \tilde{\lambda} \in (\lambda_0 - \varepsilon, \lambda_0 + \varepsilon), \qquad \Phi_{\tilde{\lambda}} \tilde{w} = \theta. \tag{4.2.82}$$

Let us assume the contrary. Then, the function $\varphi_t(w)\colon L \times [0,1] \to \tilde{V}_2$ given by

$$(w,t) \to \varphi_t(w) = w + [t(\lambda_0 + \varepsilon) + (1-t)(\lambda_0 - \varepsilon)] B_{\tilde{g}} w + \mathcal{R}_1 \circ Gw,$$

does not vanish at any point, and $\varphi_0(w) = \Phi_{\lambda_0-\varepsilon} w$, $\varphi_1(w) = \Phi_{\lambda_0+\varepsilon} w$. Hence, the mappings $\Phi_{\lambda_0-\varepsilon}$ and $\Phi_{\lambda_0+\varepsilon}$ are homotopic on L. But we have established that this is not the case. Thus, (4.2.82) holds, and the theorem is proven.

Conclusion 1 It is natural to connect the loss of stability of the plate with the least positive bifurcation point of the problem (4.2.53). Only the eigenvalues of the problem (4.2.61) can be bifurcation points of the problem (4.2.53), and every eigenvalue of the problem (4.2.61) of odd multiplicity is a bifurcation point of (4.2.53). The question whether an eigenvalue of (4.2.61) of even multiplicity is a bifurcation point of (4.2.53) demands an additional investigation.

However, below we will always connect with the loss of stability of the plate the least positive eigenvalue of the problem (4.2.61), at which the form $b_{\tilde{g}}$ is given by the expression (4.2.51), the latter being determined, in turn, by the solution of the problem (4.2.49). From the physical point of view, this means that the form $b_{\tilde{g}}$ is determined by inner forces, which depend on the function of exterior longitudinal forces Q and on the law of change of the thickness of the plate (i.e., on the function $h(x,y)$).

4.2.6 An example of nonexistence of stable solutions

Let a rectangular plate of constant thickness be clamped along the edge $x = 0$ and be subject to longitudinal compressing forces $\lambda Q_1(y)$ acting along the edge $x = a$ (see Fig. 4.2.1).

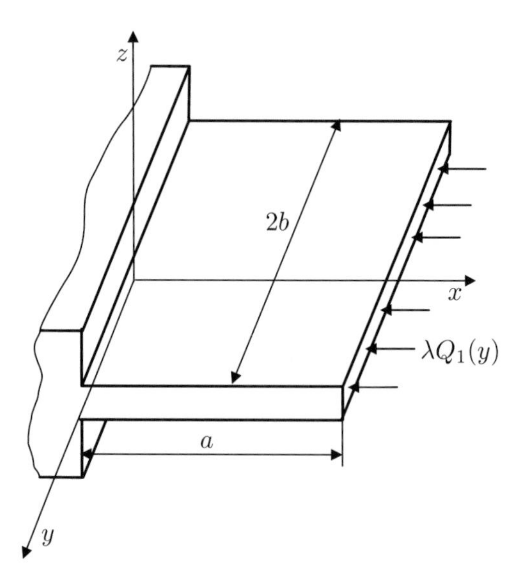

Figure 4.2.1: Rectangular plate clamped along the edge $x = 0$ and subject to compressing forces $\lambda Q_1(y)$

According to (4.2.2) and (4.2.5), the stored energy of the plate is given by

$$\Phi_\lambda(\omega) = \Phi_2(\omega) + \tfrac{1}{2}\, a^{(2)}(w,w) - \lambda \int_{-b}^{b} Q_1(y)u(a,y)\,dy, \qquad \omega = (u,v,w). \tag{4.2.83}$$

Here, the quadratic form $a^{(2)}(w,w)$ is defined by (4.2.31) and the functional $\Phi_2(\omega)$ is given by

$$\Phi_2(\omega) = \frac{1}{2}\int_0^a \int_{-b}^{b} \big[c_{11}(\varepsilon_{11}(\omega))^2 + 2c_{12}\varepsilon_{11}(\omega)\varepsilon_{22}(\omega) + c_{22}(\varepsilon_{22}(\omega))^2 + c_{33}(\varepsilon_{12}(\omega))^2\big]\,dx\,dy. \tag{4.2.84}$$

The functional (4.2.83) is considered in the space $V = V_1 \times V_2$ ($\omega = (u,v,w) \in V$, $(u,v) \in V_1$, $w \in V_2$). In the case under examination, the spaces V_1 and V_2 are the following ones

$$V_1 = \big\{\, g \,|\, g = (g_1,g_2),\ g \in \big(W_2^1(\Omega)\big)^2,\ g(0,y) = 0 \,\big\},$$
$$V_2 = \big\{\, q \,|\, q \in W_2^2(\Omega),\ q(0,y) = \frac{\partial q}{\partial x}(0,y) = 0 \,\big\},$$

where $\Omega = \big\{\,(x,y) \,|\, 0 < x < a,\ -b < y < b \,\big\}$. The space V_2 is easily verified to meet the condition (4.2.11). Introduce the function

$$\omega_c = \big(-\tfrac{2}{3}\, c^2 x^3, 0, cx^2\big), \qquad c \in \mathbb{R}. \tag{4.2.85}$$

Obviously,

$$\omega_c \in V, \qquad c \in \mathbb{R},$$

i.e.,

$$\left(-\tfrac{2}{3}c^2x^3, 0\right) \in V_1, \qquad cx^2 \in V_2, \qquad c \in \mathbb{R}.$$

Due to (4.2.1) and (4.2.85), for the strain components generated by the function ω_c, we have

$$\varepsilon_{11}(\omega_c) = 0, \qquad \varepsilon_{22}(\omega_c) = 0, \qquad \varepsilon_{12}(\omega_c) = 0. \tag{4.2.86}$$

Now, (4.2.83)–(4.2.86) yield

$$\Phi_\lambda(\omega_c) = \frac{1}{2}\, c^2 a^{(2)}(x^2, x^2) + \frac{2}{3}\,\lambda a^3 c^2 \int_{-b}^{b} Q_1(y)\, dy. \tag{4.2.87}$$

Since $\lambda Q_1(y)$ is a compressing load (see Fig. 4.2.1) and $\lambda > 0$, we get $Q_1(y) \le 0$. Hence,

$$\frac{2}{3}\, a^3 \int_{-b}^{b} Q_1(y)\, dy = k_1 < 0. \tag{4.2.88}$$

Because $a^{(2)}(w', w'')$ is a coercive form on $V_2 \times V_2$ (see Theorem 4.1.2), we conclude

$$\frac{1}{2}\, a^{(2)}(x^2, x^2) = k_2 > 0. \tag{4.2.89}$$

Assume that

$$\lambda > \frac{k_2}{|k_1|}. \tag{4.2.90}$$

Using (4.2.87)–(4.2.90), we get

$$\Phi_\lambda(\omega_c) = c^2(k_2 + \lambda k_1) < 0. \tag{4.2.91}$$

Therefore, if λ satisfies the inequality (4.2.90), then

$$\lim_{c\to\infty} \Phi_\lambda(\omega_c) = -\infty.$$

Thus, if the load is sufficiently large, that is, the inequality (4.2.90) holds, then the functional (4.2.83) is not bounded below in the space V, and there is no stable equilibrium position of the plate, although there exists at least one stationary point of the functional (4.2.83), which is $\omega = (\lambda\tilde{g}, \theta)$, where $\tilde{g}$ is the solution of the problem (4.2.49).

Suppose that

$$Q_1(y) = -c_1, \qquad c_1 = \text{const} > 0. \tag{4.2.92}$$

Let us show that then

$$\frac{k_2}{|k_1|} \ge \lambda_1. \tag{4.2.93}$$

Here, λ_1 is the least positive eigenvalue of the problem (4.2.61). Indeed, by (4.2.92), the bilinear form $b_{\tilde{g}}$ takes the form

$$b_{\tilde{g}}\left(w', w''\right) = -c_1 \int_0^a \int_{-b}^b \frac{\partial w'}{\partial x} \frac{\partial w''}{\partial x}\, dx. \tag{4.2.94}$$

(This formula is actually somewhat inaccurate, because it does not take into account the influence of the free edges of the plate. Nevertheless, for our purposes, it is precise enough if a is much less than b.)

In the case under consideration, all the eigenvalues of the problem (4.2.61) are positive and the least eigenvalue λ_1 is given by

$$\lambda_1 = \inf_{w \in V_2} \left(-\frac{a^{(2)}(w, w)}{b_{\tilde{g}}\left(w, w\right)} \right). \tag{4.2.95}$$

Making use of (4.2.89) and (4.2.94), after a simple calculation, we obtain

$$-\frac{a^{(2)}\left(x^2, x^2\right)}{b_{\tilde{g}}\left(x^2, x^2\right)} = \frac{k_2}{|k_1|}, \tag{4.2.96}$$

k_1 being defined by formulas (4.2.88) and (4.2.92). Since $w = x^2 \in V_2$, (4.2.95) and (4.2.96) imply (4.2.93).

Thus, if the inequality (4.2.90) holds, then the load exceeds the critical value, which is connected with the first positive eigenvalue of problem (4.2.61).

4.3 Model of the three-layered plate ignoring shears in the middle layer

4.3.1 Basic relations

A three-layered plate consists of two thin exterior layers, which are made of a strong material (the so-called carrier layers), and of a comparatively light, non-strong middle layer (the so-called filler), the latter ensures the joint work of the exterior layers. Consider the three-layered plate whose middle layer is of thickness $t_0(x, y)$ and two exterior layers are of thickness $h(x, y)$ (see Fig. 4.3.1).

We suppose that h is much less than t_0 ($h \ll t_0$) and that the material of the middle layer is much more flexible than the material of the exterior layers. In this case, the shearing stresses perceive mainly the middle layer, and the bending stresses perceive mainly the exterior ones.

Suppose also that, in the transversal direction, the elasticity modulus of the material of the middle layer is infinitely large. The material of the middle layer is usually light, so that the mass of the plate is concentrated in the exterior layers. This is why, in solving optimization problems for the three-layered plates, the

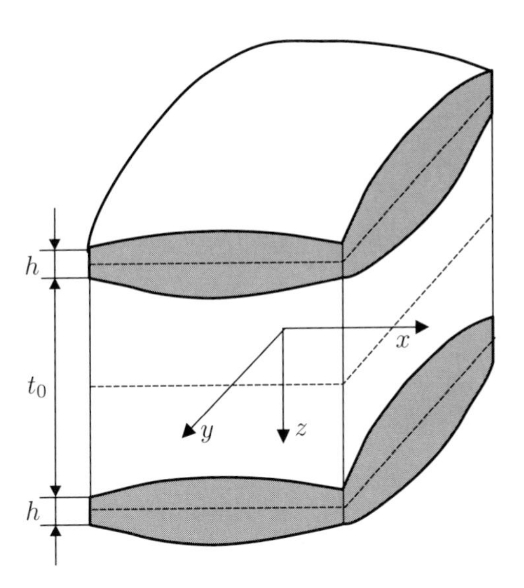

Figure 4.3.1: Three-layered plate

control is usually the function h determining the thickness of the carrier layers. In what follows, we assume that the equality

$$t_0 + h = \text{const}, \tag{4.3.1}$$

determining the parallelism of the midplanes of the carrier layers holds.

The Kirchhoff hypotheses are supposed to be fulfilled for the three-layered plate as a whole. Then, the strain components ε_1, ε_2, ε_3 are expressed by the formulas (4.1.1). The layers are assumed to be made of orthotropic materials, so that the relations (4.1.8) between stresses and strains are valid, and moreover, $E_{11} = E_{12} = E_{21} = E_{22} = G = 0$ for the inner layer, and the elasticity characteristics of the exterior layers coincide. Taking into account that $h \ll t_0$, from (4.1.3) and (4.1.8) we deduce the following relations for the bending moments M_1, M_2 and for the torque M_3:

$$\begin{aligned} M_1 = \int_{-\frac{t_0}{2}-h}^{\frac{t_0}{2}+h} z\sigma_1(z)\,dz &\approx -\frac{t_0+h}{2}\,h\sigma_1\left(-\frac{t_0+h}{2}\right) + \frac{t_0+h}{2}\,h\sigma_1\left(\frac{t_0+h}{2}\right) \\ &= D_1\chi_1 + D_2\chi_2, \end{aligned} \tag{4.3.2}$$

where

$$D_1 = \frac{E_{11}(t_0+h)^2 h}{2}, \qquad D_{12} = \frac{E_{12}(t_0+h)^2 h}{2}, \tag{4.3.3}$$

E_{11}, E_{12} are the elasticity characteristics of the exterior layers for which (4.1.9) holds true. Similarly, we have

$$M_2 = \int_{-\frac{t_0}{2}-h}^{\frac{t_0}{2}+h} z\sigma_2(z)\,dz \approx D_{21}\chi_1 + D_2\chi_2,$$
$$M_3 = \int_{-\frac{t_0}{2}-h}^{\frac{t_0}{2}+h} z\sigma_3(z)\,dz \approx D_3\chi_3. \tag{4.3.4}$$

Here

$$D_{21} = \frac{E_{21}(t_0+h)^2 h}{2} = D_{12}, \quad D_2 = \frac{E_{22}(t_0+h)^2 h}{2}, \quad D_3 = G(t_0+h)^2 h, \tag{4.3.5}$$

and E_{21}, E_{22}, G are the elasticity characteristics of the exterior layers.

The strain energy of the three-layered plate is determined by the relation (4.1.12), where D_1, D_2, D_3, and D_{12}, are expressed by the formulas (4.3.3), (4.3.5), and M_i's are described by the right-hand sides of (4.3.2) and (4.3.4).

4.3.2 Problems of the bending and of the free flexural oscillations

The bilinear form $a_h(u,v)$ for the three-layered plate is determined by (4.1.24), D_1, D_2, D_3, and D_{12} being defined by the formulas (4.1.9), (4.3.3), and (4.3.5). Suppose that

$$E_1,\ E_2,\ G \text{ are positive constants}, \tag{4.3.6}$$
$$\mu_i \text{ are constants},\ 0 \le \mu_i < 1,\ i=1,2,\quad \mu_2 E_1 = \mu_1 E_2, \tag{4.3.7}$$
$$h \in Y_p, \tag{4.3.8}$$
$$t_0 \ge c_1 > 0, \tag{4.3.9}$$
$$t_0 + h = c_2, \tag{4.3.10}$$

where

$$c_1 \text{ and } c_2 \text{ are positive numbers.} \tag{4.3.11}$$

In the same way as in Subsec. 4.1.3, one proves the following

Theorem 4.3.1 *Let the conditions* (4.3.6)–(4.3.11) *hold true, and let a set* Y_p *be determined by the formulas* (2.1.2) *and* (2.1.3). *Assume that* V *is a closed subspace of* $W_2^2(\Omega)$ *satisfying* (4.1.21). *Then, the bilinear form* $a_h(u,v)$ *defined by* (4.1.9), (4.1.24), (4.3.3), *and* (4.3.5) *is symmetric, uniformly continuous and coercive in* $h \in Y_p$ *on* $V \times V$, *i.e.,*

$$a_h(u,v) = a_h(v,u), \qquad u,v \in V, \tag{4.3.12}$$
$$|a_h(u,v)| \le c_3\|u\|_V\|v\|_V, \quad u,v \in V,\ h \in Y_p, \tag{4.3.13}$$
$$a_h(u,u) \ge c_4\|u\|_V^2, \qquad u \in V,\ h \in Y_p, \tag{4.3.14}$$

c_3 *and* c_4 *being positive numbers.*

Remark 4.3.1 The condition (4.3.10) means the parallelity of the midplanes of the exterior layers (see Fig. 4.3.1). It is easy to verify that Theorem 4.3.1 remains valid if (4.3.10) is not true.

The problem of the bending of the three-layered plate reduces to the solution of the equation (4.1.29) for the corresponding form a_h, where $f \in V^*$, f being a load. The space V for the three-layered plate is chosen just as for the one-layered plate (see Subsec. 4.1.4). In particular, for the space V of the form (4.1.31), f can be represented in the form (4.1.36).

By the Riesz theorem, or by the Lax-Milgram theorem, under the suppositions of Theorem 4.3.1, for every load $f \in V^*$, the problem (4.1.29) corresponding to the bending of the three-layered plate has a unique solution.

We proceed now to consider the problem of free oscillations of the three-layered plate. A function $U(x,y,t)$ determining the deflection of the three-layered plate, which depends on the coordinates x, y and time t, is given as the solution of the following problem

$$U \in C^2[(0,\infty);V],$$

$$\iint\limits_{\Omega} \Big[(t_0\rho_0 + 2h\rho)\frac{\partial^2 U}{\partial t^2} v + \int_{-\frac{t_0}{2}}^{\frac{t_0}{2}} \rho_0 A z^2\, dz + \int_{-h-\frac{t_0}{2}}^{-\frac{t_0}{2}} \rho A z^2\, dz$$

$$+ \int_{\frac{t_0}{2}}^{\frac{t_0}{2}+h} \rho A z^2\, dz \Big]\, dx\, dy + a_h(U,v) = 0, \qquad v \in V,\ t \in (0,\infty), \tag{4.3.15}$$

where

$$A = \frac{\partial^3 U}{\partial t^2 \partial x}\frac{\partial v}{\partial x} + \frac{\partial^3 U}{\partial t^2 \partial y}\frac{\partial v}{\partial y}. \tag{4.3.16}$$

Here, ρ_0 and ρ are the densities of the material of the interior and exterior layers.

The first term under the integral sign in (4.3.15) determines the work of the inertia forces on virtual transversal displacements, and the following terms, according to the Kirchhoff hypothesis, defines the work of the inertia forces on virtual longitudinal displacements (see Subsec. 4.1.5). By integrating in (4.3.15) in z, we arrive at the following problem

$$U \in C^2((0,\infty);V),$$

$$\iint\limits_{\Omega} \Big[(t_0\rho_0 + 2h\rho)\frac{\partial^2 U}{\partial t^2} v + \frac{1}{12}\rho_0 t_0^3 A + \Big(h^2 t_0 + \frac{1}{2}\, h t_0^2 + \frac{2}{3}\, h^3\Big)\rho A \Big]\, dx\, dy$$

$$+ a_h(U,v) = 0, \qquad v \in V,\ t \in (0,\infty). \tag{4.3.17}$$

We will search for the solution of the problem (4.3.17) in the form (4.1.50). Substituting (4.1.50) into (4.3.17) and taking into account (4.3.16), we get the problem

$$(\lambda, u) \in \mathbb{R} \times V, \qquad a_h(u,v) = \lambda b_h(u,v), \qquad v \in V, \tag{4.3.18}$$

where $\lambda = \omega^2$ and

$$
\begin{aligned}
b_h(u,v) = \iint\limits_{\Omega} \Big[&(t_0\rho_0 + 2h\rho)uv + \frac{1}{12}\rho_0 t_0^3 \left(\frac{\partial u}{\partial x}\frac{\partial v}{\partial x} + \frac{\partial u}{\partial y}\frac{\partial v}{\partial y} \right) \\
&+ \rho \left(h^2 t_0 + \frac{1}{2} h t_0^2 + \frac{2}{3} h^3 \right) \left(\frac{\partial u}{\partial x}\frac{\partial v}{\partial x} + \frac{\partial u}{\partial y}\frac{\partial v}{\partial y} \right) \Big] \, dx\, dy.
\end{aligned}
\tag{4.3.19}
$$

We suppose that

$$
\rho_0 = \text{const} > 0, \qquad \rho = \text{const} > 0. \tag{4.3.20}
$$

Theorem 4.3.2 *Let the hypotheses of Theorem* 4.3.1 *hold and let a bilinear form* $b_h(u,v)$ *be determined by the formulas* (4.3.19) *and* (4.3.20). *Then, for all* $h \in Y_p$, *the spectral problem* (4.3.18) *possesses a sequence of nonzero solutions* $\{u_i\} \subset V$ *that correspond to a sequence of eigenvalues* $\{\lambda_i\}$ *such that*

$$
\begin{gathered}
a_h(u_i, v) = \lambda_i b_h(u_i, v), \qquad v \in V, \\
0 < \lambda_1 \le \lambda_2 \le \lambda_3 \le \cdots, \qquad \lim_{i\to\infty} \lambda_i = \infty, \qquad b_h(u_i, u_j) = \delta_{ij}.
\end{gathered}
$$

Proof. It is easy to see that, under the hypotheses of Theorem 4.3.2, the bilinear form b_h is symmetric, continuous, and coercive on $W_2^1(\Omega) \times W_2^1(\Omega)$. Now, Theorem 4.3.2 is a consequence of the compactness of the embedding of V into $W_2^1(\Omega)$ and of Theorem 1.5.7.

4.4 Model of the three-layered plate accounting for shears in the middle layer

4.4.1 Basic relations

Let us consider another model of the three-layered plate which accounts for shears in the middle layer (the filler). We assume that, in the filler, a straight line perpendicular to its midsurface before deformation remains straight after the deformation, but it is no longer perpendicular to the midsurface because of shears. Thus, we accept the linear law of displacements in the filler. To the thin exterior layers we apply the straight normal (Kirchhoff) hypothesis. Hence, a normal passing through the three layers becomes a broken line after the deformation (see Fig. 4.4.1). This allows one to account for shear strains of the filler.

Denote by u, v, and w the components of the vector of displacements in the x, y, z axes of the midplane of the upper layer (see Fig. 4.3.1). Respectively, u_1, v_1, and w_1 are the components of the vector of displacements of the midplane of the lower layer. Let also w_2 be the displacement in the z axis of points of the middle layer.

The functions w, w_1, and w_2 are supposed to depend only on x and y, not on z, and moreover,

$$
w_1 \equiv w_2 \equiv w. \tag{4.4.1}
$$

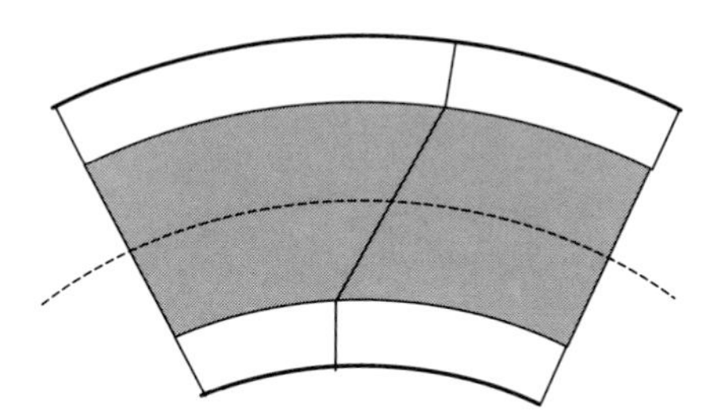

Figure 4.4.1: Three-layered plate. A normal passing through the three layers becomes a broken line after the bending

Below, we will examine problems of the bending of the three-layered plate. There one can assume that

$$u_1 = -u, \qquad v_1 = -v. \tag{4.4.2}$$

By the straight normal hypothesis (see Figs. 4.3.1 and 4.4.1), for the displacements u_{u}, v_{u}, and w_{u} of points of the upper layer in the x, y, z axes, we have

$$w_{\mathrm{u}} = w, \qquad u_{\mathrm{u}} = u - \left(z + \frac{t_0 + h}{2}\right)\frac{\partial w}{\partial x}, \qquad v_{\mathrm{u}} = v - \left(z + \frac{t_0 + h}{2}\right)\frac{\partial w}{\partial y}. \tag{4.4.3}$$

Similarly, for the displacements u_{l}, v_{l}, and w_{l} of points of the lower layer, we get

$$w_{\mathrm{l}} = w, \qquad u_{\mathrm{l}} = -u - \left(z - \frac{t_0 + h}{2}\right)\frac{\partial w}{\partial x}, \qquad v_{\mathrm{l}} = -v - \left(z - \frac{t_0 + h}{2}\right)\frac{\partial w}{\partial y}. \tag{4.4.4}$$

The displacements u_{m}, v_{m}, and w_{m} of points of the middle layer are given by

$$w_{\mathrm{m}} = w, \qquad u_m = -\frac{z}{t_0}\left(2u - h\frac{\partial w}{\partial x}\right), \qquad v_m = -\frac{z}{t_0}\left(2v - h\frac{\partial w}{\partial y}\right). \tag{4.4.5}$$

For the strain components of the layers we use the formulas

$$\varepsilon_1 = \frac{\partial \tilde{u}}{\partial x}, \qquad \varepsilon_2 = \frac{\partial \tilde{v}}{\partial y}, \qquad \varepsilon_3 = \frac{\partial \tilde{u}}{\partial y} + \frac{\partial \tilde{v}}{\partial x},$$
$$\varepsilon_4 = \frac{\partial \tilde{u}}{\partial z} + \frac{\partial w}{\partial x}, \qquad \varepsilon_5 = \frac{\partial \tilde{v}}{\partial z} + \frac{\partial w}{\partial y}, \tag{4.4.6}$$

$\tilde{u}$ and $\tilde{v}$ being the components of the vector of displacements in the corresponding layers which are defined by (4.4.3)–(4.4.5). The components ε_4 and ε_5 are taken into account only for the middle layer in which the shear strains are essential.

Let a three-layered plate be made of orthotropic materials, and let E_1, E_2, G, μ_1, and μ_2 be elasticity characteristics of the exterior layers (we assume that the upper and lower layers are made of the same material), and let G_1 be the shear modulus of the material of the middle layer.

For the stresses in the exterior layers, we have

$$\sigma_1 = E_{11}\varepsilon_1 + E_{12}\varepsilon_2, \qquad \sigma_2 = E_{21}\varepsilon_1 + E_{22}\varepsilon_2, \qquad \sigma_3 = G\varepsilon_3, \tag{4.4.7}$$

E_{11}, E_{21}, E_{12}, and E_{22} being determined by the relations (4.1.9). Substituting the values of ε_i, $i = 1, 2, 3$, from (4.4.6) into (4.4.7), taking note of the formulas (4.4.3) and (4.4.4) for the displacement components in the exterior layers, and making use of the equality (4.3.1), we obtain the following relations:

for the upper layer

$$\begin{aligned}
\sigma_1 &= E_{11}\left[\frac{\partial u}{\partial x} - \left(z + \frac{t_0+h}{2}\right)\frac{\partial^2 w}{\partial x^2}\right] + E_{12}\left[\frac{\partial v}{\partial y} - \left(z + \frac{t_0+h}{2}\right)\frac{\partial^2 w}{\partial y^2}\right],\\
\sigma_2 &= E_{21}\left[\frac{\partial u}{\partial x} - \left(z + \frac{t_0+h}{2}\right)\frac{\partial^2 w}{\partial x^2}\right] + E_{22}\left[\frac{\partial v}{\partial y} - \left(z + \frac{t_0+h}{2}\right)\frac{\partial^2 w}{\partial y^2}\right],\\
\sigma_3 &= G\left[\frac{\partial u}{\partial y} + \frac{\partial v}{\partial x} - 2\left(z + \frac{t_0+h}{2}\right)\frac{\partial^2 w}{\partial x \partial y}\right];
\end{aligned} \tag{4.4.8}$$

for the lower layer

$$\begin{aligned}
\sigma_1 &= E_{11}\left[-\frac{\partial u}{\partial x} - \left(z - \frac{t_0+h}{2}\right)\frac{\partial^2 w}{\partial x^2}\right] + E_{12}\left[-\frac{\partial v}{\partial y} - \left(z - \frac{t_0+h}{2}\right)\frac{\partial^2 w}{\partial y^2}\right],\\
\sigma_2 &= E_{21}\left[-\frac{\partial u}{\partial x} - \left(z - \frac{t_0+h}{2}\right)\frac{\partial^2 w}{\partial x^2}\right] + E_{22}\left[-\frac{\partial v}{\partial y} - \left(z - \frac{t_0+h}{2}\right)\frac{\partial^2 w}{\partial y^2}\right],\\
\sigma_3 &= -G\left[\frac{\partial u}{\partial y} + \frac{\partial v}{\partial x} + 2\left(z - \frac{t_0+h}{2}\right)\frac{\partial^2 w}{\partial x \partial y}\right].
\end{aligned} \tag{4.4.9}$$

By integrating the relations (4.4.8) in the thickness of the upper layer, we get the following formulas for the force per unit of length

$$\begin{aligned}
N_{1\mathrm{u}} &= \int_{-\frac{t_0}{2}-h}^{-\frac{t_0}{2}} \sigma_1(z)\,dz = h\left(E_{11}\frac{\partial u}{\partial x} + E_{12}\frac{\partial v}{\partial y}\right),\\
N_{2\mathrm{u}} &= \int_{-\frac{t_0}{2}-h}^{-\frac{t_0}{2}} \sigma_2(z)\,dz = h\left(E_{21}\frac{\partial u}{\partial x} + E_{22}\frac{\partial v}{\partial y}\right),\\
N_{3\mathrm{u}} &= \int_{-\frac{t_0}{2}-h}^{-\frac{t_0}{2}} \sigma_3(z)\,dz = hG\left(\frac{\partial u}{\partial y} + \frac{\partial v}{\partial x}\right).
\end{aligned} \tag{4.4.10}$$

The forces $N_{1\mathrm{l}}$, $N_{2\mathrm{l}}$, and $N_{3\mathrm{l}}$ in the lower layer are determined by the relations (4.4.10) with u and v replaced by $-u$ and $-v$, respectively, i.e.,

$$N_{1\mathrm{u}} = -N_{1\mathrm{l}}, \qquad N_{2\mathrm{u}} = -N_{2\mathrm{l}}, \qquad N_{3\mathrm{u}} = -N_{3\mathrm{l}}. \tag{4.4.11}$$

For the bending moments $M_{1\mathrm{u}}$, $M_{2\mathrm{u}}$ and for the torque $M_{3\mathrm{u}}$ in the upper layer with respect to the midplane of this layer (see Fig. 4.3.1), by (4.1.9) and (4.4.8), we get

$$\begin{aligned}
M_{1\mathrm{u}} &= \int_{-\frac{t_0}{2}-h}^{-\frac{t_0}{2}} \left(z + \frac{t_0+h}{2}\right) \sigma_1(z)\, dz = -D_1 \frac{\partial^2 w}{\partial x^2} - D_{12} \frac{\partial^2 w}{\partial y^2},\\
M_{2\mathrm{u}} &= \int_{-\frac{t_0}{2}-h}^{-\frac{t_0}{2}} \left(z + \frac{t_0+h}{2}\right) \sigma_2(z)\, dz = -D_{21} \frac{\partial^2 w}{\partial x^2} - D_2 \frac{\partial^2 w}{\partial y^2}, \qquad (4.4.12)\\
M_{3\mathrm{u}} &= \int_{-\frac{t_0}{2}-h}^{-\frac{t_0}{2}} \left(z + \frac{t_0+h}{2}\right) \sigma_3(z)\, dz = -D_3 \frac{\partial^2 w}{\partial x \partial y}.
\end{aligned}$$

Here

$$D_1 = \frac{E_1 h^3}{12(1-\mu_1\mu_2)}, \qquad D_{12} = D_{21} = \frac{E_1 \mu_2 h^3}{12(1-\mu_1\mu_2)},$$
$$D_2 = \frac{E_2 h^3}{12(1-\mu_1\mu_2)}, \qquad D_3 = \frac{1}{6} G h^3. \qquad (4.4.13)$$

Similar relations can be obtained for the bending moments M_{1l}, M_{2l} and for the torque M_{3l} in the lower layer with respect to the midplane of this layer, i.e.,

$$M_{il} = \int_{\frac{t_0}{2}}^{\frac{t_0}{2}+h} \left(z - \frac{t_0+h}{2}\right) \sigma_i(z)\, dz = M_{i\mathrm{u}}, \qquad i = 1,\, 2,\, 3. \qquad (4.4.14)$$

In the middle layer, only shearing stresses σ_4 and σ_5 are taken into account. They are given by

$$\sigma_4 = G_{\mathrm{m}} \varepsilon_4, \qquad \sigma_5 = G_{\mathrm{m}} \varepsilon_5, \qquad (4.4.15)$$

G_{m} being the shear modulus of the material of the middle layer, and ε_4, ε_5 being determined by the relations (4.4.6). From (4.4.5), (4.4.6), and (4.4.15), we get

$$\begin{aligned}
\sigma_4 &= G_{\mathrm{m}} \left(\frac{\partial u_{\mathrm{m}}}{\partial z} + \frac{\partial w}{\partial x}\right) = G_{\mathrm{m}} \left[\frac{\partial w}{\partial x} - \frac{1}{t_0}\left(2u - h\frac{\partial w}{\partial x}\right)\right],\\
\sigma_5 &= G_{\mathrm{m}} \left(\frac{\partial v_{\mathrm{m}}}{\partial z} + \frac{\partial w}{\partial y}\right) = G_{\mathrm{m}} \left[\frac{\partial w}{\partial y} - \frac{1}{t_0}\left(2v - h\frac{\partial w}{\partial y}\right)\right].
\end{aligned} \qquad (4.4.16)$$

The transversal forces in the middle layer are defined by the formulas

$$\begin{aligned}
Q_{1\mathrm{m}} &= \int_{-\frac{t_0}{2}}^{\frac{t_0}{2}} \sigma_4(z)\, dz = G_{\mathrm{m}} \left[(t_0+h)\frac{\partial w}{\partial x} - 2u\right],\\
Q_{2\mathrm{m}} &= \int_{-\frac{t_0}{2}}^{\frac{t_0}{2}} \sigma_5(z)\, dz = G_{\mathrm{m}} \left[(t_0+h)\frac{\partial w}{\partial y} - 2v\right].
\end{aligned} \qquad (4.4.17)$$

The strain energy of the three-layered plate is given by

$$\begin{aligned}\Phi = &\frac{1}{2}\iint\limits_{\Omega} dx\,dy \int_{-\frac{t_0}{2}-h}^{-\frac{t_0}{2}} (\sigma_1\varepsilon_1 + \sigma_2\varepsilon_2 + \sigma_3\varepsilon_3)\,dz \\ &+ \frac{1}{2}\iint\limits_{\Omega} dx\,dy \int_{-\frac{t_0}{2}}^{\frac{t_0}{2}} (\sigma_4\varepsilon_4 + \sigma_5\varepsilon_5)\,dz \\ &+ \frac{1}{2}\iint\limits_{\Omega} dx\,dy \int_{\frac{t_0}{2}}^{\frac{t_0}{2}+h} (\sigma_1\varepsilon_1 + \sigma_2\varepsilon_2 + \sigma_3\varepsilon_3)\,dz. \end{aligned} \tag{4.4.18}$$

Here, Ω is the domain occupied by the midplane of the plate, which is supposed to be a Lipschitz domain.

Substituting into (4.4.18) the values of σ_i from (4.4.8), (4.4.9), and (4.4.16), taking to notice (4.4.3)–(4.4.6), and denoting $\omega = (u, v, w)$, we get the following formula for the strain energy of the three-layered plate

$$\begin{aligned}\Phi(\omega) = \iint\limits_{\Omega} \Big\{ &\frac{E_1 h}{1-\mu_1\mu_2}\left[(P_1\omega)^2 + 2\mu_2(P_1\omega)(P_2\omega)\right] \\ &+ \frac{E_2 h}{1-\mu_1\mu_2}(P_2\omega)^2 + \frac{E_1 h^3}{12(1-\mu_1\mu_2)}\left[(P_4\omega)^2 + 2\mu_2(P_4\omega)(P_5\omega)\right] \\ &+ Gh(P_3\omega)^2 + \frac{E_2 h^3}{12(1-\mu_1\mu_2)}(P_5\omega)^2 + \frac{1}{3}Gh^3(P_6\omega)^2 \\ &+ \frac{1}{2}G_{\mathrm{m}} t_0\left[(P_7\omega)^2 + (P_8\omega)^2\right]\Big\}\,dx\,dy. \end{aligned} \tag{4.4.19}$$

Here, $P_i\omega$, $i = 1, 2, \ldots, 6$, are the strain components of the exterior layers, and $P_7\omega$, $P_8\omega$ are the ones of the middle layer of the plate, which are given by the formulas

$$\begin{aligned} &P_1\omega = \frac{\partial u}{\partial x}, \qquad P_2\omega = \frac{\partial v}{\partial y}, \qquad P_3\omega = \frac{\partial u}{\partial y} + \frac{\partial v}{\partial x}, \\ &P_4\omega = \frac{\partial^2 w}{\partial x^2}, \qquad P_5\omega = \frac{\partial^2 w}{\partial y^2}, \qquad P_6\omega = \frac{\partial^2 w}{\partial x \partial y}, \\ &P_7\omega = \left(1 + \frac{h}{t_0}\right)\frac{\partial w}{\partial x} - \frac{2u}{t_0}, \qquad P_8\omega = \left(1 + \frac{h}{t_0}\right)\frac{\partial w}{\partial y} - \frac{2v}{t_0}. \end{aligned} \tag{4.4.20}$$

4.4.2 Bilinear form corresponding to the three-layered plate

Introduce the notation

$$W = W_2^1(\Omega) \times W_2^1(\Omega) \times W_2^2(\Omega). \tag{4.4.21}$$

If $\omega = (u, v, w) \in W$, then $u \in W_2^1(\Omega)$, $v \in W_2^1(\Omega)$, and $w \in W_2^2(\Omega)$.

The relations (4.3.6)–(4.3.11) are supposed to be valid. Then, the formulas (4.4.20) define the operators $P_i\colon \omega \to P_i\omega$, and

$$P_i \in \mathcal{L}(W, L_2(\Omega)), \qquad i = 1, 2, \dots, 8. \tag{4.4.22}$$

The functional $\Phi(\omega)$ determined by the relations (4.4.19) and (4.4.20) is assumed to be defined in a space V which is a subspace of W and is determined according to the way of the fastening of the plate. We suppose that

$$\left.\begin{array}{l} V \text{ is a closed subspace of } W \text{ and the relations } \omega \in V, \\ P_i\omega = 0,\ i = 1, 2, \dots, 8, \text{ imply } \omega = 0. \end{array}\right\} \tag{4.4.23}$$

The norm of an element $\omega = (u, v, w) \in V$ is defined by

$$\|\omega\|_V^2 = \|u\|^2_{W_2^1(\Omega)} + \|v\|^2_{W_2^1(\Omega)} + \|w\|^2_{W_2^2(\Omega)}. \tag{4.4.24}$$

By (4.4.19), to the strain energy of the three-layered plate there corresponds the symmetric bilinear form on $V \times V$

$$\begin{aligned} a_h(\omega', \omega'') &= 2\iint_\Omega \Big\{ \frac{E_1 h}{1-\mu_1\mu_2} \big[(P_1\omega')(P_1\omega'') + \mu_2((P_1\omega')(P_2\omega'') + (P_2\omega')(P_1\omega''))\big] \\ &+ \frac{E_2 h}{1-\mu_1\mu_2}(P_2\omega')(P_2\omega'') + Gh(P_3\omega')(P_3\omega'') \\ &+ \frac{E_1 h^3}{12(1-\mu_1\mu_2)} \big[(P_4\omega')(P_4\omega'') + \mu_2((P_4\omega')(P_5\omega'') + (P_5\omega')(P_4\omega''))\big] \\ &+ \frac{E_2 h^3}{12(1-\mu_1\mu_2)}(P_5\omega')(P_5\omega'') + \frac{1}{3} Gh^3 (P_6\omega')(P_6\omega'') \\ &+ \frac{1}{2} G_{\mathrm{m}} t_0 \big[(P_7\omega')(P_7\omega'') + (P_8\omega')(P_8\omega'')\big] \Big\}\, dx\, dy, \qquad \omega', \omega'' \in V. \end{aligned} \tag{4.4.25}$$

Theorem 4.4.1 *Let Ω be a bounded Lipschitz domain in $\mathbb{R}^2$, let the conditions* (4.3.6)–(4.3.11) *hold, let a set Y_p be defined by the formulas* (2.1.2) *and* (2.1.3), *and let*

$$G_{\mathrm{m}} = \mathrm{const} > 0. \tag{4.4.26}$$

Suppose that the space V meets the condition (4.4.23) *and is equipped with the norm* (4.4.24). *Then, the bilinear form $a_h(\omega', \omega'')$ determined by the relations* (4.4.20) *and* (4.4.25), *is symmetric, uniformly continuous and uniformly coercive in $h \in Y_p$ on $V \times V$, i.e.,*

$$a_h(\omega', \omega'') = a_h(\omega'', \omega'), \qquad \omega', \omega'' \in V, \tag{4.4.27}$$

$$|a_h(\omega', \omega'')| \le c\|\omega'\|_V \|\omega''\|_V, \qquad \omega', \omega'' \in V,\ h \in Y_p, \tag{4.4.28}$$

$$a_h(\omega, \omega) \ge c_1 \|\omega\|_V^2, \qquad \omega \in V,\ h \in Y_p, \tag{4.4.29}$$

c and c_1 being positive numbers.

Proof. The equality (4.4.27) is a consequence of (4.4.25). It is easy to deduce the inequality (4.4.28) from (4.3.6)–(4.3.11), (4.4.22), and (4.4.26). Making use of the inequality $a^2+b^2 \geq 2|ab|$ which is valid for all $a, b \in \mathbb{R}$ and taking into consideration (4.3.6)–(4.3.11) together with (4.4.26), we get

$$a_h(\omega, \omega) \geq c_2 \iint_\Omega \sum_{i=1}^{8} (P_i \omega)^2 \, dx \, dy, \qquad \omega \in V, \ h \in Y_p, \tag{4.4.30}$$

c_2 being a positive number.

Let us show that the system of operators P_i, $i = 1, 2, \ldots, 8$, is W-coercive with respect to $(L_2(\Omega))^3$ (see Subsec. 1.7.1).

According to (1.7.2), the operators P_i defined by (4.4.20) can be represented as follows

$$\omega = (u, v, w) \to P_i \omega = \sum_{|k| \leq 1} g_{i1k} D^k u + \sum_{|k| \leq 1} g_{i2k} D^k v + \sum_{|k| \leq 2} g_{i3k} D^k w. \tag{4.4.31}$$

Here

$$D^k = \frac{\partial^{k_1+k_2}}{\partial x^{k_1} \partial y^{k_2}}, \qquad k = (k_1, k_2), \ |k| = k_1 + k_2,$$

and $D^k f = f$ if $|k| = 0$,

$$g_{i1k}, \ g_{i2k}, \ g_{i3k} \in L_\infty(\Omega), \qquad i = 1, 2, \ldots, 8.$$

To use Theorem 1.7.1, let us consider a rectangular matrix with the elements

$$P_{ir}(x, y, \xi) = \sum_{|k| = l_r} g_{irk} \xi_1^{k_1} \xi_2^{k_2}, \qquad i = 1, 2, \ldots, 8; \ r = 1, 2, 3, \tag{4.4.32}$$

where $l_r = 1$ for $r = 1, 2$, and $l_r = 2$ for $r = 3$ (see (4.4.21)), and $\xi = (\xi_1, \xi_2) \in \mathbb{C}^2$, $\mathbb{C}$ being the set of complex numbers. From (4.4.20), (4.4.31), and (4.4.32), we deduce

$$(P_{ir}(x, y, \xi))^{\mathrm{T}} = \begin{bmatrix} \xi_1 & 0 & \xi_2 & 0 & 0 & 0 & 0 & 0 \\ 0 & \xi_2 & \xi_1 & 0 & 0 & 0 & 0 & 0 \\ 0 & 0 & 0 & \xi_1^2 & \xi_2^2 & \xi_1 \xi_2 & 0 & 0 \end{bmatrix},$$

the index T standing for transposition.

It is easy to see that, for any $\xi \in \mathbb{C}^2$, $\xi \neq 0$, the rows of the matrix $(P_{ir}(x, y, \xi))^{\mathrm{T}}$ are linearly independent, i.e., the rank of this matrix is equal to 3. Theorem 1.7.1 implies that the system of operators P_i defined by (4.4.20) is W-coercive with respect to $(L_2(\Omega))^3$. Further, combining (4.4.23) and Theorem 1.7.3, we obtain

$$\iint_\Omega \sum_{i=1}^{8} (P_i \omega)^2 \, dx \, dy \geq c_3 \|\omega\|_V^2, \qquad \omega \in V, \tag{4.4.33}$$

where $c_3 = \text{const} > 0$.

Finally, (4.4.30) and (4.4.33) yield (4.4.29), concluding the proof.

4.4.3 Bending of the three-layered plate

Let f be a load acting on a plate, which is identified with an element of the space V^* and let V satisfy the condition (4.4.23). The problem of the bending of the plate reduces to determining a function $\tilde{\omega}$ such that

$$\tilde{\omega} \in V, \qquad a_h(\tilde{\omega}, \omega) = (f, \omega), \qquad \omega \in V. \tag{4.4.34}$$

Theorem 4.4.1 and the Riesz theorem imply the following

Theorem 4.4.2 *Under the suppositions of Theorem* 4.4.1, *for all* $h \in Y_p$ *and* $f \in V^*$, *there exists a unique function* $\tilde{\omega}$ *satisfying* (4.4.34).

To study special realizations of the space V, which are connected with the way of the fastening of the plate, we have, in view of the condition (4.4.23), to define the following subspace

$$H = \{\, \omega \,|\, \omega \in W,\ P_i\omega = 0,\ i = 1, 2, \ldots, 8 \,\}, \tag{4.4.35}$$

W being determined by (4.4.21) and P_i by (4.4.20).

From the physical point of view, the subspace H defines the set of vector functions $\omega \in W$ for which all the strain components $P_i\omega$ are equal to 0, and consequently, so is the strain energy (see (4.4.19)).

With regard to (4.4.20), the relations $P_i\omega = 0$ define a linear homogeneous system of partial differential equations. It is easy to see that the general solution of this system has the form

a) if h and t_0 are positive numbers,

$$u = c_1, \qquad v = c_2, \qquad w = \frac{2}{t_0 + h}(c_1 x + c_2 y) + c_3, \tag{4.4.36}$$

c_1, c_2, and c_3 being constants;

b) otherwise, i.e., if the thicknesses of the layers are not constant,

$$u = 0, \qquad v = 0, \qquad w = c_4, \tag{4.4.37}$$

c_4 being a constant.

Consider the case a). (4.4.36) implies that the space H is three-dimensional, and a basis in it is formed by the vector functions $\{\omega_i = (u_i, v_i, w_i\}_{i=1}^3$ given by

$$\omega_1 = \left(1,\ 0,\ \frac{2x}{t_0 + h}\right), \qquad \omega_2 = \left(0,\ 1,\ \frac{2y}{t_0 + h}\right), \qquad \omega_3 = (0, 0, 1). \tag{4.4.38}$$

Let $\omega = (u, v, w) \in H$ and $\omega(x_0, y_0) = 0$, where $(x_0, y_0) \in \overline{\Omega}$. Then, (4.4.36) implies that $c_1 = c_2 = c_3 = 0$, i.e., $\omega = 0$. Thus, if the space V corresponds to a fastening such that the displacements u, v, w vanish at least at one point, then the condition (4.4.23) is satisfied.

It should be noted that, for the functions u, v, the condition of the fastening at a single point cannot be considered, because the function $u \to u(x_0, y_0)$, $(x_0, y_0) \in \overline{\Omega}$, is not a continuous mapping of $W_2^1(\Omega)$ into $\mathbb{R}$. Thus, for the functions u, v, the condition of the fastening at a point must be replaced by that of the fastening (vanishing) of u, v on some one- or two-dimensional manifold.

Let, for example, the space V be of the form

$$V = \{\, \omega \,|\, \omega \in W,\ \omega|_{S_1} = 0 \,\}, \tag{4.4.39}$$

where

$$S_1 \subset S, \qquad \int_{S_1} ds > 0, \tag{4.4.40}$$

S being the boundary of the domain Ω.

This space satisfies the condition (4.4.23). Indeed, the space V determined by (4.4.39), is a closed subspace of W. The relations $\omega \in V$ and $P_i \omega = 0$ imply $\omega \in H$, that is,

$$\omega = c_1\omega_1 + c_2\omega_2 + c_3\omega_3. \tag{4.4.41}$$

Since $\omega|_{S_1} = 0$, we get, by the above argument, that $c_1 = c_2 = c_3 = 0$, i.e., $\omega = 0$.

Moreover, since the function $w \to w(x_0, y_0)$, where $(x_0, y_0) \in \overline{\Omega}$, is a continuous mapping of $W_2^2(\Omega)$ into $\mathbb{R}$, the space V of the form

$$V = \{\, \omega = (u, v, w) \,|\, \omega \in W,\ u|_{S_1} = v|_{S_1} = 0,\ w(x_0, y_0) = 0 \,\} \tag{4.4.42}$$

meets (4.4.23), too.

Consider now the case b) when the layers of the plate are of variable thickness. (4.4.37) yields that the subspace H is one-dimensional, and the function

$$\omega = (0, 0, 1) \tag{4.4.43}$$

forms a basis in it. Thus, for the case b), in order that the condition (4.4.23) be satisfied, it is sufficient that the space V ensures a fastening of the plate such that $w = 0$ at least at one point, i.e., the space V determined by

$$V = \{\, \omega = (u, v, w) \,|\, \omega \in W,\ w(x_0, y_0) = 0 \,\} \tag{4.4.44}$$

where $(x_0, y_0) \in \overline{\Omega}$, meets (4.4.23) in the case b).

Remark 4.4.1 From the physical point of view, for a unique solution of the problem of the bending of the three-layered plate to exist, it is necessary that the plate be fastened, i.e., it has no rigid displacements. But the space V defined by (4.4.44) does not ensure the rigid fastening of the plate, while for this space there exists a unique solution of the bending problem (the problem (4.4.34)). This is explained by the roughness of the equations of the three-layered plate, since these were defined from the hypotheses connected with displacements of points of the plate.

Nevertheless, this fact is not a seriouss obstacle in solution of applied problems; one can always choose the space V in such a way that it corresponds to a rigid fastening of the plate (in particular, this is the case for the space V from (4.4.39) if S_1 is not an intercept of a straight line).

Let a space V be defined by (4.4.39). An element $f \in V^*$ corresponding to the load can, for example, be represented in the form

$$\omega = (u, v, w), \qquad (f, \omega) = \iint_{\Omega} (f_1 u + f_2 v + f_3 w)\, dx\, dy + \int_{S_2} f_4 w\, ds. \tag{4.4.45}$$

Here f_1, f_2, and f_3 are the components of a vector function of the surface forces, f_4 is the cutting force, $S_2 = S \setminus S_1$, and

$$f_1,\ f_2,\ f_3 \in L_2(\Omega), \quad f_4 \in L_2(S_2), \quad \int_{S_2} ds > 0. \tag{4.4.46}$$

The element f determined by (4.4.45) and (4.4.46) belongs to the space V^*, and for this element, by virtue of Theorem 4.4.2, there exists a unique function $\tilde{\omega}$ solving the problem (4.4.34).

4.4.4 Natural oscillations of three-layered plate

A function $U(x, y, t) = (U_1(x, y, t), U_2(x, y, t), U_3(x, y, t))$ determining displacements of the three-layered plate depending on the coordinates x, y and time t during free oscillations is defined as the solution of the problem

$$U \in C^2((0, \infty); V),$$

$$\begin{aligned}
\iint_{\Omega} \Big[& (t_0\rho_0 + 2\rho h) \frac{\partial^2 U_3}{\partial t^2}\, w' + 2\rho h \left(\frac{\partial^2 U_1}{\partial t^2}\, u' + \frac{\partial^2 U_2}{\partial t^2}\, v' \right) \\
& + \frac{t_0\rho_0}{12} \left(2\, \frac{\partial^2 U_1}{\partial t^2} - h\, \frac{\partial^3 U_3}{\partial t^2 \partial x} \right) \left(2u' - h \frac{\partial w'}{\partial x} \right) \\
& + \frac{t_0\rho_0}{12} \left(2\, \frac{\partial^2 U_2}{\partial t^2} - h\, \frac{\partial^3 U_3}{\partial t^2 \partial y} \right) \left(2v' - h \frac{\partial w'}{\partial y} \right) \Big]\, dx\, dy + a_h(U, \omega') = 0,
\end{aligned}$$

$$t \in (0, \infty),\ \omega' \in V,\ \omega' = (u', v', w'). \tag{4.4.47}$$

Here, the first term under the integral sign corresponds to the work of the inertia forces on virtual transversal displacements, the second one corresponds to the work of the inertia forces of the exterior layers on virtual longitudinal displacements, and the third and fourth terms define the work of the inertia forces of the interior layer on virtual longitudinal displacements. Notice that the latter two terms could be obtained by using (4.4.5) and integrating the elementary work in z from $-\frac{t_0}{2}$ to $\frac{t_0}{2}$; in determining the work of the inertia forces of the exterior layers on virtual longitudinal displacements, we neglect the rotation inertia of these layers, i.e., the second terms in (4.4.3) and (4.4.4), which is justified by the fact that the thickness of the exterior layers is supposed to be small. Further, in (4.4.47), ρ and ρ_0 are the density of the material of the exterior and interior layers, respectively (recall that we suppose that both exterior layers are made of the same material); U_1 and U_2 are displacements of points of the midplane of the upper layer along the x and y axes; the corresponding displacements in the lower layer are $-U_1$ and $-U_2$.

We search for the solution of the problem (4.4.47) in the form

$$U(x,y,t) = (c_1 \cos\gamma t + c_2 \sin\gamma t)\omega(x,y), \qquad \omega \in V, \tag{4.4.48}$$

c_1 and c_2 being constants.

By substituting (4.4.48) into (4.4.47), we get the following equation for γ and ω

$$\omega \in V, \qquad \lambda \in \mathbb{R}, \qquad a_h(\omega,\omega') = \lambda b_h(\omega,\omega'), \qquad \omega' \in V. \tag{4.4.49}$$

Here

$$\lambda = \gamma^2, \tag{4.4.50}$$

$$\begin{aligned} b_h(\omega,\omega') = \iint\limits_\Omega & \Big[(t_0\rho_0 + 2\rho h)ww' + 2\rho h(uu' + vv') \\ & + \frac{t_0\rho_0}{12}\left(2u - h\frac{\partial w}{\partial x}\right)\left(2u' - h\frac{\partial w'}{\partial x}\right) \\ & + \frac{t_0\rho_0}{12}\left(2v - h\frac{\partial w}{\partial y}\right)\left(2v' - h\frac{\partial w'}{\partial y}\right)\Big]\,dx\,dy, \\ \omega = (u,v,w), & \quad \omega' = (u',v',w'). \end{aligned} \tag{4.4.51}$$

For the three-layered plates, the density of the material of the interior layer is much less than that of the exterior layers. Hence, in (4.4.51), one can often neglect the terms with the multiplier ρ_0. Then, the form b_h takes the form

$$b_h(\omega,\omega') = \iint\limits_\Omega 2\rho h(uu' + vv' + ww')\,dx\,dy. \tag{4.4.52}$$

Theorem 4.4.3 *Let the hypotheses of Theorem 4.4.1 be fulfilled and let a form b_h be defined by (4.4.51) or (4.4.52), ρ and ρ_0 being positive constants. Then, for all $h \in Y_p$, the spectral problem (4.4.49) has a sequence of nonzero solutions $\{\omega_i\}_{i=1}^\infty \subset V$ which correspond to a sequence of eigenvalues $\{\lambda_i\}_{i=1}^\infty$ such that*

$$a_h(\omega_i,\omega) = \lambda_i b_h(\omega_i,\omega), \qquad \omega \in V, \tag{4.4.53}$$

$$0 < \lambda_1 \le \lambda_2 \le \cdots, \qquad \lim_{i\to\infty} \lambda_i = \infty. \tag{4.4.54}$$

Each eigenvalue in the sequence $\{\lambda_i\}_{i=1}^\infty$ appears as many times as its multiplicity, and the multiplicity of each eigenvalue is finite.

Proof. Introduce the Hilbert space $\tilde{V}$ coinciding as a set with the space V and with the scalar product and norm generated by the form a_h. By virtue of Theorem 4.4.1, being considered as a topological space, $\tilde{V}$ coincides with V, and the norm generated by the form a_h is equivalent to the norm of the space V defined by (4.4.24). Obviously, the bilinear forms b_h from (4.4.51) and (4.4.52) are symmetric and continuous on the space $H = L_2(\Omega) \times L_2(\Omega) \times W_2^1(\Omega)$ (if $\omega = (u, v, w)$, $\omega \in H$, then $u \in L_2(\Omega)$, $v \in L_2(\Omega)$, and $w \in W_2^1(\Omega)$). Moreover, the forms (4.4.51) and (4.4.52) satisfy the estimate

$$b_h(\omega, \omega) \geq c\|\omega\|^2_{(L_2(\Omega))^3}, \qquad \omega \in H, \tag{4.4.55}$$

c being a positive number. (4.4.55) implies all the eigenvalues of the problem (4.4.49) to be positive. Now, in view of the compactness of the embedding of $\tilde{V}$ into H, the theorem follows from Theorem 1.5.8.

4.5 Basic relations of the shell theory

In this section, we briefly set forth some general facts about shells. A detailed exposition of the shell theory can be found, e.g., in Wang (1953), Timoshenko and Woinowsky-Krieger (1959), Ambartsumian (1974), Niordson (1985).

A shell is a body bounded by two surfaces, the distance between which is small as compared to the other dimensions. The set of the points equidistant from the surfaces of the shell is called the midsurface of the shell.

Define on the midsurface of the shell a curvilinear orthogonal coordinate system (α, β) such that the coordinate lines, i.e., the lines $\alpha = \text{const}$, $\beta = \text{const}$ coincide with the principal curvature lines of the midsurface (see Fig. 4.5.1).

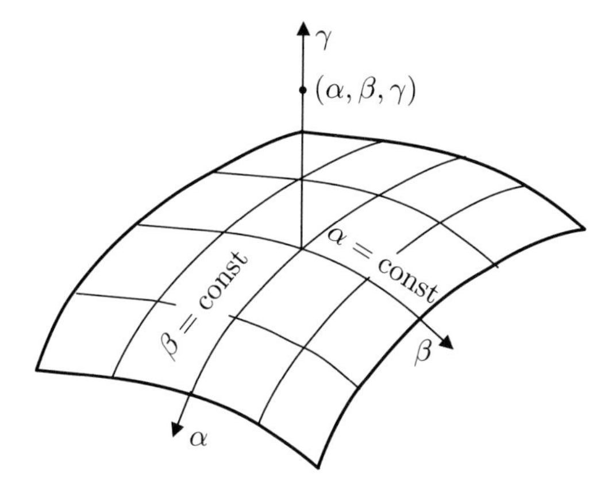

Figure 4.5.1: Curvilinear orthogonal coordinate system on the midsurface of the shell

In other words, the midsurface of the shell is a connected two-dimensional manifold with edge in $\mathbb{R}^3$, and it is defined by the parameters α, β, i.e., there exists a function f mapping a set $\Omega \subset \mathbb{R}^2$ into $\mathbb{R}^3$,

$$(\alpha,\beta) \in \Omega, \qquad (\alpha,\beta) \to f(\alpha,\beta) = \left\{ \begin{array}{c} x(\alpha,\beta) \\ y(\alpha,\beta) \\ z(\alpha,\beta) \end{array} \right\}. \tag{4.5.1}$$

Here, x, y, and z are the coordinates in $\mathbb{R}^3$ of points of the midsurface. The parameters α and β are called curvilinear coordinates. Supposing that f is a bijection of Ω onto its range $f(\Omega)$, we identify a point $(\alpha,\beta) \in \Omega$ with the point $f(\alpha,\beta)$ of the midsurface. In the chosen system of curvilinear coordinates, the midsurface is characterized by the principle curvature radii $R_1(\alpha,\beta)$, $R_2(\alpha,\beta)$, and by the coefficients of the first quadratic form $A(\alpha,\beta)$, $B(\alpha,\beta)$, by means of which the square of the length of a linear element of the midsurface is represented as follows:

$$ds^2 = A^2 d\alpha^2 + B^2 d\beta^2. \tag{4.5.2}$$

To determine location of points of the shell which are not on the midsurface, one introduces the third coordinate line γ which is normal to the midsurface and determines the distance in the normal direction between a point (α,β) of the midsurface and the point (α,β,γ) (Fig. 4.5.1).

The classical shell theory is based on the Kirchhoff hypotheses. Displacements of points of midsurface of the shell are characterized by the components u, v, w of the vector of displacements ω, i.e., $\omega = (u,v,w)$, where u, v, w are displacements along the coordinate lines α, β, γ, respectively. (More precisely, u, v, w are the components of the expansion of the vector of displacements in three mutually orthogonal unit vectors i, j, k, where i and j are tangential vectors to the coordinate lines α, β, respectively, and k is the normal vector at a given point, i, j, k being directed according to the growth of coordinates.) The components of displacements of points of the shell located at a distance γ from the midsurface are denoted by u^γ, v^γ, w^γ and determined by the components of displacements of the midsurface by the formulas

$$\begin{aligned} u^\gamma &= u + \gamma\theta_1, \qquad & v^\gamma &= v + \gamma\theta_2, \qquad w^\gamma = w, \\ \theta_1 &= -\frac{1}{A}\frac{\partial w}{\partial \alpha} + \frac{u}{R_1}, \qquad & \theta_2 &= -\frac{1}{B}\frac{\partial w}{\partial \beta} + \frac{v}{R_2}. \end{aligned} \tag{4.5.3}$$

Here $-\frac{h}{2} \le \gamma \le \frac{h}{2}$, and h – the thickness of the shell – is a function of α and β.

Similarly, the strain components in an arbitrary point of the shell are defined by those of the midsurface:

$$\varepsilon_1^\gamma = \varepsilon_1 + \gamma\chi_1, \qquad \varepsilon_2^\gamma = \varepsilon_2 + \gamma\chi_2, \qquad \varepsilon_{12}^\gamma = \varepsilon_{12} + 2\gamma\chi_{12}. \tag{4.5.4}$$

Here, ε_1, ε_2, ε_{12} are the components of the tangential strain, and χ_1, χ_2, χ_{12} are the components of the flexural and torsional strains of the midsurface, which are given by the formulas (see, e.g., Wang (1953), Novozhilov (1951))

$$\begin{aligned}
\varepsilon_1 &= \frac{1}{A}\frac{\partial u}{\partial \alpha} + \frac{1}{AB}\frac{\partial A}{\partial \beta} v + \frac{w}{R_1}, \\
\varepsilon_2 &= \frac{1}{B}\frac{\partial v}{\partial \beta} + \frac{1}{AB}\frac{\partial B}{\partial \alpha} u + \frac{w}{R_2}, \\
\varepsilon_{12} &= \frac{B}{A}\frac{\partial}{\partial \alpha}\left(\frac{v}{B}\right) + \frac{A}{B}\frac{\partial}{\partial \beta}\left(\frac{u}{A}\right), \\
\chi_1 &= -\frac{1}{A}\frac{\partial}{\partial \alpha}\left(\frac{1}{A}\frac{\partial w}{\partial \alpha} - \frac{u}{R_1}\right) - \frac{1}{AB}\frac{\partial A}{\partial \beta}\left(\frac{1}{B}\frac{\partial w}{\partial \beta} - \frac{v}{R_2}\right), \\
\chi_2 &= -\frac{1}{B}\frac{\partial}{\partial \beta}\left(\frac{1}{B}\frac{\partial w}{\partial \beta} - \frac{v}{R_2}\right) - \frac{1}{AB}\frac{\partial B}{\partial \alpha}\left(\frac{1}{A}\frac{\partial w}{\partial \alpha} - \frac{u}{R_1}\right), \\
\chi_{12} &= -\frac{1}{AB}\left(\frac{\partial^2 w}{\partial \alpha\, \partial \beta} - \frac{1}{A}\frac{\partial A}{\partial \beta}\frac{\partial w}{\partial \alpha} - \frac{1}{B}\frac{\partial B}{\partial \alpha}\frac{\partial w}{\partial \beta}\right) \\
&\quad + \frac{1}{R_1}\left(\frac{1}{B}\frac{\partial u}{\partial \beta} - \frac{1}{AB}\frac{\partial A}{\partial \beta} u\right) + \frac{1}{R_2}\left(\frac{1}{A}\frac{\partial v}{\partial \alpha} - \frac{1}{AB}\frac{\partial B}{\partial \alpha} v\right).
\end{aligned} \tag{4.5.5}$$

We assume that the shell is made of an orthotropic material, so that at every point of the shell all the three principal directions of elasticity coincide with the direction of the coordinate lines (such shells are referred to as orthotropic ones). In this case, the relations between strains and stresses take the form

$$\sigma_1^\gamma = E_{11}\varepsilon_1^\gamma + E_{12}\varepsilon_2^\gamma, \qquad \sigma_2^\gamma = E_{21}\varepsilon_1^\gamma + E_{22}\varepsilon_2^\gamma, \qquad \sigma_{12}^\gamma = G\varepsilon_{12}^\gamma, \tag{4.5.6}$$

E_{ik} being defined by (4.1.9).

The strain energy is given by

$$\Phi = \frac{1}{2}\iint\limits_{\Omega} d\alpha\, d\beta \int_{-\frac{h}{2}}^{\frac{h}{2}} (\sigma_1^\gamma \varepsilon_1^\gamma + \sigma_2^\gamma \varepsilon_2^\gamma + \sigma_{12}^\gamma \varepsilon_{12}^\gamma)\, H_1 H_2\, d\gamma. \tag{4.5.7}$$

Here, h is a function of the thickness of the shell, and H_1, H_2 are the Lamé coefficients

$$H_1 = A(1 + \frac{\gamma}{R_1}), \qquad H_2 = B(1 + \frac{\gamma}{R_2}). \tag{4.5.8}$$

By substituting into (4.5.7) the relations for the strains and stresses from (4.5.4), (4.5.6), and by integrating in γ, we get the following formula for the strain energy of the orthotropic shell

$$\begin{aligned}
\Phi = \frac{1}{2}\iint\limits_{\Omega} \big(&c_{11}\varepsilon_1^2 + 2c_{12}\varepsilon_1\varepsilon_2 + c_{22}\varepsilon_2^2 + c_{66}\varepsilon_{12}^2 \\
&+ D_{11}\chi_1^2 + 2D_{12}\chi_1\chi_2 + D_{22}\chi_2^2 + 4D_{66}\chi_{12}^2\big)\, AB\, d\alpha\, d\beta.
\end{aligned} \tag{4.5.9}$$

Here

$$c_{ik} = hE_{ik}, \qquad D_{ik} = \frac{h^3}{12}E_{ik}, \qquad E_{11} = \frac{E_1}{1-\mu_1\mu_2},$$
$$E_{22} = \frac{E_2}{1-\mu_1\mu_2}, \qquad E_{12} = \mu_2 E_{11} = \mu_1 E_{22}, \qquad E_{66} = G. \tag{4.5.10}$$

Notice that, when integrating in γ in the expressions for H_1 and H_2 (see (4.5.8)), we neglected the terms $\frac{\gamma}{R_1}$ and $\frac{\gamma}{R_2}$, since they are small as compared to 1.

4.6 Shells of revolution

4.6.1 Deformations and functional spaces

Shells whose midsurface is a surface of revolution have numerous applications in technics. Such surfaces are formed by the revolution of an arbitrary flat curve around an axis lying in the same plane. This curve is referred to as a meridian. The principal curvature lines of the shell of revolution are meridians and parallels, the latter are the lines of intersection of the surface with the planes $z = \text{const}$ (see Fig. 4.6.1). Denote by r the distance between a point of the surface and the Oz axis. Then, the equation of the meridian has the form $r = r(z)$, where $z \in [0, L]$, and the location of the meridian is determined by the angle φ which is counted from the xOz plane. Denote by X the surface of revolution that has the equation of meridian $r = r(z)$, where $0 < z < L$, i.e.,

$$X = \{\, (x, y, z) \mid 0 < z < L,\ x = r(z)\cos\varphi,\ y = r(z)\sin\varphi,\ 0 \le \varphi < 2\pi \,\}.$$

Let Ω be an open rectangle in $\mathbb{R}^2$ determined by

$$\Omega = \{\, (z, \varphi) \,|\, 0 < z < L,\ 0 < \varphi < 2\pi \,\}. \tag{4.6.1}$$

Define a mapping f of the set Ω into $\mathbb{R}^3$ by the formula

$$(z, \varphi) \to f(z, \varphi) = \begin{cases} x = r(z)\cos\varphi, \\ y = r(z)\sin\varphi, \\ z = z, \end{cases} \tag{4.6.2}$$

$r(z)$ being a smooth function.

The mapping f is a homeomorphism of the set Ω onto its range Θ – an open set in X that is the complement of the intersection of X and the half-plane

$$\{\, (x, y, z) \,|\, y = 0,\ x \ge 0,\ -\infty < z < \infty \,\},$$

z and φ being the curvilinear coordinates on X (we stress that the coordinate z in $\mathbb{R}^3$ is a usual Cartesian coordinate, while for the midsurface it is a curvilinear coordinate). The mapping (4.6.2) is a map of the set Θ.

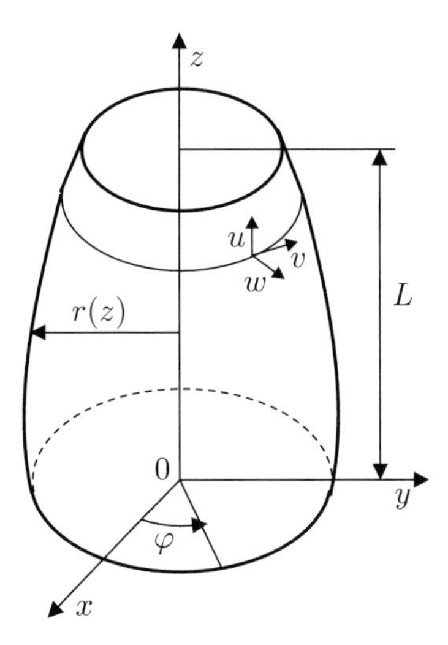

Figure 4.6.1: Shell of revolution

The coefficients of the first quadratic form and the curvature radii of the surface of revolution are given by

$$A = \left[1 + \left(\frac{dr}{dz}\right)^2\right]^{\frac{1}{2}}, \qquad B = r, \tag{4.6.3}$$

$$R_1 = -A^3 \left(\frac{d^2r}{dz^2}\right)^{-1}, \qquad R_2 = rA. \tag{4.6.4}$$

Let u, v, w stand for the components of the vector of displacements ω along the coordinate lines z, φ and the normal to the surface γ, respectively (see Fig. 4.6.1). The strain components of the midsurface from (4.5.5) in the case under investigation have the form

$$\varepsilon_1(\omega) = \frac{1}{A}\frac{\partial u}{\partial z} + \frac{w}{R_1}, \qquad \varepsilon_2(\omega) = \frac{1}{B}\frac{\partial v}{\partial \varphi} + \frac{u}{AB}\frac{\partial B}{\partial z} + \frac{w}{R_2},$$

$$\varepsilon_{12}(\omega) = \frac{1}{A}\frac{\partial v}{\partial z} + \frac{1}{B}\frac{\partial u}{\partial \varphi} - \frac{v}{AB}\frac{\partial B}{\partial z}, \qquad \chi_1(\omega) = -\frac{1}{A}\frac{\partial}{\partial z}\left(\frac{1}{A}\frac{\partial w}{\partial z} - \frac{u}{R_1}\right),$$

$$\chi_2(\omega) = -\frac{1}{B}\frac{\partial}{\partial \varphi}\left(\frac{1}{B}\frac{\partial w}{\partial \varphi} - \frac{v}{R_2}\right) - \frac{1}{AB}\frac{\partial B}{\partial z}\left(\frac{1}{A}\frac{\partial w}{\partial z} - \frac{u}{R_1}\right),$$

$$\chi_{12}(\omega) = -\frac{1}{AB}\left(\frac{\partial^2 w}{\partial z \partial \varphi} - \frac{1}{B}\frac{\partial B}{\partial z}\frac{\partial w}{\partial \varphi}\right) + \frac{1}{R_1 B}\frac{\partial u}{\partial \varphi} + \frac{1}{R_2 A}\left(\frac{\partial v}{\partial z} - \frac{1}{B}\frac{\partial B}{\partial z}v\right). \tag{4.6.5}$$

Let g be a function defined on the strip

$$Q = \{\, (z, \varphi) \,|\, 0 < z < l, \; -\infty < \varphi < \infty \,\} \tag{4.6.6}$$

and periodical in φ with period 2π, i.e.,

$$g(z, \varphi) = g(z, \varphi + 2k\pi) \tag{4.6.7}$$

for any integer k and any $z \in (0, L)$.

For $p \geq 1$, $m \in \mathbb{N}$ and Ω given by (4.6.1), define $\widetilde{W}_p^m(\Omega)$ to be the space that consists of functions g defined on the strip Q, satisfying the periodicity condition (4.6.7), and whose restriction on every bounded set Ω_1 such that $\Omega \subset \Omega_1 \subset Q$ belongs to $W_p^m(\Omega_1)$, the $\widetilde{W}_p^m(\Omega)$ norm of g is equal to the $W_p^m(\Omega)$ norm of the restriction of g on Ω.

Now let

$$W = \widetilde{W}_2^1(\Omega) \times \widetilde{W}_2^1(\Omega) \times \widetilde{W}_2^2(\Omega). \tag{4.6.8}$$

Define operators

$$P_i \in \mathcal{L}\,(W, L_2(\Omega))\,, \qquad i = 1, 2, \ldots, 6, \tag{4.6.9}$$

by the formulas

$$\begin{aligned} &P_1\omega = \varepsilon_1(\omega), && P_2\omega = \varepsilon_2(\omega), && P_3\omega = \varepsilon_{12}(\omega), \\ &P_4\omega = \chi_1(\omega), && P_5\omega = \chi_2(\omega), && P_6\omega = \chi_{12}(\omega), \end{aligned} \tag{4.6.10}$$

where $\omega = (u, v, w)$ and the right-hand sides in (4.6.10) are determined by (4.6.5). Introduce a space V such that

$$\left. \begin{array}{l} V \text{ is a closed subspace of } W \text{ and the conditions } \omega \in V, \\ P_i\omega = 0,\ i = 1, 2, \ldots, 6, \text{ imply that } \omega = 0. \end{array} \right\} \tag{4.6.11}$$

The norm of $\omega = (u, v, w) \in V$ is evidently given by the formula

$$\|\omega\|_V^2 = \|u\|^2_{W_2^1(\Omega)} + \|v\|^2_{W_2^1(\Omega)} + \|w\|^2_{W_2^2(\Omega)}. \tag{4.6.12}$$

4.6.2 The bilinear form a_h

To the strain energy of a shell of revolution there corresponds the following bilinear form on $V \times V$

$$\begin{aligned} a_h(\omega', \omega'') = \int_0^L \int_0^{2\pi} &\Big\{ c_{11}(P_1\omega')(P_1\omega'') + c_{12}\big[(P_1\omega')(P_2\omega'') + (P_2\omega')(P_1\omega'')\big] \\ &+ c_{22}(P_2\omega')(P_2\omega'') + c_{66}(P_3\omega')(P_3\omega'') + D_{11}(P_4\omega')(P_4\omega'') \\ &+ D_{12}\big[(P_4\omega')(P_5\omega'') + (P_5\omega')(P_4\omega'')\big] + D_{22}(P_5\omega')(P_5\omega'') \\ &+ 4D_{66}(P_6\omega')(P_6\omega'')\Big\}\, AB\, dz\, d\varphi. \end{aligned} \tag{4.6.13}$$

Here, $\omega', \omega'' \in V$, and the coefficients c_{ik}, D_{ik} are defined by (4.5.10). Using (4.6.10), one can easily see that

$$a_h(\omega, \omega) = 2\Phi(\omega), \qquad \omega \in V, \tag{4.6.14}$$

$\Phi(\omega)$ being the strain energy corresponding to the function of displacements ω that is determined by (4.5.9).

Theorem 4.6.1 *Let the conditions* (4.3.6)–(4.3.8) *be fulfilled, and let a set* Y_p *be defined by* (2.1.2) *and* (2.1.3). *Assume that the space* V *satisfies* (4.6.11) *and is equipped with the norm* (4.6.12). *Let also*

$$r \in C^2([0, L]), \qquad \frac{d^3 r}{dz^3} \in L_\infty(0, L), \tag{4.6.15}$$

$$r(z) \geq c_0, \qquad z \in [0, L], \; c_0 = \text{const} > 0. \tag{4.6.16}$$

Then, the bilinear form a_h *determined by the formulas* (4.5.10), (4.6.3)–(4.6.5), (4.6.10), *and* (4.6.13) *is symmetric, uniformly continuous and uniformly coercive in* $h \in Y_p$ *on* $V \times V$, *i.e.,* (4.4.27)–(4.4.29) *hold,* c *and* c_1 *being positive constants.*

Proof. The relation (4.4.27) is a consequence of (4.6.13). It is easy to verify (4.4.28). Let us prove (4.4.29). Making use of the inequality

$$a^2 + b^2 \geq 2|ab|, \qquad a, b \in \mathbb{R},$$

and taking note of (4.3.6)–(4.3.8), (4.5.10), (4.6.3), (4.6.15), and (4.6.16), we get

$$a_h(\omega, \omega) \geq c_2 \int_0^L \int_0^{2\pi} \sum_{i=1}^{6} (P_i \omega)^2 \, dz \, d\varphi, \qquad \omega \in V, \; h \in Y_p, \tag{4.6.17}$$

c_2 being a positive number.

Let us show that the system of operators P_i, $i = 1, 2, \ldots, 6$, is W-coercive with respect to $(L_2(\Omega))^3$ (see Subsec. 1.7.1), the space W being defined by (4.6.8). Analogously to the proof of Theorem 4.4.1, we introduce the matrix

$$(P_{ik}(z, \varphi, \xi)) = \begin{bmatrix} A^{-1}\xi_1 & 0 & 0 \\ 0 & B^{-1}\xi_2 & 0 \\ B^{-1}\xi_2 & A^{-1}\xi_1 & 0 \\ (AR_1)^{-1}\xi_1 & 0 & -A^{-2}\xi_1^2 \\ 0 & (BR_2)^{-1}\xi_2 & -B^{-2}\xi_2^2 \\ (R_1 B)^{-1}\xi_2 & (AR_2)^{-1}\xi_1 & -(AB)^{-1}\xi_1\xi_2 \end{bmatrix}. \tag{4.6.18}$$

Let us prove that, for all $\xi = (\xi_1, \xi_2) \in \mathbb{C}^2$, $\xi \neq 0$, and for all $(z, \varphi) \in \overline{\Omega}$, where $\overline{\Omega} = [0, L] \times [0, 2\pi]$, the rank of the matrix $(P_{ik}(z, \varphi, \xi))$ equals 3. Indeed, let $\xi = (\xi_1, \xi_2) \in \mathbb{C}^2$ and $\xi \neq 0$, so that at least one of the two components of the vector ξ does not vanish. Suppose that $\xi_1 \neq 0$. Then, the determinant of the matrix of the third order obtained from (P_{ik}) by cancelling the second, fifth, and sixth rows is equal to

$$\begin{vmatrix} A^{-1}\xi_1 & 0 & 0 \\ B^{-1}\xi_2 & A^{-1}\xi_1 & 0 \\ (AR_1)^{-1}\xi_1 & 0 & -A^{-2}\xi_1^2 \end{vmatrix} = -A^{-4}\xi_1^4 \neq 0. \tag{4.6.19}$$

Now, let $\xi_2 \neq 0$. In the same way, we can see that the determinant of the matrix obtained from (P_{ik}) by cancelling the first, fourth, and sixth rows does not vanish. Hence, the rank of the matrix $(P_{ik}(z,\varphi,\xi))$ equals 3 for all $(z,\varphi) \in \overline{\Omega}$ and $\xi \in \mathbb{C}^2$, $\xi \neq 0$. Now Theorem 1.7.1 implies the system of operators P_i to be W-coercive with respect to $(L_2(\Omega))^3$, the space W being defined by (4.6.8).

Further, due to (4.6.11), using Theorem 1.7.3, we get

$$\int_0^L \int_0^{2\pi} \sum_{i=1}^{6} (P_i\omega)^2 \, dz \, d\varphi \geq c_3 \|\omega\|_V^2, \qquad \omega \in V, \tag{4.6.20}$$

where $c_3 = \text{const} > 0$ and $\|\omega\|_V$ is defined by (4.6.12).

Finally, (4.6.17) and (4.6.20) yield

$$a_h(\omega,\omega) \geq c\|\omega\|_V^2, \qquad \omega \in V, \; h \in Y_p,$$

where $c = \text{const} > 0$, concluding the proof.

4.6.3 The subspace of functions with zero-point strain energy

In solving particular problems, the space V is determined by the way of fastening of the shell. There one faces the problem whether the condition (4.6.11) is satisfied, which in turn is connected with the problem of finding the subspace

$$H = \{\omega \,|\, \omega \in W, \; P_i\omega = 0, \; i = 1,2,\ldots,6\}. \tag{4.6.21}$$

This relation shows that, for the functions from H, all the strain components vanish, and so the strain energy equals zero. This is why H is called the space of functions with zero-point strain energy.

The relations $P_i\omega = 0$, $i = 1,2,\ldots,6$, together with (4.6.5) and (4.6.10), define a linear homogeneous system of partial differential equations.

The general solution of this system can be found by separation of variables, see Grigorenko (1973), Litvinov and Medvedev (1979). The solution implies H to be a six-dimensional space, and a basis in it is formed by the functions $\omega_i =$

(u_i, v_i, w_i) given by

$$\begin{aligned}
\omega_1 &= \left(A^{-1}\left(r - \frac{dr}{dz}\,z\right)\cos\varphi,\ z\sin\varphi,\ -A^{-1}\left(r\,\frac{dr}{dz} + z\right)\cos\varphi\right),\\
\omega_2 &= \left(A^{-1}\,\frac{dr}{dz}\cos\varphi,\ -\sin\varphi,\ A^{-1}\cos\varphi\right),\\
\omega_3 &= \left(A^{-1}\left(r - \frac{dr}{dz}\,z\right)\sin\varphi,\ -z\cos\varphi,\ -A^{-1}\left(r\,\frac{dr}{dz} + z\right)\sin\varphi\right),\\
\omega_4 &= \left(A^{-1}\,\frac{dr}{dz}\sin\varphi,\ \cos\varphi,\ A^{-1}\sin\varphi\right),\\
\omega_5 &= \left(A^{-1},\ 0,\ -A^{-1}\frac{dr}{dz}\right),\\
\omega_6 &= (0,\ r,\ 0).
\end{aligned} \tag{4.6.22}$$

Theorem 4.6.2 *Let $\omega \in H$ and let $\omega(a,\varphi_1) = \omega(a,\varphi_2) = \omega(a,\varphi_3) = 0$, where $a \in [0, L]$ and $0 \le \varphi_1 < \varphi_2 < \varphi_3 < 2\pi$. Then, $\omega = 0$.*

For the proof, see Litvinov and Medvedev (1979).

Remark 4.6.1 Theorem 4.6.2 implies that, if a closed subspace V of the space W is defined in such a way that the shell is fastened at least at three distinct points on some parallel, i.e., every vector function ω from V vanishes at these points, then the condition (4.6.11) is satisfied.

4.7 Shallow shells

A shell is said to be shallow if the rise of its midsurface does not exceed one fifth of the least linear size of its plan. For example, the shell in Fig. 4.7.1 is shallow if $\delta \le \frac{1}{5}\min(a,b)$. For a shallow shell, according to (4.5.1), one usually sets $\alpha = x$, $\beta = y$, and defines the midsurface by the relation

$$z = f(x,y). \tag{4.7.1}$$

The coefficients of the first quadratic form A and B are set equal to one. In the strain components χ_1, χ_2, χ_{12}, the terms containing u, v and their derivatives are neglected. Then, the relations (4.5.5) take the form

$$\begin{aligned}
&\varepsilon_1(\omega) = \frac{\partial u}{\partial x} + \frac{w}{R_1}, &&\varepsilon_2(\omega) = \frac{\partial v}{\partial y} + \frac{w}{R_2}, &&\varepsilon_{12}(\omega) = \frac{\partial u}{\partial y} + \frac{\partial v}{\partial x},\\
&\chi_1(\omega) = -\frac{\partial^2 w}{\partial x^2}, &&\chi_2(\omega) = -\frac{\partial^2 w}{\partial y^2}, &&\chi_{12}(\omega) = -\frac{\partial^2 w}{\partial x \partial y}.
\end{aligned} \tag{4.7.2}$$

Here

$$\frac{1}{R_1} = -\frac{\partial^2 f}{\partial x^2}, \qquad \frac{1}{R_2} = -\frac{\partial^2 f}{\partial y^2}. \tag{4.7.3}$$

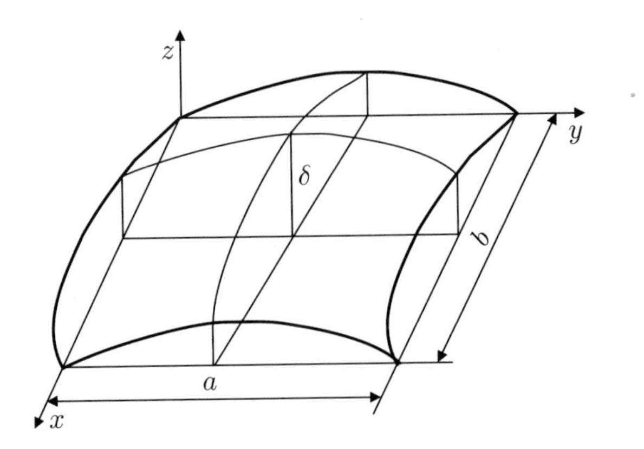

Figure 4.7.1: Shallow shell

We suppose that $R_1^{-1}, R_2^{-1} \in L_\infty(\Omega)$. However, for shallow shells, one usually sets

$$\frac{1}{R_1} = \text{const}, \qquad \frac{1}{R_2} = \text{const}. \tag{4.7.4}$$

Let $\Omega \subset \mathbb{R}^2$ be a bounded domain occupied by the projection of the midsurface of the shell onto the x, y plane.

Further, let

$$W = W_2^1(\Omega) \times W_2^1(\Omega) \times W_2^2(\Omega). \tag{4.7.5}$$

Define operators $P_i \in \mathcal{L}(W, L_2(\Omega))$ by

$$\begin{aligned} &P_1\omega = \varepsilon_1(\omega), && P_2\omega = \varepsilon_2(\omega), && P_3\omega = \varepsilon_{12}(\omega),\\ &P_4\omega = \chi_1(\omega), && P_5\omega = \chi_2(\omega), && P_6\omega = \chi_{12}(\omega), \end{aligned} \tag{4.7.6}$$

where $\omega = (u, v, w) \in W$ and the right-hand sides in (4.7.6) are determined by the formulas (4.7.2). Suppose also that

$$\left.\begin{array}{l} V \text{ is a closed subspace of } W \text{ such that the conditions}\\ \omega \in V,\ P_i\omega = 0,\ i = 1, 2, \ldots, 6, \text{ imply } \omega = 0. \end{array}\right\} \tag{4.7.7}$$

To the strain energy of the shallow shell there corresponds the following bilinear form on $V \times V$

$$\begin{aligned} a_h(\omega', \omega'') = \iint_\Omega \Big\{ & c_{11}(P_1\omega')(P_1\omega'') + c_{12}\big[(P_1\omega')(P_2\omega'') + (P_2\omega')(P_1\omega'')\big]\\ & + c_{22}(P_2\omega')(P_2\omega'') + c_{66}(P_3\omega')(P_3\omega'') + D_{11}(P_4\omega')(P_4\omega'')\\ & + D_{12}\big[(P_4\omega')(P_5\omega'') + (P_5\omega')(P_4\omega'')\big] + D_{22}(P_5\omega')(P_5\omega'')\\ & + 4D_{66}(P_6\omega')(P_6\omega'')\Big\}\, dx\, dy, \end{aligned} \tag{4.7.8}$$

c_{ik} and D_{ik} being defined by formulas (4.5.10).

Taking (4.7.6) into account, it is easy to see that

$$a_h(\omega, \omega) = 2\Phi(\omega), \qquad \omega \in V,$$

where $\Phi(\omega)$ is the strain energy corresponding to the function of displacements ω which is defined by (4.5.9) with $A = B = 1$, $\alpha = x$, $\beta = y$ and (4.7.2).

Theorem 4.7.1 *Let the conditions* (4.3.6)–(4.3.8) *be fulfilled, and let a set Y_p be defined by* (2.1.2) *and* (2.1.3). *Suppose that the space V meets* (4.7.7) *and is equipped with the norm* (4.6.12). *Let also*

$$R_1^{-1} \in L_\infty(\Omega), \qquad R_2^{-1} \in L_\infty(\Omega). \tag{4.7.9}$$

Then, the bilinear form a_h defined by (4.5.10), (4.7.2), (4.7.6), *and* (4.7.8) *is symmetric, uniformly continuous and uniformly coercive in $h \in Y_p$ on $V \times V$, i.e., the formulas* (4.4.27)–(4.4.29) *hold true, c and c_1 being positive numbers.*

Theorem 4.7.1 is proven just as Theorem 4.6.1, by using Theorems 1.7.1 and 1.7.3. So, we omit the proof. We note only that the matrix (P_{ik}) in this case has the form.

$$(P_{ik}(x, y, \xi))^{\mathrm{T}} = \begin{bmatrix} \xi_1 & 0 & \xi_2 & 0 & 0 & 0 \\ 0 & \xi_2 & \xi_1 & 0 & 0 & 0 \\ 0 & 0 & 0 & -\xi_1^2 & -\xi_2^2 & -\xi_1\xi_2 \end{bmatrix}.$$

It is easy to see that, for all $\xi \in \mathbb{C}^2$, $\xi \neq 0$, the rows of the matrix $\big(P_{ik}(x, y, \xi)\big)^{\mathrm{T}}$ are linearly independent, i.e., the rank of the matrix equals 3.

Finally, we notice that, for the shallow shell, under the condition (4.7.4), the space H given by (4.6.21), i.e., the space of functions with zero-point strain energy is five-dimensional and a basis in it is formed by the functions

$$\omega_1 = \left(R_1^{-1}x,\ R_2^{-1}y,\ -1\right), \qquad \omega_2 = \left(\frac{x^2}{2R_1} - \frac{y^2}{2R_2},\ R_2^{-1}xy,\ -x\right),$$
$$\omega_3 = \left(R_1^{-1}xy,\ \frac{y^2}{2R_2} - \frac{x^2}{2R_1},\ -y\right), \qquad \omega_4 = (1, 0, 0), \qquad \omega_5 = (0, 1, 0). \tag{4.7.10}$$

4.8 Problems of statics of shells

We proceed to study problems of stress-strain state of a shell that is subject to an exterior load. The shell is assumed to be fastened in some way, which determines the space V. The exterior load acting on the shell is identified with an element $f \in V^*$. Then, the problem of stress-strain state of the shell reduces to finding a function $\tilde{\omega} = (\tilde{u}, \tilde{v}, \tilde{w})$ such that

$$\tilde{\omega} \in V, \qquad a_h(\tilde{\omega}, \omega) = (f, \omega), \qquad \omega \in V. \tag{4.8.1}$$

By using the Riesz theorem, we conclude that, if the hypotheses of Theorem 4.6.1 are fulfilled, then the problem (4.8.1) for a shell of revolution has a unique solution for all $h \in Y_p$. Similarly, for a shallow shell, under the assumptions of Theorem 4.7.1, the problem (4.8.1) has a unique solution for all $h \in Y_p$.

As an example, let us consider a shell of revolution. Let the shell be clamped at one edge. Then, the function of displacements $\omega = (u, v, w)$ satisfies the boundary conditions

$$u(0,\varphi) = 0, \quad v(0,\varphi) = 0, \quad w(0,\varphi) = 0, \quad \frac{\partial w}{\partial z}(0,\varphi) = 0, \quad \varphi \in (0, 2\pi),$$

and the space V has the form

$$V = \left\{ \omega = (u, v, w) \,\middle|\, \omega \in W,\ \omega(0,\varphi) = 0,\ \frac{\partial w}{\partial z}(0,\varphi) = 0,\ \varphi \in (0, 2\pi) \right\}, \quad (4.8.2)$$

W defined by (4.6.8).

By the embedding theorem, or by the theorem on the trace space, V is a closed subspace of W. Now, by using Theorem 4.6.2, we deduce the space V from (4.8.2) to meet the condition (4.6.11).

Further, let the load f acting on the shell be determined by the formula

$$(f, \omega) = \int_0^L \int_0^{2\pi} (f_1 u + f_2 v + f_3 w)\, AB\, dz\, d\varphi$$
$$+ \int_0^{2\pi} \left[f_4 u(L,\varphi) + f_5 v(L,\varphi) + f_6 w(L,\varphi) + \frac{f_7}{A}\frac{\partial w}{\partial z} + \frac{f_8}{B}\frac{\partial w}{\partial \varphi} \right] B\, d\varphi, \quad (4.8.3)$$

A and B defined by (4.6.3). Here, f_1, f_2, and f_3 are the components of the force acting on the surface of the shell, f_4, f_5, and f_6 are the components of the force acting on the edge of the shell, f_7 and f_8 are the bending moment and torque. We suppose that

$$\begin{aligned} &f_i \in L_2((0,L) \times (0,2\pi)), && i = 1,2,3, \\ &f_i \in L_2(0,2\pi), && i = 4,5,\ldots,8. \end{aligned} \quad (4.8.4)$$

Then, by the embedding theorem and the theorem on the trace space, the element f defined by (4.8.3) belongs to V^*, so that under the implied conditions the problem (4.8.1) has a unique solution.

4.9 Free oscillations of a shell

Let a_h be a bilinear form on $V \times V$ generated by the strain energy of the shell; let α and β be curvilinear coordinates, i.e., the parameters determining the midsurface of the shell. Denote by U the vector function of displacements of the midsurface during free oscillations. These displacements depend on the coordinates α, β and time t, i.e.,

$$U(\alpha,\beta,t) = (U_1(\alpha,\beta,t), U_2(\alpha,\beta,t), U_3(\alpha,\beta,t)).$$

Here, $U_i(\alpha, \beta, t)$, $i = 1, 2, 3$, are the components of the expansion of the vector of displacements in the three mutually orthogonal unit vectors i, j, and k, where i and j are tangent to the coordinate lines α and β, respectively, and k is normal to the surface at a given point (see Fig. 4.5.1).

The function U is a solution of the following problem

$$U \in C^2((0, \infty); V), \tag{4.9.1}$$

$$\begin{aligned}
&\iint\limits_{\Omega} \left[\rho h \left(\frac{\partial^2 U_1}{\partial t^2}\, u' + \frac{\partial^2 U_2}{\partial t^2}\, v' + \frac{\partial^2 U_3}{\partial t^2}\, w' \right) \right. \\
&+ \frac{\rho h^3}{12} \left(\frac{1}{A} \frac{\partial^3 U_3}{\partial t^2 \partial \alpha} - \frac{1}{R_1} \frac{\partial^2 U_1}{\partial t^2} \right) \left(\frac{1}{A} \frac{\partial w'}{\partial \alpha} - \frac{u'}{R_1} \right) \\
&+ \left. \frac{\rho h^3}{12} \left(\frac{1}{B} \frac{\partial^3 U_3}{\partial t^2 \partial \beta} - \frac{1}{R_2} \frac{\partial^2 U_2}{\partial t^2} \right) \left(\frac{1}{B} \frac{\partial w'}{\partial \beta} - \frac{v'}{R_2} \right) \right] \\
&\times AB\, d\alpha\, d\beta + a_h(U, \omega') = 0, \\
&t \in (0, \infty), \; \omega' \in V, \; \omega' = (u', v', w').
\end{aligned} \tag{4.9.2}$$

Here, ρ is the density of the material of the shell and the expression under the integral sign determines the work of the inertia forces on virtual displacements. This expression is obtained from (4.5.3) just as it was done in subsecs. 4.1.5 and 4.3.2. From the physical point of view, V is the space of virtual displacements.

We search for the solution of the problem (4.9.1), (4.9.2) in the form

$$U(\alpha, \beta, t) = (c_1 \cos gt + c_2 \sin gt)\omega(\alpha, \beta), \qquad \omega = (u, v, w) \in V, \tag{4.9.3}$$

c_1 and c_2 being constants. By substituting (4.9.3) into (4.9.2), we get the following eigenvalue problem

$$\omega \in V, \qquad \lambda \in \mathbb{R}, \qquad a_h(\omega, \omega') = \lambda b_h(\omega, \omega'), \qquad \omega' \in V, \tag{4.9.4}$$

where

$$\lambda = g^2, \tag{4.9.5}$$

$$\begin{aligned}
b_h(\omega, \omega') = &\iint\limits_{\Omega} \left[\rho h\, (uu' + vv' + ww') + \frac{\rho h^3}{12} \left(\frac{1}{A} \frac{\partial w}{\partial \alpha} - \frac{u}{R_1} \right) \left(\frac{1}{A} \frac{\partial w'}{\partial \alpha} - \frac{u'}{R_1} \right) \right. \\
&+ \left. \frac{\rho h^3}{12} \left(\frac{1}{B} \frac{\partial w}{\partial \beta} - \frac{v}{R_2} \right) \left(\frac{1}{B} \frac{\partial w'}{\partial \beta} - \frac{v'}{R_2} \right) \right] AB\, d\alpha\, d\beta.
\end{aligned} \tag{4.9.6}$$

In particular, if the shell is thin, then in (4.9.6) we can neglect the terms containing h^3. Then, the form b_h takes the form

$$b_h(\omega, \omega') = \iint\limits_{\Omega} \rho h\, (uu' + vv' + ww')\, AB\, d\alpha\, d\beta. \tag{4.9.7}$$

Theorem 4.9.1 *Let a_h be a bilinear, symmetric, continuous, coercive form on $V \times V$, where V is a closed subspace of $W_2^1(\Omega) \times W_2^1(\Omega) \times W_2^2(\Omega)$ and $h \in Y_p$. Assume* *that the form* b_h *is determined either by* (4.9.6) *or by* (4.9.7), $\rho = \text{const} > 0$, *and* $A,\ B,\ A^{-1},\ B^{-1},\ R_1^{-1},\ R_2^{-1} \in L_\infty(\Omega)$. *Then, the spectral problem* (4.9.4) *has a sequence of nonzero solutions* $\{\omega_i\}_{i=1}^\infty \subset V$ *which correspond to a sequence of eigenvalues* $\{\lambda_i\}_{i=1}^\infty$ *such that*

$$a_h(\omega_i, \omega) = \lambda_i b_h(\omega_i, \omega), \qquad \omega \in V, \tag{4.9.8}$$
$$0 < \lambda_1 \le \lambda_2 \le \lambda_3 \le \cdots, \qquad \lim_{i\to\infty} \lambda_i = \infty.$$

The proof is analogous to the one of Theorem 4.4.3, and so we omit it. We notice that Theorems 4.6.1 and 4.7.1 imply the hypotheses of Theorem 4.9.1 to be fulfilled, in particular, for shallow shells and shells of revolution.

4.10 Problem of shell stability

4.10.1 On some approaches to stability problems

Basic relations of shell stability are obtained on the basis of nonlinear theory. The strain energy of a shell is determined by (4.5.9), except that additional nonlinear terms are inserted in the strain components ε_1, ε_2, ε_{12} (see Koiter (1966), Vanin et al. (1978), Grigoliuk and Kabanov (1978)).

Then, the strain energy $\Phi(\omega)$ is a functional of fourth power of $\omega \in V$, where ω is a vector function of displacements. An exterior load acting on the shell is supposed to change proportionally to a parameter $\lambda \ge 0$. Then, the stored energy of the shell is

$$\Psi_\lambda(\omega) = \Phi(\omega) - \lambda(f, \omega), \qquad \omega \in V, \tag{4.10.1}$$

where $f \in V^*$, λf being the load. When λ is sufficiently small, there usually exists a unique stationary point $\omega_\lambda \in V$ of the functional (4.10.1), at which this functional reaches its minimum on V. The stationary point ω_λ is the solution of the problem of the stress-strain state of the shell under the load λf; more exactly, ω_λ is the function of displacements of the shell under the load λf. For small λ, the second variation of the functional (4.10.1) is positive definite at the point ω_λ, i.e.,

$$\frac{d^2}{dt^2}\Psi_\lambda(\omega_\lambda + th)|_{t=0} > 0, \qquad h \in V,\ h \ne 0. \tag{4.10.2}$$

Suppose that there exists λ_0 such that at $0 < \lambda < \lambda_0$ the condition (4.10.2) is fulfilled, but for $\lambda > \lambda_0$ this is not the case. Then, the shell is considered to lose stability under the load $\lambda_0 f$. In view of this, the results on the branching of solutions of the operator equation

$$P_\lambda \omega = 0, \tag{4.10.3}$$

are applied. Here, P_λ is the Fréchet derivative of the functional (4.10.1), and λ_0 is connected with the branch point of the equation (4.10.3) (see, for example, Srubshchik (1981)).

4.10.2 Reducing of the stability problem to the eigenvalue problem

To solve the stability problem, we will use an approximate approach in which the stability problem is reduced to the eigenvalue problem.

When deflections of a shell are small as compared to its thickness, the components of the tangential strains of the midsurface have the form

$$\hat{\varepsilon}_1 = \varepsilon_1 + \frac{1}{2}\left(\frac{1}{A}\frac{\partial w}{\partial \alpha}\right)^2, \qquad \hat{\varepsilon}_2 = \varepsilon_2 + \frac{1}{2}\left(\frac{1}{B}\frac{\partial w}{\partial \beta}\right)^2,$$
$$\hat{\varepsilon}_{12} = \varepsilon_{12} + \frac{1}{AB}\frac{\partial w}{\partial \alpha}\frac{\partial w}{\partial \beta}, \tag{4.10.4}$$

where ε_1, ε_2, and ε_{12} are defined in (4.5.5), and the components of the bending and torsion strains, $\chi_1, \chi_2, \chi_{12}$, are determined by the formulas (4.5.5). The strain energy of the shell $\Phi(\omega)$ is given by (4.5.9) with ε_1, ε_2, ε_{12} replaced by $\hat{\varepsilon}_1$, $\hat{\varepsilon}_2$, $\hat{\varepsilon}_{12}$, respectively.

Let ω_λ be a solution of the problem (4.10.3), i.e., the stationary point of the functional (4.10.1), and let $g_{\omega_\lambda}(\omega', \omega'')$ be the bilinear symmetric form generated by the second variation of the functional (4.10.1) at point ω_λ, i.e.,

$$\frac{d^2}{dt^2}\Psi_\lambda(\omega_\lambda + t\omega)\big|_{t=0} = g_{\omega_\lambda}(\omega, \omega), \qquad \omega \in V. \tag{4.10.5}$$

If, for some $\lambda > 0$, there exists $\tilde{\omega}$ such that

$$\tilde{\omega} \in V, \qquad \tilde{\omega} \neq 0, \qquad g_{\omega_\lambda}(\tilde{\omega}, \omega) = 0, \qquad \omega \in V, \tag{4.10.6}$$

then the second variation of the functional (4.10.1) is not positive, i.e., the condition (4.10.2) is not satisfied.

For fixed ω', $\omega'' \in V$, the mapping $\omega \to g_\omega(\omega', \omega'')$ is a quadratic functional in V. The displacements and strains in the shell up to the loss of stability are supposed to be small. Then, in the formula for $g_{\omega_\lambda}(\omega', \omega'')$, where $\omega_\lambda = (u_\lambda, v_\lambda, w_\lambda)$, one can neglect the terms containing

$$\left(\frac{1}{A}\frac{\partial w_\lambda}{\partial \alpha}\right)^2, \qquad \left(\frac{1}{B}\frac{\partial w_\lambda}{\partial \beta}\right)^2, \qquad \frac{1}{AB}\frac{\partial w_\lambda}{\partial \alpha}\frac{\partial w_\lambda}{\partial \beta}. \tag{4.10.7}$$

The form $g_{\omega_\lambda}(\omega', \omega'')$ in which the terms containing the expressions from (4.10.7) are cancelled will be denoted by $\tilde{g}_{\omega_\lambda}(\omega', \omega'')$. Notice that $\omega \to \tilde{g}_\omega(\omega', \omega'')$ is an affine mapping of V into $\mathbb{R}$, ω' and ω'' being fixed.

Taking into account the above hypothesis about displacements and strains, we define ω_λ not as a stationary point of the functional of fourth power (4.10.1), but as a stationary point of the corresponding quadratic functional; thus

$$\omega_\lambda = \lambda\hat{\omega}, \tag{4.10.8}$$
$$\hat{\omega} \in V, \qquad a_h(\hat{\omega}, \omega) = (f, \omega), \qquad \omega \in V, \tag{4.10.9}$$

where a_h is the bilinear form generated by the quadratic strain energy and determined by the formulas (4.6.13) and (4.7.8) for a shell of revolution and for a shallow shell, respectively.

Now, instead of (4.10.6), we get the following problem: Find $\lambda \in \mathbb{R}$ for which there exists a function $\tilde{\omega}$ such that

$$\begin{aligned} &\tilde{\omega} \in V, \ \tilde{\omega} \neq 0, \\ &\tilde{g}_{\omega_\lambda}(\tilde{\omega}, \omega) = 0, \qquad \omega \in V. \end{aligned} \tag{4.10.10}$$

By (4.10.8), we get the following representation of the form $\tilde{g}_{\omega_\lambda}$:

$$\tilde{g}_{\omega_\lambda}(\omega', \omega'') = a_h(\omega', \omega'') - \lambda b_{h,\hat{\omega}}(\omega', \omega''), \tag{4.10.11}$$

and for fixed $\omega', \omega'' \in V$ the mapping $\omega \to b_{h,\omega}(\omega', \omega'')$ is a linear functional in V.

Finally, the problem of shell stability reduces to determining the least positive eigenvalue of the following problem

$$\begin{aligned} &\lambda \in \mathbb{R}, \qquad \tilde{\omega} \in V, \ \tilde{\omega} \neq 0, \\ &a_h(\tilde{\omega}, \omega) = \lambda b_{h,\hat{\omega}}(\tilde{\omega}, \omega), \qquad \omega \in V, \end{aligned} \tag{4.10.12}$$

the function $\hat{\omega}$ is the solution of the problem (4.10.9).

4.10.3 Spectral problem (4.10.12)

Under the assumption that the bilinear form a_h is continuous, symmetric, and coercive on $V \times V$, we denote by V_1 the Hilbert space that coincides as a set with V and the scalar product in which is generated by the form a_h, V_1 and V coinciding as topological spaces.

Let $B_{h,\hat{\omega}} \in \mathcal{L}(V_1, V_1)$ be the operator generated by the bilinear form $b_{h,\hat{\omega}}$ and defined by the relation

$$a_h\left(B_{h,\hat{\omega}}\omega, \omega'\right) = b_{h,\hat{\omega}}(\omega, \omega'), \qquad \omega' \in V. \tag{4.10.13}$$

Then, the problem (4.10.12) is equivalent to the following one

$$\lambda \in \mathbb{R}, \qquad \tilde{\omega} \in V_1, \ \tilde{\omega} \neq 0, \qquad \tilde{\omega} - \lambda B_{h,\hat{\omega}}\tilde{\omega} = 0. \tag{4.10.14}$$

Under additional suppositions, one can show $B_{h,\hat{\omega}}$ to be a compact mapping of V_1 into itself for the shells of revolution and shallow shells. Hence, by applying Theorem 1.5.5, we deduce the problem (4.10.14), or, equivalently, (4.10.12), has a countable set of eigenvalues $\{\lambda_i\}_{i=1}^{\infty}$. If λ_1 is the least positive eigenvalue, then $\lambda_1 f$ is the critical load under which the shell loses stability. If all the eigenvalues of the problem (4.10.12) are negative, the shell does not lose stability under all the loads λf with $0 < \lambda < \infty$.

The bilinear form $b_{h,\hat{\omega}}(\omega', \omega'')$ is defined by a rather cumbersome formula, which is not cited here. However, if the subcritical deformed state is momentless, then the form $b_{h,\hat{\omega}}$ has the essentially simpler form

$$b_{h,\hat{\omega}}(\omega', \omega'') = -\iint\limits_{\Omega} \left[\frac{T_1(h,\hat{\omega})}{A^2} \frac{\partial w'}{\partial \alpha} \frac{\partial w''}{\partial \alpha} + \frac{T_2(h,\hat{\omega})}{B^2} \frac{\partial w'}{\partial \beta} \frac{\partial w''}{\partial \beta} \right.$$

$$\left. + \frac{S(h,\hat{\omega})}{AB} \left(\frac{\partial w'}{\partial \alpha} \frac{\partial w''}{\partial \beta} + \frac{\partial w'}{\partial \beta} \frac{\partial w''}{\partial \alpha} \right) \right] AB \, d\alpha \, d\beta,$$

$$\omega' = (u', v', w'), \quad \omega'' = (u'', v'', w''). \tag{4.10.15}$$

Here, $T_1(h, \hat{\omega})$, $T_2(h, \hat{\omega})$, and $S(h, \hat{\omega})$ are the forces acting in the midsurface of the shell:

$$\begin{aligned} T_1(h,\hat{\omega}) &= c_{11}\varepsilon_1(\hat{\omega}) + c_{12}\varepsilon_2(\hat{\omega}), \\ T_2(h,\hat{\omega}) &= c_{12}\varepsilon_1(\hat{\omega}) + c_{22}\varepsilon_2(\hat{\omega}), \\ S(h,\hat{\omega}) &= c_{66}\varepsilon_{12}(\hat{\omega}), \end{aligned} \tag{4.10.16}$$

and the coefficients c_{11}, c_{12}, c_{22}, c_{66} and the components ε_1, ε_2, ε_{12} are defined by the relations (4.5.10) and (4.5.5).

Notice that, if the form $b_{h,\hat{\omega}}$ can be represented in the form (4.10.15), then, within the limit of accuracy accepted, the forces T_1, T_2, S can be determined not from the equations (4.10.9) and (4.10.16), but from the membrane (momentless) theory of shells.

Example. Consider the problem of the stability of a shell of revolution. In this case, $\alpha = z$, $\beta = \varphi$, the coefficients A and B are defined by the formulas (4.6.3), and Ω is determined by (4.6.1). Let $\lambda f \in V^*$ be an exterior load acting on the shell. If the suppositions of Theorem 4.6.1 are fulfilled, there exists a unique function $\hat{\omega}$ satisfying (4.10.9). Thus, by using the formulas (4.10.15) and (4.10.16), we define the bilinear form $b_{h,\hat{\omega}}(\omega', \omega'')$.

By using the compactness of the embedding of $W_2^2(\Omega)$ into $W_4^1(\Omega)$, one easily sees that, in the case under investigation, the operator $B_{h,\hat{\omega}}$ defined by (4.10.13) and (4.10.15) is a compact mapping of V_1 into itself (V_1 being the Hilbert space containing the elements of the space V and endowed with the scalar product generated by the form a_h). Hence, there exists a countable set of eigenvalues of the problem (4.10.14) or, equivalently, of the problem (4.10.12).

For a shallow shell, we have $A = B = 1$, $\alpha = x$, and $\beta = y$. Similarly to the above argument, by using Theorem 4.7.1, one can conclude that, for the shallow shell, there exists a countable set of eigenvalues of the problem (4.10.12).

4.11 Finite shear model of a shell

4.11.1 Strain energy of an elastic shell

So far we have considered models of plates and shells based on the Kirchhoff hypotheses. However, a great number of composite materials have a low shear stiffness. This is why shell and plate theories which account for characteristic features of these materials, the principal one being a low shear stiffness, have been developed (see, e.g., Pelekh (1973), Vanin et al. (1978), Christensen (1979), Guz et al. (1980)).

One of the most frequently used models of that kind is the finite shear model of shell, or the Timoshenko model, which is examined below. By the Timoshenko hypothesis, an element of the shell that is normal to the midsurface before deformation is no longer normal after the deformation, but rotates for some angle, remaining undistorted and of the same size. The components of displacements of points of the shell u^γ, v^γ, w^γ along the coordinate lines α, β, γ (see Fig. 4.5.1) are determined through the components of displacements of the midsurface u, v, w by the formulas

$$u^\gamma = u + \gamma\theta_1, \qquad v^\gamma = v + \gamma\theta_2, \qquad w^\gamma = w, \tag{4.11.1}$$

where θ_1 and θ_2 are the angles of rotation of the normal to the midsurface, which are supposed to be independent of the components u, v, w (compare with the formulas (4.5.3)).

Let $\omega = (u, v, w, \theta_1, \theta_2)$ be a vector function of displacements of the midsurface and the angles of the rotation of normals to it. The coordinate lines are supposed to coincide with the principal curvature lines of the midsurface. The strain components of the midsurface are expressed by the formulas (see, e.g., Guz et al. (1980))

$$\begin{gathered}
\varepsilon_1(\omega) = \frac{1}{A}\frac{\partial u}{\partial \alpha} + \frac{1}{AB}\frac{\partial A}{\partial \beta} v + \frac{w}{R_1}, \\
\varepsilon_2(\omega) = \frac{1}{B}\frac{\partial v}{\partial \beta} + \frac{1}{AB}\frac{\partial B}{\partial \alpha} u + \frac{w}{R_2}, \\
\varepsilon_{12}(\omega) = \frac{B}{A}\frac{\partial}{\partial \alpha}\left(\frac{v}{B}\right) + \frac{A}{B}\frac{\partial}{\partial \beta}\left(\frac{u}{A}\right), \\
\varepsilon_{13}(\omega) = \theta_1 + \frac{1}{A}\frac{\partial w}{\partial \alpha} - \frac{u}{R_1}, \qquad \varepsilon_{23}(\omega) = \theta_2 + \frac{1}{B}\frac{\partial w}{\partial \beta} - \frac{v}{R_2}, \\
\chi_1(\omega) = \frac{1}{A}\frac{\partial \theta_1}{\partial \alpha} + \frac{\theta_2}{AB}\frac{\partial A}{\partial \beta}, \qquad \chi_2(\omega) = \frac{1}{B}\frac{\partial \theta_2}{\partial \beta} + \frac{\theta_1}{AB}\frac{\partial B}{\partial \alpha}, \\
2\chi_{12}(\omega) = \frac{B}{A}\frac{\partial}{\partial \alpha}\left(\frac{\theta_2}{B}\right) + \frac{A}{B}\frac{\partial}{\partial \beta}\left(\frac{\theta_1}{A}\right),
\end{gathered} \tag{4.11.2}$$

where A and B are the coefficients of the first quadratic form (see (4.5.2)), R_1 and R_2 are the principal curvature radii of the midsurface of the shell.

Similarly to the above reasonings (see Section 4.5), by expressing the strain components at an arbitrary point of the shell through the strain components of the midsurface (see (4.5.4)), and by integrating in the thickness of the shell, we get the following relation for the strain energy of an orthotropic shell:

$$\begin{aligned}\Phi(\omega) = \frac{1}{2}\iint_{\Omega} \big[&c_{11}(\varepsilon_1(\omega))^2 + 2c_{12}\varepsilon_1(\omega)\varepsilon_2(\omega) + c_{22}(\varepsilon_2(\omega))^2 \\ &+ c_{33}(\varepsilon_{12}(\omega))^2 + c_{44}(\varepsilon_{13}(\omega))^2 + c_{55}(\varepsilon_{23}(\omega))^2 + D_{11}(\chi_1(\omega))^2 \\ &+ 2D_{12}\chi_1(\omega)\chi_2(\omega) + D_{22}(\chi_2(\omega))^2 + 4D_{33}(\chi_{12}(\omega))^2\big]AB\,d\alpha\,d\beta.\end{aligned} \tag{4.11.3}$$

Here,

$$\begin{aligned} c_{ik} &= hE_{ik}, \\ D_{ik} &= \frac{h^3}{12}E_{ik}, \\ E_{11} &= \frac{E_1}{1-\mu_1\mu_2}, \\ E_{22} &= \frac{E_2}{1-\mu_1\mu_2}, \\ E_{12} &= \mu_2 E_{11} = \mu_1 E_{22}, \\ E_{33} &= G_{12}, \\ E_{44} &= G_{13}, \\ E_{55} &= G_{23}, \end{aligned} \tag{4.11.4}$$

where $h = h(\alpha, \beta)$ is the function of the thickness of the shell, E_1, E_2, μ_1, μ_2, G_{12}, G_{13}, and G_{23} are elasticity constants of the material of the shell.

The functional (4.11.3) is considered in some closed subspace V of the space $\left(W_2^1(\Omega)\right)^5$, V being determined by the way of fastening of the shell. Analogously to the above, we define on $V \times V$ a bilinear symmetric form $a_h(\omega', \omega'')$. For specific kinds of shells, under natural assumptions, one easily derives the continuity of the form $a_h(\omega', \omega'')$, and, by applying Theorems 1.7.1 and 1.7.3, proves the coercivity of this form in the space V.

Remark 4.11.1 A valuable circumstance is that for the shear model $V \subset \left(W_2^1(\Omega)\right)^5$, while in the classical shell theory, which uses the Kirchhoff hypotheses, the component w of the vector function $\omega = (u, v, w)$ belongs to $W_2^2(\Omega)$. Since approximation on triangle finite elements in the space $W_2^1(\Omega)$ is easily realized, while in the space $W_2^2(\Omega)$ it leads to considerable obstacles (see, e.g., Ciarlet (1978)), this circumstance allows one to use effectively triangle finite elements for the shear model and to construct solutions of corresponding problems in the case when Ω is an arbitrary bounded polygon in $\mathbb{R}^2$.

Below we present some results for the shear model of a shallow shell.

4.11.2 Shallow shell

The bilinear form a_h and the problem of statics

For a shallow shell, one usually sets $\alpha = x$, $\beta = y$, and the midsurface of the shell is defined by a relation $z = f(x, y)$. The coefficients of the first quadratic form are set equal to 1. Then, the strain components take the form

$$\begin{aligned} &\varepsilon_1(\omega) = \frac{\partial u}{\partial x} + \frac{w}{R_1}, \qquad \varepsilon_2(\omega) = \frac{\partial v}{\partial y} + \frac{w}{R_2}, \qquad \varepsilon_{12}(\omega) = \frac{\partial u}{\partial y} + \frac{\partial v}{\partial x}, \\ &\varepsilon_{13}(\omega) = \theta_1 + \frac{\partial w}{\partial x} - \frac{u}{R_1}, \qquad \varepsilon_{23}(\omega) = \theta_2 + \frac{\partial w}{\partial y} - \frac{v}{R_2}, \\ &\chi_1(\omega) = \frac{\partial \theta_1}{\partial x}, \qquad \chi_2(\omega) = \frac{\partial \theta_2}{\partial y}, \qquad 2\chi_{12}(\omega) = \frac{\partial \theta_2}{\partial x} + \frac{\partial \theta_1}{\partial y}. \end{aligned} \tag{4.11.5}$$

Here, R_1 and R_2 are the curvature radii of the midsurface determined by the relations (4.7.3).

Let Ω be a bounded domain occupied by the projection of the midsurface of the shell on the (x, y) plane, and let $W = \big(W_2^1(\Omega)\big)^5$.

Supposing that

$$R_1^{-1},\ R_2^{-1} \in L_\infty(\Omega), \tag{4.11.6}$$

define operators $P_i \in \mathcal{L}(W, L_2(\Omega)))$ by the relations

$$\begin{aligned} &P_1\omega = \varepsilon_1(\omega), \quad P_2\omega = \varepsilon_2(\omega), \quad P_3\omega = \varepsilon_{12}(\omega), \quad P_4\omega = \varepsilon_{13}(\omega), \\ &P_5\omega = \varepsilon_{23}(\omega), \quad P_6\omega = \chi_1(\omega), \quad P_7\omega = \chi_2(\omega), \quad P_8\omega = \chi_{12}(\omega). \end{aligned} \tag{4.11.7}$$

suppose also that

$$\left. \begin{array}{l} V_1 \text{ is a closed subspace of } W = \big(W_2^1(\Omega)\big)^5 \text{ such that the} \\ \text{conditions } \omega \in V_1,\ P_i\omega = 0,\ i = 1, 2, \dots, 8, \text{ imply } \omega = 0. \end{array} \right\} \tag{4.11.8}$$

The space V_1 is equipped with the norm of the space $\big(W_2^1(\Omega)\big)^5$. With the strain energy of the shell we connect the following bilinear form a_h defined on $V_1 \times V_1$:

$$\begin{aligned} a_h(\omega', \omega'') = \iint_\Omega \Big\{ &c_{11}(P_1\omega')(P_1\omega'') + c_{12}\big[(P_1\omega')(P_2\omega'') \\ &+ (P_2\omega')(P_1\omega'')\big] + c_{22}(P_2\omega')(P_2\omega'') + c_{33}(P_3\omega')(P_3\omega'') \\ &+ c_{44}(P_4\omega')(P_4\omega'') + c_{55}(P_5\omega')(P_5\omega'') + D_{11}(P_6\omega')(P_6\omega'') \\ &+ D_{12}\big[(P_6\omega')(P_7\omega'') + (P_7\omega')(P_6\omega'')\big] + D_{22}(P_7\omega')(P_7\omega'') \\ &+ 4D_{33}(P_8\omega')(P_8\omega'')\Big\}\, dx\, dy, \qquad \omega', \omega'' \in V_1. \end{aligned} \tag{4.11.9}$$

Obviously, $a_h(\omega, \omega) = 2\Phi(\omega)$ for all $\omega \in V_1$, $\Phi(\omega)$ is the strain energy of the shell.

Theorem 4.11.1 *Let a space* V_1 *meet the condition* (4.11.8), *and let a bilinear form* a_h *be determined by the relations* (4.11.4)–(4.11.7) *and* (4.11.9), *where* E_1, E_2, G_{12}, G_{13}, *and* G_{23} *are positive numbers,* μ_1 *and* μ_2 *are constants,* $0 \le \mu_i < 1$, $i = 1,\ 2$, $\mu_1 E_2 = \mu_2 E_1$. *Then, for all* $h \in Y_p$, Y_p *being defined by the formulas* (2.1.2) *and* (2.1.3), *the bilinear form* a_h *is symmetric, continuous, and coercive on* $V_1 \times V_1$.

Proof. The symmetry and continuity of the form are obvious, and its coercivity is proven as before, with the help of Theorems 1.7.1 and 1.7.3.

The problem of stress-strain state of the shallow shell reduces to finding a function $\tilde{\omega} = (\tilde{u}, \tilde{v}, \tilde{w}, \tilde{\theta}_1, \tilde{\theta}_2)$ such that

$$\tilde{\omega} \in V_1, \qquad a_h(\tilde{\omega}, \omega) = (q, \omega), \qquad \omega \in V_1, \tag{4.11.10}$$

q being a given element of the space V_1^*, which corresponds to the load acting on the shell. The space V_1 is defined by the way of fastening of the shell.

By the Riesz theorem, under the suppositions of Theorem 4.11.1, the problem (4.11.10) has a unique solution.

Free oscillations of a shell

Denote by $U = (U_1, \ldots, U_5)$ the vector function of displacements of the midsurface and the angles of rotation of normals during free oscillations, U be a function of coordinates x, y and time t. The function U is a solution of the following problem

$$\begin{gathered}
U \in C^2((0,\infty); V_1), \\
\iint\limits_{\Omega} \Big[\rho h \Big(\frac{\partial^2 U_1}{\partial t^2}\, u' + \frac{\partial^2 U_2}{\partial t^2}\, v' + \frac{\partial^2 U_3}{\partial t^2}\, w' \Big) \\
+ \frac{\rho h^3}{12} \Big(\frac{\partial^2 U_4}{\partial t^2}\, \theta_1' + \frac{\partial^2 U_5}{\partial t^2}\, \theta_2' \Big) \Big]\, dx\, dy + a_h(U, \omega') = 0, \\
t \in (0,\infty),\ \omega' = (u', v', w', \theta_1', \theta_2') \in V_1,
\end{gathered}$$

where ρ is the density of the material of the shell, $\rho = \text{const} > 0$.

Analogously to the above, we search for the solution of the problem of free oscillations of the shell in the form

$$U(x, y, t) = (c_1 \cos gt + c_2 \sin gt)\, \omega, \qquad \omega \in V_1,$$

c_1 and c_2 being constants. As a result, we get the following eigenvalue problem

$$\begin{gathered}
\omega \in V_1,\ \omega \neq 0, \quad \lambda \in \mathbb{R}, \\
a_h(\omega, \omega') = \lambda b_h(\omega, \omega'), \qquad \omega' \in V_1,
\end{gathered} \tag{4.11.11}$$

where

$$\lambda = g^2, \qquad b_h(\omega, \omega') = \iint_\Omega \left[\rho h(uu' + vv' + ww') + \frac{\rho h^3}{12}(\theta_1\theta_1' + \theta_2\theta_2') \right] dx\, dy. \tag{4.11.12}$$

Obviously, for all $h \in Y_p$, the bilinear form b_h is continuous, symmetric, and coercive in $(L_2(\Omega))^5$. Since the embedding of $W_2^1(\Omega)$ into $L_2(\Omega)$ is compact, under the assumptions of Theorem 4.11.1, the spectral problem (4.11.11) has a sequence of nonzero solutions $\{\omega_i\}_{i=1}^\infty \subset V_1$ corresponding to a sequence of eigenvalues $\{\lambda_i\}_{i=1}^\infty$ such that

$$\begin{aligned} &a_h(\omega_i, \omega) = \lambda_i b_h(\omega_i, \omega), \qquad && \omega \in V_1, \\ &0 < \lambda_1 \le \lambda_2 \le \cdots, && \lim_{i\to\infty} \lambda_i = \infty. \end{aligned}$$

4.11.3 A relation between the Kirchhoff and Timoshenko models of shell

By comparing the relations (4.7.2) and (4.11.5), one sees that the Kirchhoff model is obtained from the Timoshenko model when

$$\theta_1 = -\frac{\partial w}{\partial x}, \qquad \theta_2 = -\frac{\partial w}{\partial y}, \tag{4.11.13}$$

and in the expressions of $\varepsilon_{13}(\omega)$ and $\varepsilon_{23}(\omega)$ the terms $\frac{u}{R_1}$ and $\frac{v}{R_2}$ are neglected. Thus, we can consider that the bilinear form (4.11.9) is deduced from (4.7.8) when, wishing to reduce the order of derivatives in the form (4.7.8), one introduces new functions θ_1 and θ_2 connected with w by (4.11.13), and solves the problem (4.8.1) by the penalty function method. In this case, G_{13} and G_{23} are considered as parameters of the penalty.

We will show that the solutions of the Tomoshenko model converge to the solution of the Kirchhoff model as G_{13} and G_{23} tend to infinity. Let, for example, the shell be clamped on a part S_1 of the boundary S. In this case, the spaces V and V_1 for the Kirchhoff and Tomoshenko models are defined by

$$\begin{aligned} V = &\{ \omega \mid \omega = (u, v, w) \in W_2^1(\Omega) \times W_2^1(\Omega) \times W_2^2(\Omega), \\ &\quad u\big|_{S_1} = v\big|_{S_1} = w\big|_{S_1} = 0,\ \frac{\partial w}{\partial \nu}\bigg|_{S_1} = 0 \}, \end{aligned} \tag{4.11.14}$$

$$\begin{aligned} V_1 = &\{ g \mid g = (u, v, w, \theta_1, \theta_2) \in \left(W_2^1(\Omega)\right)^5 \\ &\quad u\big|_{S_1} = v\big|_{S_1} = w\big|_{S_1} = 0,\ \theta_1\nu_1 + \theta_2\nu_2 = 0 \text{ on } S_1 \}, \end{aligned} \tag{4.11.15}$$

where ν_1 and ν_2 are the components of the unit outward normal ν to S_1.

We define the norm in the space V_1 as

$$\|g\|_{V_1} = \left(\|u\|^2_{W_2^1(\Omega)} + \|v\|^2_{W_2^1(\Omega)} + \|w\|^2_{W_2^1(\Omega)} + \|\theta_1\|^2_{W_2^1(\Omega)} + \|\theta_2\|^2_{W_2^1(\Omega)}\right)^{1/2}. \tag{4.11.16}$$

Suppose that the shear modula G_{13} and G_{23} depend on a parameter $\lambda \in (0,1]$. So, we denote them by G_{13}^λ, G_{23}^λ and assume that

$$\left.\begin{array}{l} \lambda \to G_{13}^\lambda,\ \lambda \to G_{23}^\lambda \text{ are positive, continuous, decreasing} \\ \text{functions given on } (0,1] \text{ such that } G_{13}^\lambda \to \infty,\ G_{23}^\lambda \to \infty \text{ as} \\ \lambda \to 0. \end{array}\right\} \tag{4.11.17}$$

Denote by $a_{h\lambda}$ the bilinear form a_h from (4.11.9) in which the terms u/R_1 and v/R_2 are omitted and G_{13}^λ, G_{23}^λ take the place of G_{13} and G_{23}. For the Timoshenko model, the problem of stress-strain state of the shallow shell is the following: Find g_λ satisfying

$$\begin{aligned} &g_\lambda = (u_\lambda, v_\lambda, w_\lambda, \theta_{\lambda 1}, \theta_{\lambda 2}) \in V_1, \\ &a_{h\lambda}(g_\lambda, g) = (q, g), \qquad g \in V_1, \end{aligned} \tag{4.11.18}$$

where q is a given element of V_1^*.

The solution of the corresponding model for the Kirchhoff model, denoted by $\tilde{\omega}$, is defined as

$$\begin{aligned} &\tilde{\omega} = (\tilde{u}, \tilde{v}, \tilde{w}) \in V, \\ &a_h(\tilde{\omega}, \omega) = (f, \omega), \qquad \omega \in V, \end{aligned} \tag{4.11.19}$$

where $f \in V^*$ and a_h is determined by (4.7.8).

Define an operator $P \in \mathcal{L}(V, V_1)$ as follows:

$$V \ni \omega = (u, v, w) \to P\omega = \left(u, v, w, -\frac{\partial w}{\partial x}, -\frac{\partial w}{\partial y}\right). \tag{4.11.20}$$

We suppose that the functionals of loading acting on the shell for the Timoshenko model $q \in V_1^*$ and for the Kirchhoff model $f \in V^*$ are equal in the following sense:

$$(f, \omega) = (q, P\omega), \qquad \omega \in V. \tag{4.11.21}$$

Theorem 4.11.2 *Suppose that the conditions of Theorems* 4.7.1 *and* 4.11.1 *hold. Let* $f \in V^*$, $q \in V_1^*$ *and let* (4.11.21) *be satisfied. Let also* (4.11.17) *be valid. Then*

$$g_\lambda \to P\tilde{\omega} \quad \text{in } (W_2^1(\Omega))^5 \text{ as } \lambda \to 0. \tag{4.11.22}$$

Proof. 1) It follows from (4.11.18) that

$$a_{h\lambda}(g_\lambda, g_\lambda) \le \|q\|_{V_1^*} \|g_\lambda\|_{V_1}. \tag{4.11.23}$$

Let $\lambda_0 \in (0,1)$. By (4.11.17) and Theorem 4.11.1 we get

$$a_{h\lambda}(\omega,\omega) \geq a_{h\lambda_0}(\omega,\omega) \geq c\|\omega\|_{V_1}^2, \qquad \omega \in V_1,\ \lambda \in (0,\lambda_0], \tag{4.11.24}$$

where c is a positive constant. The inequalities (4.11.23) and (4.11.24) give

$$\|g_\lambda\|_{V_1} \leq c^{-1}\|q\|_{V_1^*}, \qquad \lambda \in (0,\lambda_0]. \tag{4.11.25}$$

Therefore, there exists a sequence $\{\lambda_i\}_{i=1}^\infty \subset (0,\lambda_0]$, $\lim \lambda_i = 0$, such that

$$g_{\lambda_i} \to g_0 \quad \text{weakly in } V_1. \tag{4.11.26}$$

From (4.11.9), (4.11.23), and (4.11.25) we obtain

$$G_{13}^\lambda\|P_4 g_\lambda\|_{L_2(\Omega)}^2 + G_{23}^\lambda\|P_5 g_\lambda\|_{L_2(\Omega)}^2 \leq c_1, \qquad \lambda \in (0,\lambda_0]. \tag{4.11.27}$$

This inequality together with (4.11.17) yields

$$P_4 g_{\lambda_i} \to 0 \text{ and } P_5 g_{\lambda_i} \to 0 \text{ in } L_2(\Omega) \text{ as } i \to \infty. \tag{4.11.28}$$

By definition we have

$$P_4 g_{\lambda_i} = \theta_{\lambda_i 1} + \frac{\partial w_{\lambda_i}}{\partial x}, \qquad P_5 g_{\lambda_i} = \theta_{\lambda_i 2} + \frac{\partial w_{\lambda_i}}{\partial y}, \tag{4.11.29}$$

and (4.11.26), (4.11.28) imply

$$\begin{gathered} g_0 = (u_0, v_0, w_0, \theta_{01}, \theta_{02}) \in V_1 \\ \theta_{01} = -\frac{\partial w_0}{dx}, \qquad \theta_{02} = -\frac{\partial w_0}{\partial y}. \end{gathered} \tag{4.11.30}$$

Therefore, $w_0 \in W_2^2(\Omega)$ and

$$\omega_0 = (u_0, v_0, w_0) \in V, \qquad P\omega_0 = g_0. \tag{4.11.31}$$

Using (4.11.20), we get

$$P_4 P\omega = 0, \qquad P_5 P\omega = 0, \qquad \omega \in V. \tag{4.11.32}$$

We accept in (4.11.18) $\lambda = \lambda_i$, $g = P\omega$, where $\omega \in V$. Then, taking into account (4.11.21), (4.11.26), (4.7.8), (4.11.9), and (4.11.32), we pass to the limit in (4.11.18) as $i \to \infty$. As a result, we get

$$a_h(\omega_0, \omega) = (f, \omega), \qquad \omega \in V. \tag{4.11.33}$$

From the uniqueness of the solution of the problem (4.11.19), we obtain $\omega_0 = \tilde{\omega}$, and (4.11.26) is amplified in the sense

$$g_\lambda \to g_0 = P\tilde{\omega} \quad \text{weakly in } V_1 \text{ as } \lambda \to 0. \tag{4.11.34}$$

2) Let as before $\lambda \in (0,1)$. By Theorem 4.11.1, the bilinear form $a_{h\lambda_0}$ generates the norm in V_1

$$\|g\|_1 = \left(a_{h\lambda_0}(g,g)\right)^{1/2},$$

which is equivalent to the norm (4.11.16). Let us prove that

$$\lim_{\lambda\to 0} a_{h\lambda_0}(g_\lambda, g_\lambda) = a_{h\lambda_0}(g_0, g_0), \tag{4.11.35}$$

then from (4.11.34) we will get (4.11.22).

Define a functional

$$\Psi_\lambda(g) = a_{h\lambda}(g,g) - 2(q,g), \qquad g \in V_1.$$

The function g_λ is a solution of the problem (4.11.18) if and only if

$$\Psi_\lambda(g_\lambda) = \min_{g\in V_1} \Psi_\lambda(g).$$

Therefore,

$$a_{h\lambda}(g_\lambda, g_\lambda) - 2(q, g_\lambda) \le a_{h\lambda}(g_0, g_0) - 2(q, g_0), \qquad \lambda \in (0, \lambda_0]. \tag{4.11.36}$$

Since $g_0 = P\tilde{\omega}$, $\tilde{\omega} \in V$, we obtain by (4.11.32) that

$$a_{h\lambda}(g_0, g_0) = a_{h\lambda_0}(g_0, g_0), \qquad \lambda \in (0, \lambda_0]. \tag{4.11.37}$$

Passing to the limit in the left-hand side of (4.11.36), we conclude via (4.11.34) and (4.11.37) that

$$\limsup_{\lambda\to 0} a_{h\lambda}(g_\lambda, g_\lambda) \le a_{h\lambda_0}(g_0, g_0).$$

On the other hand, (4.11.17) gives

$$a_{h\lambda}(g_\lambda, g_\lambda) \ge a_{h\lambda_0}(g_\lambda, g_\lambda), \qquad \lambda \in (0, \lambda_0].$$

The two last inequalities yield

$$\limsup_{\lambda\to 0} a_{h\lambda_0}(g_\lambda, g_\lambda) \le a_{h\lambda_0}(g_0, g_0). \tag{4.11.38}$$

It follows from (4.11.34) that

$$\liminf_{\lambda\to 0} a_{h\lambda_0}(g_\lambda, g_\lambda) \ge a_{h\lambda_0}(g_0, g_0). \tag{4.11.39}$$

Now, (4.11.38) and (4.11.39) give (4.11.35), which concludes the proof.

Remark 4.11.2 Let

$$\begin{aligned} I_\lambda &= a_{h\lambda}(g_\lambda, g_\lambda) - 2(q, g_\lambda), \\ I_0 &= a_h(\tilde{\omega}, \tilde{\omega}) - 2(f, \tilde{\omega}), \end{aligned} \tag{4.11.40}$$

where g_λ and $\tilde{\omega}$ are the solutions of the problems (4.11.18) and (4.11.19), respectively. Then,

$$I_\lambda \leq I_0 \tag{4.11.41}$$

and

$$\lim_{\lambda \to 0} I_\lambda = I_0,$$
$$G_{13}^\lambda \|P_4 g_\lambda\|^2_{L_2(\Omega)} + G_{23}^\lambda \|P_5 g_\lambda\|^2_{L_2(\Omega)} \to 0 \quad \text{as } \lambda \to 0. \tag{4.11.42}$$

Indeed, since $a_{h\lambda}(g_0, g_0) = a_h(\tilde{\omega}, \tilde{\omega})$, see (4.11.34), we conclude from (4.11.21) and (4.11.36) that (4.11.41) holds.

Denote by a the bilinear form a_h defined by (4.11.9) with $G_{13} = G_{23} = 0$. We have

$$I_\lambda \geq a(g_\lambda, g_\lambda) - 2(q, g_\lambda). \tag{4.11.43}$$

It follows from (4.11.21) and (4.11.22) that

$$\lim_{\lambda \to 0} [a(g_\lambda, g_\lambda) - 2(q, g_\lambda)] = a_h(\tilde{\omega}, \tilde{\omega}) - 2(f, \tilde{\omega}) = I_0. \tag{4.11.44}$$

Now, by (4.11.41), (4.11.43), and (4.11.44), we get (4.11.42).

4.12 Laminated shells

4.12.1 The strain energy of a laminated shell

We consider a composite, laminated, thin shell fabricated of an arbitrary number of homogeneous, anisotropic laminae. We suppose that the thickness of each lamina is constant, and so the thickness of the shell is constant. Let also the coordinate surface $\gamma = 0$ coincide with the midsurface of the shell, see Fig. 4.12.1 . We apply the Kirchhoff hypotheses for the whole stack of laminae, see Section 4.5 and Ambartsumian (1974). Let $\omega = (u, v, w)$ be the vector function of displacements of points of the surface $\gamma = 0$, u, v, w being the displacements in the directions of the coordinate lines α, β, γ. Denote by u^γ, v^γ, w^γ the displacements of points situated at distance γ from the surface $\gamma = 0$. We have (see (4.5.3))

$$u^\gamma = \left(1 + \frac{\gamma}{R_1}\right) u - \frac{\gamma}{A} \frac{\partial w}{\partial \alpha}, \qquad v^\gamma = \left(1 + \frac{\gamma}{R_2}\right) v - \frac{\gamma}{B} \frac{\partial w}{\partial \beta}, \tag{4.12.1}$$
$$w^\gamma = w.$$

Here, A, B are the coefficients of the first quadratic form, R_1, R_2 the radii of the principal curvatures, the coordinate lines $\alpha = \text{const}$, $\beta = \text{const}$ coincide with lines of the principal curvatures of the surface $\gamma = 0$. The strain components are defined by the formulas (4.5.4), (4.5.5) and the stress components given by

$$\begin{aligned} \sigma_1^\gamma &= E_{11}\varepsilon_1^\gamma + E_{12}\varepsilon_2^\gamma + E_{16}\varepsilon_{12}^\gamma, \\ \sigma_2^\gamma &= E_{21}\varepsilon_1^\gamma + E_{22}\varepsilon_2^\gamma + E_{26}\varepsilon_{12}^\gamma, \\ \sigma_{12}^\gamma &= E_{61}\varepsilon_1^\gamma + E_{62}\varepsilon_2^\gamma + E_{66}\varepsilon_{12}^\gamma, \qquad E_{ij} = E_{ji}. \end{aligned} \tag{4.12.2}$$

Here, E_{ij} are elasticity coefficients, which in this case are (see Fig. 4.12.1)

$$\begin{gathered} E_{ij}(\gamma) = E_{ji}(\gamma) = E_{ij}^n, \\ \gamma \in \left(-\frac{h}{2} + \delta_{n-1},\ -\frac{h}{2} + \delta_n \right), \qquad i,j = 1,2,6,\ n = 1,2,\ldots,s, \\ \delta_0 = 0, \quad \delta_s = h, \end{gathered} \tag{4.12.3}$$

E_{ij}^n's being constants. The strain energy (see (4.5.7) and (4.5.8)) is determined by

$$\Phi = \frac{1}{2} \iint\limits_\Omega d\alpha\, d\beta \int_{-\frac{h}{2}}^{\frac{h}{2}} \left(\sigma_1^\gamma \varepsilon_1^\gamma + \sigma_2^\gamma \varepsilon_2^\gamma + \sigma_{12}^\gamma \varepsilon_{12}^\gamma \right) AB\, d\gamma. \tag{4.12.4}$$

In the expressions for H_1 and H_2 in (4.5.8), we neglect the terms $\frac{\gamma}{R_1}$ and $\frac{\gamma}{R_2}$ since they are small as compared to 1.

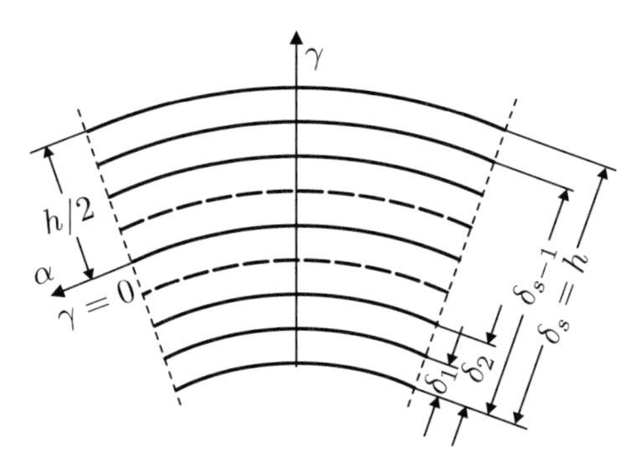

Figure 4.12.1: Laminated shell

We substitute (4.12.2) into (4.12.4). By (4.5.4) and (4.12.3), we obtain

$$\begin{aligned} \Phi = \frac{1}{2} &\iint\limits_\Omega \left(c_{11}\varepsilon_1^2 + 2c_{12}\varepsilon_1\varepsilon_2 + c_{22}\varepsilon_2^2 + c_{66}\varepsilon_{12}^2 + 2c_{16}\varepsilon_{12}\varepsilon_1 + 2c_{26}\varepsilon_{12}\varepsilon_2 \right) AB\, d\alpha\, d\beta \\ &+ \iint\limits_\Omega \big[K_{11}\varepsilon_1\chi_1 + K_{12}(\varepsilon_1\chi_2 + \varepsilon_2\chi_1) + K_{22}\varepsilon_2\chi_2 + 2K_{66}\varepsilon_{12}\chi_{12} \end{aligned}$$

$$+ K_{16}(2\varepsilon_1\chi_{12} + \varepsilon_{12}\chi_1) + K_{26}(2\varepsilon_2\chi_{12} + \varepsilon_{12}\chi_2)\big] AB\, d\alpha\, d\beta$$
$$+ \frac{1}{2} \iint_{\Omega} (D_{11}\chi_1^2 + 2D_{12}\chi_1\chi_2 + D_{22}\chi_2^2 + 4D_{66}\chi_{12}^2$$
$$+ 4D_{16}\chi_1\chi_{12} + 4D_{26}\chi_2\chi_{12}\big) AB\, d\alpha\, d\beta, \quad (4.12.5)$$

where

$$c_{ij} = \sum_{n=1}^{s} E_{ij}^n (\delta_n - \delta_{n-1}),$$
$$K_{ij} = \frac{1}{2} \sum_{n=1}^{s} E_{ij}^n \big[(\delta_n^2 - \delta_{n-1}^2) - \delta_s(\delta_n - \delta_{n-1})\big], \quad (4.12.6)$$
$$D_{ij} = \frac{1}{3} \sum_{n=1}^{s} E_{ij}^n \big[(\delta_n^3 - \delta_{n-1}^3) - \frac{3}{2}\delta_s(\delta_n^2 - \delta_{n-1}^2) + \frac{3}{4}\delta_s^2(\delta_n - \delta_{n-1})\big].$$

4.12.2 Shell of revolution

For a shell of revolution, Ω is defined by (4.6.1) and the strain components are determined by (4.6.3)–(4.6.5). We suppose that (4.6.15), (4.6.16) hold. Let

$$W = \widetilde{W}_2^1(\Omega) \times \widetilde{W}_2^1(\Omega) \times \widetilde{W}_2^2(\Omega), \quad (4.12.7)$$

where $\widetilde{W}_2^m(\Omega)$, $m = 1, 2$, is the subspace of $W_2^m(\Omega)$ consisting of periodic functions with respect to φ, see Subsec. 4.6.1. We denote by $\omega = (u, v, w)$ an element of W. Define operators $P_i \in \mathcal{L}(W, L_2(\Omega))$ by (4.6.5), (4.6.10), and associate with the strain energy the following bilinear form on W:

$$a_\delta(\omega', \omega'') = \int_0^L \int_0^{2\pi} \Big\{c_{11}(P_1\omega')(P_1\omega'') + c_{12}\big[(P_1\omega')(P_2\omega'') + (P_2\omega')(P_1\omega'')\big]$$
$$+ c_{22}(P_2\omega')(P_2\omega'') + c_{66}(P_3\omega')(P_3\omega'') + c_{16}\big[(P_1\omega')(P_3\omega'') + (P_3\omega')(P_1\omega'')\big]$$
$$+ c_{26}\big[(P_2\omega')(P_3\omega'') + (P_3\omega')(P_2\omega'')\big]\Big\} AB\, dz\, d\varphi$$
$$+ \int_0^L \int_0^{2\pi} \{K_{11}\big[(P_1\omega')(P_4\omega'') + (P_4\omega')(P_1\omega'')\big]$$
$$+ K_{12}\big[(P_1\omega')(P_5\omega'') + (P_5\omega')(P_1\omega'') + (P_2\omega')(P_4\omega'') + (P_4\omega')(P_2\omega'')\big]$$
$$+ K_{22}\big[(P_2\omega')(P_5\omega'') + (P_5\omega')(P_2\omega'')\big] + 2K_{66}\big[(P_3\omega')(P_6\omega'') + (P_6\omega')(P_3\omega'')\big]$$
$$+ K_{16}\big[2(P_1\omega')(P_6\omega'') + 2(P_6\omega')(P_1\omega'') + (P_3\omega')(P_4\omega'') + (P_4\omega')(P_3\omega'')\big]$$
$$+ K_{26}\big[2(P_2\omega')(P_6\omega'') + 2(P_6\omega')(P_2\omega'') + (P_3\omega')(P_5\omega'') + (P_5\omega')(P_3\omega'')\big]\Big\} AB\, dz\, d\varphi$$

$$+\int_0^L\int_0^{2\pi}\{D_{11}(P_4\omega')(P_4\omega'')+D_{12}\big[(P_4\omega')(P_5\omega'')+(P_5\omega')(P_4\omega'')\big]$$
$$+D_{22}(P_5\omega')(P_5\omega'')+4D_{66}(P_6\omega')(P_6\omega'')+2D_{16}\big[(P_4\omega')(P_6\omega'')+(P_6\omega')(P_4\omega'')\big]$$
$$+2D_{26}\big[(P_5\omega')(P_6\omega'')+(P_6\omega')(P_5\omega'')\big]\big\}AB\,dz\,d\varphi. \quad (4.12.8)$$

Here, $\omega',\omega''\in W$ and we denote the bilinear form by a_δ since the coefficients c_{ij}, K_{ij}, D_{ij} depend on the vector $\delta=(\delta_1,\ldots,\delta_s)$, see (4.12.6). It follows from (4.6.10), (4.12.5), and (4.12.8) that

$$a_\delta(\omega,\omega)=2\Phi(\omega),\qquad \omega\in W, \quad (4.12.9)$$

where $\Phi(\omega)$ is the strain energy for a displacement ω. We suppose that the elasticity coefficients satisfy the following conditions

$$\sum_{i,j=1,2,6}E_{ij}^n\xi_j\xi_i\ge c(\xi_1^2+\xi_2^2+\xi_6^2),\qquad (\xi_1,\xi_2,\xi_6)\in\mathbb{R}^3,$$
$$n=1,2,\ldots,s,\quad E_{ij}^n=E_{ji}^n,\quad c=\text{const}>0. \quad (4.12.10)$$

Theorem 4.12.1 *Let the conditions* (4.6.15), (4.6.16), (4.12.10) *hold. Let V be a subspace of W satisfying* (4.6.11). *Then, the bilinear form a_δ determined by* (4.6.3)–(4.6.5), (4.6.10), (4.12.6), *and* (4.12.8) *is symmetric, continuous, and coercive in V, i.e.,*

$$a_\delta(\omega',\omega'')=a_\delta(\omega'',\omega'),\qquad \omega',\omega''\in V, \quad (4.12.11)$$
$$a_\delta(\omega',\omega'')\le c_1\|\omega'\|_V\|\omega''\|_V,\qquad \omega',\omega''\in V, \quad (4.12.12)$$
$$a_\delta(\omega,\omega)\ge c_2\|\omega\|_V^2,\qquad \omega\in V. \quad (4.12.13)$$

Here, c_1, c_2 are positive constants, c_2 depending on the constant c from (4.12.10) *and on δ_s.*

Proof. By (4.5.4) and (4.6.10), we have

$$\varepsilon_1^\gamma(\omega)=P_1\omega+\gamma P_4\omega,$$
$$\varepsilon_2^\gamma(\omega)=P_2\omega+\gamma P_5\omega, \quad (4.12.14)$$
$$\varepsilon_{12}^\gamma(\omega)=P_3\omega+2\gamma P_6\omega,$$

Taking into account (4.12.2)–(4.12.4), (4.12.9), (4.12.10), (4.12.14), and denoting $\varepsilon_6^\gamma(\omega)=\varepsilon_{12}^\gamma(\omega)$, we obtain

$$a_\delta(\omega,\omega)=\int_0^L\int_0^{2\pi}AB\,dz\,d\varphi\int_{-\frac{h}{2}}^{\frac{h}{2}}\sum_{i,j=1,2,6}E_{ij}\varepsilon_i^\gamma(\omega)\varepsilon_j^\gamma(\omega)\,d\gamma$$
$$\ge c\int_0^L\int_0^{2\pi}AB\,dz\,d\varphi\int_{-\frac{h}{2}}^{\frac{h}{2}}\big[(\varepsilon_1^\gamma(\omega))^2+(\varepsilon_2^\gamma(\omega))^2+(\varepsilon_{12}^\gamma(\omega))^2\big]\,d\gamma$$
$$\ge c_3\int_0^L\int_0^{2\pi}\sum_{i=1}^6(P_i\omega)^2\,dz\,d\varphi. \quad (4.12.15)$$

This estimate together with (4.6.20) yields (4.12.13). The relations (4.12.11) and (4.12.12) are obvious.

4.12.3 Shallow shells

For these shells, the midsurface is determined by (4.7.1), the bilinear form a_δ is defined by (4.7.2), (4.7.3), (4.7.6), and (4.12.8), with $A = B = 1$, $dz = dx$, $d\varphi = dy$, and the integration over $(0, L) \times (0, 2\pi)$ is replaced with the integration over Ω, where Ω is an arbitrary bounded plane domain with a Lipschitz continuous boundary; V is defined by (4.7.5) and (4.7.7).

By analogy with the above, we prove the following

Theorem 4.12.2 *Suppose* (4.7.9) *and* (4.12.10) *hold. Then, the bilinear form* a_δ *for the shallow shell is symmetric, continuous, and coercive in* V, *i.e.,* (4.12.11)–(4.12.13) *hold.*

Chapter 5

Optimization of Deformable Solids

> " "I thought I'd try and find my way to the top of the hill –" said Alice.
> "When you say 'hill,' " the Queen interrupted, "I could show you hills, in comparison with which you'd call that a valley." "
>
> – *L. Carroll*
> "Through the Looking Glass"

5.1 Settings of optimization problems for plates and shells

5.1.1 Goal functional and a function of control

Plates and shells are main elements of many advanced structures. One of the most important characteristics of a construction is its weight, which determines the consumption of material needed for production of the construction as well as some operating features of the latter. For example, the increase of weight of an aircraft causes growth of the fuel rate in flight and degradation of some flight characteristics.

The weight of a homogeneous plate is determined by the relation

$$J = \gamma \iint\limits_{\Omega} h \, dx \, dy, \tag{5.1.1}$$

where γ is the specific gravity of the material, Ω is the domain occupied by the midplane of the plate, and h is the function of thickness. The weight of a three-layered plate is defined by the following formula (see Fig. 4.3.1)

$$J = \iint_{\Omega} (\gamma_1 t_0 + 2\gamma_2 h)\, dx\, dy, \tag{5.1.2}$$

where γ_1 and γ_2 are the specific gravities of the interior and exterior layers, respectively.

The weight of a shell is given by the formula

$$J = \iiint_{\Omega} d\alpha\, d\beta \int_{-h/2}^{h/2} \mu H_1 H_2\, d\gamma. \tag{5.1.3}$$

Here, μ is the specific gravity of the material, h is the function of thickness of the shell, α and β are orthogonal curvilinear coordinates, H_1 and H_2 are the Lamé coefficients.

Because of its great importance, in many problems of optimization of plates and shells, the weight either is considered as a goal functional, or is included into restrictions (i.e., into the set of admissible controls).

Important characteristics of a structure (plate or shell) are the frequencies of the natural oscillations and the critical load under which the construction loses stability. These characteristics can define not only the set of admissible controls, but also the goal functional. For example, studied are problems in which one maximizes the first natural frequency, the critical load under which the construction loses stability, etc.

In some problems, one considers a goal functional corresponding to optimization by some finite number of significant figures. For example, if the goal functional is of the form

$$J(h) = c_1 J_1(h) - c_2 J_2(h),$$

where h is a control, $J_1(h)$ is the weight, $J_2(h)$ is the first natural frequency, c_1 and c_2 are positive constants, then the optimization problem of finding the minimum of $J(h)$ on U_{ad} means minimization of the weight and maximization of the first natural frequency.

As control functions in optimization problems for plates and shells, one can take the function of thickness, as well as the function determining the midsurface of the shell (see the formulas (4.5.1) and (4.6.2)). Sometimes, one creates special control loads so that the external load acting on the plate or shell can also be considered as a control.

The set of admissible controls consists of a system of restrictions, which will be considered now.

5.1.2 Restrictions

Restrictions on geometry and eigenvalues

Since the thickness of a plate (shell) is bounded and positive, the function of thickness h must meet the estimates

$$\check{h} \leq h \leq \hat{h} \qquad \text{in } \Omega,$$

where $\check{h}$ and $\hat{h}$ are positive constants.

If the shape of the midsurface of the shell is a control, then one imposes restrictions on the corresponding control function. For example, for a shell of revolution, the function $r(z)$ (see the relations (4.6.2) and Fig. 4.6.1) can be a control. In this case, the condition

$$r_1 \leq r(z) \leq r_2, \qquad z \in [0, L],$$

must be fulfilled, r_1 and r_2 being positive numbers. Moreover, the function $r(z)$ should be smooth in a sense (see Theorem 4.6.1).

Restrictions on frequencies of the natural oscillation of the plate (shell) can also be imposed. In some cases, one demands that the first natural frequency be not less than a fixed value; in other cases one wishes that neighborhoods of some fixed numbers do not contain the frequencies of natural oscillations, etc.

If the plate (shell) can lose stability under the action of a given load, then one considers a restriction on stability.

Restrictions on stiffness and strength

One considers restrictions on the vector function $\omega = (u, v, w)$ of displacements of points of the midsurface of the shell. Since $u, v \in W_2^1(\Omega)$ and $w \in W_2^2(\Omega)$ (see Sections 4.6 and 4.7), the restrictions on u, v, w, e.g., for a shallow shell, may be taken in the form

$$\max_{(x,y)\in\overline{\Omega}} |w(x,y)| \leq c, \qquad \left| \iint\limits_{\Omega_i} u\, dx\, dy \right| \leq a_i,$$
$$\left| \iint\limits_{\Omega_i} v\, dx\, dy \right| \leq b_i, \qquad i = 1, 2, \dots, k. \tag{5.1.4}$$

Here, Ω_i are subdomains of the domain Ω such that $\operatorname{mes}\Omega_i > 0$, a_i, b_i, and c are positive numbers. In view of the inequalities (5.1.4), we stress that, since $w \in W_2^2(\Omega) \subset C(\overline{\Omega})$, the displacements w can be controlled at every point, whereas the functions of displacements u and v do not belong, in general, to $W_2^2(\Omega)$, so that they are controlled in the integral sense.

Restrictions on strength are also of great importance. There exist a number of different strength criteria applicable to various materials (see Goldenblat and Kopnov (1968), Wu (1974)). A generalized strength criterion for anysotropic materials has the form

$$\sum_{i,k} \Pi_{ik}\sigma_{ik} + \sum_{p,q,m,n} \Pi_{pqmn}\sigma_{pq}\sigma_{mn} + \sum_{r,s,t,l,m,n} \Pi_{rstlmn}\sigma_{rs}\sigma_{tl}\sigma_{mn} + \cdots - 1 \le 0. \qquad (5.1.5)$$

Here, σ_{ik} are the components of the stress tensor Π_{ik}, Π_{pqmn}, etc. are the components of the strength tensors of different valences.

For a number of anysotropic materials, in the criterion (5.1.5) one can consider only the first two summands, i.e., the strength criterion looks like

$$\sum_{i,k} \Pi_{ik}\sigma_{ik} + \sum_{p,q,m,n} \Pi_{pqmn}\sigma_{pq}\sigma_{mn} - 1 \le 0. \qquad (5.1.6)$$

The components of the stress tensor for the shell are determined by the formulas (4.5.4)–(4.5.6), and in this case

$$\sigma_{11} = \sigma_1^\gamma, \qquad \sigma_{22} = \sigma_2^\gamma, \qquad \sigma_{21} = \sigma_{12} = \sigma_{12}^\gamma,$$
$$\sigma_{13} = \sigma_{31} = \sigma_{23} = \sigma_{32} = \sigma_{33} = 0. \qquad (5.1.7)$$

Restrictions on the strength must be fulfilled at every point of the three-dimensional domain Q occupied by the plate or shell. Results of Chapter 4 imply that stresses in the plate (shell) are elements of the space $L_2(Q)$. Thus, in order that the relations (5.1.5) or (5.1.6) make sense at every point $(\alpha, \beta, \gamma) \in Q$ (see Fig. 4.5.1), one has to average (regularize) the stresses (see Subsec. 1.6.4). To this end, the function of stresses must be extended to a larger domain.

Let $\sigma_{ij}^{(\rho)}$ be an averaging of the function σ_{ij} relative to the α and β coordinates, ρ being the radius of the averaging kernel. $\sigma_{ij}^{(\rho)}$ are obviously functions of a point $(\alpha, \beta, \gamma) \in Q$, i.e., $\sigma_{ij}^{(\rho)} = \sigma_{ij}^{(\rho)}(\alpha, \beta, \gamma)$. According to the Kirchhoff hypotheses, the stresses along the normal to the midsurface of the shell change by affine law (see the formulas (4.5.4)–(4.5.6)), and so the conditions (5.1.5) and (5.1.6) are to be verified only on the surfaces of the plate (shell), i.e., for $\gamma = \pm\frac{h}{2}$. Thus, e.g., for the relation (5.1.6), the strength conditions take the form

$$\max_{\tilde\gamma = \frac{h}{2}, -\frac{h}{2}} \max_{(\alpha,\beta)\in\overline{\Omega}} \Bigg[\sum_{i,k=1}^{2} \Pi_{ik}\sigma_{ik}^{(\rho)}(\alpha,\beta,\tilde\gamma) + \sum_{p,q,m,n=1}^{2} \Pi_{pqmn}\sigma_{pq}^{(\rho)}(\alpha,\beta,\tilde\gamma)\sigma_{mn}^{(\rho)}(\alpha,\beta,\tilde\gamma)\Bigg] - 1 \le 0. \qquad (5.1.8)$$

We stress the operation of averaging (regularization) is legitimate because, on one hand, the relation (5.1.8) makes sense, and on the other hand, the very physical notion of a stress contains some averaging. In connection with this, one can use also integral restrictions on strength, i.e., demand that inequalities of the form (5.1.6) take place in the integral sense, on some small pieces of the surface of the shell. For example, for an isotropic shell, one can use the restrictions of the form

$$\max_{\tilde{\gamma}=\frac{h}{2},-\frac{h}{2}} \iint\limits_{\Omega_j} \sum_{i,k=1}^{2} \sigma_{ik}^2(\alpha,\beta,\tilde{\gamma}) AB\, d\alpha\, d\beta - c_j \leq 0,$$
$$j = 1, 2, \ldots, s, \quad c_j = \text{const} > 0, \tag{5.1.9}$$

A and B being coefficients of the first quadratic form (see Section 4.5). Notice that the domain Ω is divided into the subdomains Ω_j.

5.2 Approximate solution of direct and optimization problems for plates and shells

5.2.1 Direct problems and spline functions

As shown in Chapter 4, the solution of problems on the stress-strain state of a plate or shell reduces to determining a function u_h such that

$$u_h \in V, \qquad a_h(u_h, v) = (f, v), \qquad v \in V, \tag{5.2.1}$$

while the stability problems and the problems of finding frequencies of natural oscillation of plates and shells reduce to the eigenvalue problem

$$(u_i, \mu_i) \in V \times \mathbb{R}, \qquad u_i \neq 0, \qquad \mu_i a_h(u_i, v) = b_h(u_i, v), \quad v \in V. \tag{5.2.2}$$

To get approximate solutions of the problems (5.2.1), (5.2.2), one can use the Riesz and Galerkin methods (interior approximation methods) as well as exterior approximation methods, disturbed approximations, and other methods (see, e.g., Aubin (1972)).

In subsecs. 2.4.4 and 2.6.3, we considered the Riesz and Galerkin methods of approximate solution of the problems (5.2.1), (5.2.2) and established that, if a sequence of finite-dimensional subspaces $\{V_m\}_{m=1}^{\infty} \subset V$ satisfies the condition

$$\lim_{m\to\infty} \inf_{u\in V_m} \|u - v\|_V = 0, \qquad v \in V, \tag{5.2.3}$$

then the Riesz and Galerkin methods ensure convergence of approximate solutions to explicit ones.

Specific forms of the space V for different models of plates and shells were presented in Chapter 4. Below, we cite a few examples of the sequences $\{V_m\}_{m=1}^{\infty} \subset V$ meeting the condition (5.2.3).

In what follows, we will need the notion of a spline. Let there be defined a partition of an interval $[a,b]$:

$$\Delta : a = x_0 < x_1 < \cdots < x_N = b. \tag{5.2.4}$$

By $\mathrm{Sp}^{(m)}(\Delta,[a,b])$, $m \in \mathbb{N}$, we denote the space consisting of real-valued functions w such that $w \in C^{2m-2}[a,b]$ and on each interval $[x_i, x_{i+1}]$ the function w is a polynomial of degree $2m-1$.

The spaces $\mathrm{Sp}^{(m)}(\Delta,[a,b])$ for $m = 1$, 2 are of special importance for applications. They are called affine and cubic spline spaces, respectively.

Define the following subspaces of the space $\mathrm{Sp}^{(2)}(\Delta,[a,b])$:

$$\begin{aligned}
&\mathrm{Sp}_1^{(2)}(\Delta,[a,b]) = \{\, s \mid s \in \mathrm{Sp}^{(2)}(\Delta,[a,b]),\ s(x_0) = s(x_N) = 0 \,\},\\
&\mathrm{Sp}_2^{(2)}(\Delta,[a,b])\\
&\qquad = \Big\{\, s \mid s \in \mathrm{Sp}^{(2)}(\Delta,[a,b]),\ s(x_0) = s(x_N) = \frac{ds}{dx}(x_0) = \frac{ds}{dx}(x_N) = 0 \,\Big\},\\
&\mathrm{Sp}_3^{(2)}(\Delta,[a,b])\\
&\qquad = \Big\{\, s \mid s \in \mathrm{Sp}^{(2)}(\Delta,[a,b]),\ s(x_0) = s(x_N) = \frac{ds}{dx}(x_0) = 0 \,\Big\},\\
&\mathrm{Sp}_4^{(2)}(\Delta,[a,b]) = \Big\{\, s \mid s \in \mathrm{Sp}^{(2)}(\Delta,[a,b]),\ s(x_0) = \frac{ds}{dx}(x_0) = 0 \,\Big\},\\
&\mathrm{Sp}_5^{(2)}(\Delta,[a,b])\\
&\qquad = \Big\{\, s \mid s \in \mathrm{Sp}^{(2)}(\Delta,[a,b]),\ \frac{d^p s}{dx^p}(x_0) = \frac{d^p s}{dx^p}(x_N),\ p = 0,1,2 \,\Big\};
\end{aligned} \tag{5.2.5}$$

$\mathrm{Sp}_5^{(2)}(\Delta,[a,b])$ is called the periodical spline space. Bases in the spaces

$$\mathrm{Sp}^{(2)}(\Delta,[a,b]) \quad \text{and} \quad \mathrm{Sp}_i^{(2)}(\Delta,[a,b]), \qquad i = 1,2,\ldots,5,$$

are formed by corresponding fundamental splines, which are numerically constructed by using effective algorithms (see Ahlberg et al. 1967, Laurent (1972), Zavialov et al. (1980)).

5.2.2 The spaces V_m for plates

We consider the model of a plate based on the Kirchhoff hypotheses, i.e., the Kirchhoff model. In this case, $V \subset W_2^2(\Omega)$.

Suppose that Ω is a rectangular domain,

$$\Omega = \{\, (x,y) \mid a_1 < x < b_1,\ a_2 < y < b_2 \,\}, \tag{5.2.6}$$

a_i and b_i being constants. Let $\{\Delta_n = \Delta_{1n} \times \Delta_{2n}\}_{n=1}^{\infty}$ be a sequence of partitions of the rectangle $\overline{\Omega} = [a_1,b_1] \times [a_2,b_2]$, where

$$\begin{aligned}
\Delta_{1n} &: a_1 = x_{0n} < x_{1n} < \cdots < x_{N_n n} = b_1,\\
\Delta_{2n} &: a_2 = y_{0n} < y_{1n} < \cdots < y_{M_n n} = b_2.
\end{aligned} \tag{5.2.7}$$

We set

$$\pi_{1n} = \max_{0 \le i \le N_n - 1} \left(x_{(i+1)n} - x_{in} \right), \qquad \underline{\pi}_{1n} = \min_{0 \le i \le N_n - 1} \left(x_{(i+1)n} - x_{in} \right),$$

$$\pi_{2n} = \max_{0 \le i \le M_n - 1} \left(y_{(i+1)n} - y_{in} \right), \qquad \underline{\pi}_{2n} = \min_{0 \le i \le M_n - 1} \left(y_{(i+1)n} - y_{in} \right),$$

$$\pi_n = \max \left(\pi_{1n}, \pi_{2n} \right), \qquad \underline{\pi}_n = \min \left(\underline{\pi}_{1n}, \underline{\pi}_{2n} \right).$$

The notation $\Delta_n \in \mathcal{P}_\sigma(\Omega)$ will denote that $\frac{\pi_n}{\underline{\pi}_n} \le \sigma$, where $\sigma > 0$.

Consider the case when the plate is clamped on the boundary. Then $V = \overset{\circ}{W}{}_2^2(\Omega)$ (see Subsec. 4.1.4). Define spaces V_n as follows

$$V_n = \mathrm{Sp}_2^{(2)}(\Delta_{1n}, [a_1, b_1]) \otimes \mathrm{Sp}_2^{(2)}(\Delta_{2n}, [a_2, b_2]), \tag{5.2.8}$$

the symbol $\otimes$ denoting the tensor product. Notice that $V_n \subset \overset{\circ}{W}{}_2^2(\Omega)$.

If the plate is supported on the boundary, then $V = W_2^2(\Omega) \cap \overset{\circ}{W}{}_2^1(\Omega)$ (see Subsec. 4.1.4). Define in this case the spaces V_n as

$$V_n = \mathrm{Sp}_1^{(2)}(\Delta_{1n}, [a_1, b_1]) \otimes \mathrm{Sp}_1^{(2)}(\Delta_{2n}, [a_2, b_2]). \tag{5.2.9}$$

Now, $V_n \subset W_2^2(\Omega) \cap \overset{\circ}{W}{}_2^1(\Omega)$. Using results on the interpolation of splines in a rectangular domain (see Zavialov et al. (1980)) and taking into account that the set of smooth functions is dense in V, one can easily see that the spaces V_n defined by the relations (5.2.8) and (5.2.9) for the clamped and supported plates, respectively, meet the conditions (5.2.3) provided that the corresponding sequence of partitions $\{\Delta_n\}$ is such that $\Delta_n \in \mathcal{P}_\sigma(\Omega)$ for all n and $\pi_n \to 0$.

In the same way, using the tensor product of the subspaces of the form $\mathrm{Sp}_1^{(2)}$, $\mathrm{Sp}_2^{(2)}$, $\mathrm{Sp}_3^{(2)}$, and $\mathrm{Sp}_4^{(2)}$ (see (5.2.5)), one can construct a sequence of spaces satisfying (5.2.3) for the cases when, on different edges of the plate, one imposes different boundary conditions, such as those of clamping, supporting, and of free edge. A more complicated case is that when one considers different boundary conditions on the same edge. Then, one can use splines together with a method of exterior approximation (see Litvinov (1981a)).

Now, let Ω be a non-rectangular domain. If the plate is clamped or supported on the whole boundary, then one can use the "classic method" of construction of the spaces V_n, according to which these spaces are set up by using the elements of the form (see Michlin (1970))

$$u = \sum_{i=1}^{n} c_i \omega^k \varphi_i, \tag{5.2.10}$$

where c_i are constants, ω is a function which is positive in Ω and vanishes on S, the boundary of Ω, $\{\varphi_i\}_{i=1}^{\infty}$ is some system of functions, $k = 1$ in the case of supporting and $k = 2$ for clamping.

Rvachev proposed an effective method of construction of functions ω which satisfy the above conditions and given smoothness requirements for domains of a complicated shape (see Rvachev, V.L. and Rvachev, V.A. (1979)).

To obtain approximate solutions of direct problems for plates of a non-rectangular form, one can use also disturbed approximations (see Aubin (1972)). The application of the finite element method in the case when $V \subset W_2^2(\Omega)$, Ω is of a complicated form, and Ω is divided into triangles, leads to great difficulties (see, e,g., Ciarlet (1978)).

For the finite shear model of plates, $V \subset \left(W_2^1(\Omega)\right)^3$ (see Section 4.11, for a bending of the plate the vector of displacements has the form $\omega = (w, \theta_1, \theta_2)$). The approximation in a subspace of the space $\left(W_2^1(\Omega)\right)^3$ is rather easy and effectively realized by finite elements for domains of complicated forms. Hence, for such domains it is natural to use not the Kirchhoff model, but the more precise finite shear model, and to solve a corresponding problem by finite elements. (In this case, we face the situation when a refinement of the model simplifies a corresponding problem in the above mentioned sense.)

5.2.3 The spaces V_m for shells

Shallow shell

For the Kirchhoff model, V is the subspace $W_2^1(\Omega) \times W_2^1(\Omega) \times W_2^2(\Omega)$ (see Section 4.7). If Ω is a rectangular domain determined by the relation (5.2.6) and on each edge of $\overline{\Omega}$, one of the following conditions is assigned: clamp, support, or free edge, then the spaces $V_m \subset V$ satisfying the condition (5.2.3) are constructed by splines just as it was done for the plates.

For example, consider a shell clamped on the whole edge. Now, the space V has the form

$$V = \left\{ \omega = (u, v, w) \,|\, \omega \in \overset{\circ}{W}{}_2^1(\Omega) \times \overset{\circ}{W}{}_2^1(\Omega) \times \overset{\circ}{W}{}_2^2(\Omega) \right\}. \tag{5.2.11}$$

Let $\left\{\Delta_n^{(j)} = \Delta_{1n}^{(j)} \times \Delta_{2n}^{(j)}\right\}_{n=1}^{\infty}$ be a sequence of partitions of the rectangle $\overline{\Omega}$ such that $j = 1, 2, 3$ and

$$\Delta_{1n}^{(j)} : a_1 = x_{0n}^{(j)} < x_{1n}^{(j)} < \cdots < x_{N_{nj}n}^{(j)} = b_1,$$

$$\Delta_{2n}^{(j)} : a_2 = y_{0n}^{(j)} < y_{1n}^{(j)} < \cdots < y_{M_{nj}n}^{(j)} = b_2,$$

$$\pi_{1n}^{(j)} = \max_{0 \le i \le N_{nj}-1} \left(x_{(i+1)n}^{(j)} - x_{in}^{(j)} \right), \quad \underline{\pi}_{1n}^{(j)} = \min_{0 \le i \le N_{nj}-1} \left(x_{(i+1)n}^{(j)} - x_{in}^{(j)} \right),$$

$$\pi_{2n}^{(j)} = \max_{0 \le i \le M_{nj}-1} \left(y_{(i+1)n}^{(j)} - y_{in}^{(j)} \right), \quad \underline{\pi}_{2n}^{(j)} = \min_{0 \le i \le M_{nj}-1} \left(y_{(i+1)n}^{(j)} - y_{in}^{(j)} \right),$$

$$\pi_n^{(j)} = \max\left(\pi_{1n}^{(j)}, \pi_{2n}^{(j)}\right), \quad \underline{\pi}_n^{(j)} = \min\left(\underline{\pi}_{1n}^{(j)}, \underline{\pi}_{2n}^{(j)}\right). \tag{5.2.12}$$

We suppose that

$$\{\Delta_n^{(j)}\}_{n=1}^{\infty} \subset \mathcal{P}_\sigma(\Omega), \qquad \lim_{n\to\infty} \pi_n^{(j)} = 0, \qquad j = 1, 2, 3. \tag{5.2.13}$$

Define the spaces V_n in the following way:

$$V_n = V_n^{(1)} \times V_n^{(2)} \times V_n^{(3)}, \tag{5.2.14}$$

where

$$\begin{aligned} V_n^{(i)} &= \mathrm{Sp}_1^{(2)}(\Delta_{1n}^{(i)}, [a_1, b_1]) \otimes \mathrm{Sp}_1^{(2)}(\Delta_{2n}^{(i)}, [a_2, b_2]), \qquad i = 1, 2, \\ V_n^{(3)} &= \mathrm{Sp}_2^{(2)}(\Delta_{1n}^{(3)}, [a_1, b_1]) \otimes \mathrm{Sp}_2^{(2)}(\Delta_{2n}^{(3)}, [a_2, b_2]). \end{aligned} \tag{5.2.15}$$

In this case $V_n \subset V$, where V is defined by (5.2.11). Since the set of smooth functions is dense in V, making use of the error estimates of interpolation by cubic splines of two variables (see Zavialov et al. (1980)), one can show that the spaces V_n defined by the formulas (5.2.14) and (5.2.15) satisfy the condition (5.2.3) provided (5.2.13) holds.

We point out that the spaces $V_n^{(1)}$ and $V_n^{(2)}$ could be constructed as tensor products of subspaces of those affine splines that vanish at the points a_i and b_i. If $V_n^{(3)}$ is determined by the relation (5.2.15), the inclusion $V_n \subset V$ holds, and provided (5.2.13) is true, (5.2.3) is valid.

If the shell has a non-rectangular plan, i.e., Ω is a non-rectangular domain, then in order to construct the spaces V_n one can use the above stated methods for plates. Just as in the case of a plate, if Ω has a rather complicated form, it is useful to consider a finite shear model (for this model $V \subset \left(W_2^1(\Omega)\right)^5$) and to construct spaces V_n on the basis of finite elements.

Shell of revolution

For a shell of revolution, Ω is a rectangular domain defined by the relation

$$\Omega = \{\, (z, \varphi) \,|\, 0 < z < L,\ 0 < \varphi < 2\pi \,\}, \tag{5.2.16}$$

and $V \subset \widetilde{W}_2^1(\Omega) \times \widetilde{W}_2^1(\Omega) \times \widetilde{W}_2^2(\Omega)$, where $\widetilde{W}_2^l(\Omega)$, $l = 1, 2$, are the spaces of periodic functions (see Section 4.6).

For a clamped shell, the space V has the form

$$\begin{aligned} V = \Big\{ \omega = (u, v, w) \,|\, \omega \in \widetilde{W}_2^1(\Omega) \times \widetilde{W}_2^1(\Omega) \times \widetilde{W}_2^2(\Omega), \\ u\big|_{S_1} = v\big|_{S_1} = w\big|_{S_1} = \frac{\partial w}{\partial z}\Big|_{S_1} = 0 \Big\}, \end{aligned} \tag{5.2.17}$$

and

$$S_1 = \{\, (z, \varphi) \,|\, z = 0, L,\ 0 < \varphi < 2\pi \,\}. \tag{5.2.18}$$

If the shell is supported, then

$$V = \big\{\, \omega = (u, v, w) \,|\, \omega \in \widetilde{W}_2^1(\Omega) \times \widetilde{W}_2^1(\Omega) \times \widetilde{W}_2^2(\Omega),\ u\big|_{S_1} = v\big|_{S_1} = w\big|_{S_1} = 0 \,\big\}. \tag{5.2.19}$$

Let $\{\Delta_n^{(j)} = \Delta_{1n}^{(j)} \times \Delta_{2n}^{(j)}\}_{n=1}^{\infty}$ be the sequence of the partitions of the rectangle $\overline{\Omega}$, $j = 1, 2, 3$, determined by the relations (5.2.12) provided z and φ are substituted for x and y, $a_1 = 0$, $b_1 = L$, $a_2 = 0$, $b_2 = 2\pi$. For these partitions, the condition (5.2.13) is supposed to be valid.

The spaces V_n are of the form

$$V_n = V_n^{(1)} \times V_n^{(2)} \times V_n^{(3)}, \tag{5.2.20}$$

where

$$V_n^{(i)} = \mathrm{Sp}_1^{(2)}(\Delta_{1n}^{(i)}, [0, L]) \otimes \mathrm{Sp}_5^{(2)}(\Delta_{2n}^{(i)}, [0, 2\pi]), \qquad i = 1, 2,$$

$$V_n^{(3)} = \mathrm{Sp}_2^{(2)}(\Delta_{1n}^{(3)}, [0, L]) \otimes \mathrm{Sp}_5^{(2)}(\Delta_{2n}^{(3)}, [0, 2\pi]),$$

if the shell is clamped, and

$$V_n^{(i)} = \mathrm{Sp}_1^{(2)}(\Delta_{1n}^{(i)}, [0, L]) \otimes \mathrm{Sp}_5^{(2)}(\Delta_{2n}^{(i)}, [0, 2\pi]), \qquad i = 1, 2, 3,$$

if the shell is supported. Notice that, in both cases, the inclusion $V_n \subset V$ takes place and provided (5.2.13) is valid, (5.2.3) is true. Analogously to the above, in the form of tensor products of corresponding spline spaces, one can construct spaces $V_n \subset V$ satisfying the condition (5.2.3) in the case when one edge is clamped while the other edge is either supported or free.

Further, let $\{\Delta_n^{(i)}\}_{n=1}^{\infty}$ be a sequence of partitions of the segment $[0, L]$:

$$\Delta_n^{(i)} : 0 = z_{0n}^{(i)} < z_{1n}^{(i)} < \cdots < z_{N_{ni}n}^{(i)} = L, \qquad i = 1, 2, 3,$$

and

$$\pi_n^{(i)} = \max_{0 \le j \le N_{ni}-1} \left(z_{(j+1)n}^{(i)} - z_{jn}^{(i)} \right). \tag{5.2.21}$$

Denote by P_k the vector space of trigonometric polynomials of degree k, i.e., the totality of elements ψ of the form

$$\psi = a_0 + \sum_{j=1}^{k} (b_j \cos j\varphi + c_j \sin j\varphi), \tag{5.2.22}$$

a_0, b_j, and c_j being constants.

For a shell of revolution, the spaces V_n can be constructed as tensor products of spline and trigonometric polynomial spaces. Then, the spaces V_n are of form (5.2.20), and for a clamped shell $V_n^{(i)}$ are determined by

$$V_n^{(1)} = \mathrm{Sp}_1^{(2)}(\Delta_n^{(1)}, [0, L]) \otimes P_{M_n^{(1)}},$$

$$V_n^{(2)} = \mathrm{Sp}_1^{(2)}(\Delta_n^{(2)}, [0, L]) \otimes P_{M_n^{(2)}},$$

$$V_n^{(3)} = \mathrm{Sp}_2^{(2)}(\Delta_n^{(3)}, [0, L]) \otimes P_{M_n^{(3)}}, \tag{5.2.23}$$

and for a supported shell $V_n^{(i)}$ looks like

$$V_n^{(i)} = \mathrm{Sp}_1^{(2)}(\Delta_n^{(i)}, [0, L]) \otimes P_{M_n^{(i)}}, \qquad i = 1, 2, 3. \tag{5.2.24}$$

The spaces V_n defined by the formulas (5.2.20), (5.2.23), and (5.2.20), (5.2.24) for the clamped and supported shells, respectively, meet the condition of inclusion $V_n \subset V$. Then, making use of results by Litvinov (1981b), one can easily see that these spaces satisfy the condition (5.2.3) provided $M_n^{(i)} \to \infty$ as $n \to \infty$ and $\lim_{n\to\infty} \pi_n^{(i)} = 0$.

As tensor products of spline and trigonometric polynomial subspaces, one can construct spaces $\{V_n\} \subset V$ satisfying the condition (5.2.3) for other cases of fixing of a shell; for example, for the cases when one edge is clamped and the other edge is supported or free.

Calculations have shown a high efficiency of the use of the subspaces of the form (5.2.23), (5.2.24), when the thickness of the shell h varies only along the generatrix, i.e., $h = h(z)$.

5.2.4 Direct problems for nonfastened plates and shells

Above, we always considered problems in which a bilinear form a_h is coercive in V, i.e.,

$$a_h(\omega, \omega) \geq c\|\omega\|_V^2, \qquad \omega \in V, \tag{5.2.25}$$

V being a subspace of the space $W = \prod_{i=1}^k W_2^{l_i}(\Omega)$. However, in case of a nonfastened plate (shell), V is not a subspace of W, but $V = W$ and (5.2.25) does not hold. Let us present an approach to the solution of such problems, which is based upon the work by Křižek and Litvinov (1992).

From the theorems on the coerciveness of the forms a_h for plates and shells, it follows that there exists a positive constant c_1 such that

$$a_h(\omega, \omega) + \|\omega\|_{L_2(\Omega)^k}^2 \geq c_1\|\omega\|_W^2, \qquad \omega \in W. \tag{5.2.26}$$

Denote by W_1 the kernel space of the form a_h:

$$W_1 = \{\omega \mid \omega \in W,\ a_h(\omega, \omega') = 0,\ \omega' \in W\}. \tag{5.2.27}$$

W_1 is a finite-dimensional subspace of functions with zero point strain energy, see Section 4.6.3. In the theory of elasticity, W_1 is known as the subspace of rigid displacements. In the case when $\Omega \subset \mathbb{R}^2$, it is defined by (1.7.25). For the Kirchhoff model of plate, W_1 is the three-dimensional subspace with basis 1, x, y. For the models of shells, these subspaces are given by the relations (4.6.22) and (4.7.10).

Let $\{\varphi_i\}_{i=1}^N$ be a basis in W_1 and let W_2 be the orthogonal complement of W_1 to W with respect to the scalar product in $L_2(\Omega)^N$, i.e.,

$$W_2 = \{\omega \mid \omega \in W,\ (\omega, \varphi_i)_{L_2(\Omega)^N} = 0,\ i = 1, \dots, N\}. \tag{5.2.28}$$

According to (5.2.26) and Theorem1.7.3, there exists a constant c_1 such that

$$a_h(\omega, \omega) \geq c_1 \|\omega\|_W^2, \qquad \omega \in W_2. \tag{5.2.29}$$

The problem of stress-strain state of a nonfastened plate (shell) consists in finding a function $\tilde{\omega}$ such that

$$\tilde{\omega} \in W, \qquad a_h(\tilde{\omega}, \omega) = (f, \omega), \qquad \omega \in W, \tag{5.2.30}$$

where

$$f \in W^*, \qquad (f, \omega) = 0, \qquad \omega \in W_1. \tag{5.2.31}$$

The condition (5.2.31) is necessary and sufficient for the solvability of the problem (5.2.30). The solution of the latter problem is defined up to an arbitrary function of W_1, i.e., if $\tilde{\omega}$ is a solution of the problem (5.2.30), then $\omega_1 = \tilde{\omega} + \hat{\omega}$, where $\hat{\omega} \in W_1$, is also a solution of this problem. Therefore, it is an inconvenient problem from the point of view of computation. Instead of (5.2.30), we consider the following problem: Find a function $\breve{\omega}$ satisfying

$$\breve{\omega} \in W, \qquad a_h(\breve{\omega}, \omega) + r \sum_{i=1}^{N} (\breve{\omega}, \varphi_i)_{L_2(\Omega)^N} (\omega, \varphi_i)_{L_2(\Omega)^N} = (f, \omega), \qquad \omega \in W, \tag{5.2.32}$$

where r is a positive constant.

By virtue of Theorem 1.7.2 , the bilinear form $\breve{a}_h$ defined by

$$\breve{a}_h(\omega', \omega'') = a_h(\omega', \omega'') + r \sum_{i=1}^{N} (\omega', \varphi_i)_{L_2(\Omega)^N} (\omega'', \varphi_i)_{L_2(\Omega)^N}, \qquad \omega', \omega'' \in W,$$

is coercive in W for an arbitrary $r > 0$. So, there exists a unique solution $\breve{\omega}$ of the problem (5.2.32). One can verify that the function $\breve{\omega}$ is also a solution of the problem (5.2.30), i.e., $\breve{\omega} = \tilde{\omega}$, and in addition $\breve{\omega} \in W_2$. Thus, the solution $\breve{\omega}$ of the problem (5.2.32) is independent of r.

5.2.5 Solution of optimization problems

Let H be a Banach space of controls and let $\{H_n\}_{n=1}^{\infty}$ be a sequence of finite-dimensional subspaces such that

$$H_n \subset H \qquad \forall n, \tag{5.2.33}$$

$$\lim_{n \to \infty} \inf_{u \in H_n} \|u - v\|_H = 0, \qquad v \in H. \tag{5.2.34}$$

The finite-dimensional optimization problem consists in finding a function $\tilde{h}$ such that

$$\tilde{h} \in H_n \cap Q, \qquad f(\tilde{h}) = \inf_{h \in H_n \cap Q} f(h), \tag{5.2.35}$$

Q being a closed set in H.

In particular, if the control is the function of the thickness of the plate (shell), then $H = W_p^1(\Omega)$, with $p > 2$. In this case, the spaces H_n satisfying the conditions (5.2.33) and (5.2.34) can be constructed either as tensor products of one-dimensional spline spaces, or by using the finite element method. From the point of view of numerical realization, the spaces

$$H_n = \mathrm{Sp}^{(1)}(\Delta_{1n}, [a_1, b_1]) \otimes \mathrm{Sp}^{(1)}(\Delta_{2n}, [a_2, b_2]) \tag{5.2.36}$$

are the most convenient if $\overline{\Omega} = [a_1, b_1] \times [a_2, b_2]$ (for the notations, see subsecs. 5.2.1 and 5.2.2). If $\overline{\Omega}$ is a polygon, it is convenient to divide it into triangles and to choose H_n as the space of continuous functions on $\overline{\Omega}$ that are affine in each triangle, that is, to construct H_n on the basis of Courant triangles (see, e.g., Ciarlet (1978)).

However, if $\overline{\Omega}$ is a non-rectangular domain and

$$\overline{\Omega} \subset [a_1, b_1] \times [a_2, b_2],$$

then H_n can be defined as spaces of restrictions on $\overline{\Omega}$ of functions from the space defined by (5.2.36).

Let Δ_{1n} and Δ_{2n} be uniform partitions, i.e., in the relations (5.2.7)

$$x_{in} = a_1 + q_1 i, \qquad y_{jn} = a_2 + q_2 j, \qquad i = 0, 1, \dots, N_n, \ j = 0, 1, \dots, M_n,$$

where

$$q_1 = \frac{b_1 - a_1}{N_n}, \qquad q_2 = \frac{b_2 - a_2}{M_n}. \tag{5.2.37}$$

Then, in the space H_n determined by (5.2.36), a basis is formed by the following fundamental splines

$$B_{ij}(x, y) = \begin{cases} \left(1 - \left| i - \dfrac{x - a_1}{q_1} \right| \right) \left(1 - \left| j - \dfrac{y - a_2}{q_2} \right| \right), & \text{if } (x, y) \in R_{ij}, \\ 0, & \text{if } (x, y) \notin R_{ij}, \end{cases} \tag{5.2.38}$$

where

$$R_{ij} = \{\, (x, y) \mid (i - 1)q_1 \le x - a_1 \le (i + 1)q_1, \ (j - 1)q_2 \le y - a_2 \le (j + 1)q_2 \,\}$$
$$i = 0, 1, \dots, N_n; \quad j = 0, 1, \dots, M_n. \tag{5.2.39}$$

For a cylindric shell, when the control is the function of the thickness that varies only along the axial coordinate, the set of controls has the form $H = W_p^1((0, L))$, L being the length of the shell. The spaces H_n can be chosen as follows

$$H_n = \mathrm{Sp}^{(1)}(\Delta_n, [0, L]), \tag{5.2.40}$$

where

$$\Delta_n : 0 = z_0 < z_1 < \cdots < z_n = L. \tag{5.2.41}$$

If $\pi_n = \max\limits_{0 \le i \le n-1} (z_{i+1} - z_i)$ and $\pi_n \to 0$ as $n \to \infty$, the conditions (5.2.33) and (5.2.34) hold.

An arbitrary function f from the space H_n determined by the formulas (5.2.40), (5.2.41), has the form $f(z) = \sum_{i=0}^{n} f_i \varphi_i(z)$, where $f_i = f(z_i)$ and φ_i are the functions of the form

$$\varphi_i(z) = \begin{cases} \dfrac{z - z_{i-1}}{z_i - z_{i-1}}, & \text{if } z \in [z_{i-1}, z_i], \\ \dfrac{z_{i+1} - z}{z_{i+1} - z_i}, & \text{if } z \in [z_i, z_{i+1}], \\ 0, & \text{if } z \notin [z_{i-1}, z_{i+1}], \qquad i = 1, 2, \dots, n-1, \end{cases}$$

$$\varphi_0(z) = \begin{cases} \dfrac{z_1 - z}{z_1 - z_0}, & \text{if } z \in [z_0, z_1], \\ 0, & \text{if } z \notin [z_0, z_1], \end{cases}$$

$$\varphi_n(z) = \begin{cases} \dfrac{z - z_{n-1}}{z_n - z_{n-1}} & \text{if } z \in [z_{n-1}, z_n], \\ 0, & \text{if } z \notin [z_{n-1}, z_n]. \end{cases} \tag{5.2.42}$$

After defining a basis in H_n, (5.2.35) becomes a problem of nonlinear programming. Methods for solving such problems were set forth and studied by many authors; see, e.g., Céa (1971), Himmelblau (1972), Minoux (1989). Below, we present some numerical solutions of optimization problems, in which as a control we consider the functions of the thickness (first problem) and the load (second problem). For the solution of corresponding problems of nonlinear programming, different methods were applied in order to reduce the problem with restrictions to one without restrictions: the method of penalty functions for the first problem; the method of duality and the method of augmented Lagrange multipliers for the second problem. For solution of minimization problems without restrictions, we used the conjugate gradient method as well as direct methods, which do not use derivatives.

5.3 Optimization problems for plates (control by the function of the thickness)

5.3.1 Optimization under restrictions on strength

Setting of the problem

Consider the problem of optimization with respect to the weight of a plate of variable thickness, which is subject to a distributed load. Treating the function of the thickness of the plate as a control, define a set of admissible controls

$$U_{\text{ad}} = \{ h \,|\, h \in W_p^1(\Omega), \; p > 2, \; \|h\|_{W_p^1(\Omega)} \le c, \\ \check{h} \le h \le \hat{h}, \; \Psi_k(h, u_h) \le 0, \; k = 1, 2, \dots, l \}. \tag{5.3.1}$$

Here, the positive numbers $\check{h}$ and $\hat{h}$ are the lower and upper restrictions on the thickness of the plate, u_h is the solution of the problem on the stress-strain state of the plate, i.e.,

$$u_h \in V, \qquad a_h(u_h, v) = (f, v), \qquad v \in V. \tag{5.3.2}$$

The load f is supposed to be given and fixed. The bilinear form a_h is determined by the relations (4.1.13) and (4.1.24) for isotropic and orthotropic plates, respectively, where D, D_i, and D_{12} are defined by the formulas (4.1.5) and (4.1.11), $\Psi_k(h, u_h)$ are the functionals of the restrictions on the strength which may be determined by the left-hand sides of the inequalities (5.1.9) or (5.1.8) (in the latter case $l = 1$ in (5.3.1)), and for plates $\alpha = x$, $\beta = y$, $A = B = 1$. These functionals satisfy the condition (2.2.20).

As a goal functional we take the weight of the plate, which is given by the relation

$$f_1(h) = \gamma \iint_\Omega h \, dx \, dy, \tag{5.3.3}$$

γ being the specific gravity of the material. Then, results of Subsec. 2.2.2 imply that, if the set U_{ad} is not empty, there exists a function h_0 such that

$$h_0 \in U_{\text{ad}}, \qquad f_1(h_0) = \inf_{h \in U_{\text{ad}}} f_1(h). \tag{5.3.4}$$

To get an approximate solution of the problem (5.3.4), one can use the technique of Sections 2.3 and 2.4.

Optimization of a square plate

As an example, let us consider the optimization problem for a square isotropic plate which is subject to uniformly distributed load. In this case,

$$\overline{\Omega} = \{\, (x, y) \mid -a \le x \le a, \ -a \le y \le a \,\}, \tag{5.3.5}$$

the bilinear form a_h is defined by (4.1.13), and

$$(f, v) = C \iint_\Omega v \, dx \, dy, \tag{5.3.6}$$

where the constant C stands for the intensity of the distributed load.

We examined two cases of the fastening of the plate, namely, clamping and supporting. For the clamp, the space V is of the form

$$V = \Big\{\, u \mid u \in W_2^2(\Omega),\ u\big|_{x=\pm a} = u\big|_{y=\pm a} = \frac{\partial u}{\partial x}\Big|_{x=\pm a} = \frac{\partial u}{\partial y}\Big|_{y=\pm a} = 0 \,\Big\}, \tag{5.3.7}$$

and for the support

$$V = \{\, u \mid u \in W_2^2(\Omega),\ u\big|_{x=\pm a} = u\big|_{y=\pm a} = 0 \,\}. \tag{5.3.8}$$

To construct approximate solutions of the problem (5.3.2), the spaces V_n were chosen in the form (5.2.8) for the clamp, and in the form (5.2.9) for the support. The symmetry allowed us to use the product of one-dimensional splines which are symmetric with respect to 0, i.e., we applied spaces of bicubic splines symmetric with respect to the x and y axes.

To approximate the space $W_p^1(\Omega)$, we used the biaffine spline spaces H_n of the form (5.2.36). Because of the symmetry in x and y, H_n were also constructed as tensor products of spaces of splines symmetric with respect to 0.

The restrictions on the strength were taken on the basis of the Norris strength criterion:

$$\Psi_k(h, u_h) = \tilde{\sigma}_{xx}^2(h, (x,y)_k) + \tilde{\sigma}_{yy}^2(h, (x,y)_k) + \frac{\sigma_b^2}{\tau^2}\,\tilde{\sigma}_{xy}^2(h, (x,y)_k) - \sigma_b^2, \tag{5.3.9}$$
$$k = 1, 2, \dots, l.$$

Here, σ_b and τ are the tension and shear strength of the material of the plate, $\{(x,y)_k\}_{k=1}^l$ is a given set of points from $\overline{\Omega}$. Further, $\tilde{\sigma}_{xx}^2(h, (x,y)_k)$, $\tilde{\sigma}_{yy}^2(h, (x,y)_k)$, and $\tilde{\sigma}_{xy}^2(h, (x,y)_k)$ are the mean squared stresses on the area element $\Delta_k \subset \overline{\Omega}$ containing the point $(x,y)_k$:

$$\tilde{\sigma}_{xx}^2(h, (x,y)_k) = \frac{1}{S(\Delta_k)} \iint_{\Delta_k} \left[\frac{Eh}{2(1-\mu^2)}\left(\frac{\partial^2 u_h}{\partial x^2} + \mu\,\frac{\partial^2 u_h}{\partial y^2}\right)\right]^2 dx\,dy,$$
$$\tilde{\sigma}_{yy}^2(h, (x,y)_k) = \frac{1}{S(\Delta_k)} \iint_{\Delta_k} \left[\frac{Eh}{2(1-\mu^2)}\left(\frac{\partial^2 u_h}{\partial y^2} + \mu\,\frac{\partial^2 u_h}{\partial x^2}\right)\right]^2 dx\,dy,$$
$$\tilde{\sigma}_{xy}^2(h, (x,y)_k) = \frac{1}{S(\Delta_k)} \iint_{\Delta_k} \left[\frac{Eh}{2(1+\mu)}\,\frac{\partial^2 u_h}{\partial x \partial y}\right]^2 dx\,dy, \tag{5.3.10}$$

where $S(\Delta_k)$ is the area of Δ_k; E and μ are the elasticity modulus and Poisson's ratio of the material of the plate.

The finite-dimensional optimization problem corresponding to (5.3.4) consists in finding a function h_n such that

$$h_n \in H_n \cap U_{\mathrm{ad}}, \qquad f_1(h_n) = \inf_{h \in H_n \cap U_{\mathrm{ad}}} f_1(h), \tag{5.3.11}$$

f_1 being defined by the formula (5.3.3).

The problem (5.3.11) was solved by the method of penalty functions. So, we constructed a sequence $\{h_{ni}\}_{i=1}^\infty$ such that

$$h_{ni} \in H_n, \qquad f_{1i}(h_{ni}) = \inf_{h \in H_n} f_{1i}(h). \tag{5.3.12}$$

Here, f_{1i} is the augmented functional,

$$f_{1i}(h) = f_1(h) + \frac{1}{r_i}\,p(h), \tag{5.3.13}$$

where $\frac{1}{r_i}$ is the penalty coefficient, $r_i > 0$, $r_i \to 0$ as $i \to \infty$, and $p(h)$ is the penalty function which has the form

$$p(h)=c_1\left(\|h\|_{W_p^1(\Omega)}-c\right)_+^2+c_2\sum_{k=1}^{l}(\Psi_k(h,u_h))_+^2+c_3\sum_{i,j}(\check{h}-h_{ij})_+^2+c_4\sum_{i,j}(h_{ij}-\hat{h})_+^2, \tag{5.3.14}$$

where

$$(e)_+ = \begin{cases} 0, & \text{if } e < 0 \\ e, & \text{if } e \geq 0 \end{cases},$$

c_i are positive numbers, and $h_{ij} = h(x_{in}, y_{jn})$, i.e., h_{ij} are the values of the function $h(x, y)$ at the corresponding knots of the net.

It can be shown that from the sequence $\{h_{ni}\}_{i=1}^{\infty}$ one can choose a subsequence $\{h_{nj}\}_{j=1}^{\infty}$ such that $h_{nj} \to h_n$ as $j \to \infty$, where h_n is a solution of the problem (5.3.11). We would like to stress that, in optimization problems for plates and shells, the set U_{ad} is often defined inaccurately. For example, the constants $\check{h}$, $\hat{h}$, and c from (5.3.1) can take values from some domain. The same situation can be with the functionals $\Psi_k(h, u_h)$, for instance, in (5.3.9) σ_b and τ can run over some region.

Therefore, there is no need to solve the problem (5.3.12) with very small r_i. So, the main shortcoming of the method of penalty functions – the difficulty of the solution of the problem of minimization of the augmented functional under a great penalty coefficient – does not arise in the solution of many optimization problems for plates and shells.

The functionals $h \to \Psi_k(h, u_h)$ are not convex, so there exists a whole set of local minima of the problems (5.3.11) and (5.3.12). (Here, by a local minimum of the problem (5.3.12) we mean a function $\tilde{h}$ such that $f_{1i}(\tilde{h}) = \inf_{h\in Q} f_{1i}(h)$, Q being an open set in H_n containing the point $\tilde{h}$. Similarly, a function $\tilde{h}$ is a local minimum of the problem (5.3.11) if $f_1(\tilde{h}) = \inf_{h\in Q\cap U_{\text{ad}}} f_1(h)$.) In view of this, finding the global minimum, i.e., the solution of the problems (5.3.11) and (5.3.12), is usually a rather complicated task, so that one would be content with a "good local minimum."

To get the results cited below, the problem (5.3.12) was solved by the modified method of conjugate gradients, see Zangwill (1969). The local minima of the problem (5.3.12) were found for various initial approximations, the minimal one of them was determined to be an approximate solution of the problem (5.3.12).

Calculations were carried out for square plates (see (5.3.5)) for the following data: $E = 19.6 \cdot 10^{10}$ Pa, $\mu = 0.3$, $\sigma_b = 2.94 \cdot 10^8$ Pa, $\tau = 2.3 \cdot 10^8$ Pa, $a = 5$ cm, $\check{h} = 0.35$ cm, $\hat{h} = 1.5$ cm, $C = 9.8 \cdot 10^4$ Pa, C being the intensity of the load (see (5.3.6)), and the partitions of Ω for H_n were of the form

$$\Delta_{1n} : -a = x_{0n} < -\frac{a}{2} = x_{1n} < 0 = x_{2n} < \frac{a}{2} = x_{3n} < a = x_{4n},$$

$$\Delta_{2n} : -a = y_{0n} < -\frac{a}{2} = y_{1n} < 0 = y_{2n} < \frac{a}{2} = y_{3n} < a = y_{4n}. \tag{5.3.15}$$

In view of the symmetry with respect to x, y, and the diagonal of the square, the dimension of the biaffine spline space H_n corresponding to the partition (5.3.15) was equal to 6.

Below, in the table, are cited the relative values of the function of the thickness of the plate at the nodes for two essentially different optimal forms under clamped and supported edges.

Way of fastening	N_1	ξ_{22}	ξ_{23}	ξ_{24}	ξ_{33}	ξ_{34}	ξ_{44}	N_2
Clamp	I	0.35	0.97	0.61	0.72	1.00	0.70	17
	II	0.35	0.35	0.35	1.24	1.24	1.31	12
Support	I	1.35	0.49	0.78	0.95	1.25	1.06	27
	II	1.49	1.51	0.76	0.80	0.80	1.16	14

Here,

$$\xi_{ij} = \frac{\tilde{h}(x_{in}, y_{jn})}{h_0},$$

$\tilde{h}$ is the function of the thickness of the optimal plate, h_0 is the thickness of the plate with constant thickness in which stresses reach the limit values, i.e., $\max_k \Psi_k(h_0, u_{h_0}) = 0$; N_1 is the variant number, and N_2 is the gain in the weight (in percents) as compared to the plate of thickness h_0.

Figs. 5.3.1 and 5.3.2 show the distribution of thickness in the optimal plates which correspond to the variant I for the clamp (Fig. 5.3.1) and for the support (Fig. 5.3.2), because of the symmetry, only a quarter of each plate is displayed.

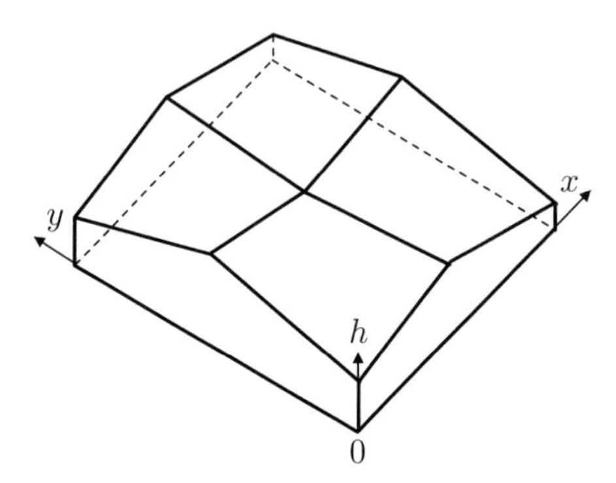

Figure 5.3.1: Distribution of thickness in the optimal plate for the clamp

Remark 5.3.1 As has been mentioned above, since the functionals $h \to \Psi_k(h, u_h)$ are not convex, there exists a whole set of the local minima of the problem (5.3.12). However, if the thickness of the plate varies within relatively small bounds, then

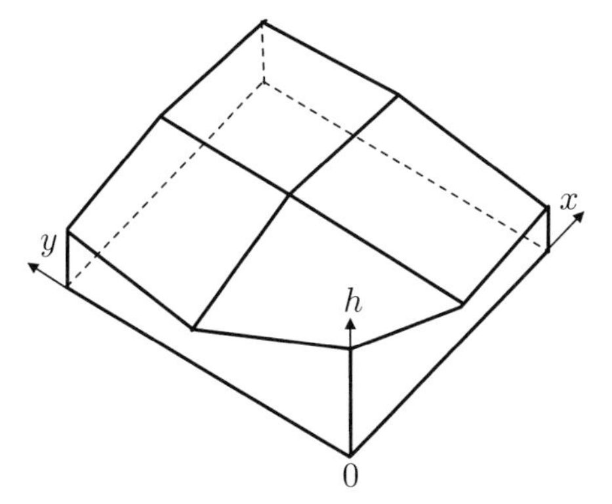

Figure 5.3.2: Distribution of thickness in the optimal plate for the support

optimization problems for such a plate can be reduced to the problem of minimization of a linear functional under linear restrictions, see Bratus (1981). In such a problem, every local minimum is the global one.

Remark 5.3.2 In the above example, the restriction $\|h\|_{W_p^1(\Omega)} \leq c$ was not used (see (5.3.1)), because for a finite-dimensional optimization problem it always holds provided c is large enough. However, if one solves the problem for various values of n, then in order to carry out the passage to the limit one has to take into account this restriction.

5.3.2 Stability optimization problem

Deformation of the plate in the (x, y) plane

Let Ω be a bounded domain occupied by the midplane of a plate, and let S be the boundary of Ω. Suppose that $S = S_1 \cup S_2 \cup S_3$, the sets S_i are disjoint, the plate is clamped on S_1, and on S_2 it is subject to a load λQ ($\lambda > 0$ being a parameter) that acts in the (x, y) plane (see Fig. 5.3.3). Thus,

$$Q = (Q_1, Q_2, Q_3), \qquad Q_1, Q_2 \in L_2(S_2), \ Q_3 = 0, \tag{5.3.16}$$

Q_1, Q_2 and Q_3 being the components of the vector function of load in the x, y, and z axes. Introduce a space V_1 by

$$V_1 = \{ g \,|\, g = (g_1, g_2), \ g \in \left(W_2^1(\Omega)\right)^2, \ g|_{S_1} = 0 \}. \tag{5.3.17}$$

V_1 being equipped with the topology of the space $\left(W_2^1(\Omega)\right)^2$ is a Banach space.

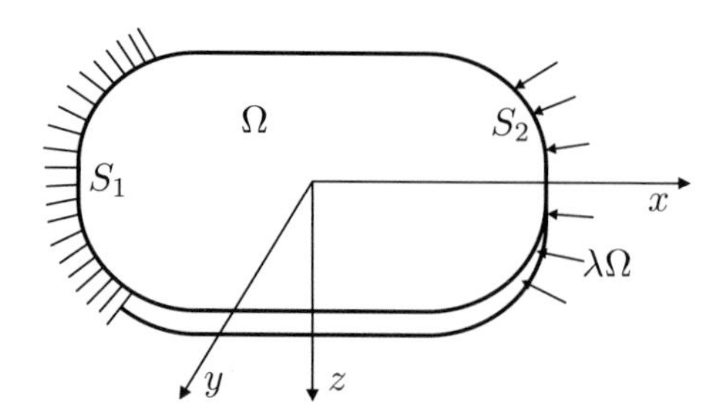

Figure 5.3.3: Plate at the buckling

The strain energy of the plate connected with deformations in the (x, y) plane generates the following bilinear form on $V_1 \times V_1$ (see Section 4.2)

$$a_h^{(1)}(q,g) = \iint_\Omega h\Big[E_{11}\frac{\partial q_1}{\partial x}\frac{\partial g_1}{\partial x} + E_{22}\frac{\partial q_2}{\partial y}\frac{\partial g_2}{\partial y} + E_{12}\left(\frac{\partial q_1}{\partial x}\frac{\partial g_2}{\partial y} + \frac{\partial q_2}{\partial y}\frac{\partial g_1}{\partial x}\right)$$

$$+ G\left(\frac{\partial q_1}{\partial y} + \frac{\partial q_2}{\partial x}\right)\left(\frac{\partial g_1}{\partial y} + \frac{\partial g_2}{\partial x}\right)\Big]\, dx\, dy, \qquad (5.3.18)$$

$$q = (q_1, q_2) \in V_1, \quad g = (g_1, g_2) \in V_1.$$

Here, h is a function of the thickness of the plate, E_{ij} and G are the elasticity constants:

$$E_{ii} = \frac{E_i}{1 - \mu_1\mu_2}, \qquad i = 1, 2, \qquad E_{12} = E_{21} = \mu_2 E_{11} = \mu_1 E_{22}. \qquad (5.3.19)$$

For all $h \in Y_p$, Y_p being defined by the relations (2.1.2) and (2.1.3), the bilinear form $a_h^{(1)}$ is symmetric, continuous, and coercive on $V_1 \times V_1$ (see Lemma 4.2.1), and hence because of (5.3.16) there exists a unique function q_h such that

$$q_h = (q_{h1}, q_{h2}) \in V_1, \qquad a_h^{(1)}(q_h, g) = \int_{S_2}(Q_1 g_1 + Q_2 g_2)\, ds, \qquad g \in V_1. \quad (5.3.20)$$

Here, q_{h1} and q_{h2} are the components of the vector function of displacements in the x and y axes of points of the midplane of the plate in the case $\lambda = 1$. These displacements cause the following forces acting in the midplane of the plate:

$$N_1(q_h) = E_{11}h\left(\frac{\partial q_{h1}}{\partial x} + \mu_2\frac{\partial q_{h2}}{\partial y}\right),$$

$$N_2(q_h) = E_{22}h\left(\frac{\partial q_{h2}}{\partial y} + \mu_1\frac{\partial q_{h1}}{\partial x}\right),$$

$$N_3(q_h) = Gh\left(\frac{\partial q_{h1}}{\partial y} + \frac{\partial q_{h2}}{\partial x}\right). \qquad (5.3.21)$$

Lemma 5.3.1 *Let the conditions* (4.1.25)–(4.1.27) *hold, and let a set* Y_p *be defined by the relations* (2.1.2) *and* (2.1.3), $p > 2$. *Assume that a form* $a_h^{(1)}$ *is determined by the formulas* (5.3.18) *and* (5.3.19), *and that* (5.3.16) *is valid. Then, the functions* $h \to N_i(q_h)$, $i = 1, 2, 3$, *defined by* (5.3.20) *and* (5.3.21) *are continuous mappings of* Y_p *equipped with the topology generated by the* $W_p^1(\Omega)$*-weak topology into* $L_2(\Omega)$.

Proof. Let $\{h_n\}_{n=1}^{\infty} \subset Y_p$ and $h_n \to h_0$ weakly in $W_p^1(\Omega)$. Then, by the embedding theorem, we have

$$h_n \to h_0 \qquad \text{in } C(\overline{\Omega}). \tag{5.3.22}$$

From here, just as it was done in the proof of Lemma 2.10.1, we conclude that

$$q_{h_n} \to q_{h_0} \qquad \text{in } V_1, \tag{5.3.23}$$

where q_{h_n} is the solution of the problem (5.3.20) for $h = h_n$, $n = 0, 1, 2, \ldots$. Now, (5.3.21)–(5.3.23) imply that

$$N_i\left(q_{h_n}\right) \to N_i\left(q_{h_0}\right) \qquad \text{in } L_2(\Omega) \text{ as } n \to \infty, \quad i = 1, 2, 3,$$

concluding the proof.

Direct stability problem

Let

$$V_2 = \left\{ f \mid f \in W_2^2(\Omega),\ f\big|_{S_1} = 0,\ \frac{\partial f}{\partial \nu}\Big|_{S_1} = 0 \right\}, \tag{5.3.24}$$

$\frac{\partial f}{\partial \nu}$ being the derivative with respect to the normal to S_1. The set V_2 being endowed with the norm of the space $W_2^2(\Omega)$ is a Banach space.

The bending energy of the plate generates the following bilinear form on $V_2 \times V_2$

$$\begin{aligned} a_h^{(2)}(w', w'') = \iint_{\Omega} \frac{h^3}{12}\Bigg[& E_{11} \frac{\partial^2 w'}{\partial x^2}\frac{\partial^2 w''}{\partial x^2} + E_{22}\frac{\partial^2 w'}{\partial y^2}\frac{\partial^2 w''}{\partial y^2} \\ + E_{12}\left(\frac{\partial^2 w'}{\partial x^2}\frac{\partial^2 w''}{\partial y^2} + \frac{\partial^2 w'}{\partial y^2}\frac{\partial^2 w''}{\partial x^2}\right) & + 4G\frac{\partial^2 w'}{\partial x \partial y}\frac{\partial^2 w''}{\partial x \partial y}\Bigg]\, dx\, dy, \qquad w', w'' \in V_2. \end{aligned} \tag{5.3.25}$$

In accordance with results of Section 4.2 (see Conclusion in Subsec. 4.2.5), one connects with the loss of stability of the plate the least positive eigenvalue of the following problem

$$\begin{gathered} (\lambda_{hi}, w_{hi}) \in \mathbb{R} \times V_2, \qquad w_{hi} \neq 0, \\ a_h^{(2)}(w_{hi}, w) + \lambda_{hi} b_{q_h}(w_{hi}, w) = 0, \qquad w \in V_2, \end{gathered} \tag{5.3.26}$$

b_{q_h} being the bilinear symmetric form on $V_2 \times V_2$ determined by the formula

$$b_{q_h}(w',w'') = \iint_{\Omega} \left[N_1(q_h) \frac{\partial w'}{\partial x} \frac{\partial w''}{\partial x} + N_2(q_h) \frac{\partial w'}{\partial y} \frac{\partial w''}{\partial y} \right.$$
$$\left. + N_3(q_h) \left(\frac{\partial w'}{\partial x} \frac{\partial w''}{\partial y} + \frac{\partial w'}{\partial y} \frac{\partial w''}{\partial x} \right) \right] dx\, dy, \qquad w', w'' \in V_2. \quad (5.3.27)$$

Taking into account the Hölder inequality, we have

$$|b_{q_h}(w',w'')| \leq 2 \sum_{i=1}^{3} \|N_i(q_h)\|_{L_2(\Omega)} \|w'\|_{W_4^1(\Omega)} \|w''\|_{W_4^1(\Omega)}, \qquad w', w'' \in V_1. \quad (5.3.28)$$

Thus, $b_{q_h} \in \mathcal{L}_2 \left(W_4^1(\Omega), W_4^1(\Omega); \mathbb{R} \right)$, i.e., b_{q_h} is a continuous form on $W_4^1(\Omega) \times W_4^1(\Omega)$. Moreover, (5.3.28) and Lemma 5.3.1 imply the following statement.

Lemma 5.3.2 *Let the hypotheses of Lemma* 5.3.1 *be satisfied and let a bilinear form* b_{q_h} *be defined by the formula* (5.3.27). *Then, for* $p > 2$, *the function* $h \to b_{q_h}$ *is a continuous mapping of* Y_p *equipped with the* $W_p^1(\Omega)$*-weak topology into* $\mathcal{L}_2 \left(W_4^1(\Omega), W_4^1(\Omega); \mathbb{R} \right)$.

Since the bilinear form $a_h^{(2)}$ is coercive in V_2 (see Theorem 4.1.2), zero cannot be an eigenvalue of the problem (5.3.26), so that we can consider the following problem, instead of (5.3.26),

$$\begin{gathered} (\mu_{hi}, w_{hi}) \in \mathbb{R} \times V_2, \qquad w_{hi} \neq 0, \\ \mu_{hi} a_h^{(2)} (w_{hi}, w) + b_{q_h} (w_{hi}, w) = 0, \qquad w \in V_2. \end{gathered} \quad (5.3.29)$$

If $\mu_{hi} \neq 0$, then

$$\mu_{hi} = \frac{1}{\lambda_{hi}}, \quad (5.3.30)$$

λ_{hi} being an eigenvalue of the problem (5.3.26). In this connection, we stress that, for some loads Q, one can get $\mu_{hi} = 0$, i.e., zero is an eigenvalue of the problem (5.3.29).

By virtue of Theorem 1.6.2, the embedding of V_2 into $W_4^1(\Omega)$ is compact. So, Theorem 1.5.8 yields that the problem (5.3.29) has a countable number of eigenvalues

$$|\mu_{h1}| \geq |\mu_{h2}| \geq \cdots, \qquad \lim_{i \to \infty} \mu_{hi} = 0. \quad (5.3.31)$$

Thus, we can define the function $h \to \mu(h) = \{\mu_{hi}\}_{i=1}^{\infty}$ which maps an element $h \in Y_p$ into $\mu_h \in \ell_{\infty,0}$. We recall that $\ell_{\infty,0}$ is the normed space of bounded number sequences converging to zero and

$$\|\mu_h\|_{\ell_{\infty,0}} = \sup_i |\mu_{hi}|. \quad (5.3.32)$$

Analogously to the proof of Lemma 2.5.1, we deduce $h \to a_h^{(2)}$ to be a continuous mapping of Y_p equipped with the $W_p^1(\Omega)$-weak topology into $\mathcal{L}_2(V_2, V_2; \mathbb{R})$ for $p > 2$. Combining this with Lemma 5.3.2 and making use of Theorem 2.5.2, we get the following statement.

Lemma 5.3.3 *Let the hypotheses of Lemma* 5.3.1 *be satisfied and let bilinear forms* $a_h^{(2)}$ *and* b_{q_h} *be defined by the formulas* (5.3.25) *and* (5.3.27). *Then, the function* $h \to \mu_h$ *is a continuous mapping of* Y_p *endowed with the* $W_p^1(\Omega)$*-weak topology into* $\ell_{\infty,0}$.

Optimization problem

Suppose we are given functionals $\Psi_k(h, \mu)$ such that

$$\left.\begin{array}{l} (h,\mu) \to \Psi_k(h,\mu) \text{ is a continuous mapping of } Y_p \times \ell_{\infty,0} \\ \text{equipped with the topology generated by the product of} \\ \text{the } W_p^1(\Omega)\text{-weak topology and } \ell_{\infty,0}\text{-strong one into } \mathbb{R}, \\ p > 2,\ k = 0,1,2,\ldots,m. \end{array}\right\} \tag{5.3.33}$$

Define a set of admissible controls as

$$U_{\text{ad}} = \{\, h \,|\, h \in W_p^1(\Omega),\ \|h\|_{W_p^1(\Omega)} \leq c,\ \check{h} \leq h \leq \hat{h},\ \Psi_k(h, \mu_h) \leq 0,\ k = 1,2,\ldots,m \,\}. \tag{5.3.34}$$

Here,

$$\left.\begin{array}{l} p > 2,\ c,\ \check{h}, \text{ and } \hat{h} \text{ are positive numbers such that} \\ e_1 < \check{h} < \hat{h} < e_2,\ e_1 \text{ and } e_2 \text{ are the positive constants from} \\ (2.1.2). \end{array}\right\} \tag{5.3.35}$$

The optimization problem consists in finding a function h_0 such that

$$h_0 \in U_{\text{ad}}, \qquad \Psi_0(h_0, \mu_{h_0}) = \inf_{h \in U_{\text{ad}}} \Psi_0(h, \mu_h). \tag{5.3.36}$$

Theorem 5.3.1 *Let the hypotheses of Lemma* 5.3.3 *be satisfied and let* (5.3.33) *hold. Assume that a nonempty set* U_{ad} *is defined by the relations* (5.3.34) *and* (5.3.35). *Then, the problem* (5.3.36) *has a solution.*

Proof. Let $\{h_n\}$ be a minimizing sequence, i.e.,

$$\{h_n\} \subset U_{\text{ad}}, \tag{5.3.37}$$

$$\lim_{n\to\infty} \Psi_0(h_n, \mu_{h_n}) = \inf_{h \in U_{\text{ad}}} \Psi_0(h, \mu_h). \tag{5.3.38}$$

Since the sequence $\{h_n\}$ is bounded in $W_p^1(\Omega)$, $p > 2$, we can choose a subsequence $\{h_m\}$ such that

$$h_m \to \tilde{h} \qquad \text{weakly in } W_p^1(\Omega), \tag{5.3.39}$$

$$h_m \to \tilde{h} \qquad \text{strongly in } C(\overline{\Omega}). \tag{5.3.40}$$

(5.3.39) and Lemma 5.3.3 yield that

$$\mu_{h_m} \to \mu_{\tilde{h}} \qquad \text{in } \ell_{\infty,0}. \tag{5.3.41}$$

Further, making use of (5.3.33) and (5.3.37)–(5.3.41), one can easily see that the function $h_0 = \tilde{h}$ is a solution of the problem (5.3.36).

An approximate solution of the problem (5.3.36) can be constructed by employing results of subsecs. 2.7.2 and 2.7.3.

Consider some realizations of the functionals Ψ_k. Introduce the notation

$$\mu_h^+ = \sup_i \mu_{hi}. \tag{5.3.42}$$

Assume that $\mu_h^+ > 0$. Then μ_h^+ is the largest positive eigenvalue of the problem (5.3.29) (see Theorem 1.5.5). The critical load under which the plate loses stability equals $\frac{Q}{\mu_h^+}$ (see Fig. 5.3.3). We point out that, if $\mu_h^+ = 0$, then the plate does not lose stability. More precisely, in this case the plate loses stability if $-Q$ is substituted for Q, i.e., the load has the opposite direction.

Suppose we need that the plate loses stability under a load that is not less than a given one. Then, the functional Ψ_1 can be taken in the following form

$$\Psi_1(h, \mu_h) = c - \frac{1}{\mu_h^+}, \tag{5.3.43}$$

c being a positive number. Of course, the plate must not be destroyed under the load cQ. Stresses in the midplane of the plate under the load cQ are given by the formulas

$$\sigma_{11,h} = \frac{cN_1(q_h)}{h}, \qquad \sigma_{22,h} = \frac{cN_2(q_h)}{h}, \qquad \sigma_{12,h} = \frac{cN_3(q_h)}{h}. \tag{5.3.44}$$

Here, $N_i(q_h)$ are determined by the relations (5.3.21), and q_h is a solution of the problem (5.3.20).

Let $\sigma_{ij,h}^{(\rho)}$ be the averaging of the stresses $\sigma_{ij,h}$ in the x and y coordinates, ρ being the radius of the averaging kernel. In view of (5.1.6), the functional Ψ_2 can be taken in the form

$$\begin{aligned} \Psi_2(h, \mu_h) = \max_{(x,y)\in\overline{\Omega}} \Bigg[& \sum_{i,j=1}^{2} \Pi_{ij}\sigma_{ij,h}^{(\rho)}(x,y) \\ & + \sum_{i,j,m,n=1}^{2} \Pi_{ijmn}\sigma_{ij,h}^{(\rho)}(x,y)\sigma_{mn,h}^{(\rho)}(x,y)\Bigg] - 1. \end{aligned} \tag{5.3.45}$$

Then, the condition $\Psi_2(h, \mu_h) \le 0$ means that the plate whose thickness is defined by the function h will not be destroyed under the load cQ.

Note that in some cases the restriction on the strength is not introduced in the formulation of the problem. Then, one has to verify whether the solution obtained satisfies this restriction.

Now, provided

$$\Psi_0(h, \mu_h) = \iint_\Omega h\, dx\, dy, \tag{5.3.46}$$

the problem (5.3.36) means that one searches for a plate of minimal weight which loses stability under a load that is not less than a given one, and meets the restriction on strength.

If we set

$$\Psi_0(h, \mu_h) = \mu_h^+, \qquad \Psi_1(h, \mu_h) = \iint_\Omega h\, dx\, dy - c_1, \tag{5.3.47}$$

c_1 being a positive constant, and assume Ψ_2 to be defined by the formula (5.3.45) in which $\sigma_{ij,h}^{(\rho)}$ is generated by the load $\frac{Q}{\mu_h^+}$, then the problem (5.3.36) corresponds to the maximization of the load that causes the plate to lose stability under the restrictions on the weight and strength.

5.3.3 Optimization of frequencies of free oscillations

In accordance with results of Subsec. 4.1.5, determination of the frequencies of free oscillations reduces to the following problem

$$\begin{gathered}(\mu_{ih}, u_{ih}) \in \mathbb{R} \times V, \\ \mu_{ih} a_h (u_{ih}, v) = b_h (u_{ih}, v), \qquad v \in V,\end{gathered} \tag{5.3.48}$$

where the bilinear forms a_h and b_h are defined by the formulas (4.1.24) and (4.1.53), and $\mu_{ih} = \frac{1}{\lambda_i}$ (see (4.1.55)). By virtue of Theorem 4.1.3, to every function $h \in Y_p$ there corresponds the element $\mu_h = (\mu_{1h}, \mu_{2h}, \dots) \in \ell_{\infty,0}$.

Lemma 5.3.4 *Let the hypotheses of Theorem* 4.1.3 *be fulfilled. Then, for* $p > 2$, *the function* $h \to \mu_h = (\mu_{1h}, \mu_{2h}, \dots)$ *is a continuous mapping of* Y_p *endowed with the* $W_p^1(\Omega)$*-weak topology into* $\ell_{\infty,0}$.

Proof. Similarly to the proof of Lemma 2.5.1, one finds that, for $p > 2$, the function $h \to a_h$ is a continuous mapping of Y_p endowed with the $W_p^1(\Omega)$-weak topology into $\mathcal{L}_2(V, V; \mathbb{R})$, and the function $h \to b_h$ is a continuous mapping of Y_p endowed with the $W_p^1(\Omega)$-weak topology into $\mathcal{L}_2(W_2^1(\Omega), W_2^1(\Omega); \mathbb{R})$. Now, Lemma 5.3.4 is a consequence of Theorem 2.5.2.

The problem of optimization of frequencies of free oscillations reduces to the problem (5.3.36) in which μ_h is a solution of the problem (5.3.48). By using Lemma 5.3.4, analogously to the proof of Theorem 5.3.1, one gets the following statement.

Theorem 5.3.2 *Let the hypotheses of Theorem* 4.1.3 *be fulfilled and the functionals* Ψ_k, $k = 0, 1, \ldots, m$, *be given and satisfy the condition* (5.3.33). *Suppose that a nonempty set* U_{ad} *is determined by* (5.3.34) *and* (5.3.35), μ_h *being defined by the solution of the problem* (5.3.48). *Then, the problem* (5.3.36) *has a solution.*

We point out that the problem of optimization of frequencies of free oscillations can correspond to the minimization of the weight of the plate under restrictions on the spectrum, as well as to the maximization of the first natural frequency λ_{1h} (i.e., the minimization of $\mu_{1h} = \frac{1}{\lambda_{1h}}$) under restrictions on the weight and spectrum. Restrictions on the spectrum can mean that the frequencies of free oscillations of the plate do not occur in some intervals, which are determined by the dynamic load. Some examples of construction of the functionals Ψ_k were presented above, in subsecs. 5.3.2 and 2.6.1.

5.3.4 Combined optimization problem and optimization for a class of loads

Assume that, in different periods of time, a plate is subject to longitudinal and transversal forces as well as certain periodic (in time) forces. In view of this, the plate should satisfy a system of restrictions on the strength, stability, and frequency of free oscillations. The goal functional can correspond to the weight of the plate, which is to be minimized. The corresponding optimization problem reduces to a combined problem (see Section 2.9). Combined problems for beams and plates were examined by Seyranian (1973, 1976, 1977). Using results of subsecs. 5.3.1–5.3.3 and Section 2.9, one can derive the existence of a solution of a combined optimization problem for plates and the convergence of the approximate solutions of this problem which are constructed by using the technique of Subsec. 2.8.2.

Taking regard of results of Section 2.10, one can optimize plates for a class of loads. Notice that problems of optimization of beams and plates for a class of loads were studied by Banichuk (1980).

5.4 Optimization problems for shells (control by functions of midsurface and thickness)

"I rode over to see her once every week for a while; and then I figured it out that if I doubled the number of trips I would see her twice as often"

– *O. Henry*
"The Pimienta Pancakes"

For optimization of shells, one can take as controls the function of the thickness of the shell and the function determining the midsurface (see (4.5.1)); for shallow shells, the control can be the shape of the plan of the shell, i.e., the shape of the projection of the midsurface of the shell on the (x, y) plane (see Fig. 4.7.1). Naturally, other kinds of controls (e.g., the optimal stiffening of the shell) are also applicable. However, in this section we study control by the shape of the midsurface and the thickness only, while in Section 5.5 we investigate the problem of control by the shape of the plan of a shallow shell.

5.4.1 Problem of optimization of a shell of revolution with respect to strength

Direct problem and continuous dependence

According to results of Sections 4.6 and 4.8, the problem on the stress-strain state of a shell of revolution reduces to finding a function of displacements $\tilde{\omega} = (\tilde{u}, \tilde{v}, \tilde{w})$ such that

$$\tilde{\omega} \in V, \qquad a_h(\tilde{\omega}, \omega) = (f, \omega), \qquad \omega \in V, \tag{5.4.1}$$

f being an element from V^* determined by the load acting on the shell, and the bilinear form a_h being defined by the formula (4.6.13).

The midsurface of the shell of revolution is determined by a function $r(z)$ which is called the meridian function (see Fig. 4.6.1). To study the optimization problem, we need the property of continuous dependence of a solution of the problem (5.4.1) of the function of the thickness of the shell $h(\varphi, z)$, of the meridian function $r(z)$, and of the load f. Therefore, let us proceed to study this property.

Define sets

$$G_1 = \{ h \mid h \in C(\overline{\Omega}),\ \check{h} \le h \le \hat{h},\ h|_{\varphi=0} = h|_{\varphi=2\pi} \}, \tag{5.4.2}$$

$$G_2 = \{ r \mid r \in C^3([0, L]),\ \check{r} \le r \le \hat{r} \}, \tag{5.4.3}$$

where

$$\check{h}, \hat{h}, \check{r}, \hat{r} \text{ are positive numbers, } \check{h} < \hat{h},\ \check{r} < \hat{r}, \tag{5.4.4}$$

and Ω is defined by (4.6.1). The sets G_1 and G_2 are equipped with the topologies generated by the topologies of the spaces $C(\overline{\Omega})$ and $C^3([0, L])$, respectively.

In view of (4.6.3)–(4.6.5), (4.6.10), and (4.6.13), the bilinear form a_h depends not only on h, but also on the function r, so that we will denote it as a_{hr}, i.e.,

$$a_h = a_{hr}. \tag{5.4.5}$$

The solution of the problem (5.4.1) depends obviously on h, r, and the load f. Hence, by (5.4.5), the problem (5.4.1) can be rewritten as follows:

$$\omega_{hrf} \in V, \qquad a_{hr}(\omega_{hrf}, \omega) = (f, \omega), \qquad \omega \in V. \tag{5.4.6}$$

Lemma 5.4.1 *Let the conditions* (4.3.6)–(4.3.8) *be satisfied, let the space V be endowed with the norm* (4.6.12) *and satisfy* (4.6.11). *Suppose that sets G_1 and G_2*

are defined by the formulas (5.4.2)–(5.4.4). *Let also* (5.4.5) *hold and let a bilinear form* a_h *be defined by the relations* (4.5.10), (4.6.3)–(4.6.5), (4.6.10), *and* (4.6.13). *Then, for any* $(h, r, f) \in G_1 \times G_2 \times V^*$, *the problem* (5.4.6) *has a unique solution* ω_{hrf}, *and the function* $(h, r, f) \to \omega_{hrf}$ *defined by this solution is a continuous mapping of* $G_1 \times G_2 \times V^*$ *into* V.

Proof. Theorem 4.6.1 implies that, for all $(h, r) \in G_1 \times G_2$, the bilinear form a_{hr} is symmetric, continuous, and coercive. Thus, by virtue of the Riesz theorem, for any $(h, r, f) \in G_1 \times G_2 \times V^*$, there exists a unique solution to the problem (5.4.6). To prove the continuity of the function $(h, r, f) \to \omega_{hrf}$, we use Theorem 2.11.1. The operators P_i from (4.6.13) determined by the formulas (4.6.3)–(4.6.5) and (4.6.10) depend obviously on r, i.e., we can write $P_i = P_i^{(r)}$. It is easy to verify that $r \to P_i^{(r)}$ is a continuous mapping of G_2 into $\mathcal{L}(W, L_2(\Omega))$, the space W being defined by (4.6.8).

The formulas (4.5.10) and (4.6.13) yield that c_{ij} and D_{lm} depend on h, i.e., $c_{ij} = c_{ij}^{(h)}$ and $D_{lm} = D_{lm}^{(h)}$. Obviously, $h \to c_{ij}^{(h)}$ and $h \to D_{lm}^{(h)}$ are continuous mappings of G_1 into $C(\overline{\Omega})$. Now, Theorem 2.11.1 yields $(h, r, f) \to \omega_{hrf}$ to be a continuous mapping of $G_1 \times G_2 \times V^*$ into V. (In this case, we set $B_1 = C(\overline{\Omega}) \times C^3([0, L])$ and $Q_1 = G_1 \times G_2$.)

Optimization problem

We will control by the function of the thickness of a shell of revolution $h(\varphi, z)$ and by the shape of its midsurface, i.e., the meridian function $r(z)$. To every pair of functions $(h, r) \in G_1 \times G_2$ we place in correspondence a load f_{hr} acting on the shell, i.e., we suppose that we are given a function $(h, r) \to f_{hr}$ such that

$$(h, r) \to f_{hr} \text{ is a continuous mapping of } G_1 \times G_2 \text{ into } V^*. \tag{5.4.7}$$

Now, by (5.4.6), the function of displacements of the midsurface of the shell $\omega_{hr} = \omega_{hrf_{hr}}$ is determined as the solution of the problem

$$\omega_{hr} \in V, \qquad a_{hr}(\omega_{hr}, \omega) = (f_{hr}, \omega), \qquad \omega \in V. \tag{5.4.8}$$

Further, suppose we are given functionals $\Psi_k(h, r, \omega)$ such that

$$\left. \begin{array}{l} (h, r, \omega) \to \Psi_k(h, r, \omega) \text{ is a continuous mapping of} \\ G_1 \times G_2 \times V \text{ into } \mathbb{R},\ k = 0, 1, 2, \ldots, m. \end{array} \right\} \tag{5.4.9}$$

Define a set of admissible controls as

$$\begin{aligned} U_{\text{ad}} = \Big\{ (h, r) \,|\, h \in W_p^1(\Omega),\ \|h\|_{W_p^1(\Omega)} \le c_h,\ \check{h} \le h \le \hat{h}, \\ h\big|_{\varphi=0} = h\big|_{\varphi=2\pi},\ r \in W_2^4((0, L)),\ \|r\|_{W_2^4((0,L))} \le c_r, \\ \check{r} \le r \le \hat{r},\ \Psi_k(h, r, \omega_{hr}) \le 0,\ k = 1, 2, \ldots, m \Big\}. \end{aligned} \tag{5.4.10}$$

Here, ω_{hr} is the solution of the problem (5.4.8) and

$$\left.\begin{array}{l}p>2,\ c_h \text{ and } c_r \text{ are positive numbers, } \check{h},\ \hat{h},\ \check{r}, \text{ and } \hat{r} \text{ are}\\ \text{the constants from (5.4.2) and (5.4.3) satisfying the}\\ \text{conditions (5.4.4).}\end{array}\right\} \tag{5.4.11}$$

The functionals Ψ_k in (5.4.10) can be considered as restrictions on the strength and stiffness of the shell (see Subsec. 5.1.2). The optimization problem consists in finding a pair (h_0, r_0) such that

$$(h_0, r_0) \in U_{\text{ad}}, \qquad \Psi_0(h_0, r_0, \omega_{h_0 r_0}) = \inf_{(h,r)\in U_{\text{ad}}} \Psi_0(h, r, \omega_{hr}). \tag{5.4.12}$$

Theorem 5.4.1 *Let the hypotheses of Lemma* 5.4.1 *be fulfilled and let the relations* (5.4.7), (5.4.9) *hold. Assume that a nonempty set* U_{ad} *is defined by the formulas* (5.4.10) *and* (5.4.11). *Then, the problem* (5.4.12) *has a solution.*

Proof. Let $\{h_n, r_n\}_{n=1}^{\infty}$ be a minimizing sequence, i.e.,

$$\{h_n, r_n\}_{n=1}^{\infty} \subset U_{\text{ad}}, \tag{5.4.13}$$

$$\lim_{n\to\infty} \Psi_0(h_n, r_n, \omega_{h_n r_n}) = \inf_{(h,r)\in U_{\text{ad}}} \Psi_0(h, r, \omega_{hr}). \tag{5.4.14}$$

Since the set U_{ad} is bounded in $W_p^1(\Omega) \times W_2^4((0, L))$, the embedding theorem implies that one can choose a subsequence $\{h_m, r_m\}_{m=1}^{\infty}$ of the sequence $\{h_n, r_n\}_{n=1}^{\infty}$ such that

$$\begin{aligned} &h_m \to \tilde{h} \quad \text{weakly in } W_p^1(\Omega), &&h_m \to \tilde{h} \quad \text{strongly in } C(\overline{\Omega}),\\ &r_m \to \tilde{r} \quad \text{weakly in } W_2^4((0, L)), &&r_m \to \tilde{r} \quad \text{strongly in } C^3([0, L]). \end{aligned} \tag{5.4.15}$$

Obviously, $\tilde{h} \in G_1$ and $\tilde{r} \in G_2$. Now, making use of Lemma 5.4.1, (5.4.7), (5.4.9), and (5.4.13)–(5.4.15), it is easy to see that the pair $(h_0 = \tilde{h}, r_0 = \tilde{r})$ is a solution of the problem (5.4.12). The theorem is proven.

Approximate solutions of the problem (5.4.12) can be constructed by using the technique of Subsec. 2.11.3. Just as above, due to results of Section 2.11, one can investigate and solve eigenvalue optimization problems which are connected with the stability and oscillations of the shell, as well as combined problems in the case when the control is the function of the thickness of the shell and the shape of the midsurface.

Optimization problems on the strength and rigidity of cylindrical shells of variable thickness that are subject to an axially symmetric load were studied by Medvedev and Totskii (1984b).

Now, let us consider an optimization problem on the stability of a cylindrical shell.

5.4.2 Optimization according to the stability of a cylindrical shell subject to a hydrostatic compressive load

Direct problem

Let a cylindrical shell be subject to a load that is uniformly distributed on the surface with intensity $-\lambda$. According to results of Subsec. 4.10.2, the problem of stability of the shell reduces to determining the least positive eigenvalue of the following problem

$$\lambda \in \mathbb{R},\ \tilde{\omega} \in V,\ \tilde{\omega} \neq 0, \qquad a_h(\tilde{\omega}, \omega) = \lambda b_{h,\hat{\omega}}(\tilde{\omega}, \omega), \quad \omega \in V. \tag{5.4.16}$$

Here, the form a_h is defined by the formulas (4.5.10), (4.6.3)–(4.6.5), (4.6.10), and (4.6.13) with $r = \text{const} = R$, where R is the radius of the shell. If the thickness of the shell is small as compared to its radius, the bilinear form $b_{h,\hat{\omega}}$ can be defined in the form (4.10.15), and the forces T_1, T_2, and S can be found from the membrane theory of shells. From this theory it follows that, in the case under investigation, $T_1 = S = 0$ and $T_2 = -R$. Then, the form $b_{h,\hat{\omega}}$ takes the form

$$b_{h,\hat{\omega}}(\omega', \omega'') = \int_0^L \int_0^{2\pi} \frac{\partial w'}{\partial \varphi} \frac{\partial w''}{\partial \varphi}\, dz\, d\varphi, \tag{5.4.17}$$

where $\omega' = (u', v', w')$ and $\omega'' = (u'', v'', w'')$.

Optimization problem

We will treat the function of the thickness of the shell h as a control and suppose h to depend only on z. Therefore, we take a set of admissible controls in the form

$$U_{\text{ad}} = \Big\{ h \,|\, h \in W_2^1((0,L)),\ \|h\|_{W_2^1((0,L))} \leq c_h,$$
$$\check{h} \leq h \leq \hat{h},\ \int_0^L h\, dz \leq c,\ f(h) \leq 0 \Big\}. \tag{5.4.18}$$

Here, f is the function of restrictions on the strength. Let λ_{1h} be the least positive eigenvalue of the problem (5.4.16), and let the goal functional f_1 have the form

$$f_1(h) = -\lambda_{1h}. \tag{5.4.19}$$

The optimization problem consists in determining a function h_0 such that

$$h_0 \in U_{\text{ad}}, \qquad f_1(h_0) = \inf_{h \in U_{\text{ad}}} f_1(h). \tag{5.4.20}$$

Since the embedding of $W_2^1((0,L))$ into $C([0,L])$ is compact, if the set U_{ad} is nonempty, then the problem (5.4.20) has a solution.

In the work by Medvedev (1980), the problem of optimization of an orthotropic cylindrical shell clamped at the edges and subject to a hydrostatic compressive load was numerically solved. The parameters of the shell were $L = 3\,\mathrm{m}$, $R = 0.75\,\mathrm{m}$, $\mu_1 = \mu_2 = 0.11$, $E_1 = E_2 = 3.7 \cdot 10^{10}\,\mathrm{Pa}$, and $G = 0.75 \cdot 10^{10}\,\mathrm{Pa}$. To construct an approximate solution of the direct problem, i.e., to calculate λ_{1h}, the spaces V_n were constructed as tensor products of splines and trigonometric polynomials (see Subsec. 5.2.3). In solution of the finite-dimensional optimization problem, the function of the thickness of the shell $z \to h(z)$ was searched for in the class of piecewise-affine functions which are continuous on $[0, L]$ and symmetric with respect to the point $z = \frac{L}{2}$. More precisely, optimal forms were searched for in the class of the functions pictured on Fig. 5.4.1 by a continuous line. The point line corresponds to the shell of a constant thickness. The values q_1 and q_2 shown in Fig. 5.4.1 were used as optimization parameters; q_3 was chosen so that the weights (volumes) of the shells of constant and variable thickness are equal, the angle α being considered to be fixed (see Fig. 5.4.1).

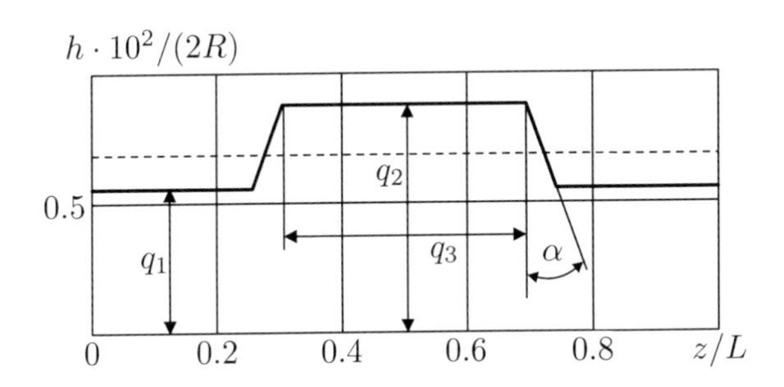

Figure 5.4.1: Function of thickness of a cylindrical shell; q_1 and q_2 are parameters of optimization

Let $q = (q_1, q_2)$. Obviously, in the case under investigation, the eigenvalues of the problem (5.4.16) depend on q. Hence, we denote them as $\lambda_i(q)$ and order them in such a way that

$$0 < \lambda_1(q) \leq \lambda_2(q) \leq \cdots . \tag{5.4.21}$$

The problem of maximization of the function $q \to \lambda_1(q)$ on the set $U_{\mathrm{ad},\check{h}}$ of the form

$$U_{\mathrm{ad},\check{h}} = \left\{ q = (q_1, q_2) \,|\, q \in \mathbb{R}^2,\ q_2 \geq q_1 \geq \check{h} \right\} \tag{5.4.22}$$

was solved. (The set of admissible controls U_{ad} depends here on $\check{h}$, so that we denoted it as $U_{\mathrm{ad},\check{h}}$.) The upper restriction on the thickness $\hat{h}$ is not included in $U_{\mathrm{ad},\check{h}}$, since $\hat{h}$ is supposed to be chosen sufficiently large, so that the optimal solution always satisfies it. We denote by $q_{\check{h}}$ a solution of the problem

$$q_{\check{h}} \in U_{\mathrm{ad},\check{h}}, \qquad \lambda_1\left(q_{\check{h}}\right) = \sup_{q \in U_{\mathrm{ad},\check{h}}} \lambda_1(q). \tag{5.4.23}$$

Fig. 5.4.2 shows the dependence $\check{h} \to \lambda_i(q_{\check{h}})$, where $\lambda_i(q_{\check{h}})$ are the eigenvalues appropriate to $q_{\check{h}}$ (see (5.4.21)). Lines 1–5 on Fig. 5.4.2 determine the values $\lambda_1(q_{\check{h}}), \ldots, \lambda_5(q_{\check{h}})$. The value $\check{h} = 10^{-2}$ m corresponds to a shell of a constant thickness. For this shell, as Fig. 5.4.2 shows, the first eigenvalue is of simple multiplicity. The decrease of $\check{h}$ causes the increase of $\lambda_1(q_{\check{h}})$ and the multiplicity of this eigenvalue.

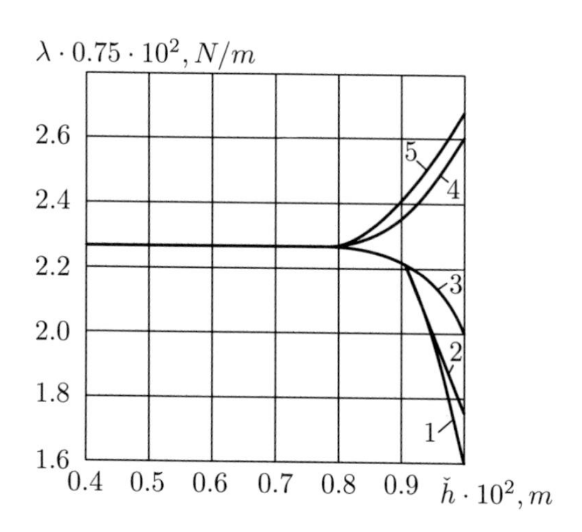

Figure 5.4.2: Eigenvalues of the optimal shells as functions of the parameter $\check{h}$

Thus, for example, for $\check{h} = 0.75 \cdot 10^{-2}$ m the multiplicity of $\lambda_1(q_{\check{h}})$ equals 5, and for the smaller values of $\check{h}$ the value of $\lambda_1(q_{\check{h}})$ is almost the same and its multiplicity does not change.

We point out that the results obtained are in agreement with the engineering approach to optimization of constructions, which demands the optimal construction to have "general" and "local" forms of loss of stability; the latter corresponds to the requirement that the multiplicity of the first eigenvalue must be greater than one.

Medvedev and Totskii (1984a) examined the problem of maximization of the first eigenvalue of the problem (5.4.16) on the set

$$U_{\mathrm{ad}} = \Big\{ h \,|\, h \in \mathrm{Sp}^{(1)}(\Delta_{38}, [0, L]), \\ h\Big(\frac{L}{2} - y\Big) = h\Big(\frac{L}{2} + y\Big) \text{ for } y \in [0, L/2],\ \check{h} \le h \le \hat{h} \Big\} \quad (5.4.24)$$

provided $b_{h,\hat{\omega}}$ is defined by (5.4.17). Here, $\mathrm{Sp}^{(1)}(\Delta_{38}, [0, L])$ is the set of piecewise-affine functions which correspond to the partition of the interval $[0, L]$ into 38 parts (see Subsec. 5.2.5), and the partition was taken uniform. As (5.4.24) shows, the optimal solution was searched for in the class of the functions symmetric with respect to the point $z = \frac{L}{2}$, and the optimization was carried out by twenty parameters. The calculation was done for the orthotropic cylindrical shell clamped on the edges in the case when $L = 3\,\mathrm{m}$, $R = 0.75\,\mathrm{m}$, $\mu_1 = \mu_2 = 0.11$, $E_1 = E_2 = 3.7 \cdot 10^{10}\, N/m^2$, $G = 0.75 \cdot 10^{10}\, N/m^2$, $\check{h} = 0.02\,\mathrm{m}$, and $\hat{h} = 0.08\,\mathrm{m}$.

In Fig. 5.4.3, we show the resulting function $h_0(z)$, which maximizes the functional $h \to \lambda_{1h}$, i.e., the first eigenvalue (the critical load) on the set U_{ad} from (5.4.24). (More precisely, h_0 is a local maximum since the function $h \to \lambda_{1h}$ is not convex.) As compared to the shell of constant thickness $\tilde{h}(z) = 0.04\,\mathrm{m}$ for all $z \in [0, L]$ and of the same weight as one of the shell with thickness $h_0(z)$, the critical load increases approximately twice, i.e., $\frac{\lambda_{1h_0}}{\lambda_{1\tilde{h}}} \approx 2$. For the optimal shell with thickness h_0, the multiplicity of the first eigenvalue λ_{1h_0} equals twelve.

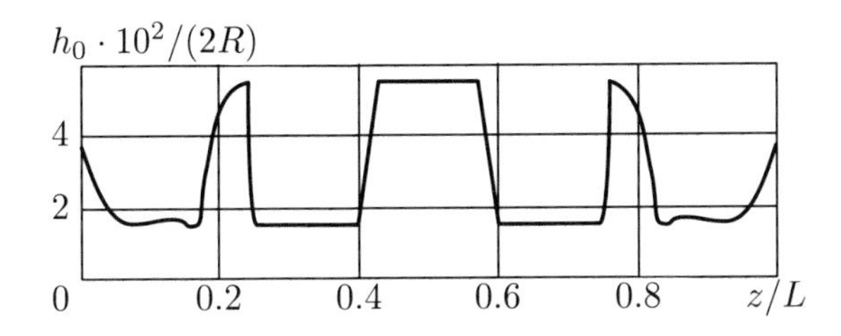

Figure 5.4.3: Function of thickness of the optimal cylindrical shell

5.5 Control by the shape of a hole and by the function of thickness for a shallow shell

5.5.1 Problem of optimization according to strength

Consider the optimization problem for a shallow shell which is described by the shear model (see Subsec. 4.11.2). Let spaces M_1 and N_1 be defined by the relations (2.12.41) with $l = 1$, and equipped with topologies as in Subsec. 2.12.4. Then the condition (2.12.10) is satisfied. To each element $q \in M_1$ we place in correspondence a two-connected domain Ω_q whose inner boundary S_1 is defined in polar coordinates by the function $\alpha \to q(\alpha)$, and whose external boundary S_2 is supposed to be fixed (Fig. 5.5.1) and determined by a function $\alpha \to q_1(\alpha)$ such that $q_1 \in C^1([0, 2\pi])$, $q_1(0) = q_1(2\pi)$, and

$$\frac{dq_1}{d\alpha}(0) = \frac{dq_1}{d\alpha}(2\pi), \qquad \min_{\alpha \in [0,2\pi]} q_1(\alpha) > r_2,$$

r_2 defined in (2.12.41).

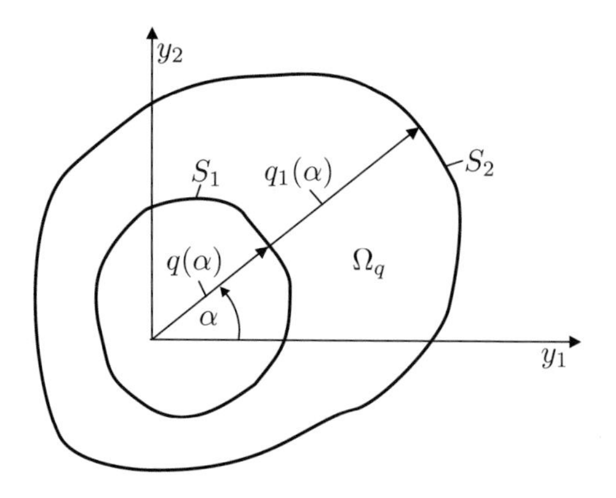

Figure 5.5.1: Two-connected domain Ω_q

By using the mapping P_q defined in Subsec. 2.12.4, one defines a diffeomorphism of the C^1 class of the set $\overline{\Omega}_q$ onto $\overline{\Omega}$, where $\overline{\Omega}$ is determined by (2.12.36).

Further, let

$$G = \{ h \,|\, h \in W_p^1(\Omega),\ e_1 \leq h \leq e_2,\ \|h\|_{W_p^1(\Omega)} \leq e_3 \}, \tag{5.5.1}$$

where e_1, e_2, and e_3 are positive numbers and $p > 2$. The set G is endowed with the topology generated by the $W_p^1(\Omega)$-weak one. Then, the condition (2.12.4) holds.

Let $V(\Omega_q)$ and $V(\Omega)$ be closed subspaces of $\left(W_2^1(\Omega_q)\right)^5$ and $\left(W_2^1(\Omega)\right)^5$, respectively, which satisfy the condition (4.11.8) and the following condition of concordance:

$$\text{the mapping } u \to u \circ P_q \text{ is an isomorphism of } V(\Omega) \text{ onto } V(\Omega_q). \tag{5.5.2}$$

Suppose we are given functions k_1 and k_2 on Ω such that $k_1 \in L_\infty(\Omega)$, $k_2 \in L_\infty(\Omega)$. Now, to every pair $(h, q) \in G \times M_1$ we place in correspondence the bilinear form a_{hq} which is generated by the strain energy of the shell. This form is defined on $V(\Omega_q) \times V(\Omega_q)$ and determined by the right-hand side of (4.11.9), provided $h \circ P_q$ is substituted for h, Ω_q is substituted for Ω, and $R_1^{-1} = k_1 \circ P_q$, $R_2^{-1} = k_2 \circ P_q$.

Define a bilinear form $P_q a_{hq}$ on $V(\Omega) \times V(\Omega)$ by the relation

$$P_q a_{hq}(u, v) = a_{hq}(u \circ P_q, v \circ P_q), \qquad u, v \in V(\Omega). \tag{5.5.3}$$

By using the rule of change of variables, one obtains the explicit form of $P_q a_{hq}$ and easily concludes that

$$\left. \begin{array}{l} (h, q) \to P_q a_{hq} \text{ is a continuous mapping of } G \times M_1 \text{ into} \\ \mathcal{L}_2(V(\Omega), V(\Omega); \mathbb{R}), \end{array} \right\} \tag{5.5.4}$$

i.e., the condition (2.12.6) is fulfilled.

Assume that we are given a continuous mapping $G \times M_1 \ni (h,q) \to B(h,q) \in (V(\Omega))^*$. From the physical point of view, for each control, i.e., for each shape of the hole which is defined by the curve S_1 (see Fig. 5.5.1), and for every function of thickness $h \circ P_q$, the mapping B defines the load $B(h,q)$ acting on the shell. (More precisely, $B(h,q)$ is the image of the load corresponding to the change of variables defined by the mapping P_q.) In particular, B can be a constant mapping, which corresponds to the case when the load, as an element of the space $(V(\Omega))^*$, does not depend on control.

Given a control $(h,q) \in G \times M_1$, determining the stress-strain state of the shell reduces to finding a function u_{hq} such that

$$u_{hq} \in V(\Omega_q), \qquad a_{hq}(u_{hq}, v) = \left(B(h,q), v \circ P_q^{-1}\right), \quad v \in V(\Omega_q). \tag{5.5.5}$$

The problem (5.5.5) is equivalent to the following one: Find a function $\tilde{u}_{hq}$ such that

$$\tilde{u}_{hq} \in V(\Omega), \qquad P_q a_{hq}(\tilde{u}_{hq}, w) = (B(h,q), w), \quad w \in V(\Omega), \tag{5.5.6}$$

in this case

$$\tilde{u}_{hq} = u_{hq} \circ P_q^{-1}. \tag{5.5.7}$$

Define a set of admissible controls as

$$U_{\text{ad}} = \left\{ (h,q) \,|\, h \in G,\ q \in N_1,\ \Psi_k(h,q,\tilde{u}_{hq}) \le 0,\ k = 1,2,\dots,l \right\}. \tag{5.5.8}$$

Here, $(h,q,u) \to \Psi_k(h,q,u)$ are continuous mappings of $G \times N_1 \times V(\Omega)$ into $\mathbb{R}$. The functionals Ψ_k can play the role of restrictions on the strength as well as those on the hole in the shell (for example, one can require that the area of the hole be not less than a given one, or that the hole contain a certain subdomain, etc.).

The set U_{ad} is supposed to be nonempty, and the goal functional is assumed to be of the form

$$J(h,q) = \int_{\Omega_q} (h \circ P_q)(y)\, dy = \int_{\Omega} h(x) \left|\det \left(P_q^{-1}\right)'(x)\right| dx. \tag{5.5.9}$$

Here, $\left(P_q^{-1}\right)'$ is the derivative (the Jacobi matrix) of the mapping P_q^{-1}.

The optimization problem of finding a pair (h_0, q_0) such that

$$(h_0, q_0) \in U_{\text{ad}}, \qquad J(h_0, q_0) = \inf_{(h,q) \in U_{\text{ad}}} J(h,q) \tag{5.5.10}$$

means that one searches for a law of the change of the thickness of the shell and a shape of the hole for which the weight of the shell is minimal provided restrictions on the strength and other requirements are fulfilled. In view of Theorem 2.12.1, under the above assumptions the problem (5.5.10) has a solution.

5.5.2 Approximate solution of the optimization and direct problems

Let $\{H_n\}$ be a sequence of finite-dimensional subspaces of $W_p^1(\Omega)$, let $\{V_n\}$ be a sequence of finite-dimensional subspaces of $\tilde{W}_2^2(0,2\pi)$, and let the limit density conditions

$$\lim_{n\to\infty} \inf_{z\in H_n} \|h - z\|_{W_p^1(\Omega)} = 0, \qquad h \in W_p^1(\Omega), \tag{5.5.11}$$

$$\lim_{n\to\infty} \inf_{\varphi\in V_n} \|q - \varphi\|_{\widetilde{W}_2^2(0,2\pi)} = 0, \qquad q \in \widetilde{W}_2^2(0,2\pi), \tag{5.5.12}$$

be fulfilled.

We recall that $\widetilde{W}_2^2(0,2\pi)$ is the subspace of periodic functions of the space $W_2^2(0,2\pi)$ (see Subsec. 2.12.4). The spaces V_n meeting the condition (5.5.12) can be constructed by trigonometric polynomials or cubic periodical splines.

Write $Q_n = H_n \times V_n$ and define an approximate solution $\left(h^{(n)}, q^{(n)}\right)$ of the problem (5.5.10) in the form

$$\left(h^{(n)}, q^{(n)}\right) \in Q_n \cap U_{\text{ad}}, \qquad J\left(h^{(n)}, q^{(n)}\right) = \inf_{(h,q)\in Q_n\cap U_{\text{ad}}} J(h,q). \tag{5.5.13}$$

For studying solvability of the problem (5.5.13) and convergence of the approximate solutions $\left(h^{(n)}, q^{(n)}\right)$ to the explicit solution (h_0, q_0), one can use results stated above (see, e.g., Section 2.3). The optimization problem (5.5.13) can be solved by the iteration method. To this end, for every iteration one has to solve the direct problem (5.5.5) in a two-connected domain Ω_q of a rather complicated shape, the domain changing at each iteration. Let us show that an approximate solution of the problem (5.5.5) can be obtained by the finite element method, by replacing the domain Ω_q with a certain two-connected domain, the connected components of the boundary of which are boundaries of polygons.

Define a mapping $n \to k(n)$ of the set of natural numbers $\mathbb{N}$ into itself such that $k(n) \to \infty$ as $n \to \infty$. To each $n \in \mathbb{N}$ we place in correspondence two closed, piecewise-straight lines S_{1n} and S_{2n} which are the boundaries of the polygons with vertices $A_{1n}^{(i)}$, $A_{2n}^{(i)}$, $i = 1, 2, \ldots, k(n)$, having the polar coordinates $(\alpha_{1n}^{(i)}, \rho_{1n}^{(i)})$, $(\alpha_{2n}^{(i)}, \rho_{2n}^{(i)})$ determined by the formulas (see Fig. 5.5.2)

$$\begin{aligned} \alpha_{1n}^{(i)} &= \frac{2\pi i}{k(n)}, \qquad \rho_{1n}^{(i)} = q\left(\frac{2\pi i}{k(n)}\right), \\ \alpha_{2n}^{(i)} &= \frac{2\pi i}{k(n)}, \qquad \rho_{2n}^{(i)} = q_1\left(\frac{2\pi i}{k(n)}\right), \qquad i = 1, 2, \ldots, k(n). \end{aligned} \tag{5.5.14}$$

If $\rho_{1n}(\alpha)$ and $\rho_{2n}(\alpha)$ are the functions determining the boundaries S_{1n} and S_{2n}, then well-known results on the estimation of the remaining term of an interpolation spline of the first power (see Zavialov et al. (1980)) imply that

$$\rho_{1n} \to q \quad \text{in } W_\infty^1((0,2\pi)), \qquad \rho_{2n} \to q_1 \quad \text{in } W_\infty^1((0,2\pi)). \tag{5.5.15}$$

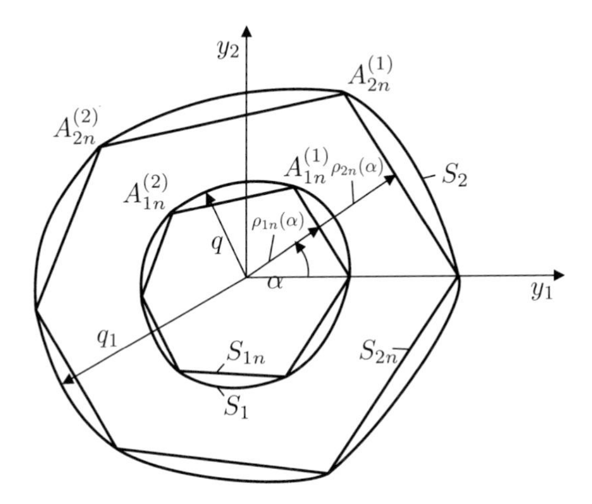

Figure 5.5.2: Domain Ω_q and polygons

Denote by Ω_n the two-connected domain bounded by the lines S_{1n} and S_{2n}. By using the functions $\rho_n = (\rho_{1n}, \rho_{2n})$ and the technique of Subsec. 2.12.4, one constructs a diffeomorphism $P_n = P_{\rho_n}$ of the W^1_∞ class of the set $\overline{\Omega}_n$ onto $\overline{\Omega}$, i.e.,

$$P_n \in W^1_\infty(\Omega_n, \Omega), \quad P_n^{-1} \in W^1_\infty(\Omega, \Omega_n). \tag{5.5.16}$$

By virtue of (5.5.16), the mappings P_n and P_n^{-1} are Lipschitz continuous, so that the mapping $u \to u \circ P_n$ is an isomorphism of $W^1_2(\Omega)$ onto $W^1_2(\Omega_n)$ (see Nečas (1967)), and because of (5.5.2) $u \to u \circ P_n$ is an isomorphism of $V(\Omega)$ onto $V(\Omega_n)$.

Now, to every n we place in correspondence the bilinear, symmetric, continuous, and coercive form a_n on $V(\Omega_n) \times V(\Omega_n)$ determined by the relation (4.11.9) provided $h \circ P_n$ is substituted for h, Ω_n is substituted for Ω, and $R_1^{-1} = k_1 \circ P_n$, $R_2^{-1} = k_2 \circ P_n$. Let also $P_n a_n$ be the image of the form a_n under the mapping P_n, given by the formula

$$P_n a_n(u, v) = a_n(u \circ P_n, v \circ P_n), \qquad u, v \in V(\Omega). \tag{5.5.17}$$

By using (5.5.15), one can verify that

$$P_n a_n \to P_q a_{hq} \qquad \text{in } \mathcal{L}_2(V(\Omega), V(\Omega); \mathbb{R}) \text{ as } n \to \infty. \tag{5.5.18}$$

Consider the following problem: Find a function $u^{(n)}$ such that

$$u^{(n)} \in V(\Omega_n), \qquad a_n\left(u^{(n)}, w\right) = \left(B(h,q), w \circ P_n^{-1}\right), \quad w \in V(\Omega_n). \tag{5.5.19}$$

By virtue of (5.5.17), for the function $u^{(n)}$ solving the problem (5.5.19), the following relation holds

$$P_n a_n \left(u^{(n)} \circ P_n^{-1}, v\right) = (B(h,q), v), \qquad v \in V(\Omega). \tag{5.5.20}$$

(5.5.6), (5.5.7), (5.5.18), (5.5.20), and Theorem 1.8.1 yield that

$$u^{(n)} \circ P_n^{-1} \to \tilde{u}_{hq} = u_{hq} \circ P_q^{-1} \qquad \text{in } V(\Omega) \text{ as } n \to \infty.$$

Since the boundary of Ω_n consists of piecewise-straight lines, by applying the finite element method one can construct approximate solutions of the problem (5.5.19) which converge in $V(\Omega_n)$ to the explicit solution $u^{(n)}$ of this problem.

5.5.3 Problem of optimization of eigenvalues

For the finite shear model of a shallow shell, the frequencies of natural oscillations are determined as solutions of the eigenvalue problem (4.11.11). According to results of Subsec. 5.5.1, to every pair $(h,q) \in G \times M_1$ there correspond a domain Ω_q and bilinear forms a_{hq} and b_{hq} defined by the right-hand sides of the relations (4.11.9) and (4.11.12), respectively, provided $h \circ P_q$ is substituted for h, Ω_q is substituted for Ω, and $R_1^{-1} = k_1 \circ P_q$, $R_2^{-1} = k_2 \circ P_q$.

By $P_q b_{hq}$ we will denote the bilinear, symmetric, continuous form on $(L_2(\Omega))^5 \times (L_2(\Omega))^5$ which is the image of the form b_{hq} under the mapping P_q:

$$P_q b_{hq}(u,v) = b_{hq}(u \circ P_q, v \circ P_q), \qquad u, v \in V(\Omega). \tag{5.5.21}$$

It is easy to verify that

$$\left.\begin{array}{l} (h,q) \to P_q b_{hq} \text{ is a continuous mapping of } G \times M_1 \text{ into} \\ \mathcal{L}_2((L_2(\Omega))^5 \times (L_2(\Omega))^5; \mathbb{R}). \end{array}\right\} \tag{5.5.22}$$

Now, consider the following problem: Given a pair $(h,q) \in G \times M_1$, find eigenvalues $\mu_{hq}^{(i)}$ of the problem

$$\begin{aligned} & u_{hq}^{(i)} \in V(\Omega_q), \quad u_{hq}^{(i)} \neq 0, \quad \mu_{hq}^{(i)} \in \mathbb{R}, \\ & \mu_{hq}^{(i)} a_{hq}(u_{hq}^{(i)}, v) = b_{hq}(u_{hq}^{(i)}, v), \qquad v \in V(\Omega_q). \end{aligned} \tag{5.5.23}$$

Results of Subsec. 4.11.2 imply the existence of a sequence of eigenvalues of the problem (5.5.23) such that $\{\mu_{hq}^{(i)}\}_{i=1}^{\infty} = \mu_{hq} \in \ell_{\infty,0}$, $\mu_{hq}^{(i)} > 0$ for any i, and

$$\mu_{hq}^{(1)} \geq \mu_{hq}^{(2)} \geq \mu_{hq}^{(3)} \geq \cdots, \qquad \lim_{i \to \infty} \mu_{hq}^{(i)} = 0.$$

Further, suppose we are given continuous mappings $\ell_{\infty,0} \ni \mu \to A_i \mu \in \mathbb{R}$, $i = 1, 2, \ldots, k$, and the set of admissible controls

$$U_{\text{ad}} = \left\{ (h,q) \,|\, (h,q) \in G \times N_1, \ A_i \mu_{hq} \leq 0, i = 1, 2, \ldots, k \right\}. \tag{5.5.24}$$

We assume that $U_{\rm ad}$ is nonempty and the goal functional $\Psi\colon (h,q) \to \Psi(h,q)$ is a continuous mapping of $G \times N_1$ into $\mathbb{R}$. Taking into account (5.5.4) and (5.5.22), from Theorem 2.12.2 we deduce that there exists a pair (h_0, q_0) such that

$$(h_0, q_0) \in U_{\rm ad}, \qquad \Psi(h_0, q_0) = \inf_{(h,q)\in U_{\rm ad}} \Psi(h,q). \tag{5.5.25}$$

5.5.4 Approximate solution of the eigenvalue problem

Let us show that to get approximate eigenvalues of the problem (5.5.23), one can replace the domain Ω_q by one with a boundary consisting of piecewise-straight lines.

Let $\{\Omega_n\}$ be a sequence of domains which are constructed as in Subsec. 5.5.2, let $\{P_n\}$ be a sequence of diffeomorphisms meeting the condition (5.5.16), and let $\{a_n\}$ be the corresponding sequence of bilinear forms on $V(\Omega_n) \times V(\Omega_n)$ defined in Subsec. 5.5.2. Further, for each n, we define a bilinear continuous form b_n on $(L_2(\Omega_n))^5 \times (L_2(\Omega_n))^5$ by the right-hand side of the relation (4.11.12), with $h \circ P_n$ substituted for h and Ω_n substituted for Ω. Denote by $P_n b_n$ the image of the form b_n under the mapping P_n:

$$P_n b_n(u,v) = b_n(u \circ P_n, v \circ P_n), \qquad u,v \in (L_2(\Omega))^5.$$

It is easy to see that

$$P_n b_n \to P_q b_{hq} \qquad \text{in } \mathcal{L}_2((L_2(\Omega))^5, (L_2(\Omega))^5; \mathbb{R}) \text{ as } n \to \infty. \tag{5.5.26}$$

Now, consider the eigenvalue problem

$$\begin{gathered} u_n^{(i)} \in V(\Omega_n), \quad u_n^{(i)} \neq 0, \quad \mu_n^{(i)} \in \mathbb{R}, \\ \mu_n^{(i)} a_n(u_n^{(i)}, v) = b_n(u_n^{(i)}, v), \qquad v \in V(\Omega_n). \end{gathered} \tag{5.5.27}$$

The problem (5.5.23) is equivalent to the following one:

$$\begin{gathered} u^{(i)} \in V(\Omega), \quad u^{(i)} \neq 0, \quad \mu_{hq}^{(i)} \in \mathbb{R}, \\ \mu_{hq}^{(i)} P_q a_{hq}\left(u^{(i)}, w\right) = P_q b_{hq}\left(u^{(i)}, w\right), \qquad w \in V(\Omega), \end{gathered}$$

where $u^{(i)} = u_{hq}^{(i)} \circ P_q^{-1}$, and problem (5.5.27) is equivalent to the problem

$$\begin{gathered} \tilde{u}_n^{(i)} \in V(\Omega), \quad \tilde{u}_n^{(i)} \neq 0, \quad \mu_n^{(i)} \in \mathbb{R}, \\ \mu_n^{(i)} P_n a_n\left(\tilde{u}_n^{(i)}, w\right) = P_n b_n\left(\tilde{u}_n^{(i)}, w\right), \qquad w \in V(\Omega), \end{gathered}$$

where $\tilde{u}_n^{(i)} = u_n^{(i)} \circ P_n^{-1}$. (5.5.18), (5.5.26), and Theorem 1.5.9 imply

$$\mu_n = \left\{\mu_n^{(i)}\right\}_{i=1}^{\infty} \to \mu_{hq} = \left\{\mu_{hq}^{(i)}\right\}_{i=1}^{\infty} \qquad \text{in } \ell_{\infty,0} \text{ as } n \to \infty.$$

Since the problem (5.5.27) is solved in a domain whose boundary consists of piecewise-straight lines, approximate values of $\mu_n^{(i)}$ can be calculated by using the finite element method, by triangulating Ω_n.

5.6 Control by the load for plates and shells

So far we have studied optimization problems for plates and shells in which as a control we took some geometric characteristics (the function of thickness, the equation defining the form of the midsurface of the shell, the form of the plan of the shell), the load being assumed to be fixed. However, if one produces a construction (here and below this word is used for a plate or a shell) having optimal values of geometric characteristics for a fixed load, then a change of the load will result in the construction's being no longer optimal. Moreover, in some cases it may happen that this construction will be even "worse than a nonoptimal one." On the other hand, the requirement that the construction be optimal for a wide class of loads which can act on it during the operation, leads to an essential complexification of design (computation) of the construction, and may considerably enlarge its weight. Besides, it is impossible to establish *a priori* all the loads that will act on the construction.

So, it seems reasonable to use an approach based on control of the stress-strain state of the construction by control loads. In the latter case, the control loads are adjusted to the basic load and change with its changes.

Static problems of control by the load, i.e., by control loads, reduce to control by the right-hand sides of elliptic systems. The problems of optimal heating and cooling of plates and shells also reduce to such problems (see, e.g., Burak et al. (1984)).

Let us consider some problems of such type.

5.6.1 General problem of control by the load

Setting for the problem

Let a construction be subject to a load $f \in V^*$. The problem on the stress-strain state of the construction reduces to determining a function of displacements ω_f such that

$$\omega_f \in V, \qquad a_h(\omega_f, \omega) = (f, \omega), \qquad \omega \in V. \tag{5.6.1}$$

Here, the function h is assumed to be fixed, while the load f runs over some set. Thus, the solution of the problem (5.6.1) is denoted by ω_f.

Now, suppose that not only the basic load f_0, but also a control load f acts on the construction. Then, in accordance with the notations accepted, the problem of the stress-strain state of the construction reduces to determining a function ω_{f_0+f} such that

$$\omega_{f_0+f} \in V, \qquad a_h\left(\omega_{f_0+f}, \omega\right) = (f_0 + f, \omega), \quad \omega \in V. \tag{5.6.2}$$

Since a_h is a bilinear form, we have

$$\omega_{f_0+f} = \omega_{f_0} + \omega_f. \tag{5.6.3}$$

Let $L \in \mathcal{L}(V, \mathcal{H})$, where $\mathcal{H}$ is a Banach space and z is a given element from $\mathcal{H}$. Define a goal functional $f \to J(f)$ as

$$J(f) = \|L\omega_{f_0+f} - z\|_{\mathcal{H}}^p . \tag{5.6.4}$$

Here, we set $p = 2$ if $\mathcal{H}$ is a Hilbert space and $p \geq 1$ otherwise.

Suppose U is a space of controls, U is a Hilbert space, and

$$V \subset U,\ V \text{ is dense in } U, \text{ the embedding of } V \text{ into } U \text{ is compact.} \tag{5.6.5}$$

Let also U_{ad} be a set of admissible controls such that

$$U_{\mathrm{ad}} \text{ is bounded and sequentially weakly closed in } U. \tag{5.6.6}$$

By identifying the space U with its dual, from (5.6.5) and Theorem 1.5.12 we deduce the embedding of U into V^* to be compact. Now, one can easily prove the existence of a function g such that

$$g \in U_{\mathrm{ad}}, \qquad J(g) = \inf_{f \in U_{\mathrm{ad}}} J(f). \tag{5.6.7}$$

Depending on the choice of the operator L and the space $\mathcal{H}$ (see (5.6.4)), the problem (5.6.7) can correspond to obtaining fields of displacements, stresses, etc., which are nearest to given ones in the mean-square or minimax sense.

If in (5.6.4) $\mathcal{H}$ is a Hilbert space, $p = 2$, and the set U_{ad} is convex, then from results of Section 3.3 one can derive necessary and sufficient optimality conditions for the problem (5.6.7), as well as utilize the technique of this section for construction of approximate solutions. In this connection, the condition (5.6.5) can be replaced by the following weaker one:

$$V \subset U,\ V \text{ is dense in } U, \text{ the embedding of } V \text{ into } U \text{ is continuous.}$$

We stress that, if J is an increasing functional, the set U_{ad} can be unbounded (see Lions (1968), Litvinov (1976)).

Let us examine some special problems.

5.6.2 Optimization problems for plates

Setting of the problem

The bilinear form a_h for a plate is defined by the formula (4.1.24) and $V \subset W_2^2(\Omega)$. Let $\mathcal{H} = L_2(\Omega)$ and let L be the operator of embedding of V into $L_2(\Omega)$. Then, the problem (5.6.4), (5.6.7) corresponds to searching for a function $g \in U_{\mathrm{ad}}$ such that the deflection function $\omega_{f_0+g} = w_{f_0+g}$ is the best mean-square approximation of a given function $z \in L_2(\Omega)$.

If the operator L acting from V into $\mathcal{H} = (L_2(\Omega))^3$ is defined by the relation

$$w_{f_0+f} \to Lw_{f_0+f} = \{M_1(w_{f_0+f}), M_2(w_{f_0+f}), M_3(w_{f_0+f})\},$$

$M_i(w_{f_0+f})$ being the bending moments and torque, determined by formulas (4.1.3) and (4.1.10), then the problem (5.6.4), (5.6.7) corresponds to searching for $g \in U_{\mathrm{ad}}$ such that the distribution of the bending moments and torque is the best mean-square approximation of a given $z = (z_1, z_2, z_3) \in (L_2(\Omega))^3$.

Let $(x = a_i, y = b_i)$ be points of Ω, $i = 1, 2, \ldots, n$, and let $\overline{\Omega}_i \subset \Omega$,

$$\overline{\Omega}_i = \{ (x, y) \mid a_i - \xi_i \le x \le a_i + \xi_i, \; b_i - \nu_i \le y \le b_i + \nu_i \}, \tag{5.6.8}$$

ξ_i and ν_i being positive numbers, $\overline{\Omega}_i$ and $\overline{\Omega}_j$ being disjoint provided $i \ne j$. Assume that

$$f = -\sum_{i=1}^{n} q_i \chi_i(x, y), \tag{5.6.9}$$

where χ_i is the characteristic function of the set $\overline{\Omega}_i$, i.e.,

$$\chi_i(x, y) = \begin{cases} 1, & \text{if } (x, y) \in \overline{\Omega}_i, \\ 0, & \text{otherwise}, \end{cases} \tag{5.6.10}$$

q_i are constants, and the vector $q = (q_1, q_2, \ldots, q_n)$ is a control. Now, if $\mathcal{H}$ is a Hilbert space, $p = 2$, then, making use of (5.6.3) and (5.6.9), we get the functional (5.6.4) to be a function of vector q which up to a constant summand looks like

$$\mathcal{I}(q) = (Aq, q)_n - 2(p, q)_n. \tag{5.6.11}$$

Here, A is an $n \times n$ matrix with elements a_{ij}, p is an n-dimensional vector with components p_i, and

$$\begin{aligned} a_{ij} &= \left(Lw_{\chi_i}, Lw_{\chi_j}\right)_{\mathcal{H}}, \qquad & i, j = 1, 2, \ldots, n, \\ p_i &= (Lw_{f_0} - z, Lw_{\chi_i})_{\mathcal{H}}, \qquad & i = 1, 2, \ldots, n, \end{aligned} \tag{5.6.12}$$

$(\cdot,\cdot)_n$ and $(\cdot,\cdot)_{\mathcal{H}}$ being the scalar products in $\mathbb{R}^n$ and $\mathcal{H}$, respectively.

By virtue of (5.6.9), the set U_{ad} is mapped onto a set $K_{\mathrm{ad}} \subset \mathbb{R}^n$, so that the problem (5.6.7) reduces to determining a vector t such that

$$t \in K_{\mathrm{ad}}, \qquad \mathcal{I}(t) = \inf_{q \in K_{\mathrm{ad}}} \mathcal{I}(q). \tag{5.6.13}$$

Let the set K_{ad} be of the form

$$K_{\mathrm{ad}} = \{ q \mid q = (q_1, q_2, \ldots, q_n) \in \mathbb{R}^n, \; q_i^- \le q_i \le q_i^+, \; i = 1, 2, \ldots, n \}. \tag{5.6.14}$$

Here, q_i^- and q_i^+ are given constants such that $q_i^+ > q_i^-$. The relation (5.6.14) corresponds to the lower and upper restrictions on control loads.

Assume that the operator L has a left inverse. Then, $\{Lw_{\chi_i}\}_{i=1}^n$ is a system of linearly independent functions in $\mathcal{H}$, and the bilinear form $q', q'' \to (Aq', q'')_n$ is symmetric, continuous, and coercive in $\mathbb{R}^n$. Now, Theorem 3.1.1 implies that problem (5.6.13), (5.6.14) has a unique solution.

Direct and dual problems

To solve the problem (5.6.13), (5.6.14), we will use duality methods (see, e.g., Séa (1971)). Let

$$M = \{\, \mu \,|\, \mu = (\mu_1, \mu_2, \dots, \mu_{2n}) \in \mathbb{R}^{2n},\ \mu_i \geq 0,\ i = 1, 2, \dots, 2n \,\}. \tag{5.6.15}$$

Consider a function $\Phi\colon \mathbb{R}^n \times M \to \mathbb{R}$ of the form

$$\Phi(q, \mu) = \mathcal{I}(q) + \sum_{i=1}^{2n} \mu_i \mathcal{I}_i(q), \tag{5.6.16}$$

where

$$\mathcal{I}_i(q) = \begin{cases} q_i - q_i^+, & \text{for } i = 1, 2, \dots, n, \\ q_{i-n}^- - q_{i-n}, & \text{for } i = n+1, \dots, 2n. \end{cases} \tag{5.6.17}$$

Taking into account (5.6.11), rewrite (5.6.16) as

$$\Phi(q, \mu) = (Aq, q)_n - 2(p, q)_n + (Bq - l, \mu)_{2n}. \tag{5.6.18}$$

Here, B is a $2n \times n$-matrix and $l \in \mathbb{R}^{2n}$. The form of the matrix B and the vector l is uniquely determined by the formulas (5.6.16), (5.6.17).

One can prove the existence of a saddle point (t, λ) of the function Φ such that

$$\begin{gathered} (t, \lambda) \in \mathbb{R}^n \times M, \\ \min_{q \in \mathbb{R}^n} \max_{\mu \in M} \Phi(q, \mu) = \Phi(t, \lambda) = \max_{\mu \in M} \min_{q \in \mathbb{R}^n} \Phi(q, \mu). \end{gathered} \tag{5.6.19}$$

The problem of determining (t, λ) from the conditions of the left-hand side of the equality (5.6.19) is referred to as a direct problem, while the problem of determining (t, λ) from the condition of the right-hand side of (5.6.19) is called a dual problem. It is easy to verify that the vector t from the pair (t, λ) is a solution of the problem (5.6.13), (5.6.14).

Consider the dual problem. Given $\mu \in M$, let q_μ be a solution of the problem of determining $\min_{q \in \mathbb{R}^n} \Phi(q, \mu)$. The vector q_μ is a solution of the equation $Aq_\mu = p - \frac{1}{2} B^* \mu$, i.e.,

$$q_\mu = A^{-1} \left(p - \frac{1}{2} B^* \mu \right). \tag{5.6.20}$$

Known results on the minimum of a quadratic functional (see Michlin (1970)) imply that

$$\Phi(q_\mu, \mu) = \min_{q \in \mathbb{R}^n} \Phi(q, \mu) = -(Aq_\mu, q_\mu)_n - (l, \mu)_{2n}.$$

Now, the dual problem takes the following form: Find a vector λ such that

$$\lambda \in M, \qquad -F(\lambda) = \max_{\mu \in M} (-F(\mu)),$$

where

$$F(\mu) = (Aq_\mu, q_\mu)_n + (l, \mu)_{2n}, \tag{5.6.21}$$

or: Find a vector λ such that

$$\lambda \in M, \qquad F(\lambda) = \min_{\mu \in M} F(\mu). \tag{5.6.22}$$

By substituting (5.6.20) into (5.6.21), we conclude that the function F up to a constant summand is determined by the relation

$$F(\mu) = \frac{1}{4}\left(BA^{-1}B^*\mu, \mu\right)_{2n} - \left(BA^{-1}p, \mu\right)_{2n} + (l, \mu)_{2n}. \tag{5.6.23}$$

By (5.6.20) and (5.6.23), we get the following relation for the gradient of the functional F

$$\operatorname{grad} F(\mu) = \frac{1}{2}BA^{-1}B^*\mu - BA^{-1}p + l = -Bq_\mu + l. \tag{5.6.24}$$

Let P_M be the orthogonal projection of $\mathbb{R}^{2n}$ onto M, which is defined by the formula

$$\mu = (\mu_1, \mu_2, \ldots, \mu_{2n}) \in \mathbb{R}^{2n}, \qquad P_M\mu = \mu^+ = \left(\mu_1^+, \mu_2^+, \ldots, \mu_{2n}^+\right),$$
$$\mu_i^+ = \begin{cases} \mu_i, & \text{if } \mu_i \geq 0, \\ 0, & \text{if } \mu_i < 0. \end{cases}$$

To solve the problem (5.6.22), (5.6.23), we apply the gradient projection method. Taking an arbitrary initial approximation $\mu^{(0)} \in M$, by using (5.6.24) and (5.6.20), we construct the sequence $\{\mu^{(n)}\} \subset M$ by the following formulas:

$$q^{(n)} = A^{-1}\left(p - \frac{1}{2}B^*\mu^{(n)}\right), \qquad \mu^{(n+1)} = P_M\left[\mu^{(n)} - \rho_n\left(l - Bq^{(n)}\right)\right], \tag{5.6.25}$$

ρ_n being a positive constant.

The sequence $\{q^{(n)}, \mu^{(n)}\}$ constructed by the algorithm (5.6.25) with a constant ρ_n satisfying the condition $0 < \rho_n < \frac{1}{d}$, where $d = \frac{1}{2}\|BA^{-1}B^*\|$, can be shown to converge to the solution (t, λ) of the problem (5.6.19).

Numerical solutions

The results below are obtained for the simply supported, rectangular plate

$$\Omega = \left\{\, (x, y) \mid -a < x < a,\ -b < y < b \right\}$$

of constant thickness $h = 0.01\,\text{m}$, with $a = b = 5h$, made of an isotropic material for which the elasticity modulus $E = 1.96 \cdot 10^{11}$ Pa and Poisson's ratio $\mu = 0.3$. There was taken $f_0 = \text{const} = 9.8 \cdot 10^4$ Pa and 16 control loads were chosen ($n = 16$)

which acted on the areas $\overline{\Omega}_i$ (see (5.6.9), (5.6.10)). The location of the first four areas $\overline{\Omega}_i$ (see (5.6.8)) was defined by the relations $a_1 = a_2 = b_1 = b_3 = 10^{-2}\,\text{m}$, $a_3 = a_4 = b_2 = b_4 = 3 \cdot 10^{-2}\,\text{m}$, other areas are located symmetrically with respect to the x and y axes, the sizes of the areas being equal; $\xi_i = \nu_i = 10^{-3}\,\text{m}$, $i = 1, 2, \ldots, 16$.

The optimization problem was solved by the algorithm (5.6.25). In the first case, the goal functional was chosen in the form

$$J(f) = \int_{-a}^{a} \int_{-b}^{b} \left[(M_1 (w_{f_0+f}))^2 + (M_2 (w_{f_0+f}))^2 + (M_3 (w_{f_0+f}))^2 \right] dx\, dy, \tag{5.6.26}$$

where M_1, M_2, and M_3 are the bending moments and torque defined by the formulas (4.1.3) and (4.1.4).

The results of calculation of the dimensionless moments $M_1^* = \frac{M_1}{4f_0a^2}$ and $M_3^* = \frac{M_3}{4f_0a^2}$ are shown on Fig. 5.6.1. Here, the dashed lines correspond to the left ordinate scale, and they show the values of M_1^* along the line $x = 0$ (Curve 1), and the values of M_3^* along the line $x = -\frac{a}{2}$ (Curve 2) provided the control loads are absent, i.e., when $q_i = 0$.

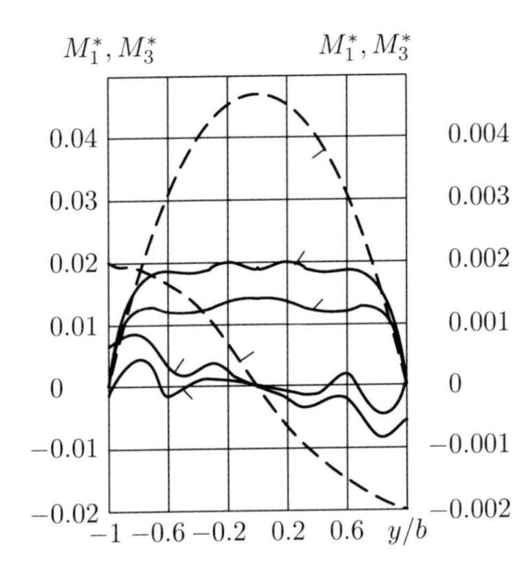

Figure 5.6.1: Bending moments and torques under the optimal control load and in the absence of control

Curves 3–6 correspond to the right ordinate scale, Curves 3 and 5 correspond to the solution of the optimization problem without restrictions, i.e., when $q_i^- = -\infty$ and $q_i^+ = \infty$ in (5.6.14), and they show the values of M_1^* and M_3^* along the lines $x = 0$ and $x = -\frac{a}{2}$, respectively. Curves 4 and 6 correspond to the solution of the optimization problem under the restrictions $0 \leq q_i \leq q_i^+$, and they show the values of the moments M_1^* and M_3^* along the same lines $x = 0$ and $x = -\frac{a}{2}$, respectively. We note that Curves 1, 3, 4 are symmetric with respect to the point $y = 0$, while Curves 2, 5, 6 are antisymmetric.

Fig. 5.6.1 shows that the values of M_1 and M_3 can be lowered by a factor 0.1 and less if the optimal values of the control loads are chosen.

In the second case, the goal functional was chosen in the form

$$J(f) = \int_{-a}^{a} \int_{-b}^{b} w_{f_0+f}^2 \, dx \, dy, \tag{5.6.27}$$

and the optimization problem was solved for the same input data and for the same location of the control loads.

The results of calculation of the function of the dimensionless deflection

$$w^* = \frac{w_{f_0+f} E h^3}{192(1 - \mu^2) f_0 a^4}$$

are shown in Fig. 5.6.2.

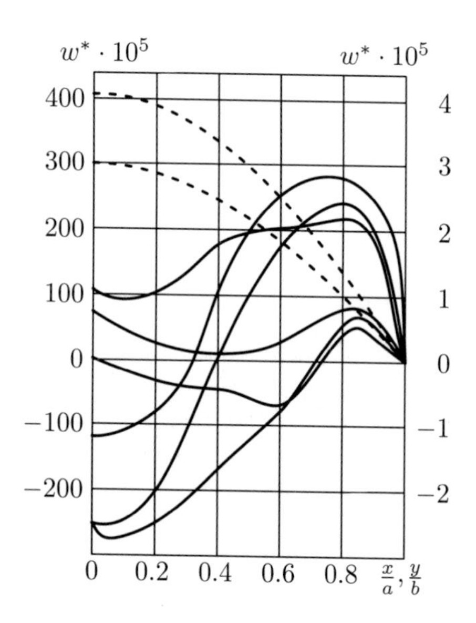

Figure 5.6.2: Deflections under the optimal control load and in the absence of control

Since the function w_{f_0+f} is symmetric with respect to the x and y axes, only the values of w^* for $0 \leq x \leq a$, $0 \leq y \leq b = a$ are presented. The dashed lines 1, 2 correspond to the left ordinate scale, and they show the values of w^* along the lines $x = 0$ and $x = \frac{a}{2}$, respectively, for the case when the control loads are absent, i.e., when all q_i are equal to zero. Curves 3–8 correspond to the right ordinate

Obviously, a^{ε}_{ijkm} is a periodic function of x with period εY, satisfying the conditions (5.7.3)–(5.7.5).

Further, let Ω be a bounded Lipshchitz domain in $\mathbb{R}^3$, let S be the boundary of Ω, $S = S_1 \cup S_2$, $S_1 \cap S_2 = \varnothing$, mes $S_1 > 0$.

Define a set H by the formula

$$H = \{u \mid u = (u_1, u_2, u_3) \in \left(W_2^1(\Omega)\right)^3, u|_{S_1} = 0\}. \tag{5.7.7}$$

The set H endowed with the scalar product and the norm of $\left(W_2^1(\Omega)\right)^3$ is a Hilbert space.

Define a bilinear form a^{ε} on $H \times H$ by the relation

$$a^{\varepsilon}(u, v) = \int_{\Omega} \sum_{i,j,k,m=1}^{3} a_{ijkm}\left(\frac{x}{\varepsilon}\right) \varepsilon_{km}(u)\varepsilon_{ij}(v)\, dx. \tag{5.7.8}$$

Suppose we are given vector functions f and F such that

$$f = (f_1, f_2, f_3) \in (L_2(\Omega))^3, \qquad F = (F_1, F_2, F_3) \in (L_2(S_2))^3. \tag{5.7.9}$$

Denote by L the linear continuous functional in H determined by the formula

$$(L, v) = \int_{\Omega} \sum_{i=1}^{3} f_i v_i\, dx + \int_{S_2} \sum_{i=1}^{3} F_i v_i\, ds. \tag{5.7.10}$$

Now, consider the following problem: Find a function u^{ε} such that

$$u^{\varepsilon} \in H, \qquad a^{\varepsilon}(u^{\varepsilon}, v) = (L, v), \quad v \in H. \tag{5.7.11}$$

Theorem 5.7.1 *Let coefficients a_{ijkm} be Y-periodic functions on $\mathbb{R}^3$, and let the conditions* (5.7.3)–(5.7.5) *be satisfied. Assume that a bilinear form a^{ε} is determined by* (5.7.8), *and an element $L \in H^*$ by the formulas* (5.7.9) *and* (5.7.10). *Then, for all $\varepsilon > 0$, the problem* (5.7.11) *has a unique solution, and the following estimate is valid*

$$\|u^{\varepsilon}\|_{(W_2^1(\Omega))^3} \leq \text{const}, \qquad \varepsilon > 0. \tag{5.7.12}$$

Proof. From (5.7.5) and (5.7.8), we conclude that

$$a^{\varepsilon}(u, u) \geq c_0 \|u\|_H^2, \qquad u \in H,\ \varepsilon > 0, \tag{5.7.13}$$

where

$$\|u\|_H^2 = \int_{\Omega} \sum_{i,j=1}^{3} (\varepsilon_{ij}(u))^2\, dx. \tag{5.7.14}$$

The Korn inequality implies that, in the space H, the formula (5.7.14) defines a norm equivalent to the norm of $\left(W_2^1(\Omega)\right)^3$. (Although in Subsec. 1.7.2 we have proved the Korn inequality in the case $\Omega \subset \mathbb{R}^2$, the same technique will give us the inequality in the case $\Omega \subset \mathbb{R}^3$.)

Now, taking into account (5.7.3) and (5.7.13), from Theorem 1.5.2 we obtain that, for any $\varepsilon > 0$, the problem (5.7.11) has a unique solution and the estimate (5.7.12) holds.

Homogenization of the structure

Define a space $W(Y)$ as follows

$$W(Y) = \{ v \,|\, v \in (\overset{\circ}{W}{}_2^1(Y))^3, \text{ the traces of } v \text{ are equal on the opposite sides of } Y \}. \quad (5.7.15)$$

Let us present another equivalent definition of this space: $W(Y)$ is the restriction on Y of the vector functions that are defined on $\mathbb{R}^3$, Y-periodical, and the restrictions of which on any bounded open set $Q \subset \mathbb{R}^3$ belong to $(W_2^1(Q))^3$.

Lemma 5.7.1 *Let K_0 be the set of constant three-dimensional vectors, and let W_1 be the orthogonal complement of K_0 to $W(Y)$ with respect to the scalar product of $(\overset{\circ}{W}{}_2^1(Y))^3$. Then, in the space W_1, the formula*

$$\|u\|_1 = \left(\int_Y \sum_{i,j=1}^{3} (\varepsilon_{ij}(u))^2 \, dx \right)^{\frac{1}{2}} \quad (5.7.16)$$

determines a norm equivalent to the norm of $(\overset{\circ}{W}{}_2^1(Y))^3$.

Proof. Let $u \in (\overset{\circ}{W}{}_2^1(Y))^3$ and let $\varepsilon_{ij}(u) = 0$, $i,j = 1,2,3$. Then, the function $u = (u_1, u_2, u_3)$ has the form (see, e.g., Nečas and Hlavaček (1970))

$$u_1 = a_1 - b_3 x_2 + b_2 x_3, \qquad u_2 = a_2 - b_1 x_3 + b_3 x_1, \qquad u_3 = a_3 - b_2 x_1 + b_1 x_2, \quad (5.7.17)$$

where a_i and b_i are constants. Moreover, if $u \in W(Y)$, then (5.7.15) and (5.7.17) yield that $u = (a_1, a_2, a_3) \in K_0$. Hence, if $v \in W_1$ and $\varepsilon_{ij}(v) = 0$, $i,j = 1,2,3$, then $v = 0$. Now, Lemma 5.7.1 is a consequence of the Korn inequality and Theorem 1.7.3.

Introduce the bilinear form

$$a(u,v) = \int_Y \sum_{i,j,k,m=1}^{3} a_{ijkm}(x) \varepsilon_{km}(u) \varepsilon_{ij}(v) \, dx. \quad (5.7.18)$$

Further, let P^{ij} be the vector function defined on Y and taking values in $\mathbb{R}^3$, the i-th component of which is equal to x_j and the other ones are equal to zero, i.e.,

$$P^{ij} = \{P_k^{ij}\}_{k=1}^3, \qquad P_k^{ij} = x_j \delta_{ki}, \quad i,j,k = 1,2,3. \quad (5.7.19)$$

The relations (5.7.3)–(5.7.5) together with Lemma 5.7.1 imply that there exists a unique function X^{ij} such that

$$X^{ij} \in W_1, \qquad a\left(X^{ij}, v\right) = a\left(P^{ij}, v\right), \quad v \in W_1, \ i,j = 1,2,3. \quad (5.7.20)$$

Theorem 5.7.2 *Let coefficients a_{ijkm} be Y-periodic functions on $\mathbb{R}^3$ satisfying the conditions* (5.7.3)–(5.7.5), *and let functions f and F meet the conditions* (5.7.9). *Then,*

$$u^\varepsilon \to u \qquad \text{weakly in } H \text{ as } \varepsilon \to 0, \tag{5.7.21}$$

where u^ε is the solution of the problem (5.7.11) *and u is the unique solution of the following problem*

$$u \in H, \qquad A(u,v) = (L,v), \quad v \in H. \tag{5.7.22}$$

The bilinear form A is defined by

$$A(u,v) = \int_\Omega \sum_{i,j,k,m=1}^{3} g_{ijkm}\varepsilon_{km}(u)\varepsilon_{ij}(v)\,dx, \tag{5.7.23}$$

$$g_{ijkm} = \frac{1}{Y_1Y_2Y_3}\, a\left(X^{ij} - P^{ij}, X^{km} - P^{km}\right), \tag{5.7.24}$$

and satisfies the conditions

$$\begin{aligned} A(u,u) &\geq c_0\|u\|_H^2, & u &\in H, \\ A(u,v) &= A(v,u), & u,v &\in H, \end{aligned} \tag{5.7.25}$$

where c_0 is the positive constant from (5.7.5).

Remark 5.7.1 Denote by A_ε and $\hat{A}$ the operators generated by the bilinear forms a^ε and A, which are defined through the relations

$$(A_\varepsilon v, w) = a^\varepsilon(v,w), \qquad v,w \in H, \tag{5.7.26}$$

$$(\hat{A}v, w) = A(v,w), \qquad v,w \in H. \tag{5.7.27}$$

Theorem 5.7.2 states that the sequence of operators $\{A_\varepsilon\}$ G-converges to the operator $\hat{A}$ as $\varepsilon \to 0$ (see Section 1.13). In the case of a Y-periodical structure being examined, the G-limit operator $\hat{A}$ has constant coefficients, which are called effective (or homogenized, or averaged) elasticity coefficients (constants).

These coefficients do not depend on the boundary conditions, on the domain Ω, and on the load $L = \{f, F\}$. Indeed, (5.7.18), (5.7.20), and (5.7.24) show that the coefficients g_{ijkm} of the operator $\hat{A}$ depend only on peculiarities of Y-periodicity, i.e., only on the coefficients $a_{ijkm}(x)$, $x \in Y$, and the domain Y.

Remark 5.7.2 Obviously, if a function X^{ij} is a solution of the problem (5.7.20), then the function $Y^{ij} = X^{ij} + C$, C being an arbitrary constant vector from K_0, is a solution of the problem

$$Y^{ij} \in W(Y), \qquad a\left(Y^{ij}, v\right) = a\left(P^{ij}, v\right), \quad v \in W(Y). \tag{5.7.28}$$

The coefficients g_{ijkm} can be determined by the formula

$$g_{ijkm} = \frac{1}{Y_1 Y_2 Y_3}\, a\left(Y^{ij} - P^{ij}, Y^{km} - P^{km}\right).$$

From the point of view of computation, the problem (5.7.28) seems to be easier than the problem (5.7.20). Apropos of the solution of the problems (5.7.20) and (5.7.28), see Subsec. 5.2.4.

Remark 5.7.3 Denote by E_1 the subspace of rigid displacements determined by (5.7.17). Let E_2 be the orthogonal complement in $\left(W_2^1(\Omega)\right)^3$ of the subspace E_1. Suppose L is an element of the dual of $\left(W_2^1(\Omega)\right)^3$, and $(L,u) = 0$ for all $u \in E_1$. Then, for $H = E_2$, the problem (5.7.11) has a unique solution and the proof of Theorem 5.7.2 stated below remains valid. Hence, the estimate (5.7.25) holds true not only for u from H defined by (5.7.7), but also for any function $u \in E_2$. Take a function $u = (u_1, u_2, u_3)$ in the form

$$u_1 = a_{11}x_1, \qquad u_2 = 2a_{21}x_1 + a_{22}x_2, \qquad u_3 = 2a_{31}x_1 + 2a_{32}x_2 + a_{33}x_3.$$

Then, by (5.7.25), taking note of (5.7.2) and (5.7.23), we get

$$\sum_{i,j,k,m=1}^{3} g_{ijkm} a_{km} a_{ij} \geq c_0 \sum_{i,j=1}^{3} a_{ij}^2, \qquad a_{ij} = a_{ji} \in \mathbb{R}.$$

Using the equalities (5.7.4), (5.7.18)–(5.7.20), and (5.7.24), one can verify that the elasticity coefficients g_{ijkm} meet the following symmetry conditions

$$g_{ijkm} = g_{jikm} = g_{ijmk} = g_{mkij}.$$

Proof of Theorem 5.7.2. 1. Introduce the notation

$$\xi_{ij}^{\varepsilon} = \sum_{k,m=1}^{3} a_{ijkm}\left(\frac{x}{\varepsilon}\right) \varepsilon_{km}\left(u^{\varepsilon}\right). \tag{5.7.29}$$

From (5.7.2), (5.7.3), and Theorem 5.7.1 (see (5.7.12)), we conclude that

$$\left\|\xi_{ij}^{\varepsilon}\right\|_{L_2(\Omega)} \leq \text{const}, \qquad \varepsilon > 0. \tag{5.7.30}$$

From here and (5.7.12), we derive the existence of subsequences $\{u^{\varepsilon}\}$ and $\{\xi_{ij}^{\varepsilon}\}$ such that

$$\varepsilon \to 0, \tag{5.7.31}$$

$$u^{\varepsilon} \to u \qquad \text{weakly in } \left(W_2^1(\Omega)\right)^3, \tag{5.7.32}$$

$$u^{\varepsilon} \to u \qquad \text{strongly in } \left(L_2(\Omega)\right)^3, \tag{5.7.33}$$

$$\xi_{ij}^{\varepsilon} \to \xi_{ij} \qquad \text{weakly in } L_2(\Omega). \tag{5.7.34}$$

Taking into account (5.7.8), (5.7.29), (5.7.34), and passing to the limit in (5.7.11), we get

$$\int_\Omega \sum_{i,j=1}^{3} \xi_{ij}\varepsilon_{ij}(v)\,dx = (L,v), \qquad v \in H. \tag{5.7.35}$$

2. Let $P = \{P_i\}_{i=1}^3$ and $P_i = \sum_{k=1}^3 c_k^i x_k$, c_k^i being constants. (5.7.3)–(5.7.5) and Lemma 5.7.1 yield the existence of a unique function ψ such that

$$\psi \in W_1, \qquad a(\psi,v) = a(P,v), \quad v \in W_1. \tag{5.7.36}$$

Introduce the following function on $\mathbb{R}^3$:

$$w^\varepsilon(x) = P(x) - \varepsilon\psi\left(\frac{x}{\varepsilon}\right), \tag{5.7.37}$$

in this case, the function $\psi(x)$ is Y-periodically extended to $\mathbb{R}^3$.

Due to the Y-periodicity, the set of functions $\{\psi(\frac{x}{\varepsilon})\}$ is bounded in $(L_2(\Omega))^3$. Hence, (5.7.31) and (5.7.37) imply that

$$w^\varepsilon \to P \qquad \text{strongly in } (L_2(\Omega))^3. \tag{5.7.38}$$

From the relations (5.7.18) and (5.7.36), taking into consideration the Y-periodicity of the functions a_{ijkm} and ψ as well as the linearity of P, we deduce that

$$\int_{\mathbb{R}^3} \sum_{i,j,k,m=1}^{3} a_{ijkm}(x)\varepsilon_{km}(P(x)-\psi(x))\varepsilon_{ij}(h)\,dx = 0, \qquad h \in \left(\mathcal{D}(\mathbb{R}^3)\right)^3.$$

From here and (5.7.37), making the change of variable, we get

$$\int_{\mathbb{R}^3} \sum_{i,j,k,m=1}^{3} a_{ijkm}\left(\frac{x}{\varepsilon}\right)\varepsilon_{km}(w^\varepsilon(x))\varepsilon_{ij}(h(x))\,dx = 0, \qquad h \in \left(\mathcal{D}(\mathbb{R}^3)\right)^3. \tag{5.7.39}$$

If $\varphi \in \mathcal{D}(\Omega)$ and $v \in H$, then (5.7.39) yields

$$a^\varepsilon(w^\varepsilon, \varphi v) = 0. \tag{5.7.40}$$

Take $v = \varphi w^\varepsilon$ in (5.7.11) and $v = u^\varepsilon$ in (5.7.40). Subtracting the second relation from the first one, we get

$$a^\varepsilon(u^\varepsilon, \varphi w^\varepsilon) - a^\varepsilon(w^\varepsilon, \varphi u^\varepsilon) = (L, \varphi w^\varepsilon). \tag{5.7.41}$$

Taking note of the notation (5.7.29), we can transform the equality (5.7.41) as follows:

$$\int_\Omega \left[\sum_{i,j=1}^{3} \xi_{ij}^\varepsilon w_i^\varepsilon \frac{\partial\varphi}{\partial x_j} - \sum_{i,j,k,m=1}^{3} a_{ijkm}\left(\frac{x}{\varepsilon}\right)\varepsilon_{km}(w^\varepsilon)u_i^\varepsilon\frac{\partial\varphi}{\partial x_j}\right]dx = \int_\Omega \sum_{i=1}^{3} f_i w_i^\varepsilon \varphi\,dx. \tag{5.7.42}$$

The functions

$$\sum_{k,m=1}^{3} a_{ijkm}\left(\frac{x}{\varepsilon}\right)\varepsilon_{km}(w^{\varepsilon})$$

are εY-periodic, so that they weakly converge in $L_2(\Omega)$ to the average on Y of the function

$$\sum_{k,m=1}^{3} a_{ijkm}(x)\varepsilon_{km}(P(x)-\psi(x)),$$

which will be denoted by $M_{ij}(P)$. Now, upon (5.7.32)–(5.7.34) and (5.7.38), passing to the limit in (5.7.42), we obtain

$$\int_{\Omega}\sum_{i,j=1}^{3}\left(\xi_{ij}P_i\frac{\partial\varphi}{\partial x_j}-M_{ij}(P)u_i\frac{\partial\varphi}{\partial x_j}\right)dx=\int_{\Omega}\sum_{i=1}^{3}f_iP_i\varphi\,dx. \tag{5.7.43}$$

Eliminating f_i from (5.7.43) by (5.7.35), we have

$$\int_{\Omega}\sum_{i,j=1}^{3}\left(\xi_{ij}P_i\frac{\partial\varphi}{\partial x_j}-M_{ij}(P)u_i\frac{\partial\varphi}{\partial x_j}\right)dx=\int_{\Omega}\sum_{i,j=1}^{3}\xi_{ij}\frac{\partial}{\partial x_j}(\varphi P_i)\,dx.$$

Hence, integration by parts gives

$$\int_{\Omega}\varphi\sum_{i,j=1}^{3}(-\xi_{ij}\varepsilon_{ij}(P)+M_{ij}(P)\varepsilon_{ij}(u))\,dx=0,\qquad\varphi\in\mathcal{D}(\Omega).$$

Thus,

$$\sum_{i,j=1}^{3}\xi_{ij}\varepsilon_{ij}(P)=\sum_{i,j=1}^{3}M_{ij}(P)\varepsilon_{ij}(u). \tag{5.7.44}$$

Let P^{rs} be a vector function such that $P^{rs}=\{P_k^{rs}\}_{k=1}^{3}$, $P_k^{rs}=x_s\delta_{kr}$. The relation (5.7.2) implies that

$$\varepsilon_{ij}\left(P^{rs}\right)=\frac{1}{2}(\delta_{ir}\delta_{js}+\delta_{is}\delta_{jr}). \tag{5.7.45}$$

Setting in (5.7.44) $P=P^{rs}$ and using (5.7.45) together with the equality $\xi_{rs}=\xi_{sr}$, we conclude

$$\xi_{rs}=\sum_{i,j=1}^{3}M_{ij}\left(P^{rs}\right)\varepsilon_{ij}(u). \tag{5.7.46}$$

Here,

$$M_{ij}\left(P^{rs}\right)=\frac{1}{Y_1Y_2Y_3}\int_{Y}\sum_{k,m=1}^{3}a_{ijkm}(x)\varepsilon_{km}\left(P^{rs}(x)-X^{rs}(x)\right)dx, \tag{5.7.47}$$

and the function X^{rs} is the solution of the following problem

$$X^{rs} \in W_1, \qquad a\left(X^{rs}, v\right) = a\left(P^{rs}, v\right), \quad v \in W_1. \tag{5.7.48}$$

By (5.7.45) and (5.7.48), the formula (5.7.47) can be rewritten as

$$M_{ij}\left(P^{rs}\right) = \frac{1}{Y_1 Y_2 Y_3} \int_Y \sum_{k,m,p,q=1}^{3} a_{pqkm}(x) \varepsilon_{km}\left(P^{rs}(x) - X^{rs}(x)\right) \times \varepsilon_{pq}\left(P^{ij}(x) - X^{ij}(x)\right) dx. \tag{5.7.49}$$

Now, writing

$$g_{rsij} = M_{ij}\left(P^{rs}\right), \tag{5.7.50}$$

from (5.7.18) and (5.7.49), we get (5.7.24). Moreover, (5.7.35) and (5.7.46) yield (5.7.22).

3. We have above defined the function u by (5.7.32) and showed that it satisfies the equation (5.7.22). The function u obviously depends on an element $L \in H^*$, so that it will be denoted as $u(L)$. Since the space H^* is separable, by the diagonal process we can choose from the sequence of bilinear forms $\{a^\varepsilon\}$ a subsequence, which still will be denoted by $\{a^\varepsilon\}$, such that

$$\begin{gathered} u^\varepsilon(Q) \to u(Q) \qquad \text{weakly in } H \text{ for all } Q \in G, \\ G \subset H^*, \ G \text{ is dense in } H^*. \end{gathered} \tag{5.7.51}$$

Here, $u^\varepsilon(Q)$ and $u(Q)$ are solutions of the problems

$$\begin{aligned} u^\varepsilon(Q) &\in H, \qquad a^\varepsilon(u^\varepsilon(Q), v) = (Q, v), \qquad v \in H, \\ u(Q) &\in H, \qquad A(u(Q), v) = (Q, v), \qquad v \in H. \end{aligned} \tag{5.7.52}$$

Thus, on the set G, which is dense in H^*, we can define a linear operator B such that

$$BQ = u(Q). \tag{5.7.53}$$

(5.7.5), (5.7.8), (5.7.14), and (5.7.26) imply

$$(A_\varepsilon v, v) \geq c_0 \|v\|_H^2, \qquad v \in H, \ \varepsilon > 0. \tag{5.7.54}$$

Hence,

$$\left(Q, A_\varepsilon^{-1} Q\right) \geq c_0 \left\|A_\varepsilon^{-1} Q\right\|_H^2, \qquad Q \in H^*, \ \varepsilon > 0, \tag{5.7.55}$$

where $A_\varepsilon^{-1} \in \mathcal{L}(H^*, H)$.

(5.7.51) and (5.7.53) yield

$$A_\varepsilon^{-1} Q \to u(Q) = BQ \qquad \text{weakly in } H \text{ for all } Q \in G. \tag{5.7.56}$$

Taking into account (5.7.51), (5.7.53), (5.7.56), we pass to the limit in (5.7.55), then we get

$$(Q, BQ) \geq c_0 \liminf \left\| A_\varepsilon^{-1} Q \right\|_H^2 \geq c_0 \|BQ\|_H^2, \qquad Q \in G. \tag{5.7.57}$$

This relation implies that $\|B\| \leq c_0^{-1}$ and the operator B can be extended by continuity to the whole space H^*. Thus, $\|B\|_{\mathcal{L}(H^*,H)} \leq c_0^{-1}$. Using (5.7.51), (5.7.53), and the inequality $\|A_\varepsilon^{-1}\|_{\mathcal{L}(H^*,H)} \leq c_0^{-1}$ for all ε (see (5.7.54) and Theorem 1.5.2), we deduce that

$$u^\varepsilon(Q) \to BQ \qquad \text{weakly in } H \text{ for all } Q \in H^*. \tag{5.7.58}$$

By the continuity, we establish that $B(Q) = u(Q)$ for all $Q \in H^*$, where $u(Q)$ is a solution of the problem (5.7.52).

Now, the inequality (5.7.57) can be strengthened as follows

$$(Q, BQ) \geq c_0 \|BQ\|_H^2, \qquad Q \in H^*. \tag{5.7.59}$$

By virtue of (5.7.3), there exists a positive number c_1 such that

$$\|A_\varepsilon\|_{\mathcal{L}(H,H^*)} \leq c_1, \qquad \text{for all } \varepsilon.$$

Hence,

$$\|A_\varepsilon^{-1} Q\|_H \geq c_1^{-1} \|Q\|_{H^*},$$

so that (5.7.55) implies

$$(Q, A_\varepsilon^{-1} Q) \geq c_0 c_1^{-2} \|Q\|_{H^*}^2, \qquad Q \in H^*,\ \varepsilon > 0.$$

This formula together with (5.7.58) yields

$$(Q, BQ) \geq c_0 c_1^{-2} \|Q\|_{H^*}^2, \qquad Q \in H^*.$$

Thus, the operator B is coercive. Hence, there exists the inverse operator $B^{-1} \in \mathcal{L}(H, H^*)$, and by (5.7.27), (5.7.52), and (5.7.53) $B^{-1} = \hat{A}$. Now, by virtue of (5.7.59), we have

$$(\hat{A}u, u) \geq c_0 \|u\|_H^2, \qquad u \in H,$$

which is equivalent to (5.7.25). Hence, the problem (5.7.22) has a unique solution and (5.7.58) holds true for any sequence of operators $\{A_\varepsilon\}$ as $\varepsilon \to 0$.

Further, let u and v be arbitrary elements from H, and $A_\varepsilon u^\varepsilon = \hat{A}u$, $A_\varepsilon v^\varepsilon = \hat{A}v$. Then, $u^\varepsilon \to u$ and $v^\varepsilon \to v$ weakly in H, and by the symmetry of the operators A_ε (see (5.7.4), (5.7.8), and (5.7.26)) we conclude that

$$(\hat{A}u, v) = \lim(\hat{A}u, v^\varepsilon) = \lim(A_\varepsilon u^\varepsilon, v^\varepsilon) = \lim(A_\varepsilon v^\varepsilon, u^\varepsilon) = \lim(\hat{A}v, u^\varepsilon) = (\hat{A}v, u).$$

Thus, the operator $\hat{A}$ is symmetric, and the theorem is proved.

Remark 5.7.4 We have above considered the approach due to Duvaut (1976) to the calculation of the G-limit operator for a periodic elastic structure. It is based on the extraction of converging subsequences and passage to the limit. Another approach to the calculation of the G-limit operator for a periodic structure is based on the two-scale expansion. The latter is widely used and has many applications, see Sanchez-Palencia (1980), Bakhvalov and Panasenko (1984), Oleinik et al. (1992). Many methods and problems of homogenization are examined by Bensoussan et al. (1978), Zhikov et al. (1993).

Remark 5.7.5 In the case of a periodical, perforated, elastic structure, the condition (5.7.5) is not satisfied, because the functions a_{ijkm} vanish in holes. However, the homogenized elastic characteristics of this structure are defined by the same formula (5.7.24) with X^{ij}'s the solutions of the problem (5.7.20). However, from the conditions $a_{ijkm} = 0$ in the holes, it follows now that zero surface forces are given on the boundary of the holes. For the proofs, see Oleinik et al. (1990).

5.7.3 Effective elasticity characteristics of granule and fiber reinforced composites

With the results of Theorem 5.7.2, one can determine the effective elasticity constants of a composite, i.e., the coefficients g_{ijkm} of the G-limit operator, see (5.7.24), assuming a corresponding structure to be Y-periodical.

For example, in the case of a granule reinforced composite, one can choose, as a model, the structure shown in Fig. 5.7.1. The dashes show the space inside of Y that is occupied by a reinforcing element (granule).

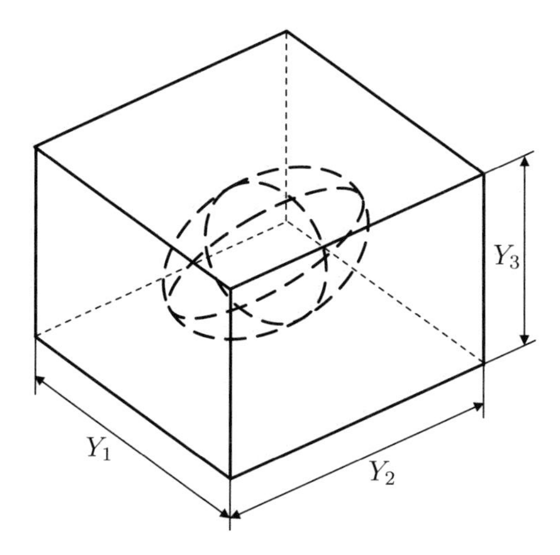

Figure 5.7.1: Structure of a granule reinforced composite

For a composite with oriented short fibers, the structure shown in Fig. 5.7.2 can be taken as a model. The dashes show the space inside of Y that is occupied by a short fiber.

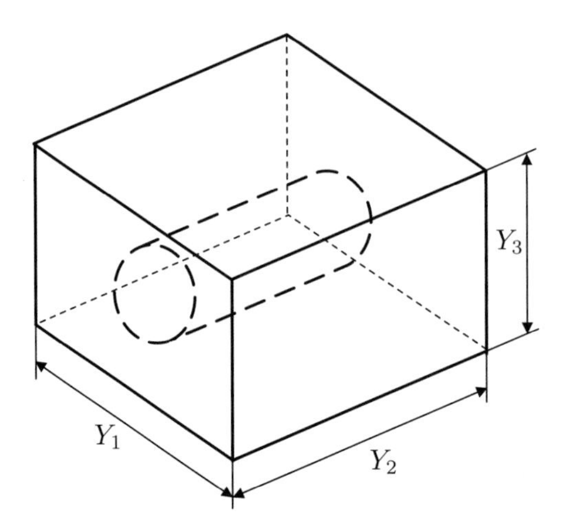

Figure 5.7.2: Structure of a composite with oriented short fibers

At last, for a unidirectional fiber reinforced composite, the model of the structure can be taken as shown in Fig. 5.7.3.

A model of unidirectional fiber reinforced composite of a more complicated structure is shown in Fig. 5.7.4.

There exist a number of distinct approaches giving the effective elasticity constants of composites, which are based upon various physical reasonings (see, e.g., Van Pho Phy (1971b), Sendeckyj (1974), Pobedria (1984)).

Given below are some simple formulas used to determine the effective elasticity constants of composite materials.

In what follows, we will refer to the stresses defined by the Hooke law (5.7.1) in which the functions a_{ijkm} are replaced by the effective elasticity constants as macroscopic, or averaged, or homogenized stresses.

In case of granule reinforced composites which contain spheroidal isotropic inclusions that are uniformly distributed in an isotropic matrix, the material remains macroscopically isotropic, i.e., the relations between the macroscopic stresses σ_{ij} and the strains $\varepsilon_{ij}(u)$ have the form

$$\sigma_{ij} = \lambda e(u)\delta_{ij} + 2G\varepsilon_{ij}(u). \tag{5.7.60}$$

Here,

$$e(u) = \sum_{i=1}^{3} \varepsilon_{ii}(u), \tag{5.7.61}$$

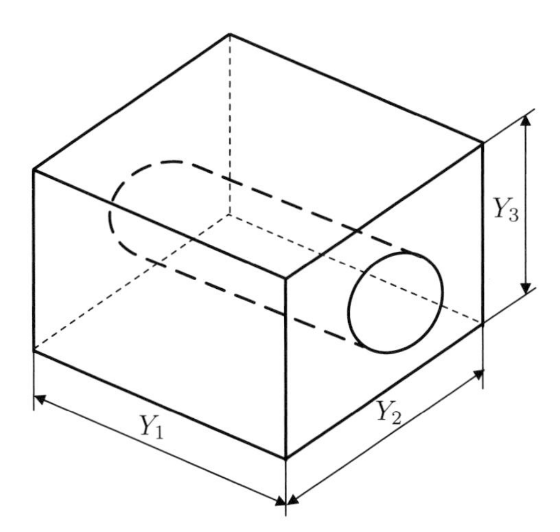

Figure 5.7.3: Structure of a unidirectional fiber reinforced composite

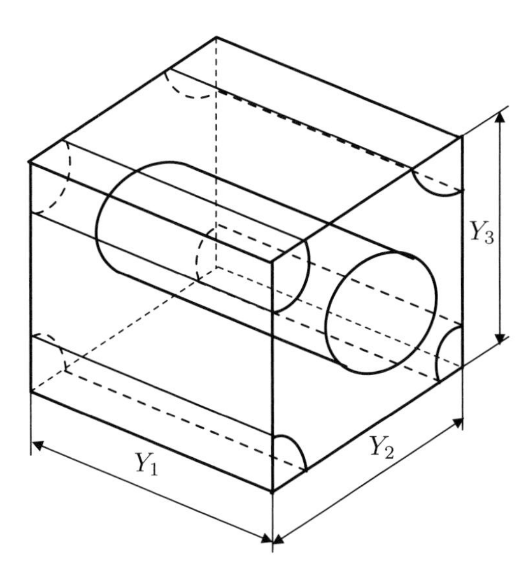

Figure 5.7.4: Complicated structure of a unidirectional fiber reinforced composite

$\varepsilon_{ij}(u)$ are defined by the formulas (5.7.2), λ and G are the effective Lamé constants, which are determined by the relations (see Sendeckyj (1974))

$$\lambda = v_1\lambda_1 + v_2\lambda_2, \qquad G = v_1 G_1 + v_2 G_2, \tag{5.7.62}$$

where λ_1, G_1 and λ_2, G_2 are the elasticity constants of the inclusion and matrix, respectively, v_1 and v_2 are the volume fractions of the inclusion and matrix,

$$v_1 + v_2 = 1. \tag{5.7.63}$$

The effective elasticity constants λ and G can also be determined by the following formulas (Sendeckyj (1974))

$$\frac{1}{\lambda} = \frac{v_1}{\lambda_1} + \frac{v_2}{\lambda_2}, \qquad \frac{1}{G} = \frac{v_1}{G_1} + \frac{v_2}{G_2}. \tag{5.7.64}$$

We point out that the relations (5.7.62) and (5.7.64) are consistent with the mixture rules.

In case of a unidirectional fiber reinforced composite (layer), the material is macroscopically orthotropic. If the x_1 axis coincides with the direction of the fibers (see Fig. 5.7.5), then the relations between the macroscopic stresses σ_{ij} and the strains $\varepsilon_{ij}(u)$ look as follows (see Sendeckyj (1974))

$$\begin{gathered}\sigma_{11} = E_{11}\varepsilon_{11}(u) + E_{12}\varepsilon_{22}(u),\\ \sigma_{22} = E_{21}\varepsilon_{11}(u) + E_{22}\varepsilon_{22}(u), \qquad \sigma_{12} = 2G\varepsilon_{12}(u).\end{gathered} \tag{5.7.65}$$

Here,

$$E_{ii} = \frac{E_i}{1-\mu_1\mu_2}, \quad i = 1,2, \qquad E_{12} = E_{21} = \mu_2 E_{11} = \mu_1 E_{22}. \tag{5.7.66}$$

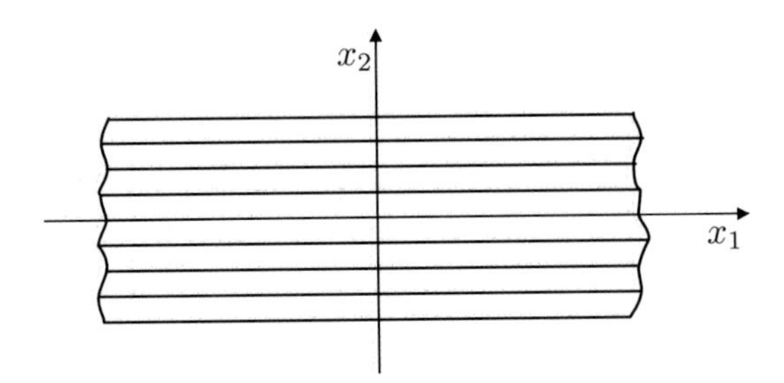

Figure 5.7.5: Unidirectional fiber reinforced composite; x_1 axis coincides with the direction of the fibers

The effective elasticity constants E_1, E_2, μ_1, μ_2, and G are defined by the formulas

$$\begin{aligned} E_1 &= v_1 E_{\mathrm f} + v_2 E_{\mathrm m}, & \mu_1 &= v_1\mu_{\mathrm f} + v_2\mu_{\mathrm m},\\ \frac{1}{E_2} &= \frac{v_1}{E_{\mathrm f}} + \frac{v_2}{E_{\mathrm m}}, & \frac{1}{G} &= \frac{v_1}{G_{\mathrm f}} + \frac{v_2}{G_{\mathrm m}},\\ \mu_1 E_2 &= \mu_2 E_1, & v_1 + v_2 &= 1. \end{aligned} \tag{5.7.67}$$

Here v_1 and v_2 are the volume fractions of the fiber and matrix, E_{f}, E_{m}, G_{f}, G_{m}, μ_{f}, and μ_{m} are the elasticity modules, shear modules, and the Poisson ratios of the fiber and matrix.

As noted by Sendeckyj (1974), the formulas (5.7.67) are in good agreement with experimental data.

From (5.7.66) and (5.7.67) one sees that, for given materials, when E_{f}, E_{m}, G_{f}, G_{m}, μ_{f}, and μ_{m} are fixed, the effective elasticity constants E_{ij}, E_i, G, and μ_i depend on the volume fraction v_1 of the fibers. Thus, further we will use the notations

$$v = v_1, \quad E_{ij}(v) = E_{ij}, \quad G(v) = G, \quad E_i(v) = E_i, \quad \mu_i(v) = \mu_i. \tag{5.7.68}$$

Notice that, due to (5.7.66), $E_{12}(v) = E_{21}(v)$, and (5.7.67) implies $v_2 = 1 - v$.

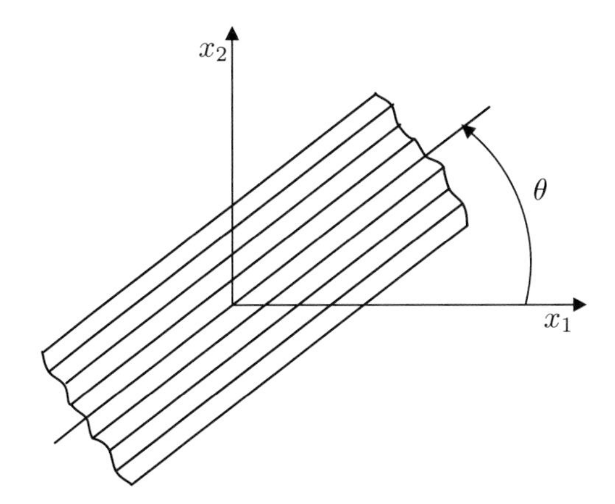

Figure 5.7.6: Composite with fibers directed at angle θ to the x_1 axis

Now, let the fibers be directed at angle θ to the x_1 axis (Fig. 5.7.6). Then, the relations between the macroscopic stresses and strains in the x_1 and x_2 coordinate axes are of the form (see Malmeister et al. (1980))

$$\begin{aligned}
\sigma_{11} &= B_{11}(\theta, v)\varepsilon_{11}(u) + B_{12}(\theta, v)\varepsilon_{22}(u) + 2B_{16}(\theta, v)\varepsilon_{12}(u), \\
\sigma_{22} &= B_{21}(\theta, v)\varepsilon_{11}(u) + B_{22}(\theta, v)\varepsilon_{22}(u) + 2B_{26}(\theta, v)\varepsilon_{12}(u), \\
\sigma_{12} &= \sigma_{21} = B_{61}(\theta, v)\varepsilon_{11}(u) + B_{62}(\theta, v)\varepsilon_{22}(u) + 2B_{66}(\theta, v)\varepsilon_{12}(u).
\end{aligned} \tag{5.7.69}$$

Here, the elasticity constants $B_{ij}(\theta, v)$ are the components of a tensor of rank 4, which are defined by the formulas

$$\begin{aligned}
&B_{11}(\theta, v) = E_{11}(v)\cos^4\theta + E_{22}(v)\sin^4\theta + (4G(v) + 2E_{12}(v))\sin^2\theta\cos^2\theta, \\
&B_{12}(\theta, v) = B_{21}(\theta, v) \\
&= (E_{11}(v) + E_{22}(v) - 4G(v))\sin^2\theta\cos^2\theta + E_{12}(v)(\cos^4\theta + \sin^4\theta),
\end{aligned}$$

$$
\begin{aligned}
B_{22}(\theta, v) &= E_{11}(v)\sin^4\theta + E_{22}(v)\cos^4\theta + (4G(v) + 2E_{12}(v))\sin^2\theta\cos^2\theta, \\
B_{16}(\theta, v) &= B_{61}(\theta, v) \\
&= \sin\theta\cos\theta[(E_{11}(v) - E_{12}(v))\cos^2\theta - (E_{22}(v) - E_{21}(v))\sin^2\theta - 2G\cos 2\theta], \\
B_{26}(\theta, v) &= B_{62}(\theta, v) \\
&= \sin\theta\cos\theta[(E_{11}(v) - E_{12}(v))\sin^2\theta - (E_{22}(v) - E_{21}(v))\cos^2\theta + 2G\cos 2\theta], \\
B_{66}(\theta, v) &= (E_{11}(v) + E_{22}(v) - 2E_{12}(v))\sin^2\theta\cos^2\theta + G(v)\cos^2 2\theta.
\end{aligned}
\tag{5.7.70}
$$

Now, let us consider a composite formed by two identical unidirectional fiber reinforced layers such that the directions of the fibers of the layers form with the x_1 axis angles θ and $-\theta$ (Fig. 5.7.7).

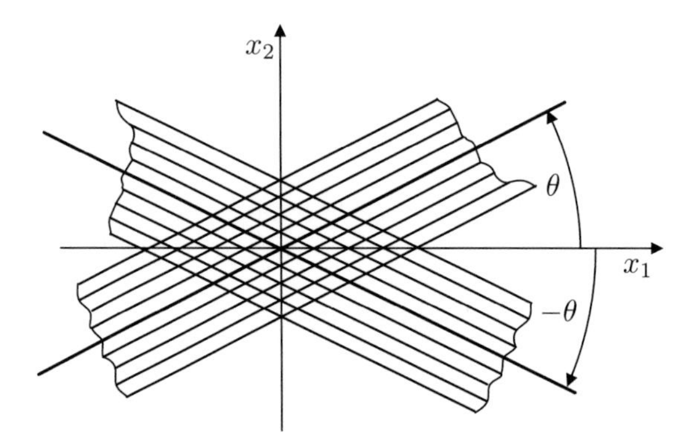

Figure 5.7.7: Composite formed by two identical fiber reinforced layers; the directions of the fibers form with the x_1 axis angles θ and $-\theta$

(5.7.69) and (5.7.70) yield this composite to be orthotropic, so that the relations between macroscopic stresses and strains in the x_1 and x_2 coordinate axes are of the following form

$$
\begin{aligned}
\sigma_{11} &= B_{11}(\theta, v)\varepsilon_{11}(u) + B_{12}(\theta, v)\varepsilon_{22}(u), \\
\sigma_{22} &= B_{21}(\theta, v)\varepsilon_{11}(u) + B_{22}(\theta, v)\varepsilon_{22}(u), \\
\sigma_{12} &= \sigma_{21} = 2B_{66}(\theta, v)\varepsilon_{12}(u),
\end{aligned}
\tag{5.7.71}
$$

where the coefficients $B_{ij}(\theta, v)$ are defined by (5.7.70).

5.7.4 Optimization of the effective elasticity constants of a composite

Setting of the optimization problem

Theorem 5.7.2 gives a method of calculation of the effective elasticity constants for a Y-periodical structure. Now, we will consider the optimization problem when a Y-periodical structure is a control.

Let the main period $Y = [0, Y_1] \times [0, Y_2] \times [0, Y_3]$ contain an inclusion. Denote by G the closed domain occupied by the inclusion, and by S its boundary.

Assume that the elasticity properties of the inclusion and matrix are characterized by the elasticity constants a^{in}_{ijkm} and a^{m}_{ijkm} (see (5.7.1)) which do not depend on the coordinates. Then, in the main period Y, the coefficients a_{ijkm} are defined by the formula

$$a_{ijkm}(x) = \begin{cases} a^{\text{in}}_{ijkm} & \text{if } x \in G, \\ a^{\text{m}}_{ijkm} & \text{if } x \in Y \setminus G. \end{cases} \tag{5.7.72}$$

G is supposed to be a star-shaped domain with respect to the point x^0 located at the center of Y, $x^0 = (\frac{Y_1}{2}, \frac{Y_2}{2}, \frac{Y_3}{2})$.

Denote by S_0 the unit sphere in $\mathbb{R}^3$ centered at x^0. Then, the boundary S of the domain G is defined by a continuous positive function $f(s)$ of a point $s \in S_0$ through the formula

$$x = x^0 + f(s)n(s), \tag{5.7.73}$$

$n(s)$ being the unit vector which is normal to the surface S_0 at the point s.

Let $C(S_0)$ be the space of real-valued continuous functions on S_0 equipped with the norm

$$\|u\|_{C(S_0)} = \max_{s \in S_0} |u(s)|. \tag{5.7.74}$$

Define a set U by the following relation

$$U = \Big\{ f \,|\, f \in C(S_0), \ \sup_{s,s' \in S_0} \frac{|f(s) - f(s')|}{\left[\sum_{i=1}^3 (x_i(s) - x_i(s'))^2\right]^{1/2}} \leq c_0, \quad f_1(s) \leq f(s) \leq f_2(s), s \in S_0 \Big\}. \tag{5.7.75}$$

Here, $x_i(s) = x_i^0 + n_i(s)$, $n_i(s)$ being the coordinate of the vector $n(s)$, and

$$\left. \begin{array}{l} f_1, f_2 \in C(S_0),\ c_1 \leq f_1(s) \leq f_2(s),\ s \in S_0,\ c_0 \text{ and } c_1 \text{ are} \\ \text{positive constants.} \end{array} \right\} \tag{5.7.76}$$

The function f_2 is assumed to satisfy the following condition

$$x^0 + f_2(s)n(s) \in Y, \qquad s \in S_0. \tag{5.7.77}$$

By the formula (5.7.73), to each function $f \in U$ there corresponds the domain $G(f)$ occupied by the inclusion:

$$G(f) = \{ x \,|\, x = x^0 + \alpha f(s)n(s), \ 0 \leq \alpha < 1, \ s \in S_0 \}. \tag{5.7.78}$$

The boundary $S(f)$ of the domain $G(f)$ is given by the formula

$$S(f) = \{ x \,|\, x = x^0 + f(s)n(s), \ s \in S_0 \}. \tag{5.7.79}$$

Further, in accordance with (5.7.72), the coefficients a^f_{ijkm} are defined on the main period Y by the relation

$$a^f_{ijkm}(x) = \begin{cases} a^{\text{in}}_{ijkm}, & \text{if } x \in G(f), \\ a^{\text{m}}_{ijkm} & \text{if } x \in Y \setminus G(f), \end{cases} \tag{5.7.80}$$

and in the space W_1 (see Lemma 5.7.1) the following bilinear form is well defined

$$a^f(u, v) = \int_Y \sum_{i,j,k,m=1}^{3} a^f_{ijkm}(x)\varepsilon_{km}(u)\varepsilon_{ij}(v)\, dx. \tag{5.7.81}$$

One can easily verify the following statement.

Lemma 5.7.2 *Let a^{in}_{ijkm} and a^{m}_{ijkm} be constants satisfying the conditions of symmetry* (5.7.4) *and positive definiteness* (5.7.5). *Assume the coefficients a^f_{ijkm} to be defined by the relations* (5.7.78) *and* (5.7.80). *Then, for any $f \in U$, the bilinear form a^f determined by the formula* (5.7.81) *is symmetric, continuous, and coercive in W_1.*

Lemma 5.7.2 implies that, for any $f \in U$, there exists a unique function $X^{ij}(f)$ such that

$$X^{ij}(f) \in W_1, \qquad a^f\left(X^{ij}(f), v\right) = a^f\left(P^{ij}, v\right), \quad v \in W_1, \ i, j = 1, 2, 3, \tag{5.7.82}$$

P^{ij} being defined by (5.7.19).

By virtue of Theorem 5.7.2, to every function $f \in U$ there correspond the effective elasticity constants $g_{ijkm}(f)$ determined by the formula

$$g_{ijkm}(f) = \frac{1}{Y_1 Y_2 Y_3} a^f\left(X^{ij}(f) - P^{ij}, X^{km}(f) - P^{km}\right), \qquad i, j, k, m = 1, 2, 3. \tag{5.7.83}$$

Now, considering the function f as a control, we define the set of admissible controls by

$$U_{\text{ad}} = \{\, f \mid f \in U, \ \Psi_k(f) \le 0, \ k = 1, 2, \ldots, l \,\}. \tag{5.7.84}$$

Here,

$$\left.\begin{array}{l} f \to \Psi_k(f) \text{ is a continuous mapping of } U \text{ equipped with} \\ \text{the topology generated by the one of } C(S_0) \text{ into } \mathbb{R}, \\ k = 1, 2, \ldots, l. \end{array}\right\} \tag{5.7.85}$$

If the functional Ψ_1 is chosen in the form

$$\Psi_1(f) = \int_{G(f)} dx - c, \tag{5.7.86}$$

c being a positive constant, then Ψ_1 defines a restriction on the volume of inclusion, which can be dictated by the weight of the composite or its cost.

Since

$$\int_{G(f)} dx = \int_Y y(x)\,dx, \qquad y(x) = \begin{cases} 1, & \text{if } x \in G(f), \\ 0, & \text{if } x \in Y \setminus G(f), \end{cases}$$

the Lebesgue theorem implies the functional (5.7.86) to meet the condition (5.7.85).

Let a goal functional be of the form

$$\Phi(f) = \sum_{i,j,k,m=1}^{3} (g_{ijkm}(f) - c_{ijkm})^2, \tag{5.7.87}$$

c_{ijkm} being given constants.

The optimization problem consists in finding a function f_0 such that

$$f_0 \in U_{\text{ad}}, \qquad \Phi(f_0) = \inf_{f \in U_{\text{ad}}} \Phi(f). \tag{5.7.88}$$

The existence theorem

Theorem 5.7.3 *Let a nonempty set U_{ad} be defined by the relations* (5.7.75), (5.7.76), (5.7.84), (5.7.85), *and let* (5.7.77) *hold. Assume that a^{in}_{ijkm} and a^{m}_{ijkm} are constants satisfying the conditions of symmetry and positive definiteness* (5.7.4), (5.7.5), *and the coefficients a^f_{ijkm} are determined by* (5.7.78), (5.7.80). *Further, let constants $g_{ijkm}(f)$ be defined by the formulas* (5.7.19), (5.7.81)–(5.7.83) *and let a goal functional be determined by* (5.7.87). *Then, the problem* (5.7.88) *has a solution.*

To prove Theorem 5.7.3, we need some auxiliary statements.

Introduce spherical coordinates r, θ, φ, which are connected with Cartesian coordinates by the formulas

$$x_1 = x_1^0 + r \sin\theta \cos\varphi, \quad x_2 = x_2^0 + r\sin\theta \sin\varphi, \quad x_3 = x_3^0 + r\cos\theta, \tag{5.7.90}$$

where $r \geq 0$, $0 \leq \theta \leq \pi$, and $0 \leq \varphi < 2\pi$. Then a point s of the unit sphere S_0 is determined by the θ and φ coordinates ($r = 1$), i.e., $s = (\theta, \varphi)$. Let

$$Q = \{\, (\theta, \varphi) \,|\, 0 < \theta < \pi,\ 0 < \varphi < 2\pi \,\}.$$

According to (5.7.79), the function $P_f \colon Q \to S(f)$ defined by the relations

$$\begin{aligned} &\theta, \varphi \to P_f(\theta, \varphi) = \{P_f^i(\theta,\varphi)\}_{i=1}^3, \\ &P_f^1(\theta,\varphi) = x_1^0 + f(\theta,\varphi)\sin\theta\cos\varphi, \\ &P_f^2(\theta,\varphi) = x_2^0 + f(\theta,\varphi)\sin\theta\sin\varphi, \\ &P_f^3(\theta,\varphi) = x_3^0 + f(\theta,\varphi)\cos\theta, \end{aligned} \tag{5.7.91}$$

is a map of the surface S_f.

First, assume that $f \in U$ and f is a continuously differentiable function on S_0. (We recall that the notion of differentiability of a function defined on a smooth manifold is introduced via local maps; see, e.g., Lions and Magenes (1972).) Then, the surface measure ds is well defined on S_f. The surface measure of $S_f \setminus P_f(Q)$ is equal to zero, and so by the theorem on the area of a parametric manifold (see, e.g., (Schwartz (1967)), we obtain

$$\int_{S_f} ds = \iint_Q D_f(\theta, \varphi)\, d\theta\, d\varphi, \tag{5.7.91}$$

where

$$D_f(\theta, \varphi) = \left[\sum_{i=1}^{3} \left(\frac{\partial P_f^i}{\partial \theta} \right)^2 \sum_{i=1}^{3} \left(\frac{\partial P_f^i}{\partial \varphi} \right)^2 - \left(\sum_{i=1}^{3} \frac{\partial P_f^i}{\partial \theta} \frac{\partial P_f^i}{\partial \varphi} \right)^2 \right]^{\frac{1}{2}}. \tag{5.7.92}$$

If f is an arbitrary function from U, then, due to (5.7.75), f satisfies the Lipschitz condition with the constant c_0. Thus, the partial derivatives of the function f exist a.e. in Q, and the following estimates are valid

$$\left| \frac{\partial f}{\partial \theta} \right| \leq c_0, \qquad \left| \frac{\partial f}{\partial \varphi} \right| \leq c_0 \quad \text{a.e. in } Q.$$

This relation together with (5.7.90) implies

$$\left| \frac{\partial P_f}{\partial \theta} \right| \leq c_2, \qquad \left| \frac{\partial P_f}{\partial \varphi} \right| \leq c_2 \quad \text{a.e. in } Q,\ f \in U. \tag{5.7.93}$$

For any $u \in \mathcal{D}(Q)$, by applying the integration by parts formula (e.g., Schwartz (1967)), we obtain

$$\iint_Q P_f \frac{\partial u}{\partial \theta}\, d\theta\, d\varphi = - \iint_Q \frac{\partial P_f}{\partial \theta} u\, d\theta\, d\varphi,$$

$$\iint_Q P_f \frac{\partial u}{\partial \varphi}\, d\theta\, d\varphi = - \iint_Q \frac{\partial P_f}{\partial \varphi} u\, d\theta\, d\varphi.$$

Since $P_f \in \left(W^1_\infty(Q)\right)^3$, there exists a sequence of smooth functions $\{P_n\}$ meeting the conditions

$$\begin{aligned} &\frac{\partial P_n}{\partial \theta} \to \frac{\partial P_f}{\partial \theta} \quad \text{a.e. in } Q, \\ &\frac{\partial P_n}{\partial \varphi} \to \frac{\partial P_f}{\partial \varphi} \quad \text{a.e. in } Q, \\ \left| \frac{\partial P_n}{\partial \theta}(\varphi, \theta) \right| \leq c_2, \qquad &\left| \frac{\partial P_n}{\partial \varphi}(\varphi, \theta) \right| \leq c_2, \qquad (\varphi, \theta) \in Q,\ \forall n. \end{aligned} \tag{5.7.94}$$

The relations (5.7.92) and (5.7.94) together with the Lebesgue theorem justify the passage to the limit as $n \to \infty$ in (5.7.91) provided P_n is substituted for P_f.

Finally, the formulas (5.7.91) and (5.7.92) define the measure of the surface S_f for any $f \in U$, which does not depend on a choice of local maps, just as in the case of a surface of the C^1 class.

Moreover, by virtue of (5.7.93), the measures of the surfaces S_f are bounded for all $f \in U$.

Thus, we have proved the following

Lemma 5.7.3 *Let a set U and a surface S_f be defined by the relations* (5.7.75), (5.7.76), *and* (5.7.79). *Then, the relations* (5.7.90)–(5.7.92) *determine the measure of the surface S_f, and there exists a number c_3 such that*

$$\int_{S_f} ds \le c_3, \qquad f \in U. \tag{5.7.95}$$

In what follows, we will need also the following lemma.

Lemma 5.7.4 *Let a set U be defined by the relations* (5.7.75) *and* (5.7.76), *and let a^{in}_{ijkm} and a^{m}_{ijkm} be constants satisfying the conditions of symmetry and positive definiteness* (5.7.4) *and* (5.7.5). *Assume that a bilinear form a^f is defined by the relations* (5.7.78), (5.7.80), *and* (5.7.81), *and a function $u(f,g)$ is a solution of the following problem*

$$u(f,g) \in W_1, \qquad a^f(u(f,g),v) = (g,v), \qquad v \in W_1, \tag{5.7.96}$$

where $g \in W_1^*$. Then, $f \to a^f$ is a continuous mapping of U, equipped with the topology induced by the $C(S_0)$-one, into $\mathcal{L}_2(W_1, W_1; \mathbb{R})$, and the function $f, g \to u(f,g)$ is a continuous mapping of $U \times W_1^*$, endowed with the topology induced by the product of the topologies of $C(S_0)$ and W_1^*, into the space W_1.

Proof. Suppose that $\{f_n\}$ is a sequence such that

$$f_n \in U \quad \forall n, \qquad f_n \to f_0 \quad \text{in } C(S_0). \tag{5.7.97}$$

Then, $f_0 \in U$ and taking into account (5.7.80), (5.7.81), and (5.7.95), for any $u, v \in W_1$, we have

$$\left| a^{f_n}(u,v) - a^{f_0}(u,v) \right| \le c_4 \|f_n - f_0\|_{C(S_0)} \|u\|_{W_1} \|v\|_{W_1}.$$

This and (5.7.97) yield

$$a^{f_n} \to a^{f_0} \qquad \text{in } \mathcal{L}_2(W_1, W_1; \mathbb{R}). \tag{5.7.98}$$

Denote by $A(f)$ the operator generated by the bilinear form a^f:

$$(A(f)u, v) = a^f(u,v), \qquad u, v \in W_1.$$

(5.7.97) and (5.7.98) imply that $f \to A(f)$ is a continuous mapping of U into $\mathcal{L}(W_1, W_1^*)$. Now, Theorem 1.8.1 implies that the function $f, g \to u(f,g)$ is a continuous mapping of $U \times W_1^*$ into W_1.

Proof of Theorem 5.7.3. Let $\{f_n\}$ be a minimizing sequence:

$$f_n \in U_{\mathrm{ad}} \quad \forall n, \qquad \lim_{n\to\infty} \Phi(f_n) = \inf_{f\in U_{\mathrm{ad}}} \Phi(f). \tag{5.7.99}$$

By virtue of (5.7.75), the functions f_n satisfy the Lipschitz condition with the constant c_0. Thus, the generalized Arzelá theorem (see, e.g., Kolmogorov and Fomin (1975)) implies that from the sequence $\{f_n\}$ one can extract a subsequence $\{f_\mu\}$ such that

$$f_\mu \to z \quad \text{in } C(S_0). \tag{5.7.100}$$

(5.7.75), (5.7.84), (5.7.85), and (5.7.100) yield that $z \in U_{\mathrm{ad}}$. Further, Lemma 5.7.4 and (5.7.100) imply

$$X^{ij}(f_\mu) \to X^{ij}(z) \qquad \text{in } W_1 \text{ as } \mu \to \infty,\ i,j = 1,2,3, \tag{5.7.101}$$

where $X^{ij}(f_\mu)$ and $X^{ij}(z)$ are the solutions of the problem (5.7.82) for f equal to f_μ and z. From Lemma 5.7.4 and (5.7.100), we deduce that

$$a^{f_\mu} \to a^z \qquad \text{in } \mathcal{L}_2(W_1, W_1; \mathbb{R}).$$

Hence, (5.7.83), and (5.7.101) imply

$$\lim_{\mu\to\infty} g_{ijkm}(f_\mu) = g_{ijkm}(z). \tag{5.7.102}$$

Now, (5.7.87), (5.7.99), and (5.7.102) yield

$$\Phi(z) = \inf_{f\in U_{\mathrm{ad}}} \Phi(f).$$

Thus, the function $f_0 = z$ is a solution of the problem (5.7.88) and the theorem is proved.

Remark 5.7.6 Denote by n the number of the effective elasticity constants g_{ijkm}, and let $g \to \Psi(g)$ is a continuous mapping of $\mathbb{R}^n$ into $\mathbb{R}$. On the set U defined by (5.7.75) and (5.7.76), introduce a function Φ_0 by $\Phi_0(f) = \Psi(g(f))$, where $g(f) = \{g_{ijkm}(f)\}$, and the functions $f \to g_{ijkm}(f)$ are determined by (5.7.83). Then, Theorem 5.7.3 remains valid provided the functional Φ_0 is substituted for the functional Φ from (5.7.87).

5.7.5 Optimization of a granule reinforced composite

Let Ω be a bounded domain in $\mathbb{R}^3$ occupied by a granule reinforced composite, and let S be the boundary of Ω, $S = S_1 \cup S_2$, $S_1 \cap S_2 = \varnothing$, $\operatorname{mes} S_1 > 0$. The composite is fastened on S_1, while the part S_2 is subject to a load.

Making the assumption that the composite contains spherical inclusions distributed in an isotropic matrix, we will use the formulas (5.7.60)–(5.7.63) to determine the macroscopic stresses in the material.

According to (5.7.60)–(5.7.63), the strain energy generates the following bilinear form

$$\begin{aligned} A(u,v) &= \int_{\Omega} \sum_{i,j=1}^{3} \sigma_{ij}(u)\varepsilon_{ij}(v)\,dx \\ &= \int_{\Omega} \left[\lambda e(u)e(v) + 2G \sum_{i,j=1}^{3} \varepsilon_{ij}(u)\varepsilon_{ij}(v) \right] dx. \end{aligned} \tag{5.7.103}$$

Here, λ and G are the effective elasticity constants. Introducing the notation $g = v_1$ and taking notice of (5.7.62) and (5.7.63), we get

$$\lambda = g\lambda_1 + (1-g)\lambda_2, \qquad G = gG_1 + (1-g)G_2. \tag{5.7.104}$$

Suppose that the volume fraction of the inclusion, g, is a function of x. This assumption requires indicating the method of calculation of the function $g(x)$ provided the structure of the composite (the distribution of the inclusion and matrix over the domain Ω) is known.

Given an averaging radius ρ that is sufficiently large as compared to the dimensions of the granules, the function $g(x)$ can be determined by

$$g(x) = \int_{\mathbb{R}^3} \omega_\rho(x-y)\,\xi(y)\,dy, \tag{5.7.105}$$

where ω_ρ is the averaging kernel (see Subsec. 1.6.4), and the function $\xi(y)$ equals one if there is an inclusion at the point y, and vanishes otherwise. The value ρ is considered to be sufficiently small as compared to the dimensions of the composite.

To calculate $g(x)$ at any point $x \in \overline{\Omega}$ by the formula (5.7.105), the function $\xi(y)$ should be extended to a domain larger than $\overline{\Omega}$. Suppose we have made such an extension. Then, the function $g(x)$ is smooth and even infinitely differentiable in $\overline{\Omega}$ if so is the function ω_ρ.

The formulas (5.7.104) determining the effective elasticity constants are obtained by averaging over a sufficiently large volume under the supposition that the inclusions are uniformly distributed over the volume of the matrix. (We stress that, in Theorem 5.7.2, the effective elasticity constants were obtained under the condition of the Y-periodicity, which in case of a granule reinforced composite corresponds to a uniform distribution of inclusions over the volume of the matrix.) Hence, in order to apply the formulas (5.7.104) in the case when g is a function, the latter must be smooth and its partial derivatives at every point must be small in absolute value, since increase of the absolute values of the derivatives leads to decrease of accuracy of the formulas (5.7.104).

Thus, in what follows, in examining the optimization problem for the granule reinforced composite, the function g is supposed to belong to the set Y_{lc} of the form

$$Y_{lc} = \{\, g \,|\, g \in W_2^l(\Omega),\ \|g\|_{W_2^l(\Omega)} \leq c_1,\ c_2 \leq g(x) \leq c_3 \,\}, \tag{5.7.106}$$

where

$$c = (c_1, c_2, c_3), \qquad c_1 > 0, \qquad 0 < c_2 < c_3 < 1, \qquad l \geq 2. \tag{5.7.107}$$

Introducing the notation $A_g = A$ and taking note of (5.7.103) and (5.7.104), we get

$$A_g(u,v) = \int_\Omega \{[g\lambda_1 + (1-g)\lambda_2]e(u)e(v) \\ +2[gG_1 + (1-g)G_2] \sum_{i,j=1}^{3} \varepsilon_{ij}(u)\varepsilon_{ij}(v)\Big\}\, dx. \tag{5.7.108}$$

The bilinear form A_g is defined on the space H determined by (5.7.7).

Theorem 5.7.4 *Let λ_1, λ_2, G_1, and G_2 be positive numbers, $g \in L_\infty(\Omega)$, $0 \leq g \leq 1$ a.e. in Ω, let a bilinear form A_g be defined by* (5.7.108)*, and let a space H be determined by* (5.7.7)*. Then, for any $L \in H^*$, there exists a unique function $u(g,L)$ such that*

$$u(g,L) \in H, \qquad A_g(u(g,L),v) = (L,v), \quad v \in H. \tag{5.7.109}$$

Proof. The bilinear form A_g is symmetric and continuous, and by the Korn inequality it is coercive in H. Hence, the existence of a unique function $u(g,L)$ satisfying the conditions (5.7.109) is implied by the Riesz theorem.

Further, assume that the composite is subject to the load $L \in H^*$ defined by the relations (5.7.9) and (5.7.10), and the function g is a control. So, the solution of the problem (5.7.109) will be denoted by u_g, i.e., $u_g = u(g,L)$.

Let there be given functionals $\Psi_k \colon Y_{lc} \times H \to \mathbb{R}$ such that

$g, u \to \Psi_k(g,u)$ is a continuous mapping of $Y_{lc} \times H$ equipped with the topology generated by the product of the $W_2^l(\Omega)$-weak topology and the H-strong topology into $\mathbb{R}$, $k = 0,1,2,\dots,p$. (5.7.110)

Now, define a set of admissible controls as

$$U_{\text{ad}} = \{\, g \,|\, g \in Y_{lc},\ \Psi_k(g,u_g) \leq 0,\ k = 1,2,\dots,p \,\}, \tag{5.7.111}$$

and let a goal functional be of the form

$$\Phi(g) = \Psi_0(g,u_g). \tag{5.7.112}$$

The optimization problem is to find a function g_0 such that

$$g_0 \in U_{\text{ad}}, \qquad \Phi(g_0) = \inf_{g \in U_{\text{ad}}} \Phi(g). \tag{5.7.113}$$

The functionals Ψ_k, $k = 1, 2, \ldots, p$, can determine restrictions on the strength, stiffness, volume fraction of inclusion in the composite, and so forth.

If the goal functional defines the mass of the composite, then it is of the form

$$\Phi(g) = (\rho_{\text{in}} - \rho_{\text{m}}) \int_\Omega g\, dx + \rho_m \int_\Omega dx. \tag{5.7.114}$$

Here, ρ_{in} and ρ_{m} are the densities of the inclusion and matrix.

If the functional Ψ_1 defines a restriction on the stiffness, it can be chosen as

$$\Psi_1(g, u_g) = A_g(u_g, u_g) - c_1, \tag{5.7.115}$$

c_1 being a positive number.

Since the embedding of Y_{lc} into $C(\overline{\Omega})$ is compact, the functional Ψ_1 from (5.7.115) can be shown to satisfy the condition (5.7.110).

If the functional Ψ_2 determines a restriction on the volume fraction of the inclusion in the composite, then it is of the form

$$\Psi_2(g, u_g) = \int_\Omega g\, dx - c_1, \qquad c_1 = \text{const} > 0.$$

Theorem 5.7.5 *Let* λ_1, λ_2, G_1, *and* G_2 *be positive numbers, let a bilinear form* A_g *be defined by* (5.7.108), *and let* $u_g = u(g, L)$ *be the solution of the problem* (5.7.109). *Assume a nonempty set* U_{ad} *is defined by the relations* (5.7.106), (5.7.107), (5.7.110), *and* (5.7.111), *and the goal functional by* (5.7.112). *Then, the problem* (5.7.113) *has a solution.*

The proof of Theorem 5.7.5 is based upon the fact that, for a three-dimensional bounded domain Ω, the embedding of $W_2^l(\Omega)$ into $C(\overline{\Omega})$ is compact for $l \geq 2$. This proof is analogous to the one of Theorem 2.2.2, and we therefore omit it.

5.7.6 Optimization of composite laminate shells

Effective relations of elasticity and the strain energy of a laminate shell

Consider a shell consisting of unidirectional fiber laminates provided the distribution of the laminates is symmetric with respect to the midsurface.

Moreover, next to every laminate the reinforcement direction of which forms an angle θ with the coordinate line, there is a layer with the reinforcement direction $-\theta$ (see Fig. 5.7.7). Here, we assume that the coordinate lines on the midsurface coincide with the principal curvature lines of this surface.

In accordance with results of Subsec. 5.7.3, two identical layers which are located at angles θ and $-\theta$ are treated as one layer in which the relations between the macroscopic stresses and strains are determined by the formulas (5.7.70) and (5.7.71).

We will use the classical shell theory, see Section 4.5. Then, (5.7.71) and (4.5.4) imply that the stresses σ_{11}, σ_{22}, σ_{12} at a point located at distance γ from the midsurface of the shell, which are denoted by σ_{11}^{γ}, σ_{22}^{γ}, σ_{12}^{γ}, are defined by the following relations

$$\begin{aligned}
\sigma_{11}^{\gamma} &= B_{11}(\theta(\gamma), v(\gamma))(\varepsilon_1 + \gamma\chi_1) + B_{12}(\theta(\gamma), v(\gamma))(\varepsilon_2 + \gamma\chi_2), \\
\sigma_{22}^{\gamma} &= B_{21}(\theta(\gamma), v(\gamma))(\varepsilon_1 + \gamma\chi_1) + B_{22}(\theta(\gamma), v(\gamma))(\varepsilon_2 + \gamma\chi_2), \\
\sigma_{12}^{\gamma} &= \sigma_{21}^{\gamma} = B_{66}(\theta(\gamma), v(\gamma))(\varepsilon_{12} + 2\gamma\chi_{12}).
\end{aligned} \tag{5.7.116}$$

Here, ε_1, ε_2, ε_{12}, χ_1, χ_2, χ_{12} are the components of the strains that are determined by the components of displacements of the midsurface of the shell through the formulas (4.5.5), and $B_{ij}(\theta(\gamma), v(\gamma))$ are defined by the relations (5.7.66)–(5.7.68), (5.7.70). (Notice that in (4.5.5) v is the component of the vector function of displacements of points (α, β) of the midsurface, while in the formulas (5.7.69), (5.7.70) v is the function of volume fraction of the fibers depending on γ.) To every $\gamma \in \left[0, \frac{h}{2}\right] \setminus \{\gamma_i\}$, where h is the thickness of the shell and $\{\gamma_i\}$ is the set of points belonging to the boundaries of the laminates on the intervals $\left[0, \frac{h}{2}\right]$, we place in correspondence $\theta(\gamma)$ – the angle of orientation of the fibers, i.e., the angle between the direction of the fibers and the α coordinate line, and the volume fraction of the fibers $v(\gamma)$.

By virtue of the above assumption on the symmetry of the distribution of the laminates with respect to the midsurface, we have

$$\theta(\gamma) = \theta(-\gamma), \qquad v(\gamma) = v(-\gamma). \tag{5.5.117}$$

The strains ε_1^{γ}, ε_2^{γ}, and $\varepsilon_{12}^{\gamma}$ at a point located at a distance γ from the midsurface are determined by the formulas (4.5.4), and in accordance with (4.5.7) and (4.5.8) the strain energy of the shell is given by

$$\Phi = \frac{1}{2} \iint\limits_{\Omega} \left[\int_{-\frac{h}{2}}^{\frac{h}{2}} (\sigma_{11}^{\gamma}\varepsilon_1^{\gamma} + \sigma_{22}^{\gamma}\varepsilon_2^{\gamma} + \sigma_{12}^{\gamma}\varepsilon_{12}^{\gamma})\, d\gamma \right] AB\, d\alpha\, d\beta. \tag{5.7.118}$$

Here, α and β are the curvilinear coordinates on the midsurface of the shell, A and B are the coefficients of the first quadratic form. In (5.7.118), we substituted A and B for H_1 and H_2, neglecting the values $\frac{\gamma}{R_1}$ and $\frac{\gamma}{R_2}$ as they are small as compared to unit, R_1 and R_2 being the principal curvature radii of the midsurface of the shell.

Substituting in (5.7.118) the expressions of σ_{ij}^{γ} from (5.7.116) and the expressions of ε_1^{γ}, ε_2^{γ}, $\varepsilon_{12}^{\gamma}$ from (4.5.4), taking note of (5.7.117), and integrating in γ, we obtain

$$\Phi = \frac{1}{2} \iint_{\Omega} \big[c_{11}(h,\theta,v)\varepsilon_1^2 + 2c_{12}(h,\theta,v)\varepsilon_1\varepsilon_2 + c_{22}(h,\theta,v)\varepsilon_2^2 + c_{66}(h,\theta,v)\varepsilon_{12}^2 \big] AB\, d\alpha\, d\beta + \frac{1}{2} \iint_{\Omega} \big[D_{11}(h,\theta,v)\chi_1^2 + 2D_{12}(h,\theta,v)\chi_1\chi_2 + D_{22}(h,\theta,v)\chi_2^2 + 4D_{66}(h,\theta,v)\chi_{12}^2 \big] AB\, d\alpha\, d\beta, \tag{5.7.119}$$

where

$$c_{ij}(h,\theta,v) = 2\int_0^{\frac{h}{2}} B_{ij}(\theta(\gamma), v(\gamma))\, d\gamma,$$
$$D_{ij}(h,\theta,v) = 2\int_0^{\frac{h}{2}} B_{ij}(\theta(\gamma), v(\gamma))\gamma^2\, d\gamma. \tag{5.7.120}$$

Thus, the strain energy of a laminate shell is defined by the formula (5.7.119) in which the elasticity constants c_{ij} and D_{ij} depend on the thickness of the shell as well as on the distributions of the angles of the orientation of the fibers and their volume fraction.

Step and continuous controls

Let the midsurface of the shell go through the boundary of a laminate, and let n be the number of the laminates which are above the midsurface. Denote by b_i the angle of the orientation of fibers, and by g_i the volume fraction of fibers in the i-th laminate, $i = 1, 2, \ldots, n$.

Further, suppose that the laminates are of the same thickness $l = \frac{h}{2n}$. Then, $\theta(\gamma)$ and $v(\gamma)$ are the step functions of the following form

$$\theta(\gamma) = b_i, \qquad v(\gamma) = g_i, \qquad \text{for } \gamma \in ((i-1)l, il),\ i = 1, 2, \ldots, n. \tag{5.7.121}$$

On the boundaries of the laminates, i.e., at the points $\gamma_k = kl$, $k = 1, 2, \ldots, n-1$, the functions θ and v are not well defined. Suppose the conditions (5.7.117) are satisfied.

Define sets G_1 and G_2 by the formulas

$$G_1 = \big\{ b \,|\, b = \{b_i\} \in \mathbb{R}^n,\ b_i \in [0, \frac{\pi}{2}],\ i = 1, 2, \ldots, n \big\},$$
$$G_2 = \big\{ g \,|\, g = \{g_i\} \in \mathbb{R}^n,\ g_i \in [e_1, e_2],\ i = 1, 2, \ldots, n \big\},$$

e_1 and e_2 being constants, $0 < e_1 < e_2 < 1$.

By (5.7.121), to every vector $b \in G_1$ there corresponds the function θ, and to every $g \in G_2$ the function v. We will denote these functions as θ_b and v_g.

Introduce mappings $P_{ij}: G_1 \times G_2 \to \mathbb{R}$ and $G_{ij}: G_1 \times G_2 \to \mathbb{R}$ by the following relations

$$P_{ij}(b,g) = c_{ij}(h, \theta_b, v_g), \qquad Q_{ij}(b,g) = D_{ij}(h, \theta_b, v_g), \tag{5.7.122}$$

where c_{ij} and D_{ij} are determined by (5.7.120).

By using (5.7.66)–(5.7.68) and (5.7.70), one can easily verify that P_{ij} and G_{ij} are continuous functions on $G_1 \times G_2$. Now, we can examine various problems of optimization of laminate shells in which optimization (control) parameters are the angles of the orientation of the fibers, the volume fractions of fibers in laminates, and the number of laminates. Since the thickness of a laminate is neglectibly small as compared to the thickness of the shell, the laminates can be considered to have zero thickness, and θ and v to be continuous functions. The thickness of the laminate shell, h, which is independent of the coordinates of points of the midsurface of the shell, can be treated as a control.

Now, define a set of controls by the formula

$$U = \left\{ t \,|\, t = (q, \theta, v),\ q \in [e_3, e_4],\ \theta, v \in C([0, e_4]),\ 0 \le \theta \le \frac{\pi}{2},\ e_1 \le v \le e_2 \right\}, \tag{5.7.123}$$

where

$$e_1, \ \ldots\ , e_4 \text{ are positive numbers, } e_1 < e_2 < 1. \tag{5.7.124}$$

For the triple $(q, \theta, v) \in U$, denote by θ_q and v_q the restrictions of the functions θ and v on the interval $[0, q]$. On the set U, let us define the functionals

$$\begin{aligned} R_{ij}(q, \theta, v) &= 2 \int_0^q B_{ij}(\theta_q(\gamma), v_q(\gamma))\, d\gamma, \\ G_{ij}(q, \theta, v) &= 2 \int_0^q B_{ij}(\theta_q(\gamma), v_q(\gamma))\gamma^2 \, d\gamma, \end{aligned} \tag{5.7.125}$$

B_{ij} being defined by the formulas (5.7.66)–(5.7.68) and (5.7.70).

(5.7.120) and (5.7.125) imply that

$$c_{ij}(2q, \theta_q, v_q) = R_{ij}(q, \theta, v), \qquad D_{ij}(2q, \theta_q, v_q) = G_{ij}(q, \theta, v). \tag{5.7.126}$$

Lemma 5.7.5 *Let a set U be defined by the relations* (5.7.123) *and* (5.7.124), *and functionals R_{ij} and G_{ij} by the formulas* (5.7.125), (5.7.66)–(5.7.68), *and* (5.7.70). *Then, R_{ij} and G_{ij} are continuous mappings of U, equipped with the topology generated by the product of the topologies of $\mathbb{R}$ and $(C([0, e_4]))^2$, into $\mathbb{R}$.*

One can easily prove Lemma 5.7.5 by using the Lebesgue theorem and the fact that the functions $\theta, v \to B_{ij}(\theta, v)$ are continuous and bounded for all $\theta \in [0, \frac{\pi}{2}]$ and $v \in [e_1, e_2]$.

Let us sum up the above arguments in the following remark.

Remark 5.7.7 To every index n – the number of the laminates above the midsurface of the shell – there correspond two sets in the n-dimensional space of step functions which determine virtual angles of the orientation of fibers and the volume fractions of fibers in the laminates. The thickness of the shell, being considered as a function of n, takes discrete values $2ln$, where l is the thickness of the laminate.

For small l and great n, one can consider the functions of orientation θ and of volume fraction of fibers v as continuous functions, then the thickness of the shell also changes continuously.

Such a passage from a discrete problem in n, with step functions of orientation and volume fraction, to a continuous problem is natural because any continuous function on an interval can be uniformly approximated by step functions, and *vice versa*, and because the functionals R_{ij} and G_{ij} are continuous with respect to the topology of uniform convergence (see Lemma 5.7.5).

However, for a given n (which can be rather great), in order to approximate a continuous function by a step function with a good accuracy, the continuous function should satisfy the Lipschitz condition. So, below, considering the optimization problem for shells, we suppose that the functions from the set of admissible controls satisfy the Lipschitz condition with some common constant for all functions.

Below, to be more specific, we will examine a cylindrical shell.

Bilinear form and the theorem on the continuity of solutions for a laminate cylindrical shell

The strain energy of a cylindrical shell is determined by the relation (5.7.119). According to the results of Section 4.6, setting $\alpha = z$, $\beta = \varphi$, we get $A = 1$, $B = r = \text{const}$ and

$$\Omega = \{\, (z, \varphi) \,|\, 0 < z < L,\ 0 < \varphi < 2\pi \,\},$$

where r is the radius of the midsurface of the shell and L is its length. The strain components are defined by the formulas (4.6.5).

Let $t = (q, \theta, v) \in U$. The strain energy, defined by (5.7.119), generates a bilinear form a_t, which, in view of (5.7.126) and (4.6.10), has the form

$$\begin{aligned} a_t(\omega', \omega'') &= \int_0^L \int_0^{2\pi} \Big\{ R_{11}(t)(P_1\omega')(P_1\omega'') + R_{12}(t)\big[(P_1\omega')(P_2\omega'') + (P_2\omega')(P_1\omega'')\big] \\ &\quad + R_{22}(t)(P_2\omega')(P_2\omega'') + R_{66}(t)(P_3\omega')(P_3\omega'') + G_{11}(t)(P_4\omega')(P_4\omega'') \\ &\quad + G_{12}(t)\big[(P_4\omega')(P_5\omega'') + (P_5\omega')(P_4\omega'')\big] + G_{22}(t)(P_5\omega')(P_5\omega'') \\ &\quad + 4G_{66}(t)(P_6\omega')(P_6\omega'') \Big\}\, r\, dz\, d\varphi. \qquad (5.7.127) \end{aligned}$$

Let V be the Hilbert space defined by the relations (4.6.8) and (4.6.11), and let $f \in V^*$. Given $t \in U$, consider the problem: Find a function $\omega(t,f)$ such that

$$\omega(t,f) \in V, \qquad a_t(\omega(t,f),\omega') = (f,\omega'), \quad \omega' \in V. \tag{5.7.128}$$

Theorem 5.7.6 *Let a set U be determined by the formulas* (5.7.123) *and* (5.7.124), *and let a bilinear form a_t, $t \in U$, be defined by the relations* (5.7.127), (5.7.66)–(5.7.68), (5.7.70), *and* (5.7.125), *provided the operators P_i in* (5.7.127) *are determined by the formulas* (4.6.5) *and* (4.6.10) *with $A = 1$ and $B = r = \text{const}$. Further, assume that the elasticity constants of the fiber and matrix meet the following conditions*:

$$E_{\rm f},\ E_{\rm m},\ G_{\rm f},\ G_{\rm m} \text{ are positive constants}, \tag{5.7.129}$$

$$\mu_{\rm f} \textit{ and } \mu_{\rm m} \text{ are constants}, \ 0 \le \mu_{\rm f} < 1,\ 0 \le \mu_{\rm m} < 1. \tag{5.7.130}$$

Then, for any $(t,f) \in U \times V^*$, there exists a unique solution of the problem (5.7.128), *and the function $t,f \to \omega(t,f)$ determined by this solution is a continuous mapping of $U \times V^*$, endowed with the topology generated by the product of the topologies of $\mathbb{R}$, $(C([0,e_4]))^2$ and V^*, into the space V.*

Proof. (5.7.70) implies that, for any $\xi_1,\xi_2,\xi_6 \in \mathbb{R}$, $\theta \in [0,\frac{\pi}{2}]$, and $v \in [e_1,e_2]$,

$$\begin{aligned} &B_{11}(\theta,v)\xi_1^2 + 2B_{12}(\theta,v)\xi_1\xi_2 + B_{22}(\theta,v)\xi_2^2 + 4B_{66}(\theta,v)\xi_6^2 \\ &\qquad = A_1(\theta,v,\xi) + A_2(\theta,v,\xi), \end{aligned} \tag{5.7.131}$$

where

$$\begin{aligned} A_1(\theta,v,\xi) &= \frac{1}{2}\big[B_{11}(\theta,v)\xi_1^2 + 2B_{12}(\theta,v)\xi_1\xi_2 + B_{22}(\theta,v)\xi_2^2 \\ &\quad + 4B_{66}(\theta,v)\xi_6^2 + 4B_{16}(\theta,v)\xi_1\xi_6 + 4B_{26}(\theta,v)\xi_2\xi_6\big], \\ A_2(\theta,v,\xi) &= \frac{1}{2}\big[B_{11}(-\theta,v)\xi_1^2 + 2B_{12}(-\theta,v)\xi_1\xi_2 + B_{22}(-\theta,v)\xi_2^2 \\ &\quad + 4B_{66}(-\theta,v)\xi_6^2 + 4B_{16}(-\theta,v)\xi_1\xi_6 + 4B_{26}(-\theta,v)\xi_2\xi_6\big]. \end{aligned} \tag{5.7.132}$$

By substituting $B_{ij}(\theta,v)$ from (5.7.70) into (5.7.132), we get

$$\begin{aligned} A_1(\theta,v,\xi) &= \frac{1}{2}\left[E_{11}(v)\lambda_1^2 + 2E_{12}(v)\lambda_1\lambda_2 + E_{22}(v)\lambda_2^2 + 4G(v)\lambda_6^2\right], \\ A_2(\theta,v,\xi) &= \frac{1}{2}\left[E_{11}(v)\beta_1^2 + 2E_{12}(v)\beta_1\beta_2 + E_{22}(v)\beta_2^2 + 4G(v)\beta_6^2\right], \end{aligned} \tag{5.7.133}$$

and

$$\lambda_1^2 + \lambda_2^2 + 2\lambda_6^2 = \beta_1^2 + \beta_2^2 + 2\beta_6^2 = \xi_1^2 + \xi_2^2 + 2\xi_6^2. \tag{5.7.134}$$

From the physical point of view, the relations (5.7.133) mean that the strain energy does not depend on a specific coordinate system. This is implied by the fact that the elasticity constants $B_{ij}(\theta, v)$, defined by (5.7.70), are tensors of rank 4, while the constants ξ_i, λ_i, β_i in the formulas (5.7.132), (5.7.133) take the part of the components of the strain tensor respectively in the original coordinate system and in the coordinate systems rotated at angles θ and $-\theta$), the rank of the strain tensor is equal to 2. The equality (5.7.134) corresponds to the fact that the sum of the squares of the strain components is invariant (see also Malmeister et al. (1980)).

(5.7.67) and (5.7.68) yield the following formulas:

$$E_1(v) = vE_{\mathrm{f}} + (1-v)E_{\mathrm{m}}, \tag{5.7.135}$$

$$E_2(v) = \frac{E_{\mathrm{f}}E_{\mathrm{m}}}{vE_{\mathrm{m}} + (1-v)E_{\mathrm{f}}}, \tag{5.7.136}$$

$$\mu_1(v) = v\mu_{\mathrm{f}} + (1-v)\mu_{\mathrm{m}}, \tag{5.7.137}$$

$$\mu_2(v) = \frac{\mu_1(v)E_2(v)}{E_1(v)}. \tag{5.7.138}$$

From (5.7.135) and (5.7.136) we obtain

$$\frac{E_2(v)}{E_1(v)} = \frac{E_{\mathrm{f}}E_{\mathrm{m}}}{E_{\mathrm{f}}E_{\mathrm{m}}(1+2v^2-2v) + (E_{\mathrm{f}}^2 + E_{\mathrm{m}}^2)(v-v^2)}.$$

This, together with the inequality $E_{\mathrm{m}}^2 + E_{\mathrm{f}}^2 \geq 2E_{\mathrm{m}}E_{\mathrm{f}}$, implies that

$$\frac{E_2(v)}{E_1(v)} \leq 1, \qquad v \in [e_1, e_2], \tag{5.7.139}$$

e_1 and e_2 being constants satisfying the condition (5.7.124).

(5.7.130) and (5.7.137) imply that

$$\max_{v\in[e_1,e_2]} \mu_1(v) < 1. \tag{5.7.140}$$

Further, (5.7.138)–(5.7.140) yield

$$\max_{v\in[e_1,e_2]} \mu_1(v)\mu_2(v) < 1. \tag{5.7.141}$$

Now, from (5.7.66)–(5.7.68), (5.7.129), (5.7.135), (5.7.136), and (5.7.141), we deduce that

$$\left.\begin{array}{l}\text{the functions } v \to E_{11}(v),\ v \to E_{22}(v), \text{ and } v \to G(v),\\ \text{being considered as mappings of } [e_1, e_2] \text{ into } \mathbb{R}, \text{ are}\\ \text{continuous and bounded below and above by positive}\\ \text{constants.}\end{array}\right\} \tag{5.7.142}$$

Let us show that

$$E_1(v) - (\mu_1(v))^2 E_2(v) \geq c_0 > 0, \qquad v \in [e_1, e_2]. \tag{5.7.143}$$

If μ_f and μ_m vanish, then (5.7.143) is a consequence of (5.7.129), (5.7.135), and (5.7.137). Therefore, we assume that at least one of the constants μ_f and μ_m is not equal to zero.

Then, from (5.7.130) and (5.7.137), we get $\mu_1(v) \geq c_1 > 0$ for any $v \in [e_1, e_2]$. Hence, (5.7.138) yields

$$E_2(v) = \frac{\mu_2(v)E_1(v)}{\mu_1(v)}.$$

From here and (5.7.135), (5.7.141), we obtain (5.7.143).

Further, (5.7.66), (5.7.141), and (5.7.143) imply

$$E_{11}(v)E_{22}(v) - (E_{12}(v))^2 \geq c_2 > 0, \qquad v \in [e_1, e_2]. \tag{5.7.144}$$

(5.7.142), (5.7.144), and the Sylvester criterion yield that the matrix $[E_{ij}]$, $i, j = 1, 2$, is positive definite for any $v \in [e_1, e_2]$, and, since $v \to E_{ij}(v)$ are continuous functions on $[e_1, e_2]$, there exists a constant $c_3 > 0$ such that

$$E_{11}(v)\lambda_1^2 + 2E_{12}(v)\lambda_1\lambda_2 + E_{22}(v)\lambda_2^2 \geq c_3 \left(\lambda_1^2 + \lambda_2^2\right),$$
$$v \in [e_1, e_2], \ \lambda_1, \lambda_2 \in \mathbb{R}.$$

Hence, (5.7.131)–(5.7.134), (5.7.142), (5.7.125), and (5.7.127) imply that

$$a_t(\omega, \omega) \geq c_4 \int_0^L \int_0^{2\pi} \sum_{i=1}^6 (P_i\omega)^2 \, dz \, d\varphi, \qquad \omega \in V, \ t \in U, \tag{5.7.145}$$

c_4 being a positive constant.

The system of operators $\{P_i\}_{i=1}^6$ is coercive in V (see the proof of Theorem 4.6.1). Hence, (5.7.145) implies the existence of a number $c_5 > 0$ such that

$$a_t(\omega, \omega) \geq c_5 \|\omega\|_V^2, \qquad \omega \in V, \ t \in U. \tag{5.7.146}$$

The bilinear form a_t is continuous and symmetric in V.

Lemma 5.7.5 yields that $t \to a_t$ is a continuous mapping of U into $\mathcal{L}_2(V, V; \mathbb{R})$. Now, Theorem 1.8.1 yields that the function $t, f \to \omega(t, f)$, where $\omega(t, f)$ is the solution of the problem (5.7.128), is a continuous mapping of $U \times V^*$ into V. The theorem is proved.

Problem of optimization of a composite laminate cylindrical shell

Let a cylindrical shell be fastened in such a way that it does not have rigid displacements, i.e., displacements without deformations.

From the mathematical point of view, this corresponds to the case when the space V defined by the way of fastening of the shell satisfies the condition (4.6.11).

In particular, if the edges are clamped, the space V is of the form

$$V = \left\{ \omega \,|\, \omega = (u, v, w) \in W, \ \omega(0, \varphi) = \omega(L, \varphi) = 0, \ \frac{\partial w}{\partial z}(0, \varphi) = \frac{\partial w}{\partial z}(L, \varphi) = 0 \right\}, \tag{5.7.148}$$

L being the length of the shell. (Notice that, in this relation, v is a component of the vector function of displacements of the midsurface of the shell, but not the function of the volume fraction of the fibers.)

For the fastened edges, we have

$$V = \{\, \omega \,|\, \omega = (u, v, w) \in W,\ \omega(0, \varphi) = \omega(L, \varphi) = 0 \,\}. \tag{5.7.148}$$

The shell is subject to a fixed load, which is identified with an element $f \in V^*$. In particular, f can be defined by the following expression

$$(f, \omega) = \int_0^L \int_0^{2\pi} (f_1 u + f_2 v + f_3 w)\, r\, dz\, d\varphi, \qquad f_i \in L_2((0, L) \times (0, 2\pi)). \tag{5.7.149}$$

We suppose the assumptions of Theorem 5.7.6 to be fulfilled. Then, for $t \in U$, there exists a unique function $\omega(t)$ such that

$$\omega(t) \in V, \qquad a_t(\omega(t), \omega') = (f, \omega'), \qquad \omega' \in V. \tag{5.7.150}$$

Here, we write $\omega(t)$ instead of $\omega(t, f)$, because the element f is fixed, while t takes values in U.

Further, let there be defined functionals Ψ_k such that

$$\left.\begin{array}{l} t, \omega \to \Psi_k(t, \omega) \text{ is a continuous mapping of } U \times V, \\ \text{equipped with the topology generated by the product of} \\ \text{the topologies of } \mathbb{R},\ (C([0, e_4]))^2, \text{ and } V, \text{ into } \mathbb{R}, \\ k = 0, 1, 2, \ldots, l. \end{array}\right\} \tag{5.7.151}$$

Define a set of the admissible controls by

$$U_{\mathrm{ad}} = \Big\{ t \,|\, t = (q, \theta, v) \in U, \quad \sup_{y, y' \in [0, e_4]} \frac{|\theta(y) - \theta(y')|}{|y - y'|} \le c_1, \quad \sup_{y, y' \in [0, e_4]} \frac{|v(y) - v(y')|}{|y - y'|} \le c_2,\ \Psi_k(t, \omega(t)) \le 0,\ k = 1, 2, \ldots, l \Big\}, \tag{5.7.152}$$

and let a goal functional be of the form

$$\Phi(t) = \Psi_0(t, \omega(t)). \tag{5.7.153}$$

The functionals Ψ_k can define restrictions on the strength and stiffness of the shell, the volume fraction of fibers in the composite, the cost of the composite, and so forth.

As a condition on the fracture of a laminate composite, one frequently uses the condition of fracture of a unidirectional fiber laminate, which is considered as a homogeneous orthotropic material. In this case, one uses generalized strength criteria of the form (5.1.6), see Malmeister et al. (1980), Teters et al. (1978).

Then, the functional of restrictions on the strength can be taken in the form

$$\Psi_1(t,\omega(t)) = \max_{(z,\varphi,\gamma)\in T}\Bigg[\sum_{i,k=1}^{2}\Pi_{ik}(v(\gamma),\theta(\gamma))\sigma_{ik}^{(\rho)}(z,\varphi,\gamma)$$

$$+\sum_{p,q,m,n=1}^{2}\Pi_{pqmn}(v(\gamma),\theta(\gamma))\sigma_{pq}^{(\rho)}(z,\varphi,\gamma)\sigma_{mn}^{(\rho)}(z,\varphi,\gamma)\Bigg]-1, \tag{5.7.154}$$

$$T=[0,L]\times[0,2\pi]\times[-q,q].$$

Here, $\sigma_{ik}^{(\rho)}$ is the averaging of the function σ_{ik}^{γ} in the z and φ coordinates (see Subsec. 5.1.2), where the function σ_{ik}^{γ} is determined by (5.7.116) via the function of displacements $\omega(t)$; $\Pi_{ik}(v,\theta)$ and $\Pi_{pqmn}(v,\theta)$ are the components of the strength tensor of a unidirectional fiber laminate which correspond to the basis rotated at angle θ with respect to the direction of the fibers of the laminate and depend on the volume fraction of fibers v.

In addition, the functional of restrictions on the strength Ψ_2 is also introduced. It is obtained from Ψ_1 by substituting $\Pi_{ik}(v,-\theta)$ and $\Pi_{pqmn}(v,-\theta)$ for $\Pi_{ik}(v,\theta)$ and $\Pi_{pqmn}(v,\theta)$.

Suppose that $v\to\Pi_{ik}(v,0)$ and $v\to\Pi_{pqmn}(v,0)$ are continuous functions on $[e_1,e_2]$. Then, making use of Theorem 5.7.6 and the properties of averaging, one can see that the functional Ψ_1 from (5.7.154) satisfies the condition (5.7.151).

As a goal functional, the mass of the shell is often taken. Then,

$$\Phi(t)=4\pi rL\int_0^q[v(\gamma)\rho_{\mathrm{f}}+(1-v(\gamma))\rho_{\mathrm{m}}]\,d\gamma. \tag{5.7.155}$$

The problem of optimization of the composite laminate shell reduces to the search for an element t_0 such that

$$t_0=(q_0,\theta_0,v_0)\in U_{\mathrm{ad}},\qquad \Phi(t_0)=\inf_{t\in U_{\mathrm{ad}}}\Phi(t). \tag{5.7.156}$$

Theorem 5.7.7 *Let the assumptions of Theorem* 5.7.6 *be fulfilled. Let a nonempty set* U_{ad} *and a goal functional* Φ *be defined by the relations* (5.7.151)–(5.7.153). *Then, the problem* (5.7.156) *has a solution.*

Proof. Let $\{t_n\}$ be a minimizing sequence:

$$t_n=(q_n,\theta_n,v_n)\in U_{\mathrm{ad}},\qquad \lim_{n\to\infty}\Phi(t_n)=\inf_{t\in U_{\mathrm{ad}}}\Phi(t). \tag{5.7.157}$$

(5.7.123), (5.7.124), and (5.7.152) yield that we can choose a subsequence $\{t_m\}$ such that

$$q_m\to q' \text{ in } \mathbb{R},\qquad \theta_m\to\theta' \text{ in } C([0,e_4]),\qquad v_m\to v' \text{ in } C([0,e_4]). \tag{5.7.158}$$

(5.7.158) and Theorem 5.7.6 imply that

$$\omega(t_m) \to \omega(t') \quad \text{in } V, \qquad t' = (q', \theta', v'). \tag{5.7.159}$$

Now, by virtue of (5.7.151) and (5.7.157)–(5.7.159), one can verify that the function $t_0 = t'$ is a solution of the problem (5.7.156). The theorem is proven.

By using the above approach, one can examine various problems of optimization of composite laminate shells under restrictions on the strength and stability.

5.7.7 Optimization of the composite structure

The set of controls and functions of effective elasticity coefficients

In Subsec. 5.7.4, we considered the problem of getting the effective elasticity constants that are most close to the given ones. There, we supposed that the composite had Y-periodic structure, and the shape of the domain G occupied by the inclusion (more exactly, the function f defining the boundary of G) considered as a control.

However, elements of constructions are, as a rule, in inhomogeneous strain-stress state. This is why in many cases it is desirable that the effective elasticity coefficients be some functions of the coordinates of points of the domain Ω occupied by the composite (by an element of a construction made of a composite material).

Now, we will suppose that the "period" Y depends on a point $y \in \overline{\Omega} \subset \mathbb{R}^3$, and the function f defining the boundary of the inclusion depends not only on a point s of the unit sphere S_0 in $\mathbb{R}^3$, but also on a point $y \in \overline{\Omega}$. Denote $Q = \overline{\Omega} \times S_0$, and define a set of controls U as follows:

$$\begin{aligned} U = \Big\{ r = (Y, f) \,\big|\, Y = \{Y_i\}_{i=1}^3 \in (C(\overline{\Omega}))^3, \ \check{Y}_i \le Y_i \le \hat{Y}_i, \\ \sup_{y, y' \in \overline{\Omega}} \frac{|Y_i(y) - Y_i(y')|}{\|y - y'\|_{\mathbb{R}^3}} \le c_i, \quad i = 1,2,3, \ f \in C(Q), \\ \check{f} \le f \le \hat{f}, \quad \sup_{y, y' \in \overline{\Omega}} \frac{|f(y,s) - f(y',s)|}{\|y - y'\|_{\mathbb{R}^3}} \le c_4, \quad s \in S_0, \\ \sup_{(y,s),\,(y',s') \in Q} \frac{|f(y,s) - f(y',s')|}{\|(y,s) - (y',s')\|_{\mathbb{R}^6}} \le c_5 \Big\}. \end{aligned} \tag{5.7.160}$$

Here, $\check{Y}_i$, $\hat{Y}_i$, $i = 1,2,3$, and $\check{f}$, $\hat{f}$ are continuous positive-number-valued functions defined in $\overline{\Omega}$ and Q, respectively, $c_1, \ldots, c_5$ are given positive constants.

For a given element $r = (Y = \{Y_i\}_{i=1}^3, f)$ from U, with every point $y \in \overline{\Omega}$ we associate the cube $\widetilde{Y}(y) = \prod_{i=1}^{3} [0, Y_i(y)]$ and the function $f(y, \cdot)\colon s \to f(y,s)$ defined on the unit sphere S_0. Put also $r(y) = (\widetilde{Y}(y), f(y, \cdot))$.

The inclusion and matrix elasticity constants $a^{\rm in}_{ijkm}$ and $a^{\rm m}_{ijkm}$ are supposed to be given and fixed. Points of a cube $\widetilde{Y}(y)$ are denoted by x.

In accordance with (5.7.78) and (5.7.80), the coefficients $a^{r(y)}_{ijkm}$ are defined on the cube $\widetilde{Y}(y)$ by

$$a^{r(y)}_{ijkm}(x) = \begin{cases} a^{\rm in}_{ijkm}, & \text{if } x \in G(r(y)), \\ a^{\rm m}_{ijkm}, & \text{if } x \in \widetilde{Y}(y) \setminus G(r(y)), \end{cases} \tag{5.7.161}$$

where

$$G(r(y)) = \bigl\{\, x \mid x = x^0(y) + \alpha f(y,s) n(s),\ 0 \le \alpha \le 1,\ s \in S_0 \,\bigr\}, \tag{5.7.162}$$

and $n(s)$ is the unit vector that is normal at the point s to the surface of the unit sphere S_{0y} centered at the point $x^0(y) = \{\frac{1}{2}Y_i(y)\}_{i=1}^3$. Notice that, on one hand, we have the fixed sphere S_0, which is supposed to be centered at the origin. On the other hand, to every $y \in \overline{\Omega}$ there corresponds the unit sphere S_{0y} centered at $x^0(y)$. Here, we suppose that the one-to-one correspondence between points of S_0 and S_{0y} is determined by the translation of the sphere S_0 without rotation. Therefore, we identify points of these spheres and denote them by a single letter s.

Denote by $W_{r(y)}$ the space W_1 defined in the cube $\widetilde{Y}(y)$, see Lemma 5.7.2. Next, according to (5.7.81), on the space $W_{r(y)}$ we define the following bilinear form:

$$a^{r(y)}(u,v) = \int_{\widetilde{Y}(y)} \sum_{i,j,k,m=1}^{3} a^{r(y)}_{ijkm}(x) \varepsilon_{km}(u) \varepsilon_{ij}(v)\, dx. \tag{5.7.163}$$

Suppose that the inclusion and matrix elasticity constants satisfy the conditions of symmetry and positive definiteness. Then, from Lemma 5.7.2, there exists a unique function $X^{ij}(r(y))$ that is the solution of the following problem:

$$\begin{gathered} X^{ij}(r(y)) \in W_{r(y)}, \\ a^{r(y)}(X^{ij}(r(y)), v) = a^{r(y)}(P^{ij}, v), \qquad v \in W_{r(y)}, \end{gathered} \tag{5.7.164}$$

where $i, j = 1,2,3$ and P^{ij} are given by (5.7.19).

Now, because of (5.7.83), the values of the effective elasticity coefficients at point $y \in \overline{\Omega}$ are given by

$$g_{ijkm}(r(y)) = \frac{a^{r(y)}\bigl(X^{ij}(r(y)) - P^{ij}, X^{km}(r(y)) - P^{km}\bigr)}{Y_1(y)Y_2(y)Y_3(y)}. \tag{5.7.165}$$

Thus, to every $r \in U$ there corresponds the functions of effective elasticity coefficients

$$g_{ijkm}(r) \colon y \to g_{ijkm}(r(y))$$

that are defined, in the domain Ω occupied by the composite and on its boundary, via the relations (5.7.82) and (5.7.83). In order to justify the use of the latter formulas, we have to suppose that the functions $y \to Y_i(y)$ and $y \to f(y,s)$ are slowly changing in $\overline{\Omega}$, i.e., the constants $c_1, \dots, c_4$ in (5.7.160) are sufficiently small.

The problem of the stress-strain state of a composite

Suppose Ω is a bounded Lipschitz domain in $\mathbb{R}^3$. By S we denote the boundary of Ω. Suppose also that $S = S_1 \cup S_2$, $S_1 \cap S_2 = \varnothing$, $\operatorname{mes} S_1 > 0$, on S_1 the composite is fastened, and surface forces act on the part S_2.

Thus, for every $r \in U$, there are functions of the elasticity coefficients $g_{ijkm}(r)$ defined in $\overline{\Omega}$. In the space H defined by (5.7.7) and equipped with the norm (5.7.14), we define a bilinear form b_r by

$$b_r(u,v) = \int_\Omega \sum_{i,j,k,m=1}^{3} g_{ijkm}(r)\varepsilon_{km}(u)\varepsilon_{ij}(v)\,dy, \qquad u,v \in H. \tag{5.7.166}$$

Let the composite be affected by a fixed load that does not depend on the control. This load is identified with an element $L \in H^*$.

Consider the problem: For a fixed $r \in U$, find a function w^r such that

$$w^r \in H, \qquad b_r(w^r, v) = (L, v), \qquad v \in H. \tag{5.7.167}$$

Theorem 5.7.8 *Let a set U be defined by* (5.7.160), $L \in H^*$, *and let a bilinear form b_r be defined by the formula* (5.7.166) *in which the functions $y \to g_{ijkm}(r(y))$ are given by the relations* (5.7.161)–(5.7.165). *Let also a^{in}_{ijkm}, a^{m}_{ijkm} be constants satisfying the conditions of symmetry and positive definiteness* (5.7.4), (5.7.5). *Then, there exists a unique solution of the problem* (5.7.167).

Proof. From Remark 5.7.3, it follows that, at every point $y \in \overline{\Omega}$, the coefficients $g_{ijkm}(r(y))$ satisfy the condition of symmetry, and the following estimate holds:

$$\sum_{i,j,k,m=1}^{3} g_{ijkm}(r(y))a_{km}a_{ij} \geq c_0 \sum_{i,j=1}^{3} a_{ij}^2, \qquad a_{ij} = a_{ji} \in \mathbb{R}, \\ r \in U,\ y \in \overline{\Omega}. \tag{5.7.168}$$

From the proof of Theorem 5.7.2 (see (5.7.48)–(5.7.50)), we conclude the existence of a constant c such that

$$\sup_{y \in \overline{\Omega}} |g_{ijkm}(r(y))| \leq c, \qquad r \in U, \tag{5.7.169}$$

for arbitrary indices $i, j, k, m = 1, 2, 3$.

So, for any $r \in U$, the bilinear form b_r is symmetric, continuous, and coercive in H. Hence, there exists a unique solution of the problem (5.7.167).

By virtue of Theorem 5.7.8, we can define the function $r \to w^r$ mapping the set U in the space H. The theorem below establishes the continuity of this function.

Theorem 5.7.9 *Under the conditions of Theorem* 5.7.8, *the function $r \to w^r$ is a continuous mapping of U (in the topology generated by the product of $(C(\overline{\Omega}))^3$ and $C(Q)$) in the space H.*

To prove Theorem 5.7.9, we need the following lemma.

Lemma 5.7.6 *Let the conditions of Theorem* 5.7.8 *be satisfied, let*

$$\{r_n = (Y_n, f_n)\}_{n=1}^{\infty}$$

be a sequence of elements of U, *and let* $Y_n \to Y_0$ *in* $(C(\overline{\Omega}))^3$, $f_n \to f_0$ *in* $C(Q)$. *Then,* $r_0 = (Y_0, f_0) \in U$ *and*

$$\lim_{n\to\infty} \sup_{y\in\overline{\Omega}} \left| g_{ijkm}(r_n(y)) - g_{ijkm}(r_0(y)) \right| = 0 \tag{5.7.170}$$

for arbitrary indices $i, j, k, m = 1, 2, 3$.

Proof. By using (5.7.160), it is not hard to see that the conditions

$$\{r_n = (Y_n, f_n)\}_{n=1}^{\infty} \subset U, \quad Y_n \to Y_0 \quad \text{in } (C(\overline{\Omega}))^3, \quad f_n \to f_0 \quad \text{in } C(Q)$$

imply that $r_0 = (Y_0, f_0) \in U$.

Let us prove the equality (5.7.170). Assume first that $Y_n = Y_0$ for each n. Since $f_n \to f_0$ in $C(Q)$, from the proof of Lemma 5.7.4, it follows that

$$\lim_{n\to\infty} \sup_{y\in\overline{\Omega}} \|a^{r_n(y)} - a^{r_0(y)}\|_{\mathcal{L}_2(W_{r_0(y)}, W_{r_0(y)}; \mathbb{R})} = 0. \tag{5.7.171}$$

Denote by $A(r_n(y))$ the operator generated by the bilinear form $a^{r_n(y)}$, $n = 0, 1, 2, \ldots$, and given by the formula

$$A(r_n(y)) \in \mathcal{L}(W_{r_0(y)}, W^*_{r_0(y)}), \quad (A(r_n(y))u, v) = a^{r_n(y)}(u, v), \qquad u, v \in W_{r_0(y)}. \tag{5.7.172}$$

From (5.7.171), we infer that

$$\lim_{n\to\infty} \sup_{y\in\overline{\Omega}} \|A(r_n(y)) - A(r_0(y))\| = 0. \tag{5.7.173}$$

As well known (see, e.g., Schwartz (1967)), if p is a linear, continuous, invertible mapping of a Banach space E into a Banach space F and if q is a linear continuous mapping of E into F satisfying the inequality $\|q\| < \|p^{-1}\|^{-1}$, then the mapping $p + q$ is invertible, and the following expansion holds:

$$(p+q)^{-1} = p^{-1} - p^{-1}qp^{-1} + p^{-1}qp^{-1}qp^{-1} - \cdots . \tag{5.7.174}$$

Thus, letting $p = A(r_0(y))$, $q = A(r_n(y)) - A(r_0(y))$ and taking (5.7.173) into account, we have that, for sufficiently large n, say $n \geq K$,

$$\begin{aligned} &\|(A(r_n(y)))^{-1} - (A(r_0(y)))^{-1}\| \\ &\quad \leq c_1 \|A(r_n(y)) - A(r_0(y))\| \, \|(A(r_0(y)))^{-1}\|^2. \end{aligned} \tag{5.7.175}$$

From the positive definiteness of the coefficients a^{in}_{ijkm} and a^{m}_{ijkm}, we conclude the existence of a constant c_2 such that

$$\sup_{y\in\overline{\Omega}} \left\| \left(A(r_0(y))\right)^{-1} \right\| \leq c_2.$$

Now, from (5.7.173) and (5.7.175), we get

$$\lim_{n\to\infty} \sup_{y\in\overline{\Omega}} \|(A(r_n(y)))^{-1} - (A(r_0(y)))^{-1}\| = 0. \tag{5.7.176}$$

By using (5.7.164), (5.7.165), (5.7.171), and (5.7.176), it is not hard to derive (5.7.170).

Assume now that the condition $Y_n = Y_0$ for all n does not hold. Then, for an arbitrary fixed point $y \in \overline{\Omega}$, to every n there corresponds the cube

$$\widetilde{Y}_n(y) = \prod_{i=1}^{3} [0, Y_{in}(y)],$$

and the bilinear forms $a^{r_n(y)}$ are defined on functions given on different cubes $\widetilde{Y}_n(y)$.

For a fixed y, introduce for each index n the variables $x^n = (x_1^n, x_2^n, x_3^n)$ that are connected with the variable $x = (x_1, x_2, x_3)$ by the equality

$$x_i^n = \frac{x_i Y_{i0}(y)}{Y_{in}(y)}.$$

In the x^n variables, the problem of finding the elasticity coefficients at any point $y \in \overline{\Omega}$ with different n reduces to the problem on the fixed domain $\prod_{i=1}^3 (0, Y_{i0}(y))$, and analogously to the above, we establish (5.7.170).

Remark 5.7.8 Analogously to the proof of Lemma 5.7.6, one can show that, for an arbitrary fixed $r \in U$, the function $y \to g_{ijkm}(r(y))$ is continuous in $\overline{\Omega}$, and therefore, in (5.7.170), the supremum can be replaced with the maximum.

Proof of Theorem 5.7.9. Let $\{r_n = (Y_n, f_n)\}_{n=1}^\infty$ be a sequence of elements of U, $r_0 = (Y_0, f_0) \in U$, $Y_n \to Y_0$ in $(C(\overline{\Omega}))^3$, $f_n \to f_0$ in $C(Q)$. Then, by Lemma 5.7.6, the condition (5.7.170) is satisfied. Hence, (5.7.166) implies

$$b_{r_n} \to b_{r_0} \quad \text{in} \quad \mathcal{L}_2(H, H; \mathbb{R}). \tag{5.7.177}$$

Denote by B_r the operator generated by the bilinear form b_r:

$$(B_r u, v) = b_r(u, v), \qquad u, v \in H.$$

By (5.7.177), we get $B_{r_n} \to B_{r_0}$ in $\mathcal{L}(H, H^*)$. From (5.7.168), the operators B_{r_n} and B_{r_0} are invertible. So, Theorem 1.8.1 gives $w^{r_n} \to w^{r_0}$ in H, concluding the proof.

Optimization problem

Suppose that we are given functionals Ψ_k such that

$$\left.\begin{array}{l}\Psi_k\colon r, w \to \Psi_k(r,w) \text{ is a continuous mapping of } U \times H \text{ (in} \\ \text{the topology generated by the product of the topologies} \\ (C(\overline{\Omega}))^3,\ C(Q), \text{ and } H) \text{ into } \mathbb{R},\ k = 0, 1, \dots, m.\end{array}\right\} \tag{5.7.178}$$

Define a set of admissible controls U_{ad} as follows:

$$U_{\mathrm{ad}} = \{\, r \mid r \in U,\ \Psi_k(r, w^r) \le 0,\ \ k = 1, 2, \dots, m \,\}, \tag{5.7.179}$$

where w^r is a solution of the problem (5.7.167).

Let also a goal functional Φ be of the form

$$\Phi(r) = \Psi_0(r, w^r). \tag{5.7.180}$$

Consider the problem of finding $\tilde{r}$ such that

$$\tilde{r} \in U_{\mathrm{ad}}, \qquad \Phi(\tilde{r}) = \inf_{r \in U_{\mathrm{ad}}} \Phi(r). \tag{5.7.181}$$

By using Theorem 5.7.9, analogously to the above (see, for example, the proof of Theorem 5.7.3), one establishes the following

Theorem 5.7.10 *Under the conditions of Theorem* 5.7.8, *let a nonempty set* U_{ad} *and a goal functional be defined by* (5.7.178)–(5.7.180). *Then, there exists a solution of the problem* (5.7.181).

We will present now some examples of Ψ_k and Φ. For $r \in U$, define a function $z_r\colon \overline{\Omega} \to \mathbb{R}$ as follows:

$$z_r(y) = \frac{1}{Y_1(y)Y_2(y)Y_3(y)} \int_{G_r(y)} dx, \tag{5.7.182}$$

where $G_r(y)$ is defined by (5.7.162).

The goal functional Φ can be taken in the form

$$\Phi(r) = \int_\Omega z_r(y)\, dy. \tag{5.7.183}$$

The formula (5.7.182) for a given control r determines the ratio of the volume of the inclusion at a point y to the volume of the cube $\widetilde{Y}(y)$. Hence, the choice of goal functional in the form (5.7.183) means minimization of the fraction of the inclusion in the volume of the composite. Usually, the specific gravity and cost of the inclusion are greater than those of the matrix, so that the minimization leads to decrease of the weight of the composite, as well as to decrease of its cost.

If, nevertheless, the specific gravity and cost of the matrix are greater than those of the inclusion, one can choose the goal functional in the form

$$\Phi(r) = -\int_\Omega z_r(y)\, dy.$$

The functionals Ψ_k, $k = 1, 2, \ldots, m$, can correspond to the restrictions on stiffness and strength.

Remark 5.7.9 There are approaches to the optimal design of composites and materials of structures that are based on the creation of a specific mixture of layers at each point of the structure. These approaches have been initiated by Olhoff (1974), Lurie and Cherkaev (1976), Armand and Lodier (1978). For further results, see Rozvany (1989), Lurie (1993), Bendsøe (1994).

5.8 Optimization of laminate composite covers according to mechanical and radio engineering characteristics

In radio-location, one faces different problems of design of the protective devices of antenna systems (radar domes), which are usually made of laminate composite materials and must satisfy some thermal, mechanical, and radio engineering demands; more exactly, they must be strong, stiff, and radioparent in a sufficiently large wave band or in some scattered wave bands, besides they should have a small weight.

Sometimes, one also faces problems of design of absorbers of radio waves, which are also made of laminate composite materials. These should be strong and stiff, and should maximally absorb radio waves, having a small weight.

Below, in Subsec. 5.8.1, we will consider the problem of propagation of radio waves through a laminate medium. In Subsec. 5.8.2, we will consider some optimization problems according to mechanical and radio engineering characteristics.

5.8.1 Propagation of electromagnetic waves through a laminated medium

We will be concerned now with a method of calculation of characteristics of the propagation of plane harmonic waves through a laminated medium. Our presentation is based on the classical work by Abelés (1950), see also Born and Wolf (1964).

Basic equations

The Maxwell equations have the following form

$$\frac{\varepsilon}{c}\frac{\partial E}{\partial t} = \operatorname{rot} H, \tag{5.8.1}$$

$$\frac{\mu}{c}\frac{\partial H}{\partial t} = -\operatorname{rot} E. \tag{5.8.2}$$

Here, $E = (E_x, E_y, E_z)$ is the vector of electric field, $H = (H_x, H_y, H_z)$ the vector of magnetic field, t the time, c the velocity of light in the vacuum, ε the dielectric permittivity, μ the magnetic permeability. We consider a laminated medium such that ε and μ are functions of z only. Moreover, μ is a real function, and ε is a real function for a nonconducting medium, and a complex function for a conducting medium; in the latter case, ε is said to be the complex dielectric permittivity.

We suppose the electromagnetic wave to be plane. Let yOz be a plane of incidence. Let TE stand for a transverse electric wave such that the electric vector is perpendicular to the yOz plane, i.e., $E = (E_x, 0, 0)$. Let also TM stand for a transverse magnetic wave for which the magnetic vector is perpendicular to the yOz plane, i.e., $H = (H_x, 0, 0)$. An arbitrary plane wave may be represented as a sum of TE and TM waves.

Consider a TE wave. We seek the solution of the equations (5.8.1), (5.8.2) in the form

$$E(x,y,z,t) = \widetilde{E}(x,y,z)\exp(-i\omega t), \qquad \widetilde{E} = (\widetilde{E}_x, \widetilde{E}_y, \widetilde{E}_z), \tag{5.8.3}$$

$$H(x,y,z,t) = \widetilde{H}(x,y,z)\exp(-i\omega t), \qquad \widetilde{H} = (\widetilde{H}_x, \widetilde{H}_y, \widetilde{H}_z), \tag{5.8.4}$$

where ω is the angular frequency. Then, the vector equations (5.8.1), (5.8.2) reduce to the following scalar equations:

$$\frac{\partial \widetilde{H}_z}{\partial y} - \frac{\partial \widetilde{H}_y}{\partial z} + \frac{i\varepsilon\omega}{c}\widetilde{E}_x = 0, \tag{5.8.5a}$$

$$\frac{\partial \widetilde{H}_x}{\partial z} - \frac{\partial \widetilde{H}_z}{\partial x} = 0, \tag{5.8.5b}$$

$$\frac{\partial \widetilde{H}_y}{\partial x} - \frac{\partial \widetilde{H}_x}{\partial y} = 0, \tag{5.8.5c}$$

$$\frac{i\omega\mu}{c}\widetilde{H}_x = 0, \tag{5.8.6a}$$

$$\frac{\partial \widetilde{E}_x}{\partial z} - \frac{i\omega\mu}{c}\widetilde{H}_y = 0, \tag{5.8.6b}$$

$$\frac{\partial \widetilde{E}_x}{\partial y} + \frac{i\omega\mu}{c}\widetilde{H}_z = 0. \tag{5.8.6c}$$

In case of a conducting medium, the complex dielectric permittivity is given by (see Born and Wolf (1964))

$$\varepsilon = \varepsilon' + i\frac{4\pi\sigma}{\omega}, \tag{5.8.6d}$$

where ε' is the dielectric permittivity, σ the conductance.

The equations (5.8.5a)–(5.8.6c) show that $\widetilde{H}_x = 0$ and $\widetilde{E}_x$, $\widetilde{H}_y$, $\widetilde{H}_z$ depend on y, z only. By (5.8.5a), (5.8.6b), and (5.8.6c), we get

$$\frac{\partial^2 \widetilde{E}_x}{\partial y^2} + \frac{\partial^2 \widetilde{E}_x}{\partial z^2} + \varepsilon\mu k^2 \widetilde{E}_x = \frac{d(\log\mu)}{dz}\frac{\partial \widetilde{E}_x}{\partial z}, \tag{5.8.7}$$

where

$$k = \frac{\omega}{c}. \tag{5.8.8}$$

We seek the solution of (5.8.7) in the form of a product of two functions, one depending on y and the other one on z:

$$\widetilde{E}_x(y,z) = Y(y)U(z). \tag{5.8.9}$$

Then, (5.8.7) gives

$$\frac{1}{Y}\frac{d^2Y}{dy^2} = -\frac{1}{U}\frac{d^2U}{dz^2} - \varepsilon\mu k^2 + \frac{d(\log\mu)}{dz}\frac{1}{U}\frac{dU}{dz}. \tag{5.8.10}$$

The left-hand side of this equation depends on y only, the right-hand side only on z, therefore

$$\frac{1}{Y}\frac{d^2Y}{dy^2} = -k^2\alpha^2, \tag{5.8.11}$$

$$\frac{d^2U}{dz^2} - \frac{d(\log\mu)}{dz}\frac{dU}{dz} + \varepsilon\mu k^2 U = k^2\alpha^2 U, \tag{5.8.12}$$

where α is a constant (possibly, complex). By (5.8.3), (5.8.9), and (5.8.11), we have

$$E_x = U(z)\exp\big[i(k\alpha y - \omega t)\big]. \tag{5.8.13}$$

It follows from (5.8.5 a), (5.8.6 b), (5.8.6 c) that H_y and H_z are given by the analogous formulas

$$H_y = V(z)\exp\big[i(k\alpha y - \omega t)\big], \tag{5.8.14}$$

$$H_z = W(z)\exp\big[i(k\alpha y - \omega t)\big], \tag{5.8.15}$$

with

$$\frac{dV}{dz} = ik(\alpha W + \varepsilon U), \tag{5.8.16}$$

$$\frac{dU}{dz} = ik\mu V, \tag{5.8.17}$$

$$\alpha U + \mu W = 0. \tag{5.8.18}$$

By substituting W from (5.8.18) into (5.8.16), we obtain the following system

$$\frac{dU}{dz} = ik\mu V, \tag{5.8.19}$$

$$\frac{dV}{dz} = ik\left(\varepsilon - \frac{\alpha^2}{\mu}\right)U. \tag{5.8.20}$$

The system (5.8.1), (5.8.2) is symmetric in the sense that, if one interchanges the positions of E and H and replaces ε by $-\mu$ and μ by $-\varepsilon$, then the system will remain invariable. Therefore, by using such interchange and replace, any result for a TM wave may be derived from a corresponding result for a TE wave. Thus, for a TM wave, we get $E_x = 0$ and

$$H_x = U(z) \exp\left[i(k\alpha y - \omega t)\right], \tag{5.8.21}$$
$$E_y = -V(z) \exp\left[i(k\alpha y - \omega t)\right], \tag{5.8.22}$$
$$E_z = -W(z) \exp\left[i(k\alpha y - \omega t)\right], \tag{5.8.23}$$

and moreover,

$$\frac{dU}{dz} = ik\varepsilon V, \tag{5.8.24}$$
$$\frac{dV}{dz} = ik\left(\mu - \frac{\alpha^2}{\varepsilon}\right)U, \tag{5.8.25}$$
$$\alpha U + \varepsilon W = 0. \tag{5.8.26}$$

Characteristic matrix, reflectivity, and transmissivity

We consider a stack of s homogeneous slabs, see Fig. 5.8.1. The first slab occupies the space from $z = 0$ to $z = z_1$, the second one from $z = z_1$ to $z = z_2$, and so forth. In this case,

$$\begin{aligned} \varepsilon(z) &= \varepsilon_n, \qquad z \in (z_{n-1}, z_n), \\ \mu(z) &= \mu_n, \qquad z \in (z_{n-1}, z_n), \ n = 1, \dots, s, \ z_0 = 0, \end{aligned} \tag{5.8.27}$$

where ε_n, μ_n are constants.

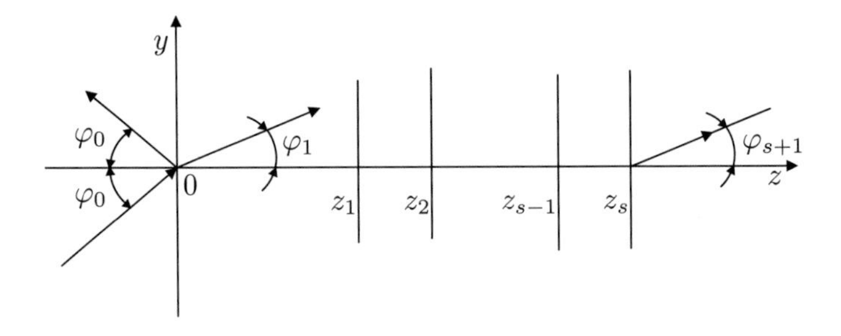

Figure 5.8.1: Stack of homogeneous slabs

We suppose that a homogeneous medium occupies the space $z < 0$; ε_0 and μ_0 are the dielectric and magnetic constants of this medium. The x component of the electric field in the medium is given by

$$E_x = E_0 \exp\left[ik\sqrt{\varepsilon_0\mu_0}\,(y\sin\varphi_0 + z\cos\varphi_0) - i\omega t\right], \qquad z < 0,$$

where φ_0 is the angle of incidence. From the continuity of the tangential component of the vector E at $z = 0$, we obtain the following relation for a TE wave:

$$E_0 \exp\left[i(k\sqrt{\varepsilon_0\mu_0}\,y\sin\varphi_0 - \omega t)\right] = U(0)\exp\left[i(k\alpha y - \omega t)\right] \qquad \forall y. \tag{5.8.28}$$

Thus,

$$\alpha = \sqrt{\varepsilon_0\mu_0}\,\sin\varphi_0. \tag{5.8.29}$$

From the continuity of the tangential component of the vector H at $z = 0$, we find that (5.8.29) holds also for a TM wave.

Now, let

$$U_n = U(z_n), \qquad V_n = V(z_n), \qquad n = 0, 1, \ldots, s-1,\ z_0 = 0. \tag{5.8.30}$$

Taking (5.8.27) into account, it is not hard to verify that, for a TE wave, the solution U, V of the problem (5.8.19), (5.8.20) with the initial data U_n, V_n is defined by

$$\begin{aligned} U(z) &= U_n F_n(z) + V_n f_n(z), \\ V(z) &= U_n G_n(z) + V_n g_n(z), \qquad z \in [z_n, z_{n+1}]. \end{aligned} \tag{5.8.31}$$

Here,

$$\begin{aligned} F_n(z) &= \cos\left[k_n(z - z_n)\right], & f_n(z) &= \frac{i}{p_{n+1}}\sin\left[k_n(z - z_n)\right], \\ G_n(z) &= ip_{n+1}\sin\left[k_n(z - z_n)\right], & g_n(z) &= \cos\left[k_n(z - z_n)\right], \end{aligned} \tag{5.8.32}$$

where

$$\begin{aligned} p_{n+1} &= \sqrt{\frac{\varepsilon_{n+1}}{\mu_{n+1}}}\cos\varphi_{n+1}, \\ k_n &= k\sqrt{\varepsilon_{n+1}\mu_{n+1}}\cos\varphi_{n+1}, \end{aligned} \tag{5.8.33}$$

and

$$\cos^2\varphi_{n+1} = 1 - \frac{\varepsilon_0\mu_0\sin^2\varphi_0}{\varepsilon_{n+1}\mu_{n+1}}. \tag{5.8.34}$$

The formula (5.8.34) follows from the condition $\alpha = \text{const}$, and moreover, we get

$$\sqrt{\varepsilon_0\mu_0}\,\sin\varphi_0 = \sqrt{\varepsilon_{n+1}\mu_{n+1}}\,\sin\varphi_{n+1}, \qquad n = 0, 1, \ldots, s, \tag{5.8.35}$$

and if ε_0, ε_{n+1} are real constants, then φ_{n+1} is the angle between the normal to the plane of constant phase and the Oz axis.

The formulas (5.8.31)–(5.8.33) are valid for a TE wave, but they also hold true for a TM wave under the condition

$$p_{n+1} = \sqrt{\frac{\mu_{n+1}}{\varepsilon_{n+1}}} \cos \varphi_{n+1}, \tag{5.8.36}$$

H_x, E_y, E_z defined by (5.8.21)–(5.8.23), (5.8.29). The formula (5.8.34) holds also for a TM wave.

Define matrices $N_n(z)$ by

$$N_n(z) = \begin{bmatrix} F_n(z) & f_n(z) \\ G_n(z) & g_n(z) \end{bmatrix}, \qquad z \in [z_n, z_{n+1}]. \tag{5.8.37}$$

By (5.8.31), (5.8.37), we have

$$\begin{bmatrix} U(z) \\ V(z) \end{bmatrix} = N_n(z) \begin{bmatrix} U_n \\ V_n \end{bmatrix}. \tag{5.8.38}$$

The matrix $N_n(z)$ is invertible as $\det N_n(z) = 1$ for an arbitrary $z \in [z_n, z_{n+1}]$; denote its inverse by $M_n(z)$,

$$M_n(z) = \begin{bmatrix} g_n(z) & -f_n(z) \\ -G_n(z) & F_n(z) \end{bmatrix}. \tag{5.8.39}$$

Then

$$\begin{bmatrix} U_n \\ V_n \end{bmatrix} = M_n(z) \begin{bmatrix} U(z) \\ V(z) \end{bmatrix}, \qquad z \in [z_n . z_{n+1}],\ n = 0, 1, \ldots, s-1, \tag{5.8.40}$$

and we have

$$\begin{bmatrix} U_0 \\ V_0 \end{bmatrix} = M_0(z_1) M_1(z_2) \cdots M_{n-1}(z_n) M_n(z) \begin{bmatrix} U(z) \\ V(z) \end{bmatrix}, \qquad z \in [z_n, z_{n+1}]. \tag{5.8.41}$$

Here, U_0, V_0 are the values of U, V at $z = 0$ (see (5.8.30)).

The reflection and transmission factors

Introduce the matrix

$$M = M_0(z_1) M_1(z_2) \cdots M_{s-1}(z_s), \qquad M = \begin{bmatrix} m_{11} & m_{12} \\ m_{21} & m_{22} \end{bmatrix}. \tag{5.8.42}$$

Let A, R, T be the amplitudes (possibly, complex) of the electric vectors of incident, reflected, and refracted (passed) waves, respectively. The boundary condition is the continuity of the tangential components of the vectors E and H at $z = 0$, see Fig. 5.8.1. Taking into account that

$$H = \sqrt{\frac{\varepsilon}{\mu}}\, e \times E, \tag{5.8.43}$$

where e is the unit vector of wave directivity, and the sign $\times$ stands for the vector product, we obtain the following relations for a TE wave:

$$\begin{aligned} U_0 &= U(0) = A + R, \qquad & U(z_s) &= T, \\ V_0 &= p_0(A - R), & V(z_s) &= p_{s+1}T, \end{aligned} \tag{5.8.44}$$

where

$$p_0 = \sqrt{\frac{\varepsilon_0}{\mu_0}} \cos\varphi_0, \qquad p_{s+1} = \sqrt{\frac{\varepsilon_{s+1}}{\mu_{s+1}}} \cos\varphi_{s+1}. \tag{5.8.45}$$

Here, ε_{s+1}, μ_{s+1} are the dielectric permittivity and the magnetic permeability of a homogeneous medium occupying the half-space $z > z_s$, $\cos\varphi_{s+1}$ defined by (5.8.34) at $n = s$.

By applying (5.8.41), (5.8.42), we obtain from (5.8.44) that

$$\begin{aligned} A + R &= (m_{11} + m_{12}p_{s+1})T, \\ p_0(A - R) &= (m_{21} + m_{22}p_{s+1})T. \end{aligned} \tag{5.8.46}$$

These equations lead us to the following expressions for the reflection and transmission factors:

$$r = \frac{R}{A} = \frac{(m_{11} + m_{12}p_{s+1})p_0 - (m_{21} + m_{22}p_{s+1})}{(m_{11} + m_{12}p_{s+1})p_0 + (m_{21} + m_{22}p_{s+1})}, \tag{5.8.47}$$

$$t = \frac{T}{A} = \frac{2p_0}{(m_{11} + m_{12}p_{s+1})p_0 + (m_{21} + m_{22}p_{s+1})}. \tag{5.8.48}$$

The reflection and transmission powers are defined by

$$\mathcal{R} = |r|^2, \qquad \mathcal{F} = \frac{p_{s+1}}{p_0}\,|t|^2. \tag{5.8.49}$$

The formulas (5.8.47)–(5.8.49) were given for a TE wave, but they are also valid for a TM wave under the conditions

$$p_0 = \sqrt{\frac{\mu_0}{\varepsilon_0}} \cos\varphi_0, \qquad p_{s+1} = \sqrt{\frac{\mu_{s+1}}{\varepsilon_{s+1}}} \cos\varphi_{s+1}. \tag{5.8.50}$$

In this case, r and t are the ratios of the amplitudes of magnetic vectors, but r is also the ratio of the amplitudes of electric vectors. For a TM wave, we denote by

t_1 the ratio of the amplitudes of the tangential components of electric vectors of transmitted and incident waves. Then, by (5.8.43), we get

$$t_1 = \sqrt{\frac{\mu_{s+1}\varepsilon_0}{\mu_0\varepsilon_{s+1}}}\,\frac{\cos\varphi_{s+1}}{\cos\varphi_0}\,t, \tag{5.8.51}$$

t defined by (5.8.48), (5.8.50). The transmission power for a TM wave may be determined by

$$\mathcal{F} = \frac{p_{s+1}}{p_0}\,|t_1|^2.$$

5.8.2 Optimization problems

We consider a cover (a radar dome, or an absorber) as a shell made of laminated composite materials. The stress-strain state of the laminated shell is defined by a function of displacements $\tilde{\omega}$ which is the solution of the following problem

$$\tilde{\omega} \in V, \qquad a_\delta(\tilde{\omega}, \omega) = (f, \omega), \qquad \omega \in V. \tag{5.8.52}$$

Here, V is a subspace of W that satisfies (4.6.11), W defined by (4.6.8) and (4.7.5) for a shell of revolution and for a shallow shell, respectively, the bilinear form a_δ for these shells determined in subsecs. 4.12.2 and 4.12.3, f is the load, $f \in V^*$.

The cover is subject to various loads. Denote by T the set of the loads on which the cover is calculated, and suppose that

$$T \text{ is a compact set in } V^*. \tag{5.8.53}$$

Introduce the set

$$B = \big\{\, \delta = (\delta_1, \delta_2, \ldots, \delta_s) \in \mathbb{R}^s \mid \delta_1 \geq a_1,\ \delta_n - \delta_{n-1} \geq a_1,\ n = 2, 3, \ldots, s,\\ \delta_s \leq a_2,\ a_1,\ a_2 \text{ are positive constants} \,\big\}. \tag{5.8.54}$$

By Theorems 4.12.1 and 4.12.2, for each $\delta \in B$ and $f \in T$ for the shell of revolution and for the shallow shell, there exists a unique solution $\tilde{\omega}$ of the problem (5.8.52); denote this solution by $\tilde{\omega}(\delta, f)$. So,

$$\tilde{\omega}(\delta, f) \in V, \qquad a_\delta(\tilde{\omega}(\delta, f), \omega) = (f, \omega), \qquad \omega \in V. \tag{5.8.55}$$

B is considered as a set of controls, i.e., the number of laminae is supposed to be fixed and equal to s. The mechanical parameters E^n_{ij} (see (4.12.6)) and the electromagnetic parameters ε_n, μ_n (see (5.8.27)) are also supposed to be fixed. For each $\delta \in B$, the state of the cover Q_δ is a set of functions $\tilde{\omega}(\delta, f)$, where f runs through T, i.e.,

$$Q_\delta = \big\{\, \tilde{\omega}(\delta, f) \,\big\}_{f \in T}. \tag{5.8.56}$$

Suppose that

$$A_i\colon \delta,\omega \to A_i(\delta,\omega) \text{ is a continuous mapping of } B\times V \text{ into } \mathbb{R},\ i=1,2,\dots,q. \tag{5.8.57}$$

Define mappings

$$B\times V^* \ni (\delta,f) \to \Gamma_i(\delta,f) = A_i(\delta,\tilde{\omega}(\delta,f)) \in \mathbb{R}, \qquad i=1,2,\dots,q. \tag{5.8.58}$$

The functionals A_i may be determined in such a way that $\{\Gamma_i(\delta,f)\}_{i=1}^q$ define stresses, displacements, and some functions of them.

We determine functionals Ψ_i on B by

$$\delta \to \Psi_i(\delta) = \sup_{f\in T} \Gamma_i(\delta,f), \qquad i=1,2,\dots,q, \tag{5.8.59}$$

$$\Psi_{q+1}(\delta) = \sum_{i=1}^{s} b_i\delta_i. \tag{5.8.60}$$

Here, b_i are positive constants that may be associated with the specific gravity or cost of the materials of laminae.

We define a set of admissible controls as follows:

$$U_{\text{ad}} = \{\,\delta\in B \mid \Psi_i(\delta) \le \tilde{c}_i,\ i=1,\dots,q+1\,\}, \tag{5.8.61}$$

where $\tilde{c}_i$ are positive constants. The relations $\Psi_i(\delta)\le\tilde{c}_i$ can be constructed in the form of restrictions on strength, stiffness, weight, cost, and so forth.

Consider some goal functionals. Suppose that the modules of the radii of the principal curvatures of the shell are $1.5 \div 2$ times greater than the wavelength

$$\lambda_0 = \frac{2\pi c}{\omega\sqrt{\varepsilon_0\mu_0}}.$$

This condition holds usually in practice. Then, we can apply the formulas (5.8.47)–(5.8.49). We suppose that the surface $\gamma = \frac{h}{2}$ is the surface of incidence of the wave. So, in (5.8.42), we should take (see Figs. 4.12.1 and 5.8.1)

$$z_n = h - \delta_{s-n}, \qquad n=1,2,\dots,s, \qquad \delta_0 = 0. \tag{5.8.62}$$

The reflection and transmission factors defined by (5.8.47), (5.8.48) depend on the vector $\delta = (\delta_1,\delta_2,\dots,\delta_s)\in B$, on the angle of incidence of the wave φ_0, and on the wavenumber k from (5.8.8). Thus, we denote the reflection and transmission factors for TE and TM waves by $r_{\text{E}}(\delta,\varphi_0,k)$, $t_{\text{E}}(\delta,\varphi_0,k)$, $r_{\text{M}}(\delta,\varphi_0,k)$, $t_{\text{M}}(\delta,\varphi_0,k)$; for these waves p_0 and p_{s+1} are defined by (5.8.45), (5.8.50).

In case of a radar dome, we should maximize the radioparency. So, we can take the following goal functional

$$\Psi_0(\delta) = -\sum_{i=1}^{n_1}\sum_{j=1}^{n_2} \{a_{ij}|t_{\text{E}}(\delta,\varphi_{0i},k_j)|^2 + b_{ij}|t_{\text{M}}(\delta,\varphi_{0i},k_j)|^2\}, \tag{5.8.63}$$

which must be minimized. Here, φ_{0i}, k_j are given values of the angles of incidence and the wave numbers, n_1, n_2 finite numbers, a_{ij}, b_{ij} weight factors, positive constants.

In case of an absorber, we should minimize the reflection and transmission powers, i.e., the following functional

$$\Psi_0(\delta) = \sum_{i=1}^{n_1}\sum_{j=1}^{n_2}\Big\{a_{ij}^{(1)}|r_{\mathrm{E}}(\delta,\varphi_{0i},k_j)|^2 + a_{ij}^{(2)}|r_{\mathrm{M}}(\delta,\varphi_{0i},k_j)|^2 + a_{ij}^{(3)}|t_{\mathrm{E}}(\delta,\varphi_{0i},k_j)|^2 + a_{ij}^{(4)}|t_{\mathrm{M}}(\delta,\varphi_{0i},k_j)|^2\Big\}, \quad (5.8.64)$$

where $a_{ij}^{(k)}$ are positive constants.

Consider the following optimization problem: Find a vector $\gamma = (\gamma_1, \gamma_2, \dots, \gamma_s)$ satisfying

$$\gamma \in U_{\mathrm{ad}}, \qquad \Psi_0(\gamma) = \inf_{\delta\in U_{\mathrm{ad}}} \Psi_0(\delta), \quad (5.8.65)$$

where Ψ_0 is given by (5.8.63) or (5.8.64).

We consider the cases when the cover is either a shell of revolution or a shallow shell.

Theorem 5.8.1 *Suppose either the conditions of Theorem* 4.12.1 *(for a shell of revolution) or of Theorem* 4.12.2 *(for a shallow shell) are satisfied. Let functionals Ψ_i and a nonempty set U_{ad} be defined by* (5.8.53), (5.8.54), (5.8.57)–(5.8.61). *Suppose a goal functional is determined by* (5.8.63) *or* (5.8.64). *Then, there exists a solution of the problem* (5.8.65).

Proof. Obviously, B is a compact set in $\mathbb{R}^s$. Therefore, there exists a sequence $\{\delta^n\}_{n=1}^{\infty}$ such that

$$\delta^n \in U_{\mathrm{ad}} \qquad \forall n, \quad (5.8.66)$$

$$\lim \Psi_0(\delta^n) = \inf_{\delta\in U_{\mathrm{ad}}} \Psi_0(\delta), \quad (5.8.67)$$

$$\delta^n \to \delta^0 \quad \text{in } \mathbb{R}^s, \qquad \delta^0 \in B. \quad (5.8.68)$$

Let $\mathcal{A}_\delta$ be the operator generated by the form a_δ, i.e.,

$$\mathcal{A}_\delta \in \mathcal{L}(V,V^*), \qquad (\mathcal{A}_\delta u, v) = a_\delta(u,v), \qquad u, v \in V. \quad (5.8.69)$$

By (5.8.68), (4.12.6), and (4.12.8), we get $\mathcal{A}_{\delta^n} \to \mathcal{A}_{\delta^0}$ in $\mathcal{L}(V,V^*)$.

Applying Theorem 1.8.1, we obtain

$$\tilde{\omega}\colon \delta, f \to \tilde{\omega}(\delta, f) \text{ is a continuous mapping of } B\times V^* \text{ into } V, \quad (5.8.70)$$

where $\tilde{\omega}(\delta, f)$ is the solution of the problem (5.8.55). Theorem 1.4.4 and (5.8.53), (5.8.57)–(5.8.59), (5.8.70) give that Ψ_i's are continuous functionals in B, $i = 1,\dots,q$. Therefore, $\delta^0 \in U_{\mathrm{ad}}$. Due to (5.8.32), (5.8.39), (5.8.42), (5.8.47), and (5.8.48), we conclude that Ψ_0 is a continuous functional on B. So, the vector $\gamma = \delta^0$ is a solution of the problem (5.8.65).

5.9 Shape optimization of a two-dimensional elastic body

Earlier, we have studied optimization problems for deformable solids in which the weak (generalized) solutions of the state equations in the form of elliptic equations were used. However, in some formulations of optimization problems for deformable solids with restrictions on strength, the stresses must be bounded at each point of a deformable solid. So, in such a case, only smooth solutions of the state equations are allowed. In the next two sections, we will consider two problems of such type for elastic bodies, using results of Section 2.13.

5.9.1 Sets of controls and domains in the optimization problem

Let M be a space of controls, which will be defined below. A domain Ω_q in $\mathbb{R}^2$ occupied by an elastic body is given for every $q \in M$. The boundary S_q of Ω_q consists of two connected components S_1 and S_{2q}, see Fig. 5.9.1. The points of S_1 are held fixed, and S_1 does not depend on a control q. Surface forces $F = (F_1, F_2)$ are given on S_{21}, which is an open set in S_{2q}, and these forces are continued onto the whole S_{2q} by zero. We suppose also that S_{21} does not depend on a control q, and $S_{2q}^{(1)} = S_{2q} \setminus S_{21}$ is the controlled part of S_{2q}, that is, $S_{2q}^{(1)}$ is the part of the boundary that must be chosen from the optimization conditions.

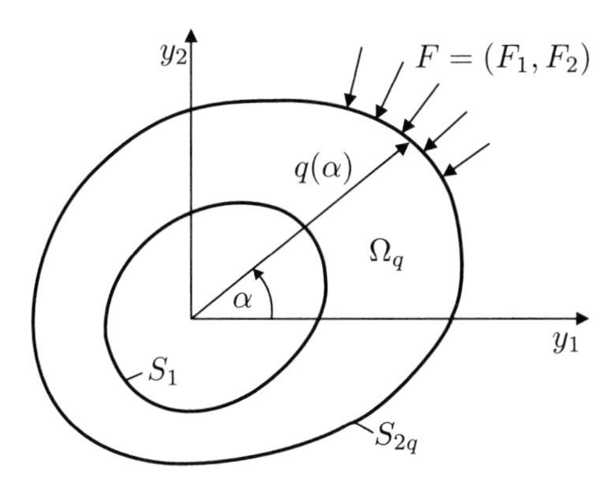

Figure 5.9.1: Domain Ω_q occupied with an elastic body

Define the space of controls in the form

$$
\begin{aligned}
M = \{\, q \mid q \in \widetilde{C}^{[l]+3}([0, 2\pi]),\ r_1 < q(\alpha) < r_2,\ \alpha \in [0, 2\pi], \\
q(\alpha) = \beta(\alpha),\ \alpha \in (\alpha_1, \alpha_2) \,\}.
\end{aligned}
\tag{5.9.1}
$$

Here, r_1, r_2 are positive constants, $r_1 < r_2$, $l > 0$, l not an integer, $[l]$ is an integer such that $l - [l] \in (0, 1)$, β is a function defined on $(\alpha_1, \alpha_2) \subset (0, 2\pi)$, and in polar coordinates r, α the part S_{21} of S_{2q} consists of points $s = (r, \alpha)$ such that $\alpha \in (\alpha_1, \alpha_2)$ and $r = \beta(\alpha)$. $\widetilde{C}^{[l]+3}([0, 2\pi])$ is the subspace of $C^{[l]+3}([0, 2\pi])$ consisting of the periodic functions. The periodicity of a function $q \in C^{[l]+3}([0, 2\pi])$ means that, if $\tilde{q}$ is the $[0, 2\pi]$-periodic continuation of q on $\mathbb{R}$, then $\tilde{q} \in C^{[l]+3}([a, b])$ for an arbitrary $[a, b] \subset \mathbb{R}$. The set M is equipped with the topology generated by that of $C^{[l]+3}([0, 2\pi])$.

For each $q \in M$, define a domain Ω_q such that its internal boundary S_1 is given in polar coordinates by a function $\gamma \in \widetilde{C}^{[l]+3}([0, 2\pi])$, and the external boundary S_{2q} by the function q.

Define the domain Ω in the form

$$\Omega = \{ x \mid x = (x_1, x_2),\ 1 < x_1^2 + x_2^2 < 4 \}. \tag{5.9.2}$$

Let E stand for the function that maps polar coordinates into Cartesian ones:

$$E\colon (r, \alpha) \to E(r, \alpha) = (y_1, y_2), \qquad y_1 = r \cos \alpha, \quad y_2 = r \sin \alpha, \tag{5.9.3}$$

and let E^{-1} be its inverse. Determine $P_q\colon \overline{\Omega}_q \to \overline{\Omega}$ by the formula

$$P_q = E \circ G_q \circ E^{-1}, \tag{5.9.4}$$

where $G_q\colon E^{-1}(\overline{\Omega}_q) \to E^{-1}(\overline{\Omega})$,

$$\begin{gathered} (r, \alpha) \to G_q(r, \alpha) = (\rho, \varphi), \\ \rho = \frac{r - 2\gamma(\alpha) + q(\alpha)}{q(\alpha) - \gamma(\alpha)}, \qquad \varphi = \alpha. \end{gathered} \tag{5.9.5}$$

The mapping $G_q^{-1}\colon E^{-1}(\overline{\Omega}) \to E^{-1}(\overline{\Omega}_q)$ has the form

$$\begin{gathered} (\rho, \varphi) \to G_q^{-1}(\rho, \varphi) = (r, \alpha), \\ r = 2\gamma(\varphi) - q(\varphi) + [q(\varphi) - \gamma(\varphi)]\rho, \qquad \alpha = \varphi. \end{gathered} \tag{5.9.6}$$

It is easily seen that the mapping P_q defined by (5.9.4), (5.9.5) is a diffeomorphism of $\overline{\Omega}_q$ onto $\overline{\Omega}$ of the $C^{[l]+3}$ class.

5.9.2 Problems of elasticity in domains

In the domain Ω_q, the operator A_q of the theory of elasticity has the form

$$A_q u = \left\{ \begin{matrix} \mu\left(\dfrac{\partial^2 u_1}{\partial y_1^2} + \dfrac{\partial^2 u_1}{\partial y_2^2}\right) + (\lambda + \mu)\left(\dfrac{\partial^2 u_1}{\partial y_1^2} + \dfrac{\partial^2 u_2}{\partial y_1 \partial y_2}\right) \\ \mu\left(\dfrac{\partial^2 u_2}{\partial y_1^2} + \dfrac{\partial^2 u_2}{\partial y_2^2}\right) + (\lambda + \mu)\left(\dfrac{\partial^2 u_1}{\partial y_1 \partial y_2} + \dfrac{\partial^2 u_2}{\partial y_2^2}\right) \end{matrix} \right\}. \tag{5.9.7}$$

Here, $u = (u_1, u_2)$ is a vector function of displacements, λ, μ are positive constants. Denote by $\varepsilon_{ij}(u)$, $\sigma_{ij}(u)$ the components of the strain and stress tensors:

$$\begin{aligned}\varepsilon_{ij}(u) &= \frac{1}{2}\left(\frac{\partial u_i}{\partial y_j} + \frac{\partial u_j}{\partial y_i}\right),\\ \sigma_{ij}(u) &= \lambda\left(\frac{\partial u_1}{\partial y_1} + \frac{\partial u_2}{\partial y_2}\right)\delta_{ij} + 2\mu\varepsilon_{ij}(u), \qquad i,j = 1,2,\end{aligned} \tag{5.9.8}$$

δ_{ij} the Kronecker delta. Define boundary operators B and B_q on S and S_{2q}, respectively, by the expressions

$$Bu = \begin{Bmatrix} u_1\big|_{S_1} \\ u_2\big|_{S_1}\end{Bmatrix}, \qquad B_q u = \begin{Bmatrix}(\sigma_{11}(u)\nu_{1q} + \sigma_{12}(u)\nu_{2q})\big|_{S_{2q}} \\ (\sigma_{21}(u)\nu_{1q} + \sigma_{22}(u)\nu_{2q})\big|_{S_{2q}}\end{Bmatrix}, \tag{5.9.9}$$

where ν_{iq} are the components of the unit outward normal to S_{2q}, $i = 1, 2$.

Theorem 5.9.1 *Let a set M be defined by (5.9.1). For each $q \in M$, determine a two-connected domain $\Omega_q \subset \mathbb{R}^2$ such that the internal and external boundaries of Ω_q are defined in polar coordinates by the functions $\gamma \in \widetilde{C}^{[l]+3}([0, 2\pi])$ and q. Then, the operator $L_q\colon u \to L_q u = (A_q u, Bu, B_q u)$ defined by (5.9.7), (5.9.9), where λ, μ are positive constants, is an isomorphism of the space $V_{lq} = C^{l+2}(\overline{\Omega}_q)^2$ onto the space $H_{lq} = C^l(\overline{\Omega}_q)^2 \times C^{l+2}(S_1)^2 \times C^{l+1}(S_{2q})^2$.*

Proof. Consider the problem

$$\begin{aligned} A_q u &= f, && \text{in } \Omega_q,\\ Bu &= g_1, && \text{on } S_1,\\ B_q u &= g_2, && \text{on } S_{2q},\end{aligned} \tag{5.9.10}$$

where $(f, g_1, g_2) \in H_{lq}$. The ellipticity of the operator A_q is implied by the Korn inequality. The ellipticity of the problem (5.9.10) follows from the ellipticity of the first and second problems of the theory of ellipticity (see Michlin (1973)). The kernel space of the operator A_q is the space of small rigid displacements, which has the form (see Subsec. 1.7.2)

$$Q = \{\, u \mid u = (u_1, u_2),\ u_1 = a_1 + a_3 y_2,\ u_2 = a_2 - a_3 y_1,\ a_1, a_2, a_3 \in \mathbb{R}\,\}. \tag{5.9.11}$$

Let $y^{(1)} = (y_1^{(1)}, y_2^{(1)})$, $y^{(2)} = (y_1^{(2)}, y_2^{(2)})$ be two different points of S_1. From the condition $Bu = 0$, we have $u(y^{(1)}) = u(y^{(2)}) = 0$, and if additionally $u \in Q$, then by (5.9.11) we get $a_1 = a_2 = a_3 = 0$. Therefore, the kernel space of the operator $L_q = (A_q, B, B_q)$ consists only of zero. For each $(f, g_1, g_2) \in H_{lq}$, there exists a solution of the problem (5.9.10). Hence, the theorem follows from Theorem 2.13.1.

5.9.3 The optimization problem

For each $q \in M$, consider the problem

$$\begin{aligned} A_q u_q &= 0, && \text{in } \Omega_q, \\ B u_q &= 0, && \text{on } S_1, \\ B_q u_q &= F, && \text{on } S_{2q}, \end{aligned} \tag{5.9.12}$$

where the operators A_q, B, B_q are defined by (5.9.7), (5.9.9). We suppose that

$$F \in C^{l+1}(S_{2q})^2, \qquad \operatorname{supp} F \subset S_{21} \subset S_{2q}, \tag{5.9.13}$$

where (see (5.9.1) and (5.9.3))

$$S_{21} = \{\, s \mid s = E(\beta(\alpha), \alpha),\ \alpha \in (\alpha_1, \alpha_2) \,\}.$$

Now, we will consider restrictions on strength. For a vector function of displacements $u = (u_1, u_2)$, the components of the stress deviator (shear stress tensor) are defined by the formula

$$\tau_{ij}(u) = \sigma_{ij}(u) - \frac{1}{2}(\sigma_{11}(u) + \sigma_{22}(u))\delta_{ij}, \qquad i, j = 1, 2,$$

and the second invariant of the stress deviator has the form

$$\mathcal{T}(u) = \sum_{i,j=1}^{2} (\tau_{ij}(u))^2 = \frac{1}{2}(\sigma_{11}(u) - \sigma_{22}(u))^2 + 2(\sigma_{12}(u))^2. \tag{5.9.14}$$

Define a functional G_1 over M by

$$G_1(q) = \max_{y \in \overline{\Omega}_q} \big[(\mathcal{T}(u_q))(y) - b\big], \tag{5.9.15}$$

where u_q is the solution of the problem (5.9.12), b a positive constant. For an isotropic material, a restriction on strength may be taken in the form $G_1(q) \leq 0$. The volume of the material is defined by

$$G_0(q) = \int_{\Omega_q} dy. \tag{5.9.16}$$

Define a set M_1 by

$$\begin{aligned} M_1 = \{\, q \,|\, q \in M,\ q \in \widetilde{C}^{l+3}([0, 2\pi]),\ \|q\|_{\widetilde{C}^{l+3}([0,2\pi])} \leq c_1, \\ r_1 + \delta \leq q(\alpha) \leq r_2 - \delta,\ \alpha \in [0, 2\pi] \,\}, \end{aligned} \tag{5.9.17}$$

where M is determined by (5.9.1), c_1, δ are positive constants, and δ is small.

We take a set of admissible controls U_{ad} in the form

$$U_{\mathrm{ad}} = \{\, q \mid q \in M_1,\ G_1(q) \le 0 \,\}. \tag{5.9.18}$$

The optimization problem consists in finding q_0 satisfying

$$q_0 \in U_{\mathrm{ad}}, \qquad G_0(q_0) = \inf_{q\in U_{\mathrm{ad}}} G_0(q). \tag{5.9.19}$$

Theorem 5.9.2 *Let operators A_q, B, B_q be defined by the formulas (5.9.7), (5.9.9), and let λ, μ be positive constants. Suppose the condition (5.9.13) holds, and functionals G_1 and G_0 over M are defined by (5.9.15), (5.9.16), where u_q is the solution of the problem (5.9.12). Let also a nonempty set U_{ad} be given by the expressions (5.9.1), (5.9.17), (5.9.18). Then, for any $l > 0$, l not an integer, there exists a solution of the problem (5.9.19).*

Proof. Define spaces V_l and H_l by

$$V_l = C^{l+2}(\overline{\Omega})^2,$$
$$H_l = C^{l}(\overline{\Omega})^2 \times C^{l+2}(S_{01})^2 \times C^{l+1}(S_{02})^2.$$

Here, Ω is the domain defined by (5.9.2), S_{01} and S_{02} are the internal and external boundaries of Ω. Just as in Subsec. 2.13.2, the problem (5.9.12) reduces to the following one: Find a function $\tilde{u}_q \in V_l$ satisfying

$$\begin{aligned} \widetilde{A}_q\tilde{u}_q &= 0, && \text{in } \Omega,\\ \widetilde{B}\tilde{u}_q &= 0, && \text{on } S_{01},\\ \widetilde{B}_q\tilde{u}_q &= F\circ P_q^{-1}, && \text{on } S_{02}. \end{aligned} \tag{5.9.20}$$

Here, the operators $\widetilde{A}_q$, $\widetilde{B}$, $\widetilde{B}_q$ are defined by

$$\begin{aligned} \widetilde{A}_q u &= (A_q(u\circ P_q))\circ P_q^{-1},\\ \widetilde{B} u &= (B(u\circ P_q))\circ P_q^{-1},\\ \widetilde{B}_q u &= (B_q(u\circ P_q))\circ P_q^{-1}. \end{aligned} \tag{5.9.21}$$

From (5.9.12), (5.9.20), and (5.9.21), it follows that $u_q = \tilde{u}_q \circ P_q$. It is easily seen that

$$\left.\begin{array}{l} q \to \widetilde{L}_q = (\widetilde{A}_q, \widetilde{B}, \widetilde{B}_q) \text{ is a continuous mapping of } M \text{ into}\\ \mathcal{L}(V_l, H_l). \end{array}\right\} \tag{5.9.22}$$

Upon (5.9.13), $q \to F\circ P_q^{-1}$ is a constant mapping, and from Theorem 2.13.3 we now obtain that

$$q \to \tilde{u}_q \text{ is a continuous mapping of } M \text{ into } V_l, \tag{5.9.23}$$

where $\tilde{u}_q$ is the solution of the problem (5.9.20). Under the change of variables corresponding to the mapping P_q, the functionals G_1 and G_0 from (5.9.15), (5.9.16) take the form

$$G_1(q) = \max_{x\in\overline{\Omega}}(\mathcal{T}(\tilde{u}_q \circ P_q))(P_q^{-1}(x)) - b, \tag{5.9.24}$$

$$G_0(q) = \int_\Omega \det\left|(P_q^{-1})'(x)\right| dx. \tag{5.9.25}$$

Here, $(\mathcal{T}(\tilde{u}_q \circ P_q))(P_q^{-1}(x))$ is the value of the function $\mathcal{T}(\tilde{u}_q \circ P_q)$ at a point $P_q^{-1}(x)$, and $(P_q^{-1})'(x)$ is the value of the Fréchet derivative of the mapping P_q^{-1} at a point x. Taking into account (5.9.23)–(5.9.25), (5.9.4)–(5.9.6), and Theorem 1.4.4, we infer that G_1 and G_0 are continuous functionals over M. As the embedding of $C^{l+3}([0,2\pi])$ into $C^{[l]+3}([0,2\pi])$ is compact, we have that M_1 is a compact set in M. Therefore, there exists a solution of the problem (5.9.19).

Remark 5.9.1 In the considered case, the function $q \to (\widetilde{A}_q, \widetilde{B}, \widetilde{B}_q)$ is a continuously Fréchet differentiable mapping of M into $\mathcal{L}(V_l, H_l)$. Thus, by using Theorem 2.13.4, we obtain that $q \to \lambda(q) = \tilde{u}_q$ is a continuously Fréchet differentiable mapping of M into V_l. However, the functional G_1 from (5.9.24) is not differentiable, because the functional $v \to \max_{x\in\overline{\Omega}} v(x)$, $v \in C(\overline{\Omega})$ is not. Nevertheless, on the set $Q = \left\{ v \mid v \in C(\overline{\Omega}),\ v(x) \geq 0,\ x \in \overline{\Omega} \right\}$, the latter functional may be approximated by the continuously Fréchet differentiable functional $v \to \|v\|_{L_p(\Omega)}$ if p is sufficiently large.

5.10 Optimization of the internal boundary of a two-dimensional elastic body

In Section 5.9, we considered the optimization problem for a two-connected elastic body in which we assigned displacements on the internal boundary and surface forces on the external boundary. In that case, there exists a unique solution of the problem (5.9.10) for any $(f, g_1, g_2) \in H_{lq}$, and owing to Theorem 5.9.1 the condition (2.13.9) is satisfied. Now, we will consider the optimization problem for a two-connected elastic body in which we specify surface forces on both internal and external boundaries. In this case, (2.13.9) is not satisfied.

Let us formulate this problem. Let M be a space of controls, and let a two-dimensional domain Ω_q occupied by an elastic body be defined for every $q \in M$. The boundary S_q of Ω_q consists of two connected components, S_{2q} and S_1 are the internal and external boundaries of Ω_q, respectively (see Fig. 5.10.1). We prescribe "self-balanced" forces F on S_1 and zero forces on S_{2q}, S_1 and F being independent of a control q, and S_{2q} being chosen from the conditions of optimization.

Let us specify the space M as follows:

$$M = \left\{ q \mid q \in \widetilde{C}^{[l]+3}([0,2\pi]),\ r_1 < q(\alpha) < r_2,\ \alpha \in [0,2\pi] \right\}, \tag{5.10.1}$$

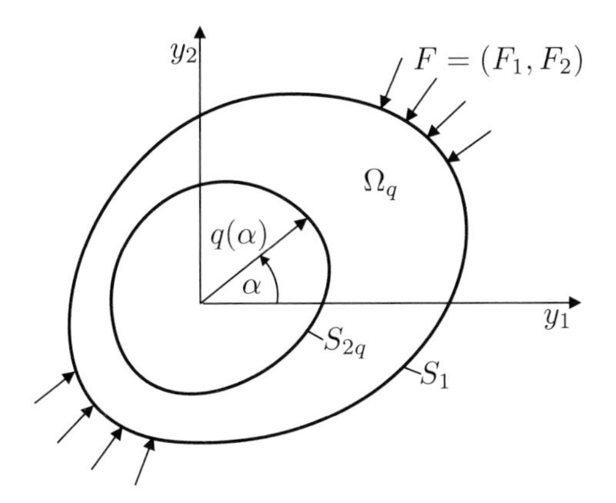

Figure 5.10.1: Domain Ω_q occupied with an elastic body; "self-balanced" forces F act on S_1

where r_1 and r_2 are positive constants. For each $q \in M$, let the two-connected domain Ω_q be such that the internal boundary S_{2q} is defined in polar coordinates by q, and the external boundary S_1 by a fixed function $\gamma \in \widetilde{C}^{[l]+3}([0, 2\pi])$. The domain Ω is defined by (5.9.2) and the mapping P_q by (5.9.4), where $G_q\colon E^{-1}(\overline{\Omega}_q) \to E^{-1}(\overline{\Omega})$ has the form

$$\begin{gathered}(r, \alpha) \to G_q(r, \alpha) = (\rho, \varphi), \\ \rho = \frac{r - 2q(\alpha) + \gamma(\alpha)}{\gamma(\alpha) - q(\alpha)}, \qquad \varphi = \alpha.\end{gathered} \tag{5.10.2}$$

The inverse mapping $G_q^{-1}\colon E^{-1}(\overline{\Omega}) \to E^{-1}(\overline{\Omega}_q)$ has the form

$$\begin{gathered}(\rho, \varphi) \to G_q^{-1}(\rho, \varphi) = (r, \alpha), \\ r = 2q(\varphi) - \gamma(\varphi) + [\gamma(\varphi) - q(\varphi)]\rho, \qquad \alpha = \varphi.\end{gathered} \tag{5.10.3}$$

The operator A_q is defined by (5.9.7) and the boundary operators B_q and B are given by the expressions

$$B_q u = \left\{ \begin{aligned} (\sigma_{11}(u)\nu_{1q} + \sigma_{12}(u)\nu_{2q})\big|_{S_{2q}} \\ (\sigma_{21}(u)\nu_{1q} + \sigma_{22}(u)\nu_{2q})\big|_{S_{2q}} \end{aligned} \right\}, \tag{5.10.4}$$

$$Bu = \left\{ \begin{aligned} (\sigma_{11}(u)\nu_1 + \sigma_{12}(u)\nu_2)\big|_{S_1} \\ (\sigma_{21}(u)\nu_1 + \sigma_{22}(u)\nu_2)\big|_{S_1} \end{aligned} \right\}. \tag{5.10.5}$$

Here, ν_{iq} and ν_i are the components of the unit outward normal to S_{2q} and S_1, respectively, $i = 1, 2$.

Define the spaces V_{lq} and H_{lq} in the form

$$\begin{aligned} V_{lq} &= C^{l+2}(\overline{\Omega}_q)^2, \\ H_{lq} &= C^l(\overline{\Omega}_q)^2 \times C^{l+1}(S_{2q})^2 \times C^{l+1}(S_1)^2, \end{aligned} \tag{5.10.6}$$

where $l > 0$, l not an integer. Obviously,

$$L_q = (A_q, B_q, B) \in \mathcal{L}(V_{lq}, H_{lq}).$$

The kernel space of the operator L_q is the three-dimensional space $\hat{V}_{lq} = Q$, where Q is defined by (5.9.11). The following functions form a basis in $\hat{V}_{lq}$:

$$\varphi_{q1} = (1, 0), \quad \varphi_{q2} = (0, 1), \quad \varphi_{q3} = (y_2, -y_1). \tag{5.10.7}$$

For given functions of volume and surface forces (f, R, F) defined on Ω_q, S_{2q}, and S_1, respectively, consider the problem: Find a function of displacements u such that

$$\begin{aligned} A_q u &= f, \quad && \text{in } \Omega_q, \\ B_q u &= R, \quad && \text{on } S_{2q}, \\ Bu &= F, \quad && \text{on } S_1. \end{aligned} \tag{5.10.8}$$

We assume that the volume and surface forces (f, R, F) are self-balanced, i.e., these are orthogonal to the space $\hat{V}_{lq}$ in the sense that

$$\begin{aligned} &\int_{\Omega_q} f_i\, dy + \int_{S_{2q}} R_i\, ds + \int_{S_1} F_i\, ds = 0, \qquad i = 1, 2, \\ &\int_{\Omega_q} (f_1 y_2 - f_2 y_1)\, dy + \int_{S_{2q}} (R_1 y_2 - R_2 y_1)\, ds \\ &\quad + \int_{S_1} (F_1 y_2 - F_2 y_1)\, ds = 0. \end{aligned} \tag{5.10.9}$$

The conditions (5.10.9) are necessary and sufficient for the existence of a solution of the problem (5.10.8), see Hlavaček and Nečas (1970). From (5.10.9), it follows that the space $\hat{H}_{lq} = H_{lq} \setminus L_q(V_{lq})$, where $L_q = (A_q, B_q, B)$, is three-dimensional, and the following functions form a basis in $\hat{H}_{lq}$:

$$\begin{aligned} \psi_{q1} &= ((1, 0), (1, 0), (1, 0)), \\ \psi_{q2} &= ((0, 1), (0, 1), (0, 1)), \\ \psi_{q3} &= ((y_2, -y_1), (y_2, -y_1), (y_2, -y_1)). \end{aligned} \tag{5.10.10}$$

Here, in the expressions for the functions ψ_{qi}, the first pair belongs to $C^l(\overline{\Omega}_q)^2$, the second one to $C^{l+1}(S_{2q})^2$, and the third one to $C^{l+1}(S_1)^2$. In this case, the equality (2.13.23) holds with $k_q = 3$, and we define the operator T_q by (2.13.24). Owing to Remark 2.13.1, we obtain the following

Theorem 5.10.1 *Let spaces* V_{lq} *and* H_{lq} *be defined by* (5.10.6), *let an operator* $L_q = (A_q, B_q, B)$ *be defined by the formulas* (5.9.7), (5.10.4), (5.10.5), *and let an operator* T_q *be given by* (2.13.24) *with* $k_q = 3$, *where* φ_{qi}, ψ_{qi} *are determined by* (5.10.7) *and* (5.10.10). *Then, the operator* $L_{q1} = L_q + T_q$ *is an isomorphism of* V_{lq} *onto* H_{lq}.

Suppose now that the surface forces F given on the external boundary satisfy the conditions

$$\begin{gathered} F \in C^{l+1}(S_1)^2, \qquad \int_{S_1} F_i\, ds = 0, \qquad i = 1, 2, \\ \int_{S_1} (F_1 y_2 - F_2 y_1)\, ds = 0. \end{gathered} \tag{5.10.11}$$

Then, due to Theorem 5.10.1, there exists a unique function $u_q \in C^{l+2}(\overline{\Omega}_q)^2$ such that

$$\begin{aligned} A_q u_q &= 0, && \text{in } \Omega_q, \\ B_q u_q &= 0, && \text{on } S_{2q}, \\ B u_q &= F, && \text{on } S_1, \\ T_q u_q &= 0, && \text{in } H_{lq}. \end{aligned} \tag{5.10.12}$$

In the same way as above, by using Theorem 5.10.1, we prove the subsequent assertion.

Theorem 5.10.2 *Let the conditions of Theorem* 5.10.1 *hold, and let sets* M, M_1 *be defined by* (5.10.1) *and* (5.9.17), *the functionals* G_1 *and* G_0 *given by* (5.9.15) *and* (5.9.16), *where* u_q *is the solution of the problem* (5.10.12). *Let also a function* F *satisfy the conditions* (5.10.11), *and let a nonempty set* U_{ad} *be given by* (5.9.18). *Then, there exists a solution of the problem* (5.9.19).

Obviously, in the considered case, the optimization problem consists in finding the shape of the internal boundary of the elastic body of minimal weight (volume) that satisfies the constraint on strength.

5.11 Optimization problems on manifolds and shape optimization of elastic solids

We have studied earlier shape optimization problems by using the following approach. Let M be a set of controls. To each $q \in M$ a domain $\Omega_q \subset \mathbb{R}^n$ is assigned, and one considers the problem of finding a function u_q defined on Ω_q that satisfies $A_q u_q = f_q$. Here, A_q is some elliptic operator acting from a space V_q to a space H_q, V_q and H_q consisting of functions defined on Ω_q and on its boundary S_q. The optimization problem is to minimize or maximize a goal functional under some restrictions. However, in the general case, the goal and restriction functionals cannot be defined on various V_q, $q \in M$. Besides, for the existence of a solution of the optimization problem, it is necessary that the goal and restriction functionals possess some continuity with respect to a control q, but it is inconvenient for one to establish the continuity when one works with various V_q's. This is why one applies a diffeomorphism P_q of the set $\overline{\Omega}_q$ onto a fixed set $\overline{\Omega}$, and after the change of variables corresponding to this diffeomorphism one arrives at the problem $A(q)u(q) = f(q)$ in the fixed domain Ω and on its boundary S, for all $q \in M$. Now, $u(q) \in V$, where V is a space of functions on $\overline{\Omega}$, and the goal and restriction functionals are given on V.

However, in some cases, the construction of the diffeomorphisms P_q for all $q \in M$, the passage to the problem $A(q)u(q) = f(q)$ in the fixed domain $\overline{\Omega}$, and the solution of the problem obtained may be difficult. In this section, we will introduce and study another approach to the shape optimization, which is based on the transition to equations on the boundary, and so on the solution of an optimization problem on manifolds. In this case, instead of the diffeomorphisms P_q, we should define maps I_q of the boundaries S_q, and the domains of the maps I_q should be the same for all $q \in M$. Denoting it by T, we obtain state equations on the fixed set $T \subset \mathbb{R}^{n-1}$, while $\Omega_q \subset \mathbb{R}^n$. Of course, such an approach may be utilized if the fundamental solution of the state equations is found.

Below, we apply this approach to the optimization of elastic solids.

5.11.1 Optimization problem for an elastic solid

Formulation of the problem

We consider a shape optimization problem for a two-dimensional elastic solid. Let M be a set of controls, and let to each $q \in M$ a bounded domain $\Omega_q \subset \mathbb{R}^2$ with a smooth boundary S_q be assigned. We suppose Ω_q is multiply-connected, and denote by S_{q0} the external boundary of Ω_q, and by S_{qi}, $1 \leq i \leq p$, the other components of S_q, where $S_{qi} \cap S_{qj} = \varnothing$ for $i \neq j$, S_{q0} envelops the others S_{qk}, and S_{qk}, $k = 1, \ldots, p$, do not envelop each other. S_{qi} is defined by a periodic function $q_i\colon (-\pi, \pi) \to \mathbb{R}^2$, $i = 0, 1, \ldots, p$, see Fig 5.11.1. So, we define a set of controls M by

$$\begin{aligned} M = \Big\{ & q = (q_0, q_1, \ldots, q_p),\ q_i = (q_{i1}, q_{i2}) \in \widetilde{C}^{m+1}([-\pi, \pi])^2, \\ & m \in \mathbb{N},\ m \geq 3,\ \left(\frac{dq_{i1}}{dt}(t)\right)^2 + \left(\frac{dq_{i2}}{dt}(t)\right)^2 > c_0 > 0, \\ & q_i(t) \in Q_i,\ t \in (-\pi, \pi],\ i = 0, 1, \ldots, p \Big\}. \end{aligned} \tag{5.11.1}$$

Here, Q_i are some open sets in $\mathbb{R}^2$ such that, for all $q \in M$, the above conditions on Ω_q are satisfied, $\widetilde{C}^{m+1}([-\pi, \pi])$ is the subspace of periodic functions of $C^{m+1}([-\pi, \pi])$.

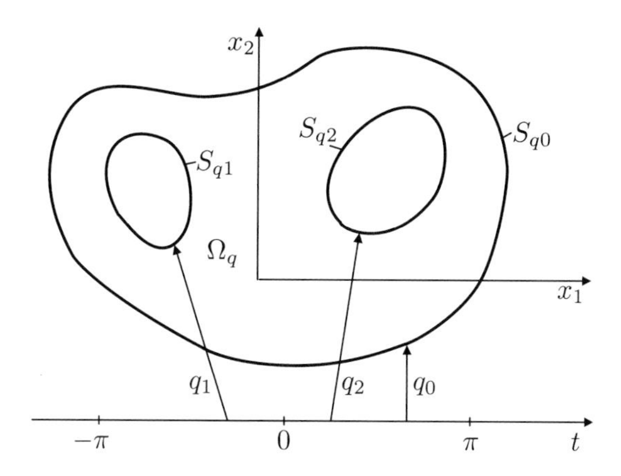

Figure 5.11.1: Multiply-connected domain Ω_q; S_{qi}'s are defined by periodic functions q_i

In the sequel, we consider all periodic functions as being given either on the unit circle or on $\mathbb{R}$ and points $t + 2\pi k$, $k \in \mathbb{Z}$, are identified. Thus, we consider periodic functions on $\mathbb{R}/2\pi\mathbb{Z}$ – the factor-group consisting of classes $\dot{t} = t + 2\pi\mathbb{Z}$ containing a point t.

The set M is provided with the topology generated by that of

$$\widetilde{C}^{m+1}([-\pi, \pi])^{2(p+1)}.$$

Notice that the mapping q_i is a homeomorphism of $(-\pi, \pi]$ onto S_{qi} for an arbitrary $q \in M$, $i = 0, 1, \ldots, p$.

The operator A_q of the theory of elasticity is defined by

$$A_q = -\mu\Delta u - (\lambda + \mu)\,\mathrm{grad}\,\mathrm{div}\,u, \qquad \text{in } \Omega_q. \tag{5.11.2}$$

Here, $u = (u_1, u_2)$ is a vector function of displacements, λ, μ are positive constants. We denote by $\varepsilon(u) = (\varepsilon_{ij}(u))$, $\sigma(u) = (\sigma_{ij}(u))$ the strain and stress tensors,

$$\begin{aligned} \varepsilon_{ij}(u) &= \frac{1}{2}\left(\frac{\partial u_i}{\partial x_j} + \frac{\partial u_j}{\partial x_i}\right), \\ \sigma_{ij}(u) &= \lambda\,\mathrm{div}\,u\delta_{ij} + 2\mu\varepsilon_{ij}(u), \qquad i, j = 1, 2, \end{aligned} \tag{5.11.3}$$

where δ_{ij} is the Kronecker delta. The traction operator (the operator of surface forces) T_q is defined on S_q by

$$\begin{aligned} T_q u &= ((T_q u)_1, (T_q u)_2), \\ (T_q u)_i &= \sigma_{ij}(u)\nu_{jq}, \qquad \text{on } S_q,\ i, j = 1, 2. \end{aligned} \tag{5.11.4}$$

Here and below, the summation over repeated indices is implied, ν_{jq} are the components of the unit outward normal ν_q to S_q. Various formulations of problems of the theory of elasticity may be considered, in particular, displacement, traction, mixed, and other ones (see, e.g., Kupradze et al. (1979)). We will be engaged in the traction formulation

$$T_q u = \mathcal{F}_q, \qquad \text{on } S_q. \tag{5.11.5}$$

So we consider the problem: Find a function u_q satisfying

$$A_q u_q = 0, \qquad \text{in } \Omega_q, \tag{5.11.6}$$

$$T_q u_q = \mathcal{F}_q, \qquad \text{on } S_q. \tag{5.11.7}$$

The case when the function of body forces, i.e., the right-hand side of (5.11.6), is not equal to zero, may be reduced to the problem (5.11.6), (5.11.7). Further, we suppose that the boundary S_{q0} is fixed, i.e.,

$$S_{q0} = S, \qquad q \in M, \tag{5.11.8}$$

the surface forces $\mathcal{F}_q$ are not equal to zero only on S, and they are fixed and self-balanced:

$$\begin{aligned} &\mathcal{F}_q\big|_{S_{qi}} = 0, \qquad i = 1, \dots, p \\ &\mathcal{F}_q\big|_S = (\mathcal{F}_1, \mathcal{F}_2) \in H^{m-\frac{3}{2}}(S)^2, \\ &\int_S \mathcal{F}_i \, ds = 0, \qquad i = 1, 2, \\ &\int_S (\mathcal{F}_1 x_2 - \mathcal{F}_2 x_1) \, ds = 0. \end{aligned} \tag{5.11.9}$$

We introduce spaces

$$\begin{aligned} V_{mq} &= H^m(\Omega_q)^2, \\ H_{mq} &= H^{m-2}(\Omega_q)^2 \times H^{m-\frac{3}{2}}(S_q)^2. \end{aligned} \tag{5.11.10}$$

Then, the operator $G_q = (A_q, T_q)$ is a linear continuous mapping of V_{mq} into H_{mq}, i.e., $G_q \in \mathcal{L}(V_{mq}, H_{mq})$, and by known results, see Theorem 2.13.1 and Section 5.10, we obtain

Theorem 5.11.1 *Let a set M be defined by* (5.11.1), *and let* (5.11.8), (5.11.9) *hold. Then, for each $q \in M$, the following representations are valid*

$$V_{mq} = \hat{V}_{mq} \oplus \check{V}_{mq}, \qquad H_{mq} = \hat{H}_{mq} \oplus \check{H}_{mq},$$

where $\hat{V}_{mq} = \ker G_q$, $\check{H}_{mq} = G_q(V_{mq})$, $G_q = (A_q, T_q)$. The subspaces $\hat{V}_{mq}$ and $\hat{H}_{mq}$ are three-dimensional, and $\varphi_1 = (1,0)$, $\varphi_2 = (0,1)$, $\varphi_3 = (x_2, -x_1)$ is a basis in $\hat{V}_{mq}$, $\psi_1 = ((1,0),(1,0))$, $\psi_2 = ((0,1),(0,1))$, $\psi_3 = ((x_2,-x_1),(x_2,-x_1))$ is a basis in $\hat{H}_{mq}$.

For each $q \in M$, there exists a unique $u_q \in \check{V}_{mq}$ satisfying (5.11.6), (5.11.7).

We introduce the following functionals on M:

$$
\begin{aligned}
\Psi_0(q) &= \int_{\Omega_q} dx, \\
\Psi_1(q) &= \max_{x\in\overline{\Omega}_q} |u_q(x)| - c_1, \qquad u_q \in \check{V}_{mq}, \\
\Psi_2(q) &= \max_{x\in\overline{\Omega}_q} \sum_{i,j=1}^{2} \Big[\sigma_{ij}(u_q)(x) - \frac{1}{2}\big(\sigma_{11}(u_q)(x) + \sigma_{22}(u_q)(x)\big)\delta_{ij}\Big]^2 - c_2, \\
\Psi_3(q) &= \int_{\Omega_q} \sigma_{ij}(u_q)\varepsilon_{ij}(u_q)\, dx - c_3,
\end{aligned}
\tag{5.11.11}
$$

where c_1, c_2, c_3 are positive constants. Note that other functionals on M may also be considered.

Now, suppose that

$$
M_1 \text{ is a compact subset of } M. \tag{5.11.12}
$$

In particular, the set M_1 may be defined by

$$
\begin{aligned}
M_1 = \big\{ & q = (q_0, q_1, \dots, q_p) \in M,\ q_i(t) \in \widetilde{Q}_i,\ t \in (-\pi, \pi], \\
& \widetilde{Q}_i \text{ are closed subsets in } Q_i, \\
& \|q_i\|_{C^{m+1+\alpha}([-\pi,\pi])} \le c,\ \alpha \in (0,1],\ i = 1, \dots, p \big\}.
\end{aligned}
$$

We recall that q_0 is considered to be fixed (see (5.11.8)), $C^{k+\alpha}([-\pi,\pi])$, where $k \in \mathbb{N}$, denotes the Hölder space with the norm

$$
\|u\|_{C^{k+\alpha}([-\pi,\pi])} = \|u\|_{C^k([-\pi,\pi])} + \sup_{t,t'\in[-\pi,\pi]} \frac{1}{|t-t'|^\alpha} \left| \frac{d^k u}{dt^k}(t) - \frac{d^k u}{dt^k}(t') \right|.
$$

Define a set of admissible controls M_{ad} as follows

$$
M_{\text{ad}} = \big\{ q \mid q \in M_1,\ \Psi_i(q) \le 0,\ i = 1, 2, 3 \big\}, \tag{5.11.13}
$$

and consider the optimization problem: Find $\tilde{q}$ satisfying

$$
\tilde{q} \in M_{\text{ad}}, \qquad \Psi_0(\tilde{q}) = \inf_{q \in M_{\text{ad}}} \Psi_0(q). \tag{5.11.14}
$$

From the physical point of view, the problem (5.11.14) corresponds to the minimization of the area (weight) of the elastic solid under restrictions on displacements, stresses, and strain energy. Other restrictions of the form $\Psi_k(q) \le 0$ may also be considered.

State equations on the boundaries and on the unit circle

Let $G(x,y) = (G_{ij}(x,y))$ be the tensor of fundamental solutions of the equation

$$Au = -\mu\Delta u - (\lambda+\mu)\,\mathrm{grad}\,\mathrm{div}\,u = 0, \qquad u = (u_1, u_2). \tag{5.11.15}$$

$G(x,y)$ is the symmetric tensor defined by

$$G_{ij}(x,y) = c_1\left(c_2\delta_{ij}\log r - \frac{(x_i-y_i)(x_j-y_j)}{r^2}\right), \qquad i,j = 1,2\ , \tag{5.11.16}$$

where

$$r = \left[\sum_{i=1}^{2}(x_i-y_i)^2\right]^{\frac{1}{2}},$$
$$c_1 = -\frac{1}{8\pi\mu(1-\sigma)}, \qquad c_2 = 3-4\sigma, \qquad \sigma = \frac{\lambda}{2(\lambda+\mu)}. \tag{5.11.17}$$

From the physical point of view, the function $G_{ij}(x,y)$ defines a displacement $u_i(x)$ engendered by the unit force $P_j(y)$ concentrated at a point y and directed along the x_j axis.

By $B_k(x,y) = (B_{ijk}(x,y))$ and $T_k(x,y) = (T_{ijk}(x,y))$ we denote the strain and stress tensors at a point x that are engendered by the force $P_k(y)$. Due to (5.11.3) and (5.11.16), we obtain

$$B_{ijk}(x,y) = \frac{c_1}{r^2}\Big[(1-2\nu)(\delta_{ik}\xi_j + \delta_{jk}\xi_i) - \delta_{ij}\xi_k + \frac{2}{r^2}\,\xi_i\xi_j\xi_k\Big], \tag{5.11.18}$$

$$T_{ijk}(x,y) = \frac{c_3}{r^2}\Big[c_4(\delta_{ik}\xi_j + \delta_{jk}\xi_i - \delta_{ij}\xi_k) + \frac{2}{r^2}\,\xi_i\,\xi_j\xi_k\Big], \tag{5.11.19}$$
$$\xi_i = x_i - y_i, \quad c_3 = -\frac{1}{4\pi(1-\sigma)}, \quad c_4 = 1-2\sigma.$$

The force $t(x) = (t_1(x), t_2(x))$ at a point x of a surface with a unit outward normal $\nu = (\nu_1, \nu_2)$ is defined by $t_i(x) = \sigma_{ij}(x)\nu_j(x)$. So, denoting by $R_{ik}(x,y)$ the value at a point x of the i-th component of the surface force generated by $P_k(y)$, we get due to (5.11.19)

$$R_{ik}(x,y) = \frac{c_3}{r^2}\left[c_4(\nu_k\xi_i - \nu_i\xi_k) + \left(c_4\delta_{ik} + \frac{2\xi_i\xi_k}{r^2}\right)\xi_j\nu_j\right]. \tag{5.11.20}$$

Let $u = (u_1, u_2)$ be a smooth function satisfying (5.11.15) in Ω_q. By Betti's formula, see Banerjee and Butterfield (1981), we obtain

$$u_i(x) = \int_{S_q}\left[(T_q u)_j(y)G_{ij}(x,y) - F_{qij}(x,y)u_j(y)\right]dS_y,$$
$$x\in\Omega_q,\ i = 1,2, \tag{5.11.21}$$

where $F_{qij}(x,y) = R_{ji}(y,x)$ for $\nu = \nu_q$, i.e.,

$$F_{qij}(x,y) = \frac{c_3}{r^2}\Big[c_4(\nu_{qi}(y)(y_j - x_j) - \nu_{qj}(y)(y_i - x_i)) + \left(c_4\delta_{ij} + \frac{2(y_i - x_i)(y_j - x_j)}{r^2}\right)(y_k - x_k)\nu_{qk}(y)\Big]. \quad (5.11.22)$$

Because of the jump relation for $F_q(x,y) = (F_{qij}(x,y))$ on S_q, see Banerjee and Butterfield (1981), the representation formula (5.11.21) yields the following expression

$$\frac{1}{2}u_i(x) = \int_{S_q}\left[(T_q u)_j(y)G_{ij}(x,y) - F_{qij}(x,y)u_j(y)\right]dS_y, \qquad x \in S_q, \quad i = 1,2. \quad (5.11.23)$$

The integral from the second addend in (5.11.23) should be understood in the sense of the Cauchy principal value.

Considering the problem (5.11.6), (5.11.7), we obtain the equation

$$N_q u(x) = u(x) + 2\int_{S_q} F_q(x,y)u(y)\,dS_y = f_q(x), \qquad x \in S_q, \quad (5.11.24)$$

where

$$f_q(x) = 2\int_{S_q} G(x,y)\mathcal{F}_q(y)\,dS_y. \quad (5.11.25)$$

The boundary S_q is defined by the function $q = (q_0, q_1, \ldots, q_p) \in M$ (see Fig. 5.11.1). Denoting $v_{qi}(t) = u(q_i(t))$, $i = 0,1,\ldots,p$, $v_q = (v_{q0},\ldots,v_{qp})$, we transform the problem (5.11.24) into the following one

$$(P(q)v_q)_i(t) = v_{qi}(t) + 2\sum_{j=0}^{p}\int_{-\pi}^{\pi} F_q(q_i(t), q_j(\tau))v_{qj}(\tau)D(q_j)(\tau)\,d\tau = g_i(t), \qquad t \in [-\pi,\pi],\ i = 0,1,\ldots,p. \quad (5.11.26)$$

Here,

$$D(q_j)(\tau) = \left[\left(\frac{dq_{j1}}{d\tau}(\tau)\right)^2 + \left(\frac{dq_{j2}}{d\tau}(\tau)\right)^2\right]^{\frac{1}{2}}, \qquad g_i(t) = f_q(q_i(t)), \quad (5.11.27)$$

and g_i is independent of q because $\mathcal{F}_q$ is not equal to zero only on S_{q0}, and S_{q0} is fixed, see (5.11.8), (5.11.9). In the case when $i = j$, the integral in (5.11.26) should be understood as the limit of the integrals

$$\int_{-\pi}^{\pi} F_{q\varepsilon}(t,\tau)v_{qi}(\tau)D(q_i)(\tau)\,d\tau, \qquad F_{q\varepsilon}(t,\tau) = \begin{cases} F_q(q_i(t), q_i(\tau)), & \text{if } |t-\tau| \geq \varepsilon \\ 0, & \text{if } |t-\tau| < \varepsilon \end{cases}$$

as ε tends to zero.

5.11.2 Spaces and operators on $\mathbb{R}/2\pi\mathbb{Z}$, auxiliary statements

Define a space W_m as follows

$$W_m = \{ u = (u_0, u_1, \dots, u_p),\ u_i = (u_{i1}, u_{i2}) \in \widetilde{H}^{m-\frac{1}{2}}(-\pi,\pi)^2,\\ i = 0, 1, \dots, p,\ m \geq 3 \}. \tag{5.11.28}$$

By $\widetilde{H}^s(-\pi,\pi)$, $s > 0$, we denote the subspace of $H^s(-\pi,\pi)$ consisting of periodic functions. The norm in W_m is defined by

$$\|v\|_m = \left(\sum_{i=0}^{p} \sum_{k=1}^{2} \|u_{ik}\|^2_{\widetilde{H}^{m-\frac{1}{2}}(-\pi,\pi)} \right)^{\frac{1}{2}}. \tag{5.11.29}$$

Here,

$$\|v\|_{\widetilde{H}^{m-\frac{1}{2}}(-\pi,\pi)} = \left[a_0^2 + \sum_{n=1}^{\infty} (a_n^2 + b_n^2) n^{2m-1} \right]^{\frac{1}{2}}, \tag{5.11.30}$$

where a_n, b_n are the Fourier coefficients of the function v, i.e.,

$$\begin{aligned} v(t) &= \frac{a_0}{2} + \sum_{n=1}^{\infty} (a_n \cos nt + b_n \sin nt), \\ a_n &= \frac{1}{\pi} \int_{-\pi}^{\pi} v(t) \cos nt\, dt, \\ b_n &= \frac{1}{\pi} \int_{-\pi}^{\pi} v(t) \sin nt\, dt. \end{aligned} \tag{5.11.31}$$

Let

$$\hat{W}_{mq} = (\gamma_q \hat{V}_{mq}) \circ q\,. \tag{5.11.32}$$

Here, $\hat{V}_{mq}$ is $\ker(A_q, T_q)$ (see Theorem 5.11.1), γ_q is the trace operator on S_q, q is the function defined in (5.11.1). It follows from Theorem 5.11.1 that $\hat{W}_{mq}$ is the three-dimensional subspace of W_m with the basis $\{\varphi_{qi}\}_{i=1}^{3}$,

$$\varphi_{qi} = (\varphi_{qi0}, \varphi_{qi1}, \dots, \varphi_{qip}), \qquad i = 1, 2, 3,\\ \varphi_{q1k} = (1, 0), \quad \varphi_{q2k} = (0, 1), \quad \varphi_{q3k} = (q_{k2}, -q_{k1}), \qquad k = 0, 1, \dots, p, \tag{5.11.33}$$

q_{ki} defined in (5.11.1).

We denote by J_q the operator of the simple layer potential

$$(J_q v)(x) = \int_{S_q} G(x,y) v(y)\, dS_y, \qquad x \in S_q, \tag{5.11.34}$$

and let

$$\begin{aligned} \psi_{qi} &= (J_q \beta_i) \circ q, && i = 1,2,3, \\ \beta_1 &= (1,0), && \text{on } S_{qi}, \\ \beta_2 &= (0,1), && \text{on } S_{qi}, \\ \beta_3 &= (x_2, -x_1), && \text{on } S_{qi}, \qquad i = 0,1,\dots,p. \end{aligned} \tag{5.11.35}$$

By (5.11.34), (5.11.35), we get

$$\begin{aligned} \psi_{qi} &= (\psi_{qi0}, \psi_{qi1}, \dots, \psi_{qip}), \qquad i = 1,2,3, \\ \psi_{qij}(t) &= \sum_{k=0}^{p} \int_{-\pi}^{\pi} G(q_j(t), q_k(\tau))\beta_i(q_k(\tau))D(q_k)(\tau)\, d\tau, \qquad j = 0,1,\dots,p, \end{aligned} \tag{5.11.36}$$

$D(q_k)$ defined by (5.11.27).

Theorem 5.11.2 *Let a set M be defined by* (5.11.1), *and let* (5.11.8), (5.11.9) *hold. Let also an operator $P(q)$ be defined by* (5.11.26). *Then, $P(q) \in \mathcal{L}(W_m)$, and for each $q \in M$ the following representations are valid*

$$W_m = \hat{W}_{mq} \oplus \check{W}_{mq} = \hat{E}_{mq} \oplus \check{E}_{mq}, \tag{5.11.37}$$

where $\hat{W}_{mq} = \ker P(q)$, $\check{E}_{mq} = P(q)(W_m)$, $\hat{E}_{mq}$ is the three-dimensional subspace of W_m with the basis $\{\psi_{qi}\}_{i=1}^{3}$ defined by (5.11.36). *There exists a unique $v_q \in \check{W}_{mq}$ satisfying* (5.11.26).

Proof. The operator J_q is a pseudodifferential operator of order -1 on S_q, and J_q is an isomorphism of $H^{m-\frac{3}{2}}(S_q)^2$ onto $H^{m-\frac{1}{2}}(S_q)^2$, see Chudinovich (1991), Wendland (1985). It is obvious that, if u_q is a solution of the problem (5.11.6), (5.11.7), then $u = \gamma_q u_q$ is a solution of the problem (5.11.24). On the contrary, if u is a solution of the problem (5.11.24), then the function

$$u_q(x) = \int_{S_q} \left[G(x,y)\mathcal{F}_q(y) - F_q(x,y)u(y) \right] dS_y, \qquad x \in \Omega_q,$$

is a solution of the problem (5.11.6), (5.11.7). The problem (5.11.26) is obtained from (5.11.24) by the change of variables corresponding to the one-to-one mapping $q \in \widetilde{C}^{m+1}(-\pi,\pi)^{2(p+1)}$. So, Theorem 5.11.2 follows from Theorem 5.11.1.

Remark 5.11.1 The space $\check{W}_{mq}$ is defined nonuniquely by (5.11.37), and so we define $\check{W}_{mq}$ so that it is orthogonal to $\hat{W}_{mq}$ with respect to the scalar product in $L_2(-\pi,\pi)^{2(p+1)}$.

Define an operator $\Gamma_{q1} \in \mathcal{L}(W_m, \hat{W}_{mq})$ by

$$\left.\begin{array}{l}\Gamma_{q1} \text{ is the orthogonal projection of } W_m \text{ onto } \hat{W}_{mq} \text{ with} \\ \text{respect to the scalar product in } L_2(-\pi,\pi)^{2(p+1)}.\end{array}\right\} \tag{5.11.38}$$

Also, define an operator $\Gamma_{q2} \in \mathcal{L}(\hat{W}_{mq}, \hat{E}_{mq})$ as follows

$$u = \sum_{i=1}^{3} c_i \varphi_{qi}, \qquad \Gamma_{q2} u = \sum_{i=1}^{3} c_i \psi_{qi}. \tag{5.11.39}$$

Theorem 5.11.3 *Let a set M and an operator P_q be defined by* (5.11.1) *and* (5.11.26), *respectively. Then, the operator $\mathcal{N}_q = P(q)+\Gamma(q)$, where $\Gamma(q) = \Gamma_{q2}\circ\Gamma_{q1}$, is an isomorphism of W_m onto itself.*

Proof. Obviously, $\Gamma(q) \in \mathcal{L}(W_m, \hat{E}_{mq})$ and $\Gamma(q)(W_m) = \hat{E}_{mq}$. Therefore, by Theorem 5.11.2, the operator $P(q)+\Gamma(q)$ is a continuous bijection from W_m onto W_m, and so by the Banach theorem $P(q)+\Gamma(q)$ is an isomorphism of W_m onto W_m.

Lemma 5.11.1 *Let a set M and an operator $\Gamma(q) = \Gamma_{q2}\circ\Gamma_{q1}$ be defined by* (5.11.1), (5.11.38), (5.11.39), *and let $q^n \to q^0$ in M (M equipped with the topology generated by the $\widetilde{C}^{m+1}(-\pi,\pi)^{2(p+1)}$ topology). Then, $\|\Gamma(q^n) - \Gamma(q^0)\|_{\mathcal{L}(W_m,W_m)} \to 0$ as $n \to \infty$.*

Proof. Let $u = (u_0, u_1, \ldots, u_p) \in W_m$. By (5.11.38), we have

$$\|\Gamma_{q^n 1} u - u\|^2_{L_2(-\pi,\pi)^{2(p+1)}} = \min_{c_i} \sum_{k=0}^{p} \|u_k - \sum_{i=1}^{3} c_i \varphi_{q^n ik}\|^2_{L_2(-\pi,\pi)^2}. \tag{5.11.40}$$

We denote by c_{in} a solution of the problem (5.11.40), i.e., c_{in} minimizes the norm in (5.11.40). Then, we get

$$c_{in} \sum_{i=1}^{3} \sum_{k=0}^{p} (\varphi_{q^n ik}, \varphi_{q^n jk})_{L_2(-\pi,\pi)^2} = \sum_{k=0}^{p} (u_k, \varphi_{q^n jk})_{L_2(-\pi,\pi)^2}, \qquad j = 1,2,3. \tag{5.11.41}$$

The matrix of the system (5.11.41) is nondegenerate because the functions $\{\varphi_{q^n i}\}_{i=1}^{3}$ form a basis in $\hat{W}_{mq^n}$. Therefore, c_{in} are defined uniquely. Let now $q^n \to q^0$ in M. Then $\varphi_{q^n ik} \to \varphi_{q^0 ik}$ in $\widetilde{C}^{m+1}(-\pi,\pi)^2$, and so $c_{in} \to c_{i0}$ as $n \to \infty$, where c_{i0} is a solution of the problem (5.11.41) for $n = 0$. As the embedding of $\widetilde{C}^{m+1}(-\pi,\pi)$ into $L_2(-\pi,\pi)$ is compact, we obtain that $c_{in} \to c_{i0}$ uniformly in $u \in K$, where K is an arbitrary bounded set in M.

It may be shown that $\psi_{q^n ij} \to \psi_{q^0 ij}$ in $\widetilde{H}^m(-\pi,\pi)^2$ as $q^n \to q^0$ in M. Therefore, $\Gamma(q^n) \to \Gamma(q^0)$ in $\mathcal{L}(W_m)$, concluding the proof of the lemma.

In the space W_m, we define an operator $\mathcal{N}(q)$ as follows

$$v = (v_0, v_1, \dots, v_p) \in W_m, \qquad \mathcal{N}(q)v = \{(\mathcal{N}(q)v)_i\}_{i=0}^{p},$$

$$(\mathcal{N}(q)v)_i(t) = \sum_{j=0}^{p} \int_{-\pi}^{\pi} F_q(q_i(t), q_j(\tau)) v_j(\tau) D(q_j)(\tau)\, d\tau, \qquad (5.11.42)$$

$$t \in [-\pi, \pi], \quad i = 0, 1, \dots, p.$$

By (5.11.26), we have

$$P(q) = I + 2\mathcal{N}(q), \qquad (5.11.43)$$

where I is the identity operator in W_m.

Lemma 5.11.2 *Let a set M be defined by* (5.11.1) *and let an operator $\mathcal{N}(q)$ be defined by* (5.11.42), (5.11.22), (5.11.27). *Then, $\mathcal{N}(q) \in \mathcal{L}(W_m)$ and the function $\mathcal{N}\colon q \to \mathcal{N}(q)$ is a continuous mapping of M into $\mathcal{L}(W_m)$.*

Proof. By (5.11.22), we have

$$F_q(q_i(t), q_j(\tau)) = (F_{qkn}(q_i(t), q_j(\tau)))_{k,n=1}^{2},$$

$$\begin{aligned} F_{qkn}(q_i(t), q_j(\tau)) = \frac{c_3}{\rho(\tau,t)} \Big\{ & c_4 \left[\nu_{qk}(q_j(\tau))(q_{jn}(\tau) - q_{in}(t)) \right. \\ & \left. - \nu_{qn}(q_j(\tau))(q_{jk}(\tau) - q_{ik}(t))\right] \\ & + \left[c_4 \delta_{kn} + \frac{2(q_{jk}(\tau) - q_{ik}(t))(q_{jn}(\tau) - q_{in}(t))}{\rho(\tau,t)} \right] \\ & \times \sum_{s=1}^{2} (q_{js}(\tau) - q_{is}(t)) \nu_{qs}(q_j(\tau)) \Big\}. \end{aligned} \qquad (5.11.44)$$

Here,

$$\rho(\tau,t) = \sum_{s=1}^{2} (q_{is}(t) - q_{js}(\tau))^2,$$

$$\nu_q(q_j(\tau)) = \left(a(\tau) \frac{dq_{j2}}{d\tau}(\tau), -a(\tau) \frac{dq_{j1}}{d\tau}(\tau) \right), \qquad (5.11.45)$$

$$a(\tau) = \left[\left(\frac{dq_{j1}}{d\tau}(\tau) \right)^2 + \left(\frac{dq_{j2}}{d\tau}(\tau) \right)^2 \right]^{-\frac{1}{2}}.$$

In the case when $i \neq j$, the elements $F_{qkn}(q_i(t), q_j(\tau))$ are nonsingular and they generate smoothing operators. So, we consider the case when $i = j$. As $q_i \in \widetilde{C}^{m+1}([-\pi, \pi])^2$, by (5.11.45) and the Taylor formula, we get

$$\left| \sum_{s=1}^{2} (q_{is}(\tau) - q_{is}(t)) \nu_{qs}(q_i(\tau)) \right| = |A(\tau, t)| \le c(t - \tau)^2.$$

Therefore, the terms containing $A(\tau,t)$ as a factor in (5.11.44) generate a nonsingular operator. The two remaining addends in (5.11.44) are singular and similar. So, it is sufficient to show that

$$\left.\begin{aligned}&\text{the operator } \mathcal{R}(q) \text{ defined by}\\ &\quad u \in \widetilde{H}^{m-\frac{1}{2}}(-\pi,\pi),\\ &(\mathcal{R}(q)u)(t)\\ &\quad = \int_{-\pi}^{\pi} \frac{1}{\rho(\tau,t)}\, \nu_{qk}(q_i(\tau))(q_{in}(\tau)-q_{in}(t))D(q_i)(\tau)u(\tau)\,d\tau\\ &\text{is a continuous mapping of } \widetilde{H}^{m-\frac{1}{2}}(-\pi,\pi) \text{ into itself, and if } q^\mu \to q^0\\ &\text{in } M, \text{ then } \mathcal{R}(q^\mu) \to \mathcal{R}(q^0) \text{ in } \mathcal{L}(\widetilde{H}^{m-\frac{1}{2}}(-\pi,\pi)).\end{aligned}\right\}. \tag{5.11.46}$$

By the Taylor formula we obtain

$$q_{in}(\tau) - q_{in}(t) = q'_{in}(t)(\tau - t) + a_q(\tau,t), \tag{5.11.47}$$

$$\rho(\tau,t) = \sum_{s=1}^{2}(q_{is}(\tau)-q_{is}(t))^2 = \sum_{s=1}^{2} q'_{is}(t)^2(\tau-t)^2 + \beta_q(\tau,t), \tag{5.11.48}$$

$$|a_q(\tau,t)| \le c|\tau-t|^2, \qquad |\beta_q(\tau,t)| \le c_1|\tau-t|^3. \tag{5.11.49}$$

Substituting (5.11.47), (5.11.48) into $\mathcal{R}(q)u$, we get

$$\mathcal{R}(q)u = \mathcal{R}_1(q)u + \mathcal{R}_2(q)u, \tag{5.11.50}$$

$$(\mathcal{R}_1(q)u)(t) = \Big(\sum_{s=1}^{2} q'_{is}(t)^2\Big)^{-1} q'_{in}(t)\int_\Gamma \frac{w_q(\tau)u(\tau)}{\tau - t}\,d\tau, \tag{5.11.51}$$

$$(\mathcal{R}_2(q)u)(t) = \int_\Gamma H(\tau,t)w_q(\tau)u(\tau)\,d\tau. \tag{5.11.52}$$

Here, Γ is the unit circle on the x_1Ox_2 plane centered at 0, the origin of the curvilinear coordinate on Γ is a point of Γ lying on Ox_1, the direction of count is counter-clockwise,

$$\begin{aligned} w_q(\tau) &= \nu_{qk}(q_i(\tau))D(q_i)(\tau),\\ H(\tau,t) &= \frac{e_q(\tau,t)}{f_q(\tau,t)},\\ e_q(\tau,t) &= \Big(\sum_{s=1}^{2} q'_{is}(t)^2\Big)a_q(\tau,t)(\tau-t) - q'_{in}(t)b_q(\tau,t),\\ f_q(\tau,t) &= \Big(\sum_{s=1}^{2} q'_{is}(t)^2\Big)^2(\tau-t)^3 + \Big(\sum_{s=1}^{2} q'_{is}(t)^2\Big)\beta_q(\tau,t)(\tau-t). \end{aligned} \tag{5.11.53}$$

Notice that the function $\tau \to \tau - t$ is discontinuous on Γ at the point $\tau_0 = t + \pi$.

Define an operator $\mathcal{R}_3$ as follows

$$(\mathcal{R}_3 u)(t) = \int_\Gamma \frac{w_q(\tau)u(\tau)}{\tau - t}\, d\tau. \tag{5.11.54}$$

Obviously, $\mathcal{R}_3 u$ is the convolution on Γ of the principal value of the distribution $\frac{1}{t}$ and $-w_q u$. We denote by $c_k(f)$ the Fourier coefficients of a function f for the complex form of the Fourier series,

$$c_k(f) = \frac{1}{2\pi} \int_{-\pi}^{\pi} f(t) e^{-ikt}\, dt.$$

Then, we have Schwartz (1961)

$$\sum_{k=-\infty}^{\infty} |c_k(\mathcal{R}_3 u)|^2 = 4\pi^2 \sum_{k=-\infty}^{\infty} |c_k(\tfrac{1}{t}) c_k(-w_q u)|^2 . \tag{5.11.55}$$

Since $|c_k(\frac{1}{t})| \le \text{const}$ for all k (see Edwards (1979)) and $w_q \in \widetilde{C}^m([-\pi, \pi])$, the Parseval equality and (5.11.55) yield

$$\|\mathcal{R}_3 u\|^2_{L_2(-\pi,\pi)} = 2\pi \sum_{k=-\infty}^{\infty} |c_k(\mathcal{R}_3 u)|^2 \le c \|u\|^2_{L_2(-\pi,\pi)}. \tag{5.11.56}$$

For $a \ne 0$, we get

$$\begin{aligned} a^{-1} &\big[(\mathcal{R}_3 u)(t+a) - (\mathcal{R}_3 u)(t)\big] \\ &= a^{-1} \int_\Gamma \left[\frac{w_q(\tau)u(\tau)}{\tau - t - a} - \frac{w_q(\tau)u(\tau)}{\tau - t} \right] d\tau \\ &= a^{-1} \int_\Gamma \frac{w_q(\tau + a)u(\tau + a) - w_q(\tau)u(\tau)}{\tau - t}\, d\tau. \end{aligned} \tag{5.11.57}$$

By (5.11.56), (5.11.57), we obtain that, if $u \in \widetilde{H}^1(-\pi, \pi)$, then

$$\left(\frac{d}{dt} (\mathcal{R}_3 u) \right)(t) = \int_\Gamma (\tau - t)^{-1} \frac{d}{d\tau} (w_q(\tau) u(\tau))\, d\tau,$$

and

$$\|\mathcal{R}_3 u\|_{H^1(-\pi,\pi)} \le c_1 \|u\|_{H^1(-\pi,\pi)}.$$

By analogy, we have

$$\|\mathcal{R}_3 u\|_{H^m(-\pi,\pi)} \le c_1 \|u\|_{H^m(-\pi,\pi)},$$

and taking into account (5.11.51), (5.11.54) we get

$$q \to \mathcal{R}_1(q) \text{ is a continuous mapping of } M \text{ into } \mathcal{L}(\widetilde{H}^m(-\pi, \pi)). \tag{5.11.58}$$

By using the interpolation (see Lions and Magenes (1972)), we obtain that (5.11.58) holds for m replaced with $m - \frac{1}{2}$. The operator $\mathcal{R}_2$ is smoothing and also satisfies (5.11.58). Now, taking into account that q is a periodic and smooth function on $\mathbb{R}$, we obtain (5.11.46), concluding the proof.

We recall that by Ω_q we denote a domain in $\mathbb{R}^2$ with a boundary $S_q = \bigcup_{i=0}^{p} S_{qi}$, S_q defined by a function $q = (q_0, q_1, \dots, q_p) \in M$.

Theorem 5.11.4 *Let a set M be defined by* (5.11.1) *and let $q^n \to q^0$ in M. Then, for each sufficiently large n, there exists a mapping P_n such that P_n is a C^m-diffeomorphism of $\overline{\Omega}_{q^0}$ onto $\overline{\Omega}_{q^n}$ and $P_n \to I$ in $C^m(\overline{\Omega}_{q^0})^2$, where I is the identity operator in $\overline{\Omega}_{q^0}$.*

Proof. First, we will prove the theorem for the case when Ω_{q^n} and Ω_{q^0} are simply connected. Let S_n and S_0 be the boundaries of Ω_{q^n} and Ω_{q^0}, respectively, let δ be a small positive number, and let

$$G_0 = \bigl\{ x \mid x \in \Omega_{q^0}, \ \inf_{y \in S_0} \|x - y\| > \delta \bigr\}. \tag{5.11.59}$$

Let also T_0 be the boundary of G_0 and $F_0 = \Omega_{q^0} \setminus \overline{G}_0$, see Fig. 5.11.2. By s we denote points of a parameterization of T_0. Then, we can consider that $s \in \mathbb{R}/2\pi\mathbb{Z}$. Outside of G_0, we define curvilinear coordinates (s, r), the r axis at a point s is normal to T_0. Let now

$$q^n \to q^0 \qquad \text{in } \widetilde{C}^{m+1}([-\pi, \pi])^2. \tag{5.11.60}$$

Then, for sufficiently large n, we have $S_n \cap T_0 = \varnothing$, and we define a function $f_n \colon \mathbb{R}/2\pi\mathbb{Z} \times [0, \delta] \to \mathbb{R}^2$ as follows

$$f_n(s, r) = (s, \beta_n(s, r)), \qquad \beta_n(s, r) = \sum_{k=1}^{m+1} a_{nk}(s) r^k, \tag{5.11.61}$$

$$\frac{\partial \beta_n}{\partial r}(s, 0) = 1, \qquad \frac{\partial^k \beta_n}{\partial r^k}(s, 0) = 0, \qquad k = 2, \dots, m, \tag{5.11.62}$$
$$Q(s, \beta_n(s, \delta)) \in S_n.$$

Here, Q is the transformation of curvilinear coordinates (s, r) into Cartesian coordinates (x_1, x_2). By (5.11.61), (5.11.62), we get

$$a_{n1}(s) = 1, \qquad a_{nk}(s) = 0, \qquad k = 2, \dots, m, \tag{5.11.63}$$

and (5.11.60) yields $a_{n(m+1)}(s) \to 0$ uniformly in s. Therefore, f_n is a one-to-one mapping. Since $S_n \in C^{m+1}$, we obtain from (5.11.62) that

$$a_{n(m+1)} \in \widetilde{C}^m([-\pi, \pi]). \tag{5.11.64}$$

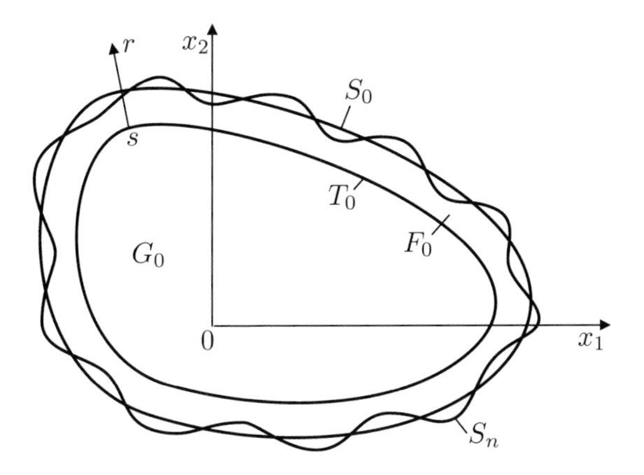

Figure 5.11.2: Domains Ω_{q_n}, Ω_{q_0} and G_0

Now, we define a mapping $P_n\colon \overline{\Omega}_{q^0} \to \mathbb{R}^2$ as follows:

$$P_n(x) = \begin{cases} x, & \text{if } x \in \overline{G}_0 \\ (Q \circ f_n \circ Q^{-1})(x), & \text{if } x \in \overline{\Omega}_{q^0} \setminus \overline{G}_0. \end{cases} \tag{5.11.65}$$

By (5.11.61)–(5.11.65), we get that P_n is a C^m-diffeomorphism of $\overline{\Omega}_{q^0}$ onto $\overline{\Omega}_{q^n}$ and $P_n \to I$ in $C^m(\overline{\Omega}_{q^0})^2$.

In the case when Ω_{q^n} and Ω_{q^0} are multiply-connected, we introduce curvilinear coordinates (s, r) in a vicinity of each component of the boundary of the domain Ω_{q^0}, the functions f_n are defined in each vicinity, and by analogy with (5.11.65), we define a C^m-diffeomorphism of $\overline{\Omega}_{q^0}$ onto $\overline{\Omega}_{q^n}$. Thus, the theorem is proven.

5.11.3 Optimization problem on $\mathbb{R}/2\pi\mathbb{Z}$

State equations and functionals

By Theorem 5.11.2, for each $q \in M$, there exists a unique v_q satisfying

$$v_q \in \check{W}_{mq}, \qquad P(q)v_q = g, \tag{5.11.66}$$

where the operator P_q and $g = (g_0, g_1, \ldots, g_p)$ are defined by (5.11.25)–(5.11.27). So, the function $M \ni q \to v_q \in W_m$ is well defined.

Theorem 5.11.5 *Let a set M be defined by* (5.11.1), *and let* (5.11.8), (5.11.9) *hold. Then, the function $q \to v_q$ defined by the solution of the problem* (5.11.66) *is a continuous mapping of M into W_m.*

Proof. We define a mapping $J\colon M \times W_m \to W_m$ by

$$J(q,u) = (P(q) + \Gamma(q))u - g, \tag{5.11.67}$$

and by λ we denote the implicit function defined by $\lambda(q) \in W_m$, $J(q,\lambda(q)) = 0$. Since $\Gamma(q)v_q = 0$ (see (5.11.38) and Remark 5.11.1), we obtain $J(q,v_q) = 0$, i.e., $\lambda(q) = v_q$. By Lemmas 5.11.1, 5.11.2 and by (5.11.43), the function $q \to P(q)+\Gamma(q)$ is a continuous mapping of M into $\mathcal{L}(W_m,)$. By Theorem 5.11.3, the operator $P(q) + \Gamma(q)$ is an isomorphism of W_m onto itself. Now, the theorem follows from the implicit function theorem (Theorem 1.9.1).

Let

$$q^n \to q^0 \qquad \text{in } M. \tag{5.11.68}$$

We denote by u^n and u^0 the solutions of the problem (5.11.6), (5.11.7) for $q = q^n$ and $q = q^0$, respectively. By (5.11.21) we have

$$u^n(x) = -\sum_{j=0}^{p} \int_{-\pi}^{\pi} F_q(x, q_j^n(\tau))v_{q^n j}(\tau)D(q_j^n)(\tau)\,d\tau + \mathcal{P}(x), \qquad x \in \Omega_{q^n}, \tag{5.11.69}$$

$$u^0(x) = -\sum_{j=0}^{p} \int_{-\pi}^{\pi} F_q(x, q_j^0(\tau))v_{q^0 j}(\tau)D(q_j^0)(\tau)\,d\tau + \mathcal{P}(x), \qquad x \in \Omega_{q^0}, \tag{5.11.70}$$

$$\mathcal{P}(x) = \int_{-\pi}^{\pi} G(x, q_0(\tau))\mathcal{F}(q_0(\tau))D(q_0)(\tau)\,d\tau. \tag{5.11.71}$$

Here, q_0 is the first component of an element $q \in M$, q_0 is fixed, $\mathcal{F} = (\mathcal{F}_1, \mathcal{F}_2)$, see (5.11.8), (5.11.9), $v_{q^n j}$, $v_{q^0 j}$ are the solutions of the problem (5.11.26) for $q = q^n$ and $q = q^0$, respectively.

By Theorem 5.11.4, for each large n, there exists a C^m-diffeomorphism P_n of $\overline{\Omega}_{q^0}$ onto Ω_{q^n}. So, we define a function $\tilde{u}^n$ as follows:

$$\tilde{u}^n(x) = u^n(P_n(x)), \qquad \tilde{u}^n \in H^m(\Omega_{q^0})^2. \tag{5.11.72}$$

Theorem 5.11.6 *Let a set M be defined by* (5.11.1), *and let* (5.11.8), (5.11.9), *and* (5.11.68) *hold. Then, $\tilde{u}^n \to u^0$ in $H^m(\Omega_{q^0})^2$, where $\tilde{u}^n$ and u^0 are defined by* (5.11.69)–(5.11.72).

Proof. Define operators U_n mapping the space $\widetilde{H}^{m-\frac{1}{2}}(-\pi,\pi)^{p+1}$ into a space of functions on Ω_{q^0} by

$$(U_n u)(x) = \sum_{j=0}^{p} \int_{-\pi}^{\pi} \frac{q_{js}^n(\tau) - P_n(x)_s}{\rho_{nj}(x,\tau)}\, u_j(\tau)\,d\tau, \tag{5.11.73}$$

$$u = (u_0, u_1, \dots, u_p) \in \widetilde{H}^{m-\frac{1}{2}}(-\pi,\pi)^{p+1}, \quad x \in \Omega_{q^0}, \quad n = 0,\, k,\, k+1,\, k+2, \dots.$$

Here, $s = 1$ or 2, P_0 is the identity operator in $\overline{\Omega}_{q^0}$, $P_n(x)_s$ is the s-th component of the vector $P_n(x)$,

$$\rho_{nj}(x,\tau) = \sum_{i=1}^{2}(q_{ji}^n(\tau) - P_n(x)_i)^2. \tag{5.11.74}$$

To prove the theorem, it is sufficient to show that

$$U_n \in \mathcal{L}(\widetilde{H}^{m-\frac{1}{2}}(-\pi,\pi)^{p+1}, H^m(\Omega_{q^0})) \tag{5.11.75}$$

and

$$\|U_n - U_0\|_{\mathcal{L}(\widetilde{H}^{m-\frac{1}{2}}(-\pi,\pi)^{p+1}, H^m(\Omega_{q^0}))} \to 0 \qquad \text{as } n \to \infty. \tag{5.11.76}$$

Let S_{q^0j} be an arbitrary component of the boundary S_{q^0}. In a vicinity of S_{q^0j} in $\overline{\Omega}_{q^0}$, we introduce curvilinear coordinates t, r, i.e., we introduce the mapping

$$[-\pi,\pi] \times [0,\delta] \ni (t,r) \to Q(t,r) = (Q_1(t,r), Q_2(t,r)) = x,$$

the function $t \to Q(t,0)$ maps $[-\pi,\pi)$ onto S_{q^0j} and Q is a C^m-diffeomorphism of $[-\pi,\pi) \times [0,\delta]$ onto $Q([-\pi,\pi) \times [0,\delta]) \subset \overline{\Omega}_{q^0}$, $Q(\cdot,r) \in \widetilde{C}^m([-\pi,\pi])$ for any fixed $r \in [0,\delta]$, and δ is a small positive constant.

Define an operator $\mathcal{M}_n$ by

$$\begin{aligned}(\mathcal{M}_n u)(t,r) &= \int_{-\pi}^{\pi} \frac{q_{js}^n(\tau) - P_n(Q(t,r))_s}{\rho_{nj}(Q(t,r),\tau)} u(\tau)\, d\tau,\\ u &\in \widetilde{H}^{m-\frac{1}{2}}(-\pi,\pi), \quad (t,r) \in \omega = (-\pi,\pi) \times (0,\delta).\end{aligned} \tag{5.11.77}$$

In order to prove (5.11.75), (5.11.76), it is sufficient to show that

$$\mathcal{M}_n \in \mathcal{L}(\widetilde{H}^{m-\frac{1}{2}}(-\pi,\pi), H^m(\omega)), \tag{5.11.78}$$

$$\|\mathcal{M}_n - \mathcal{M}_0\|_{\mathcal{L}(\widetilde{H}^{m-\frac{1}{2}}(-\pi,\pi), H^m(\omega))} \to 0 \qquad \text{as } n \to \infty. \tag{5.11.79}$$

By the Taylor formula, we obtain

$$\begin{aligned} q_{js}^n(\tau) - P_n(Q(t,r))_s &= a(q_j^n)(\tau - t) + b(q_j^n)r + \alpha_n(\tau,t,r),\\ \rho_{nj}(Q(t,r),\tau) &= e(q_j^n)(\tau-t)^2 + f(q_j^n)r^2 + \beta_n(\tau,t,r).\end{aligned} \tag{5.11.80}$$

Here,

$$\left.\begin{aligned} &a,\ b,\ e,\ f \text{ are continuous functionals on}\\ &\widetilde{C}^{m+1}([-\pi,\pi])^2,\ e(q_j^n) > 0,\ f(q_j^n) > 0 \text{ for all } n, \end{aligned}\right\} \tag{5.11.81}$$

$$\begin{aligned} |\alpha_n(\tau,t,r)| &\le c(|\tau - t|^2 + r^2),\\ |\beta_n(\tau,t,r)| &\le c(|\tau - t|^3 + r^3).\end{aligned} \tag{5.11.82}$$

By (5.11.77), (5.11.80), and (5.11.81), we get the representation

$$\mathcal{M}_n = \mathcal{M}_{n1} + \mathcal{M}_{n2}, \tag{5.11.83}$$

where

$$(\mathcal{M}_{n1}u)(t,r) = \int_{-\pi}^{\pi} \frac{a(q_j^n)(\tau - t) + b(q_j^n)r}{e(q_j^n)(\tau - t)^2 + f(q_j^n)r^2} u(\tau)\, d\tau, \tag{5.11.84}$$

and $\mathcal{M}_{n2}$ is a smoothing operator. It follows from Agmon et al. (1959) that

$$\mathcal{M}_{n1} \in \mathcal{L}(\widetilde{H}^{m-\frac{1}{2}}(-\pi,\pi), H^m(\omega)).$$

Let $\mathbb{R}_+$ stand for the set of positive numbers. Consider the function

$$\mathbb{R}_+^2 \times \mathbb{R}^2 \ni y = (y_1, y_2, y_3, y_4) \to \mathcal{H}(y) \in \mathcal{L}(\widetilde{H}^{m-\frac{1}{2}}(-\pi,\pi), H^m(\omega)),$$
$$(\mathcal{H}(y)u)(t,r) = \int_{-\pi}^{\pi} \frac{y_3(\tau - t) + y_4 r}{y_1(\tau - t)^2 + y_2 r^2} u(\tau)\, d\tau.$$

The function $\mathcal{H}$ is a continuous mapping of $\mathbb{R}_+^2 \times \mathbb{R}^2$ into $\mathcal{L}(\widetilde{H}^{m-\frac{1}{2}}(-\pi,\pi), H^m(\omega))$. So, by (5.11.68), (5.11.81), we have

$$\|\mathcal{M}_{n1} - \mathcal{M}_{01}\|_{\mathcal{L}(\widetilde{H}^{m-\frac{1}{2}}(-\pi,\pi), H^m(\omega))} \to 0 \qquad \text{as } n \to \infty.$$

Analogous results hold for the operator $\mathcal{M}_{n2}$, hence (5.11.75), (5.11.76) are fulfilled.

Theorem 5.11.7 *Let a set M be defined by* (5.11.1), *and let* (5.11.8), (5.11.9) *hold. Let also functionals Ψ_i, $i = 0,1,2,3$, be defined by* (5.11.11), *where $u_q \in \check{V}_{mq}$ is the solution of the problem* (5.11.6), (5.11.7). *Then, Ψ_i are continuous functionals on M.*

Proof. The continuity of the functional Ψ_0 is obvious. Let us prove the continuity of the functional Ψ_2. The proof for the other functionals is analogous.

Let $q_n \to q_0$ in M. By Theorem 5.11.6, we have

$$\tilde{u}^n \to u^0 \qquad \text{in } H^m(\Omega_{q^0})^2. \tag{5.11.85}$$

By the change of variables corresponding to the C^m-diffeomorphism P_n we have

$$\begin{gathered}\Psi_2(q^n) = \max_{x \in \overline{\Omega}_{q^0}} \sum_{i,j=1}^{2} \left[\tilde{\sigma}_{ij}(\tilde{u}^n)(x) - \frac{1}{2}\left(\tilde{\sigma}_{11}(\tilde{u}^n)(x) + \tilde{\sigma}_{22}(\tilde{u}^n)(x)\right)\delta_{ij} \right]^2 - c_2, \\ \tilde{\sigma}_{ij}(\tilde{u}^n)(x) = \sigma_{ij}(u^n)(P_n(x)), \quad x \in \overline{\Omega}_{q^0}.\end{gathered} \tag{5.11.86}$$

Due to (5.11.85), (5.11.86), we get $\Psi_2(q^n) \to \Psi_2(q^0)$.

Theorem 5.11.8 *Let a set M be defined by* (5.11.1), *and let* (5.11.8), (5.11.9), (5.11.12) *hold. Let also functionals Ψ_i, $i = 0, 1, 2, 3$, and let a set M_{ad} be defined by* (5.11.11), (5.11.13), *M_{ad} being nonempty. Then, there exists a solution of the problem* (5.11.14).

Proof. By (5.11.12) and Theorem 5.11.7 we get that M_{ad} is a compact set in M. The functional Ψ_0 is continuous on M, and so there exists a solution of the problem (5.11.14).

For the sensitivity analysis and for construction of approximate solutions of the problem (5.11.14), it is useful to calculate the derivative of the function $q \to v_q$, where v_q is a solution of the problem (5.11.66). So, we consider this problem.

We introduce the notation

$$\lambda(q) = v_q. \tag{5.11.87}$$

It follows from the proof of Theorem 5.11.5 that λ is the implicit function defined by $\lambda(q) \in W_m$, $J(q, \lambda(q)) = 0$, $J(q, u)$ is determined by (5.11.67). By applying (5.11.44) and the representation (5.11.50)–(5.11.53), it may be shown that $q \to P(q)$ is a continuously Fréchet differentiable mapping of M into $\mathcal{L}(W_m)$. It may also be shown that $q \to \Gamma_q$ is a continuously Fréchet differentiable mapping of M into $\mathcal{L}(W_m)$. Denote by $P'(q)$, $\Gamma'(q)$ the derivatives of these mappings at a point q. So, by applying Theorem 1.9.2, we obtain that λ is a continuously Fréchet differentiable mapping of M into W_m and its derivative at a point $q \in M$ is given by

$$\lambda'(q)h = -(P(q) + \Gamma(q))^{-1}\big[(P'(q) + \Gamma'(q))h\big]\lambda(q),$$

where $h = (h_0, h_1, \ldots, h_p)$, $h_i = (h_{i1}, h_{i2})$, $h_i \in \widetilde{C}^{m+1}([-\pi, \pi])^2$, h_0 is equal to zero since the component S_{q0} is fixed (see (5.11.8)).

5.12 Optimization of the residual stresses in an elastoplastic body

We have considered optimization problems in which the state of the object is described by linear operator equations. Now, we will be engaged in a problem of optimization of the residual stresses in a solid, supposing that the state of the solid is described by nonlinear equations.

Stresses and deformations are called residual if they stay inside of the solid after the action which induced them is removed. The fields of volume and surface forces, temperature patterns, phase transformations may be an action causing the residual stresses and deformations, which appear in almost all cold and hot workings of materials. In order to decrease stresses in a construction in the process of its operation, one tries to create, in the process of the production of the construction, the residual stresses that are equal in absolute value but oppositely directed to the stresses arising in the operation. In many cases, one aims for decreasing the residual stresses and deformations. So, various problems of modeling and optimization of them appear, see, e.g., Pozdeev et al. (1982), Grigoliuk et al. (1979), Shablii and Medynskii (1981).

Below, we consider a problem of optimization of the residual stresses and deformations in a three-dimensional elastoplastic body.

5.12.1 Force and thermal loading of a nonlinear elastoplastic body

Basic equations

Supposing that deformations are small, we accept, for the process of loading, the following relation between stresses, strains, and temperature (see Shevchenko (1970))

$$\sigma_{ij}(u,\theta) = \beta(\theta)(\varepsilon(u) - \alpha(\theta)\theta)\delta_{ij} + 2g(I(u),\theta)e_{ij}(u), \quad i,j = 1,2,3. \tag{5.12.1}$$

Here, $\sigma_{ij}(u,\theta)$ are the components of the stress tensor, which depend on a vector function of displacements (deformations) $u = (u_1, u_2, u_3)$ and temperature θ, α is the coefficient of linear expansion, β the compression modulus, α, β depending on the temperature θ,

$$\varepsilon(u) = \frac{1}{3}\sum_{i=1}^{3}\frac{\partial u_i}{\partial x_i} = \frac{1}{3}\operatorname{div} u, \tag{5.12.2}$$

$e_{ij}(u)$ are the components of the deviator of the strain tensor,

$$e_{ij}(u) = \varepsilon_{ij}(u) - \varepsilon(u)\delta_{ij}, \tag{5.12.3}$$

ε_{ij} are the components of the strain tensor

$$\varepsilon_{ij}(u) = \frac{1}{2}\left(\frac{\partial u_i}{\partial x_j} + \frac{\partial u_j}{\partial x_i}\right), \tag{5.12.4}$$

$I(u)$ is the second invariant of the deviator of the strain tensor

$$I(u) = \sum_{i,j=1}^{3}(e_{ij}(u))^2, \tag{5.12.5}$$

g is the plasticity modulus, which depends on $I(u)$ and θ.

We assume that, before power and thermal loading, $\theta(x) = 0$ for all $x \in \Omega$, where Ω is the domain occupied by the body. We could consider that, before loading, $\theta(x) = \theta_0 = \text{const}$. Then, in (5.12.1), θ is the difference between the temperature and θ_0.

We suppose that Ω is a bounded domain in $\mathbb{R}^3$ with a Lipschitz continuous boundary S, and $S = \overline{S}_1 \cup \overline{S}_2$, $S_1 \cap S_2 = \varnothing$, S_1 and S_2 open nonempty subsets of S.

If the temperature pattern θ in the medium is known, we are given volume forces $K = (K_1, K_2, K_3)$ and surface forces $F = (F_1, F_2, F_3)$ which act in Ω and on S_2, respectively, and the body is fastened at S_1, then the problem of finding the function of displacements of points of the body $u = (u_1, u_2, u_3)$ reduces to the following problem:

$$\begin{gathered}\sum_{j=1}^{3} \frac{\partial \sigma_{ij}(u(x), \theta(x))}{\partial x_j} + K_i(x) = 0, \qquad \text{in } \Omega, \\ u\big|_{S_1} = 0, \qquad \sum_{j=1}^{3} \sigma_{ij}(u, \theta)\nu_j\big|_{S_2} = F_i, \qquad i = 1, 2, 3.\end{gathered} \tag{5.12.6}$$

Here, $\nu = (\nu_1, \nu_2, \nu_3)$ is the unit Ω-outward normal vector to S_2.

Generalized solution

Define a space H as follows:

$$H = \big\{ u \mid u = (u_1, u_2, u_3) \in (W_2^1(\Omega))^3, \ u\big|_{S_1} = 0 \big\}. \tag{5.12.7}$$

Lemma 5.12.1 *In H, the norm $\|\cdot\|_H$ defined by the expression*

$$\|u\|_H = \left\{ \int_\Omega \left[(\varepsilon(u))^2 + I(u) \right] dx \right\}^{\frac{1}{2}} \tag{5.12.8}$$

is equivalent to the norm of the space $(W_2^1(\Omega))^3$.

Proof. By using (5.12.3), we have

$$\int_\Omega \sum_{i,j=1}^{3} (\varepsilon_{ij}(u))^2 \, dx = \int_\Omega \sum_{i,j=1}^{3} \left[(e_{ij}(u))^2 + 2e_{ij}(u)\varepsilon(u)\delta_{ij} + (\varepsilon(u))^2 \delta_{ij} \right] dx. \tag{5.12.9}$$

By (5.12.2)–(5.12.4), we have

$$\sum_{i,j=1}^{3} e_{ij}(u)\delta_{ij} = \sum_{i=1}^{3} \left(\frac{\partial u_i}{\partial x_i} - \frac{1}{3} \operatorname{div} u \delta_{ii} \right) = 0. \tag{5.12.10}$$

From (5.12.5), (5.12.9), and (5.12.10), it follows that

$$\int_\Omega \sum_{i,j=1}^{3} (\varepsilon_{ij}(u))^2 \, dx = \int_\Omega \left(I(u) + 3(\varepsilon(u))^2 \right) dx. \tag{5.12.11}$$

Taking into account the Korn inequality, we obtain from (5.12.11) that

$$\int_\Omega \big[(\varepsilon(u))^2 + I(u)\big]\,dx \ge c\|u\|^2_{(W_2^1(\Omega))^3}, \qquad u \in H, \tag{5.12.12}$$

where c is a positive constant. The inequality opposite to (5.12.12) is obvious, so the lemma is proven.

Suppose that the function g (see (5.12.1)) satisfies the following conditions:

I. The function $g(\xi,\eta)$ is defined and continuous on the half-plane

$$Q = \{\,(\xi,\eta) \mid \xi \in \mathbb{R}_+,\ \eta \in \mathbb{R}\,\}, \qquad \mathbb{R}_+ = \{\,y \mid y \in \mathbb{R},\ y \ge 0\,\}.$$

II. There exist positive constants a_1, a_2, a_3 such that the following estimates hold:

$$a_1 \le g(\xi,\eta) \le a_2, \qquad (\xi,\eta) \in Q, \tag{5.12.13}$$

$$\big[g(\xi_1^2,\eta)\xi_1 - g(\xi_2^2,\eta)\xi_2\big](\xi_1-\xi_2) \ge a_3(\xi_1-\xi_2)^2, \tag{5.12.14}$$

$$\xi_1,\xi_2 \in \mathbb{R}_+, \quad \eta \in \mathbb{R}.$$

From the physical point of view, the estimate (5.12.13) means that the plasticity modulus is bounded above and below by positive constants, while the estimate (5.12.14) shows that, under a simple shear, the stress increases as the strain increases.

We suppose also

$$\left.\begin{array}{l} y \to \beta(y) \text{ is a continuous mapping of } \mathbb{R} \text{ into } \mathbb{R}_+ \text{ and} \\ 0 < a_4 \le \beta(y) \le a_5 \text{ for all } y \in \mathbb{R}, \end{array}\right\} \tag{5.12.15}$$

$$\left.\begin{array}{l} y \to \alpha(y) \text{ is a continuous mapping of } \mathbb{R} \text{ into } \mathbb{R} \text{ and} \\ |\alpha(y)| \le a_6 \text{ for all } y \in \mathbb{R}, \end{array}\right\} \tag{5.12.16}$$

where a_4, a_5, a_6 are constants.

Finally, we assume that

$$K = (K_1,K_2,K_3) \in (L_2(\Omega))^3, \tag{5.12.17}$$

$$F = (F_1,F_2,F_3) \in (L_2(S_2))^3, \tag{5.12.18}$$

$$\theta \in W_2^1(\Omega). \tag{5.12.19}$$

Define an operator L_θ mapping H into H^* and an element $G_\theta \in H^*$ by

$$(L_\theta v,h) = 3\int_\Omega \beta(\theta)\varepsilon(v)\varepsilon(h)\,dx + 2\int_\Omega g(I(v),\theta)\sum_{i,j=1}^3 e_{ij}(v)e_{ij}(h)\,dx, \tag{5.12.20}$$

$$v,h \in H,$$

$$(G_\theta,h) = \int_\Omega \sum_{i=1}^3 K_i h_i\,dx + \int_{S_2}\sum_{i=1}^3 F_i h_i\,ds + 3\int_\Omega \beta(\theta)\alpha(\theta)\theta\varepsilon(h)\,dx. \tag{5.12.21}$$

A generalized solution of the problem (5.12.1), (5.12.6) is defined to be a function u satisfying

$$u \in H, \qquad (L_\theta u, h) = (G_\theta, h), \qquad h \in H. \tag{5.12.22}$$

Let us show that, if a function u is a classical solution of the problem (5.12.1), (5.12.6), then it also satisfies (5.12.22). By (5.12.4) and (5.12.6), using Green's formula, we have, for all $h \in H$,

$$\begin{aligned} -\int_\Omega \sum_{i=1}^{3} K_i h_i\, dx &= \int_\Omega \sum_{i,j=1}^{3} \frac{\partial \sigma_{ij}(u,\theta)}{\partial x_j} h_i\, dx \\ &= \int_{S_2} \sum_{i=1}^{3} F_i h_i\, ds - \int_\Omega \sum_{i,j=1}^{3} \sigma_{ij}(u,\theta)\varepsilon_{ij}(h)\, dx. \end{aligned} \tag{5.12.23}$$

Therefore, by virtue of (5.12.1), (5.12.3), and (5.12.10), we get (5.12.22). Carrying out inverse actions, one easily concludes that a smooth solution of the problem (5.12.22) is a classical solution of (5.12.1), (5.12.6).

Existence theorem

Theorem 5.12.1 *Let Conditions* I, II *be satisfied, and let* (5.12.15)–(5.12.19) *hold. Then, there exists a unique solution of the problem* (5.12.22). *Moreover, the function* $(K, F, \theta) \to u$ *is a continuous mapping of* $(L_2(\Omega))^3 \times (L_2(S_2))^3 \times W_2^1(\Omega)$ *(in the topology generated by the product of the corresponding weak topologies) into the space* H *(equipped with the strong topology).*

To prove Theorem 5.12.1, we need the following lemma.

Lemma 5.12.2 *Let the conditions of Theorem* 5.12.1 *be satisfied. Then,* L_θ *is a strictly monotone, coercive, and continuous mapping of* H *into* H^*, *more exactly, the following relations hold:*

$$\begin{aligned} (L_\theta v - L_\theta w, v - w) &\geq c \int_\Omega \left[(\varepsilon(v-w))^2 + (I^{\frac{1}{2}}(v) - I^{\frac{1}{2}}(w))^2\right] dx, \\ &v, w \in H,\ \theta \in W_2^1(\Omega),\ c = \min(2a_3, 3a_4), \end{aligned} \tag{5.12.24}$$

and the condition $(L_\theta v - L_\theta w, v - w) = 0$ *implies that* $v = w$;

$$(L_\theta v, v) \geq c_1 \|v\|_H^2, \qquad v \in H,\ \theta \in W_2^1(\Omega),\ c_1 = \min(2a_1, 3a_4), \tag{5.12.25}$$

and finally the condition $v^{(n)} \to v^{(0)}$ *in* H *yields* $L_\theta v^{(n)} \to L_\theta v^{(0)}$ *in* H^*.

Proof. By using (5.12.2), (5.12.5), (5.12.20), upon the estimate

$$\left| \sum_{i,j=1}^{3} e_{ij}(v) e_{ij}(w) \right| \leq I^{\frac{1}{2}}(v) I^{\frac{1}{2}}(w),$$

we get

$$\begin{aligned}(L_\theta v - L_\theta w, v-w) &= 3\int_\Omega \beta(\theta)(\varepsilon(v-w))^2\,dx \\ &\quad + 2\int_\Omega \Big[g(I(v),\theta)I(v) + g(I(w),\theta)I(w) - g(I(v),\theta)\sum_{i,j=1}^{3} e_{ij}(v)e_{ij}(w) \\ &\quad - g(I(w),\theta)\sum_{i,j=1}^{3} e_{ij}(w)e_{ij}(v)\Big]\,dx \\ &\geq 3\int_\Omega \beta(\theta)(e(v-w))^2\,dx \\ &\quad + 2\int_\Omega \big[g(I(v),\theta)I^{\frac12}(v) - g(I(w),\theta)I^{\frac12}(w)\big](I^{\frac12}(v) - I^{\frac12}(w))\,dx.\end{aligned} \tag{5.12.26}$$

Therefore, by using (5.12.14) and (5.12.15), we arrive at (5.12.24).

Let now

$$(L_\theta v - L_\theta w, v-w) = 0.$$

Then, because of (5.12.24), $I(v) = I(w)$ a.e. in Ω, and (5.12.20), (5.12.8), (5.12.13), (5.12.15) imply that

$$\begin{aligned}0 = (L_\theta v - L_\theta w, v-w) &= 3\int_\Omega \beta(\theta)(\varepsilon(v-w))^2\,dx \\ &\quad + 2\int_\Omega g(I(v),\theta)I(v-w)\,dx \geq c_2\|v-w\|_H^2, \qquad c_2 > 0,\end{aligned}$$

that is, $v = w$. The inequality (5.12.25) follows from (5.12.13) and (5.12.15).

It remains to show that L_θ is a continuous mapping of H into H^*. Evidently,

$$L_\theta = L_\theta^{(1)} + L_\theta^{(2)}, \tag{5.12.27}$$

where $L_\theta^{(1)}$ and $L_\theta^{(2)}$ are the mappings of H into H^* defined by the formulas

$$(L_\theta^{(1)} v, h) = 3\int_\Omega \beta(\theta)\varepsilon(v)\varepsilon(h)\,dx, \tag{5.12.28}$$

$$(L_\theta^{(2)} v, h) = 2\int_\Omega g(I(v),\theta)\sum_{i,j=1}^{3} e_{ij}(v)e_{ij}(h)\,dx. \tag{5.12.29}$$

From (5.12.2) and (5.12.15), it follows that $L_\theta^{(1)}$ is a linear continuous mapping of H into H^*. Let us show that $L_\theta^{(2)}$ is a continuous mapping. Let $v^{(n)} \to v^{(0)}$ in H and let us choose from $\{v^{(n)}\}$ a subsequence $\{v^{(m)}\}$ such that

$$I(v^{(m)}) \to I(v^{(0)}) \qquad \text{a.e. in } \Omega. \tag{5.12.30}$$

Denote

$$I^{(m)} = I(v^{(m)}), \qquad e_{ij}^{(m)} = e_{ij}(v^{(m)}), \qquad g^{(m)} = g(I(v^{(m)}), \theta), \qquad m = 0, 1, 2, \dots. \tag{5.12.31}$$

Now, taking (5.12.13) into account, we get

$$\begin{aligned} &\left|(L_\theta^{(2)} v^{(m)} - L_\theta^{(2)} v^{(0)}, h)\right| \\ &= \left| 2 \int_\Omega \left\{ \sum_{i,j=1}^{3} [g^{(m)}(e_{ij}^{(m)} - e_{ij}^{(0)}) + (g^{(m)} - g^{(0)}) e_{ij}^{(0)}] e_{ij}(h) \right\} dx \right| \\ &\leq 2 \left[a_2 \|v^{(m)} - v^{(0)}\|_H + \left(\sum_{i,j=1}^{3} \int_\Omega ((g^{(m)} - g^{(0)}) e_{ij}^{(0)})^2 \, dx \right)^{\frac{1}{2}} \right] \|h\|_H. \end{aligned} \tag{5.12.32}$$

By using Conditions I, II, the relations (5.12.30), (5.12.31), and the Lebesgue theorem, we infer that

$$\lim_{m \to \infty} \int_\Omega ((g^{(m)} - g^{(0)}) e_{ij}^{(0)})^2 \, dx = 0.$$

Hence, from (5.12.32), we get

$$L_\theta^{(2)} v^{(m)} \to L_\theta^{(2)} v^{(0)} \qquad \text{in } H^*. \tag{5.12.33}$$

From an arbitrary subsequence $\{v^{(k)}\}$ chosen from the sequence $\{v^{(n)}\}$ one can choose, in turn, a subsequence $\{v^{(m)}\}$ satisfying (5.12.33), and therefore the latter formula holds not only for the subsequence $\{v^{(m)}\}$, but for the whole sequence $\{v^{(n)}\}$, i.e., $L_\theta^{(2)} v^{(n)} \to L_\theta^{(2)} v^{(0)}$ in H^*, concluding the proof of the lemma.

Proof of Theorem 5.12.1. 1) The existence of a unique solution of the problem (5.12.22) follows from Lemma 5.12.2 and from known results on monotone operators (see, e.g., Lions (1969), Gajewski et al. (1974)). It remains to prove the continuity of the function $(K, F, \theta) \to u$ in the above mentioned sense.

Let $\{K^{(\mu)}, F^{(\mu)}, \theta_\mu\}_{\mu=1}^\infty$ be a sequence such that

$$\begin{aligned} K^{(\mu)} &\to K^{(0)} && \text{weakly in } (L_2(\Omega))^3, \\ F^{(\mu)} &\to F^{(0)} && \text{weakly in } (L_2(S_2))^3, \\ \theta_\mu &\to \theta_0 && \text{weakly in } W_2^1(\Omega). \end{aligned} \tag{5.12.34}$$

Let us choose a subsequence $\{\theta_n\}$ such that

$$\theta_n \to \theta_0 \qquad \text{strongly in } L_2(\Omega) \text{ and a.e. in } \Omega. \tag{5.12.35}$$

Upon (5.12.34), from the embedding theorem (Theorem 1.6.2), it follows that

$$\begin{aligned} K^{(n)} &\to K^{(0)} \qquad \text{strongly in } H^*, \\ F^{(n)} &\to F^{(0)} \qquad \text{strongly in } H^*. \end{aligned} \tag{5.12.36}$$

By (5.12.15), (5.12.16), we have

$$\begin{aligned} &\left| \int_\Omega \beta(\theta_n)\alpha(\theta_n)\theta_n\varepsilon(h)\,dx - \int_\Omega \beta(\theta_0)\alpha(\theta_0)\theta_0\varepsilon(h)\,dx \right| \\ &\quad = \left| \int_\Omega \big[(\theta_n - \theta_0)\beta(\theta_n)\alpha(\theta_n) \right. \\ &\qquad \left. + (\beta(\theta_n)\alpha(\theta_n) - \beta(\theta_0)\alpha(\theta_0))\theta_0\big]\varepsilon(h)\,dx \right| \\ &\quad \le (c\|\theta_n - \theta_0\|_{L_2(\Omega)} + \gamma_n)\|h\|_H, \end{aligned} \tag{5.12.37}$$

where

$$\gamma_n = \|(\beta(\theta_n)\alpha(\theta_n) - \beta(\theta_0)\alpha(\theta_0))\theta_0\|_{L_2(\Omega)}.$$

By using (5.12.15), (5.12.16), (5.12.35), and the Lebesgue theorem, we conclude

$$\lim_{n\to\infty} \gamma_n = 0. \tag{5.12.38}$$

Put

$$(G^{(n)}, h) = \int_\Omega \sum_{i=1}^{3} K_i^{(n)} h_i\,dx + \int_{S_2} \sum_{i=1}^{3} F_i^{(n)} h_i\,ds + 3\int_\Omega \beta(\theta_n)\alpha(\theta_n)\theta_n\varepsilon(h)\,dx,$$
$$n = 0, 1, 2, \dots, \qquad h \in H. \tag{5.12.39}$$

From (5.12.35)–(5.12.38), we have

$$\lim_{n\to\infty} \|G^{(n)} - G^{(0)}\|_{H^*} = 0. \tag{5.12.40}$$

Let $u^{(n)}$ be a solution of the following problem:

$$u^{(n)} \in H, \qquad (L_{\theta_n} u^{(n)}, h) = (G^{(n)}, h), \qquad h \in H, \quad n = 0, 1, 2, \dots. \tag{5.12.41}$$

We have to show that

$$\lim_{n\to\infty} \|u^{(n)} - u^{(0)}\|_H = 0. \tag{5.12.42}$$

From (5.12.25) and (5.12.41) we infer

$$c_1\|u^{(n)}\|_H^2 \le (L_{\theta_n} u^{(n)}, u^{(n)}) = (G^{(n)}, u^{(n)}) \le \|G^{(n)}\|_{H^*}\|u^{(n)}\|_H.$$

Therefore, (5.12.40) yields

$$\|u^{(n)}\|_H \leq \text{const} \qquad \forall n. \tag{5.12.43}$$

Upon (5.12.13), (5.12.15), (5.12.20), and (5.12.43), we get

$$\|L_{\theta_n} u^{(n)}\|_{H^*} \leq \text{const} \qquad \forall n. \tag{5.12.44}$$

2) Because of (5.12.43) and (5.12.44), from the sequence $\{u^{(n)}\}$ one can subtract a subsequence $\{u^{(m)}\}$ such that

$$u^{(m)} \to u \qquad \text{weakly in } H, \tag{5.12.45}$$

$$L_{\theta_m} u^{(m)} \to \chi \qquad \text{weakly in } H^*. \tag{5.12.46}$$

Passing to the limit in (5.12.41) (with $n = m$), taking note of (5.12.40) and (5.12.46), we conclude

$$(\chi, h) = (G^{(0)}, h), \qquad h \in H,$$

and so

$$\chi = G^{(0)}. \tag{5.12.47}$$

On the other hand, (5.12.40), (5.12.41), (5.12.45), and (5.12.47) imply

$$(L_{\theta_m} u^{(m)}, u^{(m)}) = (G^{(m)}, u^{(m)}) \to (G^{(0)}, u) = (\chi, u). \tag{5.12.48}$$

By (5.12.24),

$$(L_{\theta_m} u^{(m)} - L_{\theta_m} v, u^{(m)} - v) \geq 0, \qquad v \in H. \tag{5.12.49}$$

By using Conditions I, II, the relation (5.12.35), and the Lebesgue theorem, we get

$$\lim_{m\to\infty} \|g(I(v), \theta_m) e_{ij}(v) - g(I(v), \theta_0) e_{ij}(v)\|_{L_2(\Omega)} = 0. \tag{5.12.50}$$

Analogously, by using (5.12.15) and (5.12.35), we obtain

$$\lim_{m\to\infty} \|\beta(\theta_m)\varepsilon(v) - \beta(\theta_0)\varepsilon(v)\|_{L_2(\Omega)} = 0. \tag{5.12.51}$$

From (5.12.20), (5.12.45), (5.12.50), and (5.12.51)

$$\begin{aligned} \lim_{m\to\infty} (L_{\theta_m} v, v) &= (L_{\theta_0} v, v), \\ \lim_{m\to\infty} (L_{\theta_m} v, u^{(m)}) &= (L_{\theta_0} v, u). \end{aligned} \tag{5.12.52}$$

By using (5.12.45), (5.12.46), (5.12.48), and (5.12.52), we pass to the limit in (5.12.49), which gives

$$(\chi - L_{\theta_0} v, u - v) \geq 0, \qquad v \in H. \tag{5.12.53}$$

Setting in (5.12.53) $v = u - \lambda w$, $\lambda > 0$, $w \in H$, we have

$$(\chi - L_{\theta_0}(u - \lambda w), w) \geq 0.$$

Letting λ tend to zero, we conclude from Lemma 5.12.2 that

$$(\chi - L_{\theta_0} u, w) \geq 0, \qquad w \in H.$$

Hence $\chi = L_{\theta_0} u = G^{(0)}$, i.e., with the notations (5.12.41), $u = u^{(0)}$.

Since the problem (5.12.41) has a unique solution, it is easy to see that (5.12.45) holds for the whole sequence $\{u^{(n)}\}$, i.e.,

$$u^{(n)} \to u^{(0)} \qquad \text{weakly in } H. \tag{5.12.54}$$

3) Let us show that

$$\lim_{n \to \infty} \|u^{(n)}\|_H = \|u^{(0)}\|_H. \tag{5.12.55}$$

Let

$$X_n = (L_{\theta_n} u^{(n)} - L_{\theta_0} u^{(0)}, u^{(n)} - u^{(0)}). \tag{5.12.56}$$

By (5.12.41),

$$X_n = (G^{(n)} - G^{(0)}, u^{(n)} - u^{(0)}).$$

Thus, (5.12.40) and (5.12.43) give

$$\lim_{n \to \infty} X_n = 0. \tag{5.12.57}$$

Setting the notations

$$g^{(k,p)} = g(I(v^{(k)}), \theta_p), \qquad k, p = 0, 1, 2 \ldots, \tag{5.12.58}$$

by using (5.12.20), (5.12.31), and (5.12.56), we get

$$X_n = \sum_{i=1}^{4} X_n^{(i)}, \tag{5.12.59}$$

where

$$
\begin{aligned}
X_n^{(1)} &= 3\int_\Omega \beta(\theta_n)(\varepsilon(u^{(n)}-u^{(0)}))^2\,dx,\\
X_n^{(2)} &= 3\int_\Omega (\beta(\theta_n)-\beta(\theta_0))\varepsilon(u^{(0)})\varepsilon(u^{(n)}-u^{(0)})\,dx,\\
X_n^{(3)} &= 2\int_\Omega \left[\sum_{i,j=1}^{3}(g^{(n,n)}e_{ij}^{(n)}-g^{(0,n)}e_{ij}^{(0)})(e_{ij}^{(n)}-e_{ij}^{(0)})\right]dx,\\
X_n^{(4)} &= 2\int_\Omega \left[\sum_{i,j=1}^{3}(g^{(0,n)}-g^{(0,0)})e_{ij}^{(0)}(e_{ij}^{(n)}-e_{ij}^{(0)})\right]dx.
\end{aligned}
\tag{5.12.60}
$$

By using the Cauchy inequality, we get

$$
\begin{aligned}
\frac{1}{3}|X_n^{(2)}| \le {} & \left[\int_\Omega (\beta(\theta_n)-\beta(\theta_0))^2(\varepsilon(u^{(0)}))^2\,dx\right]^{\frac{1}{2}}\\
& \times\left[\int_\Omega (\varepsilon(u^{(n)}-u^{(0)}))^2\,dx\right]^{\frac{1}{2}}.
\end{aligned}
\tag{5.12.61}
$$

Utilizing (5.12.15), (5.12.35), and the Lebesgue theorem, it is not hard to see that the first multiplier on the right-hand side of (5.12.61) tends to zero as $n\to\infty$. The second multiplier in (5.12.61) is bounded for each n, by virtue of (5.12.54). Hence,

$$
\lim_{n\to\infty}|X_n^{(2)}| = 0. \tag{5.12.62}
$$

Analogously, we establish that

$$
\lim_{n\to\infty}|X_n^{(4)}| = 0.
$$

Hence, (5.12.57), (5.12.59), and (5.12.62) imply

$$
\lim_{n\to\infty}(X_n^{(1)}+X_n^{(3)}) = 0.
$$

This equality together with (5.12.24) gives

$$
\begin{aligned}
\varepsilon(u^{(n)}) &\to \varepsilon(u^{(0)}) \quad \text{in } L_2(\Omega),\\
I^{\frac{1}{2}}(u^{(n)}) &\to I^{\frac{1}{2}}(u^{(0)}) \quad \text{in } L_2(\Omega),
\end{aligned}
\tag{5.12.63}
$$

Since $y\to\|y\|_{L_2(\Omega)}$ is a continuous mapping of $L_2(\Omega)$ into $\mathbb{R}$, (5.12.8) and (5.12.63) imply (5.12.55). Now, on account of (5.12.54) and (5.12.55), we have

$$
u^{(n)}\to u^{(0)} \qquad \text{strongly in } H. \tag{5.12.64}
$$

It remains to prove that the whole sequence $\{u^\mu\}$ strongly converges to $u^{(0)}$ in H. By using the above argument and the fact that the problem (5.12.22) has a unique solution, one sees that, from an arbitrary subsequence $\{u^{(k)}\}$ extracted from the sequence $\{u^{(\mu)}\}$, one can extract, in turn, a subsequence $\{u^{(n)}\}$ for which (5.12.64) holds. Hence, $u^{(\mu)}\to u^{(0)}$ strongly in H, concluding the proof.

Continuity of the mappings

Denote by $u(K, F, \theta)$ the element u of H solving the problem (5.12.22), and define functions

$$K, F, \theta \to \sigma_{ij}(u(K, F, \theta), \theta), \qquad i, j = 1, 2, 3, \tag{5.12.65}$$

where $\sigma_{ij}(u, \theta)$'s are given by (5.12.1).

Theorem 5.12.2 *Let Conditions* I, II *and the relations* (5.12.15), (5.12.16) *hold. Then, the functions* (5.12.65) *are continuous mappings of* $(L_2(\Omega))^3 \times (L_2(S_2))^3 \times W_2^1(\Omega)$ *(in the topology generated by the product of the corresponding weak topologies) into* $L_2(\Omega)$ *endowed with the strong topology.*

Proof. Let $\{K^{(n)}\}$, $\{F^{(n)}\}$, $\{\theta_n\}$ be sequences such that

$$K^{(n)} \to K^{(0)} \qquad \text{weakly in } (L_2(\Omega))^3, \tag{5.12.66}$$

$$F^{(n)} \to F^{(0)} \qquad \text{weakly in } (L_2(S_2))^3, \tag{5.12.67}$$

$$\theta_n \to \theta_0 \qquad \text{weakly in } W_2^1(\Omega). \tag{5.12.68}$$

Introduce the notation

$$u^{(n)} = u(K^{(n)}, F^{(n)}, \theta_n), \qquad n = 0, 1, 2 \dots. \tag{5.12.69}$$

From (5.12.66)–(5.12.68) and Theorem 5.12.1, we infer

$$u^{(n)} \to u^{(0)} \qquad \text{in } H. \tag{5.12.70}$$

Because of (5.12.68) and (5.12.70), from the sequence $\{\theta_n, u^{(n)}\}$ one can subtract a subsequence $\{\theta_m, u^{(m)}\}$ such that

$$\theta_m \to \theta_0 \qquad \text{strongly in } L_2(\Omega) \text{ and a.e. in } \Omega, \tag{5.12.71}$$

$$I(u^{(m)}) \to I(u^{(0)}) \qquad \text{a.e. in } \Omega. \tag{5.12.72}$$

We have

$$\begin{aligned} &\|g(I(u^{(m)}), \theta_m)e_{ij}(u^{(m)}) - g(I(u^{(0)}), \theta_0)e_{ij}(u^{(0)})\|_{L_2(\Omega)} \\ &\quad \le \|g(I(u^{(m)}), \theta_m)(e_{ij}(u^{(m)}) - e_{ij}(u^{(0)}))\|_{L_2(\Omega)} \\ &\quad\quad + \left\|\left(g(I(u^{(m)}), \theta_m) - g(I(u^{(0)}), \theta_0)\right)e_{ij}(u^{(0)})\right\|_{L_2(\Omega)}. \end{aligned} \tag{5.12.73}$$

By (5.12.13) and (5.12.70), the first term on the right-hand side of the inequality (5.12.73) tends to zero as $m \to \infty$. From Conditions I, II and the relations (5.12.71) and (5.12.72), by virtue of the Lebesgue theorem, we get that the second term in (5.12.73) also tends to zero. So,

$$g(I(u^{(m)}), \theta_m)e_{ij}(u^{(m)}) \to g(I(u^{(0)}), \theta_0)e_{ij}(u^{(0)}) \qquad \text{in } L_2(\Omega). \tag{5.12.74}$$

Analogously, we establish that

$$\beta(\theta_m)(\varepsilon(u^{(m)}) - \alpha(\theta_m)\theta_m) \to \beta(\theta_0)(\varepsilon(u^{(0)}) - \alpha(\theta_0)\theta_0) \qquad \text{in } L_2(\Omega).$$

Hence, (5.12.1) and (5.12.74) give

$$\sigma_{ij}(u^{(m)}, \theta_m) \to \sigma_{ij}(u^{(0)}, \theta_0) \qquad \text{in } L_2(\Omega), \quad i,j = 1,2,3. \tag{5.12.75}$$

Analogously to the above, we verify that, from an arbitrary subsequence $\{u^{(\mu)}, \theta_\mu\}$ chosen from the sequence $\{u^{(n)}, \theta_n\}$, one can extract, in turn, a subsequence $\{u^{(m)}, \theta_m\}$ for which (5.12.75) holds. Thus,

$$\sigma_{ij}(u^{(n)}, \theta_n) \to \sigma_{ij}(u^{(0)}, \theta_0) \qquad \text{in } L_2(\Omega),$$

concluding the proof.

5.12.2 Residual stresses and deformations

Suppose that a nonlinear elatoplastic body is affected by fields of forces K, F and a temperature pattern θ that satisfy the conditions (5.12.17)–(5.12.19). The stresses $\sigma_{ij}(u,\theta)$ and strains $\varepsilon_{ij}(u,\theta)$ in this body are defined by the formulas (5.12.1) and (5.12.4), with u the solution of the problem (5.12.22). Let, further, there happen unloading and let K^{re}, F^{re}, θ^{re} be the fields of forces and temperature pattern at the end of unloading. We assume that the unloading is complete, that is, $K^{\text{re}} = 0$, $F^{\text{re}} = 0$, $\theta^{\text{re}} = 0$. Then, in the body under consideration, there will appear residual stresses σ_{ij}^{re}, strains $\varepsilon_{ij}^{\text{re}}$, and displacements (deformations) u^{re}, which are given by the formulas (see Shevchenko (1970))

$$\begin{aligned}
\sigma_{ij}^{\text{re}} &= \frac{g(0,0)}{g(0,\theta)}\sigma_{ij}(u,\theta) - \sigma_{ij}^{(p)},\\
\varepsilon_{ij}^{\text{re}} &= \varepsilon_{ij}(u) - \varepsilon_{ij}^{(p)},\\
u_i^{\text{re}} &= u_i - u_i^{(p)}, &&(5.12.76)\\
\sigma_{ij}^{(p)} &= \beta(0)(\varepsilon(u^{(p)}) - \alpha(0)\theta)\delta_{ij} + 2g(0,0)e_{ij}(u^{(p)}), &&(5.12.77)\\
\varepsilon_{ij}^{(p)} &= \varepsilon_{ij}(u^{(p)}), &&(5.12.78)
\end{aligned}$$

where the function $u^{(p)}$ is the solution of the problem

$$u^{(p)} \in H, \qquad (Lu^{(p)}, h) = (G_p, h), \qquad h \in H. \tag{5.12.79}$$

Here, we have used the following notations:

$$(Lw, h) = 3\int_\Omega \beta(0)\varepsilon(w)\varepsilon(h)\,dx + 2\int_\Omega g(0,0)\sum_{i,j=1}^{3} e_{ij}(w)e_{ij}(h)\,dx, \tag{5.12.80}$$

$$w, h \in H,$$

$$
\begin{aligned}
&(G_p, h) = \sum_{i=1}^{4} (G_p^{(i)}, h), \\
&(G_p^{(1)}, h) = \int_\Omega \frac{g(0,0)}{g(0,\theta)} \sum_{i=1}^{3} K_i h_i \, dx, \\
&(G_p^{(2)}, h) = -\int_\Omega \frac{g(0,0)}{(g(0,\theta))^2} \frac{\partial g}{\partial \theta}(0,\theta) \sum_{i,j=1}^{3} \sigma_{ij}(u,\theta) \frac{\partial \theta}{\partial x_i} h_j \, dx, \\
&(G_p^{(3)}, h) = \int_{S_2} \frac{g(0,0)}{g(0,\theta)} \sum_{i=1}^{3} F_i h_i \, ds, \\
&(G_p^{(4)}, h) = 3\beta(0)\alpha(0) \int_\Omega \theta \varepsilon(h) \, dx.
\end{aligned}
\tag{5.12.81}
$$

The relations (5.12.76)–(5.12.81) follow from the first theorem of unloading, which is derived under the supposition that, during the unloading, there are no additional plastic deformations (i.e., the unloading is going on elastically), and that the functions $\beta(\theta)$ and $g(0,\theta)$ are connected by the relation

$$
\beta(\theta) = \frac{2g(0,\theta)(1+\nu)}{1-2\nu},
$$

where ν is the Poisson ratio – a constant independent of the temperature θ, $0 \leq \nu < \frac{1}{2}$.

We suppose that, in addition to Conditions I, II, the function g satisfies also the following one:

$$
\left.
\begin{aligned}
&\eta \to g(0,\eta) \text{ is a continuously differentiable mapping of } \mathbb{R} \\
&\text{into } \mathbb{R} \text{ and} \\
&\qquad \left| \frac{\partial g}{\partial \eta}(0,\eta) \right| \leq a_7, \qquad \eta \in \mathbb{R}.
\end{aligned}
\right\}
\tag{5.12.82}
$$

Lemma 5.12.3 *Let Conditions* I, II *and the relations* (5.12.15)–(5.12.18), (5.12.82) *hold. Suppose that $\theta \in W_2^2(\Omega)$ and $\sigma_{ij}(u,\theta)$ are defined by the formula* (5.12.1) *in which u is the solution of the problem* (5.12.22). *Then, the element G_p defined by the formula* (5.12.81) *belongs to the space H^*, and the function $(K, F, \theta) \to G_p$ is a bilinear mapping of $(L_2(\Omega))^3 \times (L_2(S_2))^3 \times W_2^2(\Omega)$ (in the topology generated by the product of the corresponding weak topologies) in the space H^* (endowed with the strong topology).*

Proof. Let $\{K^{(n)}, F^{(n)}, \theta_n\}$ be a sequence such that

$$
K^{(n)} \to K^{(0)} \qquad \text{weakly in } (L_2(\Omega))^3, \tag{5.12.83}
$$

$$
F^{(n)} \to F^{(0)} \qquad \text{weakly in } (L_2(S_2))^3, \tag{5.12.84}
$$

$$
\theta_n \to \theta_0 \qquad \text{weakly in } W_2^2(\Omega). \tag{5.12.85}
$$

Then, by the embedding theorem, we have

$$\theta_n \to \theta_0 \qquad \text{strongly in } C(\overline{\Omega}). \tag{5.12.86}$$

For any $z \in L_2(\Omega)$,

$$\left| \int_\Omega \left(\frac{K_i^{(n)}}{g(0,\theta_n)} - \frac{K_i^{(0)}}{g(0,\theta_0)} \right) z \, dx \right| \le \left| \int_\Omega \frac{(K_i^{(n)} - K_i^{(0)}) z}{g(0,\theta_n)} \, dx \right| + \left| \int_\Omega K_i^{(0)} z \left(\frac{1}{g(0,\theta_n)} - \frac{1}{g(0,\theta_0)} \right) dx \right|. \tag{5.12.87}$$

From the Lebesgue theorem, by using Conditions I, II and the relation (5.12.86), we get

$$\frac{z}{g(0,\theta_n)} \to \frac{z}{g(0,\theta_0)} \qquad \text{strongly in } L_2(\Omega).$$

Hence, (5.12.83) yields that the first term on the right-hand side of (5.12.87) tends to zero as $n \to \infty$. Absolutely analogously, we conclude that so does the second term in (5.12.87). Thus,

$$\frac{K_i^{(n)}}{g(0,\theta_n)} \to \frac{K_i^{(0)}}{g(0,\theta_0)} \qquad \text{weakly in } L_2(\Omega). \tag{5.12.88}$$

Analogously,

$$\frac{F_i^{(n)}}{g(0,\theta_n)} \to \frac{F_i^{(0)}}{g(0,\theta_0)} \qquad \text{weakly in } L_2(S_2). \tag{5.12.89}$$

Let us denote by G_{pn} and $G_{pn}^{(i)}$ the functionals G_p and $G_p^{(i)}$, respectively, when $K = K^{(n)}$, $F = F^{(n)}$, $\theta = \theta_n$, $n = 0, 1, 2, \ldots$. From (5.12.88) and Theorem 1.5.12 it follows that $G_{pn}^{(1)} \to G_{p0}^{(1)}$ in H^*.

Let $\check{W}$ stand for the set of functions from $W_2^1(\Omega)$ vanishing on S_1, $\check{W}$ is a closed subspace of $W_2^1(\Omega)$. Denote by $H_{00}^{\frac{1}{2}}(S_2)$ the space of the traces on S_2 of the functions from $\check{W}$. This space equipped with the norm

$$\|\mu\|_{H_{00}^{\frac{1}{2}}(S_2)} = \inf \left\{ \|v\|_{W_2^1(\Omega)},\ v \in \check{W},\ v\big|_{S_2} = \mu \right\}$$

is a Hilbert space, and the following embedding takes place $H_{00}^{\frac{1}{2}}(S_2) \subset H^{\frac{1}{2}}(S_2)$, see Lions and Magenes (1972).

Let $(H_{00}^{\frac{1}{2}}(S_2))^*$ be the dual space of $H_{00}^{\frac{1}{2}}(S_2)$. The space $H_{00}^{\frac{1}{2}}(S_2)$ is compactly embedded into $L_2(S_2)$ (see Remark 1.6.1), and therefore $L_2(S_2)$ is compactly embedded into $(H_{00}^{\frac{1}{2}}(S_2))^*$ (see Theorem 1.5.12). Hence, (5.12.89) gives that $G_{pn}^{(3)} \to G_{p0}^{(3)}$ in H^*.

From (5.12.86), we conclude that $G_{pn}^{(4)} \to G_{p0}^{(4)}$ in H^*. It remains to show that

$$G_{pn}^{(2)} \to G_{p0}^{(2)} \qquad \text{in } H^*. \tag{5.12.90}$$

Since H is continuously embedded into $(L_{10/3}(\Omega))^3$, in order to prove (5.12.90) it suffices to show that

$$\frac{\sigma_{ijn}}{(g(0,\theta_n))^2}\frac{\partial g}{\partial \theta}(0,\theta_n)\frac{\partial \theta_n}{\partial x_i} \to \frac{\sigma_{ij0}}{(g(0,\theta_0))^2}\frac{\partial g}{\partial \theta}(0,\theta_0)\frac{\partial \theta_0}{\partial x_i} \qquad \text{in } L_{10/7}(\Omega). \tag{5.12.91}$$

Here, σ_{ijn} stand for the stresses σ_{ij} defined by the formula (5.12.1) in which u is the solution of the problem (5.12.22) for $K = K^{(n)}$, $F = F^{(n)}$, $\theta = \theta_n$.

By (5.12.83)–(5.12.85) and Theorem 5.12.2, we have $\sigma_{ijn} \to \sigma_{ij0}$ in $L_2(\Omega)$. Hence, taking note of Condition I and (5.12.13), (5.12.82), (5.12.86), we get

$$\frac{\sigma_{ijn}}{(g(0,\theta_n))^2}\frac{\partial g}{\partial \theta}(0,\theta_n) \to \frac{\sigma_{ij0}}{(g(0,\theta_0))^2}\frac{\partial g}{\partial \theta}(0,\theta_0) \qquad \text{in } L_2(\Omega). \tag{5.12.92}$$

Upon (5.12.85) and the embedding theorem,

$$\frac{\partial \theta_n}{\partial x_i} \to \frac{\partial \theta_0}{\partial x_i} \qquad \text{strongly in } L_5(\Omega). \tag{5.12.93}$$

Since $y', y'' \to y'y''$ is a bilinear continuous mapping of $L_2(\Omega) \times L_5(\Omega)$ into $L_{10/7}(\Omega)$, (5.12.91) follows from (5.12.92) and (5.12.93). The lemma is proved.

Theorem 5.12.3 *Let Conditions* I, II *be satisfied, let the relation* (5.12.15)–(5.12.18), (5.12.82) *hold, and let* $\theta \in W_2^2(\Omega)$. *Then, there exists a unique solution of the problem* (5.12.79), *and the function* $K, F, \theta \to u^{(p)}$ *defined by this solution is a continuous mapping of* $(L_2(\Omega))^3 \times (L_2(S_2))^3 \times W_2^2(\Omega)$ *(in the topology generated by the product of the corresponding weak topologies) into the space* H *(endowed with the strong topology).*

Proof. Evidently, the bilinear form $w, h \to (Lw, h)$ defined by (5.12.80) is symmetric and continuous on $H \times H$. By (5.12.5) and (5.12.8), this form is coercive on H. Now, the theorem follows from Lemma 5.12.3 and the Riesz theorem.

5.12.3 Temperature pattern in a medium

A stationary temperature pattern in a heated body is defined as the solution of the heat equation

$$k\sum_{i=1}^{3}\frac{\partial^2 \theta}{\partial x_i^2} = -\varphi, \qquad \text{in } \Omega, \tag{5.12.94}$$

where k is the thermal conductivity, φ is the function of density of heat sources.

We take the boundary condition in the form of the Newton law

$$\frac{\partial \theta}{\partial \nu} = -q(\theta - \theta_0), \qquad \text{on } S, \tag{5.12.95}$$

where ν is the unit outward normal to the surface, q the coefficient of the heat transfer, θ_0 the temperature of the environment.

We suppose that

$$k = \text{const} > 0, \qquad q = \text{const} > 0. \tag{5.12.96}$$

Letting $q\theta_0 = \mathcal{I}$, we write down the boundary condition (5.12.95) in the form

$$\frac{\partial \theta}{\partial \nu} + q\theta = \mathcal{I}. \tag{5.12.97}$$

Suppose S is a surface of the C^2 class and

$$\varphi \in L_2(\Omega), \qquad \mathcal{I} \in H^{\frac{1}{2}}(S). \tag{5.12.98}$$

A generalized solution of the problem (5.12.94), (5.12.97) is defined to be a function θ such that

$$\theta \in W_2^1(\Omega), \qquad a(\theta, w) = \int_\Omega \varphi w \, dx + k \int_S \mathcal{I} w \, ds, \qquad w \in W_2^1(\Omega), \tag{5.12.99}$$

where

$$a(u, v) = k \int_\Omega \sum_{i=1}^{3} \frac{\partial u}{\partial x_i} \frac{\partial v}{\partial x_i} \, dx + qk \int_S uv \, ds, \qquad u, v \in W_2^1(\Omega). \tag{5.12.100}$$

It is not hard to see that a smooth generalized solution of the problem (5.12.94), (5.12.97) is a classical solution of this problem, and vice versa, a classical solution is a generalized one.

Evidently, the bilinear form $a(u, v)$ is symmetric and continuous on $W_2^1(\Omega) \times W_2^1(\Omega)$. From the Friedrichs inequality we conclude the coercivity of this form. Therefore, the problem (5.12.99) has a unique solution. Thus, from known results on smoothness of generalized solutions of elliptic problems, see Agmon et al. (1959), Lions and Magenes (1972), we get the following

Theorem 5.12.4 *Let Ω be a bounded domain in $\mathbb{R}^3$ with the boundary of the C^2 class, and let* (5.12.96), (5.12.98) *hold. Then, there exists a unique solution of the problem* (5.12.94), (5.12.97), *and the function $\varphi, \mathcal{I} \to \theta$ is a continuous mapping of $L_2(\Omega) \times H^{\frac{1}{2}}(S)$ into $W_2^2(\Omega)$.*

By virtue of Theorem 5.12.4, we can define a linear continuous operator A mapping $L_2(\Omega) \times H^{\frac{1}{2}}(S)$ into $W_2^2(\Omega)$ such that $A(\varphi, \mathcal{I}) = \theta$, where θ is the solution of the problem (5.12.94), (5.12.97). Therefore, for any fixed $h \in (W_2^2(\Omega))^*$, the mapping $\varphi, \mathcal{I} \to (A(\varphi, \mathcal{I}), h)$ is a linear continuous functional on $L_2(\Omega) \times H^{\frac{1}{2}}(S)$. Hence, the conditions $\varphi_n \to \varphi$ weakly in $L_2(\Omega)$, $\mathcal{I}_n \to \mathcal{I}$ weakly in $H^{\frac{1}{2}}(S)$ yield

$$(A(\varphi_n, \mathcal{I}_n), h) \to (A(\varphi, \mathcal{I}), h).$$

Thus, we have

Corollary 5.12.1 *Under the conditions of Theorem* 5.12.4, *the function* $\varphi, \mathcal{I} \to \theta$ *is a continuous mapping of* $L_2(\Omega) \times H^{\frac{1}{2}}(S)$ *(in the topology generated by the product of the weak topologies of* $L_2(\Omega)$ *and* $H^{\frac{1}{2}}(S)$*) into the space* $W_2^2(\Omega)$ *endowed with the weak topology.*

5.12.4 Optimization problem

Taking note of the above results, for a given power load K, F and a given thermal load φ, $\mathcal{I}$, one can calculate the residual stresses, strains, and displacements in the following way. From the solution of the problem (5.12.94), (5.12.97), one finds the function θ. Then, one calculates the function u from the solution of the problem (5.12.22), and by using the formula (5.12.1), one derives the stresses under the load $\sigma_{ij}(u, \theta)$. Next, by (5.12.81), one finds the functional G_p, and from the solution of the problem (5.12.79) one gets the function $u^{(p)}$. Finally, by using the formulas (5.12.76)–(5.12.78), one obtains $\sigma_{ij}^{\mathrm{re}}$, $\varepsilon_{ij}^{\mathrm{re}}$, u_i^{re}.

We suppose that the functions K and F (volume and surface forces), as well as the functions φ and $\mathcal{I}$ (thermal load) are controls. Denote by t the control vector function $t = (K, F, \varphi, \mathcal{I}) \in U$, where U is the space of controls, which is taken in the form

$$U = (L_2(\Omega))^3 \times (L_2(S_2))^3 \times L_2(\Omega) \times H^{\frac{1}{2}}(S). \tag{5.12.101}$$

Evidently, the residual stresses $\sigma_{ij}^{\mathrm{re}}$, strains $\varepsilon_{ij}^{\mathrm{re}}$, and displacements u_i^{re} defined by the formula (5.12.76) depend on an element $t \in U$. This is why, in what follows, we denote them by $\sigma_{ij}^{(\mathrm{re}\,t)}$, $\varepsilon_{ij}^{(\mathrm{re}\,t)}$, $u_i^{(\mathrm{re}\,t)}$. Let $\sigma^{(\mathrm{re}\,t)}$, $\varepsilon^{(\mathrm{re}\,t)}$, $u^{(\mathrm{re}\,t)}$ stand for the tensors of residual stresses, of residual strains, and for the vector function of residual displacements, respectively. By virtue of the above results,

$$\sigma^{(\mathrm{re}\,t)} = \{\sigma_{ij}^{(\mathrm{re}\,t)}\} \in (L_2(\Omega))^6, \qquad \varepsilon^{(\mathrm{re}\,t)} = \{\varepsilon_{ij}^{(\mathrm{re}\,t)}\} \in (L_2(\Omega))^6,$$
$$u^{(\mathrm{re}\,t)} = \{u_i^{(\mathrm{re}\,t)}\} \in (W_2^1(\Omega))^3.$$

Notice that, because of the symmetricity of the tensors of stresses and strains, these are considered as elements of the space $(L_2(\Omega))^6$. Finally, by $\sigma^{(t)} = \{\sigma_{ij}^{(t)}\}$ we denote the tensor of the stresses that appear during the loading. These stresses are given by the formula (5.12.1) via the solutions of the problems (5.12.22) and (5.12.94), (5.12.97). More exactly, from the problem (5.12.94), (5.12.97), one finds the function θ, which is used in the solution of the problem (5.12.22).

Assume, next, that we are given functionals Ψ_k such that

$$\left.\begin{array}{l} t,\sigma',\varepsilon,\sigma'',u \to \Psi_k(t,\sigma',\varepsilon,\sigma'',u) \text{ is a continuous mapping} \\ \text{of } U \times (L_2(\Omega))^6 \times (L_2(\Omega))^6 \times (L_2(\Omega))^6 \times (W_2^1(\Omega))^3 \text{ (in the} \\ \text{topology generated by the product of the weak topology} \\ \text{of the space } U \text{ and the strong topology of the space} \\ (L_2(\Omega))^6 \times (L_2(\Omega))^6 \times (L_2(\Omega))^6 \times (W_2^1(\Omega))^3) \text{ into } \mathbb{R}, \\ k = 0,1,\dots,l. \end{array}\right\} \quad (5.12.102)$$

Define a set of admissible controls as follows:

$$\begin{aligned} U_{\mathrm{ad}} = \{\, t \mid t = (K,F,\varphi,\mathcal{I}) \in U,\ \|t\|_U \le c, \\ \Psi_k(t,\sigma^{(\mathrm{re}\,t)},\varepsilon^{(\mathrm{re}\,t)},\sigma^{(t)},u^{(\mathrm{re}\,t)}) \le c_k,\ k = 1,2,\dots,l\}, \end{aligned} \quad (5.12.103)$$

where c, c_k are positive constants, and let a goal functional be of the form

$$\Phi(t) = \Psi_0(t,\sigma^{(\mathrm{re}\,t)},\varepsilon^{(\mathrm{re}\,t)},\sigma^{(t)},u^{(\mathrm{re}\,t)}). \quad (5.12.104)$$

The optimization problem consists in finding a function y such that

$$y \in U_{\mathrm{ad}}, \qquad \Phi(y) = \inf_{t \in U_{\mathrm{ad}}} \Phi(t). \quad (5.12.105)$$

The functional $\Phi(t)$ can be taken, for instance, in the form

$$\Phi(t) = \int_\Omega \sum_{i,j=1}^{3} (\sigma_{ij}^{(\mathrm{re}\,t)} - a_{ij})^2\, dx, \quad (5.12.106)$$

where a_{ij} are given elements of the space $L_2(\Omega)$. Then, the problem (5.12.105) means that one must find a control for which the distribution of the residual stresses is the mean square closest to a given one. The functionals Ψ_k, $k = 1,2,\dots,l$, can be taken so that the conditions

$$\Psi_k(t,\sigma^{(\mathrm{re}\,t)},\varepsilon^{(\mathrm{re}\,t)},\sigma^{(t)},u^{(\mathrm{re}\,t)}) \le c_k$$

will define restrictions on the residual stresses, strains, and displacements, as well as restrictions on strength. The latter ones are constructed for the tensor $\sigma^{(t)}$, by using the strength criteria (see Subsec. 5.1.2), in order that the deformable body do not fail by the load.

Theorem 5.12.5 *Let Ω be a bounded domain in $\mathbb{R}^3$ with the boundary of the C^2 class, let Conditions* I, II *be satisfied for a function g, and let the relations* (5.12.15), (5.12.16), (5.12.82), (5.12.96), (5.12.102)–(5.12.104) *hold, while the set U_{ad} is nonempty. Then, there exists a solution of the problem* (5.12.105).

Proof. Let $\{t_n = (K^{(n)}, F^{(n)}, \varphi_n, \mathcal{I}_n)\}_{n=1}^{\infty}$ be a minimizing sequence, i.e.,

$$\{t_n\} \subset U_{\text{ad}}, \tag{5.12.107}$$

$$\lim_{n\to\infty} \Phi(t_n) = \inf_{t\in U_{\text{ad}}} \Phi(t). \tag{5.12.108}$$

Upon (5.12.107), the sequence $\{t_n\}$ is bounded in U. Hence, we can subtract from it a subsequence $\{t_m = (K^{(m)}, F^{(m)}, \varphi_m, \mathcal{I}_m)\}_{m=1}^{\infty}$ such that $t_m \to t_0$ weakly in U, where $t_0 = (K^{(0)}, F^{(0)}, \varphi_0, \mathcal{I}_0)$. So

$$K^{(m)} \to K^{(0)} \qquad \text{weakly in } (L_2(\Omega))^3, \tag{5.12.109}$$

$$F^{(m)} \to F^{(0)} \qquad \text{weakly in } (L_2(S_2))^3, \tag{5.12.110}$$

$$\varphi_m \to \varphi_0 \qquad \text{weakly in } L_2(\Omega), \tag{5.12.111}$$

$$\mathcal{I}_m \to \mathcal{I}_0 \qquad \text{weakly in } H^{\frac{1}{2}}(S). \tag{5.12.112}$$

Denote by θ_m the solution of the problem (5.12.94), (5.12.97) with $\varphi = \varphi_m$, $\mathcal{I} = \mathcal{I}_m$, $m = 0, 1, 2, \ldots$. From (5.12.111), (5.12.112), because of Corollary 5.12.1, it follows that

$$\theta_m \to \theta_0 \qquad \text{weakly in } W_2^2(\Omega). \tag{5.12.113}$$

Hence, by the embedding theorem,

$$\theta_m \to \theta_0 \qquad \text{in } C(\overline{\Omega}). \tag{5.12.114}$$

Let, further, $u^{(m)}$ be the solution of the problem (5.12.22) with $K = K^{(m)}$, $F = F^{(m)}$, $\theta = \theta_m$, and let $\sigma^{(t_m)} = \{\sigma_{ij}^{(t_m)}\}$ be the stress tensor defined by the formula (5.12.1) with $u = u^{(m)}$, $\theta = \theta_m$, $m = 0, 1, 2, \ldots$.

By (5.12.109), (5.12.110), (5.12.113), and Theorems 5.12.1 and 5.12.2 ,

$$u^{(m)} \to u^{(0)} \qquad \text{in } H, \tag{5.12.115}$$

$$\sigma_{ij}^{(t_m)} \to \sigma_{ij}^{(t_0)} \qquad \text{in } L_2(\Omega), \quad i,j = 1,2,3. \tag{5.12.116}$$

Let also $u^{(p,m)}$ be the solution of the problem (5.12.79) with $K = K^{(m)}$, $F = F^{(m)}$, $\theta = \theta_m$. Let $\sigma_{ij}^{(p,m)}$ stand for the function $\sigma_{ij}^{(p)}$ defined by (5.12.77) with $u^{(p)} = u^{(p,m)}$ and $\theta = \theta_m$. By (5.12.109), (5.12.110), (5.12.113), and Theorem 5.12.3, we get

$$u^{(p,m)} \to u^{(p,0)} \qquad \text{in } H, \tag{5.12.117}$$

and therefore

$$\sigma_{ij}^{(p,m)} \to \sigma_{ij}^{p,0} \qquad \text{in } L_2(\Omega), \quad i,j = 1,2,3. \tag{5.12.118}$$

By using the formulas (5.12.76) and the notations introduced, we get

$$\sigma_{ij}^{(\mathrm{re}\,t_m)} = \frac{g(0,0)}{g(0,\theta_m)}\,\sigma_{ij}^{(t_m)} - \sigma_{ij}^{(p,m)} \tag{5.12.119}$$

$$\varepsilon_{ij}^{(\mathrm{re}\,t_m)} = \varepsilon_{ij}(u^{(m)}) - \varepsilon_{ij}(u^{(p,m)}), \tag{5.12.120}$$

$$u^{(\mathrm{re}\,t_m)} = u^{(m)} - u^{(p,m)}, \tag{5.12.121}$$

where $\varepsilon_{ij}(u)$ is defined by the formula (5.12.4).

By (5.12.114), (5.12.116), (5.12.118), (5.12.119) and Conditions I, II, we have

$$\sigma_{ij}^{(\mathrm{re}\,t_m)} \to \sigma_{ij}^{(\mathrm{re}\,t_0)} \qquad \text{in } L_2(\Omega). \tag{5.12.122}$$

From (5.12.4), (5.12.115), (5.12.117), (5.12.120), (5.12.121),

$$\varepsilon_{ij}^{(\mathrm{re}\,t_m)} \to \varepsilon_{ij}^{(\mathrm{re}\,t_0)} \qquad \text{in } L_2(\Omega), \tag{5.12.123}$$

$$u^{(\mathrm{re}\,t_m)} \to u^{(\mathrm{re}\,t_0)} \qquad \text{in } H. \tag{5.12.124}$$

Taking note of (5.12.102)–(5.12.104), (5.12.107)–(5.12.112), (5.12.116), and (5.12.122)–(5.12.124), we pass to the limit, which gives

$$t_0 \in U_{\mathrm{ad}}, \qquad \Phi(t_0) = \inf_{t \in U_{\mathrm{ad}}} \Phi(t),$$

concluding the proof.

Remark 5.12.1 The problem (5.12.105) on the creation of optimal residual stresses, strains, and deformations reduces to a problem of control by the right-hand sides in the class of nonlinear elliptic problems. For approximate solution of this problem, one can use Theorem 3.2.2, as well as approaches based on extension of the set U_{ad} (see, for instance, Section 2.4).

Chapter 6

Optimization Problems for Steady Flows of Viscous and Nonlinear Viscous Fluids

" "How big and beatiful the world is!" said the toad. "But one must look round in it and not just stay in one spot all the time." And he hopped into the vegetable garden." – *H. Ch. Andersen* "The Toad"

In this chapter, we will formulate and investigate various problems of optimization of flows of nonlinear viscous, non-Newtonian fluids. The majority of real fluids are non-Newtonian and nonlinear viscous. A special case of the fluids under consideration is a viscous, Newtonian fluid, so that all the results of this chapter remain true for it.

6.1 Problem of steady flow of a nonlinear viscous fluid

6.1.1 Basic equations and assumptions

The constitutive equation of an incompressible, nonlinear viscous fluid is the following, see Litvinov (1982a),

$$\sigma_{ij}(p,v) = -p\delta_{ij} + 2\varphi(I(v))\varepsilon_{ij}(v), \qquad i,j = 1,\dots,n. \tag{6.1.1}$$

Here, $\sigma_{ij}(p,v)$ are the components of the stress tensor which depend on the functions of pressure p and velocity $v = (v_1,\dots,v_n)$, n is the dimension of the domain in which the flow is studied, δ_{ij} is the Kronecker delta, $\varepsilon_{ij}(v)$ are the components of the rate of strain tensor:

$$\varepsilon_{ij}(v) = \frac{1}{2}\left(\frac{\partial v_i}{\partial x_j} + \frac{\partial v_j}{\partial x_i}\right), \qquad i,j = 1,\dots,n, \tag{6.1.2}$$

φ is the viscosity function depending on the second invariant of the rate of strain tensor $I(v)$:

$$I(v) = \sum_{i,j=1}^{n} \big(\varepsilon_{ij}(v)\big)^2. \tag{6.1.3}$$

In the case when $\varphi = \text{const} > 0$, the formula (6.1.1) defines the constitutive equation of a viscous, Newtonian fluid. We consider flows of a nonlinear viscous fluid in a bounded domain $\Omega \subset \mathbb{R}^n$ with a Lipschitz continuous boundary S. The equations of motion and incompressibility are defined by

$$\rho\frac{\partial v_i}{\partial t} + \rho v_j\frac{\partial v_i}{\partial x_j} - \frac{\partial \sigma_{ij}(p,v)}{\partial x_j} = K_i \qquad \text{in } \Omega, \quad i = 1,\dots,n, \tag{6.1.4}$$

$$\operatorname{div} v = \sum_{i=1}^{n}\frac{\partial v_i}{\partial x_i} = 0 \qquad \text{in } \Omega. \tag{6.1.5}$$

Here and below, the summation over repeated indices is implied, ρ is the density of the fluid, K_i are the components of the vector function of volume forces $K = (K_1,\dots,K_n)$. We consider steady, slow flows, and so in the equations (6.1.4) we ignore the convection terms $\rho v_j\frac{\partial v_i}{\partial x_j}$. Thus, (6.1.1) and (6.1.4) lead to the following equations

$$\frac{\partial p}{\partial x_i} - 2\frac{\partial}{\partial x_j}\big(\varphi(I(v))\varepsilon_{ij}(v)\big) = K_i \qquad \text{in } \Omega,\ i = 1,\dots,n. \tag{6.1.6}$$

Let S_1, S_2, S_3 be open subsets of S such that $S = \overline{S}_1 \cup \overline{S}_2 \cup \overline{S}_3$, $S_i \cap S_j = \varnothing$ for $i \neq j$. We consider the mixed boundary conditions prescribing zero velocities on S_1, the surface forces $\check{F} = (\check{F}_1,\dots,\check{F}_n)$ on S_2, and the condition of filtration on S_3, see Fig. 6.1.1, i.e.,

$$v \restriction_{S_1} = 0, \tag{6.1.7}$$

$$\big[-p\delta_{ij} + 2\varphi(I(v))\varepsilon_{ij}(v)\big]\nu_j \restriction_{S_2} = \check{F}_i, \qquad i = 1,\dots,n, \tag{6.1.8}$$

$$\big[-p\delta_{ij} + 2\varphi(I(v))\varepsilon_{ij}(v)\big]\nu_j\nu_i \restriction_{S_3} - \hat{F}_i\nu_i \restriction_{S_3} = -\lambda(v_\nu^2,\cdot)v_\nu \restriction_{S_3}, \tag{6.1.9}$$

$$(v_i - v_\nu\nu_i)\restriction_{S_3} = 0, \qquad i = 1,\dots,n. \tag{6.1.10}$$

Here, ν_j are the components of the unit outward normal $\nu = (\nu_1,\dots,\nu_n)$ to S, $v_\nu = v_i\nu_i$, λ is the function of filtration, depending on v_ν^2 and a point $s \in S_3$. The function λ takes the same values when the fluid is flowing into or out of the domain Ω, so we write $\lambda(v_\nu^2, s)$, $\hat{F} = (\hat{F}_1,\dots,\hat{F}_n)$ is the function of surface forces acting on the external surface of the filtration layer S_3', see Fig. 6.1.1. We suppose that the thickness of the filtration layer δ is small and carry the function $\hat{F}$ from S_3' to S_3. If the fluid flows out into the void, or even into the air, we can take

$\hat{F} = 0$. The condition (6.1.10) means that the velocity vector is normal to S_3. The case when

$$\lambda(\alpha, s) = 0, \qquad (\alpha, s) \in \mathbb{R}_+ \times S_3,$$

where

$$\mathbb{R}_+ = \{\, \alpha \mid \alpha \in \mathbb{R},\ \alpha \geq 0 \,\},$$

corresponds to the situation when there is no filtration layer on S_3 and the normal component of the surface forces is equal to $\hat{F}_i \nu_i$. The case when $\lambda(\alpha, s) = \infty$ for each $(\alpha, s) \in \mathbb{R}_+ \times S_3$ corresponds to the solid wall and the adhesion on S_3. Indeed, in that case, by (6.1.9) and (6.1.10), we have $v = 0$ on S_3. In particular, S_3 may be an empty set, then the conditions (6.1.9), (6.1.10) are omitted.

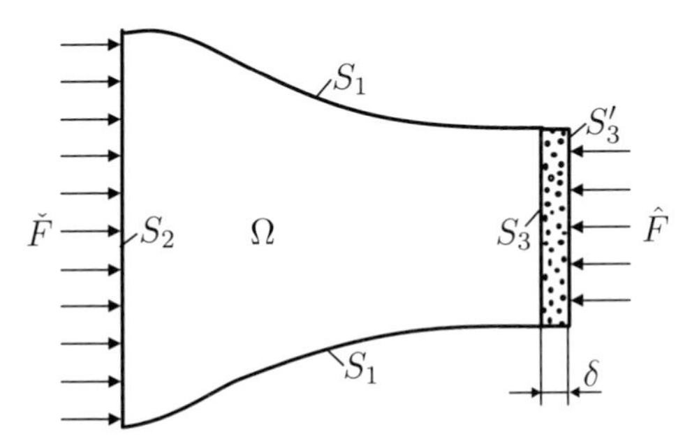

Figure 6.1.1: Domain Ω and the filtration layer on S_3

We suppose that

$$\left.\begin{array}{l}\varphi \text{ is a function continuously differentiable in } \mathbb{R}_+, \text{ and} \\ \text{there exist positive constants } a_1, \dots, a_4 \text{ such that, for an} \\ \text{arbitrary } \alpha \in \mathbb{R}_+, \text{ the following inequalities hold:}\end{array}\right\} \tag{6.1.11}$$

$$a_1 \leq \varphi(\alpha) \leq a_2, \tag{6.1.12}$$

$$\varphi(\alpha) + 2\alpha \frac{d\varphi}{d\alpha}(\alpha) \geq a_3, \tag{6.1.13}$$

$$\alpha \left| \frac{d\varphi}{d\alpha}(\alpha) \right| \leq a_4. \tag{6.1.14}$$

The physical meaning of (6.1.12) is that the function of viscosity must be bounded by two positive constants. The inequality (6.1.13) means that, in the case of a simple shearing flow, the shearing stress increases as the shearing rate increases, see Litvinov (1982a). The inequality (6.1.14) is a restriction on the increase of $\frac{d\varphi}{d\alpha}$ for large α. All these assumptions are physically natural, see Litvinov (1982a).

Concerning the function λ, we take assumptions similar to the ones concerning φ, namely,

$$\left.\begin{array}{l}\text{at each point } s \in S_3, \text{ the function } \lambda(\cdot\,, s)\colon \alpha \to \lambda(\alpha, s) \text{ is} \\ \text{continuously differentiable in } \mathbb{R}_+, \text{ and there exist positive} \\ \text{constants } b_1,\, b_2 \text{ such that, for an arbitrary} \\ (\alpha, s) \in \mathbb{R}_+ \times S_3, \text{ the following inequalities hold}\end{array}\right\} \tag{6.1.15}$$

$$0 \leq \lambda(\alpha, s) \leq b_1, \tag{6.1.16}$$

$$\lambda(\alpha, s) + 2\alpha \frac{\partial \lambda}{\partial \alpha}(\alpha, s) \geq 0, \tag{6.1.17}$$

$$\alpha \left| \frac{\partial \lambda}{\partial \alpha}(\alpha, s) \right| \leq b_2. \tag{6.1.18}$$

From the physical point of view, the estimate (6.1.17) means that the normal component of the surface force acting on the filtration layer does not decrease as the normal component of the velocity increases.

6.1.2 Formulation of the problem

We define spaces X and V as follows

$$X = \{\, u \mid u \in H^1(\Omega)^n,\ u \restriction_{S_1} = 0,\ (u - u_\nu \nu) \restriction_{S_3} = 0 \,\}, \tag{6.1.19}$$

$$V = \{\, u \mid u \in X,\ \operatorname{div} u = 0 \,\}, \tag{6.1.20}$$

where ν is the unit outward normal to S, $u_\nu = u_i \nu_i$. By virtue of Korn's inequality, the expression

$$\|u\|_X = \left(\int_\Omega I(u)\, dx \right)^{\frac{1}{2}} \tag{6.1.21}$$

defines norms in X and V that are equivalent to the norm of $H^1(\Omega)^n$, X and V being Hilbert spaces with the scalar product

$$(u, h)_X = \int_\Omega \varepsilon_{ij}(u) \varepsilon_{ij}(h)\, dx.$$

We define a function of surface forces F acting on $S_2 \cup S_3$ by

$$F(s) = \begin{cases} \check{F}(s), & \text{if } s \in S_2, \\ \hat{F}(s), & \text{if } s \in S_3. \end{cases}$$

We suppose that

$$K, F \in X^*. \tag{6.1.22}$$

If $K \in L_2(\Omega)^n$, $F \in L_2(S_2 \cup S_3)^n$, then

$$(K,h) = \int_\Omega K_i h_i \, dx, \qquad (F,h) = \int_{S_2} \check{F}_i h_i \, ds + \int_{S_3} \hat{F}_i h_i \, ds, \qquad h \in X.$$

Let us define operators $L\colon X \to X^*$ and $B \in \mathcal{L}(X, L_2(\Omega))$ as follows

$$(L(u),h) = 2\int_\Omega \varphi(I(u))\varepsilon_{ij}(u)\varepsilon_{ij}(h)\,dx + \int_{S_3} \lambda(|u|^2,\cdot)u_i h_i \, ds, \tag{6.1.23}$$

$$Bu = \operatorname{div} u, \tag{6.1.24}$$

where

$$|u| = \left(\sum_{i=1}^n u_i^2\right)^{\frac{1}{2}}.$$

By B^* we denote the operator adjoint of B, $B^* \in \mathcal{L}(L_2(\Omega), X^*)$ and

$$(Bu,\mu) = (u, B^*\mu), \qquad u \in X, \ \mu \in L_2(\Omega).$$

From (6.1.19) it follows that

$$\int_{S_3} \lambda(|u|^2,\cdot)u_i h_i \, ds = \int_{S_3} \lambda(u_\nu^2,\cdot)u_\nu h_\nu \, ds, \qquad u,h \in X. \tag{6.1.25}$$

We consider the problem: Find a pair of functions v, p satisfying

$$(v,p) \in X \times L_2(\Omega), \tag{6.1.26}$$

$$(L(v),h) - (B^*p,h) = (K+F,h), \qquad h \in X, \tag{6.1.27}$$

$$(Bv,q) = 0, \qquad q \in L_2(\Omega). \tag{6.1.28}$$

By using Green's formula, it may be shown (see Litvinov (1982a)) that, if v, p is a solution to the problem (6.1.26)–(6.1.28), then v, p is a solution to the problem (6.1.6)–(6.1.10) in the distribution sense. On the contrary, if v, p is a solution to (6.1.6)–(6.1.10) that satisfies (6.1.26), then v, p is a solution to (6.1.26)–(6.1.28).

In what follows, we will use the following

Lemma 6.1.1 *Let Ω be a bounded domain in $\mathbb{R}^n$, $n = 2, 3$, with a Lipschitz continuous boundary S, and let S_1, S_2, S_3 be open subsets of S such that $S = \overline{S}_1 \cup \overline{S}_2 \cup \overline{S}_3$, $S_i \cap S_j = \varnothing$ for $i \neq j$, S_1, S_2 being nonempty. Let spaces X and V be defined by (6.1.19), (6.1.20), and let an operator $B \in \mathcal{L}(X, L_2(\Omega))$ be defined by (6.1.24). Then, there exists a positive constant β_1 such that*

$$\inf_{\mu \in L_2(\Omega)} \sup_{v \in X} \frac{(Bv,\mu)}{\|v\|_X \|\mu\|_{L_2(\Omega)}} \geq \beta_1. \tag{6.1.29}$$

The operator B is an isomorphism of $V^{\perp}$ onto $L_2(\Omega)$, where $V^{\perp}$ is the orthogonal complement to V in X, and the operator B^* is an isomorphism of $L_2(\Omega)$ onto V^0, where

$$V^0 = \{\, f \mid f \in X^*,\ (f,u) = 0,\ u \in V \,\}, \tag{6.1.30}$$

and moreover,

$$\|B^{-1}\|_{\mathcal{L}(L_2(\Omega),\, V^{\perp})} \leq \frac{1}{\beta_1}, \tag{6.1.31}$$

$$\|(B^*)^{-1}\|_{\mathcal{L}(V^0,\, L_2(\Omega))} \leq \frac{1}{\beta_1}. \tag{6.1.32}$$

In order to prove Lemma 6.1.1, we use the following result, see Ladyzhenskaya and Solonnikov (1976), Girault and Raviart (1981).

Lemma 6.1.2 *Let Ω be a bounded domain in $\mathbb{R}^n$, $n = 2,\ 3$, with a Lipschitz continuous boundary. Then, for an arbitrary $h \in L_2(\Omega)$ such that $\int_\Omega h\,dx = 0$, there exists a function $u \in H_0^1(\Omega)^n$ for which* $\operatorname{div} u = h$.

Proof of Lemma 6.1.1. First, let us show that the operator B is an isomorphism of $V^{\perp}$ onto $L^2(\Omega)$. By Lemma 6.1.2 and Banach's theorem, it suffices to show that there exists a function w such that

$$w \in X, \qquad \operatorname{div} w = 1. \tag{6.1.33}$$

Take a function g such that

$$g \in H^1(\Omega)^n, \qquad \operatorname{supp} g \subset \Omega \cup S_2, \qquad \int_{S_2} g_i \nu_i\, ds = a \neq 0. \tag{6.1.34}$$

Then, $g \in X$ and by Green's formula and (6.1.34), we obtain

$$\int_\Omega \left(\frac{\operatorname{mes}\Omega}{a} \operatorname{div} g - 1 \right) dx = 0, \tag{6.1.35}$$

and for the function $h = \frac{\operatorname{mes}\Omega}{a} \operatorname{div} g - 1$, we have

$$h \in L_2(\Omega), \qquad \int_\Omega h\, dx = 0. \tag{6.1.36}$$

By (6.1.36) and Lemma 6.1.2, there exists a function $u \in H_0^1(\Omega)^n$ such that $\operatorname{div} u = h$, and so the function $w = \frac{\operatorname{mes}\Omega}{a} g - u$ satisfies (6.1.33).

Therefore, the operator B is an isomorphism of $V^{\perp}$ onto $L_2(\Omega)$, and there exists a positive constant β_1 such that (6.1.31) holds. The space V^0 can be identified isometrically with $(V^{\perp})^*$, and so, taking note of (6.1.31) and the following equalities:

$$(B^{-1})^* = (B^*)^{-1},$$
$$\|B^{-1}\|_{\mathcal{L}(L_2(\Omega),\, V^{\perp})} = \|(B^{-1})^*)\|_{\mathcal{L}(V^0,\, L_2(\Omega))},$$

we obtain (6.1.32). Hence,

$$\begin{aligned}\|B^*\mu\|_{V^0} &= \sup_{v\in V^\perp}\frac{(v,B^*\mu)}{\|v\|_X} = \sup_{v\in X}\frac{(v,B^*\mu)}{\|v\|_X}\\ &\geq \beta_1\|\mu\|_{L_2(\Omega)}, \qquad \mu\in L_2(\Omega),\end{aligned}$$

thus (6.1.29) holds and the lemma is proven.

Let $\{X_k\}_{k=1}^\infty$, $\{M_k\}_{k=1}^\infty$ be sequences of finite-dimensional subspaces of X and $L_2(\Omega)$, respectively, such that

$$\lim_{k\to\infty}\inf_{z\in X_k}\|u-z\|_X = 0, \qquad u\in X, \tag{6.1.37}$$

$$\lim_{k\to\infty}\inf_{y\in M_k}\|w-y\|_{L_2(\Omega)} = 0, \qquad w\in L_2(\Omega). \tag{6.1.38}$$

Define operators $B_k\in L(X_k,M_k^*)$ as follows: •

$$(B_k u,\mu) = \int_\Omega \mu\,\mathrm{div}\,u\,dx, \qquad u\in X_k,\ \mu\in M_k, \tag{6.1.39}$$

and let $B_k^*\in\mathcal{L}(M_k,X_k^*)$ be the operator adjoint of B_k:

$$(B_k u,\mu) = (u,B_k^*\mu), \qquad u\in X_k,\ \mu\in M_k.$$

We introduce spaces V_k and V_k^0 by

$$V_k = \big\{\,u \mid u\in X_k,\ (B_k u,\mu)=0,\ \mu\in M_k\,\big\}, \tag{6.1.40}$$

$$V_k^0 = \big\{\,q \mid q\in X_k^*,\ (q,u)=0,\ u\in V_k\,\big\}. \tag{6.1.41}$$

Lemma 6.1.3 *Let $\{X_k\}_{k=1}^\infty$, $\{M_k\}_{k=1}^\infty$ be sequences of finite-dimensional subspaces of X and $L_2(\Omega)$, respectively, and let there exist a positive constant β such that*

$$\inf_{\mu\in M_k}\sup_{u\in X_k}\frac{(B_k u,\mu)}{\|u\|_X\|\mu\|_{L_2(\Omega)}} \geq \beta \qquad \forall k. \tag{6.1.42}$$

Then, the operator B_k^* is an isomorphism of M_k onto V_k^0, and the operator B_k is an isomorphism of $V_k^\perp$ onto M_k^*, where $V_k^\perp$ is the orthogonal complement to V_k in X_k, and

$$\|(B_k^*)^{-1}\|_{\mathcal{L}(V_k^0,M_k)} \leq \frac{1}{\beta} \qquad \forall k, \tag{6.1.43}$$

$$\|B_k^{-1}\|_{\mathcal{L}(M_k^8,V_k^\perp)} \leq \frac{1}{\beta} \qquad \forall k. \tag{6.1.44}$$

Proof. It follows from (6.1.42) that

$$\sup_{u\in X_k}\frac{(u,B_k^*\mu)}{\|u\|_X} \geq \beta\|\mu\|_{L_2(\Omega)}, \qquad \mu\in M_k.$$

Therefore,

$$\|B_k^*\mu\|_{X_k^*} \geq \beta\|\mu\|_{L_2(\Omega)}, \qquad \mu \in M_k, \tag{6.1.45}$$

and B_k^* is an isomorphism of M_k onto its range $\mathcal{R}(B_k^*)$. By the closed range theorem, $\mathcal{R}(B_k^*) = V_k^0$. The estimate (6.1.43) follows from (6.1.45). Identifying V_k^0 with $(V_k^\perp)^*$ gives (6.1.44). The lemma is proved.

For the case when the velocity function satisfies the zero boundary condition on the whole boundary, the subspaces X_k, M_k satisfying (6.1.37), (6.1.38), and (6.1.42) are contained in Girault and Raviart (1981), Gunzburger (1986). The extensions of these subspaces corresponding to the boundary conditions (6.1.6), (6.1.9) also satisfy (6.1.37), (6.1.38), and (6.1.42).

Lemma 6.1.4 *Suppose the conditions* (6.1.11)–(6.1.18) *are satisfied and an operator L is defined by* (6.1.23). *Then, the following estimates hold:*

$$(L(u) - L(w), u - w) \geq \mu_1\|u - w\|_X^2, \qquad u, w \in X, \tag{6.1.46}$$

$$\|L(u) - L(w)\|_{X^*} \leq \mu_2\|u - w\|_X, \qquad u, w \in X, \tag{6.1.47}$$

where μ_1, μ_2 are positive constants.

Proof. Let u, w be arbitrary fixed elements of X and let

$$h = u - w. \tag{6.1.48}$$

We introduce a function ψ as follows:

$$\psi(t) = 2\int_\Omega \varphi(I(w + th))\varepsilon_{ij}(w + th)\varepsilon_{ij}(e)\,dx$$
$$+ \int_{S_3} \lambda((w + th)_\nu^2, \cdot)(w + th)_\nu e_\nu\,ds, \qquad t \in [0, 1],\ e \in X.$$

By (6.1.23) and (6.1.25), we have

$$\psi(1) = (L(u), e), \qquad \psi(0) = (L(w), e). \tag{6.1.49}$$

By using the theorem on the differentiability of a function represented as an integral, see, e.g., Schwartz (1967), we conclude that ψ is differerentiable at any point $t \in (0, 1)$. Thus, there exists $\xi \in (0, 1)$ such that

$$\psi(1) = \psi(0) + \frac{d\psi}{dt}(\xi), \tag{6.1.50}$$

where

$$\frac{d\psi}{dt}(\xi) = 2\int_\Omega \Big[\varphi(I(w + \xi h))\varepsilon_{ij}(h)\varepsilon_{ij}(e)$$
$$+ 2\,\frac{d\varphi}{d\alpha}\,(I(w + \xi h))\varepsilon_{km}(w + \xi h)\varepsilon_{ij}(w + \xi h)\varepsilon_{km}(h)\varepsilon_{ij}(e)\Big]\,dx$$
$$+ \int_{S_3} \Big[\lambda((w + \xi h)_\nu^2, \cdot)h_\nu e_\nu + 2\,\frac{\partial\lambda}{\partial\alpha}\,((w + \xi h)_\nu^2, \cdot)(w + \xi h)_\nu^2 h_\nu e_\nu\Big]\,ds. \tag{6.1.51}$$

Now, by (6.1.11), (6.1.12), (6.1.14)–(6.1.16), (6.1.18), (6.1.48)–(6.1.51), we get (6.1.47).

Define functions g_1, g_2 as follows:

$$g_1(\alpha) = \begin{cases} \dfrac{d\varphi}{d\alpha}(\alpha), & \text{if } \dfrac{d\varphi}{d\alpha}(\alpha) < 0, \\ 0, & \text{if } \dfrac{d\varphi}{d\alpha}(\alpha) \geq 0, \end{cases} \qquad \alpha \in \mathbb{R}_+, \tag{6.1.52}$$

$$g_2(\alpha, s) = \begin{cases} \dfrac{\partial\lambda}{\partial\alpha}(\alpha, s), & \text{if } \dfrac{\partial\lambda}{\partial\alpha}(\alpha, s) < 0, \\ 0, & \text{if } \dfrac{\partial\lambda}{\partial\alpha}(\alpha, s) \geq 0, \end{cases} \qquad (\alpha, s) \in \mathbb{R}_+ \times S_3. \tag{6.1.53}$$

Taking $e = h$ in (6.1.51) and applying (6.1.12), (6.1.13), (6.1.16), (6.1.17), and the estimate

$$\big(\varepsilon_{ij}(w + \xi h)\varepsilon_{ij}(h)\big)^2 \leq I(w + \xi h)I(h),$$

we get

$$\begin{aligned} \frac{d\psi}{dt}(\xi) &= 2\int_\Omega \left[\varphi(I(w+\xi h))I(h) + 2\frac{d\varphi}{d\alpha}(I(w+\xi h))(\varepsilon_{ij}(w+\xi h)\varepsilon_{ij}(h))^2\right] dx \\ &\quad + \int_{S_3} \left[\lambda((w+\xi h)_\nu^2, \cdot)h_\nu^2 + 2\frac{\partial\lambda}{\partial\alpha}((w+\xi h)_\nu^2, \cdot)(w+\xi h)_\nu^2 h_\nu^2\right] ds \\ &\geq 2\int_\Omega \left[\varphi(I(w+\xi h)) + 2g_1(I(w+\xi h))I(w+\xi h)\right] I(h)\, dx \\ &\quad + \int_{S_3} \left[\lambda((w+\xi h)_\nu^2, \cdot) + 2g_2((w+\xi h)_\nu^2, \cdot)(w+\xi h)_\nu^2\right] h_\nu^2\, ds \\ &\geq \mu_1 \|h\|_X^2, \end{aligned} \tag{6.1.54}$$

where μ_1 is a positive constant. Now, (6.1.54) together with (6.1.48), (6.1.49) for $e = h$, and (6.1.50) gives (6.1.46), which completes the proof.

6.1.3 Existence theorem

Consider another formulation of our problem: Find a function v satisfying

$$v \in V, \tag{6.1.55}$$

$$(L(v), h) = (K + F, h), \qquad h \in V. \tag{6.1.56}$$

Clearly, if v, p is a solution of the problem (6.1.26)–(6.1.28), then v is a solution of the problem (6.1.55), (6.1.56). On the contrary, if v is a solution of (6.1.55), (6.1.56), then

$$L(v) - K - F \in V^0,$$

and due to Lemma 6.1.1, there exists a unique function $p \in L_2(\Omega)$ such that

$$L(v) - K - F = B^* p.$$

Therefore, the problems (6.1.26)–(6.1.28) and (6.1.55), (6.1.56) are equivalent.

We search for an approximate solution v_k, p_k of the problem (6.1.26)–(6.1.28) in the form

$$(v_k, p_k) \in X_k \times M_k, \tag{6.1.57}$$

$$(L(v_k), h) - (B_k^* p_k, h) = (K + F, h), \qquad h \in X_k, \tag{6.1.58}$$

$$(B_k v_k, q) = 0, \qquad q \in M_k. \tag{6.1.59}$$

Theorem 6.1.1 *Suppose the conditions* (6.1.11)–(6.1.18), (6.1.22) *are satisfied. Then, there exists a unique solution* v, p *of the problem* (6.1.26)–(6.1.28). *Let also* $\{X_k\}$, $\{M_k\}$ *be sequences of finite-dimensional subspaces of* X *and* $L_2(\Omega)$ *such that* (6.1.37), (6.1.38), *and* (6.1.42) *hold and* $X_k \subset X_{k+1}$, $M_k \subset M_{k+1}$ *for each* k. *Then, for any* k, *there exists a unique solution* v_k, p_k *of the problem* (6.1.57)–(6.1.59) *and* $v_k \to v$ *in* X, $p_k \to p$ *in* $L_2(\Omega)$.

Proof. Due to the theory of monotone operators, see Gajewski et al. (1974), Vainberg (1972), Varga (1971), the existence of a unique solution of the problem (6.1.55), (6.1.56) (and hence of the problem (6.1.26)–(6.1.28)) follows from Lemma 6.1.4. We will prove the convergence of the approximate solutions v_k, p_k. The existence and uniqueness for the problem (6.1.26)–(6.1.28) will also follow from this proof.

1. From (6.1.40), (6.1.57)–(6.1.59), we get that the function v_k is a solution of the problem

$$v_k \in V_k, \qquad (L(v_k), h) = (K + F, h), \qquad h \in V_k. \tag{6.1.60}$$

By (6.1.12), (6.1.16), (6.1.21)–(6.1.23), we obtain, for an arbitrary $u \in V_k$,

$$\begin{gathered}(L(u), u) - (K + F, u) \geq 2a_1 \|u\|_X^2 - \|K + F\|_{X^*} \|u\|_X \geq 0 \\ \text{if } \|u\|_X \geq r = (2a_1)^{-1} \|K + F\|_{X^*}.\end{gathered} \tag{6.1.61}$$

So, because of a corollary of Brouwer's theorem, see Gajewski et al. (1974), Lions (1969), there exists a solution of the problem (6.1.60) and

$$\|v_k\|_X \leq r, \qquad \|L(v_k)\|_{X^*} \leq c, \tag{6.1.62}$$

the second estimate in (6.1.62) follows from (6.1.12), (6.1.16), and the embedding theorem. For an arbitrary $f \in X^*$, we denote by Gf the restriction of f to X_k, then $Gf \in X_k^*$, and by virtue of (6.1.41), (6.1.60), we conclude

$$G(L(v_k) - K - F) \in V_k^0.$$

Thus, by Lemma 6.1.3, there exists a unique $p_k \in M_k$ such that

$$B_k^* p_k = G(L(v_k) - K - F), \tag{6.1.63}$$

and also v_k, p_k is a solution of the problem (6.1.57)–(6.1.59) and

$$\|p_k\|_{L_2(\Omega)} \leq c_1. \tag{6.1.64}$$

By (6.1.62), (6.1.64), we can extract a subsequence $\{v_m, p_m\}$ such that

$$v_m \to v_0 \text{ weakly in } X, \tag{6.1.65}$$

$$p_m \to p_0 \text{ weakly in } L_2(\Omega), \tag{6.1.66}$$

$$L(v_m) \to \chi \text{ weakly in } X^*. \tag{6.1.67}$$

Let m_0 be a fixed positive integer, and let $h \in X_{m_0}$, $q \in M_{m_0}$. By (6.1.65)–(6.1.67), we pass to the limit in (6.1.58), (6.1.59), with k replaced by m, which gives

$$(\chi - B^* p_0, h) = (K + F, h), \qquad h \in X_{m_0}, \tag{6.1.68}$$

$$\int_\Omega (\operatorname{div} v_0) q \, dx = 0, \qquad q \in M_{m_0}. \tag{6.1.69}$$

Since m_0 is an arbitrary positive integer, we infer from (6.1.37), (6.1.38), (6.1.68), and (6.1.69) that

$$\chi - B^* p_0 = K + F, \tag{6.1.70}$$

$$\operatorname{div} v_0 = 0. \tag{6.1.71}$$

From Lemma 6.1.4, we get

$$(L(v_m) - L(u), v_m - u) \geq 0, \qquad u \in X, \ \forall m. \tag{6.1.72}$$

Since $(B_m^* p_m, v_m) = 0$, we get from (6.1.58), (6.1.65) that

$$(L(v_m), v_m) = (K + F, v_m) \to (K + F, v_0). \tag{6.1.73}$$

The relations (6.1.67), (6.1.70) give

$$\lim_{m \to \infty} (L(v_m), u) - (B^* p_0, u) = (K + F, u), \qquad u \in X. \tag{6.1.74}$$

By (6.1.73), (6.1.74), we pass to the limit in (6.1.72), and taking (6.1.71) into account, we get

$$(K + F - L(u) + B^* p_0, v_0 - u) \geq 0, \qquad u \in X. \tag{6.1.75}$$

Take here $u = v_0 - \gamma h$, $\gamma > 0$, $h \in X$, and let γ tend to zero. Then, due to the continuity of the mapping L (see (6.1.47)), we obtain

$$(K + F - L(v_0) + B^* p_0, h) \geq 0, \qquad h \in X.$$

Therefore, the pair $v = v_0$, $p = p_0$ is a solution of the problem (6.1.26)–(6.1.28).

2. By using (6.1.73), (6.1.74), it is easy to see that

$$(L(v_m) - L(v_0), v_m - v_0) \to 0,$$

and so, by (6.1.46), we get

$$v_m \to v_0 \qquad \text{strongly in } X. \tag{6.1.76}$$

Let us show that

$$p_m \to p_0 \qquad \text{strongly in } L_2(\Omega). \tag{6.1.77}$$

It follows from (6.1.27) and (6.1.58) that

$$\begin{aligned} (L(v_0), h) - (B^* p_0, h) &= (K + F, h), \qquad && h \in X_m, \\ (L(v_m), h) - (B^* p_m, h) &= (K + F, h), \qquad && h \in X_m. \end{aligned}$$

Thus,

$$\begin{gathered} (B^*(p_m - \mu), h) = (L(v_m) - L(v_0), h) + (B^*(p_0 - \mu), h), \\ h \in X_m, \ \mu \in M_m. \end{gathered} \tag{6.1.78}$$

The estimate (6.1.42) together with (6.1.78) yields

$$\begin{aligned} \|p_m - \mu\|_{L_2(\Omega)} &\leq \sup_{h \in X_m} \frac{(B^*(p_m - \mu), h)}{\beta \|h\|_X} \\ &\leq \beta^{-1} \|L(v_m) - L(v_0)\|_{X^*} + c\|p_0 - \mu\|_{L_2(\Omega)}, \\ &\qquad \mu \in M_m, \ c = \text{const} > 0. \end{aligned} \tag{6.1.79}$$

Hence,

$$\begin{aligned} \|p_0 - p_m\|_{L_2(\Omega)} &\leq \|p_0 - \mu\|_{L_2(\Omega)} + \|p_m - \mu\|_{L_2(\Omega)} \\ &\leq \beta^{-1} \|L(v_m) - L(v_0)\|_{X^*} + (c + 1) \inf_{\mu \in M_m} \|p_0 - \mu\|_{L_2(\Omega)}. \end{aligned} \tag{6.1.80}$$

By (6.1.47) and (6.1.76), we get $L(v_m) \to L(v_0)$ in X^*, and (6.1.77) follows from (6.1.80) and (6.1.38).

The function v_0 is a solution of the problem (6.1.55), (6.1.56). Let $v^{(1)}$, $v^{(2)}$ be two solutions of this problem, then

$$(L(v^{(1)}) - L(v^{(2)}), v^{(1)} - v^{(2)}) = 0,$$

and by (6.1.46), we get $v^{(1)} = v^{(2)}$. Hence, the solution of (6.1.55), (6.1.56) is unique, and by Lemma 6.1.1 there exists a unique p_0 such that the pair $v = v_0$, $p = p_0$ is the unique solution of (6.1.26)–(6.1.28).

From the uniqueness of the solution, we infer that $v_k \to v_0$ in X and $p_k \to p_0$ in $L_2(\Omega)$. The theorem is proved.

Remark 6.1.1 It is not hard to see that Theorem 6.1.1 remains true in the case when S_3 is an empty set. In this case,

$$X = \{ u \mid u \in H^1(\Omega)^n,\ u \restriction_{S_1} = 0 \}, \tag{6.1.81}$$

and

$$(L(u), h) = 2 \int_\Omega \varphi(I(u)) \varepsilon_{ij}(u) \varepsilon_{ij}(h)\, dx. \tag{6.1.82}$$

6.2 Theorem on continuity

It is obvious that the operator L is defined by the functions of viscosity φ and filtration λ. So, we denote it by $L_{\varphi,\lambda}$, and by (6.1.23), (6.1.25) we have

$$(L_{\varphi,\lambda}(u), h) = 2 \int_\Omega \varphi(I(u)) \varepsilon_{ij}(u) \varepsilon_{ij}(h)\, dx + \int_{S_3} \lambda(u_\nu^2, \cdot) u_\nu h_\nu\, ds, \\ u, h \in X. \tag{6.2.1}$$

The solution of the problem (6.1.26)–(6.1.28) depends on the loading K, F and φ, λ. Therefore, we denote it by $v(K, F, \varphi, \lambda)$, $p(K, F, \varphi, \lambda)$. We have

$$(v(K, F, \varphi, \lambda), p(K, F, \varphi, \lambda)) \in X \times L_2(\Omega), \tag{6.2.2}$$

$$(L_{\varphi,\lambda}(v(K, F, \varphi, \lambda)), h) - (B^* p(K, F, \varphi, \lambda), h) = (K + F, h), \qquad h \in X, \tag{6.2.3}$$

$$(Bv(K, F, \varphi, \lambda), q) = 0, \qquad q \in L_2(\Omega). \tag{6.2.4}$$

We introduce a set $\mathcal{B}_1$ as follows

$$\mathcal{B}_1 = \{ \varphi \mid \varphi \in C^1(\mathbb{R}_+),\ \varphi \text{ satisfies the conditions (6.1.12)–(6.1.14)}, \\ \text{where } a_1, \dots, a_4 \text{ are positive constants} \}. \tag{6.2.5}$$

We equip the set $\mathcal{B}_1$ with the topology generated by that of $C^1(\mathbb{R}_+)$, the norm of $C^1(\mathbb{R}_+)$ being defined by

$$\|\varphi\|_{C^1(\mathbb{R}_+)} = \sup_{\alpha \in \mathbb{R}_+} \left(|\varphi(\alpha)| + \left| \frac{d\varphi}{d\alpha}(\alpha) \right| \right). \tag{6.2.6}$$

We denote by $L_\infty(S_3; C^1(\mathbb{R}_+))$ the space of functions $\lambda \colon S_3 \to C^1(\mathbb{R}_+)$ such that the function $s \to \|\lambda(\cdot\,, s)\|_{C^1(\mathbb{R}_+)}$ is measurable and

$$\|\lambda\|_{L_\infty(S_3; C^1(\mathbb{R}_+))} = \operatorname*{sup\,ess}_{s \in S_3} \|\lambda(\cdot\,, s)\|_{C^1(\mathbb{R}_+)} < \infty. \tag{6.2.7}$$

We introduce a set $\mathcal{B}_2$ by

$$\mathcal{B}_2 = \{ \lambda \mid \lambda \in L_\infty(S_3; C^1(\mathbb{R}_+)),\ \lambda \text{ satisfies the conditions (6.1.16)–(6.1.18)}, \\ \text{where } b_1,\ b_2 \text{ are positive constants} \}. \tag{6.2.8}$$

The set $\mathcal{B}_2$ is equipped with the topology generated by that of $L_\infty(S_3; C^1(\mathbb{R}_+))$. By Theorem 6.1.1, the relations (6.2.2)–(6.2.4) define the following function G:

$$X^* \times X^* \times \mathcal{B}_1 \times \mathcal{B}_2 \ni (K, F, \varphi, \lambda) \to G(K, F, \varphi, \lambda)$$
$$= (v(K, F, \varphi, \lambda), p(K, F, \varphi, \lambda)) \in V \times L_2(\Omega). \quad (6.2.9)$$

Theorem 6.2.1 *The function G determined by* (6.2.9), (6.2.2)–(6.2.4) *is a continuous mapping of $X^* \times X^* \times \mathcal{B}_1 \times \mathcal{B}_2$ into $V \times L_2(\Omega)$.*

Proof. 1) Let $\{ K_m, F_m, \varphi_m, \lambda_m \}_{m=1}^\infty$ be a sequence such that

$$K_m \to K_0 \quad \text{in } X^*, \quad (6.2.10)$$
$$F_m \to F_0 \quad \text{in } X^*, \quad (6.2.11)$$
$$\varphi_m \to \varphi_0 \quad \text{in } \mathcal{B}_1, \quad (6.2.12)$$
$$\lambda_m \to \lambda_0 \quad \text{in } \mathcal{B}_2. \quad (6.2.13)$$

We introduce the notations

$$\begin{aligned} v_m &= v(K_m, F_m, \varphi_m, \lambda_m), \\ p_m &= p(K_m, F_m, \varphi_m, \lambda_m), \qquad m = 0, 1, 2, \ldots, \end{aligned} \quad (6.2.14)$$

where $v(K_m, F_m, \varphi_m, \lambda_m)$, $p(K_m, F_m, \varphi_m, \lambda_m)$ is the solution of the problem (6.2.2)–(6.2.4) for $K = K_m$, $F = F_m$, $\varphi = \varphi_m$, $\lambda = \lambda_m$. Then, the function v_m is the solution of the following problem:

$$(L_{\varphi_m, \lambda_m}(v_m), h) = (K_m + F_m, h), \qquad h \in V,\ m = 0, 1, 2, \ldots. \quad (6.2.15)$$

We take here $h = v_m - v_0$ and subtract (6.2.15) for $m = 0$ from (6.2.15). Then,

$$\begin{aligned} (L_{\varphi_m, \lambda_m}(v_m) - L_{\varphi_0, \lambda_0}(v_0), v_m - v_0) &= g_m, \\ g_m = (K_m - K_0 + F_m - F_0, v_m - v_0)&. \end{aligned} \quad (6.2.16)$$

Due to (6.1.12), (6.1.16), (6.2.10)–(6.2.13), we have

$$\|v_m\|_X \leq C \qquad \forall m, \quad (6.2.17)$$

and so

$$g_m \to 0. \quad (6.2.18)$$

By (6.2.1) and (6.2.16)

$$\sum_{i=1}^{4} A_{im} = g_m, \quad (6.2.19)$$

where

$$
\begin{aligned}
A_{1m} &= 2\int_\Omega \big[\varphi_m(I(v_m))\varepsilon_{ij}(v_m) - \varphi_m(I(v_0))\varepsilon_{ij}(v_0)\big]\varepsilon_{ij}(v_m - v_0)\,dx,\\
A_{2m} &= 2\int_\Omega \big[\varphi_m(I(v_0))\varepsilon_{ij}(v_0) - \varphi_0(I(v_0))\varepsilon_{ij}(v_0)\big]\varepsilon_{ij}(v_m - v_0)\,dx,\\
A_{3m} &= \int_{S_3} \big[\lambda_m(v_{m\nu}^2,\cdot)v_{m\nu} - \lambda_m(v_{0\nu}^2,\cdot)v_{0\nu}\big](v_{m\nu} - v_{0\nu})\,ds,\\
A_{4m} &= \int_{S_3} \big[\lambda_m(v_{0\nu}^2,\cdot)v_{0\nu} - \lambda_0(v_{0\nu}^2,\cdot)v_{0\nu}\big](v_{m\nu} - v_{0\nu})\,ds.
\end{aligned}
\tag{6.2.20}
$$

Lemma 6.1.4 yields

$$
A_{1m} + A_{3m} \geq \mu_1 \|v_m - v_0\|_X^2. \tag{6.2.21}
$$

We have

$$
\begin{aligned}
&A_{2m} \leq 2\alpha_m \Big|\int_\Omega \varepsilon_{ij}(v_0)\varepsilon_{ij}(v_m - v_0)\,dx\Big|, && \alpha_m = \sup_{t\in\mathbb{R}_+} |\varphi_m(t) - \varphi_0(t)|,\\
&A_{4m} \leq \beta_m \Big|\int_{S_3} v_{0\nu}(v_{m\nu} - v_{0\nu})\,ds\Big|, && \beta_m = \sup_{s\in S_3}\operatorname{ess}\sup_{t\in\mathbb{R}_+} |\lambda_m(t,s) - \lambda_0(t,s)|.
\end{aligned}
\tag{6.2.22}
$$

Thus, by (6.2.12), (6.2.13), (6.2.17), and (6.2.22), we get $A_{2m} \to 0$, $A_{4m} \to 0$, and (6.2.18), (6.2.19), (6.2.21) give

$$
v_m \to v_0 \qquad \text{in } X, \text{ i.e., in } V. \tag{6.2.23}
$$

2) By (6.2.3) and (6.2.14), we have

$$
B^*(p_m - p_0) = L_{\varphi_m,\lambda_m}(v_m) - L_{\varphi_0,\lambda_0}(v_0) - K_m + K_0 - F_m + F_0 \qquad \text{in } X^*. \tag{6.2.24}
$$

Let us show that

$$
L_{\varphi_m.\lambda_m}(v_m) \to L_{\varphi_0,\lambda_0}(v_0) \qquad \text{in } X^*. \tag{6.2.25}
$$

From (6.2.1), we get

$$
\begin{aligned}
&\big|(L_{\varphi_m,\lambda_m}(v_m) - L_{\varphi_0,\lambda_0}(v_0), h)\big|\\
&\quad\leq 2\Big\{\int_\Omega \sum_{i,j=1}^n \big[\varphi_m(I(v_m))\varepsilon_{ij}(v_m) - \varphi_0(I(v_0))\varepsilon_{ij}(v_0)\big]^2\,dx\Big\}^{\frac12}\|h\|_X\\
&\qquad + \Big\{\int_{S_3} \big[\lambda_m(v_{m\nu}^2,\cdot)v_{m\nu} - \lambda_0(v_{0\nu}^2,\cdot)v_{0\nu}\big]^2\,ds\Big\}^{\frac12}\|h_\nu\|_{L_2(S_3)}.
\end{aligned}
\tag{6.2.26}
$$

From the triangle inequality, it follows that

$$\left\{\int_{\Omega}\sum_{i,j=1}^{n}\left[\varphi_m(I(v_m))\varepsilon_{ij}(v_m)-\varphi_0(I(v_0))\varepsilon_{ij}(v_0)\right]^2dx\right\}^{\frac{1}{2}}\le B_{1m}+B_{2m}+B_{3m}, \tag{6.2.27}$$

where

$$\begin{aligned}B_{1m}&=\left\{\int_{\Omega}\sum_{i,j=1}^{n}\left[\varphi_m(I(v_m))\varepsilon_{ij}(v_m-v_0)\right]^2dx\right\}^{\frac{1}{2}},\\ B_{2m}&=\left\{\int_{\Omega}\sum_{i,j=1}^{n}\left[(\varphi_m(I(v_m))-\varphi_0(I(v_m)))\varepsilon_{ij}(v_0)\right]^2dx\right\}^{\frac{1}{2}},\\ B_{3m}&=\left\{\int_{\Omega}\sum_{i,j=1}^{n}\left[(\varphi_0(I(v_m))-\varphi_0(I(v_0)))\varepsilon_{ij}(v_0)\right]^2dx\right\}^{\frac{1}{2}}.\end{aligned} \tag{6.2.28}$$

By (6.1.12) and (6.2.12), we have

$$B_{1m}\le a_2\|v_m-v_0\|_X.$$

(6.2.23) implies $B_{1m}\to 0$, (6.2.12) gives $B_{2m}\to 0$, and by the Lebesgue theorem, $B_{3m}\to 0$.

Analogously, we get

$$\int_{S_3}\left[\lambda_m(v_{m\nu}^2,\cdot)v_{m\nu}-\lambda_0(v_{0\nu}^2,\cdot)v_{0\nu}\right]^2ds\to 0,\qquad \text{as } m\to\infty, \tag{6.2.29}$$

and (6.2.25) follows from (6.2.26). Finally, (6.2.10), (6.2.11), (6.2.24), (6.2.25), and Lemma 6.1.1 give $p_m\to p_0$ in $L_2(\Omega)$.

6.3 Continuity with respect to the shape of the domain

6.3.1 Formulation of the problem

Let Ω be a bounded domain in $\mathbb{R}^n$, $n=2,\ 3$, with a Lipschitz continuous boundary S. Let $\mathcal{M}$ be a topological space, and to each $q\in\mathcal{M}$ we assign a domain $\Omega_q\subset\mathbb{R}^n$ with a boundary S_q and a diffeomorphism P_q of the C^1 class that maps $\overline{\Omega}_q$ onto $\overline{\Omega}$:

$$P_q\in C^1(\overline{\Omega}_q,\overline{\Omega}),\qquad P_q^{-1}\in C^1(\overline{\Omega},\overline{\Omega}_q), \tag{6.3.1}$$

P_q^{-1} the diffeomorphism inverse of P_q. For each $q\in\mathcal{M}$, consider the following problem on flow of a nonlinear viscous fluid: Find a pair of functions v_q, p_q satisfying

$$\frac{\partial p_q}{\partial x_i}-2\frac{\partial}{\partial x_j}(\varphi(I(v_q))\varepsilon_{ij}(v_q))=K_{qi}\qquad \text{in } \Omega_q,\ i=1,\dots,n, \tag{6.3.2}$$

$$\operatorname{div} v_q = 0 \qquad \text{in } \Omega_q, \tag{6.3.3}$$

$$v_q \restriction_{S_{q1}} = 0, \tag{6.3.4}$$

$$\big[-p_q\delta_{ij} + 2\varphi(I(v_q))\varepsilon_{ij}(v_q)\big]\nu_{qj} \restriction_{S_{q2}} = F_{qi}, \qquad i = 1,\dots,n. \tag{6.3.5}$$

Here, S_{q1} and S_{q2} are open subsets of S_q such that $S_q = \overline{S}_{q1} \cup \overline{S}_{q2}$, $S_{q1} \cap S_{q2} = \varnothing$, ν_{qj} are the components of the unit outward normal $\nu_q = (\nu_{q1},\dots,\nu_{qn})$ to S_q, $K_q = (K_{q1},\dots,K_{qn})$, $F_q = (F_{q1},\dots,F_{qn})$ are functions of volume and surface forces given in Ω_q and S_{q2}, respectively. So, to each $q \in \mathcal{M}$ we assign the functions K_q and F_q. Let S_1, S_2 be open subsets of S – the boundary of Ω – for which $S = \overline{S}_1 \cup \overline{S}_2$, $S_1 \cap S_2 = \varnothing$.

We suppose that

$$\left.\begin{array}{l} P_q \text{ is a bijection of } S_{q1} \text{ onto } S_1 \text{ and of } S_{q2} \text{ onto } S_2 \text{ for each} \\ q \in \mathcal{M}. \end{array}\right\} \tag{6.3.6}$$

Specifically, S_{q2} and F_q may be fixed, i.e., independent of q.

Define the following spaces

$$X_q = \{\, u \mid u \in H^1(\Omega_q)^n,\ u \restriction_{S_{q1}} = 0 \,\}, \tag{6.3.7}$$

$$V_q = \{\, u \mid u \in X_q,\ \operatorname{div} u = 0 \,\}, \tag{6.3.8}$$

$$X = \{\, u \mid u \in H^1(\Omega)^n,\ u \restriction_{S_1} = 0 \,\}. \tag{6.3.9}$$

In X_q and V_q, the norm is defined by (6.1.21) with $\Omega = \Omega_q$, in X it is still defined by (6.1.21).

We suppose the function φ satisfies the conditions (6.1.11)–(6.1.14). For each $q \in \mathcal{M}$, define operators $L_q \colon X_q \to X_q^*$ and $B_q \in \mathcal{L}(X_q, L_2(\Omega_q)^*)$ by

$$(L_q(u), h) = 2\int_{\Omega_q} \varphi(I(u))\varepsilon_{ij}(u)\varepsilon_{ij}(h)\,dx, \qquad u, h \in X_q, \tag{6.3.10}$$

$$(B_q u, \mu) = \int_{\Omega_q} (\operatorname{div} u)\mu\,dx, \qquad u \in X_q,\ \mu \in L_2(\Omega_q). \tag{6.3.11}$$

We assume that

$$K_q,\ F_q \in X_q^*, \tag{6.3.12}$$

and consider the following problem: Find a pair of functions v_q, p_q satisfying

$$(v_q, p_q) \in X_q \times L_2(\Omega_q), \tag{6.3.13}$$

$$(L_q(v_q), h) - (B_q^* p_q, h) = (K_q + F_q, h), \qquad h \in X_q, \tag{6.3.14}$$

$$(B_q v_q, \mu) = 0, \qquad \mu \in L_2(\Omega_q). \tag{6.3.15}$$

The problems (6.3.2)–(6.3.5) and (6.3.13)–(6.3.15) are equivalent provided that the equalities (6.3.2), (6.3.5) are considered in the distribution sense. By

(6.3.1), (6.3.6), and Theorem 1.14.3, for each $q \in \mathcal{M}$, the function $u \to u \circ P_q$ is an isomorphism of X onto X_q and of $L_2(\Omega)$ onto $L_2(\Omega_q)$. So, the problem (6.3.13)–(6.3.15) is equivalent to the following one: Find a pair of functions $\tilde{v}_q$, $\tilde{p}_q$ such that

$$(\tilde{v}_q, \tilde{p}_q) \in X \times L_2(\Omega), \tag{6.3.16}$$

$$(L_q(\tilde{v}_q \circ P_q), h \circ P_q) - (B_q^*(\tilde{p}_q \circ P_q), h \circ P_q) = (K_q + F_q, h \circ P_q), \qquad h \in X, \tag{6.3.17}$$

$$(B_q(\tilde{v}_q \circ P_q), \mu \circ P_q) = 0, \qquad \mu \in L_2(\Omega). \tag{6.3.18}$$

By Theorem 6.1.1 and Remark 6.1.1, there exists a unique solution of the problem (6.3.13)–(6.3.15), so there exists a unique solution of the problem (6.3.16)–(6.3.18), and $\tilde{v}_q = v_q \circ P_q^{-1}$, $\tilde{p}_q = p_q \circ P_q^{-1}$.

Let us define functions Q_1 and Q_2 that map $\mathcal{M}$ into X^* by

$$(Q_1(q), h) = (K_q, h \circ P_q), \qquad (Q_2(q), h) = (F_q, h \circ P_q), \qquad h \in X. \tag{6.3.19}$$

We suppose

$$Q_1 \text{ and } Q_2 \text{ are continuous mappings of } \mathcal{M} \text{ into } X^*, \tag{6.3.20}$$

$$q \to P_q^{-1} \text{ is a continuous mapping of } \mathcal{M} \text{ into } C^1(\overline{\Omega})^n. \tag{6.3.21}$$

For $q \in \mathcal{M}$, we define operators $\widetilde{L}_q \colon X \to X^*$ and $\widetilde{B}_q \in \mathcal{L}(X, L_2(\Omega)^*)$ as follows:

$$(\widetilde{L}_q(u), h) = (L_q(u \circ P_q), h \circ P_q), \qquad u, h \in X, \tag{6.3.22}$$

$$(\widetilde{B}_q u, \mu) = (B_q(u \circ P_q), \mu \circ P_q), \qquad u \in X,\ \mu \in L_2(\Omega). \tag{6.3.23}$$

6.3.2 Lemmas on operators $\widetilde{L}_q$ and $\widetilde{B}_q$

Lemma 6.3.1 *Let a function φ satisfy the conditions* (6.1.11)–(6.1.14). *Assume that* (6.3.1), (6.3.6), (6.3.21) *hold, and the operator $\widetilde{L}_q$ is defined by* (6.3.22). *Then, for each $q \in \mathcal{M}$, there exist positive constants $\gamma_1(q)$, $\gamma_2(q)$ such that*

$$(\widetilde{L}_q(u) - \widetilde{L}_q(w), u - w) \geq \gamma_1(q) \|u - w\|_X^2, \qquad u, w \in X, \tag{6.3.24}$$

$$\|\widetilde{L}_q(u) - \widetilde{L}_q(w)\|_{X^*} \leq \gamma_2(q) \|u - w\|_X, \qquad u, w \in X, \tag{6.3.25}$$

and if $q_m \to q_0$ in $\mathcal{M}$, then there are positive constants c_1 and c_2 such that $\gamma_1(q_m) \geq c_1$, $\gamma_2(q_m) \leq c_2$ for any m.

Proof. Taking (6.3.22) into account and applying Lemma 6.1.4, we obtain

$$\begin{gathered}(\widetilde{L}_q(u) - \widetilde{L}_q(w), u - w) \geq \mu_1 \|u \circ P_q - w \circ P_q\|_{X_q}^2, \\ |(\widetilde{L}_q(u) - \widetilde{L}_q(w), h)| \leq \mu_2 \|u \circ P_q - w \circ P_q\|_{X_q} \|h \circ P_q\|_{X_q}, \\ u, w \in X,\end{gathered} \tag{6.3.26}$$

where the constants μ_1, μ_2 depend only on the function φ. Since the mapping $u \to u \circ P_q$ is an isomorphism of X onto X_q, we obtain (6.3.24), (6.3.25) from (6.3.26). The constants $\gamma_1(q)$, $\gamma_2(q)$ depend on the derivatives of the mapping P_q^{-1}. So, if $q_m \to q_0$ in $\mathcal{M}$, then due to (6.3.21) there exist positive constants c_1 and c_2 such that $\gamma_1(q_m) \geq c_1$, $\gamma_2(q_m) \leq c_2$ for each m, which completes the proof.

Lemma 6.3.2 *Let a function φ satisfy the conditions* (6.1.11)–(6.1.14). *Assume that* (6.3.1), (6.3.6), (6.3.21) *hold and*

$$q_m \to q_0 \qquad \text{in } \mathcal{M}, \tag{6.3.27}$$

$$v_m \to v_0 \qquad \text{in } X. \tag{6.3.28}$$

Then

$$\widetilde{L}_{q_m}(v_m) \to \widetilde{L}_{q_0}(v_0) \qquad \text{in } X^*. \tag{6.3.29}$$

Proof. By virtue of (6.3.10), (6.3.22), we obtain the following representation of the operator $\widetilde{L}_q$:

$$(\widetilde{L}_q(u), h) = 2 \int_\Omega \varphi((I_q(u))(x))(\varepsilon_{qij}(u))(x)(\varepsilon_{qij}(h))(x)\big|\det((P_q^{-1})'(x))\big|\, dx, \tag{6.3.30}$$

where

$$(\varepsilon_{qij}(u))(x) = \frac{1}{2}\left(\frac{\partial u_i}{\partial x_k}(x)\,\frac{\partial P_{qk}}{\partial y_j}(P_q^{-1}(x)) + \frac{\partial u_j}{\partial x_k}(x)\,\frac{\partial P_{qk}}{\partial y_i}(P_q^{-1}(x))\right), \tag{6.3.31}$$

$$(I_q(u))(x) = \sum_{i,j=1}^{n} \big[(\varepsilon_{qij}(u))(x)\big]^2, \qquad P_q = (P_{q1}, \dots, P_{qn}), \tag{6.3.32}$$

$(P_q^{-1})'(x)$ is the Fréchet derivative of the mapping P_q^{-1} at point $x \in \Omega$, and $y = (y_1, \dots, y_n) = P_q^{-1}(x) \in \Omega_q$.

Denote the components of the mapping P_q^{-1} by T_{qi}, i.e.,

$$P_q^{-1} = (T_{q1}, \dots, T_{qn}). \tag{6.3.33}$$

The formula connecting derivatives of the components of a bijection with those of its inverse gives (see (1.14.8))

$$\frac{\partial P_{qk}}{\partial y_i}(P_q^{-1}(x)) = z_q(x) a_{qik}(x), \tag{6.3.34}$$

where $z_q(x) = \big[\det((P_q^{-1})'(x))\big]^{-1}$, $a_{qik}(x)$ is the cofactor of the element $\frac{\partial T_{qi}}{\partial x_k}(x)$ of the matrix $(P_q^{-1})'(x)$.

Denote

$$\varphi_m(x) = \varphi\big((I_{q_m}(v_m))(x)\big), \qquad \widetilde{\varepsilon}_{mij}(x) = (\varepsilon_{q_m ij}(v_m))(x), \tag{6.3.35}$$

$$\mathcal{P}_m(x) = \big|\det((P_{q_m}^{-1})'(x))\big|, \qquad \lambda_{mki}(x) = z_{q_m}(x) a_{q_m ik}(x), \tag{6.3.36}$$

$$m = 0, 1, 2, \dots .$$

By (6.3.21), (6.3.27), (6.3.30)–(6.3.36), we get

$$\begin{aligned}|(\widetilde{L}_{q_m}(v_m) - \widetilde{L}_{q_0}(v_0), h)| &= 2\int_\Omega (\varphi_m \mathcal{P}_m \widetilde{\varepsilon}_{mij} - \varphi_0 \mathcal{P}_0 \widetilde{\varepsilon}_{0ij})\varepsilon_{q_m ij}(h)\,dx \\ &\quad + 2\int_\Omega \varphi_0 \mathcal{P}_0 \widetilde{\varepsilon}_{0ij}(\varepsilon_{q_m ij}(h) - \varepsilon_{q_0 ij}(h))\,dx \\ &\le (c\alpha_m + \beta_m \|v_0\|_X)\|h\|_X,\end{aligned} \tag{6.3.37}$$

where

$$\alpha_m = \left(\int_\Omega \sum_{i,j=1}^n (\varphi_m \mathcal{P}_m \widetilde{\varepsilon}_{mij} - \varphi_0 \mathcal{P}_0 \widetilde{\varepsilon}_{0ij})^2\,dx\right)^{\frac{1}{2}}, \tag{6.3.38}$$

and

$$\lim \beta_m = 0. \tag{6.3.39}$$

The triangle inequality yields

$$\alpha_m \le \alpha_{1m} + \alpha_{2m}, \tag{6.3.40}$$

where

$$\alpha_{1m} = \left(\int_\Omega (\varphi_m \mathcal{P}_m)^2 \sum_{i,j=1}^n (\widetilde{\varepsilon}_{mij} - \widetilde{\varepsilon}_{0ij})^2\,dx\right)^{\frac{1}{2}}, \tag{6.3.41}$$

$$\alpha_{2m} = \left(\int_\Omega (\varphi_m \mathcal{P}_m - \varphi_0 \mathcal{P}_0)^2 \sum_{i,j=1}^n \widetilde{\varepsilon}^2_{0ij}\,dx\right)^{\frac{1}{2}}. \tag{6.3.42}$$

By (6.3.34), (6.3.36), and the triangle inequality,

$$\begin{aligned}&\left\{\int_\Omega \left[\frac{\partial v_{mi}}{\partial x_k}(x)\,\frac{\partial P_{q_m k}}{\partial y_j}(P^{-1}_{q_m}(x)) - \frac{\partial v_{0i}}{\partial x_k}(x)\,\frac{\partial P_{q_0 k}}{\partial y_j}(P^{-1}_{q_0}(x))\right]^2 dx\right\}^{\frac{1}{2}} \\ &\quad = \left\{\int_\Omega \left[\frac{\partial v_{mi}}{\partial x_k}\lambda_{mkj} - \frac{\partial v_{0i}}{\partial x_k}\lambda_{0kj}\right]^2 dx\right\}^{\frac{1}{2}} \le \gamma_{1m} + \gamma_{2m},\end{aligned} \tag{6.3.43}$$

where v_{mi}, v_{0i} are the components of the vector functions v_m and v_0, and

$$\begin{aligned}\gamma_{1m} &= \left\{\int_\Omega \left(\frac{\partial v_{mi}}{\partial x_k} - \frac{\partial v_{0i}}{\partial x_k}\right)^2 \lambda^2_{mkj}\,dx\right\}^{\frac{1}{2}}, \\ \gamma_{2m} &= \left\{\int_\Omega \left(\frac{\partial v_{0i}}{\partial x_k}\right)^2 (\lambda_{mkj} - \lambda_{0kj})^2\,dx\right\}^{\frac{1}{2}}.\end{aligned} \tag{6.3.44}$$

The relations (6.3.21) and (6.3.27) give

$$\lambda_{mkj} \to \lambda_{0kj} \qquad \text{in } C(\overline{\Omega}) \text{ as } m \to \infty. \tag{6.3.45}$$

So, by (6.3.28), we obtain $\gamma_{1m} \to 0$, $\gamma_{2m} \to 0$, and the right-hand side of (6.3.43) also tends to zero. Now, taking into account (6.3.31), (6.3.41), and the estimate

$$\|\varphi_m \mathcal{P}_m\|_{C(\overline{\Omega})} \leq c \qquad \forall m,$$

we obtain that $\alpha_{1m} \to 0$. By (6.3.21), (6.3.27), (6.3.28), and the Lebesgue theorem, we obtain $\alpha_{2m} \to 0$. At last, (6.3.29) follows from (6.3.37), (6.3.39), and (6.3.40).

Lemma 6.3.3 *Let the conditions* (6.3.1), (6.3.6), *and* (6.3.21) *be satisfied and let an operator* $\widetilde{B}_q$ *be defined by* (6.3.11), (6.3.23). *Then, the function* $q \to \widetilde{B}_q$ *is a continuous mapping of* $\mathcal{M}$ *into* $\mathcal{L}(X, L_2(\Omega)^*)$.

Proof. Due to (6.3.11), (6.3.23), we infer the following representation of the operator $\widetilde{B}_q$:

$$(\widetilde{B}_q u, \mu) = \int_\Omega \frac{\partial u_i}{\partial x_k}(x) \, \frac{\partial P_{qk}}{\partial y_i}(P_q^{-1}(x))\mu(x) \left|\det((P_q^{-1})'(x))\right| dx, \tag{6.3.46}$$

where $y = P_q^{-1}(x) \in \Omega_q$. Suppose that

$$q_m \to q_0 \qquad \text{in } \mathcal{M}. \tag{6.3.47}$$

By using (6.3.34) and the notations (6.3.36), we get

$$\begin{aligned} \left|((\widetilde{B}_{q_m} - \widetilde{B}_{q_0})u, \mu)\right| &= \left| \int_\Omega \frac{\partial u_i}{\partial x_k}(\lambda_{mki}\mathcal{P}_m - \lambda_{0ki}\mathcal{P}_0)\mu \, dx \right| \\ &\leq c\beta_m \|u\|_X \|\mu\|_{L_2(\Omega)}, \end{aligned} \tag{6.3.48}$$

where

$$\beta_m = \max_{k,i=1,\dots,n} \|\lambda_{mki}\mathcal{P}_m - \lambda_{0ki}\mathcal{P}_0\|_{C(\overline{\Omega})}.$$

It follows from (6.3.47) and (6.3.21) that $\beta_m \to 0$, and (6.3.48) gives $\widetilde{B}_{q_m} \to \widetilde{B}_{q_0}$ in $\mathcal{L}(X, L_2(\Omega)^*)$. The lemma is proved.

6.3.3 Theorem on continuity

By applying (6.3.19), (6.3.22), and (6.3.23), we can represent the problem (6.3.16)–(6.3.18) in the following form

$$(\tilde{v}_q, \tilde{p}_q) \in X \times L_2(\Omega), \tag{6.3.49}$$

$$(\widetilde{L}_q(\tilde{v}_q), h) - (\widetilde{B}_q^* \tilde{p}_q, h) = (Q_1(q) + Q_2(q), h), \qquad h \in X, \tag{6.3.50}$$

$$(\widetilde{B}_q \tilde{v}_q, \mu) = 0, \qquad \mu \in L_2(\Omega). \tag{6.3.51}$$

Theorem 6.3.1 *Let a function φ satisfy the conditions* (6.1.11)–(6.1.14). *Assume that* (6.3.1), (6.3.6), (6.3.20), (6.3.21) *hold. Then, the function $q \to (\tilde{v}_q, \tilde{p}_q)$ is a continuous mapping of $\mathcal{M}$ into $X \times L_2(\Omega)$.*

Proof. 1) Suppose that

$$q_m \to q_0 \qquad \text{in } \mathcal{M}. \tag{6.3.52}$$

We take $q = q_m$ and $h = \tilde{v}_{q_m}$ in (6.3.17). Then, from (6.1.12), (6.3.10), and (6.3.18), we obtain

$$2a_1 \|\tilde{v}_{q_m} \circ P_{q_m}\|_{X_{q_m}} \leq \|K_{q_m} + F_{q_m}\|_{X^*_{q_m}}. \tag{6.3.53}$$

This estimate, together with (6.3.20), (6.3.21), (6.3.52), gives

$$\|\tilde{v}_{q_m}\|_X \leq c \qquad \forall m. \tag{6.3.54}$$

By (6.3.17), (6.3.52), and (6.3.53)

$$\|B^*_{q_m}(\tilde{p}_{q_m} \circ P_{q_m})\|_{X^*_{q_m}} \leq c_1 \qquad \forall m. \tag{6.3.55}$$

By Lemma 6.1.1

$$\|(B^*_{q_m})^{-1}\|_{\mathcal{L}(V^0_{q_m}, L_2(\Omega_{q_m}))} \leq \frac{1}{\beta_m}, \tag{6.3.56}$$

where

$$V^0_{q_m} = \{\, f \mid f \in X^*_{q_m},\ (f, u) = 0,\ u \in V_{q_m} \,\}, \tag{6.3.57}$$

with V_{q_m} defined by (6.3.8), and also

$$\gamma_m = \inf_{\mu \in L_2(\Omega)} \sup_{v \in X} \frac{(B_{q_m}(v \circ P_{q_m}), \mu \circ P_{q_m})}{\|v \circ P_{q_m}\|_{X_{q_m}} \|\mu \circ P_{q_m}\|_{L_2(\Omega_{q_m})}} \geq \beta_m.$$

This estimate and (6.3.21), (6.3.52) imply the existence of a positive constant $\tilde{\beta}$ such that $\gamma_m \geq \tilde{\beta}$. Thus, we can consider that $\beta_m \geq \tilde{\beta}$ for all m. Now, (6.3.55), (6.3.56), and (6.3.21) give

$$\|\tilde{p}_{q_m}\|_{L_2(\Omega)} \leq c_2 \qquad \forall m. \tag{6.3.58}$$

By (6.3.54), (6.3.58), and Lemma 6.3.1 (see (6.3.25)), we can extract from the sequence $\{\tilde{v}_{q_m}, \tilde{p}_{q_m}, \widetilde{L}_{q_m}(\tilde{v}_{q_m})\}$ a subsequence $\{\tilde{v}_{q_\nu}, \tilde{p}_{q_\nu}, \widetilde{L}_{q_\nu}(\tilde{v}_{q_\nu})\}$ such that

$$\tilde{v}_{q_\nu} \to \tilde{v}_0 \qquad \text{weakly in } X, \tag{6.3.59}$$

$$\tilde{p}_{q_\nu} \to \tilde{p}_0 \qquad \text{weakly in } L_2(\Omega), \tag{6.3.60}$$

$$\widetilde{L}_{q_\nu}(\tilde{v}_{q_\nu}) \to \chi \qquad \text{weakly in } X^*. \tag{6.3.61}$$

The relations (6.3.20) and (6.3.52) yield

$$Q_i(q_\nu) \to Q_i(q_0) \qquad \text{in } X^*, \quad i = 1, 2. \tag{6.3.62}$$

By definition, we have

$$(\widetilde{L}_{q_m}(\tilde{v}_{q_m}), h) - (\widetilde{B}^*_{q_m}\tilde{p}_{q_m}, h) = (Q_1(q_m) + Q_2(q_m), h), \qquad h \in X, \tag{6.3.63}$$

$$(\widetilde{B}_{q_m}\tilde{v}_{q_m}, \mu) = 0, \qquad \mu \in L_2(\Omega). \tag{6.3.64}$$

Applying (6.3.52), (6.3.59), (6.3.64), and Lemma 6.3.3 gives

$$\lim_{\nu\to\infty} (\widetilde{B}_{q_\nu}\tilde{v}_{q_\nu}, \mu) = (\widetilde{B}_{q_0}\tilde{v}_0, \mu) = 0, \qquad \mu \in L_2(\Omega). \tag{6.3.65}$$

By (6.3.60)–(6.3.62) and Lemma 6.3.3, we pass to the limit in (6.3.63), with m changed by ν. This gives

$$\chi - \widetilde{B}^*_{q_0}\tilde{p}_0 = Q_1(q_0) + Q_2(q_0). \tag{6.3.66}$$

Lemma 6.3.1 yields

$$(\widetilde{L}_{q_\nu}(\tilde{v}_{q_\nu}) - \widetilde{L}_{q_\nu}(u), \tilde{v}_{q_\nu} - u) \geq 0, \qquad u \in X, \ \forall \nu. \tag{6.3.67}$$

Taking into account that $(\widetilde{B}^*_{q_\nu}\tilde{p}_{q_\nu}, \tilde{v}_{q_\nu}) = 0$, by (6.3.59), (6.3.62), and (6.3.63), we get

$$(\widetilde{L}_{q_\nu}(\tilde{v}_{q_\nu}), \tilde{v}_{q_\nu}) = (Q_1(q_\nu) + Q_2(q_\nu), \tilde{v}_{q_\nu}) \to (Q_1(q_0) + Q_2(q_0), \tilde{v}_0). \tag{6.3.68}$$

The relations (6.3.61) and (6.3.66) give

$$\lim_{\nu\to\infty} (\widetilde{L}_{q_\nu}(\tilde{v}_{q_\nu}), u) - (\widetilde{B}^*_{q_0}\tilde{p}_0, u) = (Q_1(q_0) + Q_2(q_0), u), \qquad u \in X. \tag{6.3.69}$$

It follows from (6.3.52) and Lemma 6.3.2 that

$$\widetilde{L}_{q_\nu}(u) \to \widetilde{L}_{q_0}(u) \qquad \text{in } X^*, \quad u \in X. \tag{6.3.70}$$

By (6.3.59), (6.3.68)–(6.3.70), we pass to the limit in (6.3.67). By virtue of (6.3.65), we get

$$(Q_1(q_0) + Q_2(q_0) - \widetilde{L}_{q_0}(u) + \widetilde{B}^*_{q_0}\tilde{p}_0, \tilde{v}_0 - u) \geq 0, \qquad u \in X. \tag{6.3.71}$$

Take here $u = \tilde{v}_0 - \gamma h$, $\gamma > 0$, $h \in X$, and let γ tend to zero. Then, because of the continuity of the mapping $\widetilde{L}_{q_0}$ (see (6.3.25)), we obtain

$$(Q_1(q_0) + Q_2(q_0) - \widetilde{L}_{q_0}(\tilde{v}_0) + \widetilde{B}^*_{q_0}\tilde{p}_0, h) \geq 0, \qquad h \in X.$$

It follows from the latter estimate and (6.3.65) that the pair $\tilde{v}_{q_0} = \tilde{v}_0$, $\tilde{p}_{q_0} = \tilde{p}_0$ is a solution of the problem (6.3.49)–(6.3.51) for $q = q_0$.

2. By using (6.3.59), (6.3.68)–(6.3.70), is easy to see that

$$(\widetilde{L}_{q_\nu}(\tilde{v}_{q_\nu}) - \widetilde{L}_{q_\nu}(\tilde{v}_{q0}), \tilde{v}_{q_\nu} - \tilde{v}_{q0}) \to 0.$$

Thus, Lemma 6.3.1 gives $\tilde{v}_{q_\nu} \to \tilde{v}_{q0}$ strongly in X. Since the solution of the problem (6.3.16)–(6.3.18) is unique, we obtain

$$\tilde{v}_{q_m} \to \tilde{v}_{q0} \qquad \text{strongly in } X. \tag{6.3.72}$$

Subtracting the equality (6.3.63) from the one for $m = 0$ gives

$$\begin{gathered} \widetilde{B}^*_{q_m}\tilde{p}_{q_m} - \widetilde{B}^*_{q0}\tilde{p}_{q0} = \alpha_m \qquad \text{in } X^*, \\ \alpha_m = \widetilde{L}_{q_m}(\tilde{v}_{q_m}) - \widetilde{L}_{q0}(\tilde{v}_{q0}) - Q_1(q_m) - Q_2(q_m) + Q_1(q_0) + Q_2(q_0). \end{gathered} \tag{6.3.73}$$

By virtue of (6.3.20), (6.3.47), (6.3.72), and Lemma 6.3.2, we get

$$\alpha_m \to 0 \qquad \text{in } X^*. \tag{6.3.74}$$

Further,

$$\begin{gathered} \widetilde{B}^*_{q_m}\tilde{p}_{q_m} - \widetilde{B}^*_{q0}\tilde{p}_{q0} = (\widetilde{B}^*_{q_m} - \widetilde{B}^*_{q0})\tilde{p}_{q_m} + \widetilde{B}^*_{q0}(\tilde{p}_{q_m} - \tilde{p}_{q0}), \\ \|(\widetilde{B}^*_{q_m} - \widetilde{B}^*_{q0})\tilde{p}_{q_m}\|_{X^*} \leq \|\widetilde{B}^*_{q_m} - \widetilde{B}^*_{q0}\|_{\mathcal{L}(L_2(\Omega),X^*)}\|\tilde{p}_{q_m}\|_{L_2(\Omega)} = \gamma_m. \end{gathered} \tag{6.3.75}$$

Lemma 6.3.3, (6.3.52), and (6.3.58) yield $\gamma_m \to 0$. So, (6.3.73)–(6.3.75) give

$$\widetilde{B}^*_{q0}(\tilde{p}_{q_m} - \tilde{p}_{q0}) \to 0 \qquad \text{in } X^*.$$

Finally, (6.3.23) and Lemma 6.1.1 imply $\tilde{p}_{q_m} \to \tilde{p}_{q0}$ in $L_2(\Omega)$.

6.4 Control of fluid flows by perforated walls and computation of the function of filtration

The solution of the problem (6.2.2)–(6.2.4) on the flow of a nonlinear viscous, in particular, viscous fluid, depends on the functions of volume and surface forces K, F, and on the functions of viscosity φ and filtration λ. The viscosity function may be varied by the introduction of various additions in the fluid. In melted metals, the volume forces may be created by electromagnetic fields. For such problems, one should solve a coupled system of equations for the fluid and the electromagnetic field. The electrical resistance of ordinary fluids is high, and so they interact very faintly with electromagnetic fields. However, many suspensions were created which interact with them. Thus, volume forces in these suspensions may be generated by electromagnetic fields. For these problems, we have to solve a coupled system of equations for the fluid and the electromagnetic field.

So, in what follows, we do not consider the volume forces and the function of viscosity as controls. Nevertheless, they may be included in the optimization problems under consideration.

In order to control the flow of the fluid, one uses perforated walls, which may be placed at either the inlet, or outlet, or inside of the canal. In Fig. 6.4.1, a spraying device (a header) of a paper machine is schematically shown.

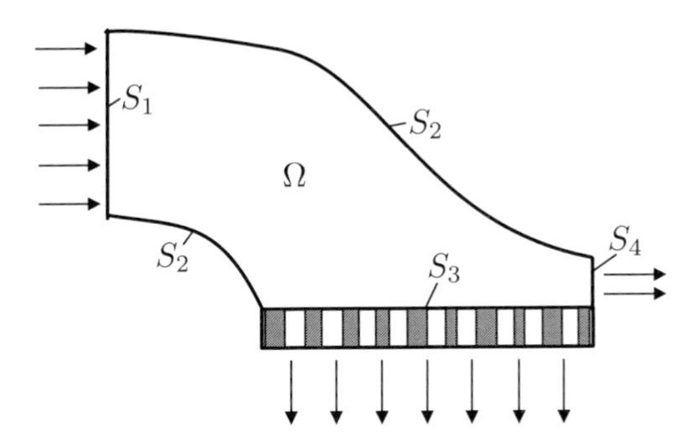

Figure 6.4.1: A scheme of a header of a paper machine

Here, we have $S = \bigcup_{i=1}^{4} \overline{S}_i$, where S is the boundary of the domain Ω defined by the shape of the header, S_1 the inflow, S_2 the hard wall, S_3, S_4 are the outflows. The canal has a perforated wall at S_3, and there is no wall at S_1 and S_4. The space of S_4 is small, and so there is a small outflow here, which is used in practice as a control parameter.

We will now show that the perforated wall may be modeled by a filtration layer.

6.4.1 The problem of flow in a circular cylinder and the function of filtration

We suppose that the holes of the perforated wall are cylinders of a circular cross-section. Let us consider the problem of flow of a nonlinear viscous fluid in a cylinder of the wall. We apply cylindrical coordinate system (r, α, z) and suppose that the velocity vector $v = (v_r, v_\alpha, v_z)$ is such that $v_\alpha = 0$ and v_r, v_z are functions of r, z only, see Fig. 6.4.2. Then, the following components of the stress tensor are not equal to zero:

$$\begin{aligned}
\sigma_{rr} &= -p + 2\varphi(I(v))\,\frac{\partial v_r}{\partial r}, \qquad & \sigma_{zz} &= -p + 2\varphi(I(v))\,\frac{\partial v_z}{\partial z}, \\
\sigma_{\alpha\alpha} &= -p + 2\varphi(I(v))\,\frac{v_r}{r}, \qquad & \sigma_{rz} &= \varphi(I(v))\Big(\frac{\partial v_z}{\partial r} + \frac{\partial v_r}{\partial z}\Big),
\end{aligned} \tag{6.4.1}$$

where

$$I(v) = \left(\frac{\partial v_r}{\partial r}\right)^2 + \left(\frac{v_r}{r}\right)^2 + \left(\frac{\partial v_z}{\partial z}\right)^2 + \frac{1}{2}\left(\frac{\partial v_z}{\partial r} + \frac{\partial v_r}{\partial z}\right)^2. \tag{6.4.2}$$

The equations of motion and incompressibility are the following:

$$\frac{\partial \sigma_{rr}}{\partial r} + \frac{\partial \sigma_{rz}}{\partial z} + \frac{\sigma_{rr} - \sigma_{\alpha\alpha}}{r} = K_r, \tag{6.4.3}$$

$$\frac{\partial \sigma_{rz}}{\partial r} + \frac{\partial \sigma_{zz}}{\partial z} + \frac{\sigma_{rz}}{r} = K_z, \tag{6.4.4}$$

$$\frac{\partial v_r}{\partial r} + \frac{\partial v_z}{\partial z} + \frac{v_r}{r} = 0. \tag{6.4.5}$$

Here, $K = (K_r, K_z)$ is the function of volume forces. In many cases, one can put $K = 0$. As the radii of the cylinders of a perforated wall are of orders smaller than the characteristic dimensions of the wall, we can consider that the velocity profile at the inflow of the cylinder is close to rectangular, i.e.,

$$v_r = 0, \qquad v_z = bf(r) \quad \text{at } z = 0, \tag{6.4.6}$$

where $b = \text{const} > 0$ and f is a function continuous on $[0, R]$, $f(r) = 1$ for an arbitrary $r \in [0, R - \delta]$, $f(R) = 0$, R being the radius of the cylinder, δ a small positive constant.

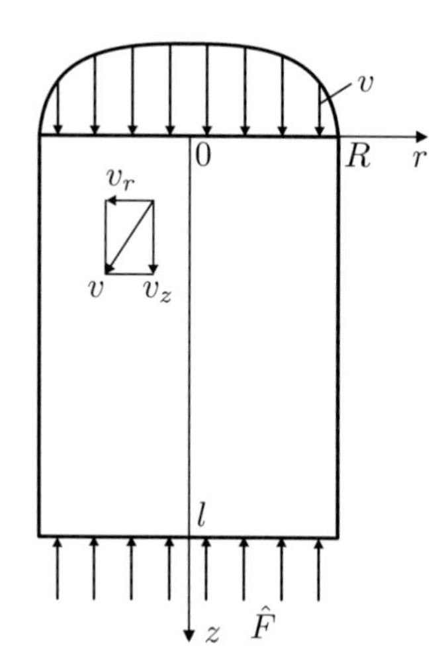

Figure 6.4.2: Notation for flow in a circular cylinder

At the outflow of the cylinder, we prescribe the following surface forces $\widehat{F} = (\widehat{F}_r, \widehat{F}_z)$:

$$\widehat{F}_r = \sigma_{rz} \restriction_{z=l} = 0, \qquad \widehat{F}_z = \sigma_{zz} \restriction_{z=l} = \xi = \text{const} \leq 0, \tag{6.4.7}$$

l being the length of the cylinder. In addition, the condition of adhesion gives

$$v = 0 \qquad \text{at } r = R. \tag{6.4.8}$$

We denote by Q the mean normal force at the inflow of the cylinder, and by $\hat{v}$ the mean velocity in the tube:

$$Q = -\frac{2}{R^2}\int_0^R \sigma_{zz}(r,0)r\,dr, \tag{6.4.9}$$

$$\hat{v} = \frac{2}{R^2}\int_0^R v_z(r,z)r\,dr. \tag{6.4.10}$$

By virtue of the condition of incompressibility (6.4.5), the latter integral is independent of z, and so, by (6.4.6), we have

$$\hat{v} = \frac{2b}{R^2}\int_0^R f(r)r\,dr. \tag{6.4.11}$$

Suppose the problem (6.4.1)–(6.4.8) is solved, its solution may be computed numerically. Then, we can compute Q and $\hat{v}$ by (6.4.9), (6.4.10). Let us define a passage factor of the tube, ν, by

$$\nu = \frac{\hat{v}}{Q+\xi}. \tag{6.4.12}$$

Obviously, ν depends on $\hat{v}$, R, l, and it is almost independent of ξ. Note that, for the case of a rectilinear flow, ν is independent of ξ, see the formulas (6.4.22), (6.4.23) below and Litvinov (1982a). Thus, we denote ν by $\nu(\hat{v}, R, l)$.

Let now G be a perforated wall which is considered as an $(n-1)$-dimensional manifold, particularly, $G = S_3$ for the header shown in Fig. 6.4.1. Let M be the number of holes (cylinders) in the wall. Assume they are numbered and $G_i \subset G$ is the region of the ith hole, $i = 1, \dots, M$, and R_i, l_i are the radius and length of the ith cylinder. We define a passage function $\tilde{\gamma}$ of the wall G as follows:

$$\mathbb{R}_+ \times G \ni (\alpha, s) \to \tilde{\gamma}(\alpha, s) = \begin{cases} \nu(\alpha^{\frac{1}{2}}, R_i, l_i), & \text{if } s \in G_i,\ i = 1, \dots, M, \\ 0, & \text{if } s \in G \setminus \bigcup_{i=1}^M G_i. \end{cases} \tag{6.4.13}$$

Comparing (6.4.12) and (6.4.13) with the boundary condition (6.1.9), we can introduce a filtration function $\tilde{\lambda}$ given on $\mathbb{R}_+ \times G$ by

$$\tilde{\lambda}(\alpha, s) = \frac{1}{\tilde{\gamma}(\alpha, s)} = \begin{cases} \dfrac{1}{\nu(\alpha^{\frac{1}{2}}, R_i, l_i)}, & \text{if } s \in G_i,\ i = 1, \dots, M, \\ \infty, & \text{if } s \in G \setminus \bigcup_{i=1}^M G_i. \end{cases} \tag{6.4.14}$$

The number M is usually very large in practice, and the radii of the cylinders are of orders smaller than the characteristic dimensions of the perforated wall. Therefore,

$\tilde{\lambda}$ is a rapidly oscillating function which can take value ∞. Such functions are inconvenient for computation. This is why we will now introduce an averaging filtration function.

An averaged passage function γ is defined by

$$\gamma(\alpha, s) = \int_G \omega_\rho(s - \tau)\tilde{\gamma}(\alpha, \tau)\, d\tau, \tag{6.4.15}$$

where ω_ρ is an averaging kernel (see Section 1.6.4). The formula (6.4.15) determines the function γ everywhere in G except for a slender band of width ρ, and it may be extended to this band. Now, define an averaging filtration function λ by

$$\lambda(\alpha, s) = \frac{1}{\gamma(\alpha, s)}. \tag{6.4.16}$$

6.4.2 The passage factor for the power model

The viscosity function for the power model is given by

$$\varphi(\alpha) = a_1 \alpha^{\frac{k-1}{2}}, \qquad \alpha \in \mathbb{R}_+, \tag{6.4.17}$$

where a_1, k are positive constants. If $k = 1$, this is just the case of a viscous, Newtonian fluid. For the power model, the problem of flow in a circular tube under the conditions that $v_r = v_\alpha = 0$, where v_α is the tangential component of v, is exactly solved, see, e.g., Astarita and Marrucci (1974), in this case, v_z is a function of r only, and p is an affine function of z, i.e.,

$$v_z(r) = \frac{k}{k+1}\left|\frac{1}{2a}\frac{dp}{dz}\right|^{\frac{1}{k}} \left[R^{\frac{k+1}{k}} - r^{\frac{k+1}{k}}\right], \tag{6.4.18}$$

$$p = p_0 + (p_1 - p_0)\frac{z}{l}, \qquad a = 2^{\frac{1-k}{2}} a_1, \tag{6.4.19}$$

where $p_0 = p(0)$, $p_1 = p(l)$. Now, $\sigma_{zz} = -p$ and (6.4.7) and (6.4.9) give $\xi = -p_1$, $Q = p_0$. Thus, (6.4.19) yields

$$\frac{dp}{dz} = -\frac{Q + \xi}{l}. \tag{6.4.20}$$

By substituting (6.4.18) into (6.4.10), we get

$$\hat{v} = \frac{k}{3k+1}\left|\frac{1}{2a}\frac{dp}{dz}\right|^{\frac{1}{k}} R^{\frac{k+1}{k}}. \tag{6.4.21}$$

From (6.4.20) and (6.4.21)

$$Q + \xi = \frac{2al}{R^{k+1}}\left|\frac{3k+1}{k}\right|^k \hat{v}^k. \tag{6.4.22}$$

By (6.4.12), (6.4.22), we obtain the following formula for the passage factor of the power model

$$\nu(\hat{v}, R, l) = \frac{\hat{v}}{Q+\xi} = \frac{R^{k+1}}{2al}\left|\frac{k}{3k+1}\right|^{k}\hat{v}^{1-k}. \tag{6.4.23}$$

By using (6.4.13), (6.4.15), (6.4.16), and (6.4.23), one can compute the filtration function for the power model.

6.4.3 Control of the surface forces at the inlet by the perforated wall

In some cases, it is necessary to obtain a required distribution of the surface forces. We will now show that, under some restrictions, the required distribution of the normal component of the surface forces at the inlet may be obtained by the filtration layer (perforated wall).

Let a fluid flow in the canal shown in Fig. 6.4.3, where Ω is the domain of the canal, S is the boundary of Ω, $S = \bigcup_{i=1}^{3} \overline{S}_i$, S_i's are open sets in S, S_1 is the hard wall, S_2 the inlet, and S_3 the outlet. The filtration layer is attached to the inlet. We suppose that the function of surface forces $\widetilde{F} = (\widetilde{F}_1, \dots, \widetilde{F}_n)$ acting on the external surface of the filtration layer S_2' is known. The functions of velocity v and pressure p satisfy the equations (6.1.6). Suppose we are given also a function $P = (P_1, \dots, P_n)$ of the surface forces at the outlet and zero velocities on the hard wall, i.e.,

$$v \restriction_{S_1} = 0, \tag{6.4.24}$$

$$\left[-p\delta_{ij} + 2\varphi(I(v))\varepsilon_{ij}(v)\right]\nu_j \restriction_{S_3} = P_i, \qquad i = 1, \dots, n, \tag{6.4.25}$$

ν_i being the components of the unit outward normal ν to S. Notice that, instead of (6.4.25), the conditions (6.1.9), (6.1.10) may be considered. In the latter case, the functions $\hat{F} = (\hat{F}_1, \dots, \hat{F}_n)$ and λ are supposed to be known.

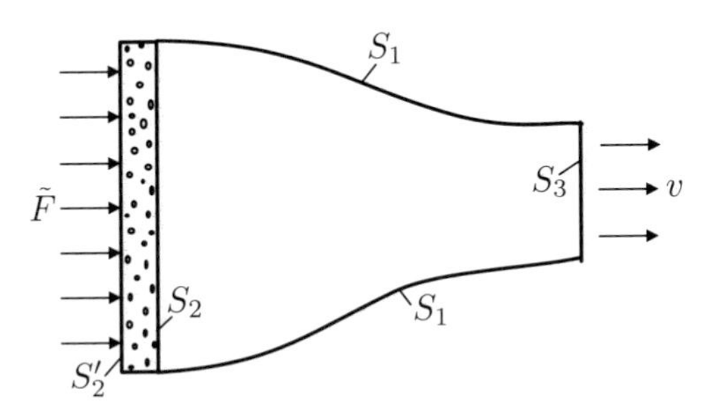

Figure 6.4.3: Domain Ω of the canal with filtration layer (perforated wall) placed at the inlet

On S_2, we have the following conditions of filtration (see (6.1.9), (6.1.10))

$$\big[-p\delta_{ij} + 2\varphi(I(v))\varepsilon_{ij}(v)\big]\nu_j\nu_i \restriction_{S_2} - \widetilde{F}_i\nu_i \restriction_{S_2} = -\tilde{\lambda}(v_\nu^2, \cdot)v_\nu \restriction_{S_2}, \tag{6.4.26}$$

$$(v_i - v_\nu \nu_i) \restriction_{S_2} = 0, \qquad i = 1, \dots, n. \tag{6.4.27}$$

Here, we suppose that the thickness of the filtration layer is small, and carry the function $\widetilde{F}$ from S_2' to S_2, see Fig. 6.4.3. We also assume that $\widetilde{F} \in C(\overline{S}_2)^n$.

Consider $\tilde{\lambda}$ as an unknown function that should be calculated so that the normal component of the function of surface forces is equal to q on S_2, i.e.,

$$\big[-p\delta_{ij} + 2\varphi(I(v))\varepsilon_{ij}(v)\big]\nu_j\nu_i \restriction_{S_2} = q, \tag{6.4.28}$$

where q is a continuous function on S_2. In this case, we suppose that

$$q(s) - \widetilde{F}_i(s)\nu_i(s) \geq \beta, \qquad s \in S_2, \tag{6.4.29}$$

β a small positive constant.

Consider the problem: Find a pair of functions v, p satisfying the equations (6.1.6) and the boundary conditions (6.4.24), (6.4.25), (6.4.27), (6.4.28). By analogy with the problem solved above, see Section 6.1, it may be shown that, if $K \in L_2(\Omega)^n$, $P \in L_2(S_3)^n$, $q \in C(\overline{S}_2)$, then there exists a unique solution of this problem such that $v \in H^1(\Omega)^n$, $p \in L_2(\Omega)$. We denote it by $v(q)$, $p(q)$. Let $v(q)_\nu$ be the normal component of $v(q)$, i.e., $v(q)_\nu = v(q)_i\nu_i$. Suppose that $v(q)_\nu < 0$ a.e. on S_2, the inlet. Now, we define a filtration function $\tilde{\lambda}\colon \mathbb{R}_+ \times S_2 \to \mathbb{R}$ at points $\big((v(q)_\nu(s))^2, s\big)$, $s \in S_2$, by

$$\tilde{\lambda}((v(q)_\nu)(s))^2, s) = -\frac{q(s) - \widetilde{F}_i(s)\nu_i(s)}{v(q)_\nu(s)}, \qquad s \in S_2\,. \tag{6.4.30}$$

It follows from (6.4.28) and (6.4.30) that the pair $v(q)$, $p(q)$ satisfies (6.4.26). Thus, if the filtration layer meets (6.4.30), then (6.4.28) is satisfied.

6.5 The flow in a canal with a perforated wall placed inside

6.5.1 Basic equations

In Section 6.1, we considered the problem of flow of a nonlinear viscous fluid in the case when the filtration layer (perforated wall) was put on the boundary of the canal. Now, we will consider the case of the filtration layer placed inside of the canal.

Let Ω be a bounded domain with a Lipschitz continuous boundary S. Suppose Ω_1 and Ω_2 are open subsets of Ω such that $\overline{\Omega} = \overline{\Omega}_1 \cup \overline{\Omega}_2$, $\Omega_1 \cap \Omega_2 = \varnothing$, and $\Gamma = \overline{\Omega}_1 \cap \overline{\Omega}_2$ is an $(n-1)$-dimensional, Lipschitz continuous manifold., see Fig. 6.5.1.

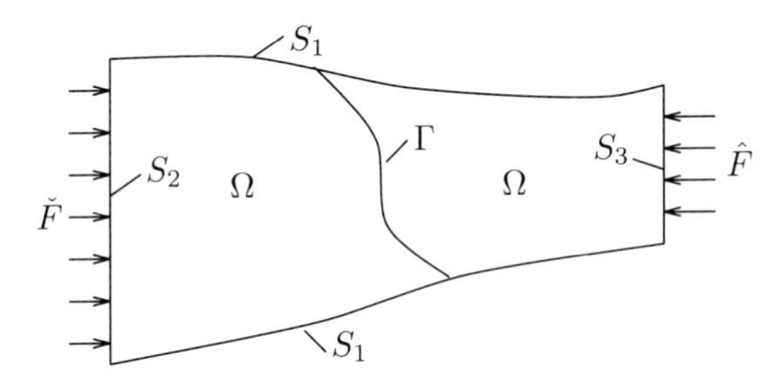

Figure 6.5.1: Notation for flow in a canal with filtration layer Γ placed inside

We assume that S_1, S_2, S_3 are open subsets of S such that $S = \bigcup_{i=1}^{3} \overline{S}_i$, $S_i \cap S_j = \varnothing$ for $i \neq j$. We consider Γ and S_3 as filtration layers and S_1 as a hard wall. Let us consider the following problem: Find a pair of functions v, p (velocity and pressure) satisfying the equations of motion and incompressibility:

$$\frac{\partial p}{\partial x_i} - 2\frac{\partial}{\partial x_j}\big(\varphi(I(v))\varepsilon_{ij}(v)\big) = K_i \qquad \text{in } \Omega_1 \cup \Omega_2, \ i = 1, \dots, n, \tag{6.5.1}$$

$$\operatorname{div} v = 0 \qquad \text{in } \Omega_1 \cup \Omega_2, \tag{6.5.2}$$

and the following boundary conditions on S_i:

$$v \restriction_{S_1} = 0, \tag{6.5.3}$$

$$\big[-p\delta_{ij} + 2\varphi(I(v))\varepsilon_{ij}(v)\big]\nu_j \restriction_{S_2} = \check{F}_i, \qquad i = 1, \dots, n, \tag{6.5.4}$$

$$\big[-p\delta_{ij} + 2\varphi(I(v))\varepsilon_{ij}(v)\big]\nu_j\nu_i \restriction_{S_3} - \hat{F}_i\nu_i \restriction_{S_3} = -\lambda(v_\nu^2, \cdot)v_\nu \restriction_{S_3}, \tag{6.5.5}$$

$$(v_i - v_\nu\nu_i) \restriction_{S_3} = 0, \qquad i = 1, \dots, n, \tag{6.5.6}$$

where $\nu = (\nu_1, \dots, \nu_n)$ is the unit outward normal to S, $v_\nu = v_k\nu_k$.

Put $v^{(i)} = v$, $p^{(i)} = p$ in Ω_i, $\nu^{(i)} = (\nu_1^{(i)}, \dots, \nu_n^{(i)})$ is the unit normal to Γ directed outwards Ω_i, $v_\nu^{(i)} = v_k^{(i)}\nu_k^{(i)}$ on Γ. The boundary conditions of filtration on Γ are the following:

$$v^{(1)} \restriction_\Gamma = v^{(2)} \restriction_\Gamma, \tag{6.5.7}$$

$$(v^{(1)} - v_\nu^{(1)}\nu^{(1)}) \restriction_\Gamma = 0, \tag{6.5.8}$$

$$\begin{aligned}&\big[-p^{(1)}\delta_{ij} + 2\varphi(I(v^{(1)}))\varepsilon_{ij}(v^{(1)})\big]\nu_j^{(1)}\nu_i^{(1)} \restriction_\Gamma \\ &+ \big[-p^{(2)}\delta_{ij} + 2\varphi(I(v^{(2)}))\varepsilon_{ij}(v^{(2)})\big]\nu_j^{(2)}\nu_i^{(1)} \restriction_\Gamma = -\tilde{\lambda}((v_\nu^{(1)})^2, \cdot)v_i^{(1)}\nu_i^{(1)}.\end{aligned} \tag{6.5.9}$$

It follows from (6.5.7) and (6.5.8) that

$$(v^{(2)} - v_\nu^{(2)}\nu^{(2)}) \restriction_\Gamma = 0.$$

6.5.2 Generalized solution of the problem

Define spaces X and V as follows:

$$X = \{ u \mid u \in H^1(\Omega)^n,\ u \restriction_{S_1} = 0,\ (u - u_\nu \nu) \restriction_{S_3} = 0,\ (u - u_\nu^{(1)} \nu^{(1)}) \restriction_\Gamma = 0 \}, \tag{6.5.10}$$

$$V = \{ u \mid u \in X,\ \operatorname{div} u = 0 \}. \tag{6.5.11}$$

Note that, in (6.5.10), the last equality is equivalent to $(u - u_\nu^{(2)} \nu^{(2)}) \restriction_\Gamma = 0$. The expression (6.1.21) defines norms in X, V that are equivalent to the norm of $H^1(\Omega)^n$. X and V are Hilbert spaces with the scalar product

$$(u, h) = \int_\Omega \varepsilon_{ij}(u) \varepsilon_{ij}(h)\, dx.$$

Introduce a set $\mathcal{B}_3$ as follows:

$$\mathcal{B}_3 = \{ \lambda \mid \lambda \in L_\infty(\Gamma; C^1(\mathbb{R}_+)),\ \lambda \text{ satisfies the conditions (6.1.16)–(6.1.18)} \\ \text{with positive constants } b_1,\ b_2 \text{ for an arbitrary } (\alpha, s) \in \mathbb{R}_+ \times \Gamma \}. \tag{6.5.12}$$

This set is equipped with the topology generated by that of $L_\infty(\Gamma; C^1(\mathbb{R}_+))$. Suppose that $\varphi \in \mathcal{B}_1$, $\lambda \in \mathcal{B}_2$, $\tilde{\lambda} \in \mathcal{B}_3$, $\mathcal{B}_1$ and $\mathcal{B}_2$ defined by (6.2.5), (6.2.8).

Define also operators $L\colon X \to X^*$ and $B \in \mathcal{L}(X, L_2(\Omega))$ by

$$(L(u), h) = 2 \int_\Omega \varphi(I(u)) \varepsilon_{ij}(u) \varepsilon_{ij}(h)\, dx + \int_\Gamma \tilde{\lambda}(|u|^2, \cdot) u_i h_i\, ds + \int_{S_3} \lambda(|u|^2, \cdot) u_i h_i\, ds, \\ Bu = \operatorname{div} u. \tag{6.5.13}$$

Here, $|u|^2 = \sum_{i=1}^n u_i^2$. We suppose

$$K = (K_1, \dots, K_n) \in X^*, \qquad \check{F} = (\check{F}_1, \dots, \check{F}_n) \in X^*, \\ \hat{F} = (\hat{F}_1, \dots, \hat{F}_n) \in X^*. \tag{6.5.14}$$

In particular, if $K \in L_2(\Omega)^n$, $\check{F} \in L_2(S_2)^n$, $\hat{F} \in L_2(S_3)^n$, then

$$(K, h) = \int_\Omega K_i h_i\, ds, \qquad (\check{F}, h) = \int_{S_2} \check{F}_i h_i\, ds, \qquad (\hat{F}, h) = \int_{S_3} \hat{F}_i h_i\, ds, \\ h \in X. \tag{6.5.15}$$

The generalized solution of the problem (6.5.1)–(6.5.9) is defined as a pair of functions v, p satisfying

$$(v, p) \in X \times L_2(\Omega), \tag{6.5.16}$$

$$(L(v), h) - (B^* p, h) = (K + \check{F} + \hat{F}, h), \qquad h \in X, \tag{6.5.17}$$

$$(Bv, q) = 0, \qquad q \in L_2(\Omega. \tag{6.5.18}$$

By using Green's formula, one can show that, if v, p is a solution to the problem (6.5.16)–(6.5.18), then v, p is a solution to the problem (6.5.1)–(6.5.9) in the distribution sense. On the contrary, if v, p is a solution to (6.5.1)–(6.5.9) that satisfies (6.5.16), then v, p is a solution to (6.5.16)–(6.5.18).

By analogy with the above, see Section 6.1, one proves the following

Theorem 6.5.1 *Suppose sets* $\mathcal{B}_1$, $\mathcal{B}_2$, $\mathcal{B}_3$ *are defined by* (6.2.5), (6.2.8), (6.5.12), *and* $\varphi \in \mathcal{B}_1$, $\lambda \in \mathcal{B}_2$, $\tilde{\lambda} \in \mathcal{B}_3$. *Assume also that operators* L *and* B *are defined by* (6.5.13) *and the functions of volume and surface forces* K, $\check{F}$, $\hat{F}$ *meet* (6.5.14). *Then, there exists a unique solution* v, p *of the problem* (6.5.16)–(6.5.18).

Let $\{X_k\}$, $\{M_k\}$ *be sequences of finite-dimensional subspaces of* X *and* $L_2(\Omega)$ *such that* (6.1.37), (6.1.38), (6.1.42) *hold, and* $X_k \subset X_{k+1}$, $M_k \subset M_{k+1}$ *for each* k. *Then, for any* k, *there exists a unique solution* v_k, p_k *of the problem* (6.1.57)–(6.1.59) *and* $v_k \to v$ *in* X, $p_k \to p$ *in* $L_2(\Omega)$.

The solution of the problem (6.5.16)–(6.5.18) depends on the functions of loading K, $\check{F}$, $\hat{F}$ and on the functions of viscosity and filtration φ, λ, $\tilde{\lambda}$, so that we denote the solution by $v(K, \check{F}, \hat{F}, \varphi, \lambda, \tilde{\lambda})$, $p(K, \check{F}, \hat{F}, \varphi, \lambda, \tilde{\lambda})$. By Theorem 6.5.1, the following function G is well defined:

$$\begin{aligned}(X^*)^3 \times \mathcal{B}_1 \times \mathcal{B}_2 \times \mathcal{B}_3 \ni (K, \check{F}, \hat{F}, \varphi, \lambda, \tilde{\lambda}) &\to G(K, \check{F}, \hat{F}, \varphi, \lambda, \tilde{\lambda}) \\ &= (v(K, \check{F}, \hat{F}, \varphi, \lambda, \tilde{\lambda}), p(K, \check{F}, \hat{F}, \varphi, \lambda, \tilde{\lambda})) \in V \times L_2(\Omega). \qquad (6.5.19)\end{aligned}$$

By analogy with the proof of Theorem 6.2.1, we get

Theorem 6.5.2 *Suppose a function* G *is given by* (6.5.19) *and sets* $\mathcal{B}_1$, $\mathcal{B}_2$, $\mathcal{B}_3$ *are defined by* (6.2.5), (6.2.8), (6.5.12) *and equipped with the topologies generated by those of* $C^1(\mathbb{R}_+)$, $L_\infty(S_3; C^1(\mathbb{R}_+))$, *and* $L_\infty(\Gamma; C^1(\mathbb{R}_+))$, *respectively. Then, the function* G *is a continuous mapping of* $(X^*)^3 \times \mathcal{B}_1 \times \mathcal{B}_2 \times \mathcal{B}_3$ *into* $V \times L_2(\Omega)$.

6.6 Optimization by the functions of surface forces and filtration

6.6.1 Formulation of the problem and the existence theorem

We again consider the case of flow studied in the previous section, i.e., the problem (6.5.16)–(6.5.18) with L defined by (6.5.13). We assume the functions of surface forces $\check{F}$, $\hat{F}$ and the functions of filtration λ, $\tilde{\lambda}$ are controls, while the functions of viscosity φ and volume forces K are fixed, and

$$\varphi \in \mathcal{B}_1, \quad K \in X^*, \qquad (6.6.1)$$

$\mathcal{B}_1$ defined by (6.2.5). As shown in Section 6.4, the functions of filtration and surface forces may be varied by changing the number of the holes in the perforated wall, their dimensions and dispositions. Thus, the number of the holes, the coordinates of the centers of the holes, and their diameters can be considered as controls. But it is much more convenient to consider λ, $\tilde{\lambda}$ as controls which are functions of points of S_3 and Γ only. Such an approach does not diminish the generality. For if λ_0, $\tilde{\lambda}_0$ are the optimal filtration functions given on S_3 and Γ, $\check{F}_0$, $\hat{F}_0$ are the optimal functions of surface forces, and v_0, p_0 is the solution of the problem (6.5.16)–(6.5.18) for $\lambda = \lambda_0$, $\tilde{\lambda} = \tilde{\lambda}_0$, $\check{F} = \check{F}_0$, $\hat{F} = \hat{F}_0$, then we can put

$$\lambda_1(|v_0(s)|^2, s) = \lambda_0(s), \qquad s \in S_3, \tag{6.6.2}$$

$$\tilde{\lambda}_1(|v_0(s)|^2, s) = \tilde{\lambda}_0(s), \qquad s \in \Gamma. \tag{6.6.3}$$

So, λ_1 and $\tilde{\lambda}_1$ are the optimal functions defined at points $(|v_0(s)|^2, s)$, $s \in S_3$ and $s \in \Gamma$. The perforated walls may be computed so as to obtain good approximations of the functions λ_1 and $\tilde{\lambda}_1$, see Section 6.4.

We introduce a set of controls by

$$\begin{aligned} N = \big\{\, T \,|\, T = (\check{F}, \hat{F}, \lambda, \tilde{\lambda}) \in H^{-\frac{1}{2}}(S_2)^n \times H^{-\frac{1}{2}}(S_3)^n \times C(\overline{S}_3) \times C(\overline{\Gamma}), \\ \lambda(s) \geq 0,\ s \in \overline{S}_3,\ \tilde{\lambda}(s) \geq 0,\ s \in \overline{\Gamma} \,\big\}. \end{aligned} \tag{6.6.4}$$

The set N is equipped with the topology generated by that of the product

$$H^{-\frac{1}{2}}(S_2)^n \times H^{-\frac{1}{2}}(S_3)^n \times C(\overline{S}_3) \times C(\overline{\Gamma}).$$

Suppose we are given functionals Ψ_i satisfying the condition

$$\Psi_i \text{ are functionals continuous on } N \times X \times L_2(\Omega),\ i = 0, 1, \ldots, m. \tag{6.6.5}$$

By Theorem 6.5.1, for each $T \in N$, there exists a unique solution of the problem (6.5.16)–(6.5.18), which will be denoted by $v(T)$, $p(T)$. On the set N, we define functionals Φ_i as follows:

$$\Phi_i(T) = \Psi_i(T, v(T), p(T)), \qquad i = 0, 1, \ldots, m. \tag{6.6.6}$$

Determine a set of admissible controls by

$$\begin{aligned} N_{\mathrm{ad}} = \big\{\, T \,|\, T = (\check{F}, \hat{F}, \lambda, \tilde{\lambda}) \in N,\ \check{F} = (\check{F}_1, \ldots, \check{F}_n) \in L_2(S_2)^n,\ \|\check{F}\|_{L_2(S_2)^n} \leq c_1, \\ \hat{F} = (\hat{F}_1, \ldots, \hat{F}_n) \in L_2(S_3)^n,\ \|\hat{F}\|_{L_2(S_3)^n} \leq c_2,\ \lambda \in W_p^l(S_3), \\ \|\lambda\|_{W_p^l(S_3)} \leq c_3,\ \tilde{\lambda} \in W_p^l(\Gamma),\ \|\tilde{\lambda}\|_{W_p^l(\Gamma)} \leq c_4,\ p > 1,\ lp > n-1, \\ \Phi_i(T) \leq 0,\ i = 1, \ldots, r,\ \Phi_i(T) = 0,\ i = r+1, \ldots, m,\ r \leq m \,\big\}. \end{aligned} \tag{6.6.7}$$

Moreover, we suppose that S_3 and Γ are $(n-1)$-dimensional manifolds of the C^l class.

The optimization problem consists in finding T_0 such that

$$T_0 = (\check{F}_0, \hat{F}_0, \lambda_0, \tilde{\lambda}_0) \in N_{\mathrm{ad}}, \qquad \Phi_0(T_0) = \inf_{T \in N_{\mathrm{ad}}} \Phi_0(T). \tag{6.6.8}$$

Theorem 6.6.1 *Suppose the conditions* (6.6.1), (6.6.5), *and* (6.6.6) *are satisfied and a nonempty set* N_{ad} *is defined by* (6.6.7). *Then, there exists a solution of the problem* (6.6.8).

Proof. It follows from Theorem 6.5.2 that the function $N \ni T \to (v(T), p(T)) \in V \times L_2(\Omega)$ is continuous. Hence, Φ_i, $i = 0, 1, \dots, m$, are functionals continuous on N, and N_{ad} is a compactum in N. So, there exists a solution of the problem (6.6.8).

Let us consider some forms of the functionals Φ_i. The goal functional can be taken in the form

$$\Phi_0(T) = \int_{S_3} \sum_{i=1}^{n} (v(T)_i - z_i)^2 \, ds, \tag{6.6.9}$$

where z_i's are given functions from $L_2(S_3)$. Hence, in this case, the velocity field at the outflow should be close to $z = (z_1, \dots, z_n)$. Particularly, for the header of the paper machine and for the extrusion head, the function z may be taken in the form

$$z_i \nu_i = \text{const} > 0, \qquad z_k - z_i \nu_i \nu_k = 0, \qquad k = 1, \dots, n. \tag{6.6.10}$$

Thus, z is normal to S_3 and z is a constant vector.

In some cases, it is necessary to obtain a velocity field that is close to some given one in the domain Ω. The goal functional is taken then in the form

$$\Phi_0(T) = \int_{\Omega} \left(\sum_{i=1}^{n} (v(T)_i - y_i)^2 \right) dx, \tag{6.6.11}$$

where y_i's are prescribed functions from $L_2(\Omega)$.

The functional of restriction Φ_1 may be defined by

$$\Phi_1(T) = \int_{\Omega} \sum_{i,j=1}^{n} \varphi(I(v(T))) \left[\varepsilon_{ij}(v(T)) \right]^2 dx - c. \tag{6.6.12}$$

The condition $\Phi_1(T) \le 0$ means now that the power must not exceed c.

Next, we are going to consider the optimization problem (6.6.8) in which the state of the system $v(T)$, $p(T)$ is calculated approximately by the Galerkin method. To this end, we will derive the Fréchet derivatives of the functionals Φ_i and necessary optimality conditions. So, let us deduce now the Fréchet derivatives of the function $T \to (v(T), p(T))$.

6.6.2 On the differentiability of the function $T \to (v(T), p(T))$

Define an operator $L_1 \colon X \to X^*$ by

$$(L_1(u), h) = 2 \int_{\Omega} \varphi(I(u)) \varepsilon_{ij}(u) \varepsilon_{ij}(h) \, dx, \qquad u, h \in X. \tag{6.6.13}$$

Let $\{X_k\}$ be a sequence of finite-dimensional subspaces of X that satisfies (6.1.37) and

$$X_k \in H^1_\infty(\Omega)^n \qquad \forall k. \tag{6.6.14}$$

Lemma 6.6.1 *Let a function φ satisfy the conditions* (6.1.11)–(6.1.14). *Let also $\{X_k\}$ be a sequence of finite-dimensional subspaces of X satisfying* (6.6.14). *Then, for any k, the operator L_1 considered as a mapping of X_k into X_k^* is continuously Fréchet differentiable in X_k, and at any point $u \in X_k$ its derivative $L_1'(u)$ is given by*

$$(L_1'(u)w,h) = 2\int_\Omega \Big[\varphi(I(u))\varepsilon_{ij}(w)\varepsilon_{ij}(h) + 2\frac{d\varphi}{d\alpha}(I(u))\varepsilon_{lm}(u)\varepsilon_{lm}(w)\varepsilon_{ij}(u)\varepsilon_{ij}(h)\Big]\,dx, \qquad w,h \in X_k. \tag{6.6.15}$$

Moreover,

$$(L_1'(u)h,h) \geq \mu\|h\|_X^2, \qquad \mu = 2\min(a_1,a_3), \tag{6.6.16}$$

where a_1, a_3 are positive constants from (6.1.12), (6.1.13).

Proof. In the subspace X_k, the norm of X is equivalent to that of $H^1_\infty(\Omega)^n$, because X_k is finite-dimensional and $X_k \subset H^1_\infty(\Omega)^n$. So, by using (6.1.11), we conclude

$$\|\varphi(I(u+w))\varepsilon_{ij}(u+w) - \varphi(I(u))\varepsilon_{ij}(u) - \varphi(I(u))\varepsilon_{ij}(w) - 2\,\frac{d\varphi}{d\alpha}(I(u))\varepsilon_{lm}(u)\varepsilon_{lm}(w)\varepsilon_{ij}(u)\|_{L_\infty(\Omega)} \leq \omega(w)\|w\|_X, \qquad u,w \in X_k, \tag{6.6.17}$$

where $\omega(w) \to 0$ as $\|w\|_X \to 0$. Thus, (6.6.15) follows from (6.6.17). By applying the estimate

$$\big(\varepsilon_{ij}(u)\varepsilon_{ij}(h)\big)^2 \leq I(u)I(h),$$

we get (6.6.16) from (6.6.15) and (6.1.12)–(6.1.14), see (6.1.54).

Remark 6.6.1 One can show that the operator L_1 considered as a mapping of X into X^* is Gâteaux differentiable in X and its Gâteaux derivative is given by (6.6.15). The estimate (6.6.16) holds also for arbitrary $u, h \in X$.

Define a space U and a set N_1 by

$$U = H^{-\frac{1}{2}}(S_2)^n \times H^{-\frac{1}{2}}(S_3)^n \times C(\overline{S}_3) \times C(\overline{\Gamma}), \tag{6.6.18}$$

$$N_1 = \big\{ T \mid T = (\check{F},\hat{F},\lambda,\tilde{\lambda}) \in U,\ \lambda(s) > -\xi,\ s \in \overline{S}_3,\ \tilde{\lambda}(s) > -\xi,\ s \in \overline{\Gamma} \big\}. \tag{6.6.19}$$

Here, ξ is a small positive constant which is defined so that

$$(L_1'(u)h,h) + \int_\Gamma \tilde{\lambda}\sum_{i=1}^n h_i^2\,ds + \int_{S_3}\lambda\sum_{i=1}^n h_i^2\,ds \geq \mu_1\|h\|_X^2,$$
$$u,h\in X,\ \lambda\in C(\overline{S}_3),\ \lambda > -\xi,\ \tilde{\lambda}\in C(\overline{\Gamma}),\ \tilde{\lambda} > -\xi,\ \mu_1 = \text{const} > 0. \tag{6.6.20}$$

By virtue of (6.6.16) and the embedding theorem, there exists a constant μ_1 for which (6.6.20) holds.

We now determine a mapping $Q\colon N_1\times X\times L_2(\Omega)\to X^*\times L_2(\Omega)^*$ by

$$T = (\check{F},\hat{F},\lambda,\tilde{\lambda})\in N_1,\qquad (u,h)\in X^2,\qquad (p,q)\in L_2(\Omega)^2,$$
$$\big(Q(T,u,p),(h,q)\big)$$
$$= \begin{cases} (L_1(u),h) + \int_\Gamma \tilde{\lambda}u_ih_i\,ds + \int_{S_3}\lambda u_ih_i\,ds - \int_\Omega p\operatorname{div}h\,dx - (K+\check{F}+\hat{F},h),\\ \int_\Omega q\operatorname{div}u\,dx. \end{cases} \tag{6.6.21}$$

Lemma 6.6.2 *Let (6.6.1) hold, let $\{X_k\}$, $\{M_k\}$ be sequences of finite-dimensional subspaces of X and $L_2(\Omega)$, respectively, let $X_k\subset H^1_\infty(\Omega)^n$ for each k, and let (6.1.42) hold. Then, the operator Q considered as a mapping of $N_1\times X_k\times M_k$ into $X_k^*\times M_k^*$ is continuously Fréchet differentiable and its derivative is given by*

$$Q'(T,u,p)(\Delta T,\Delta u,\Delta p) = \frac{\partial Q}{\partial T}(T,u,p)\Delta T + \frac{\partial Q}{\partial(u,p)}(T,u,p)(\Delta u,\Delta p),$$
$$\Delta T = (\Delta\check{F},\Delta\hat{F},\Delta\lambda,\Delta\tilde{\lambda})\in U,\ \ \Delta u\in X_k,\ \ \Delta p\in M_k,$$
$$\left(\frac{\partial Q}{\partial T}(T,u,p)\Delta T,\ (h,q)\right) = \begin{cases}\int_\Gamma \Delta\tilde{\lambda}\,u_ih_i\,ds + \int_{S_3}\Delta\lambda\,u_ih_i\,ds - (\Delta\check{F}+\Delta\hat{F},h),\\ 0,\end{cases}$$
$$\left(\frac{\partial Q}{\partial(u,p)}(T,u,p)(\Delta u,\Delta p),\ (h,q)\right)$$
$$= \begin{cases}(L_1'(u)\Delta u,h) + \int_\Gamma \tilde{\lambda}\Delta u_i\,h_i\,ds + \int_{S_3}\lambda\Delta u_i\,h_i\,ds - \int_\Omega \Delta p\operatorname{div}h\,dx,\\ \int_\Omega q\operatorname{div}\Delta u\,dx,\end{cases}$$
$$h\in X_k,\ \ q\in M_k. \tag{6.6.22}$$

At any point $(T,u,p)\in N_1\times X_k\times M_k$, the operator $\frac{\partial Q}{\partial(u,p)}(T,u,p)$ is an isomorphism of $X_k\times M_k$ onto $X_k^*\times M_k^*$.

Proof. For an arbitrary fixed $(u,p)\in X_k\times M_k$, the function $T\to Q(T,u,p)$ is a linear continuous mapping of an open subset N_1 of U into $X_k^*\times M_k^*$. So, the continuous Fréchet differentiability of the mapping $Q\colon N_1\times X_k\times M_k\to X_k^*\times M_k^*$ and the formulas (6.6.22) follow from (6.6.21) and Lemma 6.6.1.

Let (y, z) be an arbitrary point of $X_k^* \times M_k^*$. Consider the problem: Find a pair of functions w, ψ satisfying

$$\begin{aligned} &(w, \psi) \in X_k \times M_k, \\ &\frac{\partial Q}{\partial(u,p)}(T,u,p)(w,\psi) = (y,z), \end{aligned} \tag{6.6.23}$$

where $(T, u, p) \in N_1 \times X_k \times M_k$.

Introduce operators $L_2 \in \mathcal{L}(X_k, X_k^*)$ and $B_k \in \mathcal{L}(X_k, M_k^*)$ by

$$\begin{aligned} (L_2 v, h) &= (L_1'(u)v, h) + \int_\Gamma \tilde{\lambda} v_i h_i \, ds + \int_{S_3} \lambda v_i h_i \, ds, \\ (B_k v, \gamma) &= \int_\Omega \gamma \operatorname{div} v \, dx. \end{aligned} \tag{6.6.24}$$

By (6.6.22), the problem (6.6.23) may be rewritten as follows:

$$(w, \psi) \in X_k \times M_k, \tag{6.6.25}$$

$$(L_2 w, h) - (B_k^* \psi, h) = (y, h), \qquad h \in X_k, \tag{6.6.26}$$

$$(B_k w, q) = (z, q), \qquad q \in M_k. \tag{6.6.27}$$

From Lemma 6.1.3, we infer the existence of a unique $w_0 \in V_k^\perp$ such that

$$B_k w_0 = z, \qquad \|w_0\|_X \leq \beta^{-1} \|z\|_{M_k^*}. \tag{6.6.28}$$

Thus, the problem (6.6.25)–(6.6.27) reduces to the following one: Find w_1 satisfying

$$w_1 = w - w_0 \in V_k, \tag{6.6.29}$$

$$(L_2 w_1, h) = (y, h) - (L_2 w_0, h), \qquad h \in V_k. \tag{6.6.30}$$

From (6.6.20) and (6.6.24)

$$(L_2 h, h) \geq \mu_1 \|h\|_X^2, \qquad h \in X_k. \tag{6.6.31}$$

Thus, by the Lax-Milgram theorem (Theorem 1.5.2) and (6.6.28), we conclude the existence of a unique solution w_1 of the problem (6.6.29), (6.6.30) and

$$\|w_1\|_X \leq c(\|y\|_{X_k^*} + \|z\|_{M_k^*}), \tag{6.6.32}$$

where the constant c is independent of y, z, and k. Now, $L_2 w - y \in V_k^0$ (see (6.1.41)), therefore, according to Lemma 6.1.3, there exists a unique $\psi \in M_k$ such that

$$\begin{aligned} &B_k^* \psi = L_2 w - y, \\ &\|\psi\|_{L_2(\Omega)} \leq \beta^{-1} \|L_2 w - y\|_{X_k^*} \leq c_1(\|y\|_{X_k^*} + \|z\|_{M_k^*}). \end{aligned} \tag{6.6.33}$$

Thus, the operator $\dfrac{\partial Q}{\partial(u,p)}(T,u,p)$ is an isomorphism of $X_k \times M_k$ onto $X_k^* \times M_k^*$.

Theorem 6.6.2 *Let* $\{X_k\}$, $\{M_k\}$ *be sequences of finite-dimensional subspaces of* X *and* $L_2(\Omega)$, X *defined by* (6.5.10). *Let also* $X_k \subset H^1_\infty(\Omega)^n$ *for each* k *and* (6.1.42), (6.6.1) *hold. For an arbitrary fixed* k, *define a function* $\mathcal{F}\colon N_1 \to (X_k, M_k)$ *as follows:*

$$T = (\check{F}, \hat{F}, \lambda, \tilde{\lambda}) \in N_1, \qquad \mathcal{F}(T) = \big(v(T), p(T)\big) \in X_k \times M_k, \tag{6.6.34}$$

where $v(T)$, $p(T)$ *is the solution of the problem*

$$\begin{aligned}(L_1(v(T)), h) &+ \int_\Gamma \tilde{\lambda} v(T)_i h_i \, ds + \int_{S_3} \lambda v(T)_i h_i \, ds \\ &- \int_\Omega p(T) \operatorname{div} h \, dx = (K + \check{F} + \hat{F}, h), \qquad h \in X_k,\end{aligned} \tag{6.6.35}$$

$$\int_\Omega q \operatorname{div} v(T) \, dx = 0, \qquad q \in M_k. \tag{6.6.36}$$

Then, $\mathcal{F}$ *is a continuously Fréchet differentiable mapping of* N_1 *into* $X_k \times M_k$, *and its derivative is given by*

$$\mathcal{F}'(T) = -\left(\frac{\partial Q}{\partial(u,p)}(T, v(T), p(T))\right)^{-1} \circ \frac{\partial Q}{\partial T}(T, v(T), p(T)), \tag{6.6.37}$$

where $\left(\frac{\partial Q}{\partial(u,p)}(T, v(T), p(T))\right)^{-1}$ *is the inverse mapping of* $\frac{\partial Q}{\partial(u,p)}(T, v(T), p(T))$.

Proof. It follows from (6.6.21) and (6.6.34)–(6.6.36) that $\mathcal{F}$ is the implicit function defined by the equation

$$Q(T, \mathcal{F}(T)) = 0, \tag{6.6.38}$$

where Q is considered as a mapping of $N_1 \times X_k \times M_k$ into $X_k^* \times M_k^*$. So, the theorem follows from Lemma 6.6.2 and the theorem on the differentiability of an implicit function (Theorem 1.9.2).

Remark 6.6.2 In Theorem 6.6.2, by using the implicit function theorem, we have proved the Fréchet differentiability of the function $\mathcal{F}\colon T \to \mathcal{F}(T) = (v(T), p(T))$, where $v(T)$, $p(T)$ is the Galerkin approximation of the solution of the problem (6.5.16)–(6.5.18), i.e., the solution of the problem (6.6.34)–(6.6.36) for an arbitrary fixed k. The questions of both differentiability of the function $T \to (v(T), p(T))$, where $v(T)$, $p(T)$ is the solution of (6.5.16)–(6.5.18), and necessary optimality conditions for this case are open yet. Apparently, it could happen to be useful to apply the concept of an extended differentiability that is introduced and applied to the solution of optimization problems for nonlinear partial differential equations in works by Serovaiskii (1991), (1993a, b).

6.6.3 Differentiability of the functionals Φ_i and necessary optimality conditions

We suppose now that the functionals Ψ_i satisfy the following condition, which is stronger than (6.6.5),

$$\left.\begin{array}{l}\Psi_i \text{ are functionals continuously Fréchet differentiable in} \\ N_1 \times X \times L_2(\Omega),\ i = 0, 1, \dots, m.\end{array}\right\} \tag{6.6.39}$$

Theorem 6.6.3 *Let $\{X_k\}$, $\{M_k\}$ be sequences of finite-dimensional subspaces of X and $L_2(\Omega)$, X defined by (6.5.10). Let also $X_k \subset H^1_\infty(\Omega)^n$ for each k and let (6.1.42), (6.6.1) hold. Suppose that $v(t)$, $p(T)$ is the solution to the problem (6.6.34)–(6.6.36), and (6.6.39) hold. Let functionals Φ_i be defined by (6.6.6). Then, Φ_i are continuously Fréchet differentiable in N_1 and at any point $T = (\check{F}, \hat{F}, \lambda, \tilde{\lambda}) \in N_1$ the Fréchet derivative of Φ_i is given by*

$$\begin{aligned}\Phi_i'(T)\Delta T = {} & \frac{\partial \Psi_i}{\partial T}(T, v(T), p(T))\Delta T \\ & + \left((v^{(i)}, p^{(i)}), \frac{\partial Q}{\partial T}(T, v(T), p(T))\Delta T\right), \qquad \Delta T \in U,\end{aligned} \tag{6.6.40}$$

where $v^{(i)}$, $p^{(i)}$ is the solution to the problem

$$\begin{gathered}(v^{(i)}, p^{(i)}) \in X_k \times M_k, \\ \left(\frac{\partial Q}{\partial(u,p)}(T, v(T), p(T))\right)^* (v^{(i)}, p^{(i)}) = -\frac{\partial \Psi_i}{\partial(v,p)}(T, v(T), p(T)).\end{gathered} \tag{6.6.41}$$

Here, $\frac{\partial Q}{\partial T}$ and $\frac{\partial Q}{\partial(u,p)}$ are defined by (6.6.22).

Proof. By virtue of (6.6.6), (6.6.39), and Theorem 6.6.2, the functionals Φ_i are continuously Fréchet differentiable in N_1 and their derivatives are determined by

$$\begin{aligned}\Phi_i'(T)\Delta T = {} & \frac{\partial \Psi_i}{\partial T}(T, v(T), p(T))\Delta T \\ & + \frac{\partial \Psi_i}{\partial(v,p)}(T, v(T), p(T)) \circ \mathcal{F}'(T)\Delta T, \qquad \Delta T \in U.\end{aligned} \tag{6.6.42}$$

Since

$$\frac{\partial \Psi_i}{\partial(v,p)}(T, v(T), p(T)) \in (X_k \times M_k)^*, \qquad \mathcal{F}'(T)\Delta T \in X_k \times M_k,$$

we have

$$\begin{aligned}& \frac{\partial \Psi_i}{\partial(v,p)}(T, v(T), p(T)) \circ \mathcal{F}'(T)\Delta T \\ & = \left(\frac{\partial \Psi_i}{\partial(v,p)}(T, v(T), p(T)), \mathcal{F}'(T)\Delta T\right) = A.\end{aligned} \tag{6.6.43}$$

From (6.6.37), (6.6.41), and (6.6.43)

$$
\begin{aligned}
A &= -\Big(\Big(\frac{\partial Q}{\partial(u,p)}(T,v(T),p(T))\Big)^{*}(v^{(i)},p^{(i)}),\mathcal{F}'(T)\Delta T\Big) \\
&= \Big((v^{(i)},p^{(i)}),\frac{\partial Q}{\partial T}(T,v(T),p(T))\Delta T\Big).
\end{aligned}
\tag{6.6.44}
$$

Now, (6.6.40) follows from (6.6.42) and (6.6.44).

Theorem 6.6.4 *Suppose the conditions of Theorem* 6.6.3 *are satisfied, and let* T_0 *be a solution to the problem* (6.6.8) *with* $v(T)$, $p(T)$ *defined by* (6.6.34)–(6.6.36) *for an arbitrary fixed* k. *Then, there exist constants* β_i *not all equal to zero such that*

$$\sum_{i=0}^{m+4} \beta_i \Phi_i'(T_0)(T-T_0) \geq 0, \qquad T \in N, \tag{6.6.45}$$

$$\beta_i \geq 0, \qquad i = 0,1,\dots,r,m+1,\dots,m+4, \tag{6.6.46}$$

$$\beta_i \Phi_i(T_0) = 0, \qquad i = 1,\dots,r,m+1,\dots,m+4, \tag{6.6.47}$$

where $T = (\check{F},\hat{F},\lambda,\tilde{\lambda})$, $T_0 = (\check{F}_0,\hat{F}_0,\lambda_0,\tilde{\lambda}_0)$ *and*

$$
\begin{aligned}
&\Phi_{m+1}(T) = \|\check{F}\|^2_{L_2(S_2)^n} - c_1^2, \qquad &&\Phi_{m+2}(T) = \|\hat{F}\|^2_{L_2(S_3)^n} - c_2^2, \\
&\Phi_{m+3}(T) = \|\lambda\|^p_{W_p^l(S_3)} - c_3^p, \qquad &&\Phi_{m+4}(T) = \|\tilde{\lambda}\|^p_{W_p^l(\Gamma)} - c_4^p,
\end{aligned}
\tag{6.6.48}
$$

$c_1, \dots, c_4$ *being the constants from* (6.6.7).

Proof. By using the notations (6.6.48), the set N_{ad} from (6.6.7) can be represented in the form

$$
\begin{aligned}
N_{\mathrm{ad}} = \big\{ T \mid T \in N,\ &\Phi_i(T) \leq 0,\ i = 1,\dots,r,m+1,\dots,m+4, \\
&\Phi_i(T) = 0,\ i = r+1,\dots,m,\ r \leq m \big\}.
\end{aligned}
\tag{6.6.49}
$$

It follows from (6.6.4) and (6.6.18) that N is a convex set in U, N_1 is an open set in U, and $N \subset N_1$, see (6.6.19). By Theorem 6.6.3, the functionals Φ_i, $i = 0,1,\dots,m$, are continuously Fréchet differentiable in N_1, and so are the functionals $\Phi_{m+1},\dots,\Phi_{m+4}$ via Theorem 1.10.1. Finally, by applying Theorem 1.12.2, we obtain (6.6.45)–(6.6.47).

6.7 Problems of the optimal shape of a canal

We suppose that the fluid flows in a canal that occupies the following domain G, which is represented in cylindrical coordinates (r,α,z), see Fig. 6.7.1,

$$G = \{ (r,\alpha,z) \mid 0 < z < l,\ 0 \leq \alpha < 2\pi,\ 0 \leq r < \eta(\alpha,z) \}. \tag{6.7.1}$$

The boundary Γ of the canal is defined by

$$\begin{aligned} &\Gamma = \Gamma_1 \cup \Gamma_2 \cup \Gamma_3, \\ &\Gamma_1 = \{ (r,\alpha,z) \mid r = \eta(\alpha,z),\ 0 \le \alpha < 2\pi,\ 0 < z < l \}, \\ &\Gamma_2 = \{ (r,\alpha,z) \mid z = 0,\ 0 \le r \le \eta(\alpha,0),\ 0 \le \alpha < 2\pi \}, \\ &\Gamma_3 = \{ (r,\alpha,z) \mid z = l,\ 0 \le r \le \eta(\alpha,l),\ 0 \le \alpha < 2\pi \}. \end{aligned} \tag{6.7.2}$$

The cross-sections $z = 0$ and $z = l$, i.e., Γ_2 and Γ_3, are the inlet and outlet of the canal.

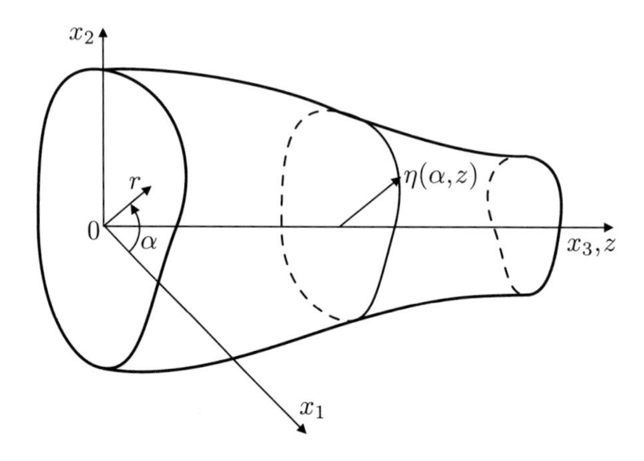

Figure 6.7.1: Domain occupied by a canal and disposition of cylindrical and Cartesian coordinate system

The following problem often appears in practice. We are given the above cross-sections, and, in addition, a function F of surface forces at the inlet may be given. The optimization problem consists in finding the length l of the canal, its lateral area, i.e., the function η, and the function of surface forces F (if it is not given) with a view to optimize, in some sense, the flow of the fluid. In what follows, we will consider these problems.

6.7.1 Set of controls and diffeomorphisms

We denote by E the transformation of cylindrical coordinates into Cartesian ones:

$$\begin{aligned} &E\colon (r,\alpha,z) \to E(r,\alpha,z) = (x_1,x_2,x_3), \\ &x_1 = r\cos\alpha, \quad x_2 = r\sin\alpha, \quad x_3 = z. \end{aligned} \tag{6.7.3}$$

The transformation E is a bijection of the set

$$W_1 = \{ (r,\alpha,z) \mid (r,\alpha,z) \in \mathbb{R}^3,\ 0 < r < \infty,\ 0 \le \varphi < 2\pi,\ z \in \mathbb{R} \} \tag{6.7.4}$$

onto the complement of the straight line

$$A = \{ x \mid x = (x_1, x_2, x_3) \in \mathbb{R}^3,\ x_1 = x_2 = 0,\ z \in \mathbb{R} \},$$

and it is a diffeomorphism of the C^∞ class of the set

$$W = \{ (r, \alpha, z) \mid (r, \alpha, z) \in \mathbb{R}^3,\ r \in (0, \infty),\ \alpha \in (0, 2\pi),\ z \in \mathbb{R} \} \tag{6.7.5}$$

onto $E(W)$. By E^{-1} we denote the inverse of E.

Let also

$$\mathcal{N} = \{ (\alpha, z) \mid \alpha \in (0, 2\pi),\ z \in (0, 1) \}. \tag{6.7.6}$$

We determine a set of controls of the form

$$\begin{aligned} \mathcal{M} = \Big\{ q \,|\, q = (f, l),\ & f \in C^1(\overline{\mathcal{N}}),\ f(0, z) = f(2\pi, z),\ \frac{\partial f}{\partial \alpha}(0, z) = \frac{\partial f}{\partial \alpha}(2\pi, z), \\ & z \in [0, 1],\ f(\alpha, z) \geq r_1,\ (\alpha, z) \in \overline{\mathcal{N}},\ f(\alpha, 0) = \xi_1(\alpha), \\ & f(\alpha, 1) = \xi_2(\alpha),\ \alpha \in [0, 2\pi],\ l \in \mathbb{R},\ l \geq l_1 \Big\}. \end{aligned} \tag{6.7.7}$$

Here, r_1, l_1 are given positive constants, ξ_1 and ξ_2 given functions. The set $\mathcal{M}$ is equipped with the topology generated by that of $C^1(\overline{\mathcal{N}}) \times \mathbb{R}$.

To each $q = (f, l) \in \mathcal{M}$, we assign the domain $\Omega_q \in \mathbb{R}^3$ of the form

$$\begin{aligned} \Omega_q = \Big\{ x \mid x = (x_1, x_2, x_3),\ & 0 < x_3 < l, \\ & 0 \leq (x_1^2 + x_2^2)^{\frac{1}{2}} < f(\alpha, \frac{x_3}{l}),\ \alpha = \arctan\ \frac{x_2}{x_1} \Big\}. \end{aligned} \tag{6.7.8}$$

By $\mathcal{B}_\gamma$ we denote the cylinder of unit length and radius γ, i.e.,

$$\mathcal{B}_\gamma = \{ x \mid x = (x_1, x_2, x_3) \in \mathbb{R}^3,\ 0 \leq x_1^2 + x_2^2 < \gamma,\ 0 < x_3 < 1 \}, \tag{6.7.9}$$

and let

$$\Omega = \mathcal{B}_{r_1}, \qquad r_2 \in \mathbb{R},\ 0 < r_2 < r_1. \tag{6.7.10}$$

To each $q \in \mathcal{M}$, we assign a mapping $P_q^{-1} \colon \overline{\Omega} \to \overline{\Omega}_q$ by

$$P_q^{-1}(x) = \begin{cases} (x_1, x_2, x_3 l), & \text{if } x \in \mathcal{B}_{r_2}, \\ (E \circ A(q) \circ E^{-1})(x), & \text{if } x \in \overline{\Omega} \setminus \mathcal{B}_{r_2}. \end{cases} \tag{6.7.11}$$

Here

$$\begin{aligned} & A(q) \colon E^{-1}(\overline{\Omega} \setminus \mathcal{B}_{r_2}) \to A_q(E^{-1}(\overline{\Omega} \setminus \mathcal{B}_{r_2})), \\ & (r, \alpha, z) \to A(q)(r, \alpha, z) = (\rho(q, r, \alpha, z), \alpha, zl), \\ & \rho(q, r, \alpha, z) = \sum_{i=0}^{2} a_i(q, \alpha, z) r^i, \end{aligned} \tag{6.7.12}$$

where

$$\begin{aligned} a_2(q,\alpha,z) &= \frac{f(\alpha,z)-r_1}{(r_1-r_2)^2}, \\ a_1(q,\alpha,z) &= 1-2r_2a_2(q,\alpha,z), \\ a_0(q,\alpha,z) &= f(\alpha,z)-r_1a_1(q,\alpha,z)-r_1^2a_2(q,\alpha,z). \end{aligned} \tag{6.7.13}$$

Lemma 6.7.1 *Let a set $\mathcal{M}$ be defined by (6.7.7) and let domains Ω_q and Ω be determined by (6.7.8) with $(f,l)=q\in\mathcal{M}$ and (6.7.9), (6.7.10). Then, for each $q\in\mathcal{M}$, the mapping P_q^{-1} given by (6.7.11)–(6.7.13) is a C^1 diffeomorphism of $\overline{\Omega}$ onto $\overline{\Omega}_q$, and its inverse P_q is a C^1 diffeomorphism of $\overline{\Omega}_q$ onto $\overline{\Omega}$.*

Proof. Applying (6.7.12) and (6.7.13) (see also Subsec. 2.12.4), one can easily verify that

$$\begin{aligned} \rho(q,r_2,\alpha,z)=r_2, \qquad \rho(q,r_1,\alpha,z)=f(\alpha,z), \\ \frac{\partial\rho}{\partial r}(q,r_2,\alpha,z)=1, \qquad (\alpha,z)\in\overline{\mathcal{N}}. \end{aligned} \tag{6.7.14}$$

Moreover,

$$\frac{\partial\rho}{\partial r}(q,r,\alpha,z)=1+2(r-r_2)a_2(q,\alpha,z). \tag{6.7.15}$$

It follows from (6.7.7) and (6.7.13) that $a_2(q,\alpha,z)\geq 0$. Thus,

$$\frac{\partial\rho}{\partial r}(q,r,\alpha,z)\geq 1, \qquad q\in\mathcal{M},\ r\in[r_2,r_1],\ (\alpha,z)\in\overline{\mathcal{N}}.$$

Therefore,

$$\det(A(q))'=\frac{\partial\rho}{\partial r}l\geq l \qquad \text{in } E^{-1}(\overline{\Omega}\setminus\mathcal{B}_{r_2}),$$

and there exists a mapping P_q that is the inverse of P_q^{-1}. Since P_q^{-1} is a continuously differentiable bijection of $\overline{\Omega}$ onto $\overline{\Omega}_q$, we obtain the inclusion $P_q\in C^1(\overline{\Omega}_q,\overline{\Omega})$ from the theorem on the differentiability of an implicit function.

Remark 6.7.1 If f is a smooth function, one can construct, in a way similar to that in Subsec. 2.12.4, a C^m diffeomorphism of $\overline{\Omega}$ onto $\overline{\Omega}_q$, where $m\geq 2$.

6.7.2 Optimization problems

For $q=(f,l)\in\mathcal{M}$, we denote by S_q the boundary of Ω_q. It follows from (6.7.8) that $S_q=S_{q1}\cup S_{q2}\cup S_{q3}$, where

$$\begin{aligned} S_{q1} &= \{\,x\mid x=(x_1,x_2,x_3)\in\mathbb{R}^3,\ (x_1^2+x_2^2)^{\frac12}=f(\alpha,\tfrac{x_3}{l}),\ 0<x_3<l\,\}, \\ S_{q2} &= \{\,x\mid x=(x_1,x_2,x_3)\in\mathbb{R}^3,\ x_3=0,\ 0\leq(x_1^2+x_2^2)^{\frac12}\leq\xi_1(\alpha)\,\}, \\ S_{q3} &= \{\,x\mid x=(x_1,x_2,x_3)\in\mathbb{R}^3,\ x_3=l,\ 0\leq(x_1^2+x_2^2)^{\frac12}\leq\xi_2(\alpha)\,\}. \end{aligned} \tag{6.7.16}$$

Here, $\alpha=\arctan\frac{x_2}{x_1}$, ξ_1, ξ_2 are the functions from (6.7.7).

We consider the following problem of flow of a nonlinear viscous fluid: Find a pair v_q, p_q satisfying

$$\frac{\partial p_q}{\partial x_i} - 2\frac{\partial}{\partial x_j}(\varphi(I(v_q))\varepsilon_{ij}(v_q)) = 0 \qquad \text{in } \Omega_q,\ i = 1,2,3, \tag{6.7.17}$$

$$\operatorname{div} v_q = 0 \qquad \text{in } \Omega_q, \tag{6.7.18}$$

$$v_q \restriction_{S_{q1}} = 0, \tag{6.7.19}$$

$$\big[-p_q\delta_{ij} + 2\varphi(I(v_q))\varepsilon_{ij}(v_q)\big]\nu_{qj} \restriction_{S_{q2}} = F_i, \qquad i = 1,2,3, \tag{6.7.20}$$

$$\big[-p_q\delta_{ij} + 2\varphi(I(v_q))\varepsilon_{ij}(v_q)\big]\nu_{qj} \restriction_{S_{q3}} = \hat{F}_i, \qquad i = 1,2,3. \tag{6.7.21}$$

Here, ν_{qj} are the components of the unit outward normal ν_q to S_q, and F_i, $\hat{F}_i$ the components of the functions F and $\hat{F}$ of surface forces at the inlet and outlet of the canal.

We note that when q passes through $\mathcal{M}$, the set S_{q3} moves along the x_3 axis without deformations. So, we consider S_{q2} and S_{q3} as fixed sets in $\mathbb{R}^2$ that are independent of q.

We define the following spaces

$$X_q = \{\, u \mid u \in H^1(\Omega_q)^3,\ u \restriction_{S_{q1}} = 0 \,\}, \tag{6.7.22}$$

$$V_q = \{\, u \mid u \in X_q,\ \operatorname{div} u = 0 \,\}, \tag{6.7.23}$$

$$X = \{\, u \mid u \in H^1(\Omega)^3,\ u \restriction_{S_1} = 0 \,\}, \tag{6.7.24}$$

where

$$S_1 = \{\, x \mid x = (x_1, x_2, x_3) \in \mathbb{R}^3,\ (x_1^2 + x_2^2)^{\frac{1}{2}} = r_1,\ x_3 \in (0,1) \,\}.$$

Suppose the function φ satisfies the conditions (6.1.11)–(6.1.14) and define operators $L_q \colon X_q \to X_q^*$ and $B_q \in \mathcal{L}(X_q, L_2(\Omega_q)^*)$ by

$$(L_q(u), h) = 2\int_{\Omega_q} \varphi(I(u))\varepsilon_{ij}(u)\varepsilon_{ij}(h)\,dx, \qquad u, h \in X_q, \tag{6.7.25}$$

$$(B_q u, \mu) = \int_{\Omega_q} (\operatorname{div} u)\mu\,dx, \qquad u \in X_q,\ \mu \in L_2(\Omega_q). \tag{6.7.26}$$

Assume that $\hat{F}$ is a fixed function and

$$\hat{F} \in H^{-\frac{1}{2}}(S_{q3})^3, \tag{6.7.27}$$

while F is a control. If the fluid flows out into the air, we can take $\hat{F} = (0, 0, p_0)$, where p_0 is a constant, the atmospheric pressure. We assume that $U = \mathcal{M} \times H^{-\frac{1}{2}}(S_{q2})^3$ is a set of controls. For $(q, F) \in U$, we consider the following problem: Find a pair of functions v_{qF}, p_{qF} satisfying

$$(v_{qF}, p_{qF}) \in X_q \times L_2(\Omega_q), \tag{6.7.28}$$

$$(L_q(v_{qF}), h) - (B_q^* p_{qF}, h) = (\hat{F} + F, h), \qquad h \in X_q, \tag{6.7.29}$$

$$(B_q v_{qF}, \mu) = 0, \qquad \mu \in L_2(\Omega_q). \tag{6.7.30}$$

The pair $v_q = v_{qF}$, $p_q = p_{qF}$ is a generalized solution to the problem (6.7.17)–(6.7.21). The existence and uniqueness of the solution to the problem (6.7.28)–(6.7.30) follow from Theorem 6.1.1. We define a goal functional in the form

$$\Psi_0(q, F) = \int_{S_{q3}} \sum_{i=1}^{3} (v_{qFi} - y_i)^2 \, ds, \tag{6.7.31}$$

where y_i are prescribed functions from $L_2(S_{q3})$ and v_{qFi} are the components of the vector function v_{qF}. In particular, for the extrusion head, one can take $y_1 = y_2 = 0$, $y_3 = c$, where c is a constant. A set of admissible controls is determined as follows

$$\begin{aligned} U_{\text{ad}} = \{ \, (q, F) \,|\, q = (f, l) \in \mathcal{M}, \ F \in L_2(S_{q2})^3, \ f \in \widetilde{W}_t^2(\mathcal{N}), \\ \|f\|_{\widetilde{W}_t^2(\mathcal{N})} \leq c_1, \ t > 2, \ \|F\|_{L_2(S_{q2})^3} \leq c_2, \ l \in [l_1, c_3] \, \}. \end{aligned} \tag{6.7.32}$$

Here, $\widetilde{W}_t^2(\mathcal{N})$ is the subspace of $W_t^2(\mathcal{N})$ consisting of periodic functions with respect to α, c_1, c_2, c_3 are positive constants, l_1 is the constant from (6.7.7).

The optimization problem consists in finding a pair q_0, F_0 satisfying

$$(q_0, F_0) \in U_{\text{ad}}, \qquad \Psi_0(q_0, F_0) = \inf_{(q,F) \in U_{\text{ad}}} \Psi_0(q, F). \tag{6.7.33}$$

Theorem 6.7.1 *Suppose a function φ satisfies the conditions* (6.1.11)–(6.1.14) *and* (6.7.27) *holds. Let a nonempty set U_{ad} and a goal functional Ψ_0 be defined by* (6.7.32) *and* (6.7.31) *with $y_i \in L_2(S_{q3})$. Then, there exists a solution of the problem* (6.7.33).

Proof. Let $\tilde{v}_{qF} = v_{qF} \circ P_q^{-1}$, $\tilde{p}_{qF} = p_{qF} \circ P_q^{-1}$, where v_{qF}, p_{qF} is the solution to the problem (6.7.28)–(6.7.30) and P_q^{-1} is defined by (6.7.11)–(6.7.13). By virtue of Lemma 6.7.1, the pair $\tilde{v}_{qF}$, $\tilde{p}_{qF}$ is the solution to the problem

$$(\tilde{v}_{qF}, \tilde{p}_{qF}) \in X \times L_2(\Omega), \tag{6.7.34}$$

$$(\widetilde{L}_q(\tilde{v}_{qF}), h) - (\widetilde{B}_q^* \tilde{p}_{qF}, h) = (Q_1(q, \hat{F}) + Q_2(q, F), h), \qquad h \in X, \tag{6.7.35}$$

$$(\widetilde{B}_q \tilde{v}_{qF}, \mu) = 0, \qquad \mu \in L_2(\Omega). \tag{6.7.36}$$

Here, the operators $\widetilde{L}_q$ and $\widetilde{B}_q$ are defined by (6.3.22) and (6.3.23), and

$$(Q_1(q, \hat{F}), h) = (\hat{F}, h \circ P_q), \qquad (Q_2(q, F), h) = (F, h \circ P_q). \tag{6.7.37}$$

As S_{q2} and S_{q3} are fixed sets in $\mathbb{R}^2$, the functions Q_1 and Q_2 are independent of q. So, from the conditions that $q_m \to q_0$ in $\mathcal{M}$, $F_m \to F_0$ in $H^{-\frac{1}{2}}(S_{q2})^3$, it follows that $Q_2(q_m, F_m) \to Q_2(q_0, F_0)$ in X^*. In addition, by (6.7.11)–(6.7.13), we get $P_{q_m}^{-1} \to P_{q_0}^{-1}$ in $C^1(\overline{\Omega})^3$. These results and Lemma 6.7.1 allow us to apply Theorem 6.3.1, which gives

$$\left.\begin{aligned} &(q, F) \to (\tilde{v}_{qF}, \tilde{p}_{qF}) \text{ is a continuous mapping of} \\ &\mathcal{M} \times H^{-\frac{1}{2}}(S_{q2})^3 \text{ into } X \times L_2(\Omega), \end{aligned}\right\} \tag{6.7.38}$$

and so Ψ_0 is a continuous functional in $\mathcal{M}\times H^{-\frac{1}{2}}(S_{q2})^3$. Because of the embedding theorem, U_{ad} is a compactum in $\mathcal{M}\times H^{-\frac{1}{2}}(S_{q2})^3$. Hence, the existence of a solution to the problem (6.7.33) follows from Theorem 1.3.9. The proof is complete.

We now consider another optimization problem. Given cross-sections at the inlet and outlet of a canal and functions of surface forces at the inlet and outlet, find the length of the canal and its lateral area which maximize the volume flow rate of the fluid. So, in this case, for each $q \in \mathcal{M}$, the functions of velocity and pressure v_q, p_q are the solution to the problem

$$(v_q, p_q) \in X_q \times L_2(\Omega_q), \tag{6.7.39}$$

$$(L_q(v_q), h) - (B_q^* p_q, h) = (\hat{F} + F, h), \qquad h \in X_q, \tag{6.7.40}$$

$$(B_q v_q, \mu) = 0, \qquad \mu \in L_2(\Omega_q). \tag{6.7.41}$$

Here, $\hat{F}$ and F are given fixed functions,

$$\hat{F} \in H^{-\frac{1}{2}}(S_{q3})^3, \qquad F \in H^{-\frac{1}{2}}(S_{q2})^3. \tag{6.7.42}$$

The goal functional is defined by

$$\Psi_0(q) = \int_{S_{q3}} v_{qi}\nu_{qi}\, ds, \tag{6.7.43}$$

where v_{qi} are the components of the velocity function v_q, ν_{qi} the components of the unit outward normal ν_q to S_{q3}. In our case, $\nu_q = (0,0,1)$ on S_{q3}. The set of admissible controls has the form

$$U_{\text{ad}} = \{\, q \mid q = (f,l) \in \mathcal{M},\ f \in \widetilde{W}_t^2(\mathcal{N}),\ \|f\|_{\widetilde{W}_t^2(\mathcal{N})} \le c_1,\ t > 2,\ l \in [l_1, c_2]\,\}. \tag{6.7.44}$$

The optimization problem consists in finding q_0 satisfying

$$q_0 \in U_{\text{ad}}, \qquad \Psi_0(q_0) = \sup_{q\in U_{\text{ad}}} \Psi_0(q). \tag{6.7.45}$$

By analogy with the proved above, we obtain

Theorem 6.7.2 *Let a function φ satisfy the conditions* (6.1.11)–(6.1.14) *and let* (6.7.42) *hold. Let v_q, p_q be the solution to the problem* (6.7.39)–(6.7.41), *and Ψ_0 be defined by* (6.7.43). *Let also a nonempty set U_{ad} be defined by* (6.7.44). *Then, there exists a solution to the problem* (6.7.45).

We note that the problem on the shape optimization of a two-dimensional canal with a view of the maximization of the volume flow rate of a viscous Newtonian fluid was studied in Litvinov (1990a).

6.8 A problem of the optimal shape of a hydrofoil

6.8.1 State equation for a moving hydrofoil

Boundary-value problem

Let a nondeformable hydrofoil move slowly in a viscous incompressible fluid. We suppose that the dimensions of the cross-sections of the hydrofoil are small with respect to its length, and the cross-sections vary slowly in the longitudinal direction. Therefore, we consider the two-dimensional problem of the flow about a hydrofoil of a constant cross-section.

We place the origin of the coordinate system in some internal point of the hydrofoil and assume that the coordinate system is rigidly attached to the hydrofoil and moves with it. In the polar coordinate system, the boundary S_q of the hydrofoil is determined by a smooth periodical function $q\colon [0,2\pi] \to \mathbb{R}$, and we designate by Q_q the domain occupied by the hydrofoil (see Fig. 6.8.1). Let Q be a bounded and sufficiently large domain in $\mathbb{R}^2$ containing the domain Q_q. In the polar coordinate system, the boundary S of Q is defined by a smooth periodical function $\gamma\colon [0,2\pi] \to \mathbb{R}$. Denote $\Omega_q = Q \setminus \overline{Q}_q$ and consider the following problem: Find a pair (v,p), where $v = (v_1, v_2)$ is the velocity and p is the pressure, satisfying the relations

$$\mu\left(\frac{\partial^2 v_1}{\partial y_1^2} + \frac{\partial^2 v_1}{\partial y_2^2}\right) - \frac{\partial p}{\partial y_1} = 0, \qquad \text{in } \Omega_q, \tag{6.8.1}$$

$$\mu\left(\frac{\partial^2 v_2}{\partial y_1^2} + \frac{\partial^2 v_2}{\partial y_2^2}\right) - \frac{\partial p}{\partial y_2} = 0, \qquad \text{in } \Omega_q, \tag{6.8.2}$$

$$\frac{\partial v_1}{\partial y_1} + \frac{\partial v_2}{\partial y_2} = 0, \qquad \text{in } \Omega_q, \tag{6.8.3}$$

$$v\big|_{S_q} = 0, \qquad v\big|_{S} = a. \tag{6.8.4}$$

Here, μ is viscosity of the fluid and $a = (a_1, a_2)$, where the functions a_i take constant values, that is,

$$a(y) = (a_1(y), a_2(y)) = c = (c_1, c_2) \in \mathbb{R}^2, \qquad y \in S. \tag{6.8.5}$$

Clearly, $-c = (-c_1, -c_2)$ is the velocity of points of the hydrofoil in the immovable fluid. We assume that the ratio $\frac{\mu}{|c|}$ is large; therefore, we use the Stokes approximation for our problem.

Such an approximation is suitable for the case when $|c|$ is small. For large c, the flow is turbulent, but the equations (6.8.1), (6.8.2) may be used if μ is considered as the turbulent viscosity. This approach traces back to Boussinesq (1877). The turbulent viscosity is far greater than the laminar one (it may be greater than the laminar viscosity by a factor 10^3–10^5), see Loitzansky (1987), Fletcher (1988). So, the ratio $\frac{\mu}{|c|}$ may be large for large $|c|$ if μ is the turbulent viscosity. For other models of turbulent flows, see Rodi (1980), Litvinov (1996).

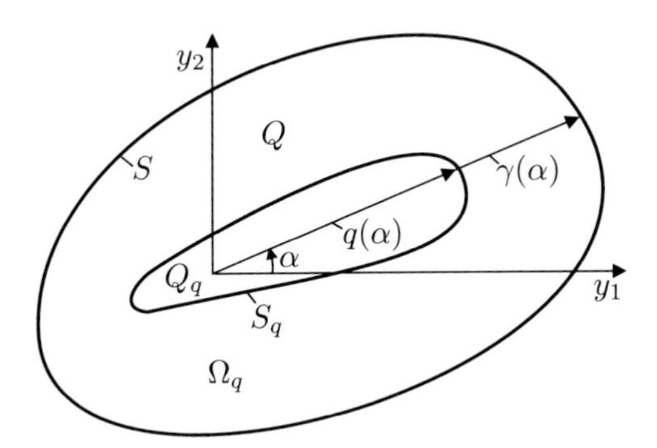

Figure 6.8.1: Domain Q_q occupied by a hydrofoil and a large domain Q containing Q_q

Set of controls and operator A_q

The problem (6.8.1)–(6.8.4) is considered in a Hölder space $C^l(\overline{\Omega}_q)$, where $l > 0$, l not an integer. The space $C^l(\overline{\Omega}_q)$ is equipped with the norm (2.13.3). We denote by $C^l(\overline{\Omega}_q)/\mathbb{R}$ the quotient space of $C^l(\overline{\Omega}_q)$ with respect to $\mathbb{R}$. In the other domains, the spaces C^l are defined analogously.

We consider the function q, determining the shape of the hydrofoil, as a control. Define a space of controls M of the form

$$M = \{\, q \mid q \in \widetilde{C}^{[l]+4}([0, 2\pi]),\ r_1 < q(\alpha) < r_2,\ \alpha \in [0, 2\pi]\,\}. \tag{6.8.6}$$

Here, r_1, r_2 are positive constants, $l > 0$, l not an integer, $\widetilde{C}^{[l]+4}([0, 2\pi])$ is the subspace of periodic functions in $C^{[l]+4}([0, 2\pi])$. We equip the set M with the topology generated by that of $C^{[l]+4}([0, 2\pi])$.

For each $q \in M$, we define in polar coordinates the boundary S_q between the domain Q_q occupied by the hydrofoil and the domain $\Omega_q = Q \setminus \overline{Q}_q$; see Fig. 6.8.1. We assume below that the domain Q is independent of the control q.

Define spaces V_{lq} and H_{lq} as follows:

$$\begin{aligned} V_{lq} &= C^{l+2}(\overline{\Omega}_q)^2 \times C^{l+1}(\overline{\Omega}_q), \\ H_{lq} &= C^l(\overline{\Omega}_q)^2 \times C^{l+1}(\overline{\Omega}_q) \times C^{l+2}(S_q)^2 \times C^{l+2}(S)^2. \end{aligned} \tag{6.8.7}$$

Also, determine an operator L_q by the formulas

$$\begin{aligned} L_q &= (A_q, B_q) \in \mathcal{L}(V_{lq}, H_{lq}), \\ (u, p) \in V_{lq}, \qquad A_q(u, p) &= \{\mu \Delta u - \operatorname{grad} p,\ \operatorname{div} u\}, \\ B_q(u, p) &= \{u|_{S_q},\ u|_S\}. \end{aligned} \tag{6.8.8}$$

Let f, ρ, α, β be functions such that

$$\begin{aligned} f = (f_1, f_2) \in C^l(\overline{\Omega}_q)^2, \qquad & \rho \in C^{l+1}(\overline{\Omega}_q), \\ \alpha = (\alpha_1, \alpha_2) \in C^{l+2}(S_q)^2, \qquad & \beta = (\beta_1, \beta_2) \in C^{l+2}(S)^2, \end{aligned} \tag{6.8.9}$$

and

$$\int_{\Omega_q} \rho \, dy = \int_{S_q} \alpha_i \nu_{qi} \, ds + \int_S \beta_i \nu_i \, ds, \tag{6.8.10}$$

where ν_{qi} and ν_i are the components of the unit outward normals ν_q and ν to S_q and S. In (6.8.10) and in the following formulas, summation over repeated indices is implied.

The theorem below follows from known results, see Solonnikov (1964, 1966).

Theorem 6.8.1 *Let Ω_q be a domain in $\mathbb{R}^2$ with two connected components of the boundary. The external boundary S and internal boundary S_q of Ω_q are defined in polar coordinates by the functions $\gamma \in \widetilde{C}^{[l]+4}([0, 2\pi])$ and $q \in M$. Let functions f, ρ, α, β satisfy the conditions* (6.8.9)–(6.8.10)*. Then, there exists a unique pair $(\tilde{v}, \tilde{p}) \in C^{l+2}(\overline{\Omega}_q)^2 \times C^{l+1}(\overline{\Omega}_q)/\mathbb{R}$ satisfying*

$$A_q(\tilde{v}, \tilde{p}) = (f, \rho), \qquad B_q(\tilde{v}, \tilde{p}) = (\alpha, \beta), \tag{6.8.11}$$

and there exists a positive constant c such that

$$\begin{aligned} \|\tilde{v}\|_{C^{l+2}(\overline{\Omega}_q)^2} + \|\tilde{p}\|_{C^{l+1}(\overline{\Omega}_q)/\mathbb{R}} \le c\big(&\|f\|_{C^l(\overline{\Omega}_q)^2} + \|\rho\|_{C^{l+1}(\overline{\Omega}_q)} \\ &+ \|\alpha\|_{C^{l+2}(S_q)^2} + \|\beta\|_{C^{l+2}(S)^2}\big). \end{aligned} \tag{6.8.12}$$

In particular, there exists a unique pair $(v, p) \in C^{l+2}(\overline{\Omega}_q)^2 \times C^{l+1}(\overline{\Omega}_q)/\mathbb{R}$ satisfying the equations (6.8.1)–(6.8.4).

Operator W_q and functionals

We denote by $\hat{V}_{lq}$ and $\check{H}_{lq}$ the kernel subspace and the range of the operator $L_q = (A_q, B_q)$,

$$\begin{aligned} \hat{V}_{lq} &= \{\, (u, p) \mid (u, p) \in V_{lq},\ L_q(u, p) = 0 \,\}, \\ \check{H}_{lq} &= L_q(V_{lq}). \end{aligned} \tag{6.8.13}$$

We denote by $\check{V}_{lq}$ and $\hat{H}_{lq}$ the complements of $\hat{V}_{lq}$ and $\check{H}_{lq}$ in V_{lq} and H_{lq}, that is,

$$V_{lq} = \hat{V}_{lq} \oplus \check{V}_{lq}, \qquad H_{lq} = \hat{H}_{lq} \oplus \check{H}_{lq}, \tag{6.8.14}$$

where the symbol $\oplus$ represents the direct sum of subspaces. From Theorem 6.8.1, it follows that $\hat{V}_{lq}$ is a one-dimensional subspace in V_{lq} and that

$$\hat{V}_{lq} = \{\, (u, p) \mid u = 0,\ p \in \mathbb{R} \,\}. \tag{6.8.15}$$

From Theorem 6.8.1 and (6.8.10), it also follows that $\hat{H}_{lq}$ is the one-dimensional subspace in H_{lq} determined by the expression

$$\hat{H}_{lq} = \{ (f, \rho, \alpha, \beta) \in H_{lq},\ f = 0,\ \rho = c,\ \alpha = -c\nu_q,\ \beta = -c\nu,\ c \in \mathbb{R} \}. \quad (6.8.16)$$

From (6.8.15)–(6.8.16), it follows that the function $\varphi = (0, 1)$, with 0 the zero element of $C^{l+2}(\overline{\Omega}_q)^2$ and 1 the unit function in Ω_q, is a basis in $\hat{V}_{lq}$, while the function $\psi = (0, 1, -\nu_q, -\nu)$ is a basis in $\hat{H}_{lq}$.

Taking into account Theorem 6.8.1, it is easy to obtain the following

Theorem 6.8.2 *Let $q \in M$ and $\gamma \in \widetilde{C}^{[l]+4}([0, 2\pi])$. Determine the operator $U_q \in \mathcal{L}(V_{lq}, H_{lq})$ of the form*

$$U_q\varphi = \psi, \qquad U_q w = 0, \qquad w \in \check{V}_{lq}. \quad (6.8.17)$$

Then, the operator $W_q = L_q + U_q$ is an isomorphism of V_{lq} onto H_{lq}.

We further denote by (v_q, p_q), with $v_q = (v_{q1}, v_{q2})$, the solution of the problem (6.8.1)–(6.8.4). The hydrodynamic lift $F_1(q)$ of the hydrofoil occupying the domain Q_q is defined by the formula

$$F_1(q) = \int_{S_q} \Big[- p_q\delta_{2j} + \mu\Big(\frac{\partial v_{q2}}{\partial y_j} + \frac{\partial v_{qj}}{\partial y_2}\Big)\Big] \nu_{qj}\, ds, \quad (6.8.18)$$

where δ_{ij} is the Kronecker delta. Because

$$\int_{S_q} \nu_{q2}\, ds = 0,$$

the integral (6.8.18) is uniquely defined for $p_q \in C^{l+1}(\overline{\Omega}_q)/\mathbb{R}$. The dissipation energy $F_0(q)$ is defined by the formula

$$F_0(q) = \frac{\mu}{2} \int_{\Omega_q} \sum_{i,j=1}^{2} \Big(\frac{\partial v_{qi}}{\partial y_j} + \frac{\partial v_{qj}}{\partial y_i}\Big)^2 dy, \quad (6.8.19)$$

and this energy is equal to the power (rate of work) needed to move the hydrofoil.

State equation in polar coordinates

We denote by E the function that transforms polar coordinates (r, α) into Cartesian coordinates (y_1, y_2),

$$E\colon (r, \alpha) \to E(r, \alpha) = (y_1, y_2), \qquad y_1 = r\cos\alpha, \quad y_2 = r\sin\alpha, \quad (6.8.20)$$

and the inverse of E is denoted by E^{-1}. We denote by Ω_{q1} the domain corresponding to Ω_q in polar coordinates (r, α) (see Fig. 6.8.1),

$$\Omega_{q1} = \{ (r, \alpha) \mid \alpha \in (0, 2\pi),\ q(\alpha) < r < \gamma(\alpha) \}. \quad (6.8.21)$$

We define sets S_{q1} and S_1 as follows:

$$S_{q1} = \{ (r,\alpha) \mid \alpha \in (0,2\pi),\ r = q(\alpha) \},$$
$$S_1 = \{ (r,\alpha) \mid \alpha \in (0,2\pi),\ r = \gamma(\alpha) \}. \tag{6.8.22}$$

These sets correspond to the internal and external boundaries of Ω_q in polar coordinates. Let $u = (u_1, u_2)$ be a velocity field in Ω_q. The function $w = (w_1, w_2)$, defined in Ω_{q1} by the formulas

$$w_1(r,\alpha) = u_1(E(r,\alpha))\cos\alpha + u_2(E(r,\alpha))\sin\alpha,$$
$$w_2(r,\alpha) = -u_1(E(r,\alpha))\sin\alpha + u_2(E(r,\alpha))\cos\alpha, \tag{6.8.23}$$

corresponds to the u in polar coordinates, with w_1, w_2 being the radial and tangential components of the velocity vector w. To a function of the pressure p in Ω_q, we place in correspondence the function $p_1 = p \circ E$ in Ω_{q1}. Then, with the replacement of the variables and functions defined by (6.8.20) and (6.8.23), the operator $L_q = (A_q, B_q)$ from (6.8.8) transforms into the operator L_{q1} of the form

$$L_{q1} = (A_{q1}, B_{q1}),$$
$$A_{q1}(w,p_1) = \begin{cases} \mu\left(\Delta_1 w_1 - \dfrac{2}{r^2}\dfrac{\partial w_2}{\partial\alpha} - \dfrac{w_1}{r^2}\right) - \dfrac{\partial p_1}{\partial r}, \\ \mu\left(\Delta_1 w_2 + \dfrac{2}{r^2}\dfrac{\partial w_1}{\partial\alpha} - \dfrac{w_2}{r^2}\right) - \dfrac{1}{r}\dfrac{\partial p_1}{\partial\alpha}, \\ \dfrac{\partial w_1}{\partial r} + \dfrac{1}{r}\dfrac{\partial w_2}{\partial\alpha} + \dfrac{w_1}{r}, \end{cases} \tag{6.8.24}$$

$$B_{q1}(w,p_1) = \{ w|_{S_{q1}}, w|_{S_1} \}. \tag{6.8.25}$$

Here, Δ_1 is the Laplace operator in polar coordinates,

$$\Delta_1 = \frac{\partial^2}{\partial r^2} + \frac{1}{r}\frac{\partial}{\partial r} + \frac{1}{r^2}\frac{\partial^2}{\partial\alpha^2}. \tag{6.8.26}$$

In the sequel, we suppose that $q \in M$, where M is defined by (6.8.6) and the function γ satisfies the conditions

$$\gamma \in \widetilde{C}^{[l]+4}([0,2\pi]), \qquad \gamma(\varphi) \geq \gamma_0 > r_2, \qquad \varphi \in (0,2\pi), \tag{6.8.27}$$

where γ_0 is a constant and r_2 is the constant from (6.8.6). By $\tilde{q}$ and $\tilde{\gamma}$, we denote the periodical extensions of the functions q and γ on $\mathbb{R}$ with the period $(0,2\pi)$, and we define the set

$$\Omega_{q2} = \{ (r,\alpha) \mid \alpha \in \mathbb{R},\ \tilde{q}(\alpha) < r < \tilde{\gamma}(\alpha) \}.$$

For $k > 0$, k not an integer, we denote by $\tilde{C}^k(\overline{\Omega}_{q1})$ the subspace of functions in $C^k(\overline{\Omega}_{q1})$ periodical with respect to α. The periodicity of a function $h \in C^k(\overline{\Omega}_{q1})$

means that, if $\tilde{h}$ is periodical with respect to the α-extension of h on Ω_{q2} with period $(0, 2\pi)$, then $\tilde{h} \in C^k(\overline{G})$, where G is an arbitrary, open, and bounded subset of Ω_{q2}.

Define functions F_1 and F_2 as follows:

$$\begin{aligned} &F_1 \colon (0, 2\pi) \to S_{q1}, \qquad \alpha \to F_1(\alpha) = (q(\alpha), \alpha), \\ &F_2 \colon (0, 2\pi) \to S_1, \qquad \alpha \to F_2(\alpha) = (\gamma(\alpha), \alpha). \end{aligned} \tag{6.8.28}$$

If u is a function given on S_{q1}, then $u \circ F_1$ is a function defined on $(0, 2\pi)$, and we determine the space $C^k(S_{q1})$, where $k \in [0, [l] + 4]$, of the form

$$C^k(S_{q1}) = \{\, u \mid u \circ F_1 \in C^k([0, 2\pi]) \,\}, \tag{6.8.29a}$$

and by analogy

$$C^k(S_1) = \{\, u \mid u \circ F_2 \in C^k([0, 2\pi]) \,\}. \tag{6.8.29b}$$

By $\widetilde{C}^k(S_{q1})$ and $\widetilde{C}^k(S_1)$, we denote the subspaces of periodic functions in $C^k(S_{q1})$ and $C^k(S_1)$. In the spaces $\widetilde{C}^k(S_{q1})$ and $\widetilde{C}^k(S_1)$, the norms are defined by

$$\|u\|_{\widetilde{C}^k(S_{q1})} = \|u \circ F_1\|_{C^k([0,2\pi])}, \qquad \|u\|_{\widetilde{C}^k(S_1)} = \|u \circ F_2\|_{C^k([0,2\pi])}. \tag{6.8.30}$$

To the spaces V_{lq} and H_{lq} from (6.8.7), we place in correspondence the following spaces:

$$\begin{aligned} V_{lq1} &= \widetilde{C}^{l+2}(\overline{\Omega}_{q1})^2 \times \widetilde{C}^{l+1}(\overline{\Omega}_{q1}), \\ H_{lq1} &= \widetilde{C}^{l}(\overline{\Omega}_{q1})^2 \times \widetilde{C}^{l+1}(\overline{\Omega}_{q1}) \times \widetilde{C}^{l+2}(S_{q1})^2 \times \widetilde{C}^{l+2}(S_1)^2. \end{aligned} \tag{6.8.31}$$

Let g, e, h, z be functions such that

$$\begin{aligned} &g = (g_1, g_2) \in \widetilde{C}^{l}(\overline{\Omega}_{q1})^2, \qquad e \in \widetilde{C}^{l+1}(\overline{\Omega}_{q1}), \\ &h = (h_1, h_2) \in \widetilde{C}^{l+2}(S_{q1})^2, \qquad z = (z_1, z_2) \in \widetilde{C}^{l+2}(S_1)^2. \end{aligned} \tag{6.8.32}$$

Consider the following problem: Find a pair (w, b) satisfying the relations

$$L_{q1}(w, b) = (g, e, h, z), \qquad (w, b) \in V_{lq1}. \tag{6.8.33}$$

From (6.8.10) and Theorem 6.8.1, it follows that the condition of solvability of the problem (6.8.32)–(6.8.33) has the form

$$\begin{aligned} \int_{\Omega_{q1}} er\, dr\, d\alpha = &\int_0^{2\pi} h_i(q(\alpha), \alpha)\nu_{q1i}(\alpha)\big[(q'(\alpha))^2 + (q(\alpha))^2\big]^{\frac{1}{2}}\, d\alpha \\ &+ \int_0^{2\pi} z_i(\gamma(\alpha), \alpha)\nu_{1i}(\alpha)\big[(\gamma'(\alpha))^2 + (\gamma(\alpha))^2\big]^{\frac{1}{2}}\, d\alpha. \end{aligned} \tag{6.8.34}$$

Here, q', γ' are the derivatives of q, γ, and ν_{q1i}, ν_{1i} are the radial ($i = 1$) and angular ($i = 2$) components of the unit outward normals to S_q and S. These components are defined by the formulas

$$\begin{aligned} \nu_{q11} &= -\cos(\arctan(\tfrac{q'}{q})), \qquad & \nu_{q12} &= \sin(\arctan(\tfrac{q'}{q})), \\ \nu_{11} &= \cos(\arctan(\tfrac{\gamma'}{\gamma})), \qquad & \nu_{12} &= -\sin(\arctan(\tfrac{\gamma'}{\gamma})). \end{aligned} \tag{6.8.35}$$

By analogy with (6.8.14), we have the following representation:

$$V_{lq1} = \hat{V}_{lq1} \oplus \check{V}_{lq1}, \qquad H_{lq1} = \hat{H}_{lq1} \oplus \check{H}_{lq1}, \tag{6.8.36}$$

where

$$\hat{V}_{lq1} = \ker L_{q1}, \qquad \check{H}_{lq1} = L_{q1}(V_{lq1}),$$

and $\check{V}_{lq1}$, $\hat{H}_{lq1}$ are the corresponding complements. From Theorem 6.8.1 and (6.8.34), it follows that $\hat{V}_{lq1}$ and $\hat{H}_{lq1}$ are one-dimensional subspaces. The function φ_q,

$$\varphi_q = (0, 1), \tag{6.8.37}$$

is the basis in $\hat{V}_{lq1}$, with 0 the zero element of $\widetilde{C}^{l+2}(\overline{\Omega}_{q1})^2$ and 1 the unit function in Ω_{q1}. The function ψ_q,

$$\psi_q = \{\, 0,\ 1,\ (-\nu_{q1i})_{i=1}^2,\ (-\nu_{1i})_{i=1}^2 \,\}, \tag{6.8.38}$$

is the basis in $\hat{H}_{lq1}$. Define an operator $U_{q1} \in \mathcal{L}(V_{lq1}, H_{lq1})$ as follows:

$$U_{q1}\varphi_q = \psi_q, \qquad U_{q1}w = 0, \qquad w \in \check{V}_{lq1}. \tag{6.8.39}$$

The operator U_{q1} is determined by the formulas

$$\begin{aligned} u \in \widetilde{C}^{l+2}(\overline{\Omega}_{q1})^2, &\qquad p \in \widetilde{C}^{l+1}(\overline{\Omega}_{q1}), \\ U_{q1}(u, p) &= c(q)\psi(q), \\ c(q) = \Big(\int_{\Omega_{q1}} r\,dr\,d\alpha \Big)^{-1} &\int_{\Omega_{q1}} pr\,dr\,d\alpha. \end{aligned} \tag{6.8.40}$$

Remark 6.8.1 By (6.8.6) and (6.8.27), the function ψ_q belongs to H_{lq1}. However, in the case when q or γ are from $\widetilde{C}^{[l]+3}([0, 2\pi])$, the function ψ_q may not belong to H_{lq1}.

From Theorem 6.8.1, we get the following theorem, which is analogous to Theorem 6.8.2.

Theorem 6.8.3 *Let the condition* (6.8.27) *hold and let $q \in M$. Let operators $L_{q1} = (A_{q1}, B_{q1})$ and U_{q1} be defined by* (6.8.24), (6.8.25), (6.8.40). *Then, L_{q1} is an isomorphism of $\check{V}_{lq1}$ onto $\check{H}_{lq1}$ and $W_{q1} = L_{q1} + U_{q1}$ is an isomorphism of V_{lq1} onto H_{lq1}.*

6.8.2 Fixed-domain problem and Fréchet differentiability of the functionals

State equations in a fixed domain

We define a domain Ω by

$$\Omega = \{\, x \mid\ x = (x_1, x_2) \in \mathbb{R}^2,\ 1 < x_1^2 + x_2^2 < 4 \,\}. \tag{6.8.41}$$

We denote by Ω_1 the domain corresponding to Ω in polar coordinates (ρ, φ) and by S_2, S_3 the sets corresponding to the internal and external boundaries of Ω,

$$\begin{aligned} \Omega_1 &= \{\, (\rho, \varphi) \mid \rho \in (1,2),\ \varphi \in (0, 2\pi) \,\}, \\ S_2 &= \{\, (\rho, \varphi) \mid \rho = 1,\ \varphi \in (0, 2\pi) \,\}, \\ S_3 &= \{(\rho, \varphi) \mid \rho = 2,\ \varphi \in (0, 2\pi) \,\}. \end{aligned} \tag{6.8.42}$$

To each $q \in M$, we assign the mapping $G_q \colon \overline{\Omega}_{q1} \to \overline{\Omega}_1$ of the form

$$\begin{gathered} (r, \alpha) \to G_q(r, \alpha) = (\rho, \varphi), \\ \rho = \frac{r - 2q(\alpha) + \gamma(\alpha)}{\gamma(\alpha) - q(\alpha)}, \qquad \varphi = \alpha. \end{gathered} \tag{6.8.43}$$

The inverse $G_q^{-1} \colon \overline{\Omega}_1 \to \bar{\Omega}_{q1}$ acts as follows:

$$\begin{gathered} (\rho, \varphi) \to G_q^{-1}(\rho, \varphi) = (r, \alpha), \\ r = 2q(\varphi) - \gamma(\varphi) + \rho[\gamma(\varphi) - q(\varphi)], \qquad \alpha = \varphi. \end{gathered} \tag{6.8.44}$$

We define the following spaces:

$$\begin{aligned} V_{l1} &= \widetilde{C}^{l+2}(\overline{\Omega}_1)^2 \times \widetilde{C}^{l+1}(\overline{\Omega}_1), \\ H_{l1} &= \widetilde{C}^{l}(\overline{\Omega}_1)^2 \times \widetilde{C}^{l+1}(\overline{\Omega}_1) \times \widetilde{C}^{l+2}(S_2)^2 \times \widetilde{C}^{l+2}(S_3)^2, \end{aligned} \tag{6.8.45}$$

which correspond to the spaces V_{lq1}, H_{lq1} from (6.8.31). The norms in $\widetilde{C}^k(S_i)$, $i = 2, 3$, $k \in [0, [l] + 4]$, are defined by formulas analogous to (6.8.30). By applying the composite function theorem (chain rule), it is easy to prove the lemma below.

Lemma 6.8.1 *Let* (6.8.27) *hold, and let a function* G_q *be defined by* (6.8.43). *Then,* G_q *is a diffeomorphism of* $\overline{\Omega}_{q1}$ *onto* $\overline{\Omega}_1$ *of the* $C^{[l]+4}$ *class, and the mapping* $u \to u \circ G_q$ *is an isomorphism of* $\widetilde{C}^k(\overline{\Omega}_1)$ *onto* $\widetilde{C}^k(\overline{\Omega}_{q1})$ *and of* $\widetilde{C}^k(S_2)$, $\widetilde{C}^k(S_3)$ *onto* $\widetilde{C}^k(S_{q1})$, $\widetilde{C}^k(S_1)$, *respectively, for* $k \in [0, [l] + 4]$.

For every $q \in M$, we define operators L_{q2}, $U_{q2} \in \mathcal{L}(V_{l1}, H_{l1})$ by the expressions

$$\begin{aligned} L_{q2} &= (A_{q2}, B_{q2}), \\ A_{q2}h &= (A_{q1}(h \circ G_q)) \circ G_q^{-1}, \\ B_{q2}h &= (B_{q1}(h \circ G_q)) \circ G_q^{-1}, \\ U_{q2}h &= (U_{q1}(h \circ G_q)) \circ G_q^{-1}, \qquad h \in V_{l1}. \end{aligned} \tag{6.8.46}$$

By (6.8.43), (6.8.44), we obtain the following formulas:

$$\left(\frac{\partial \rho}{\partial r}\right) \circ G_q^{-1} = \frac{1}{\gamma - q}, \qquad \left(\frac{\partial \rho}{\partial \alpha}\right) \circ G_q^{-1} = -\frac{1}{\gamma - q}\frac{\partial r}{\partial \varphi},$$
$$\left(\frac{\partial \varphi}{\partial r}\right) \circ G_q^{-1} = 0, \qquad \left(\frac{\partial \varphi}{\partial \alpha}\right) \circ G_q^{-1} = 1. \tag{6.8.47}$$

Let

$$w = (w_1, w_2) \in \widetilde{C}^{l+2}(\overline{\Omega}_{q1})^2, \qquad p_1 \in \widetilde{C}^{l+1}(\overline{\Omega}_{q1}),$$
$$u = (u_1, u_2) = w \circ G_q^{-1}, \qquad p = p_1 \circ G_q^{-1}.$$

From Lemma 6.8.1, it follows that $u \in \widetilde{C}^{l+2}(\overline{\Omega}_1)^2$, $p \in \widetilde{C}^{l+1}(\overline{\Omega}_1)$. We take $h = ((u_1, u_2), p)$ in (6.8.46). By applying (6.8.24), (6.8.25), (6.8.47) and the composite function theorem, we obtain the following formulas for the operators A_{q2}, B_{q2}:

$$A_{q2}(u,p) = \begin{Bmatrix} A_{q21}(u,p) \\ A_{q22}(u,p) \\ A_{q23}(u,p) \end{Bmatrix}, \qquad (u,p) \in V_{l1}, \tag{6.8.48}$$

$$B_{q2}(u,p) = \left(u\big|_{S_2}, u\big|_{S_3}\right), \tag{6.8.49}$$

where

$$\begin{aligned}
A_{q21}(u,p) = \mu\Big[& a_{11}(q)\frac{\partial^2 u_1}{\partial \rho^2} + a_{22}(q)\frac{\partial^2 u_1}{\partial \varphi^2} \\
& + a_{12}(q)\frac{\partial^2 u_1}{\partial \rho \partial \varphi} + b_1(q)\frac{\partial u_1}{\partial \rho} + b_2(q)\frac{\partial u_2}{\partial \rho} \\
& - b_3(q)\frac{\partial u_2}{\partial \varphi} + b_4(q) u_1\Big] - b_5(q)\frac{\partial p}{\partial \rho}, \\
A_{q22}(u,p) = \mu\Big[& a_{11}(q)\frac{\partial^2 u_2}{\partial \rho^2} + a_{22}(q)\frac{\partial^2 u_2}{\partial \varphi^2} \\
& + a_{12}(q)\frac{\partial^2 u_2}{\partial \rho \partial \varphi} + b_1(q)\frac{\partial u_2}{\partial \rho} - b_2(q)\frac{\partial u_1}{\partial \rho} \\
& + b_3(q)\frac{\partial u_1}{\partial \varphi} + b_4(q) u_2\Big] + b_6(q)\frac{\partial p}{\partial \rho} - b_7(q)\frac{\partial p}{\partial \varphi}, \\
A_{q23}(u,p) = b_5(q) & \frac{\partial u_1}{\partial \rho} - b_6(q)\frac{\partial u_2}{\partial \rho} + b_7(q)\left(\frac{\partial u_2}{\partial \varphi} + u_1\right).
\end{aligned} \tag{6.8.50}$$

Here, we use the following notations

$$a_{11}(q) = (\gamma - q)^{-2}\left[1 + \frac{1}{r^2}\left(\frac{\partial r}{\partial \varphi}\right)^2\right], \qquad a_{22}(q) = \frac{1}{r^2},$$

$$
\begin{aligned}
a_{12}(q) &= -\frac{2}{r^2(\gamma - q)}\frac{\partial r}{\partial \varphi}, \\
b_1(q) &= \frac{1}{r^2}\left[\frac{\partial r}{\partial \varphi}\frac{1}{(\gamma - q)^2}\left(\frac{\partial^2 r}{\partial \rho \partial \varphi} + \gamma' - q'\right)\right. \\
&\quad \left. + \frac{1}{\gamma - q}\left(r - \frac{\partial^2 r}{\partial \varphi^2}\right)\right], \\
b_2(q) &= \frac{2}{r^2}\frac{1}{\gamma - q}\frac{\partial r}{\partial \varphi}, \qquad b_3(q) = \frac{2}{r^2}, \\
b_4(q) &= -r^{-2}, \qquad b_5(q) = (\gamma - q)^{-1}, \\
b_6(q) &= \frac{1}{r}\frac{1}{\gamma - q}\frac{\partial r}{\partial \varphi}, \qquad b_7(q) = r^{-1}.
\end{aligned}
\tag{6.8.51}
$$

Note that r is defined by (6.8.44), from which one obtains the partial derivatives of r, which are present in (6.8.51).

From (6.8.40) and (6.8.46), it follows that the operator U_{q2} is defined by

$$
\begin{aligned}
U_{q2}(u,p) &= c(q)\left(\psi_q \circ G_q^{-1}\right), \qquad (u,p) \in V_{l1}, \\
c(q) &= \left(\int_{\Omega_1} r\left(\frac{\partial r}{\partial \rho}\right) d\rho\, d\varphi\right)^{-1} \int_{\Omega_1} pr\left(\frac{\partial r}{\partial \rho}\right) d\rho\, d\varphi
\end{aligned}
\tag{6.8.52}
$$

where ψ_q is determined by (6.8.35), (6.8.38). By applying Theorem 6.8.3 and Lemma 6.8.1, we obtain the following theorem.

Theorem 6.8.4 *Let (6.8.27) hold and let M be defined by (6.8.6). Then, for each $q \in M$, the operator $L_{q2} + U_{q2}$, where $L_{q2} = (A_{q2}, B_{q2})$ and U_{q2} are defined by (6.8.48)–(6.8.52), is an isomorphism of V_{l1} onto H_{l1}.*

By virtue of Theorem 6.8.4, for each $q \in M$ there exists a unique pair (u_q, g_q) satisfying the relations

$$
L_{q2}(u_q, g_q) = (0,0,0,a), \qquad U_{q2}(u_q, g_q) = 0, \qquad (u_q, g_q) \in V_{l1}, \tag{6.8.53}
$$

where a is the vector function from (6.8.4) taking constant values; see (6.8.5). Define a pair (v,p) as follows:

$$
v = (N_1(u_q \circ G_q)) \circ E^{-1}, \qquad p = g_q \circ G_q \circ E^{-1}, \tag{6.8.54}
$$

where N_1 is the following matrix:

$$
N_1 = \begin{pmatrix} \cos\alpha & -\sin\alpha \\ \sin\alpha & \cos\alpha \end{pmatrix}. \tag{6.8.55}
$$

It is easy to see that the pair (v,p) defined by (6.8.54) is a solution of the problem (6.8.1)–(6.8.4) and $(v,p) \in V_{lq}$. By Theorem 6.8.4, there exists a unique solution to the problem (6.8.53) for an arbitrary $q \in M$. So, we define a mapping $N\colon M \to V_{l1}$ such that

$$N(q) = (u_q, g_q), \qquad q \in M, \tag{6.8.56}$$

where (u_q, g_q) is the solution of the problem (6.8.53). Below, we study some properties of the mapping N.

Continuity and Fréchet differentiability of the mapping N

We now establish three lemmas, by which we prove the differentiability of the mapping N.

Lemma 6.8.2 *Let $\Omega_2 = (r_1, r_2)\times\mathbb{R}\times\Omega_1$, where r_1, r_2 are the positive constants from (6.8.6) and let a function $f\colon (x_1, x_2, \rho, \varphi) \to f(x_1, x_2, \rho, \varphi)$ be twice continuously differentiable in Ω_2. Let the function f have arbitrary continuous derivatives up to and including order four, for which the orders of the derivatives with respect to φ do not exceed two, and these derivatives may be extended by continuity onto $[r_1, r_2] \times \mathbb{R} \times \overline{\Omega}_1$. Assume that X is a bounded, open, and convex set in M, where M is given by (6.8.6) for $l \in (0, 1)$, and M, X are equipped with the topology generated by the $C^4([0, 2\pi])$ topology. Define mappings f_1 and f_2 from X into $C^2(\overline{\Omega}_1)$ as follows:*

$$\begin{aligned} f_1(q)(\rho,\varphi) &= f(q(\varphi), q'(\varphi), \rho, \varphi), \\ f_2(q)(\rho,\varphi) &= f(q(\varphi), q''(\varphi), \rho, \varphi), \qquad q \in X. \end{aligned}$$

Then, f_1 and f_2 are continuously Fréchet differentiable mappings from X into $C^2(\overline{\Omega}_1)$, and the Fréchet derivatives f_1', f_2' of these mappings at a point $q \in X$ are defined by the formulas

$$\begin{aligned} (f_1'(q)h)(\rho,\varphi) &= \frac{\partial f}{\partial x_1}(q(\varphi), q'(\varphi), \rho, \varphi)h(\varphi) \\ &\quad + \frac{\partial f}{\partial x_2}(q(\varphi), q'(\varphi), \rho, \varphi)h'(\varphi), \\ (f_2'(q)h)(\rho,\varphi) &= \frac{\partial f}{\partial x_1}(q(\varphi), q''(\varphi), \rho, \varphi)h(\varphi) \\ &\quad + \frac{\partial f}{\partial x_2}(q(\varphi), q''(\varphi), \rho, \varphi)h''(\varphi), \qquad h \in \widetilde{C}^4([0, 2\pi]). \end{aligned} \tag{6.8.57}$$

Proof. By applying the Taylor formula and by virtue of simple but cumbersome estimates, we infer that

$$\|f_i(q+h) - f_i(q) - f_i'(q)h\|_{C(\overline{\Omega}_1)} \le c_1\|h\|^2_{C^2([0,2\pi])}, \tag{6.8.58}$$

$$\left\|\frac{\partial^k}{\partial\rho^{k_1}\partial\varphi^{k-k_1}}\left[f_i(q+h) - f_i(q) - f_i'(q)h\right]\right\|_{C(\overline{\Omega}_1)} \le c_2\|h\|^2_{C^4([0,2\pi])}; \tag{6.8.59}$$

here, $i = 1,2$; $k = 1,2$; $0 \le k_1 \le k$. Also, q, $q+h \in X$, $f_i'(q)$ are defined by (6.8.57) and the constants c_1, c_2 depend on f, X and do not depend on h. Let now $\{q_n\} \subset X$, $q \in X$, and $q_n \to q$ in $C^4([0,2\pi])$. Then, the following estimate holds

$$\|(f_i'(q_n) - f_i'(q))h\|_{C^2(\overline{\Omega}_1)} \le \alpha_n \|h\|_{C^4([0,2\pi])}, \qquad h \in \widetilde{C}^4([0,2\pi]),$$
$$\lim \alpha_n = 0, \qquad i = 1,2.$$

Therefore, the function $q \to f_i'(q)$ is a continuous mapping of X into

$$\mathcal{L}(\widetilde{C}^4([0,2\pi]), C^2(\overline{\Omega}_1)).$$

By analogy with the proof of Lemma 6.8.2, we can prove the lemma below.

Lemma 6.8.3 *Let X be an arbitrary, bounded, open, and convex subset of M, where M is defined by (6.8.6) with $l \in (0,1)$. Let also $\Omega_3 = (r_1, r_2) \times \mathbb{R}$, where r_1, r_2 are the constants from (6.8.6), and let f be a function that is continuous together with the derivatives $D^k f$, $|k| \le 5$, in Ω_3, and these derivatives allow a continuous extension onto $[r_1, r_2] \times \mathbb{R}$. Define a function $f_1 \colon X \to \widetilde{C}^3([0,2\pi])$ by*

$$f_1(q)(\varphi) = f(q(\varphi), q'(\varphi)), \qquad q \in M.$$

Then, f_1 is a continuously Fréchet differentiable mapping of X into $\widetilde{C}^3([0,2\pi])$, and the Fréchet derivative f_1' of this function at a point $q \in X$ is given by

$$(f_1'(q)h)(\varphi) = \frac{\partial f}{\partial x_1}(q(\varphi), q'(\varphi))h(\varphi) + \frac{\partial f}{\partial x_2}(q(\varphi), q'(\varphi))h'(\varphi), \quad h \in \widetilde{C}^4([0,2\pi]). \tag{6.8.60}$$

Lemma 6.8.4 *Let a set M be defined by (6.8.6) and equipped with the topology generated by the topology of $C^{[l]+4}([0,2\pi])$. Let operators A_{q2}, B_{q2}, U_{q2} be defined by (6.8.48)–(6.8.52), and let (6.8.27) hold. Then, the functions $q \to L_{q2} = (A_{q2}, B_{q2})$, $q \to U_{q2}$ are continuous mappings from M into $\mathcal{L}(V_{l1}, H_{l1})$ for an arbitrary $l > 0$, l not an integer. For $l \in (0,1)$, these functions are continuously Fréchet differentiable.*

Proof. It is easily seen that

$$\left.\begin{array}{l} \text{for an arbitrary } k > 0, \text{ the multiplication of functions} \\ u, v \to uv \text{ is a bilinear continuous mapping from} \\ \widetilde{C}^k(\Omega_1) \times \widetilde{C}^k(\Omega_1) \text{ into } \widetilde{C}^k(\Omega_1). \end{array}\right\} \tag{6.8.61}$$

Let now $\{q_n\} \subset M$, $q \in M$ and $q_n \to q$ in $\widetilde{C}^{[l]+4}([0,2\pi])$. From (6.8.44), (6.8.51), (6.8.61), it follows that

$$\begin{aligned} &a_{ij}(q_n) \to a_{ij}(q), && \text{in } \widetilde{C}^{[l]+1}(\overline{\Omega}_1), && i,j = 1,2, \\ &b_i(q_n) \to b_i(q), && \text{in } \widetilde{C}^{[l]+1}(\overline{\Omega}_1), && i = 1,2,3,4, \\ &b_i(q_n) \to b_i(q), && \text{in } \widetilde{C}^{[l]+2}(\overline{\Omega}_1), && i = 5,6,7. \end{aligned} \tag{6.8.62}$$

Further, from (6.8.50) after taking into account (6.8.61) and (6.8.62), we obtain

$$\begin{aligned} A_{q_n 2i} &\to A_{q2i}, \qquad \text{in } \mathcal{L}(V_{l1}, \widetilde{C}^l(\overline{\Omega}_1)), \qquad i=1,2, \\ A_{q_n 23} &\to A_{q23}, \qquad \text{in } \mathcal{L}(V_{l1}, \widetilde{C}^{l+1}(\overline{\Omega}_1)). \end{aligned} \tag{6.8.63}$$

From (6.8.49), it follows that the operator B_{q2} is independent of q; therefore, $B_{q_n 2} = B_{q2}$, and (6.8.63) yields that $q \to L_{q2} = (A_{q2}, B_{q2})$ is a continuous mapping of M into $\mathcal{L}(V_{l1}, H_{l1})$. By applying (6.8.35), (6.8.38), (6.8.44), (6.8.52), we infer that $U_{q_n 2} \to U_{q2}$ in $\mathcal{L}(V_{l1}, H_{l1})$. Lemma 6.8.2 and the formulas (6.8.51) imply that

$$\left.\begin{array}{l} \text{for } l \in (0,1), \text{ the functions } q \to a_{ij}(q), \text{ with } i,j = 1,2, \text{ and} \\ \text{the functions } q \to b_i(q), \text{ with } i = 1, \dots, 7, \text{ are continuously} \\ \text{Fréchet differentiable mappings of } M \text{ into } \widetilde{C}^2(\overline{\Omega}_1). \end{array}\right\} \tag{6.8.64}$$

From (6.8.48)–(6.8.50), (6.8.61), (6.8.64) it follows that, for $l \in (0,1)$, the function $q \to L_{q2} = (A_{q2}, B_{q2})$ is a continuously Fréchet differentiable mapping from M into $\mathcal{L}(V_{l1}, H_{l1})$. By analogy, applying Lemma 6.8.3 and (6.8.35), (6.8.38), (6.8.44), (6.8.52), we obtain that, for $l \in (0,1)$, the function $q \to U_{q2}$ is a continuously Fréchet differentiable mapping from M into $\mathcal{L}(V_{l1}, H_{l1})$, concluding the proof.

Theorem 6.8.5 *Let a set M be defined by (6.8.6) and equipped with the topology generated by the topology of $C^{[l]+4}([0, 2\pi])$; let operators A_{q2}, B_{q2}, U_{q2} be defined by (6.8.48)–(6.8.52); let $L_{q2} = (A_{q2}, B_{q2})$ and (6.8.27) hold. Then, the function $N\colon q \to N(q) = (u_q, g_q)$, where u_q, g_q is the solution of the problem (6.8.53), is a continuous mapping of M into V_{l1}; and for $l \in (0,1)$, the function N is a continuously Fréchet differentiable mapping of M into V_{l1}; and the Fréchet derivative N' of N at a point $q \in M$ is given by*

$$N'(q)h = -(L_{q2} + U_{q2})^{-1}((L'_{q2} + U'_{q2})h)(u_q, g_q), \qquad h \in \widetilde{C}^4([0, 2\pi]). \tag{6.8.65}$$

Here, $(L_{q2} + U_{q2})^{-1}$ is the inverse of $L_{q2} + U_{q2}$ and L'_{q2}, U'_{q2} are the Fréchet derivatives of the functions $q \to L_{q2}$, $q \to U_{q2}$ at a point q.

Proof. Define a mapping $J\colon M \times V_{l1} \to H_{l1}$ as follows:

$$\begin{gathered} q \in M, \qquad (u,g) \in V_{l1}, \\ J(q,(u,g)) = (L_{q2} + U_{q2})(u,g) - (0,0,0,a). \end{gathered} \tag{6.8.66}$$

Here, a is the vector function from (6.8.4) taking constant value. It is obvious that the function $N\colon M \to V_{l1}$ introduced in (6.8.56), where u_q, g_q is the solution of the problem (6.8.53), is an implicit function defined by the mapping J, i.e.,

$$J(q, N(q)) = 0. \tag{6.8.67}$$

The existence of the function N follows from Theorem 6.8.4. To prove the continuity of N, we use the implicit function theorem (Thorem 1.9.1).

By Lemma 6.8.4, we infer that J is a continuous mapping from $M \times V_{l1}$ into H_{l1}. For an arbitrary fixed $q \in M$, the function $J(q, \cdot)\colon (u, g) \to J(q, (u, g))$ is an affine continuous mapping from V_{l1} onto H_{l1}, and the Fréchet derivative of it has the form

$$\frac{\partial J}{\partial(u,g)}(q,(u,g)) = L_{q2} + U_{q2}. \tag{6.8.68}$$

From here and Lemma 6.8.4, we obtain that $(q,(u,g)) \to \frac{\partial J}{\partial(u,g)}(q,(u,g))$ is a continuous mapping from $M \times V_{l1}$ into $\mathcal{L}(V_{l1}, H_{l1})$. From Theorem 6.8.4 and (6.8.68), it follows that $\frac{\partial J}{\partial(u,g)}(q,(u,g))$ is an isomorphism of V_{l1} onto H_{l1}. By the implicit function theorem, we now infer that N is a continuous mapping from M into V_{l1}. Taking Lemma 6.8.4 into account, one can easily see that, in case $l \in (0,1)$, the function J is a continuously Fréchet differentiable mapping from $M \times V_{l1}$ into H_{l1}. From the theorem on differentiability of an implicit function (Theorem 1.9.2), we now obtain that, for $l \in (0,1)$, the function N is a continuously Fréchet differentiable mapping from M into V_{l1}, and its Fréchet derivative is defined by (6.8.65).

Continuity and differentiability of the functionals F_0 and F_1

We now change in (6.8.18), (6.8.19) the functions $v_q = (v_{q1}, v_{q2})$, p_q and the variables y_1, y_2 for the functions $u_q = (u_{q1}, u_{q2})$, g_q and the variables ρ, φ. Taking notice of (6.8.20), (6.8.44), (6.8.54) for $v = v_q$, $p = p_q$, we obtain the following formulas for the functionals F_1 and F_0:

$$\begin{aligned} F_1(q) = \int_0^{2\pi} &\{[-g_q\delta_{1i} + 2\mu\varepsilon_{1i}(u_q)]\nu_{q1i}\sin\varphi \\ &+ [-g_q\delta_{2i} + 2\mu\varepsilon_{2i}(u_q)]\nu_{q1i}\cos\varphi\}\big|_{\rho=1} \left[(q')^2 + q^2\right]^{\frac{1}{2}}\,d\varphi, \end{aligned} \tag{6.8.69}$$

$$F_0(q) = \frac{\mu}{2}\int_1^2\int_0^{2\pi}\sum_{i,j=1}^{2}(\varepsilon_{ij}(u_q))^2 r\Big(\frac{\partial r}{\partial\rho}\Big)\,d\rho\,d\varphi. \tag{6.8.70}$$

Here, ν_{q1i} and r are defined by (6.8.35), (6.8.44), and $\varepsilon_{ij}(u_q)$ are determined by the formulas

$$\begin{aligned} \varepsilon_{11}(u_q) &= \frac{1}{\gamma - q}\frac{\partial u_{q1}}{\partial\rho}, \\ \varepsilon_{22}(u_q) &= -\frac{1}{r(\gamma-q)}\frac{\partial r}{\partial\varphi}\frac{\partial u_{q2}}{\partial\rho} + \frac{1}{r}\Big(\frac{\partial u_{q2}}{\partial\varphi} + u_{q1}\Big), \\ \varepsilon_{12}(u_q) &= \varepsilon_{21}(u_q) \\ &= \frac{1}{2}\Big[-\frac{1}{r(\gamma-q)}\frac{\partial r}{\partial\varphi}\frac{\partial u_{q1}}{\partial\rho} + \frac{1}{r}\frac{\partial u_{q1}}{\partial\varphi} + \frac{1}{\gamma-q}\frac{\partial u_{q2}}{\partial\rho} - \frac{u_{q2}}{r}\Big]. \end{aligned} \tag{6.8.71}$$

We note that $\varepsilon_{ij}(u_q)\circ G_q$ are the components of the rate of strain tensor in polar coordinates r, α (i.e., in the domain Ω_{q1}) for the velocity field $u_q \circ G_q$.

Define mappings $e_{ij}\colon M \times \widetilde{C}^{l+2}(\overline{\Omega}_1)^2 \to C^1(\overline{\Omega}_1)$, where $i,j = 1,2$, as follows:

$$q \in M, \qquad u = (u_1, u_2) \in \widetilde{C}^{l+2}(\overline{\Omega}_1)^2,$$

$$\begin{aligned}
e_{11}(q,u) &= \frac{1}{\gamma - q}\frac{\partial u_1}{\partial \rho},\\
e_{22}(q,u) &= -\frac{1}{r(\gamma - q)}\frac{\partial r}{\partial \varphi}\frac{\partial u_2}{\partial \rho} + \frac{1}{r}\Big(\frac{\partial u_2}{\partial \varphi} + u_1\Big),\\
e_{12}(q,u) &= e_{21}(q,u)\\
&= \frac{1}{2}\Big[-\frac{1}{r(\gamma - q)}\frac{\partial r}{\partial \varphi}\frac{\partial u_1}{\partial \rho} + \frac{1}{r}\frac{\partial u_1}{\partial \varphi} + \frac{1}{\gamma - q}\frac{\partial u_2}{\partial \rho} - \frac{u_2}{r}\Big]. \qquad (6.8.72)
\end{aligned}$$

From (6.8.71) and (6.8.72), it follows that

$$e_{ij}(q, u_q) = \varepsilon_{ij}(u_q). \qquad (6.8.73)$$

We now define functions $f_1\colon M \times V_{l1} \to C([0,2\pi])$ and $f_2\colon M \times V_{l1} \to C(\overline{\Omega}_1)$ as follows:

$$\begin{aligned}
&q \in M, \qquad (u,g) \in V_{l1},\\
f_1(q,(u,g)) &= \{[-g\delta_{1i} + 2\mu e_{1i}(q,u)]\nu_{q1i}\sin\varphi\\
&\qquad + [-g\delta_{2i} + 2\mu e_{2i}(q,u)]\nu_{q1i}\cos\varphi\}\big|_{\rho=1}\,[(q')^2 + q^2]^{\frac{1}{2}},\\
f_2(q,(u,g)) &= \frac{\mu}{2}\sum_{i,j=1}^{2}(e_{ij}(q,u))^2 r\frac{\partial r}{\partial \rho}. \qquad (6.8.74)
\end{aligned}$$

Here, ν_{q1i} and r are determined by (6.8.35), (6.8.44). The functionals F_1 and F_0 from (6.8.69), (6.8.70) take now the forms

$$\begin{aligned}
F_1(q) &= \int_0^{2\pi} f_1(q,(u_q,g_q))\,d\varphi,\\
F_0(q) &= \int_1^2\int_0^{2\pi} f_2(q,(u_q,g_q))\,d\rho\,d\varphi.
\end{aligned} \qquad (6.8.75)$$

Theorem 6.8.6 *Let the conditions of Theorem* 6.8.5 *hold, and let functionals* F_1 *and* F_0 *be given by* (6.8.69), (6.8.70). *Then, for arbitrary* $l > 0$, l *not an integer, the functionals* F_1 *and* F_0 *are continuously Fréchet differentiable in* M, *and their Fréchet derivatives* F_1' *and* F_0' *at a point* $q \in M$ *are defined as follows:*

$$F_1'(q)h = \int_0^{2\pi}\Big[\frac{\partial f_1}{\partial q}(q,(u_q,g_q))h + \Big(\frac{\partial f_1}{\partial(u,g)}(q,(u_q,g_q))\circ N'(q)\Big)h\Big]\,d\varphi, \quad (6.8.76)$$

$$F_0'(q)h = \int_1^2\int_0^{2\pi}\Big[\frac{\partial f_2}{\partial q}(q,(u_q,g_q))h + \Big(\frac{\partial f_2}{\partial(u,g)}(q,(u_q,g_q))\circ N'(q)\Big)h\Big]\,d\rho\,d\varphi, \qquad (6.8.77)$$

where $h \in \widetilde{C}^{[l]+4}([0,2\pi])$, $N'(q)h$ *is given by* (6.8.65), *and* $\frac{\partial f_i}{\partial q}$, $\frac{\partial f_i}{\partial (u,g)}$ *are the partial Fréchet derivatives of the functions* f_i, $i = 1,2$, *with respect to the first and second arguments.*

Proof. By analogy with the above (see the proofs of Lemmas 6.8.2 and 6.8.4), applying (6.8.35), (6.8.44), (6.8.72), and (6.8.74), we establish that f_1, f_2 are continuously Fréchet differentiable mappings from $M \times V_{l1}$ into $C([0,2\pi])$ and $C(\overline{\Omega}_1)$. Further, the functions

$$u \to \int_0^{2\pi} u\, d\varphi, \qquad u \to \int_1^2 \int_0^{2\pi} u\, d\rho\, d\varphi$$

are linear continuous mappings from $C([0,2\pi])$ and $C(\overline{\Omega}_1)$ into $\mathbb{R}$. Now, for $l \in (0,1)$, Theorem 6.8.6 and the formulas (6.8.76), (6.8.77) follow from the composite function theorem and Theorem 6.8.5. In this case, directly from the definition of Fréchet derivative, it follows that, for $l \in (0,1)$ and for arbitrary q and $q+h$ from M, the following estimates hold:

$$|F_1(q+h) - F_1(q) - F_1'(q)h| \le \alpha(h)\|h\|_{C^4([0,2\pi])},$$
$$|F_0(q+h) - F_0(q) - F_0'(q)h| \le \alpha(h)\|h\|_{C^4([0,2\pi])},$$

where $\alpha(h) \to 0$ as $h \neq 0$, $\|h\|_{C^4([0,2\pi])} \to 0$. If now $l > 1$ and $h \to 0$ in M, then all the more $\|h\|_{C^4([0,2\pi])} \to 0$. Hence, Theorem 6.8.6 is valid for an arbitrary $l > 0$, l not an integer.

6.8.3 Optimization problem

On the set M, we define a functional F_2 as follows:

$$F_2(q) = \frac{1}{2} \int_0^{2\pi} q^2\, d\varphi. \tag{6.8.78}$$

It is obvious that $F_2(q)$ is the area of a hydrofoil occupying the domain Q_q (see Fig. 6.8.1). It is easily seen that the functional F_2 is continuously Fréchet differentiable in M and its Fréchet derivative F_2' at a point $q \in M$ is given by

$$F_2'(q)h = \int_0^{2\pi} qh\, d\varphi, \qquad h \in \widetilde{C}^{[l]+4}([0,2\pi]). \tag{6.8.79}$$

We define a set of admissible controls in the form

$$\begin{aligned} M_1 = \big\{\, q \mid q \in \widetilde{C}^{l+4}([0,2\pi]),\ &\|q\|_{\widetilde{C}^{l+4}([0,2\pi])} \le c, \\ &r_3 \le q(\varphi) \le r_4,\ \varphi \in (0,2\pi),\ F_1(q) \ge c_1,\ F_2(q) = c_2 \,\big\}. \end{aligned} \tag{6.8.80}$$

Here,

$$\left.\begin{array}{l} l > 0,\ l \text{ not an integer, } c,\ c_1,\ c_2,\ r_3,\ r_4 \text{ are positive} \\ \text{constants, and } r_1 < r_3 < r_4 < r_2, \text{ where } r_1,\ r_2 \text{ are the} \\ \text{positive constants from (6.8.6).} \end{array}\right\} \tag{6.8.81}$$

Consider the following optimization problem: Find a function q_0 satisfying the relations

$$F_0(q_0) = \min_{q \in M_1} F_0(q), \qquad q_0 \in M_1. \tag{6.8.82}$$

From the physical viewpoint, the problem (6.8.82) consists in finding the shape of the hydrofoil (the function q_0) which would require minimal power to move the hydrofoil, while the constraints on the area and on the hydrodynamic lift are satisfied.

Theorem 6.8.7 *Let functionals F_0, F_1, F_2 be defined by* (6.8.18), (6.8.19), (6.8.78), *where $p_q = p$, $v_q = v = (v_1, v_2)$, and (p, v) is the solution of the problem* (6.8.1)–(6.8.4). *Let also* (6.8.27) *hold, and let a nonempty set M_1 be defined by* (6.8.80), (6.8.81). *Then, there exists a solution of the problem* (6.8.82).

Proof. As M_1 is a nonempty set, there exists a minimizing sequence $\{q_k\}_{k=1}^{\infty}$ satisfying

$$\lim F_0(q_k) = \inf_{q \in M_1} F_0(q), \qquad \{q_k\} \subset M_1. \tag{6.8.83}$$

As the embedding of $\widetilde{C}^{l+4}([0, 2\pi])$ into $\widetilde{C}^{[l]+4}([0, 2\pi])$ is compact, it is possible to extract a subsequence, still denoted by $\{q_k\}$, such that

$$q_k \to q_0, \qquad \text{in } \widetilde{C}^{[l]+4}([0, 2\pi]). \tag{6.8.84}$$

From here, by Theorem 6.8.6, we have

$$F_0(q_k) \to F_0(q_0), \qquad F_1(q_k) \to F_1(q_0). \tag{6.8.85}$$

By (6.8.84), we also have $F_2(q_k) \to F_2(q_0)$, and as $\{q_k\} \subset M_1$ by (6.8.84) we obtain

$$\|q_0\|_{\widetilde{C}^{l+4}([0,2\pi])} \leq c, \tag{6.8.86}$$

where c is the constant from (6.8.80). Now, by applying (6.8.83)–(6.8.86), we conclude that the function q_0 is a solution of the problem (6.8.82).

We now establish the necessary optimality conditions. Define a set M_2 by

$$M_2 = \big\{\, q \mid q \in \widetilde{C}^{l+4}([0, 2\pi]),\ \|q\|_{\widetilde{C}^{l+4}([0,2\pi])} \leq c,\ r_3 \leq q(\varphi) \leq r_4, \quad \varphi \in (0, 2\pi) \,\big\}. \tag{6.8.87}$$

We equip the set M_2 with the topology generated by the topology of $\widetilde{C}^{l+4}([0, 2\pi])$. By (6.8.80), we have

$$M_1 = \big\{\, q \mid q \in M_2,\ \ F_1(q) \geq c_1,\ F_2(q) = c_2 \,\big\}. \tag{6.8.88}$$

Theorem 6.8.8 *Let the conditions of Theorem* 6.8.7 *hold and let a function* q_0 *be a solution of the problem* (6.8.82). *Then, there exists constants* λ_0, λ_1, λ_2, *not all equal to zero, such that*

$$\begin{gathered}\left[\sum_{i=0}^{2} \lambda_i F_i'(q_0)\right](q - q_0) \geq 0, \qquad q \in M_2, \\ \lambda_0 \geq 0, \quad \lambda_1 \leq 0, \quad \lambda_1(F_1(q_0) - c_1) = 0.\end{gathered} \tag{6.8.89}$$

If q_0 *is an internal point of* M_2, *then*

$$\sum_{i=0}^{2} \lambda_i F_i'(q_0) = 0.$$

Proof. M_2 is a convex subset of M, and F_0, F_1, F_2 are continuously Fréchet differentiable functionals in M. Now, Theorem 6.8.8 follows from (6.8.88) and Theorem 1.12.2.

6.9 Direct and optimization problems with consideration for the inertia forces

We have above studied optimization problems for steady flows under neglect of the inertia forces. Now, we consider the general problem on steady flows of a nonlinear viscous fluid taking into account these forces. We will show that, for non-high-speed flows of high-viscous fluids, there exists a unique solution of the direct problem, and the above results on the optimization of fluid flows may be transferred to the case of consideration for the inertia forces.

It should be noted that the majority of real nonlinear viscous fluids are high-viscous, and they usually flow slowly.

6.9.1 Setting and solution of the direct problem

Equations and the definition of a generalized solution

We consider the problem of steady flows of the nonlinear viscous fluids with the constitutive equation (6.1.1) in a bounded domain $\Omega \subset \mathbb{R}^n$, $n = 2$ or 3, with a Lipschitz continuous boundary S. By (6.1.1), (6.1.4), (6.1.5), we have the following equations of motion and incompressibility:

$$\rho v_j \frac{\partial v_i}{\partial x_j} + \frac{\partial p}{\partial x_i} - 2\,\frac{\partial\big(\varphi(I(v))\varepsilon_{ij}(v)\big)}{\partial x_j} = K_i \qquad \text{in } \Omega,\ i = 1, \ldots, n, \tag{6.9.1}$$

$$\operatorname{div} v = \sum_{i=1}^{n} \frac{\partial v_i}{\partial x_i} = 0 \qquad \text{in } \Omega. \tag{6.9.2}$$

Let S_1, S_2, S_3 be open subsets of S such that $S = \overline{S}_1 \cup \overline{S}_2 \cup \overline{S}_3$, $S_i \cap S_j = \varnothing$ for $i \neq j$. We consider the mixed boundary conditions (6.1.7)–(6.1.10). The spaces X and V are the same as in (6.1.19) and (6.1.20), i.e.,

$$X = \{\, u \mid u \in H^1(\Omega)^n,\ u \restriction_{S_1} = 0,\ (u - u_\nu \nu) \restriction_{S_3} = 0 \,\}, \tag{6.9.3}$$

$$V = \{\, u \mid u \in X,\ \operatorname{div} u = 0 \,\}, \tag{6.9.4}$$

and the norm in X and V is given by (6.1.21). We suppose that the functions of volume forces K and surface forces $\check{F}$, $\hat{F}$ acting on S_2 and S_3 satisfy the conditions

$$K \in X^*, \quad \check{F} \in X^*, \quad \hat{F} \in X^*. \tag{6.9.5}$$

Suppose also that the functions φ and λ satisfy the conditions (6.1.11)–(6.1.14) and (6.1.15)–(6.1.18).

Let us define operators $L\colon X \to X^*$, $N\colon X \to X^*$, and $B \in \mathcal{L}(X, L_2(\Omega))$ as follows

$$(L(u), h) = 2 \int_\Omega \varphi(I(u)) \varepsilon_{ij}(u) \varepsilon_{ij}(h)\, dx + \int_{S_3} \lambda(|u|^2, \cdot) u_i h_i\, ds \tag{6.9.6}$$

$$(N(u), h) = \rho \int_\Omega u_j \frac{\partial u_i}{\partial x_j} h_i\, dx, \tag{6.9.7}$$

$$Bu = \operatorname{div} u. \tag{6.9.8}$$

Let also

$$M = L + N. \tag{6.9.9}$$

A generalized solution of the problem (6.9.1), (6.9.2), (6.1.7)–(6.1.10) is defined to be a pair v, p satisfying

$$(v, p) \in X \times L_2(\Omega), \tag{6.9.10}$$

$$(M(v), h) - (B^* p, h) = (K + \check{F} + \hat{F}, h), \qquad h \in X, \tag{6.9.11}$$

$$(Bv, q) = 0, \qquad q \in L_2(\Omega). \tag{6.9.12}$$

By using Green's formula, one may show that, if v, p is a solution to the problem (6.9.10)–(6.9.12), then v, p is a solution to the problem (6.9.1), (6.9.2), (6.1.7)–(6.1.10) in the distribution sense. On the contrary, if v, p is a solution to the problem (6.9.1), (6.9.2), (6.1.7)–(6.1.10) that satisfies (6.9.10), then v, p is a solution to (6.9.10)–(6.9.12).

The existence and uniqueness

Lemma 6.9.1 *Suppose that $n = 2$ or 3 and*

$$u_k \to u_0 \qquad \text{weakly in } X^*. \tag{6.9.13}$$

Then, $N(u_k) \to N(u_0)$ strongly in X^*.

Proof. Denoting

$$A_k = \rho \int_\Omega u_{0j} \left(\frac{\partial u_{ki}}{\partial x_j} - \frac{\partial u_{0i}}{\partial x_j} \right) h_i \, dx$$

and using the Hölder inequality and the embedding theorem, we obtain

$$\begin{aligned} \big|(N(u_k) - N(u_0), h)\big| &\leq \left| \rho \int_\Omega (u_{kj} - u_{0j}) \frac{\partial u_{ki}}{\partial x_j} h_i \, dx \right| + |A_k| \\ &\leq c \|u_k - u_0\|_{L_4(\Omega)^n} \|u_k\|_X \|h\|_X + |A_k|. \end{aligned} \tag{6.9.14}$$

The norm in the space $L_q(\Omega)^n$ is defined by

$$\|u\|_{L_q(\Omega)^n} = \left(\int_\Omega \sum_{i=1}^n |u_i|^q \, dx \right)^{\frac{1}{q}}, \qquad q \in [1, \infty). \tag{6.9.15}$$

Taking (6.9.3) into account and applying again Green's formula, the Hölder inequality, and the embedding theorem, we get

$$\begin{aligned} |A_k| &= \left| \rho \int_\Omega \left[\frac{\partial}{\partial x_j} (u_{oj}(u_{ki} - u_{0i}) h_i) \right. \right. \\ &\quad \left. \left. - u_{0j}(u_{ki} - u_{0i}) \frac{\partial h_i}{\partial x_j} - \frac{\partial u_{0j}}{\partial x_j} (u_{ki} - u_{0i}) h_i \right] dx \right| \\ &\leq \left| \rho \int_{S_2 \cup S_3} u_{0j}(u_{ki} - u_{oi}) h_i \nu_j \, ds \right| + c_1 \|u_0\|_X \|u_k - u_0\|_{L_4(\Omega)^n} \|h\|_X \\ &\leq \rho \sum_{i,j=1}^n \|\Gamma u_{0j}\|_{L_4(S_2 \cup S_3)} \|\Gamma(u_{ki} - u_{0i})\|_{L_2(S_2 \cup S_3)} \|\Gamma h_i\|_{L_4(S_2 \cup S_3)} \\ &\quad + c_1 \|u_0\|_X \|u_k - u_0\|_{L_4(\Omega)^n} \|h\|_X. \end{aligned} \tag{6.9.16}$$

Here, Γ is the trace operator, $\Gamma u = u \restriction_{S_2 \cup S_3}$. It follows from (6.9.13) that $u_k \to u_0$ in $L_4(\Omega)^n$ and $\Gamma u_k \to \Gamma u_0$ in $L_2(S_2 \cup S_3)^n$. So, (6.9.14) and (6.9.16) give $N(u_k) \to N(u_0)$ in X^*, which completes the proof.

Define positive constants η and γ as follows:

$$\eta = \sup_{u \in V, \, \|u\|_X \leq 1} \left| \rho \int_\Omega u_j \frac{\partial u_i}{\partial x_j} u_i \, dx \right|, \tag{6.9.17}$$

$$\gamma = \sup_{u \in V, \, \|u\|_X \leq 1} |(K + \check{F} + \hat{F}, u)|. \tag{6.9.18}$$

Lemma 6.9.2 *Suppose the conditions* (6.1.11)–(6.1.18) *are satisfied and*

$$\gamma < \frac{a_1^2}{\eta}, \tag{6.9.19}$$

where a_1 *is the positive constant from* (6.1.12). *Then, the following estimate holds:*

$$(M(u),u)-(K+\check{F}+\hat{F},u)\geq 0 \qquad \text{if } u\in V \text{ and } \|u\|_X=\frac{a_1-\sqrt{a_1^2-\eta\gamma}}{\eta}=\lambda. \tag{6.9.20}$$

Proof. By applying (6.1.12), (6.1.16), (6.9.9), (6.9.17), and (6.9.18), we get

$$\begin{aligned}(M(u),u)-(K+\check{F}+\hat{F},u)&\geq f(\|u\|_X)\\ &=2a_1\|u\|_X^2-\eta\|u\|_X^3-\gamma\|u\|_X, \qquad u\in V.\end{aligned} \tag{6.9.21}$$

Consider the quadratic equation

$$2a_1y-\eta y^2-\gamma=0. \tag{6.9.22}$$

Its roots are those of the equation $f(y)=0$ and they are equal to

$$y_1=\frac{a_1-\sqrt{a_1^2-\eta\gamma}}{\eta}, \qquad y_2=\frac{a_1+\sqrt{a_1^2-\eta\gamma}}{\eta}. \tag{6.9.23}$$

If (6.9.19) holds, then $y_1$and y_2 are real and $f(y)\geq 0$ for $y\in[y_1,y_2]$. So, (6.9.19) yields (6.9.20), and the lemma is proved.

For $l>0$, we denote by $d(0,l)$ the closed ball in V of radius l centered at 0, i.e.,

$$d(0,l)=\{\,u\mid u\in V,\ \|u\|_X\leq l\,\}. \tag{6.9.24}$$

Let a be a trilinear form in X generated by the operator N:

$$a(u,v,w)=\rho\int_\Omega u_j\,\frac{\partial v_i}{\partial x_j}\,w_i\,dx, \tag{6.9.25}$$

and

$$\gamma_1=\sup_{u,v\in d(0,1)}\big|a(u,v,v)\big|, \qquad \gamma_2=\sup_{u,v\in d(0,1)}\big|a(v,u,v)\big|. \tag{6.9.26}$$

Lemma 6.9.3 *Let the conditions* (6.1.11)–(6.1.18) *be satisfied, let* (6.9.19) *hold, and let* λ *be defined by* (6.9.20). *Let also*

$$\mu=\mu_1-(\gamma_1+\gamma_2)\lambda>0, \tag{6.9.27}$$

where μ_1 *is the positive constant from* (6.1.46). *Then, the operator* M *is strictly monotone in* $d(0,\lambda)$, *i.e.,*

$$(M(u)-M(w),u-w)\geq\mu\|u-w\|_X^2, \qquad u,w\in d(0,\lambda). \tag{6.9.28}$$

Proof. Let u, w be arbitrary functions from V and $h = u - w$. From (6.9.7) and (6.9.26), we obtain

$$\begin{aligned}\big|(N(u) - N(w), h)\big| &= \left|\rho \int_\Omega \left[u_j \frac{\partial h_i}{\partial x_j} + \frac{\partial w_i}{\partial x_j} h_j\right] h_i \, dx\right. \\ &\leq \big(\gamma_1 \|u\|_X + \gamma_2 \|w\|_X\big) \|h\|_X^2.\end{aligned} \tag{6.9.29}$$

Now, (6.1.46), (6.9.9), and (6.9.29) yield (6.9.28).

Theorem 6.9.1 *Let the conditions* (6.1.11)–(6.1.18) *be satisfied and let* (6.9.19) *hold. Then, there exists a solution v, p of the problem* (6.9.10)–(6.9.12) *such that* $\|v\|_X \leq \lambda$. *If, additionally,* (6.9.27) *holds, then there exists a unique solution of* (6.9.10)–(6.9.12) *such that* $v \in d(0, \lambda)$.

Proof. It follows from (6.9.10)–(6.9.12) that the function v is a solution of the problem

$$\begin{aligned} &v \in V, \\ &(M(v), h) = (K + \check{F} + \hat{F}, h), \qquad h \in V. \end{aligned} \tag{6.9.30}$$

Let $\{V_k\}$ be a sequence of finite-dimensional subspaces of V such that

$$\lim_{k \to \infty} \inf_{z \in V_k} \|u - z\|_X = 0, \qquad u \in V, \tag{6.9.31}$$

$$V_k \subset V_{k+1}. \tag{6.9.32}$$

We search for the Galerkin approximations v_k satisfying

$$\begin{aligned} &v_k \in V_k, \\ &(M(v_k), h) = (K + \check{F} + \hat{F}, h), \qquad h \in V_k. \end{aligned} \tag{6.9.33}$$

By virtue of Lemma 6.9.2, there exists a solution of the problem (6.9.33) and $\|v_k\|_X \leq \lambda$. Thus, we can extract a subsequence $\{v_m\}$ such that $v_m \to v_0$ weakly in V. We pass to the limit as $m \to \infty$ in (6.9.33), with k changed by m. In this case, we use the monotonicity of the operator L and the compactness of the operator N, Lemma 6.9.1. Thus, the function $v = v_0$ is a solution of the problem (6.9.30).

From Lemma 6.1.1, it follows that there exists a function $p \in L_2(\Omega)$ such that

$$B^* p = M(v) - K - \check{F} - \hat{F}. \tag{6.9.34}$$

So, the pair v, p is a solution to the problem (6.9.10)–(6.9.12). In the case when (6.9.27) holds, the operator M is strictly monotone in $d(0, \lambda)$, see Lemma 6.9.3. Therefore, in the ball $d(0, \lambda)$, there exist a unique solution of the problem (6.9.30) and a unique solution of the problem (6.9.10)–(6.9.12) such that $v \in d(0, \lambda)$.

Remark 6.9.1 Theorem 6.9.1 does not state that, under the conditions (6.1.11)–(6.1.18), (6.9.19), and (6.9.27), there exists a unique solution of the problems (6.9.30) and (6.9.10)–(6.9.12). But a unique solution such that $v \in d(0, \lambda)$ does exist.

6.9.2 Approximation of the problem (6.9.10)–(6.9.12)

Let $\{X_k\}$ and $\{M_k\}$ be sequences of finite-dimensional subspaces of X and $L_2(\Omega)$ which satisfy the conditions (6.1.37), (6.1.38), and (6.1.42). We search for an approximate solution of the problem (6.9.10)–(6.9.12) in the form

$$(v_k, p_k) \in X_k \times M_k, \tag{6.9.35}$$

$$(M(v_k), h) - (B_k^* p_k, h) = (K + \check{F} + \hat{F}, h), \qquad h \in X_k, \tag{6.9.36}$$

$$(B_k v_k, q) = 0, \qquad q \in M_k. \tag{6.9.37}$$

From the point of view of applications, in particular, of computation, the problem (6.9.35)–(6.9.37) is considerably more preferable than (6.9.33). So, we study the question of convergence of the approximations v_k, p_k.

Define constants η_1, γ_3, λ_1, γ_4, γ_5 by

$$\eta_1 = \sup_k \max_{u \in d_k} |a(u,u,u)|, \qquad \gamma_3 = \sup_k \max_{u \in d_k} |(K + \check{F} + \hat{F}, u)|,$$
$$\lambda_1 = \frac{a_1 - \sqrt{a_1^2 - \eta_1 \gamma_3}}{\eta_1}, \qquad \gamma_4 = \sup_k \max_{u,v \in d_k} |a(u,v,v)|,$$
$$\gamma_5 = \sup_k \max_{u,v \in d_k} |a(v,u,v)|, \tag{6.9.38}$$

where a_1 is the constant from (6.1.12) and

$$d_k = \{ u \mid u \in X_k, \ (B_k u, \mu) = 0, \ \mu \in M_k, \ \|u\|_X \le 1 \}. \tag{6.9.39}$$

Note that $\eta_1 \ge \eta$, $\gamma_3 \ge \gamma$, $\gamma_4 \ge \gamma_1$, $\gamma_5 \ge \gamma_2$, see (6.9.17), (6.9.18), (6.9.26).

Theorem 6.9.2 *Let the conditions* (6.1.11)–(6.1.18) *be satisfied and let* $\{X_k\}$, $\{M_k\}$ *be sequences of finite-dimensional subspaces of* X *and* $L_2(\Omega)$ *that satisfy* (6.1.37), (6.1.38), (6.1.42) *and* $X_k \subset X_{k+1}$, $M_k \subset M_{k+1}$ *for an arbitrary* k. *Let also*

$$\gamma_3 < \frac{a_1^2}{\eta_1}. \tag{6.9.40}$$

Then, for each k, *there exists a solution of the problem* (6.9.35)–(6.9.37), *and from the sequence* $\{v_k,\ p_k\}$ *one can extract a subsequence* $\{v_m,\ p_m\}$ *such that* $v_m \to v$ *in* X, $p_m \to p$ *in* $L_2(\Omega)$, *where* v, p *is a solution of the problem* (6.9.10)–(6.9.12). *If, additionally,*

$$\mu_1 - (\gamma_4 + \gamma_5)\lambda_1 > 0, \tag{6.9.41}$$

then for each k *there exists a unique solution of the problem* (6.9.35)–(6.9.37) *such that* $\|v_k\|_X \le \lambda_1$.

Proof of this theorem is analogous to that of Theorem 6.1.1, so we give only a sketch of it.

By using (6.9.40) and (6.1.42), we argue the solvability of the problem (6.9.35)–(6.9.37) for each k, and the boundedness of the solutions. By selecting appropriate subsequences, we pass to the limit, using the monotonicity of the operator L and the compactness of the operator N, see Lemma 6.9.1. If (6.9.41) holds, then for each k the operator $L = M + N$ is strictly monotone in the ball

$$d_{k\lambda_1} = \{\, u \mid u \in X_k,\ (B_k u, q) = 0\ q \in M_k,\ \|u\|_X \le \lambda_1 \,\}. \tag{6.9.42}$$

In this case, $\|v_k\|_X \le \lambda_1$ and there exists a unique solution of the problem (6.9.35)–(6.9.37) such that $v_k \in d_{k\lambda_1}$.

6.9.3 Some remarks on models, optimization problems, and existence results

For small data, more exactly, for the case when (6.9.19) holds, we have established above the results on the existence of a solution of the general problem on steady flows of the nonlinear viscous fluid with the mixed boundary conditions (6.1.7)–(6.1.10), when zero velocities, a function of surface forces, and the conditions of filtration are prescribed on different parts of the boundary. If, additionally, (6.9.27) holds, a unique solution of this problem exists in a small ball.

It should be mentioned that, for large data, i.e., for large velocities, our problem, as well as the corresponding problem for the Navier-Stokes equations, which are obtained from (6.9.1) when $\varphi = \text{const}$, is ill-posed. Moreover, the equations for the nonlinear viscous fluid and the Navier-Stokes equations do not describe the main phenomena observed at high-speed and turbulent flows of real fluids, see Litvinov (1996), and they are not suitable for such flows. This is why we only considered flows with small velocities, which are well described by our equations. Notice also that not only the equations (6.9.1), but also those (6.1.6) may be used for such flows.

The optimization problems for the Navier-Stokes equations without assumptions on the smallness of velocities and with a function of volume forces considered as a control were first set and studied by Fursikov (1980, 1982, 1983a). For further investigation of optimization problems for the Navier-Stokes equations, see Abergel and Temam (1990), Sritharan (1991, 1992), Gunzburger et al. (1991, 1992), Casas (1993).

By using the continuity and strict monotonicity of the operator M in a small ball of the space V, for small data, one can carry over, by analogy, all the results of Sections 6.2 and 6.3 on the continuity of the function control-solution of the direct problem to the case of consideration for the inertia forces, provided that the viscosity of the fluid is large, while the velocities are not large.

Bibliography

Abelés, F. (1950). Recherches sur la propagation des ondes électromagnétiques sinusoidales dans les milieux stratifiés. Application aux couches minces. *Annales de Physique*, Ser. 12, **5**, 596–782

Abergel, F. and Temam, R. (1990). On some control problems in fluid mechanics. *Theoret. Comput. Fluid Dynamics.* **1**, 303–325

Adams, R.A. (1975). *Sobolev Spaces.* Academic Press. New York

Agmon, S., Douglis, A., and Nirenberg, L. (1959). Estimates near the boundary for solutions of elliptic partial differential equations satisfying general boundary conditions I. *Com. Pure Appl. Math.* **12**, 623–727

Agmon, S., Douglis, A., and Nirenberg, L. (1964). Estimates near the boundary for solutions of elliptic partial differential equations satisfying general boundary conditions II. *Com. Pure Appl. Math.* **17**, 35–92

Ahlberg, J.H., Nilson, E.N., and Walsh, J.L. (1967). *The Theory of Splines and Their Applications.* Academic Press. New York

Allaire, G. (1994). Explicit lamination parameters for three-dimensional shape optimization. *Control and Cybernetics* **23**, 309–326

Allaire, G. and Kohn, R. (1993). Optimal bounds on the effective behavior of a mixture of two well-ordered elastic materials. *Quart. Appl. Math.* **51**, 643–671

Ambartsumian, S.M. (1974). *General Theory of Anisotropic Shells.* Nauka. Moscow (in Russian)

Armand J.-L. (1972). *Applications of the Theory of Optimal Control of Distributed Parameter Systems to Structural Optimization.* NASA Report CR-2044

Armand, J.-L. and Lodier, B. (1978). Optimal design of bending elements. *Int. J. Num. Meth. Engng.* **13**, 373–384

Astarita, G. and Marrucci, G. (1974). *Principles of Non-Newtonian Fluid Mechanics.* McGraw-Hill. London

Aubin, J.-P. (1972). *Approximation of Elliptic Boundary Value Problems.* Wiley-Interscience. New York

Baiocchi, C. and Capelo, A. (1984). *Variational and Quasivariational Inequalities – Applications to Free Boundary Problems.* John Wiley and Sons. New York

Bakhvalov, N.S. (1980). Homogenization and perturbation problems. In *Computing Methods in Applied Science and Engineering*, pp. 645–658. North-Holland. Amsterdam

Bakhvalov N.S. and Panasenko, G.P. (1984). *Homogenization of Processes in Periodic Media.* Nauka. Moscow (in Russian)

Banerjee, P.K. and Butterfield, R. (1981). *Boundary Element Methods in Engineering Scince.* McGraw-Hill. London

Banichuk, N.V. (1980). *Optimization of the Shapes of Elastic Bodies.* Nauka. Moscow (in Russian); English transl.: *Problems and Methods of Optimal Structural Design.* Plenum Press. 1983

Banichuk, N.V., Kartvelishvili, V.M., and Mironov, A.A. (1977). Numerical solutions of two-dimensional optimization problems for elastic plates. *Mechanics of Solids* **12**, 65–74

Bendsøe, M.P. (1994). Methods for optimization of structural topology, shape and material. Preprint, Mathematical Institute, Technical University of Denmark

Bendsøe, M. and Guedes, J. (1994). Some computational aspects of using extremal material properties in the optimal design of shape, topology and material. *Control and Cybernetics* **23**, 327–349

Bendsøe, M. and Kikuchi, N. (1988). Generating optimal topologies in structural design using a homogenization method. *Comp. Meth. Appl. Mech. Engrg.* **71**, 197–224

Bendsøe, M.P, and Rodrigues, H.C. (1990). On topology and boundary variations in shape optimization. *Control and Cybernetics* **19**, no. 3–4, 9–26

Bensoussan, A., Lions, J.L., and Papanicolau, G. (1978). *Asymptotic analysis for Periodic Structures.* North Holland. Amsterdam

Bernadou, M., Palma, F.J., and Rousselet, B. (1991). Shape optimization of an elastic thin shell under various criteria. *Structural Optimization* **3**, 7–21

Besov, O.V., Il'in, V.P., and Nikol'skii, S.M. (1975). *Integral Representation of Functions and Embedding Theorems.* Nauka. Moscow (in Russian); English transl.: Vol. 1, 2, Wiley, 1979

Bolotin, V.V. and Novikov, Yu.N. (1980). *Mechanics of Multilayer Constructions.* Mashinostroenie. Moscow (in Russian)

Born, M. and Wolf, E. (1964). *Principles of Optics.* Pergamon Press. Oxford

Boussinesq, J. (1877). Essai sur la théorie des eaux courantes. In *Mémoires Présentées par Diverses Savants à l'Acad. d. Sci.*, Vol. 23. Paris

Bourbaki, N. (1955). *Éléments de Mathématique, Livre V, Espaces Vectoriels Topologiques.* Hermann. Paris

Bourbaki, N. (1960). *Éléments de Mathématique, Livre III, Topologie Générale.* Hermann. Paris

Bratus, A.S. (1981). Asymptotic solutions in problems of optimal control by coefficients of elliptic operators. *DAN SSSR* **259**, 1035–1038 (in Russian)

Bratus, A.S. and Seiranian, A.P. (1983a). Two multiple eigenvalues in optimization problems. *DAN SSSR* **272**, 275–278 (in Russian)

Bratus, A.S. and Seiranian, A.P. (1983b). Bimodal solutions in problems of optimization of eigenvalues. *Prikl. Matem. Mekh.* **47**, 546–554 (in Russian)

Brezis, H. (1983). *Analyse Fonctionnelle, Théorie et Applications.* Masson. Paris; English transl.: Springer-Verlag. Heidelberg, 1987

Burak, Ya.I. and Budz, S.F. (1974). Determination of the optimal rate of heating of a thin spherical shell. *Prikl. Mekh.* **10**, no. 2, 14–20 (in Russian)

Burak, Ya.I. and Domanskii, P.P. (1982). Optimization of dynamic effects in shells of revolution under axisymmetric force load. *Soviet Appl. Mechan.* **18**, no. 2, 92–98

Burak, Ya.I., Zozuliak, Yu.D., and Gera, B.V. (1984). *Optimization of Transients for Thermoelastic Shells.* Naukova Dumka. Kiev (in Russian)

Burczyński, T. and Fedeliński, P. (1990). Shape sensitivity analysis and optimal design of static and vibrating systems using the boundary element method. *Control and Cybernetics* **19**, no. 3–4, 47–71

Calderon, A.P. (1961). Lebesgue spaces of differentiable functions and distributions. In *Proc. Sympos. Pure Math.* Vol. 4, pp. 33–49. Providence

Casas, E. (1990). Optimality conditions and numerial approximations for some optimal design problems. *Control and Cybernetics* **19**, no. 3–4, 73–91

Casas, E. (1993). Some optimal control problems of turbulent flows. In *Optimal Design and Control of Structures under Statical and Vibration Response. Advanced Tempus Course. Lecture Notes, Part 1*, pp. 1–17, Banach Centre, Poland

Céa, J. (1971). *Optimisation. Théorie et Algorithmes.* Dunod. Paris

Céa, J. (1978). Numerical search method for an optimal domain. In *Numerical Methods in Mathematical Physics, Geophysics, and Optimal Control* (Sympos. Novosibirsk (1976), Marchuk, G.I. and Lions, J.L., eds., pp. 64–74. Nauka. Novosibirsk (in Russian)

Céa, J. and Malanowski, K. (1970). An example of a max-min problem in partial differential equations. *SIAM J. Control* **8**, 305–316

Chenais, D. (1987). Optimal design of midsurface of shells: differentiability and sensitivity computation. *Appl. Math. and Optimization* **16**, 93–133

Chenais, D. (1994). Shape optimization of shells., *Control and Cybernetics* **23**, 351–382

Cherkaev, A. V. (1994). Relaxation of problems of optimal structural design. *Int. J. Solids and Structures* **31**, 2251–2280

Christensen, R.M. (1979). *Mechanics of Composite Materials.* Wiley-Interscience. New York

Chudinovich, I.Yu. (1991). *Methods of Boundary Equations in Problems of the Elastic Medium Dynamics.* Kharkov University. Kharkov (in Russian)

Ciarlet, P. (1978). *The Finite Element Method for Elliptic Problems.* North-Holland. Amsterdam

Donnell, L.H. (1976). *Beams, Plates, and Shells.* McGraw-Hill. New York

Dunford, N. and Schwartz, J.T. (1958). *Linear Operators. Part I: General Theory.* Interscience Publishers. New York

Duvaut, G. (1976). Etude de matériaux composites élastiques à structure périodique. Homogénéisation. In *Proc. Congress of Theoretical and Applied Mechanics, Delft (1976)*, Koiter, ed. North-Holland. Amsterdam

Duvaut, G. and Lions, J.L. (1972). *Les Inéquations en Mécanique et en Physique.* Dunod. Paris

Edwards, R.E. (1979). *Fourier Series – A Modern Introduction, Vol. 1.* Springer-Verlag. New York, *Vol. 2.* Springer-Verlag. New York, 1982

Ekeland, I. and Temam, R. (1976). *Convex Analysis and Variational Problems.* North-Holland. Amsterdam

Fedorenko, R.P. (1978). *Approximate Solution of Optimal Control Problems.* Nauka. Moscow (in Russian)

Fernández Cara, E. (1989). Optimal design in fluid mechanics. In *Control of Partial Differential Equations. Proc. of IFIP Conference in Santiago de Compostela (1987)*, A. Bermúdez, ed. Lecture Notes in Control and Information Sciences, Vol. 114, pp. 120–131. Springer-Verlag. Berlin/New York

Fichera, G. (1972). *Existence Theorems in Elasticity.* Springer-Verlag. Berlin

Fikhtengolts, G.M. (1966). *A Course of Differential and Integral Calculus, Vol. 1.* Nauka. Moscow (in Russian)

Fil'shtinskii, L.A. (1964). Stresses and displacements in an elastic sheet weakened by a doubly periodic set of equal circular holes. *J. Appl. Math. Mech.* **28**, 530–543

Fix, G.J. and Strang, G. (1973). *An Analysis of the Finite Element Method.* Prentice Hall. Englewood Cliffs, N.J.

Fletcher, C.A.I. (1988). *Computational Techniques for Fluid Dynamics, Vol. 2.* Springer-Verlag. Berlin

Flügge, W. (1957). *Statik und Dynamik der Schalen.* Springer-Verlag. Berlin

Fučik, S. Kratochvil, A., and Nečas, J. (1973). Kachanov-Galerkin method. *Comment. Math. Univ. Carolinae* **14**, 651–659

Fursikov, A.V. (1980). On some control problems and results concerning the unique solvability of a mixed boundary value problem for the three-dimensional Navier-Stokes and Euler systems. *Soviet Math. Dokl.* **21**, 889–893

Fursikov, A.V. (1982). Control problems and theorems concerning the unique solvability of a mixed boundary value problem for the three-dimensional Navier-Stokes and Euler equations. *Math. USSR Sbornik* **43**, 251–273

Fursikov, A.V. (1983a). Properties of solutions of some extremal problems connected with the Navier-Stokes system. *Math. USSR Sbornik* **46**, 323–351

Fursikov, A.V. (1983b). Some questions of the theory of optimal control over nonlinear systems with distributed parameters. *Trudy Seminara Petrovskogo*, no. 1, 167–189 (in Russian)

Gajewski, H., Gröger, K., and Zacharias, K. (1974). *Nichtlineare Operatorgleichungen und Operatordifferentialgleichungen.* Akademie-Verlag. Berlin

Gibiansky, L.V. and Cherkaev, A.V. (1994). Microstructures of composites of extremal rigidity and exact estimates of provided energy density. In *Topics in the Mathematical Modelling of Composite Materials*, Kohn, R., ed. Birkhäuser. Boston

Girault, V. and Raviart, P.-A. (1981). *Finite Element Approximations of the Navier-Stokes Equations.* Lect. Notes Math., Vol. 749. Springer-Verlag. Berlin/ New York

Glowinski, R., Lions, J.L., and Trêmolières, R. (1976). *Analyse Numérique des Inéquations Variationnelles, Tome 1, Tome 2.* Dunod. Paris

Goldenblat, I.I. and Kopnov, V.A. (1968). *Criteria of the Strength and Plasticity of Structural Materials.* Mashinostroenie. Moscow (in Russian)

Gould, S.H. (1966). *Variational Methods for Eigenvalue Problems.* Oxford University Press. London

Grigoliuk, E.I. and Kabanov, V.V. (1978). *Stability of Shells.* Nauka. Moscow (in Russian)

Grigoliuk, E.I, Podstrigach, Ya.S., and Burak, Ya.I. (1979). *Optimization of the Heating of Plates and Shells.* Naukova Dumka. Kiev (in Russian)

Grigorenko, Ya. M. (1973). *Isotropic and Anisotropic Laminated Shells of Revolution which Possess Variable Stiffness.* Naukova Dumka. Kiev (in Russian)

Gunzburger, M. (1986). Mathematical aspects of finite element methods for incompressible viscous flows. In *Finite Element Theory and Applications. Proc. of ICASE Conf., Hampton, Virginia (1986)*, pp. 124–150. Springer-Verlag. Berlin/New York

Gunzburger, M. Hou, L., and Svobodny, T. (1991). Analysis and finite element approximation of optimal control problems for the stationary Navier-Stokes equations with Dirichlet conditions. *Math. Mod. Numer. Anal.* **25**, 711-748

Gunzburger, M., Hou, L., and Svobodny, T. (1992). Boundary velocity control of incompressible flow with an application to viscous drag reduction. *SIAM J. Control Optim.* **30**, 167–181

Guz, A.N. and Babich, I.Yu. (1980). *Three-Dimensional Theory of Stability of Bars, Plates, and Shells.* Vyshcha Shkola. Kiev (in Russian)

Guz, A.N., Chernyshenko, I.S., Chekhov, Val.N., Chekhov, Vik.N., and Shnerenko, K.I. (1980). *Theory of Thin Shells Weakened by Holes.* Naukova Dumka, Kiev (in Russian)

Haftka, R.T. and Gurdal, Z. (1992). *Elements of Structural Optimization* (Third revised edition). Kluwer. Dordrecht/Boston/London

Haslinger, J. (1993). Fictitious domain approach in optimal shape design problems. In *Optimal Design and Control of Structures under Statical and Vibration Response, Advanced Tempus Course. Lecture Notes, Part 1*, pp. 1–8. Banach Centre. Poland

Haslinger, J. and Neittaanmäki, P. (1988). *Finite Element Approximation for Optimal Shape Design. Theory and Applications.* John Wiley and Sons. New York

Haug, E.J. and Arora, J.S. (1979). *Applied Optimal Design. Mechnical and Structural Systems.* John Wiley and Sons. New York

Haug, E.J., Choi, K.K., and Komkov, V. (1986). *Design Sensitivity Analysis of Structural Systems.* Academic Press. New York

Himmelblau, D. (1972). *Applied Nonlinear Programming.* McGraw-Hill. London

Hlaváček, I. and Nečas, J. (1970). On inequalities of Korn's type. *Arch. Rat. Mech. and Anal.* **36**, 305–334

Hlaváček, I. and Nečas, J. (1982). Optimization of the domain in elliptic unilateral boundary value problems by finite element method. *R.A.I.R.O. Numerical Analysis* **16**, 351–371

Horoshun, L.P. (1978). Methods of theory of random functions in problems of macroscopic properties of micrononhomogeneous medium. *Prikl. Mekh.* **14**, no. 2, 3–17 (in Russian)

Ioffe, A. and Tikhomirov, V. (1979). *Extremal Problems.* North-Holland. Amsterdam

Ivanov, L.A., Kotko, L.A., and Krein, S.G. (1977). Boundary value problems in variable domains. In Dif. Uravn. i Primen. Trudy Sem. Processy Optimal. Upravl., Vyp. 19 (in Russian)

Kantorovich, L.V. and Akilov, G.P. (1977), *Functional Analysis* (2nd rev. ed.). Nauka. Moscow (in Russian); English transl.: Pergamon Press. 1982

Kartvelishvili, V.M. and Kobelev, V.V. (1984). Analytical solutions to problems of optimal reinforcement for multilayer plates made of composite materials. *Mekh. Komposits. Material.* **19**, 571–581 (in Russian)

Kato, T. (1976). *Perturbation Theory for Linear Operators.* Springer-Verlag. New York

Kohn, R.V. (1989). Recent progress in the mathematical modelling of composite materials. In *Composite Material Response: Constitutive Relations and Damage Mechanisms*, Sih, G.C. et al. eds., pp. 155–177. Elsevier. London

Kohn, R.V. and Strang, G. (1986). Optimal design and relaxation of variational problems. *Comm. Pure Appl. Math.* **39**, 113–137, 139–182, 353–377

Koiter, W.T. (1966). On the nonlinear theory of thin elastic shells, Parts I, II, III. *Proc. Koninkl. Nederl. Akad. Wet.* **B69**, 1–17, 18–32, 33–54

Kolmogorov, A.N. and Fomin, S.V. (1975). *Introductory Real Analysis.* Dover. New York

Komkov, V. (1972). *Optimal Control Theory for the Damping of Vibrations of Simple Elastic Systems.* Springer-Verlag. Berlin

Krasnoselskii, M.A. (1956). *Topological Methods in the Theory of Nonlinear Integral Equations.* Gostekhizdat. Moscow (in Russian)

Krasnoselskii, M.A., Vainiko, G.M., Zabreiko, P.P., Rutitskii, Ya.B., and Stetsenko, B.Ya. (1969). *Approximate Solution of Operator Equations.* Nauka. Moscow (in Russian)

Krein, S.G. (1968). The behavior of solutions of elliptic problems under variation of the domain. *Studia Math.* **31**, 411–424 (in Russian)

Křižek, M. and Litvinov, W.G. (1994). On the methods of penalty functions and Lagrange's multipliers in the abstract Neumann problem. *Z. Angew. Math. Mech.* **74**, 216–218

Krys'ko, V.A. and Pavlov, S.P. (1982). Problem of optimal control of the natural frequency of inhomogeneous shells. *Soviet Appl. Mech.* **18**, 319–325

Kupradze, V.D., Gegelia, T.G., Bashelishvili, M.O., and Burchuladze, T.V. (1979). *Three-Dimensional Problems of the Mathematical Theory of Elasticity.* North-Holland. Amsterdam

Kwak, B.M. and Choi, J.H. (1987). Shape design sensitivity analysis using boundary integral equation for potential problems. In *Computer Aided Optimal Design: Structural and Mechanical Systems.* Springer-Verlag. Berlin

Ladyzhenskaya, O.A. (1969). *The Mathematical Theory of Viscous Incompressible Flow.* Gordon and Breach. New York

Ladyzhenskaya, O.A. and Solonnikov, V.A. (1976). Some problems of vector analysis and generalized formulations of boundary value problems for the Navier-Stokes equations. *Zap. Nauchn. Sem. LOMI*, **59**, 81–116 (in Russian); English transl.: in *J. Soviet Math.* **10** (1978), no. 2

Ladyzhenskaya, O.A. and Uraltseva, N.N. (1973). *Linear and Quasilinear Elliptic Equations.* Nauka. Moscow (in Russian)

Laurent, P.-J. (1972). *Approximation et Optimisation.* Hermann. Paris

Lavrentiev, M.A. and Shabat, B.V. (1973). *Methods of the Theory of Functions of a Complex Variable.* Nauka. Moscow (in Russian)

Lekszycki, T. (1990). Eigenvalues optimization – New view about the old problem. *Control and Cybernetics* **19**, no. 3–4, 189–201

Lewinski, T. (1992). Homogenizing stiffnesses of plates with periodic structures. *Int. J. Solids Struct.* **29**, 309–326

Lions, J.L. (1968). *Contrôle Optimal de Systèmes Gouvernés par des Équations aux Dérivées Partielles.* Dunod Gauthier-Villars. Paris

Lions, J.L. (1969). *Quelques Méthodes de Résolution des Problèmes aux Limites non Linéaires.* Dunod Gauthier-Villars. Paris

Lions, J.L. (1983). *Contrôle des Systèmes Distribués Singuliers.* Gauthier-Villars. Paris

Lions, J.L. and Magenes, E. (1972). *Non-Homogeneous Boundary Value Problems and Applications, Vol. 1.* Springer-Verlag. New York

Litvinov, W.G. (1976). Some inverse problems for bended plates. *Prikl. Mat. Mekh.* **40**, 682–691 (in Russian)

Litvinov, W.G. (1981a). A modification of the Ritz method for variational equations and applications to the problems with mixed boundary conditions. *Diff. Uravn.* **17**, 519–526 (in Russian)

Litvinov, W.G. (1981b). Approximation of functions by a tensor product of splines and trigonometric polynomials. *Ukrain. Mat. Zh.* **33**, 252–257 (in Russian)

Litvinov, W.G. (1982a). *Motion of Nonlinear Viscous Fluid.* Nauka. Moscow (in Russian)

Litvinov, W.G. (1982b). Optimal control by coefficients for elliptic systems. *Diff. Uravn.* **18**, 1036–1047 (in Russian)

Litvinov, W.G. (1987). *Optimization in Elliptic Boundary Value Problems and Applications to Mechanics.* Nauka. Moscow (in Russian)

Litvinov, W.G. (1990a). Control of the shape of the domain in elliptic systems and choice of the optimal domain in the Stokes problem. *Math. USSR Sbornik* **67**, 165–175

Litvinov, W.G. (1990b). Domain shape optimization in mechanics. *Control and Cybernetics* **19**, no. 3–4, 203–220

Litvinov, W.G. (1994). Optimization problems on manifolds and the shape optimization of elastic solids. *Control and Cybernetics* **23**, 495–512

Litvinov, W.G. (1995). On the optimal shape of a hydrofoil. *J. Optim. Theory and Appl.* **85**, 325–345

Litvinov, W.G. (1996). Some models and problems for laminar and turbulent flows of viscous and nonlinear viscous fluids. *J. Math. and Phys. Sciences.* **30**, 101–157

Litvinov, W.G. and Medvedev, N.G. (1979). Some questions of the stability of shells of revolution. *Mat. Fiz.* **26**, 101–109 (in Russian)

Litvinov, W.G. and Panteleev, A.D. (1980). Optimization problems for plates of variable thickness. *Mech. Solids* **15**, 140–146

Litvinov, W.G. and Rubezhanskii, Yu.I. (1978). Methods of the theory of duality in problems of optimal loading of thin plates. *Prikl. Mekh.* **14**, no. 2, 73–79 (in Russian)

Litvinov, W.G. and Rubezhankii, Yu.I. (1982). Problems of control by the right-hand sides of elliptic systems and their application to control of the stress-strain state in shells. *J. Appl. Math. Mech.* **46**, 256–260

Litvinov, W.G. and Rubezhankii, Yu.I. (1990). On the convergence of solutions of problems of the Timoshenko theory of shells to the solution of the Kirchhoff-Love theory. *Matem. Fiz. Nelin. Mekh.* **13**, 46–51 (in Russian)

Loitzansky, L.G. (1987). *Mechanics of Fluid and Gas.* Nauka. Moscow (in Russian)

Lovišek, J. (1991). Optimization of the thickness of layers of sandwich conical shells. *Appl. Math. Optim.* **24**, 1–33

Lurie, K.A. (1993). *Applied Optimal Control.* Plenum. New York

Lurie, K.A. and Cherkaev, A.V. (1976). On application of Prager's theorem to the problems of optimal design of thin plates. *Izvest. Akad. Nauk SSSR MTT* **6**, 157–159 (in Russian)

Mäkelä, M.M. and Neittaanmäki, P. (1995). *Nonsmooth Optimization Analysis and Algorithms with Applications to Optimal Control.* World Scientific. Singapore/River Edge, NJ

Malmeister, A.K., Tamuzh, V.P., and Teters, G.A. (1980). *Strength of Polymeric and Composite Materials.* Zinatne. Riga (in Russian)

Marcellini, P. (1979). Convergence of second order linear elliptic operators. *Boll. Un. Mat. Ital.* **16-B**, 278–290

Marchenko, V.A. and Khruslov, E.Ya. (1974). *Boundary Value problems in Domains with Fine-Grained Boundary.* Naukova Dumka. Kiev (in Russian)

Marino, A. and Spagnolo, S. (1969). Un tipo di approssimazione dell' operatore $\sum_{i,j} D_i(a_{ij}D_j)$ con operatori $\sum_j D_j(bD_j)$. *Ann. Scuola Norm. Sup. Pisa, Cl. Sci.* **23**, 657–673

Masson, J. (1985). *Method of Functional Analysis for Application in Solid Mechanics.* Elsevier. New York

Maz'ja, V.G. (1985). *Sobolev Spaces.* Springer-Verlag. Berlin

Mazur, S. (1933). Über schwache Konvergenz in den Räumen (L^p). *Studia Math.* **4**, 128–133

Medvedev, N.G. (1980). Some spectral peculiarity of problems of optimization of stability of shells of variable thickness. *Dokl. AN Ukrainy*, Ser. A, no. 9, 59–63 (in Russian)

Medvedev, N.G. and Totskii, N.P. (1984a). On multiplicity of eigenvalues in problems of optimization of stability of cylindric shells of variable thickness. *Prikl. Mekh.* **20**, no. 6, 113–116 (in Russian)

Medvedev, N.G. and Totskii, N.P. (1984b). Optimization of cylindric shells of variable thickness at axially symmetric load. *Prikl. Mekh.* **20**, no. 9, 53–57 (in Russian)

Michlin, S.G. (1962). *Multi-Dimensional Singular Integral Equations.* Nauka. Moscow (in Russian)

Michlin, S.G. (1970). *Variational Methods in Mathematical Physics.* Nauka. Moscow (in Russian)

Michlin, S.G. (1973). Spectrum of the bunch of operators of the elasticity theory. *Usp. Mat. Nauk* **28**, no. 3, 43–82 (in Russian)

Michlin, S.G. and Prössdorf, S. (1980). *Singuläre Integraloperatoren.* Akademie-Verlag. Berlin

Mikhailov, V. (1980). *Equations aux Dérivées Partielles.* Mir. Moscow

Minoux, M. (1989). *Programmation Mathématique Théorie et Algorithmes.* Dunod. Paris

Mota Soares, C.A., Infante Barbosa, J., and Mota Soares, C.M. (1994). Axisymmetric thin shell structures sizing and shape optimization. *Control and Cybernetics* **23**, 513–551

Mroz, Z. and Rozvany, G.I.N. (1975). Optimal design of structures with variable support conditions. *J. Optim. Theory Appl.* **15**, 85–101

Muschelishvili, N.I. (1963). *Some Basic Problems of the Mathematical Theory of Elasticity.* Noordhoff. Groningen

Murat, F. and Simon, J. (1976). Étude de problèmes d'optimal design. In *Optimization Techniques: Modeling and Optimization in the Service of Man, Part 2. Proc. of the Seventh IFIP Conf., Nice (1975)*, J. Céa, ed. Lect. Notes in Computer Sci., Vol. 41, pp. 54–62. Springer-Verlag. Berlin

Natanson, I.P. (1974). *Theory of Functions of Real Variable.* Nauka. Moscow (in Russian)

Nečas, J. (1967). *Les Méthodes Directes en Théorie des Equations Elliptiques.* Masson. Paris

Nečas, J. and Hlavaček, I. (1981). *Mathematical Theory of Elastic and Elastico-Plastic Bodies – An Introduction.* Elsevier. Amsterdam

Neittaanmäki, P. and Tiba, D. (1994). *Optimal Control of Nonlinear Parabolic Systems. Theory, Algorithms, and Application.* Marcel Dekker. New York

Niordson, F.I. (1985). *Shell Theory.* North-Holland. Amsterdam

Novozhilov, V.V. (1951). *Theory of Thin Shells.* Sudpromgiz. Leningrad (in Russian)

Obraztsov, I.F., Vasiliev, V.V., and Bunakov, V.A. (1976). *Optimal Reinforcement of Shells of Revolution Made of Composite Materials.* Mashinostroenie. Moscow (in Russian)

Oleinik, O.A., Iosiphian, G.A., and Shamaev, A.S. (1990). *Mathematical Problems of the Theory of Strongly Nonhomogeneous Elastic Medium.* Izd. MGU. Moscow (in Russian)

Oleinik, O.A., Shamaev, A.S., and Iosiphian, G.A. (1992). *Mathematical Problems in Elasticity and Homogenization.* North-Holland. Amsterdam

Olhoff, N. (1974). Optimal design of vibrating rectangular plates. *Int. J. Solids and Structures* **10**, 93–109

Osipov, Yu.S. and Suetov, A.P. (1990). Existence of optimal domain for elliptic problems with Dirichlet boundary condition. Preprint, Acad. Sci. Inst. Math. Mech., Sverdlovsk

Panagiotopoulos, P.D. (1985). *Inequality Problems in Mechanics and Applications.* Birkhäuser. Boston

Panteleev, A.D. (1991). A problem of design of three-layered shells of revolutions according to mechanical and radioegineering characteristics. *Izv. AN SSR, Mekh. Tverd. Tela*, no. 6, 112–116 (in Russian)

Pelekh, B.L. (1973). *Theory of Thin Shells with Finite Shear Stiffness.* Naukova Dumka. Kiev (in Russian)

Petukhov, L.V. (1986). Optimization problems with unknown boundaries in the theory of elasticity. *Izv. AN SSSR, Prikl. Mat. Mekh.* **50**, 231–236 (in Russian)

Pironneau, O. (1974). On the optimum design in fluid mechanics. *J. Fluid Mech.* **64**, 97–110

Pironneau, O. (1984). *Optimal Shape Design for Elliptic Systems.* Springer-Verlag. New York

Pobedria, B.E. (1984). *Mechanics of Composite Materials.* Izd. MGU, Moscow (in Russian)

Podstrigach, Ya.S., Burak, Ya.I., Shelepets, V.I., Budz, S.F., and Piontkovskii, A.B. (1980). *Optimization and Control in the Production of Electro-Vacuum Devices.* Naukova Dumka. Kiev (in Russian)

Pozdeev, A.A., Niashin, Yu.I., and Trusov, P.V. (1982), *Residual Stresses – Theory and Applications*, Nauka, Moscow (in Russian)

Pshenichny, B.N. (1980). *Convex Analysis and Extremum Problems.* Nauka, Moscow (in Russian)

Pshenichny, B.N. (1983). *Method of Linearization.* Nauka. Moscow (in Russian)

Pshenichny, B.N. and Danilin, Yu.M. (1978). *Numerical Methods in Extremal Problems.* Mir. Moscow

Rao, M. and Sokolowski, J. (1989). Shape sensitivity analysis of state constrained optimal control problems for distributed parameter systems. In Lect. Notes in Control and Information Sciences, Vol. 114, pp. 236–245. Springer-Verlag. Berlin

Raitum, U.E. (1989). *Problems of Optimal Control for Elliptic Equations.* Zinatne. Riga (in Russian)

Rempel, S. and Schulze, B.-W. (1982). *Index Theory of Elliptic Boundary Problems.* Akademie-Verlag. Berlin

Riesz, F. and Sz.-Nagy, B. (1972). *Leçons d'Analyse Fonctionnelle.* Akadémiai Kiadó. Budapest

Rodi, W. (1980). *Turbulence Models and Their Application in Hydraulics.* I.A.H.R. Delft

Rojtberg, Ja.A. (1964). Elliptic problems with nonhomogeneous boundary conditions and the local raise of smoothness up to the boundary of generalized solutions. *DAN SSSR* **157**, 798–801 (in Russian)

Rojtberg, Ja.A. (1975). Theorems of the complete set of isomorphisms for elliptic systems in the sense of Douglis-Nirenberg. *Ukrain. Mat. Zh.* **27**, 544–548 (in Russian)

Rojtberg, Ja.A. and Sheftel, Z.G. (1979). Optimal control of systems described by general elliptic problems. *Diferents. Uravn.* **15**, 1075–1087 (in Russian)

Rozvany, G.I.N. (1989). *Structural Design via Optimality Criteria.* Kluwer. Dordrecht

Rubezhanskii, Yu.I. (1979). Some inverse problems for cylindrical shells. *Prikl. Mekh.* **15**, no. 9, 32–36 (in Russian)

Rudin, W. (1973). *Functional Analysis.* McGraw-Hill. New York

Rvachov, V.L. and Rvachov, V.A. (1979). *Nonclassical Methods of the Approximation Theory of Boundary Value Problems.* Naukova Dumka. Kiev (in Russian)

Sanchez-Palencia, E. (1980). *Non-Homogeneous Media and Vibration Theory.* Springer-Verlag. New York

Schwartz, L. (1961). *Méthodes Mathématiques pour les Sciences Physiques.* Hermann. Paris

Schwartz, L. (1966). *Théorie des Distributions I, II.* Hermann. Paris

Schwartz, L. (1967). *Analyse Mathématique.* Hermann. Paris

Sendeckyj, G.P. (1974). Elastic properties of composites. In *Mechanics of Composite Materials, Vol. 2,* Sendeckyj, G.P., ed. Academic Press. New York

Serovaiskii, S.Ya. (1991). Necessary and sufficient optimality conditions for a system described by a nonlinear elliptic equation. *Sibir. Mat. Zh.* **32**, 141–150 (in Russian)

Serovaiskii, S.Ya. (1993a). Optimization in a nonlinear elliptic system with control in coefficients. *Mat. Zametki* **54**, no. 2, 85–95 (in Russian)

Serovaiskii, S.Ya. (1993b). Differentiation of an inverse function in nonnormed spaces. *Funk. Anal. Prilozh.*, no. 4, 84–87 (in Russian)

Seyranian, A.P. (1973). Elastic plates and shells of minimum weight with several bending loads. *Mechanics of Solids* **8**, no. 5, 83–89

Seyranian, A.P. (1976). Optimal beam design with limitations on natural vibration frequency and the critical force of stability. *Mechanics of Solids* **11**, no. 1, 133–138

Seyranian, A.P. (1977). Quasioptimal solutions of a problem on optimal design with various restrictions. *Soviet Appl. Mech.* **13**, 544–550

Seyranian, A.P. (1991). Sensitivity Analysis of Multiple Eigenvalues. Report no. 431, The Technical University of Denmark

Shablii, O.N. and Medynskii, Ya.R. (1981). To the problem of creation of necessary residual stresses in a circular disk. *Prikl. Mekh.* **17**, no. 6, 90–93 (in Russian)

Shevchenko, Yu.N. (1970). *Thermoplasticity under Variable Loading.* Naukova Dumka. Kiev (in Russian)

Simon, J. (1980). Differentiation with respect to the domain in boundary value problems. *Numer. Func. Anal. and Optimiz.* **2**, 649–687

Simon, J. (1989). Diferenciación de problemas de contorno respecto del dominio. Lectures in the University of Sevilla

Simon, J. (1990). Domain variation for drag in Stokes flow. In *Proc. IFIP Conf. in Shanghaï*, Li Xunjing, ed. Lect. Notes in Control and Iformation Sciences. Springer-Verlag. Berlin

Smirnov, V.I. (1959). *A Course of Higher Mathematics, Vol. 5.* Gos. Izd. Fiz. Mat. Lit. Moscow (in Russian)

Sobolev, S.L. (1963). *Applications of Functional Analysis in Mathematical Physics.* Transl. Amer. Math. Soc. Mathematical Monograph **7**

Sokolowski, J. (1987). Sensitivity analysis of control constrained optimal control problems for distributed parameter systems. *SIAM J. Control and Optimization* **25**, 1542–1556

Sokolowski, J. (1990). Sensitivity analysis of shape optimization problems. *Control and Cybernetics* **19**, 271–286

Sokolowski, J. and Zolesio, J.-P. (1992). *Introduction to Shape Optimization. Shape Sensitivity Analysis.* Springer-Verlag. New York

Solonnikov, V.A. (1964). On general boundary value problems for systems elliptic in the sense of Douglis-Nirenberg, I. *Izv. AN SSSR Ser. Matem.* **28**, 665–706 (in Russian)

Solonnikov, V.A. (1966). On general boundary value problems for systems elliptic in the sense of Douglis-Nirenberg, II. *Trudy MIAN SSSR* **92**, no. 4, 233–297 (in Russian)

Sritharan, S. (1991). Dynamic programming of the Navier-Stokes equations. *Systems and Control Letters* **16**, 299–307

Sritharan, S. (1992). An optimal control problem in exterior hydrodynamics. *Proc. Royal Soc. of Edinburgh* **121A**, 5–32

Srubshchik, L.S. (1981). *Buckling and Post-Buckling Behaviour of Shells.* Izdat. Rostov Univers. Rostov-Don (in Russian)

Suetov, A.P. (1994). A shape optimization algorithm for an elliptic system. *Control and Cybernetics* **23**, 565–573

Sveshnikov, A.G. and Tikhonravov, A.V. (1989). Mathematical methods in problems of analysis and design of layered medium. *Matem. Modelir.* **1**, no. 7, 13–38 (in Russian)

Tartar, L. (1975). Problèmes de contrôle des coefficients dans des équations aux dérivées partielles. In Lect. Notes Econ. and Math. Syst., Vol. 107, pp. 420–426

Teters, G.A., Rikards, R.B., and Narysberg, B.L. (1978). *Optimization of Laminate Composite Shells.* Zinatne. Riga (in Russian)

Thelen, A. (1989). *Design of Optical Interference Coatings.* McGraw-Hill. New York

Tikhonov, A.N., Tikhonravov, A.V., and Trubetskov, M.K. (1993). Second order optimization in problems of design of multilayered coverings. *Zhur. Vych. Mat. i Mat. Fiz.* **33**, 1518–1536 (in Russian)

Timoshenko, S. and Woinowsky-Krieger, S. (1959). *Theory of Plates and Shells.* McGraw-Hill. New York

Triebel, H. (1978). *Interpolation Theory, Function Spaces, Differential Operators.* Veb. Deutscher Verlag der Wissenschaften. Berlin

Troitskii, V.A. and Petukhov, L.V. (1982). *Shape Optimization of Elastic Solids.* Nauka. Moscow (in Russian)

Tsai, S.W. and Hahn, H.T. (1980). *Introduction to Composite Materials.* Technomic. Lancaster

Vainberg, M.M. (1972). *Variational Method and Method of Monotone Operators.* Nauka. Moscow (in Russian)

Vanin, G.A., Semeniuk, N.P., and Emelianov, R.F. (1978). *Stability of Shells of Reinforced Materials.* Naukova Dumka. Kiev (in Russian)

Van Pho Phy, G.A. (1971a). *Constructions of Reinforced Materials.* Tekhnika. Kiev (in Russian)

Van Pho Phy, G.A. (1971b). *Theory of Reinforced Materials with Coatings.* Naukova Dumka. Kiev (in Russian)

Varga, R.S. (1971). *Functional Analysis and Approximation Theory in Numerical Analysis.* Soc. for Industrial and Applied Mathematics. Philadelphia

Vigak, V.M. (1979). *Optimal Control of Nonstationary Temperature Condition.* Naukova Dumka. Kiev (in Russian)

Vigak, V.M. (1988). *Optimal Control of Temperature Stresses and Displacements.* Naukova Dumka. Kiev (in Russian)

Vigdergauz, S.B. (1977). On a case of the inverse problem of the two-dimensional theory of elasticity. *J. Appl. Math. Mech.* **41**, 927–933

Vigdergauz, S.B. (1983). Inverse problem of three-dimensional elasticity. *Mechanics of Solids* **18**, no. 2, 83–86

Vorovich, I.I. (1989). *Mathematical Problems of the Nonlinear Theory of Shallow Shells.* Nauka. Moscow (in Russian)

Wang, C.-T. (1953). *Applied Elasticity.* McGraw-Hill. New York

Washizu, K. (1982). *Variational Methods in Elasticity and Plasticity.* Pergamon Press. Oxford/New York

Wendland, W.L. (1985). On some aspects of boundary element methods for elliptic problems. In *The Mathematics of Finite Elements and Applications, V*, pp. 193–227. Academic Press. London

Wu, E.M. (1974). Phenomenological criteria of fracture of anisotropic medium. In *Mechanics of Composite Materials, Vol. 2*, Sendeckyj, G.P. ed. Academic Press. New York

Yosida, K. (1971). *Functional Analysis.* Springer-Verlag. Berlin

Zangwill, W.I. (1969). *Nonlinear Programming – A Unified Approach.* Prentice-Hall. Englewood Cliffs

Zavialov, Yu.S., Kvasov, B.I., and Miroshnichenko, V.L, (1980). *Methods of Spline Functions.* Nauka. Moscow (in Russian)

Zhikov, V.V., Kozlov, S.M., and Oleinik, O.A. (1993). *Homogenization of Differential Operators.* Nauka. Moscow (in Russian)

Index

averaging of a function, 40

ball
  closed, 5
  open, 5
base
  of neighborhoods, 9
base of neighborhoods, 4
bending moment, 211, 212, 268
bifurcation, 235
  point, 235
boundary of a set, 4

chain rule, 52
closure of a set, 4
coefficients of the first quadratic form, 258
completion of the set of real numbers, 3
component
  of the flexural strain, 258
  of the tangential strain, 258
  of the torsional strain, 258
computation of eigenvalues, 127
condition
  complementing boundary, 169
  limit density, 103, 112
  natural, 220
  of ellipticity, 168
  of optimality necessary, 90, 95, 135
  of stability, 220
  of the free edge, 220
  of transversality, 220
  optimality, 192
  supplementary, 168
condition of filtration, 432
cone, 6
constitutive equation of a nonlinear viscous fluid, 431
continuity modulus, 93
continuity of the spectrum, 118
control
  discontinuous, 97
  nonregular, 86
  nonsmooth, 97
  regular, 88
convergence, 4
  strong, 9
  weak, 8
curvilinear coordinates, 260
cylindrical stiffness of a plate, 211

distance, 5, 7
distribution, 38
domain in $\mathbb{R}^n$, 36

eigenfunction, 20
  approximate, 127
eigenspace, 20
eigenvalue, 20
  approximate, 127
  its multiplicity, 20
elasticity characteristics, 243
element of the best approximation, 112
energy
  stored, 223
  strain, 211

finite-dimensional approximation, 122

form
  bilinear, 11
  coercive, 12, 81
  symmetric, 12
Fréchet derivative, 51
free oscillations of a shell, 268
function
  continuous, 11
  generalized, 37
  lower semicontinuous, 11
  test, 37
  upper semicontinuous, 11

G-closedness of linear operators, 72
G-convergence of linear operators, 72
Gâteaux derivative, 53

homeomorphism, 10

image, 10
implicit function, 52
index of elliptic operator, 173
inequality
  Korn, 48
  Schwartz, 17
  variational, 201
infimum, 3
interior of a set, 4
isomorphism, 10

Kirchhoff hypotheses, 209, 258
Kronecker delta, 19

Lagrange principle, 70
lemma
  inner point, 183
limit
  lower, 3
  upper, 3

majorant, 2
mapping, 2
  k-linear, 11
  bijective, 2
  bilinear, 11
  bounded, 8
  continuous, 10
  convex, 7
  Fréchet differentiable, 51
    continuously, 51
  injective, 2
  sequentially continuous, 10
  surjective, 2
  uniformly continuous, 14
material
  orthotropic, 259
maximum, 2
maximum function, 14, 16
  discrete, 14
midsurface
  of a shell, 257
minimizing element
  its characterization, 181
minimum, 2
  local, 70, 303
minorant, 2
model
  finite shear, of a shell, 274
multi-index, 36

neighborhood of a point, 4
norm, 7
  equivalent, 7
  Euclidean, 8

observation, 191
operator
  coercive, 81
  compact, 20
  nonlinear compact, 234
  selfadjoint, 20
optimization
  domain shape, 173

orthogonal
complement, 19
elements, 18
subspaces, 18

plate
isotropic, 212
orthotropic, 211, 214
three-layered, 242
problem
combined, 142, 157
control
basic, 154
general, 149
eigenvalue, 117
eigevalue optimizaton, 162
finite-dimensional, 103, 105
finite-dimensional regular, 113
minimax control, 198
of bending of a plate, 215
of free oscillations of a plate, 221
spectral, 117
product
of sets, 2
of topologies, 5
prototype, 10

regularization of a function, 40
Riesz
method, 112
operator, 112

scalar product, 17
scale of Hilbert spaces, 35
segment, 6
sequence
Cauchy, 6
fundamental, 6
sequence weakly fundamental, 9
set
bounded, 2
bounded above, 2
bounded below, 2
closed, 4
compact, 13
convex, 6
dense, 4
open, 5
relatively compact, 13
sequentially $*$-weakly closed, 102
sequentially weakly closed, 9, 110
weakly closed, 9
shell
laminated, 282
of revolution, 260
shallow, 276
space
Banach, 7
dual, 8
Hausdorff, 4
Hilbert, 17
metric, 5
complete, 6
sequentially weakly complete, 9
Sobolev, 39
topological, 4
compact, 13
metrizable, 6
separable, 4
sequentially compact, 13
vector, 6
normed, 6
spline
fundamental, 299
strain component, 210, 211
stress component, 210
subspace of functions with
zero-point strain energy, 264
supremum, 3
system
ellpitic in the sense of
Douglis-Nirenberg, 168
system of operators
W-coercive, 45
coercive, 44

theorem
Calderon, 39

embedding, 39
Lax-Milgram, 18
Lebesgue, 37
on a composite function, 52
on equivalent norms, 41
on the invariance of Sobolev spaces, 76
on trace space, 43
Riesz, 17
three-layered plate
free oscillations, 245
natural oscillations, 255
topology, 4
induced, 5
weak, 8
torque, 211, 212, 268
trace of a function, 43